Multivariate Statistics

This book explores multivariate statistics from both traditional and modern perspectives. The first section covers core topics like multivariate normality, MANOVA, discrimination, PCA, and canonical correlation analysis. The second section includes modern concepts such as gradient boosting, random forests, variable importance, and causal inference.

A key theme is leveraging classical multivariate statistics to explain advanced topics and prepare for contemporary methods. For example, linear models provide a foundation for understanding regularization with AIC and BIC, leading to a deeper analysis of regularization through generalization error and the VC theorem. Discriminant analysis introduces the weighted Bayes rule, which leads into modern classification techniques for class-imbalanced machine learning problems. Steepest descent serves as a precursor to matching pursuit and gradient boosting. Axis-aligned trees like CART, a classical tool, set the stage for more recent methods like super greedy trees.

Another central theme is training error. Introductory courses often caution that reducing training error too aggressively can lead to overfitting. At the same time, training error, also referred to as empirical risk, is a foundational concept in statistical learning theory. In regression, training error corresponds to the residual sum of squares, and minimizing it results in the least squares solution, which can lead to overfitting. Regardless of this concern, empirical risk plays a pivotal role in evaluating the potential for effective learning. The principle of empirical risk minimization demonstrates that minimizing training error can be advantageous when paired with regularization. This idea is further examined through techniques such as penalization, matching pursuit, gradient boosting, and super greedy tree constructions.

Key Features:

- Covers both classical and contemporary multivariate statistics.
- Each chapter includes a carefully selected set of exercises that vary in degree of difficulty and are both applied and theoretical.
- The book can also serve as a reference for researchers due to the diverse topics covered, including new material on super greedy trees, rule-based variable selection, and machine learning for causal inference.
- Extensive treatment on trees that provides a comprehensive and unified approach to understanding trees in terms of partitions and empirical risk minimization.
- New content on random forests, including random forest quantile classifiers for class-imbalanced problems, multivariate random forests, subsampling for confidence regions, super greedy forests. An entire chapter is dedicated to random survival forests, featuring new material on random hazard forests extending survival forests to time-varying covariates.

Dr. Hemant Ishwaran's work focuses on advancing machine learning techniques for applications in public health, medicine, and informatics. His contributions include the development of open-source tools, such as R packages for his pioneering methods, including the widely-used random survival forests—a significant extension of the random forest algorithm in machine learning. His collaborations with healthcare experts have resulted in precision models for cardiovascular disease (CVD), heart transplantation, cancer staging, and resistance to gene cancer therapy.

Multivariate Statistics

Classical Foundations and Modern Machine Learning

Hemant Ishwaran

CRC Press
Taylor & Francis Group
Boca Raton London New York

CRC Press is an imprint of the
Taylor & Francis Group, an **informa** business
A CHAPMAN & HALL BOOK

Designed cover image: © Hemant Ishwaran

First edition published 2025
by CRC Press
2385 NW Executive Center Drive, Suite 320, Boca Raton FL 33431

and by CRC Press
4 Park Square, Milton Park, Abingdon, Oxon, OX14 4RN

CRC Press is an imprint of Taylor & Francis Group, LLC

ISBN: 978-1-032-75879-4 (hbk)
ISBN: 978-1-032-79750-2 (pbk)
ISBN: 978-1-003-49367-9 (ebk)

DOI: 10.1201/9781003493679

Typeset in Utopia Regular font
by KnowledgeWorks Global Ltd.

Publisher's note: This book has been prepared from camera-ready copy provided by the authors.

Contents

Preface

This book originated from notes compiled over many years from a graduate course I taught at the University of Miami, primarily attended by second- and third-year Ph.D. biostatistics students. The course was designed to cover multivariate statistics from both classical and contemporary perspectives. The first part covers classical topics such as multivariate normality, MANOVA, discrimination, PCA, and canonical correlation analysis. The second part includes contemporary material like gradient boosting, random forests, variable importance, and causal inference.

A recurring theme is to build on classical multivariate statistics to explain difficult concepts and prepare for contemporary material. For example:

- *Linear models* (Chapter 4) are used to understand regularization with AIC and BIC, leading to a deeper treatment of regularization with generalization error and the VC theorem (Chapter 7).
- *Discriminant analysis* (Chapter 6) leads to the weighted Bayes rule, paving the way for contemporary classification techniques for class-imbalanced problems in machine learning (Chapter 16.12).
- *Steepest descent* (Chapter 11) serves as a foundation for matching pursuit and gradient boosting (Chapter 12).
- *Axis-aligned trees* like CART, a classical tool, pave the way for super greedy trees, their modern counterparts (Chapter 15).

Another major theme is training error. Although we are taught in introductory courses that reducing training error leads to overfitting, we also know that training error is another name for empirical risk. In regression, training error is the residual sum of squares, and reducing this quantity gives us least squares. Surely, this cannot be the best way to proceed? However, empirical risk plays a central role in the theory of statistical learning and is fundamental to determining whether coherent learning can take place. The importance of the empirical risk minimization principle is discussed in detail in Chapter 7. This shows that reducing training error is not detrimental as long as regularization is in place. This concept is further explored in discussions on penalization, matching pursuit, gradient boosting, and super greedy tree constructions.

Strategies for teaching

The book was intended for a 13-week, one-semester course, but given the amount of material, it could be split into two separate courses. The first semester would cover classical material, similar to a traditional multivariate course, while the second semester would cover contemporary material, akin to a statistical learning or machine learning course.

For the first semester, focusing on multivariate statistics with some machine learning, the syllabus would be:

- 1.1*, 1.2*
- 2.1, 2.3–2.8, 2.9*, 2.11*, 2.13*, 2.17*

- 3, 5, 6, 8, 9
- 4^*, (7.1–7.3, 7.9, $7.10^\dagger)^*$, 7.8, 7.11, 10^*

Optional material is indicated by * and difficult material by †. Chapters 4 and most of 7, which are optional, cover regularization, overfitting, generalization error, and cross-validation. Covering as much of this as possible in the first semester will facilitate the second semester.

For the second semester, focusing on machine learning, the syllabus would be:

- 11, 12
- $13^{*\dagger}$, 14^* (14.1^*, $14.2^{*\dagger}$, 14.3^*)
- 15.1–15.10, $15.11^\dagger$, 15.12–15.14
- 16, 17.1–17.4, 17.6 (condensed)
- 19–19.5, 19.6^*, 19.7^*, $19.8^{*\dagger}$
- 20.1–20.3, $20.4^{*\dagger}$

Chapter 12, one of the first topics covered, discusses gradient boosting and assumes students understand generalization error, cross-validation, and have been exposed to the lasso (Chapters 4.9, 4.10, and 7). Chapters 19 and 20 cover random survival forests and causal inference. Both are difficult chapters, and it may only be possible to cover one of them.

For instructors who want to cover both sets of topics, material can be selected from each list. The following selections have worked for me:

- 2.1, 2.3–2.8, 2.9^*, 2.11^*, 2.13^*, 2.17^*
- 3, 4, 6
- 7.1–7.3, 7.8, $7.9^\dagger$, 7.10^*, 7.11
- 8, 11, 12
- $13.5^{*\dagger}$ (and condensed necessary material from 13)
- 15.1–15.10, 15.12–15.14
- 16.1–16.8, 16.12 (condensed)
- 17.1–17.4, 17.6 (condensed)
- 20.1–20.3

Exercises

At the end of every chapter, instructors will find a carefully selected set of exercises that vary in degree of difficulty and are both applied and theoretical. For applied problems, I balanced the need to use readily available data, often well-known benchmark datasets available in R packages, against the need for students to work with more challenging real-world data. In the latter cases, I avoided using web links to point to this data since this will be generally unreliable. Instead, I refer to them by their well-known names. An example is the `METABRIC` dataset, which is a breast cancer gene expression profile data used in later parts of the book. A quick online search for this dataset with some keywords should quickly reveal many sources where this data can be downloaded, such as the Kaggle website. Real-world data like this will generally require some effort to download and preprocess correctly. In these cases, the steps needed have been outlined in the exercise. For example, the `METABRIC` dataset contains missing data that can be imputed using the random forest missing data algorithms introduced in Chapter 16. This might make some of these exercises very long and time-consuming for students. Instructors could break these into several smaller exercises or download and preprocess the data themselves to help out.

Prerequisites

Although the intended audience for these notes was Ph.D. biostatistics students, the material can be used for other students as well. However, I recommend prerequisites of advanced mathematical statistics, intermediate probability, and generalized linear models. If students have been exposed to asymptotic theory, more of the book will be accessible to them. Chapter 2 introduces the notation and covers most of the theory needed to understand the entire book. Instructors should consider this to gauge the prerequisites. Some of the material might seem overly theoretical, but it can be easily toned down by cutting back on asymptotic theory. To help gauge the difficulty, advanced chapters have been highlighted with a curved bend icon to indicate difficult material.

Other uses for the book

The book can also serve as a reference for researchers due to the diverse topics covered, including new material on super greedy trees, rule-based variable selection, and machine learning techniques for causal inference. The chapter on trees provides a comprehensive and unified approach to understanding trees in terms of partitions and empirical risk minimization. There is also new content on random forests, including random forest quantile classifiers for class-imbalanced problems, multivariate random forests, subsampling for confidence regions, and super greedy forests. An entire chapter is dedicated to random survival forests, featuring new material on random hazard forests that extend survival forests to time-varying covariates.

Given the detailed chapters on trees and forests, this book could be used for a standalone short course on random forests and trees, covering Chapters 15, 16, 17, 18.1*–18.3*, 19–19.6, and 19.7*–19.9*. Selected material from Chapters 7.6 and 7.7 could also be included. Chapter 17 includes variable selection for random forests and rule-based procedures, making it suitable for a short course focused on variable selection for machine learning.

Author

Motivated by an early passion for mathematics and computer science, Hemant Ishwaran pursued his undergraduate studies in Mathematics and Statistics at the University of Toronto. Inspired by this, he pursued his graduate work in statistics, receiving an M.Sc. in Applied Statistics from Oxford University and a Ph.D. in Statistics from Yale University. Currently, he holds the position of Professor of Biostatistics at the Miller School of Medicine, University of Miami.

Dr. Ishwaran's work focuses on innovating machine learning methods in public health, medicine, and informatics. His contributions include developing open-source software, such as R-packages for methods he has pioneered, including the highly popular random survival forests, an important extension of the random forest method used in machine learning. His collaborations with healthcare professionals have led to precision models for cardiovascular disease (CVD) and heart transplantation, cancer staging, and resistance to gene cancer therapy.

As an Expert Panel Member for the American Joint Committee on Cancer (AJCC), Dr. Ishwaran developed a machine learning procedure for cancer staging. This procedure is now used by the AJCC and published in the AJCC Cancer Staging Manuals, playing a critical role in staging esophageal cancer patients in the U.S.

Dr. Ishwaran also continues his involvement with the International Society for Diseases of the Esophagus (ISDE), contributing to new data platforms aimed at enhancing the understanding of survival for esophageal cancer patients. Furthermore, he has been actively engaged with the American Association for Thoracic Surgery (AATS), where he has developed machine learning algorithms and procedures to assess the quality of cardiothoracic outcomes.

Dr. Ishwaran's hobbies include running, ice hockey, collecting vinyl records, and listening to high-fidelity music.

1

Introduction

1.1 Why Multivariate Methods are Needed

If you think about it, multivariate data is everywhere in real life. This is because the processes we encounter daily are rarely simple and therefore inherently involve many variables. Consider, for example, the weather: it's influenced by temperature, humidity, pressure, wind speed, and many other factors. Similarly, our health is affected by diet, exercise, genetics, environment, and countless other variables.

However, making sense of this data presents significant challenges. Each variable, when analyzed individually, has a compelling story to tell, but this story might differ from what another variable tells us. Multivariate data is also often messy. It can include a mix of continuous variables (like temperature), discrete variables (like the number of rainy days), and ordinal categorical data (like the severity of a storm rated on a scale). Additionally, in many modern settings, it is common for the number of variables to be larger than the sample size. This high-dimensional data requires sophisticated methods to uncover meaningful patterns.

To effectively analyze multivariate data, we have to use tools that synthesize information from all variables together. These multivariate tools, such as principal component analysis, penalization, and multivariate regression, help reveal the underlying structure and relationships between variables. By considering the data holistically, they help provide a more accurate and comprehensive understanding. While it might be tempting to stick with familiar univariate analyses, which seem simpler and easier to interpret, this can obscure the real story and lead us astray.

1.2 Cancer Staging

We illustrate this last point using a real-life example related to cancer staging for the esophagus. Cancer staging assesses the anatomic extent or severity of individual cancers in relation to their survival [4]. The anatomic extent of disease remains the most critical factor in determining the prognosis of cancer patients and serves as the foundation for therapeutic decision-making. Ideally, cancer staging should: (i) categorize cancer characteristics such that patient survival differs between groups (distinctiveness); (ii) consistently decline with increasing stage group (monotonicity); and (iii) exhibit similar survival within groups (homogeneity). Cancer is classified into five stages. Conventionally, Stage 0 is designated for noninvasive cancer, while Stage IV is for cancer that has metastasized to distant sites. This leaves Stages I, II, and III as the primary targets for stage groupings, which ideally should be equally spaced in terms of survival outcomes.

Esophageal cancer staging has traditionally relied solely on the anatomic extent of the disease, using an orderly, progressive grouping of TNM cancer classifications. TNM represents

DOI: 10.1201/9781003493679-1

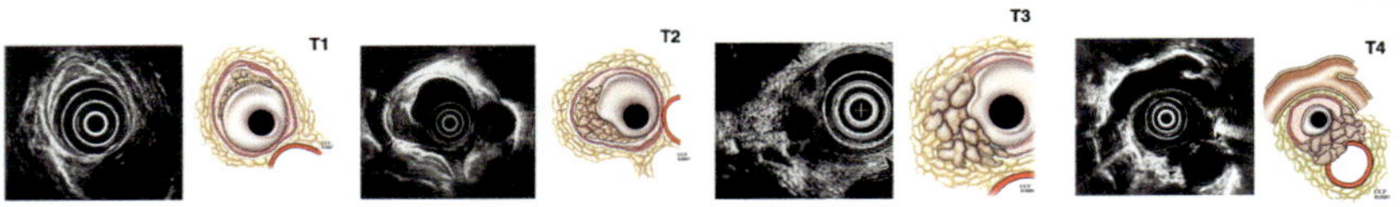

FIGURE 1.1
Esophagus: T1 through T4 as seen by imaging.

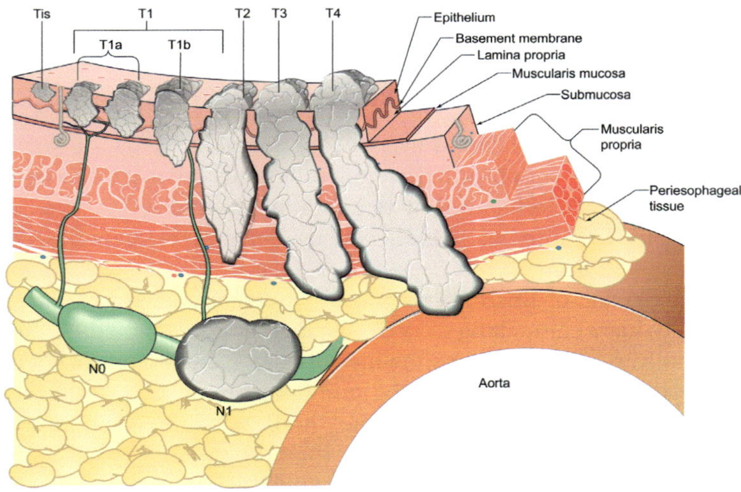

FIGURE 1.2
Anatomic cancer classification is by the depth of cancer invasion (T) and regional lymph node classification (N), defined by the absence (N0) or presence (N1) of cancer-positive lymph nodes. Distant metastasis (M) not illustrated.

three anatomic features of esophageal cancer: the measured depth of tumor (cancer) invasion into the esophageal wall and surrounding tissues (T; see Figures 1.1 and 1.2), the presence of cancer metastases in regional lymph nodes along the esophagus (N), and the presence of cancer metastases at distant sites (M). There are five subclassifications of T (Tis, T1, T2, T3, and T4), two of N (absence N0 or presence N1 of cancer-positive nodes), and two of M (absence M0 or presence M1 of distant metastases) [1].

Table 1.1 describes the previous staging system used for esophageal cancer and reflects the orderly grouping of TNM cancer classifications of the traditional approach. Line 1 describes Stage 0 defined as TisN0M0. This classification describes patients with a minimally invasive tumor (Tis), no nodal involvement (N0), and no metastatic involvement (M0). This stage represents the best scenario for a patient. Line 2 is Stage I, classified as T1N0M0. Here, there is still no metastatic or nodal involvement, but the tumor has grown deeper to T1. This is worse than Tis but still relatively favorable. Following this, cancer stages worsen systematically in terms of T, N, and M. Stages advance from IIA and IIB to III in terms of T and N, while Stage IV (IVA, IVB), reserved for the sickest patients, describes those with metastatic cancer regardless of T or N.

While this approach might seem intuitive, several problems arise. First, using only TNM classifications is simplistic because they reflect only the anatomical extent of cancer. Other important cancer characteristics include the location of the cancer along the esophagus [2],

TABLE 1.1
Previous stage groupings of esophageal cancer [1].

Stage groupings	T	N	M
0	Tis	N0	M0
I	T1	N0	M0
IIA	T2-3	N0	M0
IIB	T1-2	N1	M0
III	T3	N1	M0
	T4	Any N	M0
IV	Any T	Any N	M1
IVA	Any T	Any N	M1a
IVB	Any T	Any N	M1b

histopathologic cell type (squamous cell carcinoma vs. adenocarcinoma) [7, 10], and histologic grade (G), an indicator of biological activity [7]. For N, an increasing number of cancer-positive regional lymph nodes is associated with non-linear decreasing survival [6, 9]. The unique lymphatic anatomy of the esophagus allows cancer to spread to regional lymph nodes with minimal invasion, resulting in an interaction of T with N [5].

These complications challenge the orderly, progressive stage grouping seen in Table 1.1. To improve the prognosis of esophageal cancer patients, a refined data-driven staging system was required. This led to the Worldwide Esophageal Cancer Collaboration (WECC), a global effort to pool data and provide robust evidence for new stage groupings.

1.3 The Dangers of Inference Based on Univariable Analyses

The primary outcome used by WECC was time to all-cause death measured from date of surgery. Figure 1.3 displays survival curves estimated by the Kaplan-Meier method, and these estimates are accompanied by 68% confidence limits, equivalent to ± 1 standard error (Chapter 19 reviews survival analysis and discusses the Kaplan-Meier estimator).

The top plots are Kaplan-Meier survival curves stratified by pT; the bottom figures are stratified by nodal involvement (pN). In both cases, the two criteria of stage grouping (i) distinctiveness and (ii) monotonicity are clearly satisfied, but homogeneity cannot be. Stage grouping defined using only pT or pN separately cannot be homogeneous because they ignore the interaction between pT and pN on survival. The complex nature of esophageal cancer suggests many other variables may also be at play.

Thus, while univariable analyses identify covariates with monotonic survival behavior, they do not capture the simultaneous effects on survival. For a coherent and biologically meaningful analysis, we turn to multivariate methods. Chapter 19 discusses a machine learning method called random survival forests, which was used for multivariable stage grouping for this data [3, 8]. As expected, this analysis revealed a complex interplay between cancer variables and survival.

To give some idea of this complexity, Figure 1.4 shows variable importance values calculated using VarPro, a rule-based variable selection method discussed in Chapter 17.6. Red bars show signal variables (those affecting survival), and blue bars show noisy variables (those not affecting survival). We observe many signal variables. TNM variables are significant, as expected, but patient characteristics like age and bilirubin are also significant. There's also a strong hospital (site) effect. These factors need consideration when forming stage groupings. More details on

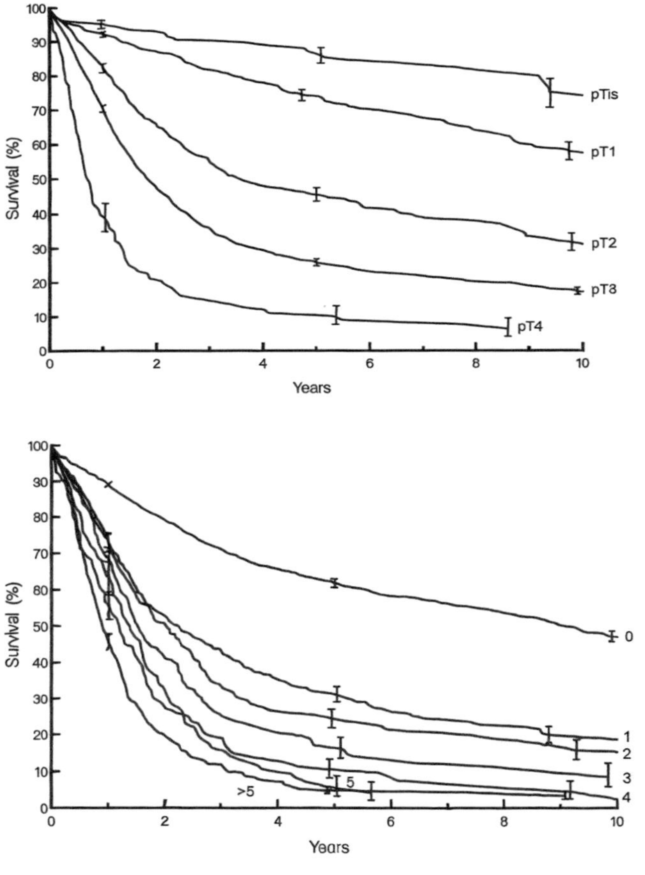

FIGURE 1.3
Kaplan-Meier survival curves for WECC esophageal cancer data stratified by T (top), N (bottom).

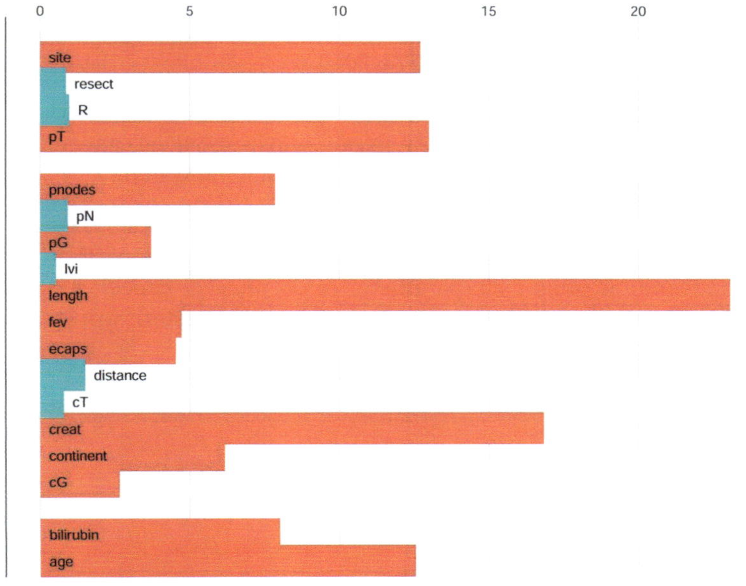

FIGURE 1.4
Variable selection for WECC esophageal cancer data using random survival forests.

incorporating complex survival characteristics into stage groupings can be found in Ishwaran et al. [3] and Rice et al. [8]. In Chapter 19 we will revisit this example when we discuss random survival forests.

Bibliography

[1] American Joint Committee on Cancer. *AJCC Cancer Staging Manual. 6th ed.*, New York: Springer-Verlag, 2002.

[2] Goan Y.G., Chang H.C., Hsu H.K. and Chou Y.P. An audit of surgical outcomes of esophageal squamous cell carcinoma. *Eur. J. Cardiothorac Surg.*, 31:536–544, 2007.

[3] Ishwaran H., Blackstone E.H., Hansen. C.A. and Rice T.W. A novel approach to cancer staging: application to esophageal cancer. *Biostatistics*, 10:603–620, 2009.

[4] National Cancer Institute Fact Sheet. *Cancer Staging* 1-5, 2002. `http://www.cancer.gov/cancertopics/factsheet/detection/staging`.

[5] Rice T.W., Zuccaro G., Jr., Adelstein D.J., Rybicki L.A., Blackstone E.H. and Goldblum J.R. Esophageal carcinoma: depth of tumor invasion is predictive of regional lymph node status. *Ann. Thorac. Surg.*, 65:787–792, 1998.

[6] Rice T.W., Blackstone E.H., Rybicki L.A., Adelstein D.J., Murthy S.C., DeCamp M.M., et al. Refining esophageal cancer staging. *J. Thorac. Cardiovasc. Surg.*, 125:1103–1113, 2003.

[7] Rice T.W., Mason D.P., Murthy S.C., Zuccaro G., Jr., Adelstein D.J., Rybicki L.A., et al. T2N0M0 esophageal cancer. *J. Thorac. Cardiovasc. Surg.*, 133:317–324, 2007.

[8] Rice T.W., Rusch V.W., Ishwaran H., and Blackstone E.H. Cancer of the esophagus and esophagogastric junction: data-driven staging for the 7th edition of the AJCC/UICC cancer staging manuals. *Cancer,* 116(16):3763–3773, 2010.

[9] Rizk N., Venkatraman E., Park B., Flores R., Bains M.S. and Rusch V. The prognostic importance of the number of involved lymph nodes in esophageal cancer: implications for revisions of the American Joint Committee on Cancer staging system. *J. Thorac. Cardiovasc. Surg.*, 132:1374–1381, 2006.

[10] Siewert J.R., Stein H.J., Feith M., Bruecher B.L., Bartels H. and Fink U. Histologic tumor type is an independent prognostic parameter in esophageal cancer: lessons from more than 1,000 consecutive resections at a single center in the western world. *Ann. Surg.*, 234: 360–369, 2001.

2

Properties of Random Vectors and Background Material

This chapter collects together properties of random vectors and provides background material for the book. Now is a good time to scan this material to get a sense of what the book will cover. At the very least, please review the first few sections (Sections 2.1, 2.3, 2.4), which describe the notation used throughout and basic facts about means and covariances of random vectors. Once understood, readers can skip to other chapters but feel free to return when needed.

2.1 Notation

We adopt the standard convention of indicating random variables and vectors by upper case notation and observed values by lower case notation. Thus $\mathbf{X} = (X_1, \ldots, X_p)^T$ is a p-dimensional random vector, whereas $\mathbf{x} = (x_1, \ldots, x_p)^T$ is the observed value of $\mathbf{X}$; i.e., $\mathbf{X} = \mathbf{x}$. Subscripts are used to indicate sampled values. Thus $\mathbf{X}_1, \ldots, \mathbf{X}_n$ is a random sample of n random vectors, while $\mathbf{x}_1, \ldots, \mathbf{x}_n$ are the n observed values. We write $\mathbf{X}_i = (X_{i1}, \ldots, X_{ip})^T$, where X_{ij} is the measurement for the ith individual on the jth coordinate.

In later sections, when discussing variable selection with a focus on a specific variable, the notation changes slightly. We write $\mathbf{X} = (X^{(1)}, \ldots, X^{(p)})^T$, where $X^{(j)}$ is the jth coordinate or variable of $\mathbf{X}$.

2.2 Densities, Integration, Measures

Here we discuss densities, measures, and integration, starting with a brief review of measure theoretic concepts. These concepts allow us to rigorously discuss certain material in this chapter but can be skipped by readers not interested in these technical aspects.

We assume that random values are constructed over a common probability space $(\Omega, \mathscr{A}, \mathbb{P})$. A probability space is a measurable space $(\Omega, \mathscr{A})$ where Ω is a set and $\mathscr{A}$ is a σ-algebra of subsets of Ω. A probability measure $\mathbb{P}$ is a measure over $(\Omega, \mathscr{A})$ such that $\mathbb{P}(\Omega) = 1$ and a measure is a nonnegative and countably additive function on $\mathscr{A}$. A σ-algebra of Ω is a set that contains Ω and is closed under complements and countable unions.

Formally, random variables and random vectors are defined as follows. If $(\Omega, \mathscr{A}, \mathbb{P})$ is a probability space and $(\mathscr{X}, \mathscr{B})$ is a measurable space, then any measurable function ζ from Ω to $\mathscr{X}$ is called a random element. Thus a random element is nothing more than a measurable function $\zeta : \Omega \to \mathscr{X}$. Generally speaking, Ω is taken to be some abstract space and $\mathscr{X}$ is the space

DOI: 10.1201/9781003493679-2

where the actual measurements of ζ take place. When $\mathscr{X} \subseteq \mathbb{R}$ and $\zeta(\omega) = X(\omega) := X$, then X is called a random variable; thus X is real-valued. Otherwise, if $\mathscr{X} \subseteq \mathbb{R}^p$ and $\zeta(\omega) = \mathbf{X}(\omega) := \mathbf{X}$, then $\mathbf{X}$ is called a random vector; thus $\mathbf{X}$ is a real-valued vector. In the latter case the underlying measure space is assumed to be $(\mathbb{R}^p, \mathscr{B}(\mathbb{R}^p))$ where $\mathscr{B}(\mathbb{R}^p)$ are the Borel sets on $\mathbb{R}^p$.

Events of interest for a random vector $\mathbf{X}$ are described in terms of sets $B \in \mathscr{B}$ over $\mathscr{X}$ but the actual probability calculation is implicitly done over the original probability space $(\Omega, \mathscr{A}, \mathbb{P})$. Thus $\mathbb{P}(B)$ for $B \in \mathscr{B}$ is taken to mean

$$\mathbb{P}(B) = \mathbb{P}\{\mathbf{X} \in B\} = \mathbb{P}\{\omega \in \Omega : \mathbf{X}(\omega) \in B\}. \tag{2.1}$$

The measure $\mathbb{P}$ is called the law (or distribution) of $\mathbf{X}$. By (2.1), the law of $\mathbf{X}$ equals the image measure $\mathbb{P} \circ \mathbf{X}^{-1}$ over $\mathscr{B}$. This is called the induced measure for $\mathbf{X}$ where often the distinction between the two spaces $\mathscr{X}$ and Ω is ignored.

Let $(\Omega, \mathscr{A})$ be a measurable space and let λ be a σ-finite measure on it. The measure λ is called σ-finite if Ω is the countable union of measurable sets with finite measure. Notice that a probability measure is always σ-finite as it is a finite measure. An example of a σ-finite measure is Lebesgue measure on the real line. This is because the real line can be described as a countable union of finite intervals and a finite interval has finite Lebesgue measure. Let μ be another σ-finite measure on $(\Omega, \mathscr{A})$. We say that μ is absolutely continuous with respect to λ, written as $\mu < \lambda$, if $\mu(A) = 0$ whenever $\lambda(A) = 0$. The following result shows that μ can always be described in terms of a density taken with respect to λ. Note that a.e. $[\lambda]$ denotes almost everywhere with respect to the measure λ (the term almost all, abbreviated as a.a., is also sometimes used in place of a.e.).

Theorem 2.1. *(Radon-Nikodym theorem) Let λ and μ be σ-finite measures on the measurable space $(\Omega, \mathscr{A})$ such that $\mu < \lambda$. Then there exists a λ-integrable real-valued function $f : \Omega \to \mathbb{R}$ such that*

$$\mu(A) = \int_A f \, d\lambda, \quad \textit{for all } A \in \mathscr{A}.$$

Futhermore, f is unique a.e. $[\lambda]$; i.e. if $\mu(A) = \int_A h \, d\lambda$ for every $A \in \mathscr{A}$ then $h = f$ a.e. $[\lambda]$.

The function f is called the Radon-Nikodym density (or derivative) of μ with respect to λ and is often written as $f = d\mu/d\lambda$. If $f \geq 0$ a.e. $[\lambda]$ and $\int f \, d\lambda = 1$, then μ is a probability measure. In this case, f corresponds to the usual notion of a probability density function. Specifically, the density is the derivative of the measure μ with respect to λ.

The actual choice of the dominating measure often does not matter since we have a type of cancellation that occurs with respect to this value. In the following, notice that using λ or λ^* for the dominating measure yields the same value, since

$$\int_A f \, d\mu = \int_A f (d\mu/d\lambda) d\lambda = \int_A f (d\mu/d\lambda^*) d\lambda^*.$$

In much of this book, we assume that the distribution $\mathbb{P}$ of our random vector $\mathbf{X} \in \mathbb{R}^p$ is absolutely continuous with respect to Lebesgue measure. This assumption will often be implied to avoid excessive technicality and to keep the focus on the important multivariate concepts. When we state that ***X** is absolutely continuous* without specifying the measure, it means that the density $f_{\mathbf{X}}$ for $\mathbf{X}$ is taken with respect to Lebesgue measure on $\mathbb{R}^p$. The probability of a set A is then

$$\mathbb{P}(A) = \int_A f_{\mathbf{X}}(\mathbf{x}) \, d\mathbf{x} = \int_A f_{\mathbf{X}}(x_1, \ldots, x_p) \, dx_1 \cdots dx_p$$

and the density is positive everywhere and integrates to one

$$f_{\mathbf{X}} > 0, \quad \int_{\mathbb{R}^p} f_{\mathbf{X}}(\mathbf{x}) \, d\mathbf{x} = 1.$$

A conditional density can be defined as follows. If $\mathbf{X}$ is partitioned into separate coordinate vectors $\mathbf{X}_1, \mathbf{X}_2$ then the joint density of $\mathbf{X}_1$ given $\mathbf{X}_2 = \mathbf{x}_2$ is

$$f_{\mathbf{X}_1|\mathbf{X}_2}(\mathbf{x}_1|\mathbf{x}_2) = \frac{f_{\mathbf{X}}(\mathbf{x}_1, \mathbf{x}_2)}{f_{\mathbf{X}_2}(\mathbf{x}_2)}$$

where $f_{\mathbf{X}_2}(\mathbf{x}_2) = \int f_{\mathbf{X}}(\mathbf{x}_1, \mathbf{x}_2)\, d\mathbf{x}_1$ is the marginal density of $\mathbf{X}_2$.

The expectation of a function is taken with respect to the underlying distribution and density. The expectation of $h : \mathscr{X} \to \mathbb{R}$ for a random vector $\mathbf{X}$ is

$$\mathbb{E}(h(\mathbf{X})) = \int h(\mathbf{x})\, d\mathbb{P}(\mathbf{x}), \quad \left(\text{also written as} \int h(\mathbf{x})\, \mathbb{P}(d\mathbf{x})\right).$$

This can be written in terms of the density. If $\mathbf{X}$ is absolutely continuous over $\mathbb{R}^p$,

$$\mathbb{E}(\mathbf{h}(\mathbf{X})) = \int_{\mathbb{R}^p} h(\mathbf{x}) f_{\mathbf{X}}(\mathbf{x})\, d\mathbf{x}.$$

2.3 Mean and Covariance

Two important values for summarizing a univariate distribution are the center (mean) and spread (variance). These values have extensions to the multivariate setting. Let $\mathbf{X} = (X_1, \ldots, X_p)^T$ be a p-dimensional absolutely continuous random vector. We write $\boldsymbol{\mu} = (\mu_1, \ldots, \mu_p)^T$ for its mean $\boldsymbol{\mu} = \mathbb{E}(\mathbf{X})$, where

$$\boldsymbol{\mu} = \begin{pmatrix} \mathbb{E}(X_1) \\ \vdots \\ \mathbb{E}(X_p) \end{pmatrix}_{p\times 1}$$

and $\boldsymbol{\Sigma} = (\sigma_{ij})$ for its variance-covariance matrix $\mathrm{Var}(\mathbf{X}) = \mathbb{E}(\mathbf{X} - \boldsymbol{\mu})(\mathbf{X} - \boldsymbol{\mu})^T$, where

$$\boldsymbol{\Sigma} = \begin{pmatrix} \mathbb{E}(X_1 - \mu_1)^2 & \cdots & \mathbb{E}(X_1 - \mu_1)(X_p - \mu_p) \\ \vdots & \ddots & \vdots \\ \mathbb{E}(X_p - \mu_p)(X_1 - \mu_1) & \cdots & \mathbb{E}(X_p - \mu_p)^2 \end{pmatrix}_{p\times p}.$$

Notice that $\boldsymbol{\mu}$ and $\boldsymbol{\Sigma}$ are defined by stacking values in a vector and matrix format where each entry is obtained by calculating an expectation with respect to the underlying multivariate density $f_{\mathbf{X}}$. In the case of $\boldsymbol{\mu}$, coordinate j is the expected value of X_j calculated by the marginal density of X_j

$$\begin{aligned} \mu_j = E(X_j) &= \int_{\mathbb{R}} x_j f_{X_j}(x)\, dx \\ f_{X_j}(x) &= \int_{\mathbb{R}^{p-1}} f_{\mathbf{X}}(x_1, \ldots, x_p)\, dx_1 \cdots dx_{j-1} dx_{j+1} \cdots dx_p. \end{aligned}$$

While for $\boldsymbol{\Sigma}$, entry σ_{ij} is calculated by taking the expectation of $(X_i - \mu_i)(X_j - \mu_j)$ with respect to the joint density for $(X_i, X_j)^T$

$$\begin{aligned} \sigma_{ij} = \mathbb{E}(X_i - \mu_i)(X_j - \mu_j) &= \int_{\mathbb{R}^2} (x_i - \mu_i)(x_j - \mu_j) f_{X_i, X_j}(x_i, x_j)\, dx_i dx_j \\ f_{X_i, X_j}(x_i, x_j) &= \int_{\mathbb{R}^{p-2}} f_{\mathbf{X}}(x_1, \ldots, x_p) \prod_{l \neq i,j} dx_l. \end{aligned}$$

Useful properties of expectation that allow for manipulations of mean vectors and covariance matrices are as follows ($\mathbf{a} \in \mathbb{R}^p$, $b \in \mathbb{R}$, $\mathbf{b} \in \mathbb{R}^q$, and $\mathbf{A}_{q\times p}$ are fixed values):

1. $\mathbb{E}(\mathbf{a}^T\mathbf{X} + b) = \mathbf{a}^T\boldsymbol{\mu} + b$.
2. $\mathbb{E}(\mathbf{AX} + \mathbf{b}) = \mathbf{A}\boldsymbol{\mu} + \mathbf{b}$.
3. $\boldsymbol{\Sigma} = \mathbb{E}(\mathbf{XX}^T) - \boldsymbol{\mu}\boldsymbol{\mu}^T$.
4. $\mathrm{Var}(\mathbf{a}^T\mathbf{X} + b) = \mathrm{Var}(\mathbf{a}^T\mathbf{X}) = \mathbf{a}^T\boldsymbol{\Sigma}\mathbf{a}$.
5. $\mathrm{Var}(\mathbf{AX} + \mathbf{b}) = \mathrm{Var}(\mathbf{AX}) = \mathbf{A}\boldsymbol{\Sigma}\mathbf{A}^T$.
6. $\mathrm{Cov}(\mathbf{X}, \mathbf{Y}) = \mathbb{E}\left[(\mathbf{X} - \boldsymbol{\mu}_X)(\mathbf{Y} - \boldsymbol{\mu}_Y)^T\right] = \mathbb{E}(\mathbf{XY}^T) - \boldsymbol{\mu}_X\boldsymbol{\mu}_Y^T$.
7. $\mathrm{Cov}(\mathbf{X}, \mathbf{X}) = \mathrm{Var}(\mathbf{X})$.
8. $\mathrm{Cov}(\mathbf{X}, \mathbf{Y}) = \mathrm{Cov}(\mathbf{Y}, \mathbf{X})^T$.
9. $\mathrm{Cov}(\sum_i \mathbf{X}_i, \sum_j \mathbf{Y}_j) = \sum_{i,j} \mathrm{Cov}(\mathbf{X}_i, \mathbf{Y}_j)$.
10. $\mathrm{Cov}(\mathbf{AX}, \mathbf{BY}) = \mathbf{A}\mathrm{Cov}(\mathbf{X}, \mathbf{Y})\mathbf{B}^T$.

The multivariate sample estimators for the mean and covariance matrix are extensions of their univariate counterparts. The sample mean is

$$\overline{\mathbf{X}} = \begin{pmatrix} \overline{X}_1 \\ \vdots \\ \overline{X}_p \end{pmatrix}_{p\times 1}$$

where $\overline{X}_j = \sum_{i=1}^n X_{ij}/n$. The sample variance-covariance matrix is

$$\mathbf{S} = \begin{pmatrix} S_{11} & \cdots & S_{1p} \\ \vdots & \ddots & \vdots \\ S_{p1} & \cdots & S_{pp} \end{pmatrix}_{p\times p}$$

where $S_{jk} = \sum_{i=1}^n (X_{ij} - \overline{X}_j)(X_{ik} - \overline{X}_k)/n$. It is sometimes convenient to write $\mathbf{S}$ in terms of $\overline{\mathbf{X}}$. It is easily verified that

$$\mathbf{S} = \frac{1}{n}\sum_{i=1}^n (\mathbf{X}_i - \overline{\mathbf{X}})(\mathbf{X}_i - \overline{\mathbf{X}})^T = \frac{1}{n}\sum_{i=1}^n \mathbf{X}_i\mathbf{X}_i^T - \overline{\mathbf{X}}\overline{\mathbf{X}}^T.$$

Finally, we mention a useful approach involving conditioning for calculating the mean and covariance matrix of a random vector $\mathbf{X}$. This follows from the Tower property. Let $\mathbf{Y}$ be a random vector (not necessarily of the same dimension as $\mathbf{X}$), then

$$\mathbb{E}(\mathbf{X}) = \mathbb{E}_{\mathbf{Y}}\{\mathbb{E}[\mathbf{X}|\mathbf{Y}]\}. \tag{2.2}$$

This states that the mean of $\mathbf{X}$ can be calculated by integrating the conditional mean $\mathbb{E}[\mathbf{X}|\mathbf{Y}]$ with respect to the distribution of $\mathbf{Y}$. Also

$$\mathbf{V}(\mathbf{X}) = \mathbb{E}_{\mathbf{Y}}\{\mathbf{V}[\mathbf{X}|\mathbf{Y}]\} + \mathbf{V}_{\mathbf{Y}}\{\mathbb{E}[\mathbf{X}|\mathbf{Y}]\}. \tag{2.3}$$

Thus, the variance of $\mathbf{X}$ equals the mean of the conditional variance plus the variance of the conditional mean.

2.4 The Data Matrix

It is often convenient to express the data in the form of a matrix referred to as the *data matrix*. For the sample $\mathbf{X}_1,\ldots,\mathbf{X}_n$, let $\mathbf{D}$ be the $n \times p$ data matrix defined by

$$\mathbf{D} = \begin{pmatrix} \mathbf{X}_1^T \\ \vdots \\ \mathbf{X}_n^T \end{pmatrix}_{n\times p} = \left[\mathbf{X}_{(1)},\ldots,\mathbf{X}_{(p)}\right]_{n\times p}$$

where the jth column $\mathbf{X}_{(j)}$ is the n-dimensional vector of the jth coordinate over the sample, $\mathbf{X}_{(j)} = (X_{1j},\ldots,X_{nj})^T$. The $p \times p$ sum-of-squares matrix $\mathbf{D}^T\mathbf{D}$ (also called the Gram matrix of $\mathbf{D}$) can be written as

$$\mathbf{D}^T\mathbf{D} = \sum_{i=1}^{n} \mathbf{X}_i\mathbf{X}_i^T.$$

From this, notice that the sample covariance matrix can be written in terms of the sum-of-squares matrix:

$$\mathbf{S} = \frac{1}{n}\mathbf{D}^T\mathbf{D} - \overline{\mathbf{X}}\,\overline{\mathbf{X}}^T.$$

2.5 Inner Products and Norms

The notion of distance and length plays an important role in multivariate statistics. The inner (or scalar or dot) product between two vectors $\mathbf{x},\mathbf{y} \in \mathbb{R}^p$ is defined by

$$(\mathbf{x},\mathbf{y}) = \mathbf{x}\cdot\mathbf{y} = \mathbf{x}^T\mathbf{y} = \sum_{j=1}^{p} x_j y_j.$$

The norm of $\mathbf{x}$ equals

$$||\mathbf{x}||_2 = \sqrt{\mathbf{x}\cdot\mathbf{x}} = \sqrt{\mathbf{x}^T\mathbf{x}} = \sqrt{\sum_{j=1}^{p} x_j^2}.$$

This is called the L_2-norm or Euclidean distance and is a special case of a norm.

In general, we can define a norm for a vector space as follows. Let $\mathcal{V}$ be a vector space (this means $\mathcal{V}$ is a set closed under finite vector addition and scalar multiplication). A norm $||\cdot||$ is a function from $\mathcal{V}$ to $[0,\infty)$ satisfying the following three conditions:

1. $||x|| = 0$ if and only if $x = 0$ for all $x \in \mathcal{V}$.
2. $||cx|| = |c|\cdot||x||$ for all $c \in \mathbb{R}$ and all $x \in \mathcal{V}$.
3. $||x + y|| \le ||x|| + ||y||$ for all $x, y \in \mathcal{V}$.

This is definition applies to all types of vector spaces. An example is the L_p-space of functions for $1 \le p < \infty$ over $\mathbb{R}$ containing all functions $\mathbb{E}(|f(X)|^p) < \infty$ (as a special case: random variables X satisfying $\mathbb{E}|X|^p < \infty$). The L_p-norm is denoted by $||\cdot||_p$ and is defined by

$$||X||_p = (\mathbb{E}|X|^p)^{1/p}.$$

When $p = 2$ this yields the L_2 space and

$$||X||_2 = (\mathbb{E}|X|^2)^{1/2}.$$

The L_p-space can be formally defined over an arbitrary measurable space for an arbitrary measure μ.

Definition 2.1. Let $(\mathscr{X}, \mathscr{B}, \mu)$ be a measurable space and $0 < p < \infty$. Then $L_p(\mathscr{X}, \mathscr{B}, \mu) := L_p(\mathscr{X}, \mathscr{B}, \mu, \mathbb{R})$ equals the set of all real-valued measurable functions f on $\mathscr{X}$ such that $\int |f| d\mu < \infty$. For $1 \le p < \infty$, define $||f||_p = (\int |f|^p d\mu < \infty)^{1/p}$ to be the L_p-norm of f. □

L_p-spaces are also defined for $p = \infty$. A measurable function f is called essentially bounded if $|f| \le M$ almost everywhere for some $M < \infty$. The L_∞-space is defined to be the set of all essentially real-valued bounded functions with norm defined by $||f||_\infty = \inf\{M : |f| \le M \text{ almost everywhere}\}$.

There is an ordering of L_p-spaces, a type of inclusion property. This applies to finite measure spaces and therefore as a special case to probability measures.

Theorem 2.2. *Let $(\mathscr{X}, \mathscr{B}, \mathbb{P})$ be a probability space. Then for $1 \le q < p \le \infty$, $L_p(\mathscr{X}, \mathscr{B}, \mathbb{P}) \subset L_q(\mathscr{X}, \mathscr{B}, \mathbb{P})$.*

The inner product also applies to general vector spaces (in fact, a norm can always be defined by an inner product, $||x||^2 = (x, x)$). In general, the inner product between two elements $x, y \in \mathscr{V}$ is denoted by (x, y) and satisifies the following three conditions for any $u, v, w \in \mathscr{V}$:

1. $(v, v) = 0$ if and only if $v = 0$.
2. $(v, w) = (w, v)$.
3. $(au + bv, w) = a(u, w) + b(v, w)$ for all $a, b \in \mathbb{R}$.

When $\mathscr{V}$ is combined with an inner product $(\cdot, \cdot)$, $\mathscr{V}$ is called an inner product space. An important aspect of inner product spaces is that this allows for a notion of orthogonality. Two vectors $u, v \in \mathscr{V}$ are said to be orthogonal to one another if and only if $(u, v) = 0$. Orthogonality of two vectors implies that the length squared of their sum equals the sum of their lengths squared (where length is defined by the norm induced by the inner product):

$$||u + v||^2 = ||u||^2 + ||v||^2.$$

As mentioned earlier, an example of an inner product is the dot product on $\mathbb{R}^n$ defined by $(\mathbf{x}, \mathbf{y}) = \mathbf{x}^T\mathbf{y}$, which combined with the vector space on $\mathbb{R}^n$ yields the Euclidean inner product space. Another example is the inner product space of L_2 real valued functions with respect to Lebesgue measure where the inner product for two elements $f, g \in L_2$ is defined by

$$(f, g) = \int f(x)g(x)\, dx.$$

Similarly, if the underlying measure is a probability measure $\mathbb{P}$, then for $f, g \in L_2(\mathscr{X}, \mathscr{B}, \mathbb{P}, \mathbb{R})$,

$$(f, g) = \int f(x)g(x)\, d\mathbb{P}(x).$$

2.6 Linear Subspaces on $\mathbb{R}^n$

The following are some basic results pertaining to linear subspaces on $\mathbb{R}^n$.

1. Vectors $\mathbf{v}_1, \ldots, \mathbf{v}_p \in \mathbb{R}^n$ are said to be linearly independent if $\lambda_1\mathbf{v}_1 + \cdots + \lambda_p\mathbf{v}_p = \mathbf{0}$ if and only if $\lambda_1 = \cdots = \lambda_p = 0$.

2. The span of a set of vectors $S = \{\mathbf{v}_1, \ldots, \mathbf{v}_p\}$ is defined by

$$\text{span}(S) = \text{span}\{\mathbf{v}_1, \ldots, \mathbf{v}_p\} = \left\{ \sum_{k=1}^{p} \lambda_k \mathbf{v}_k : \lambda_1, \ldots, \lambda_p \in \mathbb{R} \right\}.$$

3. Let $\mathscr{H}$ be a linear subspace of $\mathbb{R}^n$. A basis of $\mathscr{H}$ is a maximal linearly independent set of variables.

   (a) Every basis of $\mathscr{H}$ contains the same number of vectors called the dimension of $\mathscr{H}$, $\dim(\mathscr{H})$.

   (b) If $\mathbf{v}_1, \ldots, \mathbf{v}_p$ is a basis of $\mathscr{H}$ then every $\mathbf{z} \in \mathscr{H}$ can be written as a linear combination of the form $\mathbf{z} = \lambda_1\mathbf{v}_1 + \cdots + \lambda_p\mathbf{v}_p$ for some $\lambda_1, \ldots, \lambda_p$.

   (c) We write $\mathscr{H} := \text{span}\{\mathbf{v}_1, \ldots, \mathbf{v}_p\}$.

4. The (column) rank of a matrix $\mathbf{A}_{n\times p}$ is the maximum number of linearly independent columns. If $\mathbf{A} = (\mathbf{v}_1, \ldots, \mathbf{v}_p)$ then

$$\mathscr{H} := \text{column space}(\mathbf{A}) = \text{span}\{\mathbf{v}_1, \ldots, \mathbf{v}_p\}, \quad \dim(\mathscr{H}) = q \le p.$$

For convenience we remind the reader of the definition of a projection.

1. The projection of $\mathbf{x}$ onto $\mathbf{y}$ equals $\dfrac{\mathbf{x}\cdot\mathbf{y}}{\mathbf{y}\cdot\mathbf{y}}\mathbf{y} = \dfrac{\mathbf{x}^T\mathbf{y}}{\mathbf{y}^T\mathbf{y}}\mathbf{y} = \text{proj}_{\mathbf{y}}\mathbf{x}$.

2. The angle θ between two vectors $\mathbf{x}$ and $\mathbf{y}$ satisfies the law of cosines

$$\cos(\theta) = \frac{\mathbf{x}\cdot\mathbf{y}}{||\mathbf{x}||\cdot||\mathbf{y}||}.$$

   If $\mathbf{x}$ and $\mathbf{y}$ are unit vectors (length 1)

$$\cos(\theta) = \mathbf{x}\cdot\mathbf{y}.$$

3. If $\mathbf{x}\cdot\mathbf{y} = 0$ then $\theta = 90$ degrees (hence $\mathbf{x} \perp \mathbf{y}$).

Finally, we recall the definition of an orthonormal basis and some useful facts.

1. $\{\mathbf{v}_1, \ldots, \mathbf{v}_n\}$ is called an orthonormal basis of $\mathbb{R}^n$ if:

   (a) $\mathbf{v}_i^T\mathbf{v}_j = \begin{cases} 1 & \text{if } i = j \\ 0 & \text{otherwise.} \end{cases}$

   (b) $\text{span}\{\mathbf{v}_1, \ldots, \mathbf{v}_n\} = \mathbb{R}^n$.

   If all properties hold but $\mathbf{v}_i^t\mathbf{v}_i \neq 1$ (vector lengths are not constrained to be 1), then $\{\mathbf{v}_1, \ldots, \mathbf{v}_n\}$ is called an orthogonal basis.

2. Given an orthogonal basis $\mathbf{b} = \{\mathbf{v}_1, \ldots, \mathbf{v}_n\}$ of $\mathbb{R}^n$, any vector $\mathbf{z} \in \mathbb{R}^n$ can be rewritten as

$$\mathbf{z} = \text{proj}_{<\mathbf{b}>}\mathbf{z} = \sum_{i=1}^{n} \text{proj}_{\mathbf{v}_i}\mathbf{z} = \sum_{i=1}^{n} \left(\frac{\mathbf{z}^T\mathbf{v}_i}{\mathbf{v}_i^T\mathbf{v}_i}\right)\mathbf{v}_i =^* \sum_{i=1}^{n} (\mathbf{z}^T\mathbf{v}_i)\mathbf{v}_i,$$

   where $(\mathbf{z}^T\mathbf{v}_i)$ is the amount of $\mathbf{z}$ projected onto $\mathbf{v}_i$. Note the last identity $*$ holds if $\mathbf{b}$ is orthonormal.

3. The Gram-Schmidt algorithm for constructing an orthogonal basis $\mathbf{b}_j = \{\mathbf{v}_1, \ldots, \mathbf{v}_j\}$ spanning $\mathcal{H}_j = <\mathbf{x}_1, \ldots, \mathbf{x}_j>$, for $j = 1, \ldots, p$, is:

$$\begin{array}{lllll} \mathbf{v}_1 & = & \mathbf{x}_1 & & \\ \mathbf{v}_2 & = & \mathbf{x}_2 - \text{proj}_{\mathcal{H}_1}\mathbf{x}_2 & = & \mathbf{x}_2 - \text{proj}_{\mathbf{v}_1}\mathbf{x}_2 \\ \mathbf{v}_3 & = & \mathbf{x}_3 - \text{proj}_{\mathcal{H}_2}\mathbf{x}_3 & = & \mathbf{x}_3 - \text{proj}_{\mathbf{v}_1}\mathbf{x}_3 - \text{proj}_{\mathbf{v}_2}\mathbf{x}_3 \\ \vdots & & \vdots & & \vdots \end{array}$$

2.7 Matrix Algebra

1. *Quadratic form.* A quadratic in $\mathbf{x} \in \mathbb{R}^p$ is a function expressible as

$$\begin{aligned} Q(\mathbf{x}) &= \mathbf{x}^T\mathbf{A}\mathbf{x} = \sum_{i=1}^{p}\sum_{j=1}^{p} a_{ij}x_ix_j \\ &= \left(a_{11}x_1^2 + \cdots + a_{pp}x_2^2\right) + \left(2a_{12}x_1x_2 + \cdots + 2a_{p-1,p}x_{p-1}x_p\right) \end{aligned}$$

where $\mathbf{A}_{p\times p}$ is a square symmetric matrix.

2. *Positive definite.* A square symmetric matrix $\mathbf{A}_{p\times p}$ is called positive definite if $Q(\mathbf{x}) = \mathbf{x}^T\mathbf{A}\mathbf{x} > 0$ for all $\mathbf{x} \neq \mathbf{0}$. Write $\mathbf{A} > 0$ in this case. Call $\mathbf{A}$ positive semidefinite if $Q(\mathbf{x}) \geq 0$ for all $\mathbf{x}$ and $Q(\mathbf{x}) = 0$ for some $\mathbf{x} \neq \mathbf{0}$. Denote this by $\mathbf{A} \geq 0$.

   In geometric terms, positive semidefiniteness says that for every $\mathbf{x}$, the angle between $\mathbf{x}$ and $\mathbf{A}\mathbf{x}$ does not exceed $\pi/2$, since

$$(\mathbf{A}\mathbf{x}, \mathbf{x}) = ||\mathbf{A}\mathbf{x}|| \cdot ||\mathbf{x}|| \cdot \cos(\theta),$$

   and thus $\cos(\theta) \geq 0$.

3. *Transpose.* The transpose of a matrix $\mathbf{A} = (a_{jk})_{p\times n}$ is defined by $\mathbf{A}^T = (a_{kj})_{n\times p}$.

   (a) $\left(\mathbf{A}^T\right)^T = \mathbf{A}$.

   (b) $(\mathbf{A} + \mathbf{B})^T = \mathbf{A}^T + \mathbf{B}^T$.

   (c) $(\mathbf{A}\mathbf{B})^T = \mathbf{B}^T\mathbf{A}^T$.

4. *Trace.* The trace function for a matrix $\mathbf{A}$ is defined by $\text{tr}(\mathbf{A}) = \sum_j a_{jj}$. For $\mathbf{A}_{p\times p}, \mathbf{B}_{p\times p}, \mathbf{C}_{p\times n}, \mathbf{D}_{n\times p}$ and $\alpha \in \mathbb{R}$:

   (a) $\text{tr}(\alpha) = \alpha$ for $\alpha \in \mathbb{R}$.

   (b) $\text{tr}(\mathbf{A} \pm \mathbf{B}) = \text{tr}(\mathbf{A}) \pm \text{tr}(\mathbf{B})$.

   (c) If $\mathbf{A}$ and $\mathbf{B}$ are positive semidefinite, $\text{tr}(\mathbf{A}\mathbf{B}) \leq \text{tr}(\mathbf{A})\text{tr}(\mathbf{B})$.

   (d) $\text{tr}(\alpha\mathbf{A}) = \alpha\text{tr}(\mathbf{A})$ for $\alpha \in \mathbb{R}$.

   (e) $\text{tr}(\mathbf{C}\mathbf{D}) = \text{tr}(\mathbf{D}\mathbf{C}) = \sum_{i,j} c_{ij}d_{ji}$.

   (f) $\text{tr}(\mathbf{C}^T\mathbf{C}) = \text{tr}(\mathbf{C}\mathbf{C}^T) = \sum_{i,j} c_{ij}^2$.

   (g) $\sum_{i=1}^n \mathbf{x}_i^T\mathbf{A}\mathbf{x}_i = \text{tr}(\mathbf{A}\mathbf{T})$ where $\mathbf{T} = \sum_{i=1}^n \mathbf{x}_i\mathbf{x}_i^T$.

Illustration: Let $\mathbf{e} \in \mathbb{R}^n$ be a random vector with zero mean and covariance matrix $\sigma^2 \mathbf{I}_n$ and let $\mathbf{D}_{n\times p}$ be the data matrix for the fixed vectors $\mathbf{X}_1, \ldots, \mathbf{X}_n$. Define

$$\hat{\sigma}_n^2 = (\mathbf{e}^T\mathbf{e} - \mathbf{e}^T\mathbf{H}\mathbf{e})/(n-p), \quad \text{where } \mathbf{H} = \mathbf{D}(\mathbf{D}^T\mathbf{D})^{-1}\mathbf{D}^T.$$

Using properties of the trace and our previous discussions regarding properties of expectation, we show that $\hat{\sigma}_n^2$ is an unbiased estimator of σ^2. Observe that

$$\begin{aligned}
\mathbb{E}\left[\mathbf{e}^T\mathbf{e} - \mathbf{e}^T\mathbf{H}\mathbf{e}\right] &= \mathbb{E}\left[\mathrm{tr}(\mathbf{e}^T\mathbf{e})\right] - \mathbb{E}\left[\mathrm{tr}\left(\mathbf{e}^T\mathbf{H}\mathbf{e}\right)\right] \\
&= \mathrm{tr}\left[\mathbb{E}(\mathbf{e}\mathbf{e}^T)\right] - \mathrm{tr}\left(\mathbb{E}\left[\mathbf{H}\mathbf{e}\mathbf{e}^T\right]\right) \\
&= n\sigma^2 - \mathrm{tr}\left(\mathbf{H}\mathbb{E}\left[\mathbf{e}\mathbf{e}^T\right]\right) \\
&= n\sigma^2 - \sigma^2\mathrm{tr}\,(\mathbf{H}) \\
&= n\sigma^2 - p\sigma^2.
\end{aligned}$$

Therefore it follows that $\mathbb{E}(\hat{\sigma}_n^2) = \sigma^2$. Note that the second last line uses the identity $\mathrm{tr}(\mathbf{H}) = \mathrm{tr}(\mathbf{I}) = p$ (another proof is to use $H_{ii} = \mathbf{X}_i(\mathbf{D}^T\mathbf{D})^{-1}\mathbf{X}_i^T$ combined with property (g) where H_{ii} are the so-called leverage points from regression analysis).

5. *Idempotent matrix.* A square matrix $\mathbf{A}_{p\times p}$ is idempotent if $\mathbf{A}^2 = \mathbf{A}$.

6. *Rank.* The rank for a matrix $\mathbf{A}_{n\times p}$, denoted by $\mathrm{rank}(\mathbf{A})$, is defined as the maximum number of linearly independent rows (columns) of $\mathbf{A}$.

   (a) $0 \le \mathrm{rank}(\mathbf{A}) \le \min(n, p)$.

   (b) $\mathrm{rank}(\mathbf{A}) = \mathrm{rank}(\mathbf{A}^T)$.

   (c) $\mathrm{rank}(\mathbf{A}+\mathbf{B}) \le \mathrm{rank}(\mathbf{A}) + \mathrm{rank}(\mathbf{B})$.

   (d) $\mathrm{rank}(\mathbf{AB}) \le \min(\mathrm{rank}(\mathbf{A}), \mathrm{rank}(\mathbf{B}))$.

   (e) $\mathrm{rank}(\mathbf{A}^T\mathbf{A}) = \mathrm{rank}(\mathbf{A}\mathbf{A}^T) = \mathrm{rank}(\mathbf{A})$.

7. *Determinant.* The determinant of a square matrix $\mathbf{A} = (a_{jk})_{p\times p}$, written $|\mathbf{A}|$, is defined as

$$|\mathbf{A}| = \sum_{\pi} (-1)^{|\pi|} a_{1\pi(1)} \cdots a_{p\pi(p)}$$

   where the summation is over all permutations π of $\{1, \ldots, p\}$ and $|\pi| = \pm 1$ depending on whether π can be written as an even $(+1)$ or odd (-1) number of transpositions. Equivalently, let A_{ij} be the (i, j)th cofactor of $\mathbf{A}$ defined as $(-1)^{i+j}$ times the determinant of $\mathbf{A}$ after removing removing row i and column j. Then

$$|\mathbf{A}| = \sum_{i=1}^{p} a_{ij}A_{ij} = \sum_{j=1}^{p} a_{ij}A_{ij}, \quad \text{for any } 1 \le i, j \le p.$$

   The determinant satisfies the following properties:

   (a) If $\mathbf{A}$ is triangular (or diagonal), $|\mathbf{A}| = \prod_j a_{jj}$.

   (b) $|\alpha\mathbf{A}| = \alpha^p|\mathbf{A}|$ for $\alpha \in \mathbb{R}$.

   (c) $|\mathbf{AB}| = |\mathbf{A}||\mathbf{B}|$.

   (d) If $|\mathbf{A}| \ne 0$, then $\mathbf{A}$ is non-singular (i.e., has a unique inverse).

   (e) $|\mathbf{A}+\mathbf{B}| \ne |\mathbf{A}| + |\mathbf{B}|$.

   (f) $|\mathbf{A}^T| = |\mathbf{A}|$.

8. *Non-singularity.* If $\mathbf{A} > 0$, then $\mathbf{A}$ is non-singular, $|\mathbf{A}| > 0$.

9. *Eigenvalues and eigenvectors.* For a square matrix $\mathbf{A}_{p\times p}$ if $q(\lambda) = |\mathbf{A} - \lambda\mathbf{I}|$, then the p roots of $q(\cdot)$ are called the eigenvalues of $\mathbf{A}$.

   (a) If $\lambda_1, \ldots, \lambda_p$ are the eigenvalues of $\mathbf{A}$ then

   $$q(\lambda) = \prod_{j=1}^{p} (\lambda_j - \lambda).$$

   We call $q(\lambda)$ the *characteristic polynomial* of $\mathbf{A}$.

   (b) A vector $\mathbf{v}$ is called an eigenvector for the eigenvalue λ if if $\mathbf{A}\mathbf{v} = \lambda\mathbf{v}$. There can be infinitely many vectors $\mathbf{v}$ satisfying $\mathbf{A}\mathbf{v} = \lambda\mathbf{v}$ and in fact $\{\mathbf{v} \in \mathbb{R}^P : \mathbf{A}\mathbf{v} = \lambda\mathbf{v}\}$ is a subspace of $\mathbb{R}^p$.

   (c) If $\mathbf{v}$ is an eigenvector such that $||\mathbf{v}||^2 = 1$, then $\mathbf{v}$ is called a unit eigenvector.

   (d) $|\mathbf{A}| = \prod_{j=1}^{p} \lambda_j$.

   (e) $\mathrm{tr}(\mathbf{A}) = \sum_{j=1}^{p} \lambda_j$.

   (f) If $\mathbf{A} > 0$, then $\lambda_1, \ldots, \lambda_p > 0$. If $\mathbf{A} \geq 0$, then $\lambda_1, \ldots, \lambda_p \geq 0$.

10. *Inverse.* The inverse of a square matrix $\mathbf{A}_{p\times p}$ with $|\mathbf{A}| \neq 0$ is the unique matrix $\mathbf{A}^{-1}$ satisfying $\mathbf{A}\mathbf{A}^{-1} = \mathbf{A}^{-1}\mathbf{A} = \mathbf{I}$.

    (a) $(\alpha\mathbf{A})^{-1} = \alpha^{-1}\mathbf{A}^{-1}$.

    (b) $(\mathbf{A}\mathbf{B})^{-1} = \mathbf{B}^{-1}\mathbf{A}^{-1}$.

    (c) $(\mathbf{A} + \mathbf{B})^{-1} \neq \mathbf{A}^{-1} + \mathbf{B}^{-1}$.

11. *Invertibility.* If $\mathbf{A} > 0$, then $\mathbf{A}$ is invertible and $\mathbf{A}^{-1} > 0$.

12. *Orthonormal matrix.* A square matrix $\mathbf{A}_{p\times p}$ is orthonormal if $\mathbf{A}^T\mathbf{A} = \mathbf{A}\mathbf{A}^T = \mathbf{I}$.

    (a) $\mathbf{A}^{-1} = \mathbf{A}^T$.

    (b) $|\mathbf{A}| = \pm 1$.

    (c) $\mathbf{A}$ is column orthonormal: $\mathbf{A}^T\mathbf{A} = \mathbf{I}$. Thus if we write $\mathbf{A}$ in "column format", $\mathbf{A} = [\mathbf{a}_{(1)}, \ldots, \mathbf{a}_{(p)}]$, then $\mathbf{a}_{(j)}^T\mathbf{a}_{(k)} = 0$ if $j \neq k$ and $\mathbf{a}_{(j)}^T\mathbf{a}_{(j)} = 1$.

    (d) $\mathbf{A}$ is row orthonormal: $\mathbf{A}\mathbf{A}^T = \mathbf{I}$. Thus if we write $\mathbf{A}$ in "row format",

    $$\mathbf{A} = \begin{pmatrix} \mathbf{a}_1^T \\ \vdots \\ \mathbf{a}_p^T \end{pmatrix}_{p\times p},$$

    then $\mathbf{a}_j^T\mathbf{a}_k = 0$ if $j \neq k$ and $\mathbf{a}_j^T\mathbf{a}_j = 1$.

13. *Equicorrelation matrix.* The covariance matrix $\boldsymbol{\Sigma}_{p\times p}$ is called an equicorrelation matrix if all its diagonal values (variance) are 1 and all its off-diagonal values (correlation) are $-1 < \rho < 1$. Let $\mathbf{1} = (1, \ldots, 1)^T$ be the p-dimensional vector of ones. Then $\boldsymbol{\Sigma} = (1-\rho)\mathbf{I} + \rho\mathbf{1}\mathbf{1}^T$.

    (a) Let $\mathbf{J} = \mathbf{1}\mathbf{1}^T$. Then $\mathbf{J}$ has rank one and eigenvector $\mathbf{1}$ with eigenvalue p. Let $\mathbf{v}_1, \ldots, \mathbf{v}_{p-1}$ be orthogonal vectors chosen to be orthogonal to $\mathbf{1}$. Let $\mathbf{v}_p = \mathbf{1}$. Then $\mathbf{v}_1, \ldots, \mathbf{v}_p$ are eigenvectors for $\mathbf{I}$ and consequently they are eigenvectors for $\boldsymbol{\Sigma}$ with eigenvalues $\lambda_j = (1-\rho)$ for $j = 1, \ldots, p-1$ and $\lambda_p = (1-\rho) + p\rho = 1 + (p-1)\rho$.

(b) $|\boldsymbol{\Sigma}| = \prod_{j=1}^{p} \lambda_j = (1-\rho)^{p-1}\left[1+(p-1)\rho\right]$.

(c) Using $\mathbf{JJ} = p\mathbf{J}$ and $\mathbf{IJ} = \mathbf{J}$ it can be directly verified that

$$\boldsymbol{\Sigma}^{-1} = \frac{1}{1-\rho}\left[\mathbf{I} - \frac{\rho}{1+(p-1)\rho}\mathbf{J}\right], \quad -\frac{1}{p-1} < \rho < 1.$$

(The above can also be proved using the Sherman-Morrison identity given below).

14. *Sherman-Morrison-Woodbury formula.* Let $\mathbf{A}_{n\times n}$ be a square invertible matrix and let $\mathbf{u}, \mathbf{v} \in \mathbb{R}^n$ be any two vectors. Then $\mathbf{A} + \mathbf{uv}^T$ is invertible if and only if $1 + \mathbf{v}^T\mathbf{A}^{-1}\mathbf{u} \neq 0$ in which case

$$\left(\mathbf{A} + \mathbf{uv}^T\right)^{-1} = \mathbf{A}^{-1} - \frac{\mathbf{A}^{-1}\mathbf{uv}^T\mathbf{A}^{-1}}{1 + \mathbf{v}^T\mathbf{A}^{-1}\mathbf{u}}.$$

15. *Woodbury formula.* For any matrices $\mathbf{A}_{p\times p}$, $\mathbf{B}_{p\times m}$, $\mathbf{C}_{m\times m}$, and $\mathbf{D}_{m\times p}$, and assuming the necessary inverses exist,

$$(\mathbf{A} + \mathbf{BCD})^{-1} = \mathbf{A}^{-1} - \mathbf{A}^{-1}\mathbf{B}(\mathbf{C}^{-1} + \mathbf{DA}^{-1}\mathbf{B})^{-1}\mathbf{DA}^{-1}.$$

16. *Frobenius norm.* The Frobenius norm $||\cdot||_F$ for a matrix $\mathbf{A} = (a_{ij})_{m\times n}$ is defined by

$$||\mathbf{A}||_F = \sqrt{\sum_{i,j} a_{ij}^2}.$$

(a) $||\mathbf{A}||_F^2 = \text{tr}(\mathbf{A}^T\mathbf{A}) = \text{tr}(\mathbf{AA}^T) = \sum_{k=1}^{r} \lambda_k$ where $\{\lambda_k\}_1^r$ are the non-zero eigenvalues of $\mathbf{A}^T\mathbf{A}$ and $r = \text{rank}(\mathbf{A}) \leq \min(m, n)$.

(b) $||\mathbf{A}||_F \geq 0$ and $||\mathbf{A}||_F = 0$ if and only if $\mathbf{A} = \mathbf{0}$.

(c) $||\alpha\mathbf{A}||_F = |\alpha| \cdot ||\mathbf{A}||_F$ for all $\alpha \in \mathbb{R}$.

(d) $||\mathbf{A} + \mathbf{B}||_F \leq ||\mathbf{A}||_F + ||\mathbf{B}||_F$ for $m \times n$ matrices $\mathbf{A}, \mathbf{B}$.

(e) $||\mathbf{AB}||_F \leq ||\mathbf{A}||_F \cdot ||\mathbf{B}||_F$ for $m \times n$ matrices $\mathbf{A}, \mathbf{B}$.

17. *Induced matrix norm.* An induced matrix norm can be defined by any vector norm $||\mathbf{x}||$ where $\mathbf{x} \in \mathbb{R}^n$ by $||\mathbf{A}|| = \sup_{||\mathbf{x}||=1}\{||\mathbf{Ax}||\}$. In particular, properties (b)-(e) from above hold for this norm.

2.8 Inversion, Generalized Inverse

Often in linear algebra, we need to solve for $\mathbf{x}$ in a system of linear equations:

$$\mathbf{Ax} = \mathbf{z} \tag{2.4}$$

where $\mathbf{A}_{m\times p}$ and $\mathbf{z}_{m\times 1}$ are known quantities. A classical example is the normal equations in linear regression. In least squares, we observe a vector of responses $\mathbf{y}$ such that $\mathbb{E}(\mathbf{Y}|\mathbf{D}) = \mathbf{D}\boldsymbol{\beta}$, where $\boldsymbol{\beta}$ is the unknown regression vector and $\mathbf{D}$ is a fixed $n \times p$ design matrix. The goal is to find the vector $\boldsymbol{\beta} \in \mathbb{R}^p$ that minimizes

$$\|\mathbf{y} - \mathbf{D}\boldsymbol{\beta}\|^2.$$

The solution to this is called the Ordinary Least Squares (OLS) estimator, which corresponds to the $\boldsymbol{\beta}$ that solves the normal equations:

$$\mathbf{D}^T\mathbf{D}\boldsymbol{\beta} = \mathbf{D}^T\mathbf{y}, \tag{2.5}$$

which is of the form (2.4) with $\mathbf{A} = \mathbf{D}^T\mathbf{D}$ and $\mathbf{z} = \mathbf{D}^T\mathbf{y}$. Solving for $\boldsymbol{\beta}$ by inverting $\mathbf{D}^T\mathbf{D}$ is tempting but should be avoided. Instead, we use numerically stable procedures like the QR decomposition.

Theorem 2.3. *For each matrix $\mathbf{A}_{m\times p}$ there exists an orthonormal matrix $\mathbf{Q}_{m\times m}$ ($\mathbf{Q}^T\mathbf{Q} = \mathbf{I}$) and an upper triangular matrix $\mathbf{R}_{m\times p}$ with nonnegative diagonal elements such that $\mathbf{A} = \mathbf{QR}$.*

To solve for $\boldsymbol{\beta}$ in (2.5), let $\mathbf{A} = \mathbf{D}^T\mathbf{D}$ and $\mathbf{z} = \mathbf{D}^T\mathbf{y}$. Obtain the QR decomposition $\mathbf{A} = \mathbf{QR}$ and multiply throughout by $\mathbf{Q}^T$ to obtain

$$\mathbf{Q}^T\mathbf{A}\boldsymbol{\beta} = \mathbf{Q}^T(\mathbf{QR})\boldsymbol{\beta} = \mathbf{R}\boldsymbol{\beta} = \mathbf{Q}^T\mathbf{z}.$$

The final system $\mathbf{R}\boldsymbol{\beta} = \mathbf{Q}^T\mathbf{z}$ involves an upper triangular matrix and can be quickly solved using back substitution. See Exercise 2.10.

There is no unique OLS solution to the normal equations (2.5) when $p > n$. In this case, we use the minimum least squares (MLS) estimator [6]. This estimator depends on the Moore-Penrose generalized inverse of a matrix, which has some remarkable properties.

Definition 2.2. The generalized inverse of a matrix $\mathbf{A}_{n\times p}$ is the unique matrix $\mathbf{A}^+_{p\times n}$ that satisfies the following conditions:

$$\mathbf{A}\mathbf{A}^+\mathbf{A} = \mathbf{A}, \quad \mathbf{A}^+\mathbf{A}\mathbf{A}^+ = \mathbf{A}^+, \quad (\mathbf{A}\mathbf{A}^+)^T = \mathbf{A}\mathbf{A}^+.$$

If $\mathbf{A}$ is non-singular than $\mathbf{A}^+ = \mathbf{A}^{-1}$. □

The generalized inverse is a very useful solution and has the so-called best approximation property. To state this property, recall that the Frobenius norm $||\mathbf{A}||_F$ for $\mathbf{A}$ is defined by $||\mathbf{A}||_F^2 = \sum_{i,j} a_{ij}^2$. Consider the following definition.

Definition 2.3. We say that $\mathbf{B}^*$ is the best approximate solution to $f(\mathbf{B}) = \mathbf{C}$, if for every $\mathbf{B} \neq \mathbf{B}^*$ one of the following holds:

(i) $||f(\mathbf{B}) - \mathbf{C}||_F^2 > ||f(\mathbf{B}^*) - \mathbf{C}||_F^2$.

(ii) $||f(\mathbf{B}) - \mathbf{C}||_F^2 = ||f(\mathbf{B}^*) - \mathbf{C}||_F^2$ and $||\mathbf{B}||_F^2 > ||\mathbf{B}^*||_F^2$.

□

The following result states the best approximation property for the generalized inverse.

Theorem 2.4. *$\mathbf{B}^* = \mathbf{A}^+\mathbf{C}$ is the unique best approximation for $\mathbf{AB} = \mathbf{C}$.*

This property leads to many useful results, including the following corollary for the best approximation to the system of linear equations (2.4).

Corollary 2.5. *$\mathbf{x}^* = \mathbf{A}^+\mathbf{z}$ is the unique best approximation to the equation $\mathbf{Ax} = \mathbf{z}$ for any $\mathbf{x} \in \mathbb{R}^p$.*

The minimum least squares estimator follow as a special case by setting $\mathbf{A} = \mathbf{D}$ and $\mathbf{z} = \mathbf{y}$.

Corollary 2.6. *The minimum least squares estimator $\boldsymbol{\beta}^* = \mathbf{D}^+\mathbf{y}$ is the unique best approximation to $\mathbf{D}\boldsymbol{\beta} = \mathbf{y}$.*

Another corollary is that $\mathbf{A}^+$ is the best approximate solution to inverting $\mathbf{A}$.

Corollary 2.7. *The best approximate solution to $\mathbf{AB} = \mathbf{I}$ is $\mathbf{B} = \mathbf{A}^+$.*

2.9 Differentiability in More than One Dimension

A function $\mathbf{f}: \mathbb{R}^n \to \mathbb{R}^m$ is called differentiable at $\mathbf{x}_0 \in \mathbb{R}^n$ if

$$\mathbf{f}(\mathbf{x}) = \mathbf{f}(\mathbf{x}_0) + \mathbf{J}(\mathbf{x} - \mathbf{x}_0) + \mathbf{r}(\mathbf{x}, \mathbf{x}_0), \tag{2.6}$$

where the matrix $\mathbf{J}_{m \times n}$ depends on $\mathbf{x}_0$ but not $\mathbf{x}$ and the remainder term $\mathbf{r}: \mathbb{R}^n \to \mathbb{R}^m$ satisfies

$$\frac{\mathbf{r}(\mathbf{x}, \mathbf{x}_0)}{||\mathbf{x} - \mathbf{x}_0||} \to \mathbf{0}, \quad \text{as } \mathbf{x} \to \mathbf{x}_0.$$

Because $\mathbf{J}(\mathbf{x} - \mathbf{x}_0)$ is a $m \times 1$ vector, what this says is that

$$\mathbf{f}(\mathbf{x}) = \mathbf{f}(\mathbf{x}_0) + \{\text{linear function of } \mathbf{x} \text{ into } \mathbb{R}^m\} + o(||\mathbf{x} - \mathbf{x}_0||).$$

The last term, which is the remainder term, is called "little-o," which is a type of calculus for recording the rate of decrease of a sequence. This concept will be reviewed in Chapter 2.10.

To find conditions ensuring differentiability, it is helpful to study a property implied by it. The derivative of f in the direction $\mathbf{u}$ is defined by

$$D_{\mathbf{u}}\mathbf{f}(\mathbf{x}) = \lim_{\delta \to 0} \frac{\mathbf{f}(\mathbf{x} + \delta\mathbf{u}) - \mathbf{f}(\mathbf{x})}{\delta ||\mathbf{u}||}.$$

The directional derivative is related to the Jacobian, since by the definition of differentiability

$$D_{\mathbf{u}}\mathbf{f}(\mathbf{x}) = \lim_{\delta \to 0} \left\{ \frac{\mathbf{J}\delta\mathbf{u} + o(||\delta\mathbf{u}||)}{\delta ||\mathbf{u}||} \right\} = \mathbf{J}\frac{\mathbf{u}}{||\mathbf{u}||}.$$

In particular, if we set $\mathbf{u} = \mathbf{e}_j$ where $\mathbf{e}_j$ is the jth Euclidean basis function (equal to one in coordinate j and zero elsewhere),

$$\begin{aligned}
\mathbf{f}_j(\mathbf{x}) &= D_{\mathbf{e}_j}\mathbf{f}(\mathbf{x}) \\
&= \lim_{\delta \to 0} \frac{\mathbf{f}(x_1, \ldots, x_{j-1}, x_j + \delta, x_{j+1}, \ldots, x_n) - \mathbf{f}(\mathbf{x})}{\delta} \\
&= \frac{\partial \mathbf{f}}{\partial x_j} \\
&= j\text{th column of } \mathbf{J} \quad \left(\text{i.e. } \mathbf{J}\frac{\mathbf{e}_j}{||\mathbf{e}_j||} \right).
\end{aligned}$$

Setting $\mathbf{f}(\mathbf{x}) = (f^{(1)}(\mathbf{x}), \ldots, f^{(m)}(\mathbf{x}))^T$, we find

$$\frac{\partial \mathbf{f}}{\partial x_j} = \left(\frac{\partial f^{(1)}}{\partial x_j}, \ldots, \frac{\partial f^{(m)}}{\partial x_j} \right)^T := \left(f_j^{(1)}, \ldots, f_j^{(m)} \right)^T.$$

Therefore, the matrix $\mathbf{J} = \mathbf{J}(\mathbf{f})$ appearing in (2.6), which we can call the Jacobian matrix of $\mathbf{f}$, is given by

$$\mathbf{J}(\mathbf{f}) = \begin{pmatrix} f_1^{(1)} & \cdots & f_n^{(1)} \\ \vdots & \ddots & \vdots \\ f_1^{(m)} & \cdots & f_n^{(m)} \end{pmatrix}.$$

The most common case of differentiability occurs when $f(\mathbf{x}) = f(x_1, \ldots, x_n)$ is real valued, $f: \mathbb{R}^n \to \mathbb{R}$. Then

$$\mathbf{D}_{\mathbf{u}} f = (\nabla f)^T \frac{\mathbf{u}}{||\mathbf{u}||}$$

where

$$\nabla f = \left(\frac{\partial f}{\partial x_1}, \dots, \frac{\partial f}{\partial x_n}\right)^T := (f_1, \dots, f_n)^T$$

is called the *gradient* for f. Differentiability for f then means that

$$\begin{aligned} f(\mathbf{x}+\mathbf{h}) - f(\mathbf{x}) &= \mathbf{h}^T \nabla f(\mathbf{x}) + o(||\mathbf{h}|) \\ &= h_1 \frac{\partial f}{\partial x_1} + \cdots + h_n \frac{\partial f}{\partial x_n} + o(||\mathbf{h}||). \end{aligned}$$

Theorem 2.8. *Let* $\boldsymbol{f} = (f^{(1)}(\boldsymbol{x}), \dots, f^{(m)}(\boldsymbol{x}))^T : \mathbb{R}^n \to \mathbb{R}^m$ *be defined in a ball around* $\boldsymbol{x}_0$. *Then* $\boldsymbol{f}$ *is differentiable at* $\boldsymbol{x}_0$ *if and only if* $f^{(1)}, \dots, f^{(m)}$ *are. A sufficient condition for* $\boldsymbol{f}$ *to be differentiable at* $\boldsymbol{x}_0$ *is that all partial derivatives* $\{f_j^{(i)}\}$ *are continuous at* $\boldsymbol{x}_0$.

Theorem 2.9. *If* $\mathbf{g}: \mathbb{R}^p \to \mathbb{R}^n$ *is differentiable at* $\boldsymbol{x}_0$ *and* $\boldsymbol{f}: \mathbb{R}^n \to \mathbb{R}^m$ *is differentiable at* $\mathbf{g}(\boldsymbol{x}_0)$, *then* $\boldsymbol{f} \circ \mathbf{g}$ *is differentiable at* $\boldsymbol{x}_0$ *and*

$$(\boldsymbol{f} \circ \mathbf{g})_j(\boldsymbol{x}_0) = \frac{\partial \boldsymbol{f} \circ \mathbf{g}(\boldsymbol{x}_0)}{\partial x_j} = \sum_{i=1}^{n} \frac{\partial \boldsymbol{f}(\mathbf{g}(\boldsymbol{x}_0))}{\partial x_i} \mathbf{g}_j^{(i)}(\boldsymbol{x}_0), \quad j = 1, \dots, p.$$

In other words, $\boldsymbol{J}(\boldsymbol{f} \circ \mathbf{g}) = \boldsymbol{J}(\boldsymbol{f})\boldsymbol{J}(\mathbf{g})$.

Higher order partial derivatives of a real-valued function $f : \mathbb{R}^n \to \mathbb{R}$ are defined as follows. The second-order partial derivative is by definition

$$f_{ij}(\mathbf{x}) = \frac{\partial^2 f(\mathbf{x})}{\partial x_i \partial x_j} = \lim_{\delta \to 0} \frac{f_i(\mathbf{x} + \delta \mathbf{e}_j) - f_i(\mathbf{x})}{\delta},$$

where $f_i = \partial f / \partial x_i$. If the partial derivatives are continuous then the second-order partial derivatives are the same in each direction. Thus if f_{ij} is continuous, then f_{ji} exists and $f_{ij} = f_{ji}$.

Third and higher-order derivatives are defined analogously:

$$f_{i_1,\dots,i_m}(\mathbf{x}) = \frac{\partial^m f(\mathbf{x})}{\partial x_{i_1} \cdots \partial x_{i_m}} = \frac{\partial f_{i_1,\dots,i_{m-1}}(\mathbf{x})}{\partial x_{i_m}}.$$

2.9.1 Taylor's series

Making use of the above definitions, we state a multivariable Taylor's theorem.

Theorem 2.10. ***(Taylor's theorem in*** $\mathbb{R}^n$***)*** *Let* $f : \mathbb{R}^n \to \mathbb{R}$ *be a real-valued function such that its partial derivatives of order* $1, \dots, m$ *exist and are continuous. For each* $\boldsymbol{x} \in \mathbb{R}^n$ *there exists a* $\theta := \theta(\boldsymbol{x})$ *where* $0 < \theta < 1$ *such that*

$$\begin{aligned} f(\boldsymbol{x}+\boldsymbol{h}) &= f(\boldsymbol{x}) + \sum_{i=1}^{n} h_i f_i(\boldsymbol{x}) + \frac{1}{2!} \sum_{i_1=1}^{n} \sum_{i_2=1}^{n} h_{i_1} h_{i_2} f_{i_1 i_2}(\boldsymbol{x}) + \cdots \\ &\quad + \frac{1}{(m-1)!} \sum_{i_1=1}^{n} \sum_{i_2=1}^{n} \cdots \sum_{i_{m-1}=1}^{n} h_{i_1} h_{i_2} \cdots h_{i_{m-1}} f_{i_1 i_2 \cdots i_{m-1}}(\boldsymbol{x}) \\ &\quad + \frac{1}{m!} \sum_{i_1=1}^{n} \sum_{i_2=1}^{n} \dots \sum_{i_m=1}^{n} h_{i_1} h_{i_2} \cdots h_{i_m} f_{i_1 i_2 \dots i_m}(\boldsymbol{x} + \theta \boldsymbol{h}). \end{aligned}$$

Theorem 2.11. ***(Taylor's theorem in*** $\mathbb{R}^n$ ***with little-o term)*** *Let* $f : \mathbb{R}^n \to \mathbb{R}$ *be a real-valued function such that its first and second-order partial derivatives exist. Then the Taylor's series with little-o remainder term is*

$$f(\boldsymbol{x}+\boldsymbol{h}) = f(\boldsymbol{x}) + \boldsymbol{h}^T \nabla f(\boldsymbol{x}) + \frac{1}{2} \boldsymbol{h}^T \boldsymbol{H}(\boldsymbol{x}) \boldsymbol{h} + o(||\boldsymbol{h}||^2),$$

where $\boldsymbol{H} = \{f_{ij}\}$ is an $n \times n$ matrix of second-order derivatives called the Hessian matrix. Observe that the expansion requires f to be differentiable to the second order. However, the little-o term can be matched to the order of differentiability m by replacing the last term in Theorem 2.10 with a remainder term of order $o(||\boldsymbol{h}||^m)$. For example, in the case that f is only differentiable to order one, the expansion becomes

$$f(\boldsymbol{x}+\boldsymbol{h}) = f(\boldsymbol{x}) + \boldsymbol{h}^T \nabla f(\boldsymbol{x}) + o(||\boldsymbol{h}||).$$

2.9.2 Derivatives involving vectors and matrices

We end this section by listing some useful derivative identities involving vectors and matrices.

1. $\dfrac{\partial \mathbf{a}^T\mathbf{z}}{\partial \mathbf{z}} = \mathbf{a}$.
2. $\dfrac{\partial \mathbf{z}^T\mathbf{z}}{\partial \mathbf{z}} = 2\mathbf{z}$, $\dfrac{\partial \mathbf{z}^T\mathbf{A}\mathbf{z}}{\partial \mathbf{z}} = 2\mathbf{A}\mathbf{z}$ (if $\mathbf{A}$ is symmetric), $\dfrac{\partial \mathbf{z}^T\mathbf{A}\mathbf{y}}{\partial \mathbf{z}} = \mathbf{A}\mathbf{y}$.
3. $\dfrac{\partial |\mathbf{A}|}{\partial a_{ij}} = A_{ij}$ if all elements of $\mathbf{A}_{p\times p}$ are distinct, where A_{ij} is the (i, j)th cofactor of $\mathbf{A}$ defined as $(-1)^{i+j}$ times the determinant of $\mathbf{A}$ after removing removing row i and column j.
4. $\dfrac{\partial \mathrm{tr}(\mathbf{AB})}{\partial \mathbf{A}} = \mathbf{B}^T$ if all elements of $\mathbf{A}$ are distinct.
5. $\dfrac{\partial \mathbf{A}^{-1}}{\partial a_{ij}} = -\mathbf{A}^{-1}\mathbf{A}_{ij}\mathbf{A}^{-1}$ where $\mathbf{A}_{ij}$ denotes the matrix with one in entry (i, j) and zeroes elsewhere.

2.9.3 Global minimum for least squares

We use the least squares problem to connect several of the concepts introduced earlier. Recall that least squares is the optimization problem where the goal is to minimize $f(\boldsymbol{\beta}) = \|\mathbf{y} - \mathbf{D}\boldsymbol{\beta}\|^2$, where $\mathbf{D}$ is an $n \times p$ design matrix, $\mathbf{y} \in \mathbb{R}^n$ is the observed response, and $\boldsymbol{\beta}$ is the regression vector.

We can find the solution by solving for the roots of the gradient (we will show why this is a valid approach below). Thus we want to find $\boldsymbol{\beta}$ such that $\nabla f(\boldsymbol{\beta}) = \mathbf{0}$. Applying our rules for derivatives, we have

$$\begin{aligned}
\frac{\partial f(\boldsymbol{\beta})}{\partial \boldsymbol{\beta}} &= \frac{\partial}{\partial \boldsymbol{\beta}} ||\mathbf{y} - \mathbf{D}\boldsymbol{\beta}||_2^2 \\
&= \frac{\partial}{\partial \boldsymbol{\beta}} \left\{ (\mathbf{y} - \mathbf{D}\boldsymbol{\beta})^T (\mathbf{y} - \mathbf{D}\boldsymbol{\beta}) \right\} \\
&= \frac{\partial}{\partial \boldsymbol{\beta}} \left\{ \mathbf{y}^T\mathbf{y} - 2\boldsymbol{\beta}^T\mathbf{D}^T\mathbf{y} + \boldsymbol{\beta}^T\mathbf{D}^T\mathbf{D}\boldsymbol{\beta} \right\} \\
&= \frac{\partial}{\partial \boldsymbol{\beta}} \mathbf{y}^T\mathbf{y} - 2\frac{\partial}{\partial \boldsymbol{\beta}} \boldsymbol{\beta}^T\mathbf{D}^T\mathbf{y} + \frac{\partial}{\partial \boldsymbol{\beta}} \boldsymbol{\beta}^T\mathbf{D}^T\mathbf{D}\boldsymbol{\beta} \\
&= -2\mathbf{D}^T\mathbf{y} + 2\mathbf{D}^T\mathbf{D}\boldsymbol{\beta}.
\end{aligned}$$

Solving $\nabla f(\boldsymbol{\beta}) = \mathbf{0}$, we obtain $\mathbf{D}^T\mathbf{D}\boldsymbol{\beta} = \mathbf{D}^T\mathbf{y}$ (c.f. (2.5)) which are the normal equations and yield the solution $\hat{\boldsymbol{\beta}} = (\mathbf{D}^T\mathbf{D})^{-1}\mathbf{D}^T\mathbf{y}$ if $\mathbf{D}$ has full rank.

Now to show that this is the global minimum, apply a Taylor's expansion using Theorem 2.11 to see that

$$\begin{aligned}
f(\boldsymbol{\beta}) &= f(\hat{\boldsymbol{\beta}}) + (\boldsymbol{\beta} - \hat{\boldsymbol{\beta}})^T \nabla f(\hat{\boldsymbol{\beta}}) + (\hat{\boldsymbol{\beta}} - \boldsymbol{\beta})^T \mathbf{H}(\hat{\boldsymbol{\beta}})(\hat{\boldsymbol{\beta}} - \boldsymbol{\beta}) + o(||\hat{\boldsymbol{\beta}} - \boldsymbol{\beta}||^2) \\
&= f(\hat{\boldsymbol{\beta}}) + (\hat{\boldsymbol{\beta}} - \boldsymbol{\beta})^T \mathbf{H}(\hat{\boldsymbol{\beta}})(\hat{\boldsymbol{\beta}} - \boldsymbol{\beta}) + o(||\hat{\boldsymbol{\beta}} - \boldsymbol{\beta}||^2).
\end{aligned}$$

The Hessian equals $\mathbf{H}(\hat{\boldsymbol{\beta}}) = 2\mathbf{D}^T\mathbf{D}$ which is positive definite if $\mathbf{D}$ is full rank. In this case f is convex and $\hat{\boldsymbol{\beta}}$ is the unique minimum.

One technical point we want to emphasize is the remainder term in the above expression. Although we used little-o notation for this term, in the context of least squares, the estimator $\hat{\boldsymbol{\beta}}$ is random due to measurement error in the linear model. Therefore, this term does not converge deterministically as indicated by little-o notation but is actually stochastic, and the convergence is called "little-op," indicated by o_p. This and other topics related to stochastic convergence will be covered in Chapter 2.11.

2.10 Convergence for Real-valued Sequences

This section reviews convergence for real-valued sequences. Throughout we use the notation $b_n \gg a_n$ to mean $b_n / a_n \to \infty$. For simplicity results are mostly stated for real values but they have obvious extensions to vectors.

First, we recall the Cauchy-convergence for a sequence.

Theorem 2.12. ***(Cauchy convergence criterion)*** *The limit of the sequence* $\{a_n\}_1^\infty$ *exists and is finite if and only if for every* $\varepsilon > 0$ *there is a positive integer* $N_0 := N_0(\varepsilon)$ *such that* $|x_m - x_n| < \varepsilon$ *for all* $m, n > N_0$.

Sequences converge under fairly simple conditions. One of the simplest conditions is that the sequence should be bounded and monotone.

Theorem 2.13. ***(Monotone convergence)*** *If* $a_n \le a_{n+1}$ *for all* n, *then* $a_n \to \infty$ *or* $a_n \le M$ *for all* n *for some* $M < \infty$. *In the latter case,* $\lim_{n\to\infty} a_n = \sup\{a_n, n \ge 1\}$.

A consequence of the monotone convergence theorem is the ratio test.

Theorem 2.14. ***(Ratio test)*** *Suppose that* $|a_{n+1}| \le L|a_n|$ *for sufficiently large* n. *Then* $\{a_n\}_1^\infty$ *converges if* $L < 1$. *In particular if*

$$L = \lim_{n\to\infty} \frac{|a_{n+1}|}{|a_n|}$$

then $\{a_n\}_1^\infty$ *converges if* $L < 1$ *and diverges if* $L > 1$.

2.10.1 Little-o and big-O notation for real-valued sequences

Given that a sequence converges, it is useful to summarize the rate at which it converges. This is done using "little-o" and "big-O" notation. We will deal only with the real-valued case, but the extension to vector-valued sequences is straightforward by using the Euclidean distance $||\cdot||_2$ as a norm.

Definition 2.4. For a sequence $\{a_n\}_1^\infty$ and a nonnegative sequence $\{b_n\}_1^\infty$, we say $a_n = o(b_n)$ if $|a_n|/b_n \to 0$ as $n \to \infty$. Thus, $b_n \gg a_n$. □

Definition 2.5. In particular, $a_n = o(1)$ if $a_n \to 0$ as $n \to \infty$. □

Definition 2.6. For a sequence $\{a_n\}_1^\infty$ and a nonnegative sequence $\{b_n\}_1^\infty$, we say $a_n = O(b_n)$ if $|a_n|/b_n$ is bounded, i.e., if there exists a finite constant M and a finite integer n_0 such that $|a_n| \le M b_n$ for all $n \ge n_0$. □

Definition 2.7. In particular, $a_n = O(1)$ if $|a_n|$ is bounded, i.e., if there exists a finite constant M and a finite integer n_0 such that $|a_n| \le M$ for all $n \ge n_0$. □

2.11 Convergence of Random Vectors

This section reviews various modes of convergence and formally describes stochastic order notation. As before, we assume that random variables and random vectors are constructed over a common probability space $(\Omega, \mathscr{A}, \mathbb{P})$.

2.11.1 Different modes of stochastic convergence

Convergence in distribution. A sequence of random variables $\{X_n\}_1^\infty$ is said to converge in distribution to a random variable X, if $\mathbb{P}\{X_n \le x\} \to \mathbb{P}\{X \le x\}$ for every x at which the limit distribution function is continuous. Alternative names are weak convergence and convergence in law. Symbolically, convergence in distribution is denoted by $X_n \overset{\text{d}}{\rightsquigarrow} X$ (other notation is $X_n \overset{\text{d}}{\to} X$ or $X_n \rightsquigarrow X$). In general, a sequence of random vectors $\{\mathbf{X}_n\}_1^\infty$ converges in distribution to a random vector $\mathbf{X}$, written $\mathbf{X}_n \overset{\text{d}}{\rightsquigarrow} \mathbf{X}$, if $\mathbb{E}(h(\mathbf{X}_n)) \to \mathbb{E}(h(\mathbf{X}))$ for all bounded continuous functions h.

Remark 2.1. Convergence in distribution is the one case where its not required for random variables and vectors to be defined on a common probability space. In particular, when we say $\mathbf{X}_n$ is converging in distribution to $\mathbf{X}$, this does not mean $\mathbf{X}_n$ is converging to $\mathbf{X}$ which is a stronger form of convergence to be discussed next, but that $\mathbf{X}_n$ is converging to some distribution and that distribution is the same as $\mathbf{X}$. □

Convergence in probability. A sequence of random variables $\{X_n\}_1^\infty$ is said to converge to X in probability, in symbols $X_n \overset{\text{p}}{\to} X$, if for every $\varepsilon > 0$, $\mathbb{P}\{|X_n - X| \le \varepsilon\} \to 1$ as $n \to \infty$, or equivalently, $\mathbb{P}\{|X_n - X| \ge \varepsilon\} \to 0$ as $n \to \infty$. The result extends to random vectors by using Euclidean distance. Equivalently, the sequence of vectors $\{\mathbf{X}_n\}_1^\infty$ converges to $\mathbf{X}$ in probability if and only if each of its univariate coordinates converges in probability. Convergence in probability implies convergence in distribution.

Almost sure convergence (convergence with probability 1). A sequence of random vectors $\{\mathbf{X}_n\}_1^\infty$ is said to converge almost surely to $\mathbf{X}$ (also stated as $\mathbf{X}_n$ converges to $\mathbf{X}$ with probability 1), written $\mathbf{X}_n \overset{\text{a.s.}}{\to} \mathbf{X}$, if

$$\mathbb{P}\left\{\lim_{n\to\infty} \mathbf{X}_n = \mathbf{X}\right\} = 1.$$

This means you are virtually guaranteed that $\mathbf{X}_n$ approaches the values of $\mathbf{X}$ in the sense that events for which $\mathbf{X}_n$ do not converge to $\mathbf{X}$ have probability 0. Using the probability space $(\Omega, \mathscr{A}, \mathbb{P})$ and the concept of the random vector as a function from Ω to $\mathbb{R}^k$, this is equivalent to the statement

$$\mathbb{P}\left\{\omega \in \Omega : \lim_{n\to\infty} \mathbf{X}_n(\omega) = \mathbf{X}(\omega)\right\} = 1.$$

Almost sure convergence for a sequence of vectors is equivalent to almost sure convergence of each of the univariate coordinates.

Almost sure convergence is a very strong form of convergence; however, it is important to remember convergence holds only for $\omega \in \Omega$ in the support of $\mathbb{P}$. The following is an example of a sequence of random variables that converges almost surely but not everywhere. The space is $\Omega = [0,1]$ and the underlying measure $\mathbb{P}$ is uniform over Ω (i.e. Lebesgue measure). Define

$$X_n(\omega) = \begin{cases} n & \text{if } 0 \le \omega < 1/n \\ 0 & \text{if } 1/n < \omega \le 1. \end{cases}$$

We have $X_n \overset{\text{a.s.}}{\to} 0$, since $X_n(\omega) \to 0$ for each $0 < \omega \le 1$ and $\mathbb{P}(0,1] = 1$. However, X_n does not converge everywhere, since $X_n(0) = n \to \infty$.

L_p-convergence. We say a random variable $X \in L_p$ for $1 \le p < \infty$ if $\mathbb{E}|X|^p < \infty$. The most important case is when $p = 2$, which is the L_2-space where the inner product X, Y of two variables equals their correlation. A sequence of variables $\{X_n\}_1^\infty$ converges in L_p to a variable X, written $X_n \overset{L_p}{\to} X$, if $\mathbb{E}(|X_n - X|^p) \to 0$. Convergence in L_p implies convergence in probability.

2.11.2 Relationship between the different modes of stochastic convergence

Convergence in distribution is weaker than convergence in probability, which in turn is weaker than almost sure convergence. The following is a list of useful results connecting the three types of convergence. Let X, Y, X_n, Y_n for $n = 1, 2, \ldots$ be random variables constructed over the same probability space $(\Omega, \mathscr{A}, \mathbb{P})$.

1. If $X_n \overset{\text{a.s.}}{\to} X$ then $X_n \overset{\text{p}}{\to} X$.
2. If $X_n \overset{\text{p}}{\to} X$ then $X_n \overset{\text{d}}{\rightsquigarrow} X$.
3. $X_n \overset{\text{p}}{\to} c$ where c is a real-valued constant if and only if $X_n \overset{\text{d}}{\rightsquigarrow} c$.
4. If $X_n \overset{\text{d}}{\rightsquigarrow} X$ and $|X_n - Y_n| \overset{\text{p}}{\to} 0$ then $Y_n \overset{\text{d}}{\rightsquigarrow} X$.
5. If $X_n \overset{\text{d}}{\rightsquigarrow} X$ and $Y_n \overset{\text{p}}{\to} c$ where c is a constant then $(X_n, Y_n) \overset{\text{d}}{\rightsquigarrow} (X, c)$.
6. If $X_n \overset{\text{p}}{\to} X$ and $Y_n \overset{\text{p}}{\to} Y$ then $(X_n, Y_n) \overset{\text{p}}{\to} (X, Y)$.

The following result describes an interesting connection between convergence in probability and almost sure convergence.

Theorem 2.15. *Suppose $\{X_n\}_1^\infty$ are real-valued random variables.*

(a) Cauchy Criterion: $\{X_n\}_1^\infty$ converges in probability if and only if $\{X_n\}_1^\infty$ is Cauchy in probability; i.e., $\sup_{m \ge n} \mathbb{P}\{|X_m - X_n| \ge \varepsilon\} \to 0$.

(b) $X_n \overset{\text{p}}{\to} X$ if and only if each subsequence $\{X_{n_j}\}$ contains a further subsequence $\{X_{n_{j_k}}\}$ which converges almost surely to X.

(c) $X_n \overset{a.s.}{\to} X$ if and only if $\sup_{j \ge n} |X_j - X| \overset{\text{p}}{\to} 0$.

2.11.3 Stochastic order notation

For simplicity, results here are presented for random variables. However, extensions to random vectors are straightforward (e.g., replace absolute values $|\cdot|$ with norms $||\cdot||$).

Definition 2.8. For a sequence of random variables $\{X_n\}_1^\infty$ and a nonnegative sequence $\{b_n\}_1^\infty$, we say $X_n = O_p(b_n)$, i.e., the sequence is stochastically bounded in probability, if given $\varepsilon > 0$ there exists a finite constant $M = M(\varepsilon)$ such that $\mathbb{P}\{|X_n| \le M b_n\} \ge 1 - \varepsilon$, eventually. □

Definition 2.9. In particular, $X_n = O_p(1)$, if given $\varepsilon > 0$, there exists a finite constant $M = M(\varepsilon)$ such that $\mathbb{P}\{|X_n| \le M\} \ge 1 - \varepsilon$, eventually. □

Definition 2.10. For a sequence of random variables $\{X_n\}_1^\infty$ and a nonnegative sequence $\{b_n\}_1^\infty$, we say $X_n = o_p(b_n)$ if $|X_n|/b_n \overset{\text{p}}{\to} 0$ as $n \to \infty$. In other words, if $X_n = o_p(b_n)$ then for every $\varepsilon > 0$, $\mathbb{P}\{|X_n| \ge \varepsilon b_n\} \to 0$ as $n \to \infty$. □

Definition 2.11. In particular, $\{X_n\}_1^\infty$ converges to 0 in probability if $X_n = o_p(1)$. In other words, if $X_n = o_p(1)$ then for every $\varepsilon > 0$, $\mathbb{P}\{|X_n| \geq \varepsilon\} \to 0$ as $n \to \infty$. □

Definition 2.12. A sequence of random variables $\{X_n\}_1^\infty$ is said to be order $o^*(b_n)$, where $\{b_n\}_1^\infty$ is a nonnegative sequence, written $X_n = o^*(b_n)$, if $X_n/b_n \to 0$ with probability 1 as $n \to \infty$. In particular, $X_n = o^*(1)$ if $X_n \to 0$ with probability 1; in other words $X_n \overset{\text{a.s.}}{\to} 0$. □

2.12 Convergence Theorems

Here we list some key convergence theorems.

Theorem 2.16. ***(Kolmogorov strong law of large numbers)*** *Let $\{X_n\}_1^\infty$ be i.i.d. random variables. Then $\sum_{i=1}^n X_i/n \overset{a.s.}{\to} c < \infty$ where c is a constant if and only if $\mathbb{E}(X)$ exists and $\mathbb{E}(X) = c$.*

Theorem 2.17. ***(Srong law of large numbers for vectors)*** *The strong law of large numbers applies to random vectors if and only if the strong law applies to each of the coordinates.*

Theorem 2.18. ***(Kolmogorov three series theorem)*** *Let $\{X_n\}_1^\infty$ be independent random variables. Then $\sum_{i=1}^n X_i$ converges almost surely if and only if*

(a) $\sum_{n=1}^\infty \mathbb{P}\{|X_n| > 1\} < \infty$.

(b) $\sum_{n=1}^\infty \mathbb{E}[|X_n| 1_{\{|X_n| \leq 1\}}] < \infty$.

(c) $\sum_{n=1}^\infty Var(X_n 1_{\{|X_n| \leq 1\}}) < \infty$.

Theorem 2.19. ***(Law of the iterated logarithm)*** *Let $\{X_n\}_1^\infty$ be i.i.d. random variables with mean 0 and variance σ^2. Let $S_n = \sum_{i=1}^n X_i$, then*

$$\limsup_{n\to\infty} \frac{|S_n|}{\sqrt{2n\log\log n}} = \sigma^2.$$

The law of the iterated logarithm [1] is the rate condition for a sum of i.i.d. random variables which stradles the rates for the law of large numbers and the central limit theorem. The law of large numbers tells us that scaling S_n by n^{-1} causes the sum to converge to zero almost surely, while the central limit theorem states that scaling S_n by $n^{-1/2}$ causes the sum to converge to a limiting distribution. Thus in the latter case, we do not have convergence almost surely (or even in probability). The question is, what is the smallest scaling factor leading to a breakdown in almost sure convergence? The law of the iterated logarithm shows that the rate $\sqrt{2n\log\log n}$ is this lower bound. In fact, it is the point where convergence in probability and almost sure convergence deviates:

$$\frac{S_n}{\sqrt{2n\log\log n}} \overset{\text{p}}{\to} 0, \qquad \frac{S_n}{\sqrt{2n\log\log n}} \overset{\text{a.s.}}{\not\to} 0, \qquad \text{as } n \to \infty.$$

Theorem 2.20. ***(Continuous mapping theorem)*** *Let $h : \mathbb{R}^k \to \mathbb{R}^m$ be continuous at every point of a set C such that $\mathbb{P}\{\boldsymbol{X} \in C\} = 1$. Then*

(a) If $\boldsymbol{X}_n \overset{\text{d}}{\rightsquigarrow} \boldsymbol{X}$, then $h(\boldsymbol{X}_n) \overset{\text{d}}{\rightsquigarrow} h(\boldsymbol{X})$.

(b) If $\boldsymbol{X}_n \overset{\text{p}}{\to} \boldsymbol{X}$, then $h(\boldsymbol{X}_n) \overset{\text{p}}{\to} h(\boldsymbol{X})$.

(c) If $\boldsymbol{X}_n \overset{a.s.}{\to} \boldsymbol{X}$, then $h(\boldsymbol{X}_n) \overset{a.s.}{\to} h(\boldsymbol{X})$.

Theorem 2.21. ***(Slutsky's theorem)*** *If $\boldsymbol{X}_n \overset{\text{d}}{\rightsquigarrow} \boldsymbol{X}$ and $\boldsymbol{Y}_n \overset{\text{p}}{\to} \boldsymbol{y}$ (a constant), then*

(a) $\boldsymbol{X}_n + \boldsymbol{Y}_n \overset{\text{d}}{\rightsquigarrow} \boldsymbol{X} + \boldsymbol{y}$.

(b) $\boldsymbol{X}_n \boldsymbol{Y}_n \overset{\text{d}}{\rightsquigarrow} \boldsymbol{y}\boldsymbol{X}$.

(c) $\boldsymbol{X}_n \boldsymbol{Y}_n \overset{\text{p}}{\to} \boldsymbol{0}$, *if* $\boldsymbol{y} = \boldsymbol{0}$.

Theorem 2.22. ***(Lévy continuity theorem)*** *Let $\boldsymbol{X}_n$ and $\boldsymbol{X}$ be random vectors on $\mathbb{R}^k$. Then $\boldsymbol{X}_n \overset{\text{d}}{\rightsquigarrow} \boldsymbol{X}$ if and only if $\mathbb{E}(e^{i\boldsymbol{t}^T\boldsymbol{X}_n}) \to \mathbb{E}(e^{i\boldsymbol{t}^T\boldsymbol{X}})$ for every $\boldsymbol{t} \in \mathbb{R}^k$. Moreover, if $\mathbb{E}(e^{i\boldsymbol{t}^T\boldsymbol{X}_n})$ converges pointwise to $\phi(\boldsymbol{t})$ that is continuous at zero, then ϕ is a characteristic function of a random vector $\boldsymbol{X}$ and $\boldsymbol{X}_n \overset{\text{d}}{\rightsquigarrow} \boldsymbol{X}$.*

Remark 2.2. Lévy's continuity theorem implies that weak convergence of a random vector is equivalent to weak convergence of linear combinations of random variables: $\mathbf{X}_n \overset{\text{d}}{\rightsquigarrow} \mathbf{X}$ iff $\ell^T\mathbf{X}_n \overset{\text{d}}{\rightsquigarrow} \ell^T\mathbf{X}$ for all $\ell \in \mathbb{R}^k$. This is known as the Cramér-Wold device. It reduces higher-dimensional problems to the one-dimensional case. □

Theorem 2.23. ***(Lebesgue dominated convergence theorem)*** *Suppose that $\{f_n\}_1^\infty$ is a sequence of measurable functions such that $f_n \to f$ pointwise almost everywhere as $n \to \infty$ and $|f_n| \le g$ for all n, where g is integrable. Then f is integrable, and*

$$\lim_{n\to\infty} \int f_n d\mu = \int f d\mu.$$

The Lebesgue-dominated convergence theorem is often used to assert convergence for the expected value of a sequence random variables. The following is a restatement of the theorem for this purpose. The result is also slightly generalized by requiring the weaker condition of convergence in probability.

Theorem 2.24. ***(Lebesgue dominated convergence theorem for random variables)*** *If $X_n \overset{\text{p}}{\to} X$ and if there exists a dominating random variable $\zeta \in L_1$ such that $|X_n| \le \zeta$ and $E\zeta < \infty$, then $\mathbb{E}(X_n) \to \mathbb{E}(X)$.*

Theorem 2.25. ***(Scheffé's theorem)*** *Let $\{X, X_n, n \ge 1\}$ be a sequence of random variables with densities $\{f, f_n, n \ge 1\}$ such that $f_n \to f$ almost everywhere. Then*

$$\int |f_n(x) - f(x)|\, dx \to 0.$$

In particular this implies $X_n \overset{\text{d}}{\rightsquigarrow} X$.

Remark 2.3. The result as stated is with respect to Lebesgue measure but can be extended to densities with respect to other dominating measures. For example, using counting measure for countable spaces we have the following result establishing the equivalent between convergence in distribution and convergence in probability. Let $\{X_n\}_1^\infty$ be discrete random variables taking values over a countable set $S = \{a_1, a_2, \ldots\}$. Let X be another random variable over S. Then $\mathbb{P}\{X_n = a_k\} \to \mathbb{P}\{X = a_k\}$ for every k implies

$$\sum_{k=1}^{\infty} |\mathbb{P}\{X_n = a_k\} - \mathbb{P}\{X = a_k\}| \to 0.$$

However, since $\mathbb{P}\{X_n = a_k\} \to \mathbb{P}\{X = a_k\}$ for every k is equivalent to $X_n \overset{\text{d}}{\rightsquigarrow} X$ for a countable space, we have

$$X_n \overset{\text{d}}{\rightsquigarrow} X \quad \text{if and only if} \quad \sum_{k=1}^{\infty} |\mathbb{P}\{X_n = a_k\} - \mathbb{P}\{X = a_k\}| \to 0.$$

□

Corollary 2.26. *For a countable probability space $(\Omega, \mathscr{A}, \mathbb{P})$, convergence in distribution and convergence in probability are equivalent.*

2.13 The Empirical Measure

The empirical measure is a key quantity in the theory of empirical processes. In later chapters, we will use it in the study of learning theory and when discussing bootstrapping and subsampling of data. Therefore, it is helpful to define it and state one of its well-known consistency properties (more development will follow in later chapters).

Let $\mathbf{X}_1, \ldots, \mathbf{X}_n$ be a sequence of independent and identically distributed random vectors over $(\mathbb{R}^p, \mathscr{B})$. The empirical measure denoted by $\mathbb{P}_n$ is a measure defined over $(\mathbb{R}^p, \mathscr{B})$ such that for each $B \in \mathscr{B}$

$$\mathbb{P}_n(B) = \frac{1}{n} \sum_{i=1}^{n} 1_{\{\mathbf{X}_i \in B\}}.$$

Observe that $\mathbb{P}_n$ puts equal mass at each of the observations $\mathbf{X}_1, \ldots, \mathbf{X}_n$. Thus $\mathbb{P}_n(B)$ amounts to calculating the relative frequency of the sample whose values are in the set B. In a similar fashion, we can define the expectation of a real-valued function $g : \mathbb{R}^p \to \mathbb{R}$ under the empirical measure

$$\mathbb{P}_n g = \frac{1}{n} \sum_{i=1}^{n} g(\mathbf{X}_i).$$

Notice if $\mathbb{P}g < \infty$, then $\mathbb{P}_n g \overset{\text{a.s.}}{\to} \mathbb{P}g$ by the strong law of large numbers. Thus we can see that $\mathbb{P}_n g$ is a consistent estimator of $\mathbb{P}g$ which holds without knowledge of $\mathbb{P}$. This seems to suggest that $\mathbb{P}_n$ might be a good way to approximate $\mathbb{P}$ in general. Since in many problems, we may not know what $\mathbb{P}$ is, then this would indeed be a valuable property. The following famous result due to Glivenko and Cantelli shows $\mathbb{P}_n$ is able to uniformly approximate $\mathbb{P}$.

Theorem 2.27. ***(Glivenko-Cantelli theorem)*** *For every distribution $\mathbb{P}$ on the real line,*

$$\sup_t |\mathbb{P}_n(-\infty, t] - \mathbb{P}(-\infty, t]| \overset{a.s.}{\to} 0.$$

Theorem 2.27 applies to random variables ($p = 1$), and the notation $\mathbb{P}(-\infty, t]$ means $\mathbb{P}\{X \in (-\infty, t]\} = \mathbb{P}\{X \le t\}$. Since this equals the cumulative distribution function (cdf) for X evaluated at t, the result asserts that the empirical cdf uniformly approximates the true cdf.

However, we can also interpret $\mathbb{P}(-\infty, t]$ as the expected value of the function $g_t(X) = 1_{\{X \in (-\infty, t]\}}$. Thus, another way to view Theorem 2.27 is that it shows $\mathbb{P}_n$ uniformly approximates $\mathbb{P}$ over a specific class of functions. Let $\mathscr{G} = \{g_t : t \in \mathbb{R}\}$ be the class of all g_t functions. Then Glivenko-Cantelli shows that

$$\sup_{g \in \mathscr{G}} |\mathbb{P}_n g - \mathbb{P}g| \overset{\text{a.s.}}{\to} 0.$$

2.14 The Borel-Cantelli Lemma

Let $\{A_n\}_1^\infty$ be an infinite sequence of events over a probability space $(\Omega, \mathscr{A}, \mathbb{P})$. We say that events in the sequence occur *infinitely often* (i.o.) if there is an infinite number of events $A_{n_1}, A_{n_2}, A_{n_3}, \ldots$ that are true. An event A is said to be "true" if there exists a probability experiment that produces an outcome $\omega \in \Omega$ for which $\omega \in A$. Let $\{A_n\, i.o.\}$ denote the event that an infinite number of events occurs:

$$\{A_n\, i.o.\} = \{\omega \in \Omega : \omega \in A_{n_1}, A_{n_2}, A_{n_3}, \ldots\} = \{\omega \in \Omega : \omega \in A_n \text{ for infinitely many } n\}.$$

It is a basic fact of set theory and probability that the following are true:

1. If $\{A_n\, i.o.\}$ is true, $\bigcup_{i=n}^{\infty} A_i$ holds for all $n = 1, 2, \ldots$.
2. If $\{A_n\, i.o.\}$ is false, $\bigcup_{i=n}^{\infty} A_i$ must be false for some $n \in \{1, 2, \ldots\}$.
3. $\limsup_{n\to\infty} A_n = \bigcap_{n=1}^{\infty} \bigcup_{i=n}^{\infty} A_i = \{A_n\, i.o.\}$.
4. $\bigcup_{i=n}^{\infty} A_i \downarrow \{A_n\, i.o.\}$.
5. $\lim_{n\to\infty} \mathbb{P}\left[\bigcup_{i=n}^{\infty} A_i\right] = \mathbb{P}\{A_n\, i.o.\}$.

We say that the events in the sequence occur *finitely often* (f.o.) if only a finite number of sets are true:

$$\{A_n\, f.o\} = \{A_n\, i.o.\}^c = \{\omega \in \Omega : \omega \in A_n \text{ for only a finite number of } n\}.$$

From the above it follows that:

1. If $\{A_n\, f.o.\}$ is true $\bigcup_{i=n}^{\infty} A_i$ must be false for some $n \in \{1, 2, \ldots\}$.
2. If $\{A_n\, f.o.\}$ is true $\bigcap_{i=n}^{\infty} A_i^c$ must be true for some $n \in \{1, 2, \ldots\}$.
3. $\liminf_{n\to\infty} A_n = \bigcup_{n=1}^{\infty} \bigcap_{i=n}^{\infty} A_i = \{A_n\, f.o.\}$.
4. $\bigcap_{i=n}^{\infty} A_i \uparrow \{A_n\, f.o.\}$.

These facts are useful for understanding the Borel-Cantelli lemma, a key result used in probability theory and asymptotics.

Lemma 2.28. ***(Borel-Cantelli lemma)*** *Let $\{A_n\}_1^{\infty}$ be an infinite sequence of events over a probability space $(\Omega, \mathscr{A}, \mathbb{P})$.*

1. *If $\sum_{n=1}^{\infty} \mathbb{P}(A_n) < \infty$, then $\mathbb{P}\{A_n\, i.o.\} = 0$ (i.e. $\mathbb{P}\{A_n\, f.o.\} = 1$).*
2. *If $\sum_{n=1}^{\infty} \mathbb{P}(A_n) = \infty$ and the events are mutually exclusive (i.e., independent), then $\mathbb{P}\{A_n\, i.o.\} = 1$.*

As an illustration, consider the following coin-flipping experiment. Suppose that we flip a sequence of coins where A_n denotes the event that coin n is a head. Suppose that $\mathbb{P}(A_n) = 1/n$. Then

$$\sum_{n=1}^{\infty} \mathbb{P}(A_n) = \sum_{n=1}^{\infty} \frac{1}{n} = \infty.$$

Thus, if the coin flips are independent, the second part of the Borel-Cantelli lemma tells us that the coin-tossing experiment will result in an infinite number of heads with probability one. Indeed, a direct calculation reveals why this must be the case:

$$\mathbb{E}[\text{number of heads}] = \sum_{n=1}^{\infty} \mathbb{E}(1_{A_n}) = \sum_{n=1}^{\infty} \mathbb{P}(A_n) = \infty.$$

On the other hand, if $\mathbb{P}(A_n)$ decreases rapidly enough so that $\sum_{i=1}^{\infty} \mathbb{P}(A_n) < \infty$ (for example, if $\mathbb{P}(A_n) = 1/n^2$), then only a finite number of coins will be heads with probability one. Notice this result holds regardless of whether the coin tosses are independent or not. Thus, whether the number of heads is infinite or finite ultimately depends on the summability of the event probabilities. This fact is often called the Borel zero-one law, which can be stated as follows.

Corollary 2.29. *If* $\{A_n\}_1^\infty$ *are independent events,* $\mathbb{P}\{A_n\, i.o.\} = 0$ *or* $\mathbb{P}\{A_n\, i.o.\} = 1$ *according to* $\sum_{n=1}^\infty \mathbb{P}(A_n) < \infty$ *or* $\sum_{n=1}^\infty \mathbb{P}(A_n) = \infty$.

An important application of the Borel-Cantelli lemma is in proving almost sure convergence. The following theorem is instrumental in this.

Theorem 2.30. *Let* $\{X_n\}_1^\infty$ *be a sequence of random variables on a common probability space* $(\Omega, \mathscr{A}, \mathbb{P})$*. Then* $X_n \overset{a.s.}{\to} 0$ *if and only if*

$$\mathbb{P}\left\{|X_n| > \frac{1}{k}, i.o.\right\} = 0, \quad k = 1, 2, \ldots. \tag{2.7}$$

Proof. Let $A = \bigcup_{k=1}^\infty A_k$, where $A_k = \{|X_n| > 1/k, i.o.\}$. If $\omega \notin A$, then $|X_n(\omega)| > 1/k$ for only finitely many n for every positive integer k, implying $X_n(\omega) \to 0$ and thereby $X_n \overset{\text{a.s.}}{\to} 0$ since $\mathbb{P}(A) \le \sum_{k=1}^\infty \mathbb{P}(A_k) = 0$ (and thus $\mathbb{P}(A^c) = 1$) assuming (2.7) holds. Conversely if $X_n(\omega) \to 0$, then $\omega \notin A_k$ for every positive integer k, and thus $\omega \notin A$. Hence $A^c = \{\omega : X_n(\omega) \to 0\}$, which implies that $\mathbb{P}(A) = 0$, or equivalently that $\mathbb{P}(A_k) = 0$ for every k.

□

Theorem 2.30 equates almost sure convergence to verifying (2.7), thus making it possible to establish almost sure convergence by showing the summability of event probabilities by using the first half of Borel-Cantelli. This technique is an important tool commonly used in asymptotic theory.

We illustrate this approach by proving a strong law of large numbers for binary variables. This result only requires independence and can easily be extended to arbitrary bounded variables. Let $X_1, X_2, \ldots, X_n$ be independent Bernoulli random variables. Let $S_n = \sum_{i=1}^n X_i$ and $\mu_n = \sum_{i=1}^n \mathbb{P}\{X_i = 1\}$. By Hoeffding's inequality (see Theorem 2.41),

$$\mathbb{P}\{|S_n - \mu_n| \ge t\} \le 2\exp\left(-\frac{2t^2}{n}\right).$$

Setting $t = \varepsilon n$, we have

$$\mathbb{P}\{|\overline{X}_n - \overline{\mu}_n| \ge \varepsilon\} \le 2\exp\left(-2n\varepsilon^2\right)$$

where $\overline{X}_n = S_n/n$ and $\overline{\mu}_n = \mu_n/n$. Observe that for every $\varepsilon > 0$

$$\sum_{n=1}^\infty \mathbb{P}\{|\overline{X}_n - \overline{\mu}_n| \ge \varepsilon\} \le 2\sum_{n=1}^\infty \exp\left(-2n\varepsilon^2\right) < \infty.$$

By the first half of the Borel-Cantelli lemma this shows (2.7) holds for the sequence $\overline{X}_n - \overline{\mu}_n$ for any $\varepsilon = 1/k$ for an integer $k \ge 1$, from which it follows from Theorem 2.30 that $\overline{X}_n - \overline{\mu}_n \overset{\text{a.s.}}{\to} 0$. In fact, using a slight modification of this argument, we have established the following general result.

Theorem 2.31. *Let* $X_1, X_2, \ldots$ *be independent random variables bounded by real numbers* A, B *such that* $-\infty < -|A| < a_i \le X_i \le b_i \le |B| < \infty$ *for each* $i = 1, \ldots, n$*. Then* $\overline{X}_n - \overline{\mu}_n \overset{a.s.}{\to} 0$*, where* $\overline{X}_n = \sum_{i=1}^n X_i/n$ *and* $\overline{\mu}_n = \sum_{i=1}^n \mathbb{E}[X_i]/n$.

2.15 The Skorohod Representation Theorem

Recall $X_n \overset{\text{a.s.}}{\to} X$ implies $X_n \overset{\text{d}}{\rightsquigarrow} X$ but the converse is not true in general. Despite the fact convergence in distribution does not imply almost sure convergence, the Skorohod representation theorem provides a partial converse. The result is used as a conceptual tool for proving weak convergence results by taking advantage of continuity arguments.

Theorem 2.32. ***(Skorohod representation theorem)*** *Let* $\{X, X_n : n = 1,2,\ldots\}$ *be random variables defined over a probability space* $(\Omega, \mathscr{A}, \mathbb{P})$ *such that* $X_n \overset{\text{d}}{\rightsquigarrow} X$. *Then there exists random variables* $\{X^*, X_n^* : n = 1,2,\ldots\}$ *defined over the Lebesgue probability space* $([0,1], \mathscr{A}[0,1], \lambda =$ *Lebesgue measure) such that*

$$X \overset{d}{=} X^*, X_n \overset{d}{=} X_n^* \text{ and } X_n^* \overset{a.s.}{\to} X^*,$$

where "d" and "a.s." are taken with respect to the Lebesgue probability space.

The proof rests on the probability integral transform which is a technique used to simulate random variables. Let F_n be the cumulative distribution function for X_n. The probability integral transform states that if U is uniformly distributed over $[0,1]$, then $F_n^{-1}(U)$ is distributed as X_n. The Skorohod construction is as follows:

$$X_n^* := F_n^{-1}(U) \overset{d}{=} X_n$$
$$X^* := F^{-1}(U) \overset{d}{=} X.$$

Using $X_n \overset{\text{d}}{\rightsquigarrow} X$ it follows that $F_n \to F$ and therefore $X_n^* \to X^*$. An interesting observation is that the construction uses a random draw from only variable, U. This is what makes the convergence $X_n^* \to X^*$ almost sure.

As a first illustration of Theorem 2.32, we prove the continuous mapping theorem result that $X_n \overset{\text{d}}{\rightsquigarrow} X$ implies $h(X_n) \overset{\text{d}}{\rightsquigarrow} h(X)$ for $h : \mathbb{R} \to \mathbb{R}$ a bounded function continuous almost everywhere $\mathbb{P}$ (Theorem 2.20 (a)).

Proof. Construct $\{X^*, X_n^* : n = 1,2,\ldots\}$ over $[0,1]$ where $X \overset{d}{=} X^*$ and $X_n \overset{d}{=} X_n^*$. Because $X_n^* \overset{\text{a.s.}}{\to} X^*$ with respect to Lebesgue measure, we have $X_n^*(t) \to X^*(t)$ for a.a. $t \in [0,1]$. Let $C(h)$ denote the continuity points of h. Then,

$$\begin{aligned}
&\lambda\{t \in [0,1] : h(X_n^*(t)) \to h(X^*(t))\} \\
&\quad \geq \lambda\{t \in [0,1] : X^*(t) \in C(h)\} \\
&\quad = \mathbb{P}\{\omega : X(\omega) \in C(h)\} = 1,
\end{aligned}$$

which implies $h(X_n^*) \overset{\text{a.s.}}{\to} h(X^*)$. Because almost sure convergence implies convergence in distribution, $h(X_n^*) \overset{\text{d}}{\rightsquigarrow} h(X^*)$. Thus, we have shown:

$$h(X_n) \overset{d}{=} h(X_n^*) \overset{\text{d}}{\rightsquigarrow} h(X) \overset{d}{=} h(X^*).$$

□

The following example highlights the remarkable effectiveness of the Skorohod method. With minimal effort, we can demonstrate the well-known "Delta Method," which is used to calculate the asymptotic variance for a normal limit under a continuous map.

Theorem 2.33. *Let* $\overline{X}_n = \sum_{i=1}^n X_i/n$ *where* X_i *are i.i.d. with mean* $\mathbb{E}(X_1) = \mu$ *and variance* $Var(X_1) = \sigma^2$. *If* ϕ *is a continuous function with non-zero derivative* $\phi'(\mu)$ *at* μ, *then*

$$\sqrt{n}\left(\phi(\overline{X}_n) - \phi(\mu)\right) \overset{\text{d}}{\rightsquigarrow} N\left(0, \sigma^2 \phi'(\mu)^2\right).$$

Thus the asymptotic variance for $\phi(\overline{X}_n)$ *is* $\sigma^2 \phi'(\mu)^2$.

Proof. By the central limit theorem

$$Z_n := \sqrt{n}\left(\frac{\overline{X}_n - \mu}{\sigma}\right) \overset{\mathrm{d}}{\rightsquigarrow} Z \overset{d}{=} \mathrm{N}(0,1).$$

By the Skorohod construction, there exists random variables Z^*, Z_n^* over $[0,1]$ such that $Z \overset{d}{=} Z^*$, $Z_n \overset{d}{=} Z_n^*$ and $Z_n^* \overset{\text{a.s.}}{\to} Z^*$. Let $\overline{X}_n^* = \mu + \sigma Z_n^*/\sqrt{n}$. Then $\overline{X}_n \overset{d}{=} \overline{X}_n^*$. By the definition of the derivative, and using $\sigma Z_n^*/\sqrt{n} \overset{\text{a.s.}}{\to} 0$,

$$\begin{aligned}
\sqrt{n}\left(\frac{\phi(\overline{X}_n) - \phi(\mu)}{\sigma\phi'(\mu)}\right) &\overset{d}{=} \sqrt{n}\left(\frac{\phi\left(\mu + \sigma Z_n^*/\sqrt{n}\right) - \phi(\mu)}{\sigma\phi'(\mu)}\right) \\
&= \frac{\phi(\mu + \sigma Z_n^*/\sqrt{n}) - \phi(\mu)}{\sigma Z_n^*/\sqrt{n}} \cdot \frac{\sigma Z_n^*}{\sigma\phi'(\mu)} \\
&\overset{\text{a.s.}}{\to} \phi'(\mu) \cdot \frac{\sigma Z^*}{\sigma\phi'(\mu)} = Z^* \overset{d}{=} Z \overset{d}{=} \mathrm{N}(0,1).
\end{aligned}$$

□

2.16 Manipulating o_p and O_p Terms

Here we list various rules for manipulating stochastic order notation.

Proposition 2.34. $o_p(1) \times o_p(1) = o_p(1)$.

Proof. Let $X_n = o_p(1)$ and $Y_n = o_p(1)$. Then

$$\begin{aligned}
\mathbb{P}\{|X_n Y_n| > \epsilon\} &= \mathbb{P}\{|X_n Y_n| > \epsilon, |Y_n| \le 1\} + \mathbb{P}\{|X_n Y_n| > \varepsilon, |Y_n| > 1\} \\
&\le \mathbb{P}\{|X_n| > \varepsilon\} + \mathbb{P}\{|Y_n| > 1 \to 0.
\end{aligned}$$

□

Proposition 2.35. $o_p(1) \times O_p(1) = o_p(1)$.

Proof. Let $X_n = o_p(1)$ and $Y_n = O_p(1)$. Then for any $\varepsilon > 0$, there exists a finite constant M such that

$$\begin{aligned}
\mathbb{P}\{|X_n Y_n| > \varepsilon\} &= \mathbb{P}\{|X_n Y_n| > \varepsilon, |Y_n| \le M\} + \mathbb{P}\{|X_n Y_n| > \varepsilon, |Y_n| > M\} \\
&\le \mathbb{P}\{|X_n| > \varepsilon/M\} + \mathbb{P}\{|Y_n| > M\} \le \varepsilon, \qquad \text{eventually.}
\end{aligned}$$

□

Proposition 2.36. $O_p(1) \times O_p(1) = O_p(1)$.

Proof. Let $X_n = O_p(1)$ and $Y_n = O_p(1)$. Then for $\varepsilon > 0$ there exists finite constants M and M_2 such that

$$\begin{aligned}
\mathbb{P}\{|X_n Y_n| > M\} &= \mathbb{P}\{|X_n Y_n| > M, |Y_n| \le M_2\} + \mathbb{P}\{|X_n Y_n| > \varepsilon, |Y_n| > M_2\} \\
&\le \mathbb{P}\{|X_n| > M/M_2\} + \mathbb{P}\{|Y_n| > M_2\} \le \varepsilon, \qquad \text{eventually.}
\end{aligned}$$

□

Proposition 2.37. $o_p(1) + o_p(1) = o_p(1)$.

Proposition 2.38. $o_p(1) + O_p(1) = O_p(1)$.

Proposition 2.39. $O_p(1) + O_p(1) = O_p(1)$.

2.17 Inequalities

The following are standard inequalities used in many calculations.

1. The HM-GM-AM-QM (harmonic-geometric-arithmetic-quadratic mean) inequality. For any set of positive real numbers $x_1, \ldots, x_n$

$$0 < \frac{n}{\sum_{i=1}^n x_i^{-1}} \leq \left(\prod_{i=1}^n x_i\right)^{1/n} \leq \frac{\sum_{i=1}^n x_i}{n} \leq \sqrt{\frac{\sum_{i=1}^n x_i^2}{n}}.$$

   Note that the far right inequality is an example of the Cauchy-Schwarz inequalty discussed next.

2. Cauchy-Schwarz inequality. Suppose we have two random variables $X, Y \in L_2$. Then

$$|\mathbb{E}(XY)| \leq \mathbb{E}|XY| \leq \sqrt{\mathbb{E}(X^2)}\sqrt{\mathbb{E}(Y^2)}$$

   In integral format this can be written

$$\left|\int f(x)g(x)dx\right| \leq \int |f(x)g(x)|dx \leq \sqrt{\int f(x)^2 dx}\sqrt{\int g(x)^2 dx}.$$

   For two vectors $\mathbf{u}, \mathbf{v}$ with inner product $(\mathbf{u}, \mathbf{v})$ and norm $||\cdot||$ this can be written as

$$|(\mathbf{u}, \mathbf{v})| \leq |||\mathbf{u}|| \cdot ||\mathbf{v}||.$$

   For example, for the L_2 norm on $\mathbb{R}^n$, this yields

$$\left|\sum_{i=1}^n u_i v_i\right| \leq \sum_{i=1}^n |u_i v_i| \leq \sqrt{\sum_{i=1}^n u_i^2}\sqrt{\sum_{i=1}^n v_i^2}.$$

   Note that equality holds if and only if $\mathbf{u}$ is a scalar multiple of $\mathbf{v}$.

3. Holder's inequality. Suppose p, q are real values satisfying

$$p > 1, q > 1, \frac{1}{p} + \frac{1}{q} = 1$$

   and that we have two random variables X, Y such that $E|X|^p < \infty$ and $E|Y|^q < \infty$ (i.e., $X \in L_p$ and $Y \in L_q$). Then

$$|\mathbb{E}(XY)| \leq \mathbb{E}|XY| \leq (\mathbb{E}|X|^p)^{1/p}(\mathbb{E}|Y|^q)^{1/q}.$$

   Notice that Schwartz inequality is the special case $p = q = 2$.

4. Minkowski's inequality. Assume $X, Y \in L_p$ where $1 \leq p < \infty$. Then $X + Y \in L_p$ and

$$||X + Y||_p \leq ||X||_p + ||Y||_P.$$

2.18 Concentration Inequalities

Here we present some well known inequalities, some of which will be used in later sections.

2.18.1 Chebychev's inequality

Chebychev's inequality provides an upper bound to the probability of a random variable exceeding its mean. For a random variable X, with mean $\mathbb{E}(X) = \mu$ and variance $\mathbb{E}(X-\mu)^2 = \sigma^2$, Chebychev's inequality asserts that for any $t > 0$

$$\mathbb{P}\{|X-\mu| \geq t\} \leq \frac{\sigma^2}{t^2}.$$

This shows when $t \gg \sigma$, the probability of X exceeding its mean by more than t will be small. This type of inequality is referred to as a large deviations bound or *concentration inequality*.

However, Chebychev's inequality is often too crude for most applications. Usually, we need a probability of large deviations to be exponentially small. An example of such an inequality is the extension to tails of binomial distributions due to Chernoff.

Theorem 2.40. ***(Chernoff)*** *Let $X_1, X_2, \ldots, X_n$ be independent $\{0,1\}$ random variables such that $\mathbb{P}\{X_i = 1\} = p = 1 - \mathbb{P}\{X_i = 0\}$ for each $i = 1, \ldots, n$. If $S_n = \sum_{i=1}^n X_i$, then for any $t > 0$*

$$\mathbb{P}\{|S_n - np| \geq nt\} \leq 2\exp\left(-2nt^2\right).$$

Example 2.1. The superiority of Chernoff's bound to Chebychev's inequality is immediately apparent. In the binomial case, $\mu = np$ and $\sigma^2 = np(1-p)$. Chebychev's inequality gives

$$\mathbb{P}\{|S_n - np| \geq nt\} \leq \frac{np(1-p)}{(nt)^2} = \frac{p(1-p)}{nt^2}.$$

This is much larger than Chernoff's bound; for example, if $p = 1/2$ the bound is $1/(4nt^2)$ compared with $2\exp(-2nt^2)$. □

2.18.2 Hoeffding's inequality

Hoeffding's inequality [4] generalizes the Chernoff bound to the case of arbitrarily bounded random variables.

Theorem 2.41. ***(Hoeffding)*** *Let $X_1, X_2, \ldots, X_n$ be independent random variables such that $a_i \leq X_i \leq b_i$ for each i. If $S_n = \sum_{i=1}^n X_i$ and $\mu_n = \mathbb{E}(S_n)$, then for any $t > 0$*

$$\mathbb{P}\{|S_n - \mu_n| \geq t\} \leq 2\exp\left(-\frac{2t^2}{\sum_{i=1}^n (b_i - a_i)^2}\right).$$

Example 2.2. Chernoff's bound follows from the Hoeffding inequality when $a_i = 0$ and $b_i = 1$. Since $\mu_n = np$,

$$\mathbb{P}\{|S_n - \mu_n| \geq nt\} = \mathbb{P}\{|S_n - np| \geq nt\} \leq 2\exp\left(-\frac{2n^2t^2}{\sum_{i=1}^n 1}\right) = 2\exp\left(-2nt^2\right).$$

□

Example 2.3. For i.i.d. data we obtain an exponential bound for the sample average

$$\mathbb{P}\{|\overline{X} - \mu| \geq t\} = \mathbb{P}\{|S_n - n\mu| \geq nt\} \leq 2\exp\left(-\frac{2n^2t^2}{n(b-a)}\right) = 2\exp\left(-\frac{2nt^2}{(b-a)}\right).$$

□

2.18.3 Kolmogorov's inequality

Another useful inequality are maximal-inequalities for bounding partial sums of independent random variables. The following bound is due to Kolmogorov.

Theorem 2.42. ***(Kolmogorov)*** *Let $X_1, X_2, \dots, X_n$ be independent random variables with $\mathbb{E}(X_i) = 0$ for each $i = 1, \dots, n$, then for any $C > 0$*

$$\mathbb{P}\left\{\max_{1 \le m \le n}\left|\sum_{1 \le i \le m} X_i\right| \ge C\right\} \le \frac{1}{C^2}\sum_{1 \le i \le n} \mathbb{E}(X_i^2).$$

2.18.4 McDiarmid's inequality

In statistical theory, it is often useful to have large deviation bounds for functions that can depend upon independent random variables, rather than bounds on sums of independent variables. McDiarmid's inequality [5] is a concentration inequality developed for this purpose.

Theorem 2.43. ***(McDiarmid)*** *Let $X_1, X_2, \dots, X_n$ be independent random variables with X_i taking values in the space $\mathscr{X}_i$. Suppose that the real-valued function ϕ defined on $\mathscr{X} = \mathscr{X}_1 \times \cdots \times \mathscr{X}_n$ satisfies for each $i = 1, \dots, n$ and all $(x_1, \dots, x_n) \in \mathscr{X}$ and all $x_i' \in \mathscr{X}_i$*

$$\left|\phi(x_1, \dots, x_{i-1}, x_i, x_{i+1}, \dots, x_n) - \phi(x_1, \dots, x_{i-1}, x_i', x_{i+1}, \dots, x_n)\right| \le c_i.$$

Then if $\mu = \mathbb{E}[\phi(X_1, \dots, X_n)]$, for each $t > 0$

$$\mathbb{P}\{\phi(X_1, \dots, X_n) - \mu \ge t\} \le \exp\left(-\frac{2t^2}{\sum_{i=1}^n c_i^2}\right).$$

In particular, applying the inequality to $-\phi$, we have

$$\mathbb{P}\{|\phi(X_1, \dots, X_n) - \mu| \ge t\} \le 2\exp\left(-\frac{2t^2}{\sum_{i=1}^n c_i^2}\right).$$

Example 2.4. Setting $\phi(X_1, \dots, X_n) = \sum_{i=1}^n X_i = S_n$, we obtain Hoeffding's inequality as a special case. Notice

$$\begin{aligned}&\left|\phi(x_1, \dots, x_{i-1}, x_i, x_{i+1}, \dots, x_n) - \phi(x_1, \dots, x_{i-1}, x_i', x_{i+1}, \dots, x_n)\right| \\ &\quad = |x_i - x_i'| \le b_i - a_i,\end{aligned}$$

where the bound on the right holds if X_i is bounded such that $a_i \le X_i \le b_i$. Set $c_i = b_i - a_i$ to deduce Hoeffding's inequality. □

Example 2.5. Let $\mathscr{F}$ be a class of bounded functions over $\mathscr{X}$ such that each $f \in \mathscr{F}$ satisfies $f : \mathscr{X} \to [a, b]$. Define

$$\phi(X_1, \dots, X_n) = \sup_{f \in \mathscr{F}} |\mathbb{E}_n f - \mathbb{E} f|$$

where $\mathbb{E}_n f = \sum_{i=1}^n f(X_i)/n$ is the sample average and $\mathbb{E} f = \mathbb{E} f(X_1)$ is the population average for $X_1, \dots, X_n$, an i.i.d. sequence. By the boundedness of f

$$\begin{aligned}&\left|\phi(x_1, \dots, x_{i-1}, x_i, x_{i+1}, \dots, x_n) - \phi(x_1, \dots, x_{i-1}, x_i', x_{i+1}, \dots, x_n)\right| \\ &\quad = \left|\sup_{f \in \mathscr{F}} |\mathbb{E}_n f - \mathbb{E} f| - \sup_{f \in \mathscr{F}}\left|\mathbb{E}_n f - \mathbb{E} f + \frac{1}{n}\left[f(x_i') - f(x_i)\right]\right|\right| \\ &\quad \le \frac{1}{n}\sup_{f \in \mathscr{F}} |f(x_i') - f(x_i)| \\ &\quad \le \frac{b - a}{n}.\end{aligned}$$

Hence setting $c_i = (b-a)/n$,

$$\mathbb{P}\left\{\left|\sup_{f\in\mathscr{F}}|\mathbb{E}_n f - \mathbb{E}f| - \mathbb{E}\left[\sup_{f\in\mathscr{F}}|\mathbb{E}_n f - \mathbb{E}f|\right]\right| \geq t\right\} \leq 2\exp\left(-\frac{2nt^2}{(b-a)^2}\right).$$

This is type of empirical process result. □

Example 2.6. Let $(\mathbf{X}, Y), (\mathbf{X}_1, Y_1), \ldots, (\mathbf{X}_n, Y_n)$ be i.i.d. values from some common distribution $\mathbb{P}$, where it is assumed that

$$Y_i = f(\mathbf{X}_i) + \varepsilon_i, \qquad i = 1, \ldots, n,$$

such that $\mathbb{E}(Y_i|\mathbf{X}_i) = f(\mathbf{X}_i)$. The goal is to estimate $f(\mathbf{X})$.

Given the observed data $\mathscr{L} = \{(\mathbf{x}_1, y_1), \ldots, (\mathbf{x}_n, y_n)\}$, consider a local weighted estimator $\hat{f}(\mathbf{x})$ defined as follows

$$\hat{f}(\mathbf{x}) = \sum_{i=1}^{n} W_{\mathscr{L}}(\mathbf{x}, \mathbf{x}_i) y_i,$$

where $0 \leq W_{\mathscr{L}}(\mathbf{x}, \mathbf{x}_i) \leq 1$ are convex weights satisfying $\sum_{i=1}^{n} W_{\mathscr{L}}(\mathbf{x}, \mathbf{x}_i) = 1$. An example of a local weighted estimator is a random forest regression ensemble.

To apply McDiarmid's inequality, we define ϕ as follows:

$$\phi((\mathbf{x}_1, y_1), \ldots, (\mathbf{x}_n, y_n)) = \sum_{i=1}^{n} W_{\mathscr{L}}(\mathbf{x}, \mathbf{x}_i) y_i,$$

where $\mathbf{x}$ is some arbitrary value. To derive the concentration inequality, we will assume $\hat{f}$ satisfies a property we call k-robustness. To define this term, let us lighten the notational load by denoting $\mathbf{z}_i = (\mathbf{x}_i, y_i)$.

Definition 2.13. The local weighted estimator $\hat{f}$ is said to satisfy k-robustness if there exists a $1 \leq k \leq n$ such that for each $\mathbf{x}$, and each data set $\mathscr{L}_i' = \{\mathbf{z}_1', \ldots, \mathbf{z}_n'\}$ that differs from $\mathscr{L}$ at a single coordinate $1 \leq i \leq n$ (i.e, $\mathbf{z}_i' \neq \mathbf{z}_i$ and $\mathbf{z}_j' = \mathbf{z}_j$ for $j \neq i$), we have

$$\sum_{j=1}^{n} |W_{\mathscr{L}}(\mathbf{x}, \mathbf{x}_j) - W_{\mathscr{L}_i'}(\mathbf{x}, \mathbf{x}_j')| \leq \frac{k}{n}, \qquad W_{\mathscr{L}_i'}(\mathbf{x}, \mathbf{x}_i') \leq \frac{k}{n}.$$

□

We also require the observed responses to be bounded: $|y_i| \leq C$ for $i = 1, \ldots, n$. Then for each $1 \leq i \leq n$

$$\begin{aligned}
&\left|\phi(\mathbf{z}_1, \ldots, \mathbf{z}_{i-1}, \mathbf{z}_i, \mathbf{z}_{i+1}, \ldots, \mathbf{z}_n) - \phi(\mathbf{z}_1, \ldots, \mathbf{z}_{i-1}, \mathbf{z}_i', \mathbf{z}_{i+1}, \ldots, \mathbf{z}_n)\right| \\
&\quad = \left|\sum_{j\neq i}\left[W_{\mathscr{L}}(\mathbf{x}, \mathbf{x}_j) - W_{\mathscr{L}_i'}(\mathbf{x}, \mathbf{x}_j)\right] y_j + \left[W_{\mathscr{L}}(\mathbf{x}, \mathbf{x}_i) y_i - W_{\mathscr{L}_i'}(\mathbf{x}, \mathbf{x}_i') y_i'\right]\right| \\
&\quad \leq \sum_{j\neq i} |W_{\mathscr{L}}(\mathbf{x}, \mathbf{x}_j) - W_{\mathscr{L}_i'}(\mathbf{x}, \mathbf{x}_j)||y_j| + |W_{\mathscr{L}}(\mathbf{x}, \mathbf{x}_i) y_i - W_{\mathscr{L}_i'}(\mathbf{x}, \mathbf{x}_i') y_i'| \\
&\quad \leq C\frac{k}{n} + |W_{\mathscr{L}}(\mathbf{x}, \mathbf{x}_i) y_i - W_{\mathscr{L}_i'}(\mathbf{x}, \mathbf{x}_i') y_i| + |W_{\mathscr{L}_i'}(\mathbf{x}, \mathbf{x}_i')[y_i - y_i']| \\
&\quad \leq 4C\frac{k}{n}.
\end{aligned}$$

Apply McDiarmid's inequality with $c_i = 4Ck/n$ to deduce

$$\mathbb{P}\left\{\left|\hat{f}(\mathbf{x}) - \mathbb{E}[\hat{f}(\mathbf{x})]\right| \geq t\right\} \leq \exp\left(-\frac{nt^2}{8C^2k^2}\right).$$

Note the right-hand side remains exponentially small as long as $n/k^2 \to \infty$. □

2.19 *U*-Statistics

We cover some basic theory for U-statistics. For a more detailed presentation see Hoeffding [2, 3]. Let $X_1, X_2, \ldots$ be i.i.d. random variables and let $f : \mathbb{R}^m \to \mathbb{R}$ be a real-valued function. For $n \ge m$, define

$$\overline{f}_{n,m} = \frac{1}{n(n-1)\cdots(n-m+1)} \sum_{S_m} f(X_{i_1}, \ldots, X_{i_m})$$

where the sum is over S_m equal to all $\binom{n}{m}$ combinations of m distinct integers $\{i_1, \ldots, i_m\}$ selected from $\{1, \ldots, n\}$. Call the random variable $\overline{f}_{n,m}$ a U-statistic of order m.

Without loss of generality, it can be assumed that f is a symmetric function. This is because we can redefine $\overline{f}_{n,m}$ by replacing each term $f(X_{i_1}, \ldots, X_{i_m})$ in the above sum with the mean of f evaluated over the $m!$ permutations of $\{X_{i_1}, \ldots, X_{i_m}\}$. From this, it also follows that $\overline{f}_{n,m}$ is an unbiased estimator for $\mathbb{E}[f(X_1, \ldots, X_m)]$.

Here are some examples of U-statistics.

1. If $f(x) = x$ then $\overline{f}_{n,1} = \overline{X}$.
2. If $f(x_1, x_2) = |x_1 - x_2|$ then $\overline{f}_{n,2} = 2/(n(n-1)) \sum_{i>j} |X_i - X_j|$.
3. If $f(x_1, x_2) = (x_1 - x_2)^2/2$ then $\overline{f}_{n,2} = 1/(n-1) \sum_i (X_i - \overline{X})^2$.

Theorem 2.44. *If* $\mathbb{E}|f(X_1, \ldots, X_m)| < \infty$, *then*

$$\overline{f}_{n,m} \to \mathbb{E}\left[f(X_1, \ldots, X_m)\right] \quad \textit{almost surely.}$$

Theorem 2.44 proves many interesting results. Our first example, $f(x) = x$, implies the strong law of large numbers. Our third example, $f(x_1, x_2) = (x_1 - x_2)^2/2$, shows that the sample variance estimator is almost surely consistent for the population variance under the assumption of a finite second moment.

Another useful bound due to Hoeffding is as follows. This is similar to Theorem 2.41 but applies to U-statistics.

Theorem 2.45. *Let* $\theta = \mathbb{E}[f(X_1, \ldots, X_m)]$ *and* $\sigma^2 = Var(f(X_1, \ldots, X_m))$. *If* $a \le f(X_1, \ldots, X_m) \le b$, *then for each* $t > 0$ *and* $n \ge m$

$$\mathbb{P}\{\overline{f}_{n,m} - \theta \ge t\} \le \exp\left(-\frac{2nt^2}{m(b-a)^2}\right)$$

and

$$\mathbb{P}\{\overline{f}_{n,m} - \theta \ge t\} \le \exp\left(-\frac{nt^2}{2m(\sigma^2 + t(b-\theta)/3)}\right).$$

Remark 2.4. When $m = 1$ the first inequality implies Theorem 2.41. □

Exercises

Exercise 2.1. Use Holder's inequality to prove Theorem 2.2.

Exercise 2.2. Prove all properties stated in Chapter 2.7.

Exercise 2.3. Prove (2.2) and (2.3).

Exercise 2.4. Suppose $\mathbf{X} = (X_1, \ldots, X_p)^T$ is a random vector with mean $\boldsymbol{\mu}$ and covariance matrix $\boldsymbol{\Sigma}$ and let $\mathbf{a} \in \mathbb{R}^p$ be a fixed vector. If r_j is the correlation between X_j and $\mathbf{a}^T\mathbf{X}_j$, show that $\mathbf{r} = (r_1, \ldots, r_p)^T$ can be written as $\mathbf{r} = (s\mathbf{D})^{-1/2}\boldsymbol{\Sigma}\mathbf{a}$ where $s = \mathbf{a}^T\boldsymbol{\Sigma}\mathbf{a}$ and $\mathbf{D} = \text{diag}(\boldsymbol{\Sigma})$.

Exercise 2.5. Prove Theorem 2.4.

Exercise 2.6. Write a program to calculate the generalized inverse $\mathbf{A}^+$ of any matrix $\mathbf{A}$. Illustrate the program on an interesting example and directly verify that the three properties of a generalized inverse hold in your example.

Exercise 2.7. Prove the properties for derivatives stated in Chapter 2.9.

Exercise 2.8. Prove the following properties for the gradient:

(a) $\nabla(\alpha f(\mathbf{x}) + \beta g(\mathbf{x})) = \alpha\nabla(f(\mathbf{x})) + \beta\nabla(g(\mathbf{x}))$.

(b) $\nabla(f(\mathbf{x})g(\mathbf{x})) = g(\mathbf{x})\nabla(f(\mathbf{x})) + f(\mathbf{x})\nabla(g(\mathbf{x}))$.

(c) $\nabla(g(\mathbf{x})^{-1}) = g(\mathbf{x})^{-2}\nabla(g(\mathbf{x}))$.

Exercise 2.9. Let $\mathbf{f}, \mathbf{g}$ be differentiable functions from $\mathbb{R}^n$ to $\mathbb{R}^m$. Prove the following properties for the directional derivative $\mathbf{D_u}$.

(a) $D_{\mathbf{u}}(\mathbf{f} + \mathbf{g}) = D_{\mathbf{u}}(\mathbf{f}) + D_{\mathbf{u}}(\mathbf{g})$.

(b) $D_u(\mathbf{f} \circ \mathbf{g}) = \mathbf{f} \circ D_{\mathbf{u}}\mathbf{g} + (D_{\mathbf{u}}\mathbf{f}) \circ \mathbf{g}$.

Exercise 2.10. The QR decomposition can be used to efficiently solve a system of linear equations. To solve for $\mathbf{x}$ in $\mathbf{Ax} = \mathbf{z}$, for $\mathbf{A}$ a square $p \times p$ matrix, we determine the QR decomposition $\mathbf{A} = \mathbf{QR}$, substitute, and with some rearrangement obtain $\mathbf{Rx} = \mathbf{Q}^T\mathbf{z}$. Let $\mathbf{c} = \mathbf{Q}^T\mathbf{z}$. Show that $\mathbf{x}$ can be solved efficiently in $\mathbf{Rx} = \mathbf{c}$ by using back substitution. As a hint, notice that $\mathbf{Rx} = \mathbf{c}$ implies

$$\begin{pmatrix} r_{11} & r_{12} & \cdots & r_{1p} \\ & r_{22} & \cdots & r_{2p} \\ & & \ddots & \vdots \\ & & & r_{pp} \end{pmatrix} \begin{pmatrix} x_1 \\ x_2 \\ \vdots \\ x_p \end{pmatrix} = \begin{pmatrix} c_1 \\ c_2 \\ \vdots \\ c_p \end{pmatrix}.$$

In particular, the last row yields $r_{p,p}x_p = c_p$, which immediately yields $x_p = c_p / r_{p,p}$. Substituting this into the second last row, and solving, and repeating recursively, yields the back substitution method. State this solution.

Exercise 2.11. Find an underdetermined data set ($p > n$) and calculate the minimum least squares estimator described in Corollary 2.6 using the generalized inverse.

Exercise 2.12. Devise an underdetermined linear model ($p > n$) and solve for $\boldsymbol{\beta}$ by using the QR decomposition. Now use the generalized inverse to obtain the minimum least squares estimator for $\boldsymbol{\beta}$. Which estimator is better (hint: consider the best-approximation property and the L_2-length of each solution).

Exercise 2.13. Let $\boldsymbol{\beta}^+$ be the minimum least squares estimator described in Chapter 2.8. Describe the Taylor's series for $f(\boldsymbol{\beta}) = ||\mathbf{Y} - \mathbf{D}\boldsymbol{\beta}||_F^2$ pivoted around $\boldsymbol{\beta} = \boldsymbol{\beta}^+$ to two terms where $\mathbf{D}$ is the design matrix from a linear regression for $\mathbf{Y}$.

(a) What does the gradient equal?

(b) What is the Hessian matrix in this example?

(c) What properties of f can you infer from the gradient and Hessian in this example?

Exercise 2.14. Let $\{\mathbf{X}_n\}_1^\infty$ be a sequence of random vectors converging to $\mathbf{X}$. Give examples where:

(a) $\mathbf{X}_n \overset{d}{\rightsquigarrow} \mathbf{X}$.

(b) $\mathbf{X}_n \overset{p}{\rightarrow} \mathbf{X}$.

(c) $\mathbf{X}_n \overset{a.s.}{\rightarrow} \mathbf{X}$.

Exercise 2.15. Prove convergence in distribution to a constant implies convergence in probability.

Exercise 2.16. Prove convergence in probability implies convergence in distribution.

Exercise 2.17. Prove convergence almost surely implies convergence in probability.

Exercise 2.18. Prove convergence in L_p implies convergence in probability.

Exercise 2.19. Let $\{X_n\}_1^\infty$ be a sequence of Bernoulli random variables such that $\mathbb{P}\{X_n = 1\} = 1/n$. Show that $X_n \overset{p}{\rightarrow} 0$. However, using the Borel-Canelli lemma prove $X_n \overset{a.s.}{\nrightarrow} 0$ if X_n are independent.

Exercise 2.20. Given an example where $\mathbf{X}_n \overset{p}{\rightarrow} \mathbf{X}$ but $\mathbf{X}_n \overset{a.s.}{\nrightarrow} \mathbf{X}$.

Exercise 2.21. Suppose that X, Y, X_n, Y_n for $n = 1, 2, \ldots$ are random variables constructed over the same probability space $(\Omega, \mathscr{A}, \mathbb{P})$. Prove or disprove the following assertions:

(a) If $X_n \overset{a.s.}{\rightarrow} 0$ and $Y_n \overset{d}{\rightsquigarrow} Y$ then $X_n Y_n \overset{p}{\rightarrow} 0$.

(b) If $X_n \overset{a.s.}{\rightarrow} 0$ and $Y_n \overset{p}{\rightarrow} Y$ then $X_n Y_n \overset{d}{\rightsquigarrow} 0$.

(c) If $X_n \overset{a.s.}{\rightarrow} X$ and $Y_n \overset{p}{\rightarrow} 0$ then $X_n Y_n \overset{a.s.}{\rightarrow} 0$.

(d) If $X_n \overset{a.s.}{\rightarrow} 0$ and $Y_n \overset{p}{\rightarrow} 0$ then $X_n Y_n \overset{a.s.}{\rightarrow} 0$.

Bonus question: If $X_n Y_n \overset{a.s.}{\rightarrow} 0$ what necessary conditions are required for the convergence of X_n and Y_n?

Exercise 2.22. Prove the remaining propositions from Chapter 2.16.

Exercise 2.23. Let $X, X_1, X_2, \ldots$ be independent and identically distributed random variables with $\mathbb{E}(X) = 0$ and $\mathbb{E}(X^2) < \infty$. Use the continuous mapping theorem to prove

$$T_n = \sqrt{n} \cdot \frac{\overline{X}_n}{s_n} \overset{d}{\rightsquigarrow} \mathrm{N}(0,1)$$

where $\overline{X}_n = \sum_{i=1}^n X_i / n$ and $s_n^2 = \sum_{i=1}^n (X_i - \overline{X}_n)^2/(n-1)$. (Hint: start by proving $s_n^2 \overset{p}{\rightarrow} \mathrm{Var}(X)$.)

Bibliography

[1] Hartman P. and Wintner A. On the law of the iterated logarithm. *American Journal of Mathematics*, 63:169–176, 1941.

[2] Hoeffding W. A class of statistics with asymptotically normal distribution. *Annals of Mathematical Statistics*, 19:293–325, 1948.

[3] Hoeffding W. The strong law of large numbers for U-statistics. *Technical Report, North Carolina State University, Department of Statistics*, 1961.

[4] Hoeffding W. Probability inequalities for sums of bounded random variables. *Journal of the American Statistical Association,* 58(301):13–30, 1963.

[5] McDiarmid C. On the method of bounded differences. *Surveys in Combinatorics,* Invited Papers at the Twelfth British Combinatorial Conference: 148–188, 1989.

[6] Penrose R. A generalized inverse for matrices. *Proceedings - Cambridge Philosophical Society,* 51:406–413, 1955.

3

Multivariate Normal Distribution

3.1 Introduction

A multivariate normal (or Gaussian) distribution is a generalization of the one-dimensional (univariate) normal distribution to higher dimensions. It describes a random vector whose components are jointly normally distributed. The multivariate normal distribution is a cornerstone of multivariate statistical analysis. Its mathematical properties and ability to describe the joint behavior of multiple random variables make it a critical tool for developing and applying statistical methods across a wide range of disciplines. Understanding the multivariate normal distribution is essential for effectively modeling, analyzing, and interpreting multivariate data.

This chapter reviews the properties of the multivariate normal distribution and provides various applications of its use, including the famous regression to the mean phenomenon, Stein's lemma, normal mixture models, Bayesian multivariate normal hierarchical models, and unbiased estimating equations with Godambe information. The latter illustrates how the multivariate normal distribution often arises in asymptotic theory. Readers will also find an appendix at the end that conveniently lists several classical statistics related to the univariate and multivariate normal distributions.

3.2 Properties of the Multivariate Normal

A p-dimensional random vector $\mathbf{X} = (X_1, \ldots, X_p)^T$ is called a multivariate normal vector if has the density

$$f_{\mathbf{X}}(\mathbf{x}|\boldsymbol{\mu}, \boldsymbol{\Sigma}) = |2\pi\boldsymbol{\Sigma}|^{-1/2} \exp\left(-\frac{1}{2}(\mathbf{x}-\boldsymbol{\mu})^T \boldsymbol{\Sigma}^{-1}(\mathbf{x}-\boldsymbol{\mu})\right).$$

In this case, we say $\mathbf{X}$ is distributed as a multivariate normal and write this as $\mathbf{X} \sim \mathrm{N}_p(\boldsymbol{\mu}, \boldsymbol{\Sigma})$, where $\boldsymbol{\mu}$ is the mean and $\boldsymbol{\Sigma} > 0$ is the covariance matrix. Below we list several key properties of the multivariate normal to be used throughout the book.

1. *Standard multivariate normal:* When $\boldsymbol{\mu} = \mathbf{0}$ and $\boldsymbol{\Sigma} = \mathbf{I}$, then $\mathbf{X}$ is called a standard multivariate normal, in which case it is denoted by $\mathbf{Z} = (Z_1, \ldots, Z_p)^T$. Its density equals

$$f_{\mathbf{Z}}(\mathbf{z}) = (2\pi)^{-p/2} \exp\left(-\frac{1}{2}\mathbf{z}^T\mathbf{z}\right) = (2\pi)^{-p/2} \exp\left(-\frac{1}{2}\sum_{j=1}^{p} z_j^2\right).$$

   Note that $Z_k \overset{\text{iid}}{\sim} \mathrm{N}(0, 1)$ for $k = 1, \ldots, p$ and $\mathbf{Z}^T\mathbf{Z} \sim \chi_p^2$.

DOI: 10.1201/9781003493679-3

2. *Bivariate multivariate normal:* When $p = 2$ the density for $\mathbf{X}$ is $f(x_1, x_2)$, which can be written

$$\left(2\pi\sqrt{\sigma_{11}\sigma_{22}(1-\rho)^2}\right)^{-1}$$
$$\times \exp\left(-\frac{1}{2(1-\rho^2)}\left[\frac{(x_1-\mu_1)^2}{\sigma_{11}} + \frac{(x_2-\mu_2)^2}{\sigma_{22}} - 2\rho\frac{(x_1-\mu_1)(x_2-\mu_2)}{\sqrt{\sigma_{11}\sigma_{22}}}\right]\right).$$

3. *Characteristic function:* Let $\mathbf{X} \sim \mathrm{N}_p(\boldsymbol{\mu}, \boldsymbol{\Sigma})$ then the characteristic function for $\mathbf{X}$ evaluated at $\mathbf{t} \in \mathbb{R}^p$ is

$$\psi(\mathbf{t}) = \mathbb{E}(\exp(i\mathbf{t}^T\mathbf{X})) = \exp\left[i\mathbf{t}^T\boldsymbol{\mu} - \frac{1}{2}\mathbf{t}^T\boldsymbol{\Sigma}\mathbf{t}\right].$$

4. *Cramér–Wold device (see Theorem 2.22):* $\mathbf{X}$ is multivariate normal if and only if $\ell^T\mathbf{X}$ is univariate normal for each fixed $\ell \in \mathbb{R}^p$.

5. *Cramér–Wold properties:* The following useful properties for $\mathbf{X} \sim \mathrm{N}_p(\boldsymbol{\mu}, \boldsymbol{\Sigma})$ follow directly from the Cramér–Wold Device and by basic properties for the expectation and variance of a random vector (Chapter 2):

   (a) $\ell^T\mathbf{X} \sim \mathrm{N}(\ell^T\boldsymbol{\mu}, \ell^T\boldsymbol{\Sigma}\ell)$, for each $\ell \in \mathbb{R}^p$.

   (b) $\mathbf{AX} + \mathbf{b} \sim \mathrm{N}_q(\mathbf{A}\boldsymbol{\mu} + \mathbf{b}, \mathbf{A}\boldsymbol{\Sigma}\mathbf{A}^T)$, where $\mathbf{A}_{q\times p}$ and $\mathbf{b}_{q\times 1}$ are fixed.

   (c) If $\mathbf{X}$ is partitioned as $(\mathbf{X}_1, \mathbf{X}_2)$, where $\mathbf{X}_1 \in \mathbb{R}^q$ and $\mathbf{X}_2 \in \mathbb{R}^{p-q}$, then

   $$(\mathbf{X}_1 | \mathbf{X}_2 = \mathbf{x}_2) \sim \mathrm{N}_q(\boldsymbol{\mu}^*, \boldsymbol{\Sigma}^*),$$

   where $\boldsymbol{\mu}^* = \boldsymbol{\mu}_1 + \boldsymbol{\Sigma}_{12}\boldsymbol{\Sigma}_{22}^{-1}(\mathbf{x}_2 - \boldsymbol{\mu}_2)$ and $\boldsymbol{\Sigma}^* = \boldsymbol{\Sigma}_{11} - \boldsymbol{\Sigma}_{12}\boldsymbol{\Sigma}_{22}^{-1}\boldsymbol{\Sigma}_{21}$.

6. *Independence and zero correlation are equivalent:* The multivariate normal is one of the few distributions where independence can be inferred directly from correlation. By 5(c) if $\boldsymbol{\Sigma}_{12}$ is zero then the distribution of $\mathbf{X}_1|\mathbf{X}_2$ does not depend on $\mathbf{X}_2$ which shows that $\mathbf{X}_1$ and $\mathbf{X}_2$ are independent. In other words, if $\mathbf{X}_1$ and $\mathbf{X}_2$ are uncorrelated, then $\mathbf{X}_1$ and $\mathbf{X}_2$ are independent if $\mathbf{X}$ is multivariate normal.

7. *Simulation:* A multivariate normal can always be simulated from a standard normal:

$$\mathbf{X} \sim \mathrm{N}_p(\boldsymbol{\mu}, \boldsymbol{\Sigma}) \quad \text{if} \quad \mathbf{X} = \boldsymbol{\Sigma}^{1/2}\mathbf{Z} + \boldsymbol{\mu}.$$

   This follows directly from a change of variables. Let $u(\mathbf{x}) = \boldsymbol{\Sigma}^{-1/2}(\mathbf{x} - \boldsymbol{\mu})$ be the linear map from $\mathbf{x}$ to $\mathbf{z}$, then

$$\begin{aligned} f_{\mathbf{X}}(\mathbf{x})\,d\mathbf{x} &= f_{\mathbf{Z}}(u(\mathbf{x}))\left|\frac{\partial}{\partial\mathbf{x}}u(\mathbf{x})\right|^+ d\mathbf{x} \\ &= f_{\mathbf{Z}}\left(\boldsymbol{\Sigma}^{-1/2}(\mathbf{x} - \boldsymbol{\mu})\right)|\boldsymbol{\Sigma}^{-1/2}|\,d\mathbf{x} \end{aligned}$$

   where $|\cdot|^+$ is used for the positive value of the determinant. This gives the desired density.

8. *Mahalanobis distance:* The Mahalanobis distance, $\sqrt{D(\mathbf{X})}$, is defined by

$$D(\mathbf{X}) = (\mathbf{X} - \boldsymbol{\mu})^T\boldsymbol{\Sigma}^{-1}(\mathbf{X} - \boldsymbol{\mu}).$$

   The Mahalanobis distance is particularly useful in multivariate statistics because it accounts for the correlations between variables. In essence, it measures how many standard deviations a point is from the mean of the distribution, considering the shape of the distribution as defined by the covariance matrix. The distance can be used to construct various test-statistics. For example, $D(\mathbf{X}) \sim \chi_p^2$ by (7) assuming normality, and therefore

$$\mathbb{P}\left\{\mathbf{x}: D(\mathbf{x}) \leq \chi_p^2(\alpha)\right\} = 1-\alpha,$$

where $\chi_p^2(\alpha)$ is the $100(1-\alpha)$ percentile for a χ_p^2.

9. *Contours:* The probability density contours for a $\mathrm{N}_p(\boldsymbol{\mu}, \boldsymbol{\Sigma})$ are p-dimensional ellipsoids centered at $\boldsymbol{\mu}$ with axes along the eigenvectors of $\boldsymbol{\Sigma}$.

10. *Multivariate central limit theorem:* Suppose that $\mathbf{X}_1, \ldots, \mathbf{X}_n$ is an i.i.d. sequence with $\mathbb{E}(\mathbf{X}_j) = \boldsymbol{\mu}$ and $\mathrm{Var}(\mathbf{X}_j) = \boldsymbol{\Sigma}$. Then
$$\sqrt{n}(\overline{\mathbf{X}} - \boldsymbol{\mu}) \overset{\mathrm{d}}{\rightsquigarrow} \mathrm{N}_p(\mathbf{0}, \boldsymbol{\Sigma}).$$

11. *Density for a sample:* Suppose that $\mathbf{X}_1, \ldots, \mathbf{X}_n$ are independent $\mathrm{N}_p(\boldsymbol{\mu}, \boldsymbol{\Sigma})$ vectors. Then the joint density equals
$$\begin{aligned} f(\mathbf{x}_1, \ldots, \mathbf{x}_n) &\\ &= |2\pi\boldsymbol{\Sigma}|^{-n/2} \exp\left(-\frac{1}{2}\sum_{i=1}^{n}(\mathbf{x}_i - \boldsymbol{\mu})^T \boldsymbol{\Sigma}^{-1}(\mathbf{x}_i - \boldsymbol{\mu})\right) \\ &= |2\pi\boldsymbol{\Sigma}|^{-n/2} \exp\left(-\frac{n}{2}\mathrm{trace}\left[\boldsymbol{\Sigma}^{-1}\mathbf{S}\right] - \frac{n}{2}(\overline{\mathbf{x}} - \boldsymbol{\mu})^T \boldsymbol{\Sigma}^{-1}(\overline{\mathbf{x}} - \boldsymbol{\mu})\right), \end{aligned}$$
where $\mathbf{S}$ is the sample variance (see Chapter 2).

12. *MLE for $\boldsymbol{\mu}$ and $\boldsymbol{\Sigma}$:* By the previous result, the log-likelihood for the sample in terms of $\boldsymbol{\mu}$ and $\boldsymbol{\Sigma}$ is
$$l(\boldsymbol{\mu}, \boldsymbol{\Sigma}) = -\frac{n}{2}\log|2\pi\boldsymbol{\Sigma}| - \frac{n}{2}\mathrm{trace}\left[\boldsymbol{\Sigma}^{-1}\mathbf{S}\right] - \frac{n}{2}(\overline{\mathbf{x}} - \boldsymbol{\mu})^T \boldsymbol{\Sigma}^{-1}(\overline{\mathbf{x}} - \boldsymbol{\mu}).$$
Since $\boldsymbol{\Sigma}^{-1} > 0$, the quadratic $(\overline{\mathbf{x}} - \boldsymbol{\mu})^T \boldsymbol{\Sigma}^{-1}(\overline{\mathbf{x}} - \boldsymbol{\mu})$ is non-negative. Therefore, the likelihood is maximized for $\boldsymbol{\mu}$ at the MLE value $\hat{\boldsymbol{\mu}} = \overline{\mathbf{x}}$. To determine the MLE for $\boldsymbol{\Sigma}$, let $\mathbf{A} > 0$ be any $p \times p$ matrix and define
$$\psi(\boldsymbol{\Sigma}) = |\boldsymbol{\Sigma}|^{-n/2} \exp\left(-\frac{1}{2}\mathrm{trace}\left[\boldsymbol{\Sigma}^{-1}\mathbf{A}\right]\right).$$
Then it can be shown that $\psi(\boldsymbol{\Sigma})$ is maximized over $\boldsymbol{\Sigma} > 0$ by $\boldsymbol{\Sigma} = n^{-1}\mathbf{A}$. Using this, it follows that the MLE for $\boldsymbol{\Sigma}$ is $\hat{\boldsymbol{\Sigma}} = \mathbf{S}$.

3.3 Interesting Properties of the Multivariate Normal

This section collects together some interesting and perhaps surprising results due to the properties of the multivariate normal.

3.3.1 Regression to the mean

Regression toward the mean occurs when the correlation between two variables is less than one. This phenomenon was discovered by Francis Galton in his work on correlation. Galton observed that the average height of parents and the height of their children were not identical but were modulated by the correlation between the two measurements. Table 3.1 describes the data used by Galton. The height of children is given in the columns, and the average height of their parents is given in the rows. The table shows the frequencies of these measurements. Taller parents tend to have slightly shorter children, and shorter parents tend to have slightly taller children. This tilting away from the $y = x$ axis is called regression to the mean.

TABLE 3.1
Galton's data used to study regression to the mean. Displayed are frequencies of a child's height compared to a parent's height.

Average Height of Parents	Height of Child														Total Number of Children
	< 61.7	62.2	63.2	64.2	65.2	66.2	67.2	68.2	69.2	70.2	71.2	72.2	73.2	> 73.7	
> 73.0												1	3		4
72.5								1	2	1	2	7	2	4	19
71.5					1	3	4	3	5	10	4	9	2	2	43
70.5	1		1		1	1	3	12	18	14	7	4	3	3	68
69.5			1	16	4	17	27	20	33	25	20	11	4	5	183
68.5	1		7	11	16	25	31	34	48	21	18	4	3		219
67.5		3	5	14	15	36	38	28	38	19	11	4			211
66.5		3	3	5	2	17	17	14	13	4					78
65.5	1		9	5	7	11	11	7	7	5	2	1			66
64.5	1	1	4	4	1	5	5		2						23
< 64.0	1		2	4	1	2	2	1	1						14
Total	5	7	32	59	48	117	138	120	167	99	64	41	17	14	928

In fact we can show that regression to the mean is a statistical phenomenon. One way to explain this is by the following regression argument. Assume that (X, Y) satisfy a simple slope-intercept relationship

$$Y_i = \alpha + \beta X_i + \varepsilon_i, \quad i = 1, \ldots, n.$$

The least squares estimators are

$$\hat{\alpha} = \overline{Y} - \hat{\beta}\overline{X}, \quad \hat{\beta} = \frac{\sum_{i=1}^{n}(X_i - \overline{X})(Y_i - \overline{Y})}{\sum_{i=1}^{n}(X_i - \overline{X})^2} = r_{XY}\frac{S_{YY}^{1/2}}{S_{XX}^{1/2}}$$

The predicted value is $\hat{Y}_i = \hat{\alpha} + \hat{\beta}X_i$. Rearranged, this gives

$$\frac{\hat{Y}_i - \overline{Y}}{S_{YY}^{1/2}} = r_{YX}\frac{X_i - \overline{X}}{S_{XX}^{1/2}}.$$

Thus, an increase in one standardized value of X corresponds to an increase of a scale factor r_{YX} in the standardized value of $\hat{Y}$, where r_{YX} is the correlation. For example, if X is the average height of a child's parent, the child's height Y will be less by a factor of r_{YX} than their parent's height. This often comes as a surprise, because the smaller the correlation, the more Y is regressed to the mean $\bar{Y}$, which might be non-intuitive to the layperson.

Another way to understand regression to the mean is that it is a property of bivariate normality. To explain this, we use the conditional distribution result stated earlier for the multivariate normal. Assuming that the data is bivariate normal,

$$\begin{pmatrix} Y \\ X \end{pmatrix} \sim \mathrm{N}_p\left(\begin{pmatrix} \mu_Y \\ \mu_X \end{pmatrix}, \begin{bmatrix} S_{YY} & S_{YX} \\ S_{XY} & S_{YY} \end{bmatrix}\right),$$

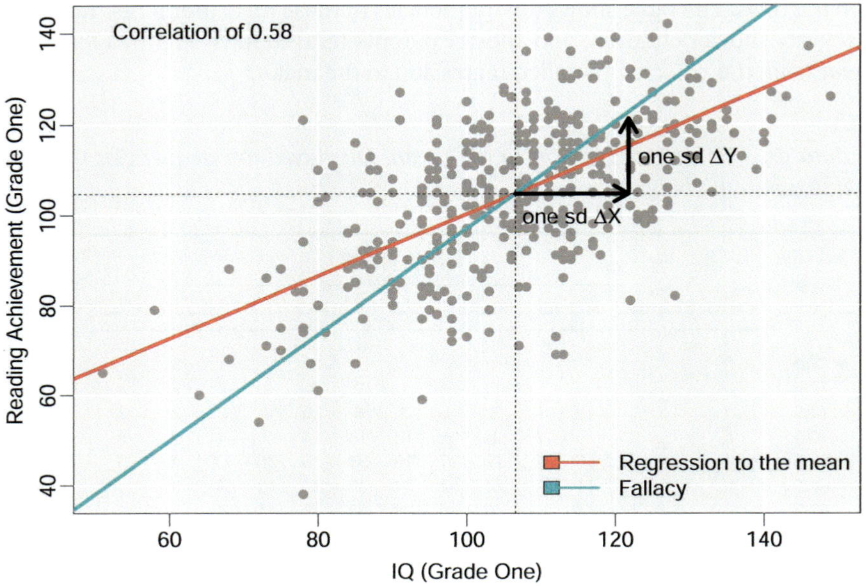

FIGURE 3.1
Regression to the mean for Connecticut Longitudinal Study (Chapter 3.5).

then by the previous result, we have $(Y|X = x) \sim \mathrm{N}(\mu^*, S^*)$ where

$$\begin{aligned} \mu^* &= \mu_Y + S_{YX}S_{XX}^{-1}(x - \mu_X) \\ S^* &= S_{YY} - S_{YX}S_{XX}^{-1}S_{XY}. \end{aligned}$$

This implies

$$\frac{\mu^* - \mu_Y}{S_{YY}^{1/2}} = \frac{\mathbb{E}(Y|X = x) - \mu_Y}{S_{YY}^{1/2}} = r_{YX}\frac{(x - \mu_X)}{S_{XX}^{1/2}}.$$

Therefore, as before, this shows an increase in the standardized value of X corresponds to an r_{YX} increase in the standardized value of the conditional mean $\mu^* = \mathbb{E}(Y|X = x)$. Figure 3.1 displays IQ and reading achievement scores for children in grade 1 from the Connecticut Longitudinal Study (see Chapter 3.5). As these scores are by definition standardized, the regression to the mean effect is clearly seen.

3.3.2 Correlation in regression models

The following example shows how correlation can play a deceptive role in determining effect size. Consider the following model

$$Y|\mathbf{X} \sim \mathrm{N}(F(\mathbf{X}), \sigma^2), \tag{3.1}$$

where

$$\begin{aligned} F(\mathbf{X}) = \; & 5X_1 + 5X_2 + 2X_3 + 0 \cdot X_4 - 5X_5 - 5X_6 \\ & -2X_7 + 0 \cdot X_8 + 0 \cdot X_9 + 0 \cdot X_{10} + 0 \cdot X_{11} + 0 \cdot X_{12} \end{aligned} \tag{3.2}$$

and $\mathbf{X} \sim \mathrm{N}_{12}(\mathbf{0}, \mathbf{\Sigma})$, where

$$\mathbf{\Sigma} = \begin{bmatrix} 1 & \rho & \rho & \rho & 0 & 0 & 0 & 0 & 0 & 0 & 0 & 0 \\ \rho & 1 & \rho & \rho & 0 & 0 & 0 & 0 & 0 & 0 & 0 & 0 \\ \rho & \rho & 1 & \rho & 0 & 0 & 0 & 0 & 0 & 0 & 0 & 0 \\ \rho & \rho & \rho & 1 & 0 & 0 & 0 & 0 & 0 & 0 & 0 & 0 \\ 0 & 0 & 0 & 0 & 1 & 0 & 0 & 0 & 0 & 0 & 0 & 0 \\ 0 & 0 & 0 & 0 & 0 & 1 & 0 & 0 & 0 & 0 & 0 & 0 \\ 0 & 0 & 0 & 0 & 0 & 0 & 1 & 0 & 0 & 0 & 0 & 0 \\ 0 & 0 & 0 & 0 & 0 & 0 & 0 & 1 & 0 & 0 & 0 & 0 \\ 0 & 0 & 0 & 0 & 0 & 0 & 0 & 0 & 1 & 0 & 0 & 0 \\ 0 & 0 & 0 & 0 & 0 & 0 & 0 & 0 & 0 & 1 & 0 & 0 \\ 0 & 0 & 0 & 0 & 0 & 0 & 0 & 0 & 0 & 0 & 1 & 0 \\ 0 & 0 & 0 & 0 & 0 & 0 & 0 & 0 & 0 & 0 & 0 & 1 \end{bmatrix}_{12\times 12}. \tag{3.3}$$

Thus the first four attributes X_1, X_2, X_3, X_4 are block-correlated with correlation $\rho > 0$ while the remaining attributes are uncorrelated with themselves and the first four attributes.

The model described by (3.1)–(3.3) is a linear model with conditional expectation $\mathbb{E}(Y|\mathbf{X}) = F(\mathbf{X})$, where by properties of the multivariate normal, the response and random feature vector $\mathbf{X}$ are related by

$$\mathrm{Cor}(Y, \mathbf{X})^T = \frac{\mathbf{\Sigma}\boldsymbol{\beta}}{\mathrm{Var}(Y)^{1/2}}.$$

We use $\boldsymbol{\beta} = (5, 5, 2, 0, -5, -5, -2, 0, 0, 0, 0, 0)^T$ to denote the vector of coefficients for the features. In particular, if we consider X_4, notice that $\mathrm{Cor}(Y, X_4) = 12\rho$ even though X_4 has a zero coefficient in the regression model. This type of "spurious" correlation can have unintended consequences when performing inference as it can lead to misinterpreted effect sizes.

As an illustration, data was simulated from a linear model ($n = 500$, $\rho = 0.9$). The data was fit using least squares. Then, variable X_3 was removed, and the data was refit using least squares. The results are given in Table 3.2. In the full model, the least squares estimator correctly identifies the true signal variables X_1, X_2, X_3, X_5, X_6, and X_7. However, in the misspecified model, X_4 is incorrectly identified as significant. Removing X_3 allows X_4 to act as a proxy for X_3 due to their high pairwise correlation. Therefore, in the presence of correlation, the estimated effect correctly reflects the true causal variables only when the model is correctly specified.

This same issue occurs with marginal plots. It is common (but naive) to plot the response against a variable to determine potential relationships. Plotting Y against X_4 in this example would suggest that X_4 is causal due to its correlation with X_3. The only way to prevent this is to *adjust for all signal variables*, as was done in the full model least squares analysis. A related discussion of this problem from the perspective of causal inference is given in Chapter 20.5. Chapter 20.5.2 illustrates the difference between naive marginal plots and partial dependence causal plots.

3.3.3 Singular covariance

In our examples, we have assumed a positive definite covariance matrix, $\mathbf{\Sigma} > 0$. But what happens if $\mathbf{\Sigma}$ is singular? In this case, the multivariate normal distribution $\mathrm{N}_p(\boldsymbol{\mu}, \mathbf{\Sigma})$ is no longer defined on $\mathbb{R}^p$ but concentrates on a translate of a subspace

$$\{\mathbf{x} \in \mathbb{R}^p : (\mathbf{x} - \boldsymbol{\mu})^T \mathbf{\Sigma}(\mathbf{x} - \boldsymbol{\mu}) = 0\} = \{\boldsymbol{\mu} + \mathbf{y} : \mathbf{A}\mathbf{y} = \mathbf{0}\} \tag{3.4}$$

TABLE 3.2
Least squares applied to a normal linear model with correlated features. Only $X_1, X_2, X_3, X_5, X_6, X_7$ are true causal (signal) variables.

	Full Model			Misspecified Model		
	$\hat{\beta}$	t-statistic	p-value	$\hat{\beta}$	t-statistic	p-value
Intercept	-0.069	-1.476	0.141	-0.116	-2.041	0.042
X_1	4.776	36.738	0.000	5.572	37.892	0.000
X_2	5.038	39.334	0.000	5.573	36.745	0.000
X_3	2.097	15.767	0.000			
X_4	0.204	1.607	0.109	0.830	5.614	0.000
X_5	-4.991	-107.495	0.000	-4.983	-87.417	0.000
X_6	-5.027	-107.108	0.000	-4.976	-86.555	0.000
X_7	-1.924	-41.202	0.000	-1.930	-33.663	0.000
X_8	0.060	1.281	0.201	0.073	1.269	0.205
X_9	-0.020	-0.431	0.666	0.065	1.139	0.255
X_{10}	0.066	1.431	0.153	0.026	0.466	0.641
X_{11}	0.057	1.230	0.219	-0.003	-0.051	0.959
X_{12}	-0.047	-1.008	0.314	-0.055	-0.971	0.332

where $\mathbf{\Sigma} = \mathbf{A}^T\mathbf{A}$. Here we have used the assumption that $\mathbf{\Sigma}$ can be written as a product of two matrices, which is an application of the spectral decomposition theorem; see Chapter 8.2.

3.3.4 Dimension free behavior

The next result bounds the behavior of the maximum coordinate of a multivariate normal. A curious result is that this bound is dimension free. To state the result, let $\mathbf{X} = (X_1, \dots, X_p)^T$ be a multivariate normal with mean zero. Let $\sigma_k^2 = \mathbb{E}(X_k^2) > 0$ and define $\tilde{\sigma} = \max_{1\le k\le p} \sigma_k$. It can be shown [2]

$$\mathbb{P}\left\{\left|\max_{1\le k\le p} X_k - \mathbb{E}\left(\max_{1\le k\le p} X_k\right)\right| \ge t\right\} \le 2\exp\left(\frac{-t^2}{2\tilde{\sigma}^2}\right), \quad t > 0. \tag{3.5}$$

This is a concentration inequality which states that the maximum coordinate remains close to the mean of the maximum coordinate. The most interesting thing is that the bound does not depend on p and even holds when $p = \infty$. See Chernozhukov, Chetverikov and Kato [1] for related results. Exercise 3.14 provides an illustrative example of the above bound.

3.3.5 Stein's lemma

The following lemma is due to Stein [9].

Lemma 3.1. *If $\boldsymbol{X} \sim N_p(\boldsymbol{\mu}, \boldsymbol{I})$, then for any real-valued function $g : \mathbb{R}^p \to \mathbb{R}$ with gradient ∇g such that $\mathbb{E}||\nabla g(\boldsymbol{X})|| < \infty$,*

$$\mathbb{E}\left[(\boldsymbol{X} - \boldsymbol{\mu})g(\boldsymbol{X})\right] = \mathbb{E}(\nabla g(\boldsymbol{X})).$$

Stein's lemma leads to several important results in statistics and probability. One such result is a useful characterization of the principal Hessian, originally provided by Li [5]. Consider a regression model $Y = F(\mathbf{X}) + \varepsilon$, where ε is independent of $\mathbf{X}$ and $\mathbb{E}(\varepsilon) = 0$. The principal Hessian, $\overline{\mathbf{H}}_x$, is defined as the mean value of the Hessian of F under the data distribution, i.e., $\overline{\mathbf{H}}_x = \mathbb{E}(\mathbf{H}(\mathbf{X}))$, where recall that the Hessian of a multivariable function $F(\mathbf{x})$ is (Chapter 2.9)

$$\mathbf{H}(\mathbf{x}) = \begin{pmatrix} \frac{\partial^2 F(\mathbf{x})}{\partial x_1^2} & \frac{\partial^2 F(\mathbf{x})}{\partial x_1 \partial x_2} & \cdots & \frac{\partial^2 F(\mathbf{x})}{\partial x_1 \partial x_p} \\ \frac{\partial^2 F(\mathbf{x})}{\partial x_2 \partial x_1} & \frac{\partial^2 F(\mathbf{x})}{\partial x_2^2} & \cdots & \frac{\partial^2 F(\mathbf{x})}{\partial x_2 \partial x_p} \\ \vdots & \vdots & \ddots & \vdots \\ \frac{\partial^2 F(\mathbf{x})}{\partial x_p \partial x_1} & \frac{\partial^2 F(\mathbf{x})}{\partial x_p \partial x_2} & \cdots & \frac{\partial^2 F(\mathbf{x})}{\partial x_p^2} \end{pmatrix}_{p \times p}.$$

Under this framework the principal Hessian has an alternate representation that involves the covariances of $\mathbf{X}$ and $\mathbf{Y}$, therefore making it relatively easy to calculate $\overline{\mathbf{H}}_x$. We offer a slightly different characterization and proof of this result compared to Li [5].

Lemma 3.2. *Suppose that* $\boldsymbol{X} \sim N_p(\boldsymbol{\mu}_x, \boldsymbol{\Sigma}_x)$ *where* $\boldsymbol{\Sigma}_x > 0$ *and* $F(\boldsymbol{x}) = \mathbb{E}(Y|\boldsymbol{X} = \boldsymbol{x})$ *is second-order differentiable in* $\boldsymbol{x}$ *with all continuous second-order partial derivatives that are integrable. Then the principal Hessian satisfies the following identity*

$$\overline{\boldsymbol{H}}_x = \boldsymbol{\Sigma}_x^{-1} \boldsymbol{\Sigma}_{yxx} \boldsymbol{\Sigma}_x^{-1} \tag{3.6}$$

where $\boldsymbol{\Sigma}_{yxx} = \mathbb{E}\left[(Y - \mu_y)(\boldsymbol{X} - \boldsymbol{\mu}_x)(\boldsymbol{X} - \boldsymbol{\mu}_x)^T\right]$ *and* $\mu_y = \mathbb{E}(Y) = \mathbb{E}(F(\boldsymbol{X}))$.

Proof. Let $\mathbf{Z} = \boldsymbol{\Sigma}_x^{-1/2}(\mathbf{X} - \boldsymbol{\mu}_x) = (Z_1, \ldots, Z_p)^T$ and define $g_j(\mathbf{z}) = z_j F(\mathbf{z})$. Then since $\mathbf{Z}$ is a standard multivariate normal, applying Lemma 3.1 and using the chain rule,

$$\mathbb{E}(\mathbf{Z} Z_j F(\mathbf{Z})) = \mathbb{E}(\mathbf{Z} g_j(\mathbf{Z})) = \mathbb{E}(\nabla g_j(\mathbf{Z})) = \mathbb{E}(F(\mathbf{Z}))\mathbf{1}_j + \mathbb{E}(Z_j \nabla F(\mathbf{Z}))$$

where $\mathbf{1}_j$ is the jth Euclidean basis vector. Repeating this for coordinates $j = 1, \ldots, p$ and column binding the resulting vectors into a $p \times p$ matrix, we have

$$\mathbb{E}(\mathbf{Z}\mathbf{Z}^T F(\mathbf{Z})) = \mathbb{E}(F(\mathbf{Z}))\mathbf{I} + \mathbb{E}(\mathbf{Z}(\nabla F(\mathbf{Z}))^T). \tag{3.7}$$

Now define $\tilde{g}_j(\mathbf{z}) = \partial F(z)/\partial z_j$ to be the jth partial derivative of F with respect to z_j. Applying Lemma 3.1 once more, we have

$$\mathbb{E}(\mathbf{Z}\tilde{g}_j(\mathbf{Z})) = \mathbb{E}(\nabla \tilde{g}_j(\mathbf{Z})) = \mathbb{E}(\nabla_j^2 F(\mathbf{Z})).$$

Stacking the resulting vectors,

$$\mathbb{E}(\mathbf{Z}(\nabla F(\mathbf{Z}))^T) = \mathbb{E}(\mathbf{H}(\mathbf{Z})) := \overline{\mathbf{H}}_z,$$

and hence with some rearrangement of (3.7) using $\mathbb{E}(\mathbf{Z}\mathbf{Z}^T) = \mathbf{I}$,

$$\mathbb{E}\left[\mathbf{Z}\mathbf{Z}^T(F(\mathbf{Z}) - \mathbb{E}(F(\mathbf{Z}))\right] = \overline{\mathbf{H}}_z.$$

By the chain rule, $\overline{\mathbf{H}}_z = \boldsymbol{\Sigma}_x^{1/2} \overline{\mathbf{H}}_x \boldsymbol{\Sigma}_x^{1/2}$. Therefore

$$\mathbb{E}\left[(\mathbf{X} - \boldsymbol{\mu}_x)(\mathbf{X} - \boldsymbol{\mu}_x)^T(F(\mathbf{Z}) - \mathbb{E}(F(\mathbf{Z}))\right] = \boldsymbol{\Sigma}_x \overline{\mathbf{H}}_x \boldsymbol{\Sigma}_x.$$

Because $\mu_y = \mathbb{E}(F(\mathbf{Z}))$ and writing $Y = F(\mathbf{Z}) + \varepsilon$, it follows that (3.6) holds by the independence of ε and $\mathbf{X}$ and $\mathbb{E}(\varepsilon) = 0$. □

Lemma 3.2 has many useful consequences. For example, Li [5] used it to characterize the principal Hessian direction (pHd) and illustrated how this could be used for dimension reduction. Another application, which we pursue here, is to show that the result can be used for interaction detection in regression models.

The idea will be to replace Y with a suitable fitted polynomial from which we can directly infer the interactions of F. First, observe that adding or subtracting a linear function of **X** from Y will not change the Hessian matrix. Hence, instead of Y in Lemma 3.2, we can use the residual $r = Y - h(\mathbf{X})$ from a linear least squares fit, where $h(\mathbf{X})$ is the least squares model. We can take this idea a step further. Suppose instead of a linear model, we fit a least squares model using a polynomial. Then consider replacing Y with the polynomial regression fit. The following result, due to Li [5], shows that the resulting Hessian remains unchanged as long as the polynomial degree does not exceed the rank of **X**.

Corollary 3.3. *Suppose that* **X** *is normally distributed and consider a polynomial regression of* Y *onto* **X**:

$$h(\boldsymbol{x}) = \underset{Q(\boldsymbol{x})}{\arg\min}\, \mathbb{E}\,(Y - Q(\boldsymbol{x}))^2,$$

where $Q(\boldsymbol{x})$ *is a polynomial function of total degrees* $q \le p$. *Then the principal Hessian for the fitted polynomial* h *is the same as the principal Hessian for* Y.

Corollary 3.3 allows us replace Y with any fitted polynomial of degree less than $p+1$. Consider, therefore, fitting a polynomial of degree $q = 2$:

$$h(\mathbf{x}) = \beta_0 + \sum_{j=1}^{p} x_j\beta_j + \sum_{k,l} x_k x_l \beta_{k,l}.$$

Since h is quadratic, its Hessian is $\{\beta_{k,l}\}_{k,l}$. By Corollary 3.3, this must be identical to the principal Hessian, therefore $\overline{\mathbf{H}}_x = \{\beta_{k,l}\}_{k,l}$. Furthermore, because the (k,l) entries of $\overline{\mathbf{H}}_x$ can be interpreted as the interaction effect between the k and l feature attributes, this shows that the regression parameters $\beta_{k,l}$ used in the polynomial regression model h can be used for interaction detection in the original regression model F.

This idea can be further refined to improve computational efficiency. Let $\mathbf{Z} = \boldsymbol{\Sigma}_x^{-1/2}(\mathbf{X} - \boldsymbol{\mu}_x)$. With some rearrangement, Lemma 3.2 implies that:

$$\mathbb{E}\left[\mathbf{Z}\mathbf{Z}^T(Y - \mathbb{E}[F(\mathbf{Z})])\right] = \boldsymbol{\Sigma}_x^{1/2}\overline{\mathbf{H}}_x\boldsymbol{\Sigma}_x^{1/2}.$$

Let $\tilde{h}$ be the least squares solution obtained by regressing Y onto **Z** using a polynomial of degree 2. Denote this solution by

$$\tilde{h}(\mathbf{z}) = \tilde{\beta}_0 + \sum_{j=1}^{p} z_j\tilde{\beta}_j + \sum_{k,l} z_k z_l \tilde{\beta}_{k,l}.$$

By Corollary 3.3, it follows that

$$\mathbb{E}\left[\mathbf{Z}\mathbf{Z}^T(Y - \mathbb{E}[F(\mathbf{Z})])\right] = \mathbb{E}\left[\mathbf{Z}\mathbf{Z}^T(\tilde{h}(\mathbf{Z}) - \mathbb{E}[\tilde{h}(\mathbf{Z})])\right] = \boldsymbol{\Sigma}_x^{1/2}\overline{\mathbf{H}}_x\boldsymbol{\Sigma}_x^{1/2}.$$

It is easy to see that $\tilde{\beta}_0 = \mathbb{E}(\tilde{h}(\mathbf{Z}))$. Therefore,

$$\mathbb{E}\left[\mathbf{Z}\mathbf{Z}^T\left(\sum_{j=1}^{p} Z_j\tilde{\beta}_j + \sum_{k,l} Z_k Z_l \tilde{\beta}_{k,l}\right)\right] = \boldsymbol{\Sigma}_x^{1/2}\overline{\mathbf{H}}_x\boldsymbol{\Sigma}_x^{1/2}.$$

The matrix in square brackets on the left is comprised of products of coordinates of **Z** of the form $Z_m Z_n Z_j$ and $Z_m Z_n Z_k Z_l$. But since the coordinates of **Z** are independent standard normal variables, only second and fourth moments, equal to 1 and 3 are nonzero, which occurs for entries $Z_m Z_n Z_k Z_l$ when $m = n$ and $k = l$ and never for entries $Z_m Z_n Z_j$. Thus, the left-hand side is a diagonal matrix $\mathbf{D}$ with entries

$$d_{m,m} = 3\tilde{\beta}_{m,m} + \sum_{k \neq m} \tilde{\beta}_{k,k}, \quad m = 1, \ldots, p.$$

Combining this with the above, we have $\boldsymbol{\Sigma}_x^{-1/2} \mathbf{D} \boldsymbol{\Sigma}_x^{-1/2} = \overline{\mathbf{H}}_x$.

Therefore, the principal Hessian values can be obtained from $\boldsymbol{\Sigma}_x^{-1/2} \mathbf{D} \boldsymbol{\Sigma}_x^{-1/2}$ by substituting the estimated sample covariance matrix for $\boldsymbol{\Sigma}_x$ and estimating $\mathbf{D}$ using regression parameters from a regression of y onto normalized $\mathbf{z}$. The potential advantage of this approach is that the least squares regression requires fewer parameters. This is because $\mathbf{D}$ only depends on values $\{\tilde{\beta}_{k,k}\}_{k=1}^{p}$, and therefore it would suffice to consider a regression using the reduced polynomial

$$\tilde{h}^* (\mathbf{z}) = \tilde{\beta}_0 + \sum_{k=1}^{p} z_k^2 \tilde{\beta}_{k,k}.$$

This could be beneficial in high dimensional problems since it reduces the dimension from $1 + p + p(p-1)/2$ parameters using regression with $\mathbf{X}$ to $1 + p$ parameters using regression with $\mathbf{Z}$.

3.3.6 Multivariate normal mixture model

The multivariate normal distribution is one of the few distributions where independence is equivalent to being uncorrelated. However, this property breaks down when mixing over the normal density, as shown in the following example.

Let $\mathbf{X} \in \mathbb{R}^p$ be a continuous random vector with density from a finite normal mixture density. Let $\phi(\cdot; \boldsymbol{\mu}, \boldsymbol{\Sigma})$ be the density for a multivariate normal $\mathrm{N}_p(\boldsymbol{\mu}, \boldsymbol{\Sigma})$. Then $\mathbf{X}$ is said to have a finite multivariate normal mixture distribution if it has density

$$f_{\mathbf{X}}(\mathbf{x}) = \sum_{l=1}^{L} \pi_l \phi(\mathbf{x}; \boldsymbol{\mu}_l, \boldsymbol{\Sigma}_l) \tag{3.8}$$

where $0 < \pi_l < 1$ are mixing weights satisfying $\sum_{l=1}^{L} \pi_l = 1$.

We will work out the variance for $\mathbf{X}$ and use this to demonstrate our point about dependence and correlation. First, observe that $\mathbf{X}$ can be equivalently described as the conditional model

$$\mathbf{X}|\zeta \sim \mathrm{N}_p(\boldsymbol{\mu}_\zeta, \boldsymbol{\Sigma}_\zeta), \quad \mathbb{P}\{\zeta = l\} = \pi_l, \tag{3.9}$$

where in the above, $\zeta \in \{1, \ldots, L\}$ is a latent variable recording the normal mixture component.

Using the conditional property for the mean (2.2) and variance (2.3), conditioning on ζ, we have

$$\begin{aligned}
\mathrm{Var}(\mathbf{X}) &= \mathbb{E}_\zeta\{\mathrm{Var}(\mathbf{X}|\zeta)\} + \mathrm{Var}_\zeta\{\mathbb{E}(\mathbf{X}|\zeta)\} \\
&= \mathbb{E}_\zeta\{\boldsymbol{\Sigma}_\zeta\} + \mathrm{Var}_\zeta\{\boldsymbol{\mu}_\zeta\} \\
&= \sum_{l=1}^{L} \pi_l \boldsymbol{\Sigma}_l + \sum_{l=1}^{L} \pi_l (\boldsymbol{\mu}_l - \boldsymbol{\mu})(\boldsymbol{\mu}_l - \boldsymbol{\mu})^T
\end{aligned}$$

where $\boldsymbol{\mu} = \sum_l \pi_l \boldsymbol{\mu}_l$. It will be assumed that $\boldsymbol{\mu} = \mathbf{0}$, therefore

$$\mathrm{Var}(\mathbf{X}) = \sum_{l=1}^{L} \pi_l \boldsymbol{\Sigma}_l + \sum_{l=1}^{L} \pi_l \boldsymbol{\mu}_l \boldsymbol{\mu}_l^T.$$

For notational simplicity we write $\boldsymbol{\Sigma} = \mathrm{Var}(\mathbf{X})$.

3.3.7 Example

Let $p = 3$ and assume that

$$\boldsymbol{\mu}_l = \begin{pmatrix} \mu_{l1} \\ \mu_{l2} \\ 0 \end{pmatrix}, \quad \boldsymbol{\Sigma}_l = \begin{pmatrix} 1 & \rho_{l12} & 0 \\ \rho_{l12} & 1 & 0 \\ 0 & 0 & 1 \end{pmatrix}, \quad -1 < \rho_{l12} < 1.$$

Therefore if $\boldsymbol{\mu} = \sum_l \pi_l \boldsymbol{\mu}_l = \mathbf{0}$,

$$\boldsymbol{\Sigma} = \begin{pmatrix} 1 + \sum_l \pi_l \mu_{l1}^2 & \sum_l \pi_l \rho_{l12} + \sum_l \pi_l \mu_{l1}\mu_{l2} & 0 \\ \sum_l \pi_l \rho_{l12} + \sum_l \pi_l \mu_{l1}\mu_{l2} & 1 + \sum_l \pi_l \mu_{l2}^2 & 0 \\ 0 & 0 & 1 \end{pmatrix}.$$

Even though the mixture components are correlated along the first two coordinates due to the way $\boldsymbol{\Sigma}_l$ is defined, this does not automatically guarantee that $\mathbf{X}$ is correlated. In fact, if the following conditions are met, then $\boldsymbol{\Sigma}$ is diagonal and therefore $\mathbf{X}$ is uncorrelated:

$$\sum_l \pi_l \mu_{l1} = 0, \quad \sum_l \pi_l \mu_{l2} = 0, \quad \sum_l \pi_l \rho_{l12} + \sum_l \pi_l \mu_{l1}\mu_{l2} = 0. \tag{3.10}$$

As this is a system of homogeneous equations it can be solved by the trivial solution of zero. This is undesirable since this means $\rho_{l12} = 0$ which is not interesting since obviously $\mathbf{X}$ would be uncorrelated in such a case. However, a non-trivial solution can be obtained by setting $\rho_{l12} = -\mu_{l1}\mu_{l2}$. Then (3.10) is non-trivially satisfied if

$$\sum_l \pi_l \mu_{l1} = 0, \quad \sum_l \pi_l \mu_{l2} = 0, \quad -1 < \mu_{l1}, \mu_{l2} < 1, \quad \mu_{l1}\mu_{l2} \neq 0, \quad l = 1, \ldots, L.$$

Here is one solution. Take $\pi_l = 1/L$ and let $S_1 = \{\mu_{l1}^*\}_{l=1}^L$ and $S_2 = \{\mu_{l2}^*\}_{l=1}^L$ be any two sets of non-zero numbers, each summing to zero. Then normalize the numbers in S_1 and S_2 so that each value lies in $(-1, 1)$ (for example by dividing by the largest absolute value in each set plus some small number greater than zero). This gives the values for μ_{l1} and μ_{l2}. Setting $\rho_{l12} = -\mu_{l1}\mu_{l2}$ we obtain (3.10).

Under such a specification, this shows even if the coordinates are correlated within a mixture component, the observed $\mathbf{X}$ may be uncorrelated. In our example the variance of $\mathbf{X}$ is

$$\boldsymbol{\Sigma} = \begin{pmatrix} \sigma_1 & 0 & 0 \\ 0 & \sigma_2 & 0 \\ 0 & 0 & 1 \end{pmatrix}, \quad \sigma_1, \sigma_2 > 1.$$

Therefore given a sample $\mathbf{X}_1, \ldots, \mathbf{X}_n$ from $f_{\mathbf{X}}$, the covariance matrix of the sample would show a higher variance for coordinates X_1 and X_2 but we would not be alerted to their underlying correlation in the mixture components.

An illustrating example is provided in Figure 3.2. Here 1000 values are sampled from a $p = 3$ multivariate normal mixture with $L = 2$ components using parameters set as above. In the figure, the color (green and red) indicates the mixture component for a sampled data point. Correlations for coordinates within a mixture are given in the upper diagonals and show that X_1 and X_2 are highly correlated within both mixture components. However, X_1 and X_2 have near-zero correlation for the mixed random vector (these correlations are given in black).

3.4 Unbiased Estimating Equations: Godambe Information

The multivariate normal distribution often arises in asymptotic theory, such as in the case of unbiased estimating equations and the Godambe information. Here we outline this scenario

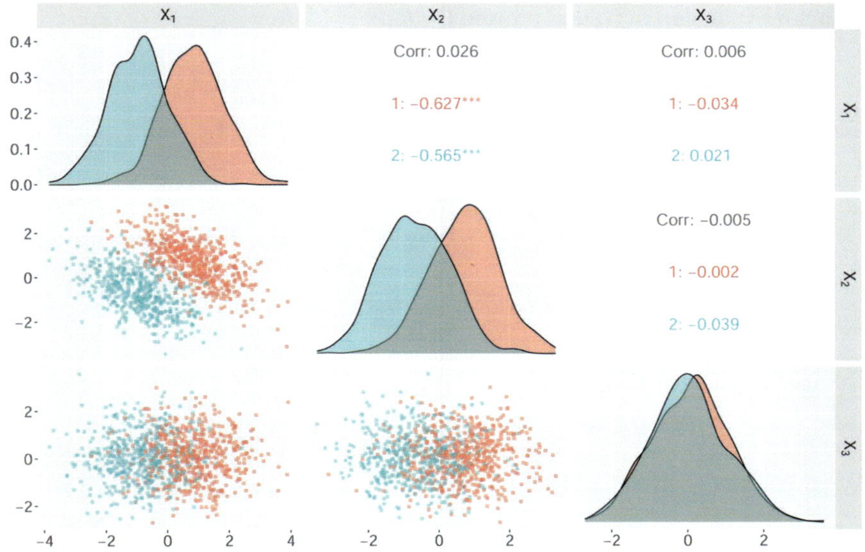

FIGURE 3.2
Multivariate $p = 3$ normal mixture model with $L = 2$ components (red and green colors identify the component data comes from). Even though normal mixture components are correlated (see correlation values in color), the mixed random vector has coordinates that are uncorrelated (see correlation values in black).

and indicate its connection to maximum likelihood estimation. We will draw upon some results from Chapter 2 and the properties of the multivariate normal for our discussion.

Let $\mathbf{X}_1, \ldots, \mathbf{X}_n$ be an i.i.d. sample from a common distribution $\mathbb{P}$. The goal is to estimate a parameter vector $\boldsymbol{\theta} \in \Theta \subseteq \mathbb{R}^p$ using the estimator $\hat{\boldsymbol{\theta}}_n$ satisfying

$$\mathbf{G}_n(\hat{\boldsymbol{\theta}}_n) = \mathbf{0}, \quad \text{where } \mathbf{G}_n(\boldsymbol{\theta}) = \frac{1}{n}\sum_{i=1}^{n} \mathbf{g}(\mathbf{X}_i, \boldsymbol{\theta}), \tag{3.11}$$

where $\mathbf{g} : \mathbb{R}^p \to \mathbb{R}^p$ is a multivariate function. We say (3.11) are *unbiased estimating equations* [4] if

$$\mathbb{E}\left[\mathbf{g}(\mathbf{X}, \boldsymbol{\theta})\right] = \mathbf{0}, \quad \text{for all } \boldsymbol{\theta} \in \Theta.$$

The above is a generalization of maximum likelihood estimation. That is, let $f(\mathbf{X}, \boldsymbol{\theta})$ be the density and define the log-likelihood to be

$$l_n(\boldsymbol{\theta}) = \sum_{i=1}^{n} \log f(\mathbf{X}_i, \boldsymbol{\theta}).$$

Define $\mathbf{g}(\mathbf{X}, \cdot) \in \mathbb{R}^p$ to be the gradient

$$\mathbf{g}(\mathbf{X}, \boldsymbol{\theta}) = \frac{\partial \log f(\mathbf{X}, \boldsymbol{\theta})}{\partial \boldsymbol{\theta}} := \boldsymbol{\nabla} \log f(\mathbf{X}, \boldsymbol{\theta})$$

where $\boldsymbol{\nabla}$ denotes the gradient operator with respect to $\boldsymbol{\theta}$. The function $\mathbf{g}$ is called the *score function* in maximum likelihood estimation. Then $\mathbf{G}_n(\boldsymbol{\theta}) = \boldsymbol{\nabla} l_n(\boldsymbol{\theta})/n = \sum_{i=1}^{n} \boldsymbol{\nabla} \log f(\mathbf{X}_i, \boldsymbol{\theta})/n$ and it is well known that

$$\mathbb{E}\left[\boldsymbol{\nabla} \log f(\mathbf{X}, \boldsymbol{\theta})\right] = \mathbf{0}$$

under standard regularity conditions.

Let $\boldsymbol{\theta}_0$ be the true parameter. We assume regularity conditions hold for (3.11) that yield the following Taylor expansion of $\mathbf{G}_n(\hat{\boldsymbol{\theta}}_n)$ around $\mathbf{G}_n(\boldsymbol{\theta}_0)$:

$$\mathbf{0} = \mathbf{G}_n(\hat{\boldsymbol{\theta}}_n) = \mathbf{G}_n(\boldsymbol{\theta}_0) + [\nabla \mathbf{G}_n(\boldsymbol{\theta}_0)]\left(\hat{\boldsymbol{\theta}}_n - \boldsymbol{\theta}_0\right) + o_p(n^{-1/2}). \tag{3.12}$$

Notice that $\mathbf{g}(\mathbf{X}_i, \boldsymbol{\theta})$ are i.i.d. with mean zero and therefore by the multivariate central limit theorem

$$n^{1/2}\mathbf{G}_n(\boldsymbol{\theta}_0) \overset{\text{d}}{\rightsquigarrow} \mathrm{N}_p(\mathbf{0}, \mathbf{V})$$

where $\mathbf{V} = \mathrm{Var}(\mathbf{g}(\mathbf{X}, \boldsymbol{\theta}_0))$. Assuming $\mathbf{U}_{p\times p} = -\mathbb{E}(\nabla \mathbf{g}(\mathbf{X}, \theta_0))$ exists, then by the strong law of large numbers

$$-\nabla \mathbf{G}_n(\boldsymbol{\theta}_0) = -\frac{1}{n}\sum_{i=1}^{n} \nabla \mathbf{g}(\mathbf{X}_i, \boldsymbol{\theta}_0) \overset{\text{a.s.}}{\to} \mathbf{U}.$$

Thus we can rewrite (3.12) as

$$n^{1/2}\left(\hat{\boldsymbol{\theta}}_n - \boldsymbol{\theta}_0\right) = -\left[\nabla \mathbf{G}_n(\boldsymbol{\theta}_0)\right]^{-1}\left[n^{1/2}\mathbf{G}_n(\boldsymbol{\theta}_0)\right] + o_p(1).$$

By Slutsky's theorem (Theorem 2.21) and the continuous mapping theorem (Theorem 2.20),

$$n^{1/2}\left(\hat{\boldsymbol{\theta}}_n - \boldsymbol{\theta}_0\right) \overset{\text{d}}{\rightsquigarrow} \mathbf{U}^{-1}\mathbf{Z},$$

where $\mathbf{Z} \sim \mathrm{N}_p(\mathbf{0}, \mathbf{V})$, and hence by the properties of the multivariate normal

$$n^{1/2}\left(\hat{\boldsymbol{\theta}}_n - \boldsymbol{\theta}_0\right) \overset{\text{d}}{\rightsquigarrow} \mathrm{N}_p(\mathbf{0}, \mathbf{U}^{-1}\mathbf{V}(\mathbf{U}^{-1})^T).$$

The asymptotic variance $\mathbf{U}^{-1}\mathbf{V}(\mathbf{U}^{-1})^T$ is called the *Godambe information* [4].

The asymptotic variance has a simplification in terms of the *Fisher information* in the case of maximum likelihood estimation. Fisher information is a concept from statistical theory that measures the amount of information $\mathbf{X}$ carries about the unknown parameter $\boldsymbol{\theta}_0$. In essence, it quantifies the precision with which $\boldsymbol{\theta}$ can be estimated. Mathematically, the Fisher information for $\boldsymbol{\theta}$ is defined as

$$\mathbf{I}(\boldsymbol{\theta}) = \mathbb{E}\left(\left[\nabla \log f(\mathbf{X}_i, \boldsymbol{\theta})\right]\left[\nabla \log f(\mathbf{X}_i, \boldsymbol{\theta})\right]^T\right).$$

Assuming standard regularity conditions, the following identities hold for the Fisher information:

$$\begin{aligned}
\mathbf{V} &= \mathrm{Var}\left(\nabla \log f(\mathbf{X}_i, \boldsymbol{\theta}_0)\right) \\
&= \mathbb{E}\left(\left[\nabla \log f(\mathbf{X}_i, \boldsymbol{\theta}_0)\right]\left[\nabla \log f(\mathbf{X}_i, \boldsymbol{\theta}_0)\right]^T\right) := \mathbf{I}(\boldsymbol{\theta}_0) \\
&= -\mathbb{E}\left[\frac{\partial^2 \log f(\mathbf{X}, \boldsymbol{\theta})}{\partial \boldsymbol{\theta} \partial \boldsymbol{\theta}^T}\bigg|_{\boldsymbol{\theta}_0}\right] \\
&= -\mathbb{E}(\nabla \mathbf{g}(\mathbf{X}, \boldsymbol{\theta}_0)) \\
&= \mathbf{U}.
\end{aligned}$$

Therefore, we have $\mathbf{U} = \mathbf{V}$ and

$$\mathbf{U}^{-1}\mathbf{V}(\mathbf{U}^{-1})^T = \mathbf{U}^{-1} = \mathbf{V}^{-1} := \mathbf{I}(\boldsymbol{\theta}_0)^{-1}.$$

This shows that the inverse of the Fisher information is the limiting asymptotic variance.

Note that the Fisher information depends on the true parameter $\boldsymbol{\theta}_0$, so it has to be approximated in practice. The maximum likelihood estimator is used for this purpose, giving the estimator $\hat{\mathbf{U}}_n$:

$$\hat{\mathbf{U}}_n = -\frac{1}{n}\sum_{i=1}^{n} \mathbf{g}(\mathbf{X}_i, \hat{\boldsymbol{\theta}}_n) \overset{\text{a.s.}}{\to} -\mathbb{E}(\nabla \mathbf{g}(\mathbf{X}, \boldsymbol{\theta}_0)) = \mathbf{U} = \mathbf{V}$$

where $\mathbf{g}(\mathbf{X},\boldsymbol{\theta}) = \nabla \log f(\mathbf{X},\boldsymbol{\theta})$, and the limit holds by the strong law of large numbers and the continuous mapping theorem.

Maximum likelihood estimation is a special case and in general the asymptotic variance $\mathbf{U}^{-1}\mathbf{V}(\mathbf{U}^{-1})^T$ may not simplify. In this case, $\mathbf{V}$ is estimated by

$$\hat{\mathbf{V}}_n = -\frac{1}{n}\sum_{i=1}^{n}\left[\nabla \log f(\mathbf{X}_i,\hat{\boldsymbol{\theta}}_n)\right]\left[\nabla \log f(\mathbf{X}_i,\hat{\boldsymbol{\theta}}_n)\right]^T \overset{\text{a.s.}}{\to} \mathbf{V}$$

and the asymptotic variance is estimated by

$$\hat{\mathbf{U}}_n^{-1}\hat{\mathbf{V}}_n(\hat{\mathbf{U}}_n^{-1})^T \overset{\text{a.s.}}{\to} \mathbf{U}^{-1}\mathbf{V}(\mathbf{U}^{-1})^T.$$

This estimator is called the *sandwich estimator* in the generalized estimating equations literature.

3.5 Reading Disability: Connecticut Longitudinal Study

In the next few sections, we illustrate the usefulness of the multivariate normal distribution through a detailed case study of childhood dyslexia. This topic is important as there is still no consensus on the conceptual and operational definitions of reading disability in children [3, 7, 8]. We will use the discrepancy model and discrepancy score (DS), which consider reading disability as a discrepancy between a child's ability and actual achievement [6, 12]. The properties of the multivariate normal distribution will help us define and study the DS.

The data comes from the Connecticut Longitudinal Study, a survey of children attending Connecticut public kindergartens from 1983–1984. A total of $n = 445$ children participated. Children's intellectual ability was assessed by the verbal IQ Wechsler Intelligence Scales for Children-Revised (WISC-R; [10]). Reading achievement was assessed by the Woodcock-Johnson Psychoeducational Test Battery [11]. The WISC-R was administered in grades one, three, five, and seven, while the Woodcock-Johnson was administered in grades one through eight. There were 407 children with complete IQ records and 403 with complete achievement records; these complete records are used for our analysis.

Figure 3.3 shows bivariate contour plots of the data. The blue points are the observed data, and the black elliptical regions represent the 50th, 90th, 95th, and 99th percentile contours from a bivariate normal distribution (with mean and covariance estimated from the sample). The elliptical nature of the data suggests it closely matches a multivariate normal distribution, which is a key assumption in our analysis. We will formally verify this assumption after reviewing different methods for testing multivariate normality.

3.6 Testing Multivariate Normality

Many statistical tests rely on normality or approximate normality for their validity. For this reason, it is useful to have a set of tools for verifying this assumption. Using properties of the multivariate normal distribution, we list a few simple but effective methods for this purpose.

1. *Mahalanobis distance.* If $\mathbf{X} \sim \mathrm{N}_p(\mu,\boldsymbol{\Sigma})$, then the Mahalanobis distance squared $D(\mathbf{X}) = (\mathbf{X}-\boldsymbol{\mu})^T\boldsymbol{\Sigma}^{-1}(\mathbf{X}-\boldsymbol{\mu})$ has a χ^2_p-distribution. This distance can be used as a univariate statistic to

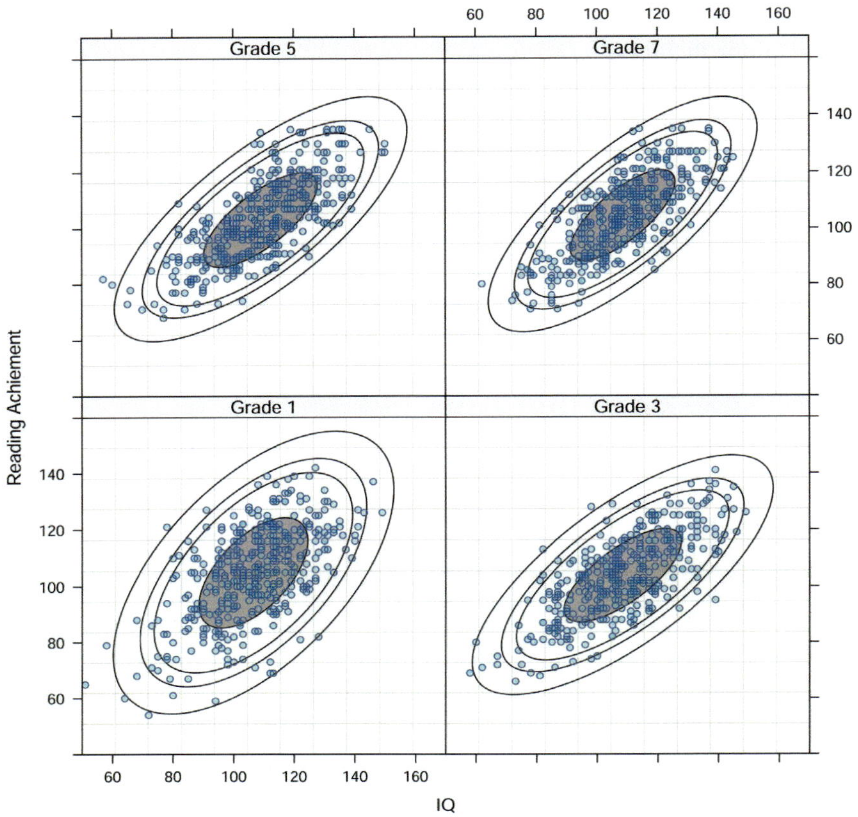

FIGURE 3.3
Bivariate normal contour plots (50, 90, 95, 99%) for Verbal IQ and Woodcock-Johnson reading achievement scores from the Connecticut Longitudinal Study.

assess normality. The population mean $\boldsymbol{\mu}$ and population variance $\boldsymbol{\Sigma}$ are estimated by $\hat{\boldsymbol{\mu}} = \overline{\mathbf{X}}$ and $\hat{\boldsymbol{\Sigma}} = \mathbf{S}$. Then $D(\mathbf{X})$ is approximated by

$$\hat{D}(\mathbf{X}) = (\mathbf{X} - \hat{\boldsymbol{\mu}})^T \hat{\boldsymbol{\Sigma}}^{-1} (\mathbf{X} - \hat{\boldsymbol{\mu}})$$

which is approximately χ^2_p-distributed under the assumption of normality.

2. *Q-Q plot.* A quantile-quantile plot (Q-Q plot) is a graphical method for comparing two probability distributions by plotting their quantiles against each other. A point (x, y) on the plot corresponds to one of the quantiles of the first distribution (y-coordinate) plotted against the same quantile of the second distribution (x-coordinate). Therefore, the line represents a parametric curve parameterized by the quantile. For assessing normality, we use the sample values $\hat{D}(\mathbf{X}_1), \ldots, \hat{D}(\mathbf{X}_n)$ for y. The x values are the quantiles for the theoretical target distribution, which, for $\hat{D}(\mathbf{X})$, are the quantiles for a χ^2_p-distribution. If the sampled values y follow the x target distribution, then the points on the Q-Q plot will approximately lie on the line $y = x$.

3. *Order statistics.* If the univariate data is $X_1, \ldots, X_n$, we transform the data to $X_i^* = (X_i - \overline{x})/s_n$, where $\overline{x}$ is the sample mean and s_n^2 is the unbiased estimator for the population variance. To test normality we order the transformed values $X^*_{(1)} \le X^*_{(2)} \le \cdots \le X^*_{(n)}$ and perform the following.

(a) Plot $(\mu_{(i)}, X^*_{(i)})$ for $i = 1, \dots, n$ where $\mu_{(i)}$ is the mean for the ith order statistic from an i.i.d. sample of $\mathrm{N}(0,1)$ variables.

(b) Compute the correlation between $\mu_{(i)}$ and $X^*_{(i)}$. We call this the r_Q statistic.

There are many other ways to use order statistics to test normality. A popular statistic is the Shapiro-Wilk's statistic.

4. *Kolmogorov-Smirnov distance.* This statistic measures the distance between the empirical distribution of the sample versus a theoretical distribution, and is defined as

$$K_n = \sup_x |\hat{F}_n(x) - F_0(x)|,$$

where $\hat{F}_n$ is the empirical distribution function and F_0 is the target theoretical distribution function. We reject the null that the sample is from F_0 if $\sqrt{n}K_n \geq c_\alpha$ where c_α is the cutoff value for a level of significance $0 < \alpha < 1$ determined by

$$\mathbb{P}\{\sqrt{n}K_n \leq t\} \asymp 1 - 2\sum_{j=1}^{\infty}(-1)^{j-1}\exp(-2j^2t^2).$$

This test can be used to assess whether $\hat{D}(\mathbf{X})$ has a χ^2_p-distribution.

3.7 Reading Disability: Using the Discrepancy Score

Returning to the Connecticut Longitudinal Study, we define the discrepancy score (DS) as the difference between reading achievement and the expected achievement measure given the IQ. Let I_j denote the WISC-R Verbal IQ and ACH_j denote the Woodcock-Johnson reading achievement score for grade j. The discrepancy score for grade j is defined as

$$\mathrm{DS}_j = \mathrm{ACH}_j - \mathbb{E}(\mathrm{ACH}_j | \mathrm{I}_j), \quad j = 1,3,5,7. \tag{3.13}$$

Our analysis relies on normality, so we test this assumption using the previously described methods. Define $\mathbf{X}$ as the vector of IQ and achievement scores from grades 1 through 8,

$$\mathbf{X} = (\mathrm{I}_1, \mathrm{I}_3, \mathrm{I}_5, \mathrm{I}_7, \mathrm{ACH}_1, \dots, \mathrm{ACH}_8)^T.$$

Results for the normality tests are given in Figure 3.4. Panel (A) is the histogram and density plot for $\hat{D}(\mathbf{X})$. The skewed right tail is typical of a χ^2-density. Panel (B) is a Q-Q plot where the y-axis are quantiles for the observed $\hat{D}(\mathbf{X})$ which are plotted against the theoretical quantiles of a χ^2_{12}-distribution on the x-axis. As seen, the points closely approximate the $y = x$ line. The r_Q values for each of the $\mathbf{X}$ coordinates are given in (C) and have noticeably large values. Panel (D) are the estimated distribution and theoretical distribution for $\hat{D}(\mathbf{X})$ under the null. The two distribution functions lie nearly on top of one another. Although the p-value is 0.03 (suggesting rejecting the null of equality), the statistic can be overly sensitive, especially to small differences around the center of the distribution.

Thus, the analysis provides strong evidence of normality. Therefore, for each grade j we can assume

$$\begin{pmatrix} \mathrm{ACH}_j \\ \mathrm{I}_j \end{pmatrix} \sim \mathrm{N}_2\left(\begin{pmatrix} \mu^{\mathrm{A}}_j \\ \mu^{\mathrm{I}}_j \end{pmatrix}, \begin{pmatrix} \sigma^{\mathrm{AA}}_j & \sigma^{\mathrm{AI}}_j \\ \sigma^{\mathrm{IA}}_j & \sigma^{\mathrm{II}}_j \end{pmatrix}\right).$$

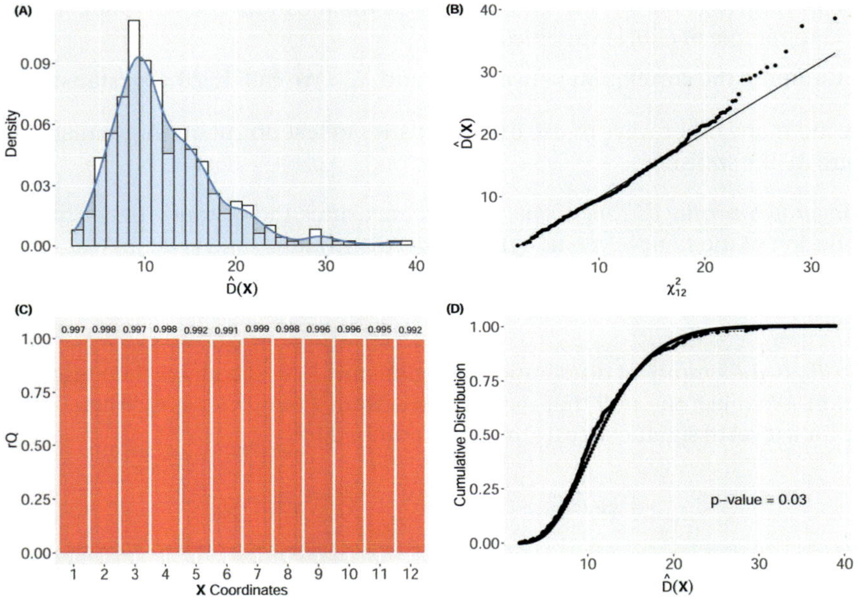

FIGURE 3.4
Testing multivariate normality of IQ and achievement scores from the Connecticut Longitudinal Study.

By the conditional distribution for a bivariate normal,

$$\mathbb{E}(\text{ACH}_j|\text{I}_j) = \mu_j^{\text{A}} + \frac{\sigma_j^{\text{AI}}}{\sigma_j^{\text{II}}}\left(\text{I}_j - \mu_j^{\text{I}}\right).$$

Substituting into (3.13), we obtain

$$\text{DS}_j = \left(\text{ACH}_j - \mu_j^{\text{A}}\right) - \frac{\sigma_j^{\text{AI}}}{\sigma_j^{\text{II}}}\left(\text{I}_j - \mu_j^{\text{I}}\right).$$

Notice that DS_j is a linear function of ACH_j and I_j. Therefore by the Cramér–Wold Device, and our assumption of bivariate normality, it follows that DS_j must be normally distributed. Thus, its distribution is fully specified by its mean and variance. To determine its mean, notice that

$$\mathbb{E}(\text{DS}_j) = \mathbb{E}\left(\text{ACH}_j - \mu_j^{\text{A}}\right) - \frac{\sigma_j^{\text{AI}}}{\sigma_j^{\text{II}}}\mathbb{E}\left(\text{I}_j - \mu_j^{\text{I}}\right)$$

which must be zero. For its variance

$$\begin{aligned}
\text{Var}(\text{DS}_j) &= \text{Var}(\text{ACH}_j) - 2\frac{\sigma_j^{\text{AI}}}{\sigma_j^{\text{II}}}\text{Cov}(\text{ACH}_j, \text{I}_j) + \left(\frac{\sigma_j^{\text{AI}}}{\sigma_j^{\text{II}}}\right)^2 \text{Var}(\text{I}_j) \\
&= \sigma_j^{\text{AA}} - 2\frac{\left(\sigma_j^{\text{AI}}\right)^2}{\sigma_j^{\text{II}}} + \frac{\left(\sigma_j^{\text{AI}}\right)^2}{\sigma_j^{\text{II}}} \\
&= \sigma_j^{\text{AA}} - \frac{\left(\sigma_j^{\text{AI}}\right)^2}{\sigma_j^{\text{II}}} = \sigma_j^{\text{AA}} - \rho_j^2\sigma_j^{\text{AA}}, \quad \text{where } \rho_j = \frac{\sigma_j^{\text{AI}}}{\sqrt{\sigma_j^{\text{AA}}}\sqrt{\sigma_j^{\text{II}}}}.
\end{aligned}$$

Therefore, the DS has a normal distribution specified by

$$\mathrm{DS}_j \sim \mathrm{N}\left(0, \sigma_j^{\mathrm{AA}}(1-\rho_j^2)\right).$$

Standardizing this provides a method for assessing reading disability

$$\mathrm{DS}_j^* = \frac{\mathrm{DS}_j}{\sqrt{\sigma_j^{\mathrm{AA}}(1-\rho_j^2)}}.$$

For example, a standardized score less than -1.65 represents the 5% tail of the DS continuum and evidence of learning disability.

To pursue this further, we investigate the lower tails for the standardized scores DS_j^* for each grade $j = 1,3,5,7$ under varying cut-off values. Let $C < 0$ be a cutoff value for defining reading disability. For child i in grade j, define $B_{ij} = 1$ if i has a standardized DS less than or equal to C; otherwise, define $B_{ij} = 0$. Thus, the number of reading-disabled children in grade j is $B = \sum_{i=1}^{n} B_{ij}$. Assuming children are independent,

$$B = \sum_{i=1}^{n} B_{ij} \sim \mathrm{Binom}(n, p), \quad \text{where } p = \mathbb{P}\{\mathrm{DS}_j^* \le C\} = \mathbb{P}\{\mathrm{N}(0,1) \le C\}.$$

By the central limit theorem, $B \asymp \mathrm{N}(np, np(1-p))$, and therefore with 95% certainty, of the n children in a grade,

$$np \pm 1.96\sqrt{np(1-p)}$$

will be classified as disabled under the cutoff C.

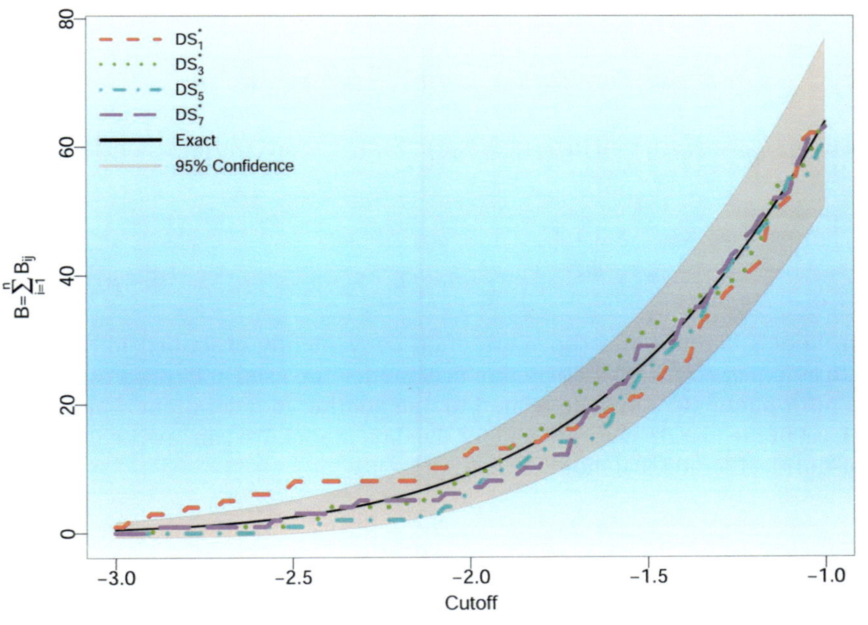

FIGURE 3.5
Standardized discrepancy score, DS_j^* for $j = 1,3,5,7$: comparison of empirical to theoretical behavior under the assumption of normality.

Figure 3.5 tests this. The dashed lines display $B := B(C)$ for each grade as a function of C over a range of cutoff values. The expected number of children with standardized scores below C is

np, displayed with a solid black line. The filled-in colored region represents the 95% confidence limits. The values for B are well within this confidence band. Only for very extreme tail behavior ($C < -2.5$) is there a difference, which occurs for grade 1. For this grade, the observed number of disabilities exceeds the theoretical value. This is interesting because higher variability could be associated with early childhood development. Considering Figure 3.3, the variance for grade 1 measurements is higher than for other grades, supporting this hypothesis.

3.8 Bayesian Multivariate Normal Model

In the Bayesian approach, one specifies a prior density over the parameters of interest. Then using the likelihood (the density of the data), the prior is "updated" yielding the posterior and from this one obtains inference for the parameters or what is called posterior inference. In the multivariate normal model, the two parameters of interest are the mean $\boldsymbol{\mu}$ and the covariance matrix $\boldsymbol{\Sigma}$. Here we look at how Bayesian posterior inference applies to these parameters.

3.8.1 Known variance

We begin by first considering the simplified setting where $\boldsymbol{\Sigma}$ is known. For the prior for $\boldsymbol{\mu}$ we assume a multivariate normal distribution with mean $\boldsymbol{\mu}_0$ and covariance matrix $\boldsymbol{\Sigma}_0$

$$(\boldsymbol{\mu}|\boldsymbol{\Sigma}_0) \sim \mathrm{N}_p(\boldsymbol{\mu}_0, \boldsymbol{\Sigma}_0).$$

The prior density for $\boldsymbol{\mu}$ is

$$f(\boldsymbol{\mu}|\boldsymbol{\mu}_0, \boldsymbol{\Sigma}_0) = |2\pi\boldsymbol{\Sigma}_0|^{-1/2} \exp\left(-\frac{1}{2}(\boldsymbol{\mu}-\boldsymbol{\mu}_0)^T \boldsymbol{\Sigma}_0^{-1}(\boldsymbol{\mu}-\boldsymbol{\mu}_0)\right).$$

The data $(\mathbf{X}_i)_{1\le i\le n}$ are assumed to be i.i.d. $\mathrm{N}_p(\boldsymbol{\mu}, \boldsymbol{\Sigma})$. Therefore, the joint density for the observed data is

$$\begin{aligned} &f(\mathbf{x}_1, \mathbf{x}_2, \ldots, \mathbf{x}_n|\boldsymbol{\mu}, \boldsymbol{\Sigma}) \\ &\quad = |2\pi\boldsymbol{\Sigma}|^{-n/2} \exp\left(-\frac{n}{2}\mathrm{trace}\left[\boldsymbol{\Sigma}^{-1}\mathbf{S}\right] - \frac{n}{2}(\overline{\mathbf{x}}-\boldsymbol{\mu})^T \boldsymbol{\Sigma}^{-1}(\overline{\mathbf{x}}-\boldsymbol{\mu})\right). \end{aligned}$$

Bayesian models like the one above are often conveniently described in a compact format referred to as a *hierarchical model.* Hierarchical models are used in Bayesian applications where multiple prior parameters may be related to one another in a conditional structured manner. The top level of the hierarchy is reserved for the data model. We can rewrite the above model as a normal-normal hierarchical model as follows

$$\begin{aligned} (\mathbf{X}_i|\boldsymbol{\mu}, \boldsymbol{\Sigma}) &\overset{\text{iid}}{\sim} \mathrm{N}_p(\boldsymbol{\mu}, \boldsymbol{\Sigma}), \quad i = 1, \ldots, n, \\ (\boldsymbol{\mu}|\boldsymbol{\mu}_0, \boldsymbol{\Sigma}_0) &\sim \mathrm{N}_p(\boldsymbol{\mu}_0, \boldsymbol{\Sigma}_0). \end{aligned}$$

The posterior density for $\boldsymbol{\mu}$ equals

$$f(\boldsymbol{\mu}|\boldsymbol{\Sigma}, \mathbf{x}_1, \ldots, \mathbf{x}_n) = \frac{f(\mathbf{x}_1, \mathbf{x}_2, \ldots, \mathbf{x}_n|\boldsymbol{\mu}, \boldsymbol{\Sigma}) f(\boldsymbol{\mu}|\boldsymbol{\mu}_0, \boldsymbol{\Sigma}_0)}{f(\mathbf{x}_1, \mathbf{x}_2, \ldots, \mathbf{x}_n|\boldsymbol{\Sigma})},$$

where

$$f(\mathbf{x}_1, \mathbf{x}_2, \ldots, \mathbf{x}_n|\boldsymbol{\Sigma}) = \int_{\mathbb{R}^p} f(\mathbf{x}_1, \mathbf{x}_2, \ldots, \mathbf{x}_n|\boldsymbol{\mu}, \boldsymbol{\Sigma}) f(\boldsymbol{\mu}|\boldsymbol{\mu}_0, \boldsymbol{\Sigma}_0)\, d\boldsymbol{\mu}. \tag{3.14}$$

The density (3.14) is the normalizing constant for the posterior density of $\boldsymbol{\mu}$. However, we do not need to calculate (3.14) because we use a normal prior density, which is a conjugate prior in this problem. A conjugate prior is a prior that yields a posterior density in the same family type. In our normal-normal model, the posterior is multivariate normal. Knowing this, to identify the posterior distribution, we only need to determine its value up to a constant (not depending on $\boldsymbol{\mu}$) because we already know that the posterior must be normally distributed. The following calculations are sufficient for this purpose:

$$\begin{aligned} & f(\boldsymbol{\mu}|\boldsymbol{\Sigma},\mathbf{x}_1,\ldots,\mathbf{x}_n) \\ & \quad \propto f(\mathbf{x}_1,\mathbf{x}_2,\ldots,\mathbf{x}_n|\boldsymbol{\mu},\boldsymbol{\Sigma}) f(\boldsymbol{\mu}|\boldsymbol{\mu}_0,\boldsymbol{\Sigma}_0) \\ & \quad \propto \exp\left(-\frac{n}{2}(\overline{\mathbf{x}}-\boldsymbol{\mu})^T\boldsymbol{\Sigma}^{-1}(\overline{\mathbf{x}}-\boldsymbol{\mu})\right)\exp\left(-\frac{1}{2}(\boldsymbol{\mu}-\boldsymbol{\mu}_0)^T\boldsymbol{\Sigma}_0^{-1}(\boldsymbol{\mu}-\boldsymbol{\mu}_0)\right) \\ & \quad \propto \exp\left(-\frac{1}{2}(\boldsymbol{\mu}-\boldsymbol{\mu}^*)^T\boldsymbol{\Sigma}^{*-1}(\boldsymbol{\mu}-\boldsymbol{\mu}^*)\right) \end{aligned}$$

where

$$\begin{aligned} \boldsymbol{\mu}^* &= \boldsymbol{\Sigma}^*\left(n\boldsymbol{\Sigma}^{-1}\overline{\mathbf{x}}+\boldsymbol{\Sigma}_0^{-1}\boldsymbol{\mu}_0\right) \\ \boldsymbol{\Sigma}^* &= (n\boldsymbol{\Sigma}^{-1}+\boldsymbol{\Sigma}_0^{-1})^{-1}. \end{aligned}$$

(The last line in the multiline equation follows by completing the quadratic in $\boldsymbol{\mu}$ and discarding a constant not depending on $\boldsymbol{\mu}$.) Using the fact that the posterior must be multivariate normal, we can infer that the posterior mean for $\boldsymbol{\mu}$ is $\boldsymbol{\mu}^*$ and that its posterior variance is $\boldsymbol{\Sigma}^*$. Therefore,

$$(\boldsymbol{\mu}|\boldsymbol{\Sigma},\mathbf{x}_1,\ldots,\mathbf{x}_n) \sim \mathrm{N}_p(\boldsymbol{\mu}^*,\boldsymbol{\Sigma}^*).$$

Observe that $\boldsymbol{\mu}^*$ is a weighted average of $\overline{\mathbf{x}}$ and the prior mean $\boldsymbol{\mu}_0$, where the weights depend on $n\boldsymbol{\Sigma}^{-1}$ and $\boldsymbol{\Sigma}_0^{-1}$. For example, under a flat (diffuse) prior for $\boldsymbol{\mu}$, we have $\boldsymbol{\mu}^* = \overline{\mathbf{x}}$. In fact, letting $\boldsymbol{\Sigma}_0^{-1} \to \mathbf{0}$ (a diffuse prior), we have $\boldsymbol{\Sigma}^* \to \boldsymbol{\Sigma}/n$ and $\boldsymbol{\mu}^* \to \overline{\mathbf{x}}$. This yields the limit

$$(\boldsymbol{\mu}|\boldsymbol{\Sigma},\mathbf{x}_1,\ldots,\mathbf{x}_n) \sim \mathrm{N}_p(\overline{\mathbf{x}},\boldsymbol{\Sigma}/n).$$

3.8.2 Unknown variance

When $\boldsymbol{\Sigma}$ is unknown we require priors for both $\boldsymbol{\mu}$ and $\boldsymbol{\Sigma}$. For the multivariate normal model, the conjugate prior for $(\boldsymbol{\mu},\boldsymbol{\Sigma})$ is the normal-inverse Wishart distribution. Notationally it will be more convenient to work with $\mathbf{B}=\boldsymbol{\Sigma}^{-1}$ in place of $\boldsymbol{\Sigma}$ as the Bayesian parameter. The normal-Wishart distribution for $(\boldsymbol{\mu},\mathbf{B})$ with parameters $(\boldsymbol{\mu}_0,\kappa_0,\boldsymbol{\Sigma}_0,\upsilon_0)$ has density

$$f(\boldsymbol{\mu},\mathbf{B}|\boldsymbol{\mu}_0,\kappa_0,\boldsymbol{\Sigma}_0,\upsilon_0) = f(\boldsymbol{\mu}|\mathbf{B},\boldsymbol{\mu}_0,\kappa_0) f(\mathbf{B}|\boldsymbol{\Sigma}_0,\upsilon_0)$$

where $\boldsymbol{\mu}$ has a conditional multivariate normal density $\mathrm{N}_p(\boldsymbol{\mu}_0,(\mathbf{B}\kappa_0)^{-1})$,

$$f(\boldsymbol{\mu}|\mathbf{B},\boldsymbol{\mu}_0,\kappa_0) = \frac{|\kappa_0\mathbf{B}|^{1/2}}{(2\pi)^{p/2}}\exp\left(-\frac{\kappa_0}{2}(\boldsymbol{\mu}-\boldsymbol{\mu}_0)^T\mathbf{B}(\boldsymbol{\mu}-\boldsymbol{\mu}_0)\right),$$

and $\mathbf{B}$ has a Wishart$(\boldsymbol{\Sigma}_0,\upsilon_0)$ distribution (see Appendix 3.E) with density

$$f(\mathbf{B}|\boldsymbol{\Sigma}_0,\upsilon_0) \propto |\mathbf{B}|^{(\upsilon_0-p-1)/2}\exp\left(-\frac{1}{2}\mathrm{trace}[\mathbf{B}\boldsymbol{\Sigma}_0^{-1}]\right).$$

Due to the conditional structure of the distribution for $(\boldsymbol{\mu},\mathbf{B})$, we can write the Bayesian model as the following hierarchical model

$$\begin{aligned} (\mathbf{X}_i|\boldsymbol{\mu},\mathbf{B}) &\overset{\text{iid}}{\sim} \mathrm{N}_p(\boldsymbol{\mu},\mathbf{B}^{-1}), \quad i=1,\ldots,n, \\ (\boldsymbol{\mu}|\mathbf{B},\boldsymbol{\mu}_0,\kappa_0) &\sim \mathrm{N}_p(\boldsymbol{\mu}_0,(\mathbf{B}\kappa_0)^{-1}) \\ (\mathbf{B}|\boldsymbol{\Sigma}_0,\upsilon_0) &\sim \mathrm{Wishart}_p(\boldsymbol{\Sigma}_0,\upsilon_0). \end{aligned}$$

The posterior density satisfies

$$\begin{aligned}
&f(\boldsymbol{\mu},\mathbf{B}|\boldsymbol{\mu}_0,\kappa_0,\boldsymbol{\Sigma}_0,\upsilon_0,\mathbf{x}_1,\ldots,\mathbf{x}_n)\\
&\quad\propto f(\mathbf{x}_1,\mathbf{x}_2,\ldots,\mathbf{x}_n|\boldsymbol{\mu},\mathbf{B})f(\boldsymbol{\mu}|\mathbf{B},\boldsymbol{\mu}_0,\kappa_0)f(\mathbf{B}|\boldsymbol{\Sigma}_0,\upsilon_0)\\
&\quad\propto |\mathbf{B}|^{(n+\upsilon_0-p)/2}\exp\left(-\frac{n}{2}\text{trace}\,[\mathbf{BS}]-\frac{1}{2}\text{trace}[\mathbf{B}\boldsymbol{\Sigma}_0^{-1}]\right)\\
&\qquad\times\exp\left(-\frac{n}{2}(\overline{\mathbf{x}}-\boldsymbol{\mu})^T\mathbf{B}(\overline{\mathbf{x}}-\boldsymbol{\mu})-\frac{\kappa_0}{2}(\boldsymbol{\mu}-\boldsymbol{\mu}_0)^T\mathbf{B}(\boldsymbol{\mu}-\boldsymbol{\mu}_0)\right)\\
&\quad\propto |\mathbf{B}|^{(\upsilon_0^*-p)/2}\exp\left(-\frac{1}{2}\text{trace}\left[\mathbf{B}\boldsymbol{\Sigma}_0^{*-1}\right]\right)\exp\left(-\frac{\kappa_0^*}{2}(\boldsymbol{\mu}-\boldsymbol{\mu}_0^*)^T\mathbf{B}(\boldsymbol{\mu}-\boldsymbol{\mu}_0^*)\right)
\end{aligned}$$

where

$$\begin{aligned}
\boldsymbol{\mu}_0^* &= \frac{n}{\kappa_0+n}\overline{\mathbf{x}}+\frac{\kappa_0}{\kappa_0+n}\boldsymbol{\mu}_0\\
\kappa_0^* &= \kappa_0+n\\
\boldsymbol{\Sigma}_0^* &= \left(\mathbf{S}+\boldsymbol{\Sigma}_0^{-1}+\frac{\kappa_0 n}{\kappa_0+n}(\overline{\mathbf{x}}-\boldsymbol{\mu}_0)(\overline{\mathbf{x}}-\boldsymbol{\mu}_0)^T\right)^{-1}\\
\upsilon_0^* &= \upsilon_0+n.
\end{aligned}$$

Determining $\boldsymbol{\Sigma}_0^*$ used the following fact

$$-\kappa_0^*\boldsymbol{\mu}_0^{*T}\mathbf{B}\boldsymbol{\mu}_0^*+n\overline{\mathbf{x}}^T\mathbf{B}\overline{\mathbf{x}}+\kappa_0\boldsymbol{\mu}_0^T\mathbf{B}\boldsymbol{\mu}_0=\text{trace}\left[\frac{\kappa_0 n}{\kappa_0+n}\mathbf{B}(\overline{\mathbf{x}}-\boldsymbol{\mu}_0)(\overline{\mathbf{x}}-\boldsymbol{\mu}_0)^T\right].$$

Therefore, the posterior distribution for $(\boldsymbol{\mu},\mathbf{B})$ is normal-Wishart with parameters $(\boldsymbol{\mu}_0^*,\kappa_0^*,\boldsymbol{\Sigma}_0^*,\upsilon_0^*)$.

Example 3.1. We analyzed data on fertility and birth rates in the United States for U.S. residents and non-residents from 2007–2010, obtained from the Center for Disease Control (CDC). Birth rates were calculated as the number of births divided by the total population. Fertility rates were calculated as the number of births divided by the number of females aged 15–44 years old. Rates were provided for both state and county, based on the mother's place of residence recorded on the birth certificate. There were $n=50$ states and the District of Columbia, with the number of counties per state ranging up to 36 (California being the largest).

The rates for each state and district are shown in Figure 3.6, where the size of each point is proportional to the number of counties within a state. All rates are stated per 1000 individuals. Figure 3.6 shows that fertility rates are generally proportional to birth rates. The largest rates are observed for Utah and the smallest for Rhode Island. The District of Columbia, with only one observation, appears to be an outlier with a surprisingly small fertility rate given its relatively large birth rate; this is likely a spurious effect due to the small sample size.

Let $\mathbf{X}_{ij}\in\mathbb{R}^2$ denote the fertility and birth rate for county $j=1,\ldots,n_i$ within state $i=1,\ldots,n$. We use a normal model with a state-specific normal-Wishart prior for the mean and covariance. The model is

$$\begin{aligned}
(\mathbf{X}_{ij}|\boldsymbol{\mu}_i,\mathbf{B}_i) &\overset{\text{iid}}{\sim} \text{N}_2(\boldsymbol{\mu}_i,\mathbf{B}_i^{-1}),\quad i=1,\ldots,n,\ j=1,\ldots,n_i\\
(\boldsymbol{\mu}_i|\mathbf{B}_i,\boldsymbol{\mu}_0,\kappa_0) &\sim \text{N}_2(\boldsymbol{\mu}_0,(\mathbf{B}_i\kappa_0)^{-1})\\
(\mathbf{B}_i|\boldsymbol{\Sigma}_0,\upsilon_0) &\sim \text{Wishart}_2(\boldsymbol{\Sigma}_0,\upsilon_0).
\end{aligned}$$

Notice this takes advantage of the nested hierarchical structure of the data. Counties reside within states which are model independent of other states. This is one of the strengths of Bayesian hierarchical models. Starting with a simple model, like the normal-normal-Wishart hierarchy, it is a simple matter to extend this to address nested data.

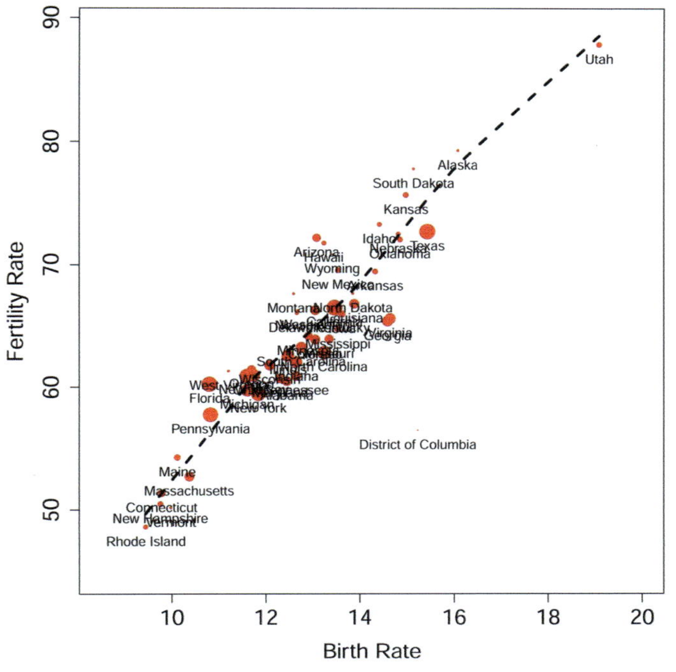

FIGURE 3.6
Fertility rate versus birth rate (per 1000) in the U.S. from 2007 to 2010. Size of circle is proportional to the number of counties within a state.

The posterior derivation is straightforward due to conjugacy. By our previous calculations, the posterior for each state must be normal-Wishart:

$$(\boldsymbol{\mu}_i, \mathbf{B}_i | \boldsymbol{\mu}_0, \kappa_0, \boldsymbol{\Sigma}_0, \upsilon_0, \mathbf{x}_{i1}, \ldots, \mathbf{x}_{in_i}) \sim \text{normal-Wishart}_2(\boldsymbol{\mu}^*_{i0}, \kappa^*_{i0}, \boldsymbol{\Sigma}^*_{i0}, \upsilon^*_{i0})$$

where

$$\begin{aligned}
\boldsymbol{\mu}^*_{i0} &= \frac{n_i}{\kappa_0 + n_i}\overline{\mathbf{x}}_i + \frac{\kappa_0}{\kappa_0 + n_i}\boldsymbol{\mu}_0 \\
\kappa^*_{i0} &= \kappa_0 + n_i \\
\boldsymbol{\Sigma}^*_{i0} &= \left(\mathbf{S}_i + \boldsymbol{\Sigma}_0^{-1} + \frac{\kappa_0 n_i}{\kappa_0 + n_i}(\overline{\mathbf{x}}_i - \boldsymbol{\mu}_0)(\overline{\mathbf{x}}_i - \boldsymbol{\mu}_0)^T\right)^{-1} \\
\upsilon^*_{i0} &= \upsilon_0 + n_i.
\end{aligned}$$

Having a closed-form expression like this simplifies posterior inference. Then we can calculate values by analytical derivation or by directly sampling the posterior, which is the method we use here.

The prior used the following parameters

$$\boldsymbol{\mu}_0 = \overline{\mathbf{x}} = \frac{1}{n}\sum_{i=1}^{n}\sum_{j=1}^{n_i}\frac{\mathbf{x}_{ij}}{n_i}, \quad \boldsymbol{\Sigma}_0 = \mathbf{I}, \quad \upsilon_0 = 1.$$

We varied κ_0 using $\kappa_0 = 0.01, 1, 2, 20$. Small values of κ_0 elicit a diffuse prior, large values result in an informative prior. For each κ_0 we sampled 1000 values from the posterior. Each value was sampled using the two-stage sampling algorithm:

1. Draw $(\mathbf{B}_i | \boldsymbol{\Sigma}^*_{i0}, \upsilon^*_{i0})$ from a $\text{Wishart}_2(\boldsymbol{\Sigma}^*_{i0}, \upsilon^*_{i0})$.
2. Using the draw for $\mathbf{B}_i$, draw $(\boldsymbol{\mu}_i | \mathbf{B}_i, \boldsymbol{\mu}^*_{i0}, \kappa^*_{i0})$ from a $\text{N}_2(\boldsymbol{\mu}^*_{i0}, (\mathbf{B}_i \kappa^*_{i0})^{-1})$.

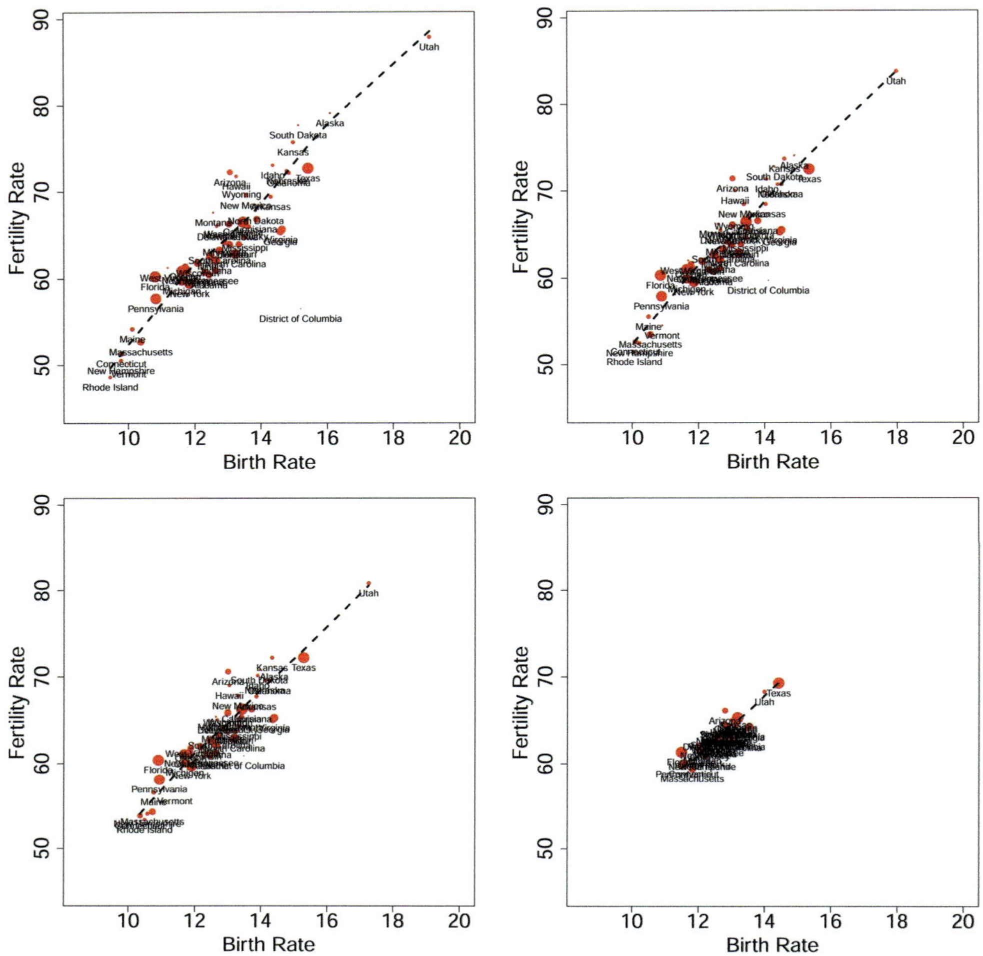

FIGURE 3.7
Posterior median for fertility and birth rates under a normal-Wishart $(\boldsymbol{\mu}_0, \kappa_0, \boldsymbol{\Sigma}_0, v_0)$ *prior where* $\boldsymbol{\mu}_0 = \overline{\mathbf{x}}$, $\boldsymbol{\Sigma}_0 = \boldsymbol{I}$, $v_0 = 1$, *and* $\kappa_0 = 0.01, 1, 2, 20$ *(left to right, top to bottom).*

This gives a draw from a normal-Wishart$(\boldsymbol{\mu}^*_{i0}, \kappa^*_{i0}, \boldsymbol{\Sigma}^*_{i0}, v^*_{i0})$. The posterior median was obtained from the sampled values and these values are shown in Figure 3.7. When $\kappa_0 = 0.01$ (top left) the values closely approximate those of Figure 3.6 but as κ_0 increases (left to right, top to bottom), the values converge to the prior $\boldsymbol{\mu}_0 = \overline{\mathbf{x}}$. This shows that an overly informative prior can "swamp" the data and results in a posterior that closely resembles the original prior. □

Exercises

Exercise 3.1. Simulate a dataset with $n = 1000$ observations from a multivariate normal $\mathbf{X} = (X_1, X_2, X_3)^T$ with the following parameters:

$$\boldsymbol{\mu} = \begin{pmatrix} 3 \\ -1 \\ 2 \end{pmatrix}, \quad \boldsymbol{\Sigma} = \begin{pmatrix} 4 & 2 & 0 \\ 2 & 3 & 1 \\ 0 & 1 & 2 \end{pmatrix}.$$

1. Add random noise to each observation by adding a vector of independent t_υ random variables. Test the normality of the data using different values $\upsilon = 1, 5, 10, 20, 100$.
2. Estimate the mean vector and covariance matrix of the generated data and compare this to the true parameters.
3. Create a binary response variable Y based on a linear combination of the original variables where $Y = 1$ if $X_1 + X_2 > 1$, otherwise $Y = 0$. What is the distribution of Y?
4. Let $Y = 2X_1 - X_2 + 0.5X_3 + \epsilon$ where $\epsilon \sim \mathrm{N}(0,1)$. What is the distribution of Y?

Exercise 3.2. Let (X, Y) have a bivariate normal distribution with parameters $\mu_X = 415$, $\sigma_X^2 = 611$, $\mu_Y = 347$, $\sigma_Y^2 = 689$, and $\rho = -0.25$. Find:

1. $\mathbb{P}\{309.2 < Y < 380.6\}$.
2. $\mathbb{E}(Y|X = x)$.
3. $\mathrm{Var}(Y|X = x)$.
4. $\mathbb{P}\{309.2 < Y < 380.6 \mid X = 385.1\}$.
5. $\mathrm{Cov}(\mathbf{AZ}, \mathbf{BZ})$, where

$$\mathbf{Z} = \begin{pmatrix} X \\ Y \end{pmatrix}, \qquad \mathbf{A} = \begin{pmatrix} 1 & 2 \\ 2 & 5 \end{pmatrix}, \qquad \mathbf{B} = \begin{pmatrix} 4 & 0 \\ 0 & 5 \end{pmatrix}.$$

Exercise 3.3. Consider a particle of mass m confined in a three-dimensional harmonic oscillator potential with angular frequency ω. The wave function for the ground state of this system is given by:

$$\Psi(\mathbf{r}) = \left(\frac{m\omega}{\pi\hbar}\right)^{3/4} \exp\left(-\frac{m\omega}{2\hbar}(x^2 + y^2 + z^2)\right)$$

where $\mathbf{r} = (x, y, z)$ is the position vector and $\hbar$ is the reduced Planck constant.

1. Show that the probability density function $f(\mathbf{r}) = |\Psi(\mathbf{r})|^2$ is a multivariate normal density.
2. Calculate the expectation values of the position coordinates x, y, and z.
3. Derive the covariance matrix for the position vector $\mathbf{r}$.
4. The wave function $\Phi(\mathbf{p})$ in momentum space equals the Fourier transform of the wave function $\Psi(\mathbf{r})$,

$$\Phi(\mathbf{p}) = \frac{1}{(2\pi\hbar)^{3/2}} \int \Psi(\mathbf{r}) e^{-i\mathbf{p}\cdot\mathbf{r}/\hbar} \mathbf{r}.$$

   Show that $\Phi(\mathbf{p})$ has a multivariate normal distribution. Identify the mean vector and covariance matrix in momentum space.
5. Calculate the standard deviations σ_x and σ_p and show their product satisfies the uncertainty principle, $\sigma_x \sigma_p \geq \hbar$.

Exercise 3.4. Use the Cramér–Wold Device to prove properties 5 (a)-(d) of Chapter 3.2. You can assume basic properties for the expectation and variance of a random vector. [Hint for (d): Show that $\mathbf{A}(\mathbf{X} - \boldsymbol{\mu}) \sim \mathrm{N}_p(\mathbf{0}, \mathbf{A\Sigma A}^T)$ where

$$\mathbf{A} = \begin{bmatrix} \mathbf{I} & -\boldsymbol{\Sigma}_{12}\boldsymbol{\Sigma}_{22}^{-1} \\ \mathbf{0} & \mathbf{I} \end{bmatrix}_{p \times p}$$

and that

$$\mathbf{A\Sigma A}^T = \begin{bmatrix} \boldsymbol{\Sigma}^* & \mathbf{0} \\ \mathbf{0} & \boldsymbol{\Sigma}_{22} \end{bmatrix}_{p \times p}.$$

The result follows by considering the distribution of the first q coordinates of $\mathbf{A}(\mathbf{X} - \boldsymbol{\mu})$, where q is the dimension of $\mathbf{X}_1$.]

Exercise 3.5. Suppose that $\mathbf{X}_1, \mathbf{X}_2, \mathbf{X}_3$ are i.i.d. $\mathrm{N}_p(\boldsymbol{\mu}, \boldsymbol{\Sigma})$. Derive the conditional distribution of $\mathbf{Y}_1$ given $\mathbf{Y}_2$ and $\mathbf{Y}_1$ given $\mathbf{Y}_2, \mathbf{Y}_3$ if $\mathbf{Y}_1 = \mathbf{X}_1 + \mathbf{X}_2$, $\mathbf{Y}_2 = \mathbf{X}_2 + \mathbf{X}_3$, and $\mathbf{Y}_3 = \mathbf{X}_1 + \mathbf{X}_3$.

Exercise 3.6. If $\mathbf{X} \sim \mathrm{N}_p(\boldsymbol{\mu}, \boldsymbol{\Sigma})$ and $\mathbf{Q}\boldsymbol{\Sigma}\mathbf{Q}^T$ $(q \times q)$ is non-singular, then show that the conditional distribution of $\mathbf{X}$ given $\mathbf{QX} = \mathbf{q}$ is normal with mean

$$\boldsymbol{\mu} + \boldsymbol{\Sigma}\mathbf{Q}^T(\mathbf{Q}\boldsymbol{\Sigma}\mathbf{Q}^T)^{-1}(\mathbf{q} - \mathbf{Q}\boldsymbol{\mu})$$

and (singular) covariance matrix $\boldsymbol{\Sigma} - \boldsymbol{\Sigma}\mathbf{Q}^T(\mathbf{Q}\boldsymbol{\Sigma}\mathbf{Q}^T)^{-1}\mathbf{Q}\boldsymbol{\Sigma}$.

Exercise 3.7. If $\mathbf{X} \sim \mathrm{N}_p(\mathbf{0}, \sigma^2\mathbf{I})$, show that $\mathbf{AX}$ and $(\mathbf{I} - \mathbf{A}^+\mathbf{A})\mathbf{X}$ are independent and that each are normally distributed, where $\mathbf{A}^+$ is the generalized inverse of $\mathbf{A}$.

Exercise 3.8. For the $p = 2$ dimensional case, derive the distribution for a degenerate bivariate normal distribution using (3.4).

Exercise 3.9. Write out the log-likelihood for a multivariate normal sample and then taking the derivative for $\boldsymbol{\mu}$ and $\boldsymbol{\Sigma}$ directly solve for the MLE by setting the derivative to zero.

Exercise 3.10. Let $\mathbf{D}$ be the $n \times p$ data matrix with rows $\mathbf{X}_1, \ldots, \mathbf{X}_n$ which are i.i.d. $\mathrm{N}_p(\mathbf{0}, \mathbf{I}_p)$. A function h is called scale invariant if $h(\mathbf{D}) = h(\alpha\mathbf{D})$ for all $\alpha > 0$. Show that $||\mathbf{D}||_F^2$ is statistically independent of $h(\mathbf{D})$ for any scale-invariant function h of $\mathbf{D}$ (hint: use the fact that $||\mathbf{D}||_F^2 = \mathrm{trace}(\mathbf{D}^T\mathbf{D})$ and that the rows of $\mathbf{D}$ are radially symmetric; i.e., this means that the distribution of $\mathbf{D}$ depends only on the distances from the origin and not on the direction).

Exercise 3.11. (Spherical symmetry). If $\mathbf{X} \sim \mathrm{N}_p(\boldsymbol{\mu}, \boldsymbol{\Sigma})$, show that $\mathbf{X}$ and $\boldsymbol{\Gamma}\mathbf{X}$ have the same distribution for all orthogonal matrices $\boldsymbol{\Gamma}$ if and only if $\boldsymbol{\mu} = \mathbf{0}$ and $\boldsymbol{\Sigma} = \sigma^2\mathbf{I}$ for some $\sigma^2 > 0$.

Exercise 3.12. Let $\mathbf{X} \sim \mathrm{N}_p(\boldsymbol{\mu}, \boldsymbol{\Sigma})$ and let $\mathbf{b} \in \mathbb{R}^p$.

1. Show that if $\mathbf{b}$ is a fixed vector,
$$Z = \frac{\mathbf{b}^T(\mathbf{X} - \boldsymbol{\mu})}{(\mathbf{b}^T\boldsymbol{\Sigma}\mathbf{b})^{1/2}} \sim \mathrm{N}(0, 1).$$
2. Show if $\mathbf{b}$ is a random vector independent of $\mathbf{X}$ such that $\mathbb{P}\{\mathbf{b}^T\boldsymbol{\Sigma}\mathbf{b} = 0\} = 0$, then $Z \sim N(0, 1)$ and is independent of $\mathbf{b}$.
3. Hence show that if $\mathbf{X} = (X_1, X_2, X_3)^T$ such that $\mathbf{X} \sim \mathrm{N}_3(\mathbf{0}, \mathbf{I})$, then
$$\frac{X_1 \exp(X_3) + X_2 \log|X_3|}{\left(\exp(2X_3) + (\log|X_3|)^2\right)^{1/2}} \sim \mathrm{N}(0, 1).$$

Exercise 3.13. Consider the following model

$$Y|\mathbf{X} \sim \mathrm{N}(\mathbf{X}^T\boldsymbol{\beta}, \sigma^2),$$

where $\boldsymbol{\beta}$ is some unknown parameter vector and $\mathbf{X} \sim \mathrm{N}_p(\mathbf{0}, \boldsymbol{\Sigma})$.

1. Show that $\mathrm{Cor}(Y, \mathbf{X})^T = \dfrac{\mathrm{Diag}(\boldsymbol{\Sigma})^{-1/2}\boldsymbol{\Sigma}\boldsymbol{\beta}}{\mathrm{Var}(Y)^{1/2}}$.
2. Work out an explicit representation for $\mathrm{Var}(Y)$ in terms of $\boldsymbol{\beta}$ and $\mathbf{R}$ where $\mathbf{R}$ is the correlation matrix for $\mathbf{X}$.

Exercise 3.14. Consider a high-dimensional dataset representing the returns of p assets in a financial portfolio. The returns of these assets over a given time period are modeled as a multivariate normal distribution with mean zero. Let $\mathbf{R} = (R_1, R_2, \ldots, R_p)^T$ be the vector of asset returns. Assume that the variance of each asset return is given by $\sigma_k^2 = \mathbb{E}(R_k^2)$, and define $\tilde{\sigma} = \max_{1 \le k \le p} \sigma_k$.

1. Explain the meaning of the term $\mathbb{E}(\max_{1\le k\le p} R_k)$.
2. Discuss why the maximum return might be of interest to a risk-averse investor.
3. Using the concentration inequality (3.5), derive an upper bound for the probability that the maximum return exceeds its expectation by a certain amount t.
4. Suppose you have a portfolio of $p = 1000$ assets with variances σ_k^2 uniformly distributed between 0.01 and 0.02. Calculate $\tilde{\sigma}$.
5. For $t = 0.2$, compute the upper bound for the probability that the maximum return exceeds its expectation by at least 0.2. Interpret your results and discuss how this bound can be useful for portfolio risk management.

Exercise 3.15. Let $\mathbf{X} \sim \mathrm{N}_p(\boldsymbol{\mu}, \mathbf{I})$ and define

$$\theta_1(k) = \mathbb{E}\left[(\mathbf{X}-\boldsymbol{\mu})\left(\sum_{i=1}^{p} X_i^k\right)\right], \quad \theta_2(k) = \mathbb{E}\left[(\mathbf{X}-\boldsymbol{\mu})\left(\sum_{i=1}^{p} X_i\right)^k\right].$$

Determine $\{\theta_1(k), \theta_2(k)\}$ for $k = 0, 1, 2, \ldots$.

Exercise 3.16. Give a proof of Corollary 3.3. Make sure to state your assumptions carefully.

Exercise 3.17. Prove that the distribution for $\mathbf{X}$ specified by (3.9) for the finite normal mixture model is identical to the distribution of $\mathbf{X}$ specified by (3.8).

Exercise 3.18. You are an analyst at an investment firm. Your task is to analyze and optimize a portfolio of assets. You are given the daily returns of five different stocks over the past two years. Assume that the daily returns of these stocks are jointly normally distributed. The steps below guide you through the analysis.

1. Generate synthetic data for daily returns of five stocks. Assume 500 trading days (approximately two years), use a multivariate normal distribution to generate the data using the following parameters for the mean vector and covariance matrix:

$$\boldsymbol{\mu} = \begin{pmatrix} 0.0005 \\ 0.0007 \\ 0.0003 \\ 0.0004 \\ 0.0006 \end{pmatrix}, \quad \boldsymbol{\Sigma} = \begin{pmatrix} 0.0004 & 0.0001 & 0.00005 & 0.00002 & 0.00003 \\ 0.0001 & 0.0003 & 0.00007 & 0.00003 & 0.00004 \\ 0.00005 & 0.00007 & 0.00025 & 0.00004 & 0.00006 \\ 0.00002 & 0.00003 & 0.00004 & 0.0002 & 0.00005 \\ 0.00003 & 0.00004 & 0.00006 & 0.00005 & 0.00035 \end{pmatrix}.$$

2. Estimate the mean vector and covariance matrix of the daily returns from the generated data.
3. Perform a multivariate normality test to check if the data follows a multivariate normal distribution.
4. Formulate the problem of finding the optimal portfolio weights $\mathbf{w}$ that maximizes the Sharpe ratio. The Sharpe ratio is defined as:

$$S(\mathbf{w}) = \frac{\mathbf{w}^T\boldsymbol{\mu} - r_f}{\sqrt{\mathbf{w}^T\boldsymbol{\Sigma}\mathbf{w}}},$$

where r_f is the risk-free rate, assumed here to be $r_f = 0.0001$. Solve the optimization problem to find the optimal weights.
5. Visualize the efficient frontier, which plots the expected portfolio return against the portfolio risk (standard deviation).

Exercise 3.19. In this exercise, you will participate in a "hypothetical Kaggle competition" focused on detecting anomalies in financial data. You are provided a dataset of daily returns for 10 different stocks over a period of 5 years (approximately 1250 trading days). The dataset includes the following columns:

- *Date*: The trading date.
- $Stock_1, Stock_2, \ldots Stock_{10}$: Daily returns for each of the 10 stocks.

Download the dataset from the provided link (instructor provides data for the exercise, otherwise the student should simulate the data themselves). The goal will be to identify days when the returns significantly deviate from the expected multivariate normal distribution.

1. Load the dataset, check for missing values, standardize the data and perform a multivariate normality test to ensure the data is suitably normally distributed.
2. Define an appropriate threshold for anomaly detection based on the Mahalanobis distance
$$\hat{D}(\mathbf{x}) = (\mathbf{x} - \hat{\boldsymbol{\mu}})^T \hat{\boldsymbol{\Sigma}}^{-1} (\mathbf{x} - \hat{\boldsymbol{\mu}}).$$
3. Identify potential anomalies and report your results.

Exercise 3.20. Under what conditions does the finite normal mixture model (3.9) have the same distribution as a multivariate normal? Provide a proof for your answer.

Exercise 3.21. Consider the following Bayesian normal-normal hierarchy:

$$\begin{aligned}(\mathbf{Y}|\mathbf{X}, \boldsymbol{\beta}, \sigma^2) &\sim \mathrm{N}_n(\mathbf{X}\boldsymbol{\beta}, \sigma^2 \mathbf{I}_n)\\ (\boldsymbol{\beta}|\boldsymbol{\Gamma}) &\sim \mathrm{N}_p(\mathbf{0}, \boldsymbol{\Gamma}).\end{aligned}$$

Using conjugacy show that the posterior distribution for $\boldsymbol{\beta}$ is multivariate normal with distribution

$$(\boldsymbol{\beta}|\mathbf{Y}, \boldsymbol{\Gamma}, \sigma^2) \sim N_p(\boldsymbol{\mu}_{\Gamma}, \sigma^2 \boldsymbol{\Sigma}_{\Gamma})$$

where $\boldsymbol{\mu}_{\Gamma} = \boldsymbol{\Sigma}_{\Gamma}\mathbf{X}^T\mathbf{Y}$ is the posterior mean of $\boldsymbol{\beta}$ and $\boldsymbol{\Sigma}_{\Gamma} = (\mathbf{X}^T\mathbf{X} + \sigma^2\boldsymbol{\Gamma}^{-1})^{-1}$.

Exercise 3.22. Describe a non-trivial setting where the goal is to estimate a parameter $\boldsymbol{\theta} \in \mathbb{R}^p$ using an estimator $\hat{\boldsymbol{\theta}}_n$ with unbiased estimating equations of the form (3.11). Describe the limiting distribution of your estimator and its Godambe information. Give closed form expressions whenever possible.

Exercise 3.23. In Example 3.1 how does the posterior for $\boldsymbol{\mu}_i$ depend on υ_0? What is the limit if $\upsilon_0 \to \infty$?

Exercise 3.24. Consider the following normal-normal hierarchical model

$$\begin{aligned}(\mathbf{X}|\boldsymbol{\mu}, \boldsymbol{\Sigma}) &\sim \mathrm{N}_p(\boldsymbol{\mu}, \boldsymbol{\Sigma})\\ (\boldsymbol{\mu}|\boldsymbol{\mu}_0, \boldsymbol{\Sigma}_0) &\sim \mathrm{N}_p(\boldsymbol{\mu}_0, \boldsymbol{\Sigma}_0).\end{aligned}$$

Calculate the mean and covariance of $\mathbf{X}$ ($\boldsymbol{\mu}_0$ and $\boldsymbol{\Sigma}_0$ are assumed fixed).

Bibliography

[1] Chernozhukov V., Chetverikov D. and Kato K. Gaussian approximations and multiplier bootstrap for maxima of sums of high-dimensional random vectors. *The Annals of Statistics*, 41:2786–2819, 2013.

[2] Chernozhukov V., Chetverikov D. and Kato K. Comparison and anti-concentration bounds for maxima of Gaussian random vectors. *Probability Theory and Related Fields*, 162:47–70, 2015.

[3] Fletcher J.M., Esp K. A., Francis D.J., Davidson K.C., Rourke B.P. and Shaywitz S.E. Comparisons of cut-off and regression-based definitions of reading disabilities. *Journal of Learning Disabilities,* 22:334-338, 1989.

[4] Godambe V.P. An optimum property of regular maximum likelihood estimation. *The Annals of Mathematical Statistics,* 31: 1208–1211, 1960.

[5] Li K.-C. On principal Hessian directions for data visualization and dimension reduction: Another application of Stein's lemma. *Journal of the American Statistical Association,* 87 (420):1025–1039, 1992.

[6] Reynolds C.R. Critical measurement issues in learning disability. *Journal of Special Education,* 18:451–476, 1984.

[7] Shaywitz B.A., Fletcher J.M., Holahan J.M. and Shaywitz S.E. Discrepancy compared to low achievement definitions of reading disability: results from the Connecticut Longitudinal Study. *Journal of Learning Disabilities,* 25:639–648, 1992.

[8] Shaywitz S.E., Escobar M.D., Shaywitz B.A., Fletcher J.M. and Makuch R. Evidence that dyslexia may represent the lower tail of a normal distribution of reading disability. *New England Journal of Medicine,* 326:145–150, 1992.

[9] Stein C.M. Estimation of the mean of a multivariate normal distribution. *The Annals of Statistics,* 1135–1151, 1981.

[10] Wechsler D. *Wechsler Intelligence Scale for Children—Revised.* New York: Psychological corporation, 1974.

[11] Woodcock R.W. and Johnson M.B. *Woodcock-Johnson psychoeducational battery.* Boston: Teaching Resources, 1977.

[12] Yule W., Rutter M., Berger M. and Thompson J. Over- and under-achievement in reading: distribution in the general population *British Journal of Educational Psychology,* 44:1–12, 1974.

APPENDIX

3.A The Normal Random Variable

1. X is said to be normal with mean $\mathbb{E}(X) = \mu$ and variance $\mathrm{Var}(X) = \sigma^2$ if it has the density

$$f_X(x) = \frac{1}{\sigma\sqrt{2\pi}} \exp\left(-\frac{1}{2\sigma^2}(x-\mu)^2\right).$$

   This is written as $X \sim \mathrm{N}(\mu, \sigma^2)$. When $\mu = 0$ and $\sigma^2 = 1$, then X is called a standard normal, which is denoted by Z.

2. Normal random variables are closed under linear transformations. If $X \sim \mathrm{N}(\mu, \sigma^2)$:

   (a) $aX + b \sim \mathrm{N}(a\mu + b, a^2\sigma^2)$.

   (b) $Z = (X - \mu)/\sigma \sim \mathrm{N}(0, 1)$.

3. The 68%–95%–99% rule:

$$\begin{array}{rcccl} \mathbb{P}\{\mu-\sigma \le X \le \mu+\sigma\} & = & \mathbb{P}\{-1 \le Z \le 1\} & \approx & 0.68 \\ \mathbb{P}\{\mu-2\sigma \le X \le \mu+2\sigma\} & = & \mathbb{P}\{-2 \le Z \le 2\} & \approx & 0.95 \\ \mathbb{P}\{\mu-3\sigma \le X \le \mu+3\sigma\} & = & \mathbb{P}\{-3 \le Z \le 3\} & \approx & 0.99. \end{array}$$

4. The 95th and 97.5 percentile for the standard normal is 1.645 and 1.96. This is written as $z_{0.05} = 1.645$ and $z_{0.025} = 1.96$.

5. The cumulative distribution function for X is

$$F_X(x) = \mathbb{P}\{X \le x\}.$$

For the standard normal this is written as $\Phi(z)$. For example, $\Phi(0) = 0.5$ and $\Phi(1.96) = 0.975$.

6. If $X_1, \ldots, X_n$ are independent $\mathrm{N}(\mu_i, \sigma_i^2)$, then

$$a_1 X_1 + \cdots + a_n X_n \sim \mathrm{N}(\mu, \sigma^2), \quad \mu = \sum_{i=1}^{n} a_i \mu_i, \text{ and } \sigma^2 = \sum_{i=1}^{n} a_i^2 \sigma_i^2.$$

e.g, if $X_1 \sim \mathrm{N}(3,2)$ and $X_2 \sim \mathrm{N}(4,5)$, then $X_1 + 2X_2 \sim \mathrm{N}(11, 22)$.

3.B The t_n Random Variable

1. The t_n random variable with n degrees of freedom is continuous with symmetric density centered at zero. Its density looks normal but with fatter tails. As $n \to \infty$, its distribution converges to a standard normal. For example, $t_n(0.025)$ (the 97.5 percentile) is equal to 12.706, 2.228, 2.086, 2.042, 1.984, 1.965, and 1.962 when $n = 1, 10, 20, 30, 100, 500$, and 1000.

2. Formally, t_n is defined by

$$t_n = \frac{Z}{\sqrt{\chi_n^2 / n}},$$

where $Z \sim \mathrm{N}(0,1)$ is independent of χ_n^2 (a chi-squared variable with n degrees of freedom).

3. One of its most important use is constructing confidence intervals for the mean for data sampled from a normal population. In particular, if $X_1, \ldots, X_n$ are independent $\mathrm{N}(\mu, \sigma^2)$,

$$t_{n-1} = \frac{\sqrt{n}(\overline{X} - \mu)}{S},$$

where $\overline{X} = \sum X_i / n$ and $S^2 = \sum (X_i - \overline{X})^2 / (n-1)$. Note that $\overline{X}$ and S^2 are independent.

3.C The χ_n^2 Random Variable

1. If $Z_1, \ldots, Z_n$ are independent $\mathrm{N}(0,1)$,

$$\chi_n^2 = Z_1^2 + \cdots + Z_n^2$$

is a chi-squared variable with n degrees of freedom.

2. Another construction is possible from non-standardized normal variables. If $X_1,\ldots,X_n$ are independent $\mathrm{N}(\mu,\sigma^2)$,

$$\sum_{i=1}^{n}(X_i-\overline{X})^2 \sim \sigma^2\chi^2_{n-1}.$$

3.D The F_{n_1,n_2} Random Variable

1. The F_{n_1,n_2} random variable with n_1 degrees of freedom in the numerator and n_2 degrees of freedom in the denominator is defined by

$$F_{n_1,n_2}=\frac{\chi^2_{n_1}/n_1}{\chi^2_{n_2}/n_2}$$

where $\chi^2_{n_1}$ is independent of $\chi^2_{n_2}$.

2. $F_{1,n_2}=t^2_{n_2}$ when $n_1=1$.

3.E The Wishart Distribution

Let $\mathbf{X}_1,\ldots,\mathbf{X}_n$ be i.i.d. $\mathrm{N}_p(\mathbf{0},\boldsymbol{\Sigma})$ vectors where $\boldsymbol{\Sigma}>0$. Then,

$$\mathbf{W}=\sum_{i=1}^{n}\mathbf{X}_i\mathbf{X}_i^T \sim \mathrm{Wishart}_p(\boldsymbol{\Sigma},n)$$

is a random matrix with the Wishart distribution. The density for $\mathbf{W}$ is

$$f(\mathbf{W})=\frac{|\mathbf{W}|^{(n-p-1)/2}\exp\left(-\frac{1}{2}\mathrm{trace}[\mathbf{W}\boldsymbol{\Sigma}^{-1}]\right)}{2^{pn/2}\pi^{p(p-1)/4}|\boldsymbol{\Sigma}|^{n/2}\prod_{k=1}^{p}\Gamma((n+1-k)/2)}.$$

Some useful properties and facts:

1. If $\mathbf{A}\sim\mathrm{Wishart}_p(\boldsymbol{\Sigma},n_1)$ independently of $\mathbf{B}\sim\mathrm{Wishart}_p(\boldsymbol{\Sigma},n_2)$, then

$$\mathbf{A}+\mathbf{B}\sim\mathrm{Wishart}_p(\boldsymbol{\Sigma},n_1+n_2).$$

2. If $\mathbf{A}\sim\mathrm{Wishart}_p(\boldsymbol{\Sigma},n)$ and $\mathbf{B}_{q\times p}$ is a fixed matrix, then

$$\mathbf{BAB}^T\sim\mathrm{Wishart}_q(\mathbf{B}\boldsymbol{\Sigma}\mathbf{B}^T,n).$$

3. $f(\mathbf{W}|\boldsymbol{\Sigma},n)\propto\exp\left(-\frac{1}{2}\mathrm{trace}\left[\boldsymbol{\Sigma}^{-1}\mathbf{W}\right]\right)$.

4. Suppose that $\mathbf{X}_1,\ldots,\mathbf{X}_n$ are i.i.d. $\mathrm{N}_p(\boldsymbol{\mu},\boldsymbol{\Sigma})$. Then $\overline{\mathbf{X}}\sim\mathrm{N}_p(\boldsymbol{\mu},\boldsymbol{\Sigma}/n)$ is independent of

$$n\mathbf{S}=\sum_{i=1}^{n}(\mathbf{X}_i-\overline{\mathbf{X}})(\mathbf{X}_i-\overline{\mathbf{X}})^T\sim\mathrm{Wishart}_p(\boldsymbol{\Sigma},n-1).$$

3.F The Hotelling T^2 Distribution

Let $\mathbf{Z} \sim \mathrm{N}_p(\mathbf{0},\mathbf{I})$ be a standard normal independent of $\mathbf{W} \sim \mathrm{Wishart}_p(\mathbf{I}, m)$. Then,

$$m\mathbf{Z}^T\mathbf{W}^{-1}\mathbf{Z} \sim T^2(p,m)$$

has the Hotelling T^2 distribution.

Some useful properties and facts:

1. $T^2(p,m) \stackrel{d}{=} \frac{pm}{m-p+1}F_{p,m-p+1}$.
2. Suppose that $\mathbf{X}_1,\ldots,\mathbf{X}_n$ are i.i.d. $\mathrm{N}_p(\boldsymbol{\mu},\boldsymbol{\Sigma})$. Then
$$T^2 = n(\overline{\mathbf{X}}-\boldsymbol{\mu})^T\mathbf{S}^{-1}(\overline{\mathbf{X}}-\boldsymbol{\mu}) \sim T^2(p,n-1).$$
Put another way,
$$\frac{1}{n}T^2 \stackrel{d}{=} \frac{p(n-1)}{n(n-p)}F_{p,n-p},$$
so that
$$\left\{\boldsymbol{\mu} : (\boldsymbol{\mu}-\overline{\mathbf{x}})^T\mathbf{s}^{-1}(\boldsymbol{\mu}-\overline{\mathbf{x}}) \le \frac{p(n-1)}{n(n-p)}F_{p,n-p}(\alpha)\right\}$$
is a $100(1-\alpha)\%$ elliptical confidence region for $\boldsymbol{\mu}$.
3. If $\mathbf{X} \sim \mathrm{N}_p(\boldsymbol{\mu},\boldsymbol{\Sigma})$ is independent of $\mathbf{W} \sim \mathrm{Wishart}_p(\boldsymbol{\Sigma}, m)$, then
$$m(\mathbf{X}-\boldsymbol{\mu})^T\mathbf{W}^{-1}(\mathbf{X}-\boldsymbol{\mu}) \sim T^2(p,m).$$
4. Let $\mathbf{Y} = \mathbf{AX}+\mathbf{b}$, for fixed non-singular $\mathbf{A}_{p\times p}$ and fixed $\mathbf{b}_{p\times 1}$. Then $T^2(\mathbf{Y}) = T^2(\mathbf{X})$. Thus, the Hotelling T^2 is invariant under non-singular affine transformations.
5. $T^2(p,n) \to \chi^2_p$, as $n \to \infty$.

3.G Wilks' Lambda Distribution

Suppose that $\mathbf{A} \sim \mathrm{Wishart}_p(\mathbf{I}, m_1)$ independently of $\mathbf{B} \sim \mathrm{Wishart}_p(\mathbf{I}, m_2)$, then

$$\frac{|\mathbf{A}|}{|\mathbf{A}+\mathbf{B}|} \sim \Lambda(p, m_1, m_2),$$

has the Wilks' lambda distribution.

Some useful properties and facts:

1. Bartlett's Approximation:
$$-[m_1 - \frac{1}{2}(p-m_2+1)]\log\Lambda(p,m_1,m_2) \to \chi^2_{m_2 p}, \quad \text{as } m_1 \to \infty.$$
2. A special case relating to Hotelling's T^2:
$$\frac{1-\Lambda(p,m,1)}{\Lambda(p,m,1)} \sim \frac{p}{m-p+1}F_{p,m-p+1}.$$

4

Linear Regression

4.1 Introduction

This chapter introduces the linear regression model, a fundamental tool in statistical analysis and predictive modeling. The initial sections will cover the basics of the linear model, laying the groundwork for more advanced topics. The real focus of this chapter, however, is the concept of regularization, a set of techniques used to address overfitting in predictive machine learning models. Due to its simplicity, the linear model serves as an excellent tool for illustrating these nuanced points.

We will explore various aspects of least squares, restricted least squares, final prediction error, mean-squared error, AIC, and BIC. Furthermore, we will delve into penalization methods such as bridge estimation and lasso, demonstrating how these concepts relate to regularization. By the end of this chapter, you will have a comprehensive understanding of how regularization techniques can enhance the performance and reliability of linear regression models.

Throughout the chapter, we denote the learning data by $\{(\mathbf{X}_i, Y_i)\}_1^n$, where Y_i is the response and $\mathbf{X}_i \in \mathbb{R}^p$ is the p-dimensional covariate for observation i. Throughout (unless indicated otherwise), we treat $\mathbf{X}_i$ as being "fixed", meaning we condition on $\mathbf{X}_i$ as if it were known, and write $\mathbf{x}_i = (x_{i1}, \ldots, x_{ip})^T$ for this conditioned value of $\mathbf{X}_i$. The linear model is taken to be

$$Y_i = \mathbf{x}_i^T \boldsymbol{\beta} + \varepsilon_i, \quad i = 1, \ldots, n, \tag{4.1}$$

where $\boldsymbol{\beta} = (\beta_1, \ldots, \beta_p)^T \in \mathbb{R}^p$ is the unknown regression vector. The ε_i denote i.i.d. measurement error assumed to be independent of $\mathbf{X}_i$ and satisfying $\mathbb{E}(\varepsilon_i|\mathbf{X}_i) = 0$ and $\mathbb{E}(\varepsilon_i^2|\mathbf{X}_i) = \sigma^2$. The regression model can also be described in matrix format

$$\mathbf{Y} = \mathbf{D}\boldsymbol{\beta} + \boldsymbol{\epsilon} \tag{4.2}$$

where $\mathbf{Y} = (Y_1, \ldots, Y_n)^T$, $\boldsymbol{\epsilon} = (\varepsilon_1, \ldots, \varepsilon_n)^T$, and $\mathbf{D}$ denotes the $n \times p$ design matrix whose ith row entry is $\mathbf{x}_i^T$.

4.2 Least Squares

We will use basic properties of linear subspaces to present a geometric view of least squares. This approach will help us develop theoretical results for least squares and derive useful properties. These results will also be applied in our later discussions on regularization.

Formally, the ordinary least squares (OLS) estimator is the solution to the L_2-optimization problem

$$\hat{\boldsymbol{\beta}} = \underset{\boldsymbol{\beta} \in \mathbb{R}^p}{\arg\min} ||\mathbf{Y} - \mathbf{D}\boldsymbol{\beta}||^2.$$

Even though an estimator for this problem always exists it may not be unique.

DOI: 10.1201/9781003493679-4

We can observe this and other properties for the OLS by the following. First observe that (4.2) can be rewritten as $\mathbf{Y} = \boldsymbol{\mu} + \boldsymbol{\epsilon}$, where $\boldsymbol{\mu} = \mathbf{D}\boldsymbol{\beta}$. Let $\mathbf{X}_{(k)} \in \mathbb{R}^n$ denote the kth column of the design matrix $\mathbf{D}$. Notice that

$$\boldsymbol{\mu} = \sum_{k=1}^{p} \beta_k \mathbf{X}_{(k)}$$

is a linear combination of the columns of $\mathbf{D}$ and therefore $\boldsymbol{\mu}$ spans the model space $\mathscr{H} := \text{span}\{\mathbf{X}_{(1)},\ldots,\mathbf{X}_{(p)}\}$. Hence determining the OLS amounts to identifying the vector $\boldsymbol{\mu} \in \mathscr{H}$ with smallest squared residual length $||\mathbf{Y} - \boldsymbol{\mu}||^2$.

In order to describe $\boldsymbol{\mu}$, define $\{\mathbf{v}_1,\ldots,\mathbf{v}_n\}$ to be an orthonormal basis spanning $\mathbb{R}^n$. Let the rank for $\mathbf{D}$ equal $1 \le r \le \min(n,p)$ were we can assume that $\{\mathbf{v}_i\}_{i=1}^n$ are chosen such that the first r values span the model space:

$$\text{span}\{\mathbf{v}_1,\ldots,\mathbf{v}_r\} = \mathscr{H}.$$

Recall from Chapter 2.6 that any vector $\mathbf{z} \in \mathbb{R}^n$ can be rewritten as $\mathbf{z} = \sum_{i=1}^n (\mathbf{z}^T\mathbf{v}_i)\mathbf{v}_i$. Let $Z_i = \mathbf{Y}^T\mathbf{v}_i$ and $\eta_i = \boldsymbol{\mu}^T\mathbf{v}_i$ be the inner products (the projection values) from projecting $\mathbf{Y}$ and $\boldsymbol{\mu}$ onto $\mathbf{v}_i$. Then

$$\mathbf{Y} = \sum_{i=1}^{n} Z_i\mathbf{v}_i, \quad \boldsymbol{\mu} = \sum_{i=1}^{n} \eta_i\mathbf{v}_i = \sum_{i=1}^{r} \eta_i\mathbf{v}_i.$$

The second identity for $\boldsymbol{\mu}$ follows because $\eta_{r+1} = \cdots = \eta_n = 0$ (convince yourself why this is the case). The following results follow directly from the above.

Theorem 4.1. *Let $\hat{\boldsymbol{\mu}} = \sum_{i=1}^r Z_i\mathbf{v}_i$ be the projection of $\mathbf{Y}$ onto $\mathscr{H}$. Then,*

(i) $\mathbf{D}\hat{\boldsymbol{\beta}} = \hat{\boldsymbol{\mu}}$.

(ii) If $r = p$ then $\boldsymbol{\beta}$ is identifiable and $\hat{\boldsymbol{\beta}}$ is the unique value

$$\hat{\boldsymbol{\beta}} = (\mathbf{D}^T\mathbf{D})^{-1}\mathbf{D}^T\mathbf{Y} = (\mathbf{D}^T\mathbf{D})^{-1}\mathbf{D}^T\hat{\boldsymbol{\mu}}.$$

Result (i) is often referred to as the normal equations.

Proof of Theorem 4.1. To prove part (i), observe that for any $\boldsymbol{\zeta} \in \mathscr{H}$,

$$\mathbf{Y} - \boldsymbol{\zeta} = \sum_{i=1}^{n} Z_i\mathbf{v}_i - \sum_{i=1}^{r} (\boldsymbol{\zeta}^T\mathbf{v}_i)\mathbf{v}_i.$$

Therefore

$$\left\|\mathbf{Y} - \boldsymbol{\zeta}\right\|^2 = \left\|\sum_{i=1}^{r} (Z_i - \boldsymbol{\zeta}^T\mathbf{v}_i)\mathbf{v}_i\right\|^2 + \left\|\sum_{i=r+1}^{n} Z_i\mathbf{v}_i\right\|^2.$$

The left-hand side is minimized by setting $(\boldsymbol{\zeta}^T\mathbf{v}_i) = Z_i$ for $i \le r$. Thus, $\mathbf{D}\hat{\boldsymbol{\beta}} = \hat{\boldsymbol{\mu}} = \sum_{i=1}^r Z_i\mathbf{v}_i$. For part (ii), note that if $r = p$ then $\mathbf{D}^T\mathbf{D}$ is non-singular. From part (i) we have $\mathbf{D}^T\mathbf{D}\hat{\boldsymbol{\beta}} = \mathbf{D}^T\hat{\boldsymbol{\mu}}$. Therefore

$$\begin{aligned}
\hat{\boldsymbol{\beta}} &= (\mathbf{D}^T\mathbf{D})^{-1}\mathbf{D}^T\hat{\boldsymbol{\mu}} \\
&= (\mathbf{D}^T\mathbf{D})^{-1}\mathbf{D}^T\left(\sum_{i=1}^{r} Z_i\mathbf{v}_i\right) \\
&= (\mathbf{D}^T\mathbf{D})^{-1}\mathbf{D}^T\left(\sum_{i=1}^{n} Z_i\mathbf{v}_i\right) \quad \text{(because } \mathbf{v}_1,\ldots,\mathbf{v}_r \text{ spans } \mathscr{H}\text{)} \\
&= (\mathbf{D}^T\mathbf{D})^{-1}\mathbf{D}^T\mathbf{Y}.
\end{aligned}$$

□

4.2.1 Slope-intercept model

Let's look at a simple toy example using a slope-intercept model to illustrate the above results,

$$Y_i = \alpha + x_i\beta + \varepsilon_i, \qquad i = 1,\ldots,n. \tag{4.3}$$

Although it is straightforward to directly calculate the OLS and its predicted value for (4.3), we will derive these quantities from Theorem 4.1. This approach will yield some surprisingly simple results, such as the distribution for R^2 goodness of fit.

Let $\mathbf{1}$ denote the n-dimensional vector $(1,\ldots,1)^T$ and let $\mathbf{x} = (x_1,\ldots,x_n)^T$. We assume that $\{\mathbf{1},\mathbf{x}\}$ are linearly independent which is satisfied if $S_{xx} \neq 0$, where $S_{xx} = \sum_i (x_i - \overline{x})^2$ and $\overline{x} = \sum_i x_i/n$. Let $\mathbf{v}_1 = \mathbf{1}/\sqrt{n}$ and $\mathbf{v}_2 = (\mathbf{x} - \overline{x}\mathbf{1})/\sqrt{S_{xx}}$. The assumption of linear independence ensures that $\mathbf{v}_1$ and $\mathbf{v}_2$ are orthonormal vectors satisfying

$$\text{span}\{\mathbf{v}_1,\mathbf{v}_2\} = \mathscr{H} := \text{span}\{\mathbf{1},\mathbf{x}\}.$$

Note that

$$Z_1 = \mathbf{Y}^T\mathbf{v}_1 = \sqrt{n}\overline{Y}, \qquad Z_2 = \mathbf{Y}^T\mathbf{v}_2 = S_{xy}/\sqrt{S_{xx}}$$

where $\overline{Y} = \sum_i Y_i/n$ and $S_{xy} = \sum_i Y_i(x_i - \overline{x})$. Therefore by Theorem 4.1(i), the predicted value of $\mathbf{Y}$ is

$$\hat{\mathbf{Y}} = Z_1\mathbf{v}_1 + Z_2\mathbf{v}_2 = \overline{Y}\mathbf{1} + \frac{S_{xy}}{S_{xx}}(\mathbf{x} - \overline{x}\mathbf{1}).$$

Denote the OLS by $(\hat{\alpha},\hat{\boldsymbol{\beta}})$. Since we must have

$$\hat{\mathbf{Y}} = \hat{\alpha}\mathbf{1} + \hat{\beta}\mathbf{x} = \left(\overline{Y} - \overline{x}\frac{S_{xy}}{S_{xx}}\right)\mathbf{1} + \frac{S_{xy}}{S_{xx}}\mathbf{x},$$

we can immediately deduce that $\hat{\alpha} = \overline{Y} - \hat{\beta}\overline{x}$ and $\hat{\beta} = S_{xy}/S_{xx}$.

This might seem like a lot of effort to derive a simple result. However, let's consider a more complex case. Take the R^2 goodness of fit for the model, defined as

$$R^2 = 1 - \frac{||\mathbf{Y} - \hat{\mathbf{Y}}||^2}{||\mathbf{Y} - \overline{Y}\mathbf{1}||^2}.$$

Assuming normality, we can derive the distribution of R^2 for (4.3) using a geometric argument. In fact, this result holds for linear models in general, which is not easily demonstrated using conventional approaches.

Assume $\varepsilon_1,\ldots,\varepsilon_n$ are i.i.d. $\mathrm{N}(0,\sigma^2)$. Choose $\mathbf{v}_3,\ldots,\mathbf{v}_n$ to be any set of orthonormal vectors that are orthogonal to $\mathbf{v}_1$ and $\mathbf{v}_2$. Then $\{\mathbf{v}_1,\ldots,\mathbf{v}_n\}$ is an orthonormal basis spanning $\mathbb{R}^n$. In particular, notice that $\mathbf{Y} = \sum_{i=1}^n Z_i\mathbf{v}_i$. Hence, using $\overline{Y}\mathbf{1} = Z_1\mathbf{v}_1$, we have

$$\begin{aligned} R^2 &= \frac{||\mathbf{Y} - \overline{Y}\mathbf{1}||^2 - ||\mathbf{Y} - \hat{\mathbf{Y}}||^2}{||\mathbf{Y} - \overline{Y}\mathbf{1}||^2} \\ &= \frac{||\sum_{i=2}^n Z_i\mathbf{v}_i||^2 - ||\sum_{i=3}^n Z_i\mathbf{v}_i||^2}{||\sum_{i=2}^n Z_i\mathbf{v}_i||^2} \\ &= \frac{\sum_{i=2}^n Z_i^2 - \sum_{i=3}^n Z_i^2}{\sum_{i=2}^n Z_i^2} \\ &= \frac{Z_2^2}{Z_2^2 + \sum_{i=3}^n Z_i^2}. \end{aligned} \tag{4.4}$$

In order to work out this distribution, observe that

$$Z_i = \mathbf{Y}^T \mathbf{v}_i = \boldsymbol{\mu}^T \mathbf{v}_i + \boldsymbol{\epsilon}^T \mathbf{v}_i. \tag{4.5}$$

Since $\boldsymbol{\mu} \in \mathscr{H} = \text{span}\{\mathbf{v}_1, \mathbf{v}_2\}$, we have $\boldsymbol{\mu}^T \mathbf{v}_i = 0$ for $i \geq 3$. This implies, due to normality, that Z_i^2 are i.i.d. $\sigma^2\chi_1^2$ for $i \geq 3$. However, $\boldsymbol{\mu}^T \mathbf{v}_i$ may not be zero for $i = 1, 2$, so Z_1^2 and Z_2^2 have a non-central chi-squared distribution. Because the expression for R^2 depends on Z_2, its distribution will depend on unknown parameters, making it cumbersome to work with. To resolve this, we can restrict the model space to eliminate this dependence. For instance, under the null hypothesis $H_0 : \beta = 0$, we have $\boldsymbol{\mu}^T \mathbf{v}_2 = 0$ and $Z_2 \sim \text{N}(0, \sigma^2)$. Therefore, $Z_2^2, \ldots, Z_n^2$ are i.i.d. $\sigma^2\chi_1^2$, and from the last line in (4.4), we obtain $R^2 \sim \text{Beta}(1/2, (n-2)/2)$ under the null.

As a simple illustration, let's consider the statistical significance of R^2 using the results mentioned above. Table 4.1 shows the value of R^2 required to achieve 1%, 5%, and 10% levels of significance under the null hypothesis $H_0 : \beta = 0$. The results here are for the slope-intercept model (4.3), but extensions to the general regression model are straightforward (see Exercise 4.6). Significance values were obtained from a $\text{Beta}(1/2, (n-2)/2)$ distribution and are given for different sample sizes n.

TABLE 4.1
Statistical significance of R^2 for slope-intecept model $Y = \alpha + \beta x + \varepsilon$.

n	10%	5%	1%
3	0.98	0.99	1.00
4	0.81	0.90	0.98
5	0.65	0.77	0.92
10	0.30	0.40	0.58
50	0.06	0.08	0.13
100	0.03	0.04	0.07

Remark 4.1. Table 4.1 highlights a common issue with statistical significance. When n is small, it is nearly impossible to obtain a significant result unless R^2 is very close to 1. Conversely, for large n, even the smallest R^2 will reject the null hypothesis. Therefore, testing the null hypothesis can only tell us if $\beta = 0$ but may not provide a good assessment of the true effect size. □

4.2.2 Hypothesis testing

In the previous example, we considered hypothesis testing for a test statistic derived under the null hypothesis. We demonstrated that if the slope is zero in a slope-intercept model, then the R^2 goodness of fit test follows a beta distribution.

This same geometric approach can be extended to other types of hypothesis tests (Chapter 5 provides illustrations of hypothesis testing with multivariate linear models). Consider the following hypothesis testing framework. Suppose that $\hat{\boldsymbol{\mu}}_0$ is the least squares predictor for $\mathbf{Y}$ under the null hypothesis

$$H_0\colon \mathscr{H}_0 = \text{span}\{\mathbf{v}_1, \ldots, \mathbf{v}_q\} \quad \text{for some } 1 \leq q < r.$$

We would like to test H_0 against the unconstrained hypothesis

$$H\colon \mathscr{H} = \text{span}\{\mathbf{v}_1, \ldots, \mathbf{v}_q, \mathbf{v}_{q+1}, \ldots, \mathbf{v}_r\}.$$

Denoting $\hat{\boldsymbol{\mu}}$ for the least squares predictor of $\mathbf{Y}$ under H, we have the following result.

Theorem 4.2. *Assume that* $\varepsilon_i \overset{\text{iid}}{\sim} N(0,\sigma^2)$. *To test* H_0 *against* H, *let*

$$L_n = \frac{n-r}{r-q}\,\frac{||\mathbf{Y}-\hat{\boldsymbol{\mu}}_0||^2 - ||\mathbf{Y}-\hat{\boldsymbol{\mu}}||^2}{||\mathbf{Y}-\hat{\boldsymbol{\mu}}||^2}$$

(note that L_n *is equivalent to the likelihood ratio test). We have:*

(i) If H_0 *is true,* $L_n \sim F_{r-q,n-r}$.

(ii) In general, L_n *has a non-central F-distribution,* $L_n \sim F_{r-q,n-r}(\delta^2)$, *where*

$$\delta^2 = \frac{||\boldsymbol{\mu}-\boldsymbol{\mu}_0||^2}{\sigma^2} = \frac{\sum_{i=q+1}^{r}\eta_i^2}{\sigma^2}$$

and $\boldsymbol{\mu}_0$ *is the projection of* $\boldsymbol{\mu}$ *onto* $\mathscr{H}_0$. *Furthermore,* $||\mathbf{Y}-\hat{\boldsymbol{\mu}}_0||^2 \sim \sigma^2\chi^2_{n-q}(\delta^2)$.

Proof of Theorem 4.2. By definition, we have

$$\hat{\boldsymbol{\mu}}_0 = \sum_{i=1}^{q} Z_i\mathbf{v}_i, \quad \hat{\boldsymbol{\mu}} = \sum_{i=1}^{r} Z_i\mathbf{v}_i, \quad \mathbf{Y} = \sum_{i=1}^{n} Z_i\mathbf{v}_i.$$

Therefore,

$$\begin{aligned} L_n &= \frac{n-r}{r-q}\,\frac{||\sum_{i=q+1}^{n} Z_i\mathbf{v}_i||^2 - ||\sum_{i=r+1}^{n} Z_i\mathbf{v}_i||^2}{||\sum_{i=r+1}^{n} Z_i\mathbf{v}_i||^2} \\ &= \frac{n-r}{r-q}\,\frac{||\sum_{i=q+1}^{r} Z_i\mathbf{v}_i||^2}{||\sum_{i=r+1}^{n} Z_i\mathbf{v}_i||^2} = \frac{n-r}{r-q}\,\frac{\sum_{i=q+1}^{r} Z_i^2}{\sum_{i=r+1}^{n} Z_i^2}. \end{aligned}$$

Under H_0, $\boldsymbol{\mu}^T\mathbf{v}_i = 0$ if $i \geq q+1$. Deduce that $\{Z_{q+1}^2,\ldots,Z_n^2\}$ are i.i.d. $\sigma^2\chi_1^2$ using (4.5). Part (i) follows. Part (ii) follows using a similar argument. The sum in the denominator of L_n is the same, equaling in distribution $\sum_{i=r+1}^{n}\zeta_i^2$ where ζ_i are i.i.d. $N(0,\sigma^2)$. However, now the sum in the numerator is

$$\sum_{i=q+1}^{r} Z_i^2 = \sum_{i=q+1}^{r} (\boldsymbol{\mu}^T\mathbf{v}_i + \boldsymbol{\epsilon}^T\mathbf{v}_i)^2 \overset{d}{=} \sum_{i=q+1}^{r} (\eta_i+\zeta_i)^2.$$

Deduce that $\sum_{i=q+1}^{r} Z_i^2 \sim \sigma^2\chi^2_{n-q}(\delta^2)$ where $\delta^2 = \sigma^{-2}\sum_{i=q+1}^{r}\eta_i^2$. □

Remark 4.2. Part (ii) indicates how to construct an unbiased estimator for σ^2. Namely,

$$\hat{\sigma}^2 = \frac{1}{n-r}||\mathbf{Y}-\hat{\boldsymbol{\mu}}||^2 \sim \frac{\sigma^2}{n-r}\chi^2_{n-r}.$$

Hence, $\mathbb{E}(\hat{\sigma}^2) = \sigma^2$ and $\hat{\sigma}^2$ is unbiased. However, the estimator

$$\hat{\sigma}_0^2 = \frac{1}{n-q}||\mathbf{Y}-\hat{\boldsymbol{\mu}}_0||^2$$

is proportional to a non-central χ^2_{n-q} variable with non-centrality parameter

$$\delta^2 = \frac{\sum_{i=q+1}^{q}\eta_i^2}{\sigma^2} \neq 0.$$

In particular, $\hat{\sigma}_0^2$ is unbiased for σ^2 only under the null H_0. □

4.3 Regularization and Overfitting

Building on our understanding of linear regression, we now turn our attention to the concept of regularization. Often in statistical analyses, we aim to select the "best" model m from a space $\mathcal{M}$ using some selection criteria. However, achieving this in practice can be challenging, especially when the space size $M = |\mathcal{M}|$ is large. For example, with p variables, the number of possible models is $M = 2^p$, which is impractical even for a moderately sized p. The dimension $p_m = |m|$ for a model must also be considered in the selection criteria, since including more variables will fit the training data better but may not perform well on test data.

Figure 4.1 demonstrates why residual sums of squares (RSS), also known as goodness of fit, is not a suitable criterion for model selection. The following definitions are used:

$$
\begin{array}{lcll}
\hat{\boldsymbol{\mu}} & = & \mathbf{D}\hat{\boldsymbol{\beta}} & \text{OLS predictor} \\
\hat{\boldsymbol{\mu}}_m & = & \mathbf{D}_m\hat{\boldsymbol{\beta}}_m & \text{restricted OLS predictor} \\
\hat{S} & = & ||\mathbf{Y}-\hat{\boldsymbol{\mu}}||^2 & \text{RSS for } \hat{\boldsymbol{\mu}} \\
\hat{S}_m & = & ||\mathbf{Y}-\hat{\boldsymbol{\mu}}_m||^2 & \text{RSS for } \hat{\boldsymbol{\mu}}_m \\
\mathbf{D}_m & = & n \times p_m \text{ design matrix for } m & \\
\hat{\boldsymbol{\beta}}_m & = & \text{restricted OLS estimator} & \\
p_m & = & \text{size of model } m. &
\end{array}
$$

The notation m denotes the linear regression model that uses a subset $\mathcal{S}_m \subseteq \{1,\ldots,p\}$ of the p variables for model fitting. The predictor for m uses the design matrix $\mathbf{D}_m$ with columns selected from $\mathcal{S}_m$ and is denoted by $\hat{\boldsymbol{\mu}}_m \in \mathbb{R}^n$ and called the *restricted OLS* predictor. The *full model OLS* predictor uses the design matrix $\mathbf{D}$ with all variables and is denoted by $\hat{\boldsymbol{\mu}} \in \mathbb{R}^n$. The notation

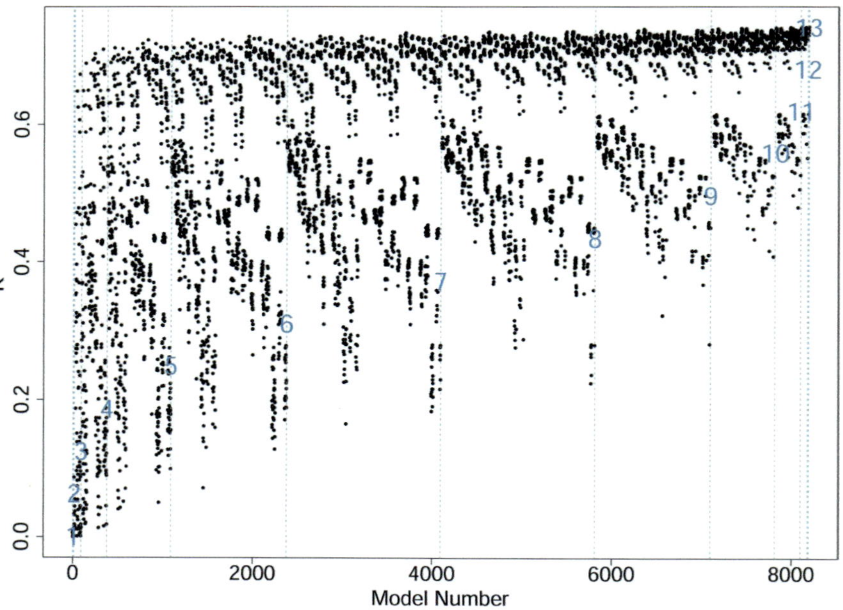

FIGURE 4.1
Illustration of the pitfalls in using R^2 as a model selection criterion using body fat data [9]. The goodness of fit for each model is displayed stratified in terms of model size, with model size increasing from left to right displayed by large blue text.

p_m denotes the size of m, which equals $|\mathscr{S}_m|$ and therefore is the number of columns of the design matrix $\mathbf{D}_m$. However, due to the possibility of linear dependence, for technical reasons, we define the true dimension $p_m^* \le p_m$ to be the dimension of the span of the design matrix $\mathbf{D}_m$. The dimension of the OLS is denoted by $p^* \le p$.

Figure 4.1 plots the goodness of fit $R_m^2 = 1 - \hat{S}_m/\hat{S}_0$, where $\hat{S}_0 = ||\mathbf{Y} - \overline{Y}\mathbf{1}||^2$, for each model m from a dataset of body fat measurements [9]. The data consists of measurements from $n = 251$ individuals, with $p = 13$ variables including age, weight, height, and various body circumference measurements, with body fat as the response variable. There are a total of $2^{13} - 1 = 8,192$ models, making it feasible to calculate R^2 for the entire space of models.

The blue text indicates model size; starting from models of size $p_m = 1$ (a total of 13) and ending with the full model of size $p_m = 13$, corresponding to the OLS. The OLS which uses all variables has the largest R^2 because it achieves the smallest RSS. This can be easily seen by the following argument. Let $\{\mathbf{v}_1, \dots, \mathbf{v}_n\}$ be an orthonormal basis such that

$$\text{span}\{\mathbf{v}_1, \dots, \mathbf{v}_{p_m^*}\} = \text{column span}\{\mathbf{D}_m\}, \quad \text{span}\{\mathbf{v}_1, \dots, \mathbf{v}_{p^*}\} = \text{column span}\{\mathbf{D}\}.$$

Then,

$$\begin{aligned}
\hat{S}_m &= ||\mathbf{Y} - \hat{\boldsymbol{\mu}}_m||^2 \\
&= \left\| \sum_{i=1}^{n} Z_i \mathbf{v}_i - \sum_{i=1}^{p_m^*} Z_i \mathbf{v}_i \right\|^2 \\
&= \sum_{i=p_m^*+1}^{n} Z_i^2 \\
&\ge \sum_{i=p^*+1}^{n} Z_i^2 = ||\mathbf{Y} - \hat{\boldsymbol{\mu}}||^2 = \hat{S},
\end{aligned}$$

which shows that the OLS achieves the largest R^2. This argument also shows that goodness of fit increases with increasing model size, which can also be seen in the figure. However, we also observe a diminishing effect, where we find smaller models with R^2 nearly as good as the OLS. Thus, a concern with R^2 is that it *does not adjust for model size* to allow favoring smaller models with high goodness of fit performance.

A more subtle issue involves the *crowding* of the model space. Even if we identify an appropriate model size, such as $p_m = 7$, there are still many models with the same dimension and nearly identical goodness of fit. Therefore, accounting for model size alone may not be sufficient for good model selection. Figure 4.2 presents the results in different ways, and crowding remains apparent.

This example illustrates that several issues are at play if we want to be successful in optimal model selection. Below, we highlight three criteria for achieving appropriate regularization. We will first discuss point 1, adjusting for model size in the selection criteria, and demonstrate its importance by showing that OLS is not optimal under mean-squared error (MSE). The remaining points, related to challenges of crowding, will be addressed when we discuss bridge estimators.

Regularization requires:

1. *Adjusting for the model size.*
2. *Determining the best model with the same dimension.*
3. *Dealing with model spaces too large to compute every model.*

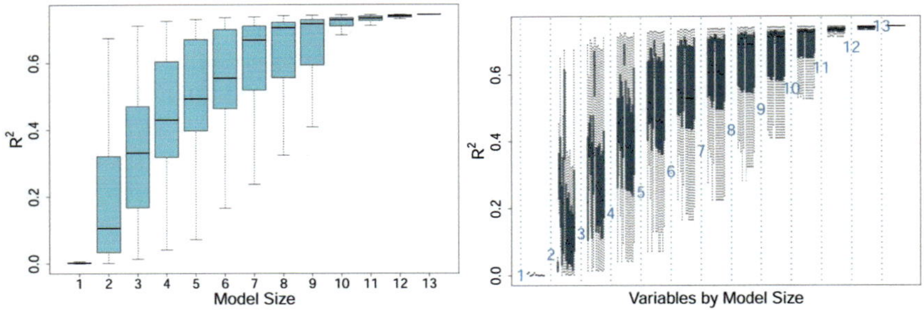

FIGURE 4.2
R^2 values for body fat displayed somewhat differently. The left panel is in terms of different model sizes. The right panel is in terms of each of the $p = 13$ variables where each each boxplot are models containing a specific variable, where size of model has been stratified.

4.3.1 The OLS is not regularized and overfits

To show overfitting by OLS, we compare its MSE, $\mathbb{E}||\hat{\boldsymbol{\mu}} - \boldsymbol{\mu}||^2$, to the MSE for restricted OLS, $\mathbb{E}||\hat{\boldsymbol{\mu}}_m - \boldsymbol{\mu}||^2$. MSE is a well-known metric for measuring performance, and we will explain why it is well-suited for regularization. We will show that OLS is generally suboptimal under this metric compared to restricted OLS.

Let $\boldsymbol{\mu} = \mathbb{E}(\mathbf{Y})$ and $\boldsymbol{\mu}_m = \mathbb{E}(\hat{\boldsymbol{\mu}}_m)$. Using the previous notation, let $Z_i = \mathbf{Y}^T\mathbf{v}_i, \eta_i = \boldsymbol{\mu}^T\mathbf{v}_i$ and $\zeta_i = \boldsymbol{\epsilon}^T\mathbf{v}_i$. Then

$$\hat{\boldsymbol{\mu}}_m = \sum_{k=1}^{p_m^*} Z_k\mathbf{v}_k = \sum_{k=1}^{p_m^*} \eta_k\mathbf{v}_k + \sum_{k=1}^{p_m^*} \zeta_k\mathbf{v}_k, \quad \boldsymbol{\mu}_m = \sum_{k=1}^{p_m^*} \eta_k\mathbf{v}_k, \quad \boldsymbol{\mu} = \sum_{k=1}^{p^*} \eta_k\mathbf{v}_k,$$

and hence,

$$\begin{aligned} ||\hat{\boldsymbol{\mu}}_m - \boldsymbol{\mu}||^2 &= ||\hat{\boldsymbol{\mu}}_m - \boldsymbol{\mu}_m + \boldsymbol{\mu}_m - \boldsymbol{\mu}||^2 \\ &= \left\| \sum_{k=1}^{p_m^*} (Z_k - \eta_k)\mathbf{v}_k - \sum_{k=p_m^*+1}^{p^*} \eta_k\mathbf{v}_k \right\|^2 \\ &= \sum_{k=1}^{p_m^*} \zeta_k^2 + \sum_{k=p_m^*+1}^{p^*} \eta_k^2 \\ &= \sum_{k=1}^{p_m^*} \zeta_k^2 + ||\boldsymbol{\mu}_m - \boldsymbol{\mu}||^2. \end{aligned} \tag{4.6}$$

It follows that the MSE for $\hat{\boldsymbol{\mu}}_m$ is

$$\mathbb{E}||\hat{\boldsymbol{\mu}}_m - \boldsymbol{\mu}||^2 = p_m^*\sigma^2 + ||\boldsymbol{\mu}_m - \boldsymbol{\mu}||^2. \tag{4.7}$$

The right-hand side of (4.7) can be interpreted as a MSE decomposition. Recall that, in general, the MSE for an estimator $\hat{\theta}$ of a parameter θ can be decomposed into variance and bias-squared terms:

$$\underbrace{\mathbb{E}\left[(\hat{\theta} - \theta)^2\right]}_{\text{MSE}} = \underbrace{\mathbb{E}\left[(\hat{\theta} - \mathbb{E}(\hat{\theta}))^2\right]}_{\text{variance}} + \underbrace{\left[\mathbb{E}(\hat{\theta}) - \theta\right]^2}_{\text{bias-squared}}.$$

It can be shown that (4.7) has this same type of decomposition:

$$\mathbb{E}\left[||\hat{\boldsymbol{\mu}}_m - \boldsymbol{\mu}||^2\right] = \underbrace{p_m^*\sigma^2}_{\text{variance}} + \underbrace{||\boldsymbol{\mu}_m - \boldsymbol{\mu}||^2}_{\text{bias-squared}}. \tag{4.8}$$

Some work is required to show this rigorously, which we leave as an exercise for readers (Exercise 4.8; see also Chapter 7.10.1). Therefore, we can see that MSE is an appropriate metric for assessing performance because it balances variance against bias.

Now to show why the OLS overfits, let $\mathscr{S}_0 \subseteq \{1,\ldots,p\}$ be the set of indices corresponding to the non-zero β coefficients. The set $\mathscr{S}_0$ corresponds to the true model, denoted by m_0. If m is a model such that $\mathscr{S}_0 \subseteq \mathscr{S}_m$, then m includes the entire set of useful variables, and possibly some irrelevant variables. If m is a model like this, then $\eta_k = 0$ for $i > p_m^*$, and by (4.7),

$$\mathbb{E}\left[||\hat{\boldsymbol{\mu}}_m - \boldsymbol{\mu}||^2\right] = \underbrace{p_m^*\sigma^2}_{\text{variance}} + \underbrace{||\boldsymbol{\mu}_m - \boldsymbol{\mu}||^2}_{\text{bias-squared}=0} = p_m^*\sigma^2. \tag{4.9}$$

Thus, m has zero bias; however, by including irrelevant variables, the MSE for m increases due to excess variance. Therefore, among all m such that $\mathscr{S}_0 \subseteq \mathscr{S}_m$, the best model is $m_0 := \mathscr{S}_0$ with MSE $p_0\sigma^2$, where $p_0 = p_0^*$ is the true (effective) dimension. This is the optimal model because it is the smallest model with zero bias. On the other hand, the OLS is the worst model since it incurs the largest excess variance among the unbiased models. By (4.9), its MSE is $p^*\sigma^2$, which can be substantially larger than the optimal value $p_0\sigma^2$ in high dimensions.

4.3.2 Keypoints

This analysis reveals several key points about regularization which we summarize.

1. Mean-squared error (MSE) is an appropriate selection criteria. Because MSE balances variance against bias, it balances model size against goodness of fit, which is a desired property. The MSE is the gold standard for regularization but it requires knowing the true distribution, therefore it must be approximated in practice.
2. Least squares is an unregularized procedure. This is because it does not adjust for model size.
3. The full model OLS is an unregularized estimator because it is an estimator from an unregularized procedure (least squares). OLS overfits in high dimensional settings in terms of MSE due to excess variance from including non-essential variables.
4. Restricted least squares is a regularized procedure because it adjusts for model size. The restricted OLS with the smallest MSE has the optimality property that it is the smallest model with zero bias.

Methods for approximating the MSE raised in the first point will be discussed shortly. The next section builds on the issue of optimality raised in point 4.

4.4 Oracle Property

The previous discussion indicated that the restricted OLS estimator containing only the non-zero coefficients $\mathscr{S}_0$ represents the ideal estimator. We formalize this optimality in a related way in terms of the so-called *oracle property* [7].

Let $\hat{\boldsymbol{\beta}} = (\hat{\beta}_1,\ldots,\hat{\beta}_p)^T$ be an estimator for $\boldsymbol{\beta}$. Without loss of generality, assume $\boldsymbol{\beta}$ has been reordered so that its non-zero coefficients are the first p_0 coefficients

$$\boldsymbol{\beta} = (\boldsymbol{\beta}_0, 0,\ldots,0)^T, \quad \boldsymbol{\beta}_0 \in \mathbb{R}^{p_0}$$

Let m_0 denote the true model corresponding to the first p_0 coordinates, i.e., $\boldsymbol{\beta}_0$.

The estimator $\hat{\boldsymbol{\beta}}$ is called an oracle estimator if

1. $\mathbb{P}\{$model selected by $\hat{\boldsymbol{\beta}}$ equals $m_0\} \to 1$

2. $\sqrt{n}\left((\hat{\beta}_1,\ldots,\hat{\beta}_{p_0})^T - \boldsymbol{\beta}_0\right) \overset{\mathrm{d}}{\rightsquigarrow} \mathrm{N}_{p_0}\left(\mathbf{0},\sigma^2\mathbf{Q}_{11}^{-1}\right)$, where

$$\frac{1}{n}\mathbf{D}^T\mathbf{D} \to \mathbf{Q} = \begin{pmatrix} \mathbf{Q}_{11} & \mathbf{Q}_{12} \\ \mathbf{Q}_{21} & \mathbf{Q}_{22} \end{pmatrix}_{p\times p}.$$

Note that $\mathbf{Q}_{11}^{-1}$ is not necessarily the same as the upper diagonal matrix of $\mathbf{Q}^{-1}$, where the former reflects the efficiency of the restricted OLS for the true coefficients, while the latter is the efficiency obtained using the full model OLS. Thus, the oracle property requires an estimator to perform asymptotically as well as the estimator calculated using only coordinates with non-zero coefficients. This is a very stringent requirement which makes the property more of hypothetical construct than something achievable in practice. It is often used as an acid test for gauging optimality of a regularized method. We will revisit the oracle property when we introduce the lasso.

4.5 Final Prediction Error

From the previous discussion, we can see that the restricted OLS estimator with the smallest MSE corresponding to the true model m_0 satisfies the oracle property. However, determining this value is hypothetical since it requires knowing the true MSE. To address this, researchers instead turn to information-based metrics to identify good performing models.

One of the earliest proposals came from Akaike [1] who introduced the Akaike criteria, often simply referred to as AIC. For the regression model (4.1), the AIC for model m is defined as

$$\mathrm{AIC}(m) = n\log(\hat{S}_m/n) + 2p_m.$$

The selected model $\hat{m}$ is defined to be the model minimizing the AIC:

$$\hat{m} = \arg\min_{m\in\mathcal{M}}\{\mathrm{AIC}(m)\}.$$

Compared to RSS, the AIC prevents overfitting by applying the term p_m which penalizes larger sized models.

Related to AIC is Mallows' C_P statistic [12]. For a given model m, the C_P statistic is defined as

$$C_m = \frac{\hat{S}_m}{\hat{\sigma}^2} + 2p_m - n,$$

where $\hat{\sigma}^2$ is the unbiased estimator for σ^2 based on the full model OLS. Mallows recommended that any model with a C_P statistic satisfying $C_m < p_m$ should be carefully investigated as a potential optimal model. However, he advised against using the C_P statistic for model selection. In practice, it is often used in this way. Thus, the C_P-optimized model is found by minimizing

$$C_m^* = \hat{S}_m + 2\hat{\sigma}^2 p_m.$$

Nishii [13] showed that the C_P statistic and AIC are asymptotically equivalent.

Schwartz's criteria [16], referred to as BIC, is another popular model selection criteria. BIC uses a logarithmic penalty in n and therefore is a heavier penalty that tends to choose smaller and more parsimonious models. The BIC value is defined as

$$\text{BIC}(m) = n\log(\hat{S}_m/n) + \log(n)p_m.$$

We can consider all of the above as examples of final prediction error (FPE) metrics. The FPE is trying to estimate the model-fitting error when the trained model is used on test data. Shibata [18] presented a generalization of the FPE, the generalized FPE, defined as

$$C(m,\lambda) = \hat{S}_m + \lambda\hat{\sigma}^2 p_m.$$

Sometimes the FPE is written in a logarithmic format so as to bring it more in alignment with AIC and BIC, thus the generalized FPE is $n\log(\hat{S}_m/n) + \lambda\hat{\sigma}^2 p_m$. There is an enormous literature on model selection criteria. For convenience, we summarize some of the better known procedures in Table 4.2.

TABLE 4.2
Various model selection criteria.

$n\log(\hat{S}_m/n) + 2p_m$	Akaike [2]
$n\log(\hat{S}_m/n) + \log\log(n)p_m$	Hannan and Quinn [8]
$\hat{S}_m/\hat{\sigma}^2 + 2p_m - n$	Mallows [12]
$n\log(\hat{S}_m/n) + \log(n)p_m$	Schwarz [16]
$\hat{S}_m + \lambda\hat{\sigma}^2 p_m$	Shibata [18]

4.6 AIC versus BIC

The two most popular information metrics are BIC and AIC. Considering Table 4.2, we see that AIC and BIC appear to be on opposite extremes. BIC uses a logarithmic penalty that tends to choose smaller and more parsimonious models than AIC. The two metrics come from different philosophies. AIC is motivated by minimizing Kullback-Leibler divergence of the true and estimated models, whereas BIC selects the model maximizing the posterior model probability. These differences are important and result in different asymptotic properties for the two procedures.

BIC is generally consistent if the model space $\mathcal{M}$ is finite and contains the true model, m_0 [13]. As before, let $\mathcal{S}_m \subseteq \{1,\ldots,p\}$ record the indices of the coordinate contained by model m. The true model m_0 has indices $\mathcal{S}_0$. Let $\mathcal{M}_U = \{\mathcal{S}_m : \mathcal{S}_0 \nsubseteq \mathcal{S}_m\}$ be the set of *underfit models*. These models contain some of the true non-zero coefficients of the true model, but not all of them. Let $\mathcal{M}_O = \{\mathcal{S}_m : \mathcal{S}_0 \subseteq \mathcal{S}_m\}$ be the set of potentially *overfit models*. These models contain all the non-zero coefficients of the true model but may also contains irrelevant variables (note that the true model is contained in $\mathcal{M}_O$).

Under regularity conditions, the following holds for the estimator $\hat{\mu}_{\text{BIC}}$ which selects models using the BIC criterion (see Theorem 4 of Nishii [13]):

(a) $\mathbb{P}\{\hat{\mu}_{\text{BIC}} \in \mathcal{M}_U\} = o(n^{-r})$ for any $r > 0$.
(b) $\mathbb{P}\{\hat{\mu}_{\text{BIC}} \in \mathcal{M}_O - \{m_0\}\} = o(1)$.
(c) $\mathbb{P}\{\hat{\mu}_{\text{BIC}} = m_0\} \to 1$.
(d) $\text{MSE}(\hat{\mu}_{\text{BIC}}) \to p_0\sigma^2$.

Parts (a) and (b) show that BIC will not select underfit or overfit models, and the probability of underfitting in (a) decreases rapidly. Part (c) shows that BIC is consistent (converges with probability one); (d) shows its MSE converges to the optimal value, which is the asymptotic MSE for the oracle estimator (see Exercise 4.9).

Thus, BIC is model consistent if the true model is contained within a finite dimensional model space. On the other hand, AIC is risk consistent in problems where the space of models increases even if there is no true model; see Theorem 2.2 of Shibata [17]. Yang [21] proved the following related result. Let $\mathscr{K}$ be an integer set and let $\mathscr{F}_k = \{f_k(\mathbf{x},\boldsymbol{\theta}_k) : \boldsymbol{\theta}_k \in \Theta_k\}$ for $k \in \mathscr{K}$ be a class of parametric linear regression functions where $\boldsymbol{\theta}_k$ has dimension p_k. Let $\hat{\delta} \in \mathscr{K}$ be a model selection criteria that selects a model $k \in \mathscr{K}$ and estimates $\boldsymbol{\theta}_k$ by the least squares estimator $\hat{\boldsymbol{\theta}}_k$. The risk for $\hat{\delta}$ for a regression function f is defined by

$$R_n(f,\hat{\delta}) = \frac{1}{n}\sum_{i=1}^{n}\mathbb{E}\big(f(\mathbf{x}_i) - f_{\hat{\delta}}(\mathbf{x}_i,\hat{\boldsymbol{\theta}}_{\hat{\delta}})\big)^2.$$

Consider the class of models $\mathscr{F} = \cup\mathscr{F}_k$ where the number of models of size p_k, $n_k := |\mathscr{F}_k|$, satisfies $n_k \le \exp(cp_k)$ for some constant c. Let $\hat{\delta}_{\text{AIC}}$ be the AIC model selection criteria. Then if model k^* is the index of the true model in $\mathscr{F}$, there exists a constant $C := C(c) > 0$, depending only on c, such that

$$\sup_{f\in\mathscr{F}_{k^*}} R_n(f,\hat{\delta}_{\text{AIC}}) \le \frac{Cp_{k^*}}{n}. \tag{4.10}$$

This shows in the worst-case scenario that the risk for AIC converges at a $1/n$ parametric rate, therefore, AIC is minimax-rate optimal if the true model is in the candidate class.

4.6.1 A simple illustrative example

Let us consider a simple regression example to help illustrate the previous ideas (we borrow this example from Yang [22]). Assume a simple slope-intercept model: $Y_i = \alpha + x_i\beta + \varepsilon_i$, for $i = 1,\dots,n$, where ε_i are i.i.d. $\mathrm{N}(0,\sigma^2)$ errors. The variance σ^2 is unknown and the covariates satisfy $\overline{x} = 0$ and $\sum_{i=1}^n x_i^2 = n$. The goal is to determine whether $\beta = 0$.

Using a classical hypothesis approach, we test the null hypothesis $H_0 : \beta = 0$ versus the alternative hypothesis $H_1 : \beta \neq 0$ by using a t-test. The null is rejected in favor of the alternative if

$$\frac{n^{1/2}|\hat{\beta}|}{\hat{\sigma}} \ge t_{n-2,\alpha_0/2}, \tag{4.11}$$

where α_0 is some preset type-I error level. Here $\hat{\alpha} = \overline{Y}$ and $\hat{\beta} = n^{-1}\sum_{i=1}^n x_iY_i$ are the OLS estimates and

$$\hat{\sigma}^2 = \frac{1}{n-2}\sum_{i=1}^{n}(Y_i - \hat{\alpha} - \hat{\beta}x_i)^2$$

is the unbiased estimator for σ^2.

In this simplified setting, the hypothesis test can be seen to be related to FPE. There are two models: $\mathscr{M}_0$ containing only the intercept, and $\mathscr{M}_1$ containing both the intercept and the slope (i.e., the full-model). The FPE for model $\mathscr{M}_j$, $j = 0,1$, is

$$n\log(\hat{S}_j/n) + \lambda_n p_j \tag{4.12}$$

where $\hat{S}_j$ is the RSS for model $\mathscr{M}_j$, λ_n is the penalty term, and p_j is the size of the model ($p_0 = 0$ and $p_1 = 1$).

By FPE, we choose the model that minimizes (4.12). Thus, we select $\mathscr{M}_1$ if and only if

$$\frac{\hat{S}_0}{\hat{S}_1} \ge \exp\Big(\lambda_n/n\Big).$$

Notice that

$$\sum_{i=1}^{n}(Y_i-\hat{\alpha}-\hat{\beta}x_i)^2=\sum_{i=1}^{n}(Y_i-\hat{\alpha})^2-2\hat{\beta}\sum_{i=1}^{n}(Y_i-\overline{Y})x_i+\hat{\beta}^2\sum_{i=1}^{n}x_i^2=\sum_{i=1}^{n}(Y_i-\hat{\alpha})^2-n\hat{\beta}^2.$$

Therefore, $\hat{S}_1=\hat{S}_0-n\hat{\beta}^2$. Also because $\hat{S}_1=(n-2)\hat{\sigma}^2$,

$$\frac{\hat{S}_0}{\hat{S}_1}=1+\frac{n\hat{\beta}^2}{(n-2)\hat{\sigma}^2}.$$

Hence, we select the full-model if and only if

$$\frac{n^{1/2}|\hat{\beta}|}{\hat{\sigma}}\geq\sqrt{\left[\exp\left(\lambda_n/n\right)-1\right](n-2)}.$$

For example, under the AIC penalty $\lambda_n=2$, this becomes

$$\frac{n^{1/2}|\hat{\beta}|}{\hat{\sigma}}\geq\sqrt{\left[\exp\left(2/n\right)-1\right](n-2)}\approx\sqrt{2},$$

while for the BIC penalty $\lambda_n=\log(n)$, we have

$$\frac{n^{1/2}|\hat{\beta}|}{\hat{\sigma}}\geq\sqrt{\left(n^{1/n}-1\right)(n-2)}\approx\sqrt{\log(n)}.$$

4.6.1.1 Thresholding and consistency

We can see that the full model is selected if and only if

$$\frac{n^{1/2}|\hat{\beta}|}{\hat{\sigma}}\geq a_n \tag{4.13}$$

for a sequence $a_n>0$ related to the FPE penalty term. But this is actually nothing more than a generalization of the hypothesis test (4.11). We have simply replaced the $t_{n-2,\alpha_0/2}$ cutoff with a more general threshold value, a_n.

If $a_n\to\infty$ such that $a_n/\sqrt{n}\to 0$, then it is not hard to show that (4.13) leads to a consistent model selection procedure. More precisely, let $\hat{\delta}(a_n)$ be the decision rule which selects model $\mathcal{M}_1$ if (4.13) holds, otherwise $\mathcal{M}_0$ is selected. Let $\mathcal{M}_{\text{true}}$ be the true model. If $a_n=o(\sqrt{n})$, then

$$\mathbb{P}\{\hat{\delta}(a_n)=\mathcal{M}_{\text{true}}\}\to 1. \tag{4.14}$$

In particular, this shows that BIC leads to a consistent procedure.

But what about AIC and the hypothesis test (4.11)? Are they also consistent? In fact, they are not. One can show that if a_n remains bounded and $\mathcal{M}_{\text{true}}=\mathcal{M}_0$

$$\limsup_{n\to\infty}\mathbb{P}_0\{\hat{\delta}(a_n)=\mathcal{M}_{\text{true}}\}<1.$$

Thus, a bounded penalty precludes consistency if the true model is the null model. On the other hand, if a_n is bounded but $\mathcal{M}_{\text{true}}=\mathcal{M}_1$:

$$\mathbb{P}_1\{\hat{\delta}(a_n)=\mathcal{M}_{\text{true}}\}\to 1.$$

So all three procedures, BIC, hypothesis testing and AIC, have similar properties under the full-model, but not under the null model.

4.6.1.2 Risk consistency

The previous result suggests that BIC, and other thresholding procedures which use heavy and increasing penalties, should be preferred to fixed penalty methods like AIC when the aim is model selection. However, there is a price to be paid for a heavier penalty in terms of risk performance. The problem is that while BIC is pointwise consistent, it is not uniformly consistent, and this translates into poor uniform MSE performance.

To see this, let $\hat{Y}(x_0)$ be the predictor for Y using the model selection procedure $\hat{\delta}(a_n)$,

$$\hat{Y}(x_0) = \hat{\alpha} + \hat{\beta}\, x_0\, 1_{\{\hat{\delta}(a_n)=\mathcal{M}_1\}}.$$

Let $Y(x_0) = \alpha + \beta x_0$ be the true value of Y at x_0. The risk for $\hat{\delta}(a_n)$ at x_0 is

$$\begin{aligned} R_n(x_0, a_n) &= \mathbb{E}_\beta (Y(x_0) - \hat{Y}(x_0))^2 \\ &= \frac{\sigma^2}{n} + x_0^2\, \mathbb{E}_\beta \Big(\hat{\beta} 1_{\{\hat{\delta}(a_n)=\mathcal{M}_1\}} - \beta\Big)^2. \end{aligned}$$

For convenience we will work with the scaled risk $R_n^* = nR_n$. Consider the uniform behavior of the risk over bounded β. For concreteness take $x_0 = 1$. We have

$$\sup_{|\beta|\le C} R_n^*(x_0, a_n) \ge \sup_{\beta\in\mathcal{B}_n} n\mathbb{E}_\beta \Big(\hat{\beta} 1_{\{\hat{\delta}(a_n)=\mathcal{M}_1\}} - \beta\Big)^2,$$

where $\mathcal{B}_n = \Big\{\beta : |\beta| \le \min(\hat{\sigma} a_n/\sqrt{n}, C)\Big\}$. Also,

$$n\mathbb{E}_\beta \Big(\hat{\beta} 1_{\{\hat{\delta}(a_n)=\mathcal{M}_1\}} - \beta\Big)^2 = n\mathbb{E}_\beta \Big((\hat{\beta}-\beta)^2 1_{\{\hat{\delta}(a_n)=\mathcal{M}_1\}}\Big) + n\beta^2 \mathbb{P}_\beta\{\hat{\delta}(a_n) \ne \mathcal{M}_1\}.$$

If $\beta_n = \min(\hat{\sigma} a_n/\sqrt{n}, C)$ one can show with a little work that if $a_n \to \infty$

$$\mathbb{P}_{\beta_n}\{\hat{\delta}(a_n) \ne \mathcal{M}_1\} \to 1. \tag{4.15}$$

This shows that using a heavy penalty does not guarantee uniform model selection consistency (compare (4.15) with (4.14)). From this we obtain

$$\sup_{|\beta|\le C} R_n^*(x_0, a_n) \ge n\beta_n^2 \mathbb{P}\{\hat{\delta}(a_n) \ne \mathcal{M}_{\text{true}}\} \to \infty.$$

Therefore, the scaled uniform risk is unbounded even over β values less than a fixed constant value. Compare this to (4.10) which has a uniformly bounded scaled risk.

4.6.2 Theoretical conclusions

Combining the previous discussion with published results mentioned earlier, we can summarize the differences between BIC and AIC as follows.

Take away message about AIC and BIC:

1. *BIC is model consistent when a true model exists, but its heavy penalty can cause problems in small samples. This penalty can lead to underestimating the model, resulting in an estimator with high bias and poor mean squared error (MSE) performance when there is uncertainty in model determination.*
2. *AIC tends to select larger models due to its lighter penalty, making it difficult to choose among overfit models that include the true model. This results in a larger model with low bias, which generally means AIC has good risk performance. However, this same property prevents AIC from being pointwise model selection consistent, as shown in our example.*

4.6.3 Empirical example

So far we have looked at asymptotic results, now we consider a simulation to see how AIC and BIC perform in finite samples. Data was simulated from a linear model with $n = 200$, $p = 13$ where the true number of features was $p_0 = 3$ and

$$Y_i = x_{i1}\beta + x_{i2}\beta + x_{i3}\beta + \varepsilon_i, \quad i = 1, \ldots, n.$$

The ε_i are i.i.d. $\mathrm{N}(0,1)$ and the $\mathbf{X}$ features are independently drawn from a multivariate normal $\mathrm{N}_p(\mathbf{0}, \mathbf{\Sigma})$ where $\mathbf{\Sigma}$ is an eqicorrelation matrix with correlation parameter ρ. Four different experiments were considered as summarized by Table 4.3.

TABLE 4.3
Linear model simulation used to investigate finite sample behavior of AIC and BIC.

Experiment	$\mathbf{X}$	β
Strong Orthogonal	$\rho = 0$	3
Weak Orthogonal	$\rho = 0$	0.25
Strong Correlated	$\rho = 0.7$	3
Weak Correlated	$\rho = 0.7$	0.25

The simulations were repeated 1000 times. Figure 4.3 displays results for the two orthogonal ($\rho = 0$) simulations where on the vertical axis the Monte Carlo averaged values of AIC (in red) and BIC (in green) are recorded for each of the 2^{p-1} possible models. Left is the experiment with $\beta = 3$ (strong signal); right is $\beta = .25$ (weak signal). Notice how each figure has 4 "clouds" (groups) of points for both AIC and BIC. Going from bottom to top the clouds correspond to models with: all signal variables; one signal variable missing; two signal variables missing; and all signal variables missing. This shows that FPE increases when a model excludes signal variables: the more signal variables missing, the larger the FPE.

A noticeable difference between the two figures is the penalization effect seen for the weak signal experiment. In particular, FPE increases in the weak signal experiment as dimension increases going from left to right, whereas no such effect is observed in the strong signal experiment. This is because in the latter case the signal is so strong that the dominating term for the FPE is the RSS and penalization has little to no effect on this.

Results for the two correlated experiments are given in Figure 4.4. We observe some differences here. In the weak signal case, the results are similar to the weak orthogonal case,

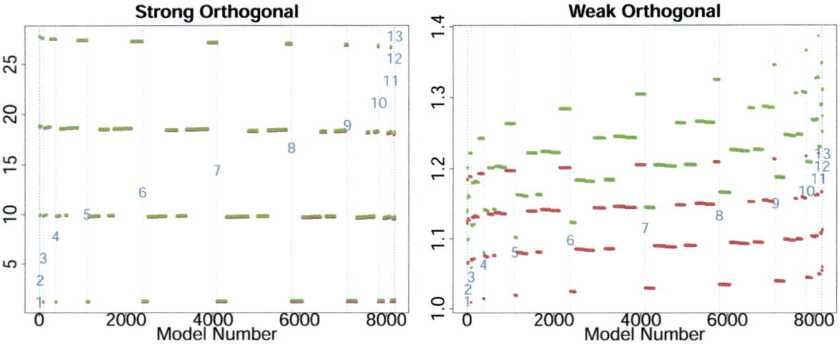

FIGURE 4.3
AIC (red) and BIC (green) for all subsets from linear regression simulation with strong/weak uncorrelated features. Blue integers signify model size.

but the effect of penalization is more dramatic. This is because the RSS is lower due to features being correlated, which increases the impact of the penalization on FPE. In the strong signal experiment, we observe a surprising change: FPE *decreases* as dimension increases. This effect is seen for the underfit models, which are misspecified models that exclude one or more signal variables. The effect of model misspecification on RSS is amplified when features are correlated. For a linear multivariate normal model as in our simulation, it was shown in Chapter 3.3 that $\text{Cor}(Y, \mathbf{X})^T = \text{Var}(Y)^{-1/2}\boldsymbol{\Sigma}\boldsymbol{\beta}$. Therefore, Y is correlated with non-signal variables, meaning that for underfit models, a non-signal variable can contribute to reducing RSS. This effect is amplified with larger underfit models when the signal is strong, as there is more opportunity to decrease the RSS. This explains why FPE decreases with dimension in the correlated strong experiment but not in the correlated weak setting.

Considering the size of the models selected by the two procedures (Figure 4.5) we find that AIC and BIC's performance are independent of correlation when signal is strong. Otherwise, both methods tend to select smaller models when $\mathbf{X}$ is correlated. Overall, BIC selects smaller models and chooses the true model size $p_0 = 3$ more often than AIC. This is due to BIC's model selection consistency property as we have discussed. On the other hand, BIC also underfits as seen in the

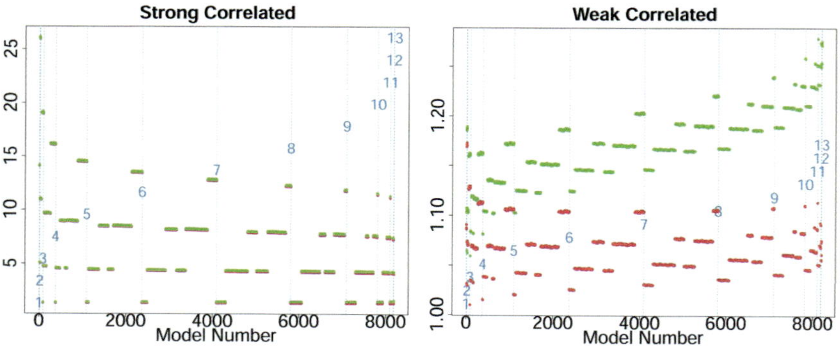

FIGURE 4.4
AIC (red) and BIC (green) for all subsets from linear regression simulation with strong/weak correlated features. Blue integers signify model size.

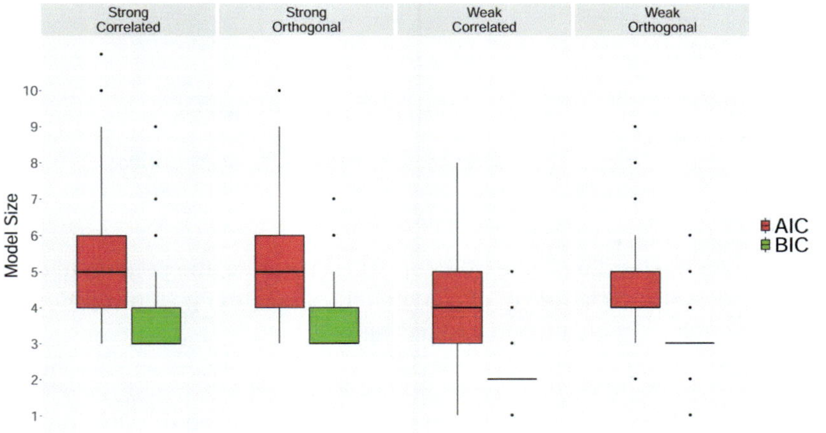

FIGURE 4.5
Size of models selected by AIC and BIC for all subsets from linear regression simulation.

weakly correlated experiment. This results in bias which degrades MSE performance; thus, in this experiment AIC might be regarded as being superior even though it tends to overfit.

4.7 Prediction Error

FPE and MSE are interconnected concepts. Both actually have the same goal in mind, which is assessing how well a procedure performs over new data. That is to say, both are metrics designed to estimate *prediction error.* In this section we will define prediction error and discuss how FPE methods like AIC are related to this through the connection to MSE.

For this discussion, we take the opportunity to present a more general definition of prediction error and return later specifically to discuss the linear model. Denote the learning data by $\mathscr{L} = \{(\mathbf{X}_i, Y_i)\}_1^n$ where $(\mathbf{X}_i, Y_i)$ are i.i.d. values from $\mathbb{P}$. As before, $\mathbf{X}_i$ are features for predicting Y_i, however, we remove our previous assumption that $\mathbf{X}_i$ is fixed. Denote a generic new set of values by $(\mathbf{X}, Y)$ which are assumed to be distributed as $\mathbb{P}$.

Let $\hat{\mu}(\mathbf{X}) := \hat{\mu}(\mathbf{X}, \mathscr{L})$ be an estimator for the unknown regression function $\mu(\mathbf{X}) = \mathbb{E}(Y|\mathbf{X})$. Notice that $\hat{\mu}$ is constructed using only the learning data $\mathscr{L}$. This makes it natural to consider *conditional prediction error* defined as the integrated squared error for $\hat{\mu}$ conditioned on $\mathscr{L}$. Formally the conditional prediction error is defined as

$$\mathrm{err}(\hat{\mu}) = \mathbb{E}_{\mathbf{X},Y}\left\{(Y - \hat{\mu}(\mathbf{X}))^2|\mathscr{L}\right\} := \mathbb{E}_{\mathbf{X},Y}\left[Y - \hat{\mu}(\mathbf{X})\right]^2.$$

The *(unconditional) prediction error*, or *generalization error*, is the expectation of the conditional error over all learning sets of size n with the same distribution as $\mathscr{L}$. Let Err denote this value. Then,

$$\mathrm{Err}(\hat{\mu}) = \mathbb{E}_{\mathscr{L}}\mathbb{E}_{\mathbf{X},Y}\left[Y - \hat{\mu}(\mathbf{X})\right]^2.$$

Now notice that

$$\begin{aligned}\mathrm{Err}(\hat{\mu}) &= \mathbb{E}_{\mathscr{L}}\mathbb{E}_{\mathbf{X},Y}\left[(Y-\mu(\mathbf{X})+\mu(\mathbf{X})-\hat{\mu}(\mathbf{X}))^2\right]\\ &= \mathbb{E}_{\mathbf{X},Y}\left[Y-\mu(\mathbf{X})\right]^2+\mathbb{E}_{\mathscr{L},\mathbf{X}}\left[\mu(\mathbf{X})-\hat{\mu}(\mathbf{X})\right]^2\\ &\quad +2\mathbb{E}_{\mathbf{X}}\left\{\mathbb{E}_{Y|\mathbf{X}}\left[Y-\mu(\mathbf{X})\right]\mathbb{E}_{\mathscr{L}}\left[\mu(\mathbf{X})-\hat{\mu}(\mathbf{X})\right]\right\}\\ &= \sigma^2+\mathbb{E}_{\mathscr{L},\mathbf{X}}\left[\hat{\mu}(\mathbf{X})-\mu(\mathbf{X})\right]^2. \qquad (4.16)\end{aligned}$$

The second term on the last line is the MSE for $\hat{\mu}(\mathbf{X})$. Therefore, the prediction error equals the MSE plus the underlying variance σ^2 of the model. Hence, the goal of minimizing MSE and minimizing prediction error are the same in regression.

Now let us return to the linear regression setting. We show the C_P statistic can be used as an unbiased estimator for the MSE. Because C_P is closely related to AIC and since MSE is related to prediction error, this demonstrates that both metrics are designed to select procedures on the basis of prediction error. The analysis given below also reveals exactly why AIC uses the penalty $2\sigma^2 p_m$.

As before, we assume a fixed design matrix. Let $\hat{\boldsymbol{\mu}}_m$ be the restricted OLS predictor for model m. Then

$$\begin{aligned}||\mathbf{Y}-\hat{\boldsymbol{\mu}}_m||^2 &= \left\|\sum_{k=1}^{n} Z_k\mathbf{v}_k-\sum_{k=1}^{p_m} Z_k\mathbf{v}_k\right\|^2\\ &= \sum_{k=p_m+1}^{n} Z_k^2\\ &= \sum_{k=p_m+1}^{n}\eta_k^2+\sum_{k=p_m+1}^{n}\zeta_k^2+2\sum_{k=p_m+1}^{n}\eta_k\zeta_k\\ &= ||\hat{\boldsymbol{\mu}}_m-\boldsymbol{\mu}||^2-\sum_{k=1}^{p_m}\zeta_k^2+\sum_{k=p_m+1}^{n}\zeta_k^2+2\sum_{k=p_m+1}^{n}\eta_k\zeta_k \quad \text{(by (4.6))}.\end{aligned}$$

Taking expectations,

$$\begin{aligned}\mathbb{E}||\mathbf{Y}-\hat{\boldsymbol{\mu}}_m||^2 &= \mathbb{E}||\hat{\boldsymbol{\mu}}_m-\boldsymbol{\mu}||^2+(n-2p_m)\sigma^2\\ &= n\mathbb{E}\left[\hat{\mu}_m(\mathbf{X})-\mu(\mathbf{X})\right]^2+(n-2p_m)\sigma^2.\end{aligned}$$

For any estimator $\hat{\sigma}^2 \xrightarrow{\text{p}} \sigma^2 > 0$,

$$\begin{aligned}\mathbb{E}(C_m) &= \mathbb{E}\left[\frac{\hat{S}_m}{\hat{\sigma}^2}+2p_m-n\right]\\ &= \mathbb{E}\left(\frac{||\mathbf{Y}-\hat{\boldsymbol{\mu}}_m||^2}{\hat{\sigma}^2}\right)+2p_m-n\\ &= \frac{1}{\sigma^2}\mathbb{E}||\mathbf{Y}-\hat{\boldsymbol{\mu}}_m||^2+2p_m-n+o_p(1)\\ &= \frac{n}{\sigma^2}\mathbb{E}\left[\hat{\mu}_m(\mathbf{X})-\mu(\mathbf{X})\right]^2+o_p(1).\end{aligned}$$

Therefore, $\sigma^2 C_m/n$ is asymptotically unbiased for the MSE. Now replacing the FPE criterion with the slightly modified value

$$\mathrm{FPE}^*(m)=\hat{S}_m+2\sigma^2 p_m,$$

it follows that the average of $\mathrm{FPE}^*(m)$ (which is a modified AIC) is

$$\mathbb{E}[\mathrm{FPE}^*(m)]=n\mathbb{E}\left[\hat{\mu}_m(\mathbf{X})-\mu(\mathbf{X})\right]^2+\text{constant}.$$

Asymptotically, we observe that minimizing AIC is, on average, equivalent to minimizing the MSE, which serves as a measure of risk in the model. Additionally, we see that adding the penalty $2\sigma^2 p_m$ to the RSS provides the exact amount of penalization needed to correct the bias.

4.8 Penalization by Bridge Estimation

Model selection criteria such as in Table 4.2 avoid overfitting by imposing a penalty on the size of the model. The larger the model size, the larger the penalty. The basic principle is always the same: only the size of the penalty changes.

This addresses the first point needed for appropriate regularization. Now, we turn to the remaining points: selecting optimal models within the same dimension and managing large model spaces. For the latter, we must acknowledge that FPE may not be practical due to the vast number of models that need to be evaluated. The regression framework involves $M = 2^p - 1$ models, which quickly becomes infeasible as p increases.

To tackle this, we adopt a new principle. Instead of penalizing the size of the model estimator, we penalize the size of its estimated coefficients. This approach is accommodated by the class of *bridge estimators*, which includes a range of useful penalty estimators.

For a given $\lambda \geq 0$, a bridge estimator with bridge parameter $\gamma > 0$ is the estimator defined as the solution to the constrained L_2-optimization problem:

$$\hat{\boldsymbol{\beta}}_\gamma = \underset{\boldsymbol{\beta} \in \mathbb{R}^p}{\arg\min} \left\{ ||\mathbf{Y} - \mathbf{D}\boldsymbol{\beta}||^2 + \lambda \sum_{k=1}^{p} |\beta_k|^\gamma \right\}. \tag{4.17}$$

Setting $\lambda = 0$ gives the OLS estimator which is unregularized since it involves no penalty. Otherwise, setting $\lambda > 0$ results in a penalized estimator. Below we summarize examples of some well-known classes of bridge estimators:

1. $\gamma = 2$ gives the ridge regression estimator of Hoerl and Kennard [10].
2. $0 < \gamma \leq 1$ results in bridge estimators with the sparsity property that their coefficient estimates can shrink exactly to zero.
3. The lasso [19] is a special case of (2) with $\gamma = 1$, known as L_1-penalization. The lasso is particularly appealing due to its computational efficiency and its membership in the sparse family ($0 < \gamma \leq 1$). This property ensures that the estimated lasso solution includes zero values, making it an effective tool for model selection by identifying models with non-zero coefficients. Additionally, the L_1-penalization used by the lasso has certain unique properties that will be discussed in detail.
4. As a limiting case $\gamma \downarrow 0$, bridge estimators include penalization methods based on cardinality similar to FPE,
$$\lim_{\gamma \downarrow 0} \sum_{k=1}^{p} |\beta_k|^\gamma \to \sum_{k=1}^{p} 1_{\{\beta_k \neq 0\}}.$$

In many cases the bridge optimization (4.17) can be solved efficiently (ridge regression and lasso are examples). This assumes a fixed penalty parameter λ but this value can be tuned relatively quickly using cross-validation (see Chapter 7.9). Thus, the final selected model using bridge estimation can often be computed efficiently without the problems of exhaustive enumeration as required by FPE.

4.9 Lasso

Unlike standard methods, the lasso solution often contains many zeroes, a property resulting from L_1-penalization. This ability to shrink coefficient estimates to zero will be explained by contrasting the lasso with the ridge estimator.

According to (4.17), the lasso is the solution to

$$\hat{\boldsymbol{\beta}}_{\text{lasso}}(\lambda) = \underset{\boldsymbol{\beta}\in\mathbb{R}^p}{\arg\min}\left\{||\mathbf{Y}-\mathbf{D}\boldsymbol{\beta}||^2 + \lambda\sum_{k=1}^{p}|\beta_k|\right\}, \tag{4.18}$$

where $\lambda \geq 0$ is the lasso regularization parameter. Solving (4.18) is equivalent to solving the optimization problem

$$\text{minimize } ||\mathbf{Y}-\mathbf{D}\boldsymbol{\beta}||^2, \quad \text{subject to } \sum_{k=1}^{p}|\beta_k| \leq C.$$

This shows that the lasso is the solution to the least-squares problem subject to an L_1-constraint.

The ridge estimator obtained by setting $\gamma = 2$ in (4.17) is defined by

$$\hat{\boldsymbol{\beta}}_{\text{ridge}}(\lambda) = \underset{\boldsymbol{\beta}\in\mathbb{R}^p}{\arg\min}\left\{||\mathbf{Y}-\mathbf{D}\boldsymbol{\beta}||^2 + \lambda\sum_{k=1}^{p}\beta_k^2\right\}, \tag{4.19}$$

where $\lambda \geq 0$ is the ridge parameter. Equivalently, this is the solution to the least-squares problem subject to an L_2-constraint:

$$\text{minimize } ||\mathbf{Y}-\mathbf{D}\boldsymbol{\beta}||^2, \quad \text{subject to } \sum_{k=1}^{p}\beta_k^2 \leq t.$$

Now to see why lasso has a different behavior than ridge, let $\hat{\boldsymbol{\beta}}$ denote the OLS. It is easily verified that

$$||\mathbf{Y}-\mathbf{D}\boldsymbol{\beta}||^2 = (\boldsymbol{\beta}-\hat{\boldsymbol{\beta}})^T\mathbf{D}^T\mathbf{D}(\boldsymbol{\beta}-\hat{\boldsymbol{\beta}}) + \text{constant}. \tag{4.20}$$

The first term on the right-hand side is an ellipsoid centered at the OLS. Therefore, the ridge estimator (4.19) is the solution point where the OLS-centered ellipsoid first touches the spherical constraint region (see Figure 4.6), whereas the lasso (4.18) is the solution where the ellipsoid first touches the cubic constraint region (see Figure 4.7). The cubical nature of the constraint region forces some coordinates of the lasso to zero. This contrasts with the ridge estimator, which cannot shrink values to zero due to the spherical constraint.

This ability to shrink coefficients to zero makes L_1-penalization desirable in many problems, especially those with a large number of parameters. Examples are underdetermined systems of equations (i.e., systems with fewer equations than unknowns), such as signal processing, inverse problems, and genomic data analysis. Donoho [5] quantified the sparsity capability of lasso for solving such systems. Define $\tau = p/n$ as the complexity. Donoho [5] showed that there exists a $\rho = \rho(\tau) > 0$ such that for most large $\mathbf{D}_{n\times p}$ matrices, if convex optimization is used to find the L_1-minimal solution and it has fewer than ρn nonzeros, then this is the unique sparsest solution. If the result does not have fewer than ρn nonzeros, then no solution with fewer than ρn nonzero coefficients exists. Thus, if a sparse solution with significantly fewer than n nonzeros exists, it may be found by the lasso.

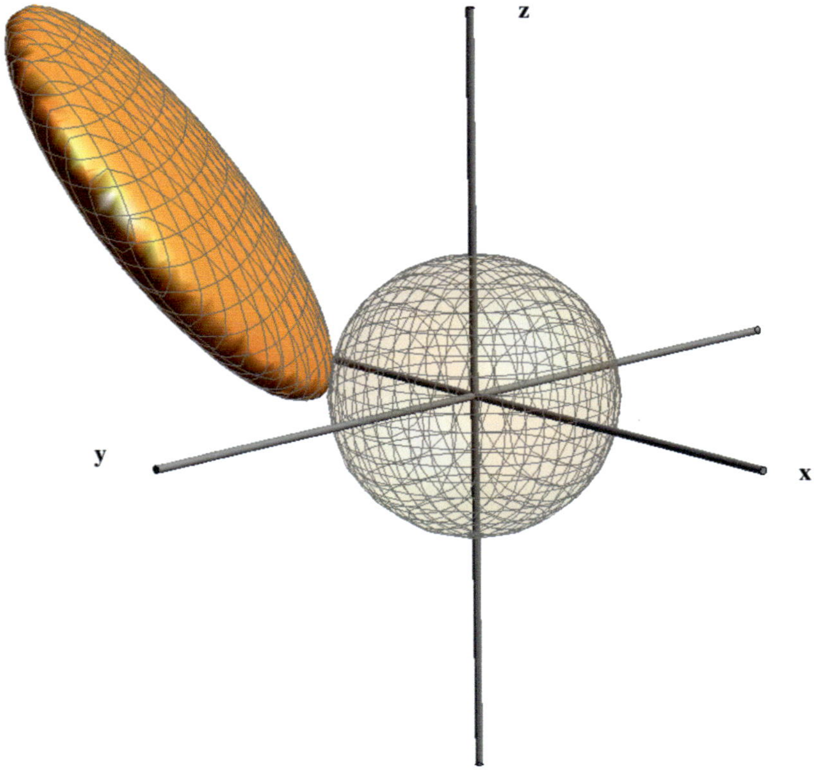

FIGURE 4.6
The ridge solution ($p = 3$ dimensions) is the point where the ellipsoid centered at the OLS touches the constraint region $x^2 + y^2 + z^2 \leq t$.

Below we list some basic facts about the uniqueness of the lasso [20].

1. There is either a unique lasso solution or an uncountably infinite number of solutions.
2. Every lasso solution $\hat{\boldsymbol{\beta}}_{\text{lasso}}$ gives the same predictor $\mathbf{X}\hat{\boldsymbol{\beta}}_{\text{lasso}}$.
3. If $\lambda > 0$, every lasso solution $\hat{\boldsymbol{\beta}}_{\text{lasso}}$ has the same L_1-norm, $||\hat{\boldsymbol{\beta}}_{\text{lasso}}||_1$.

As will be shown, another important property of the lasso is its prediction-loss property, which means it can achieve high prediction accuracy even when p is very large. This is again due to the lasso's ability to shrink coefficients to zero.

4.9.1 Considering the oracle property

We have mentioned some positive properties of the lasso, such as its predictive capabilities, but we should also consider how well the lasso performs in terms of model selection and estimation. As we have seen in our comparison of BIC and AIC, the properties of prediction and estimation are not necessarily the same. In particular, the question of whether the lasso posesses the oracle

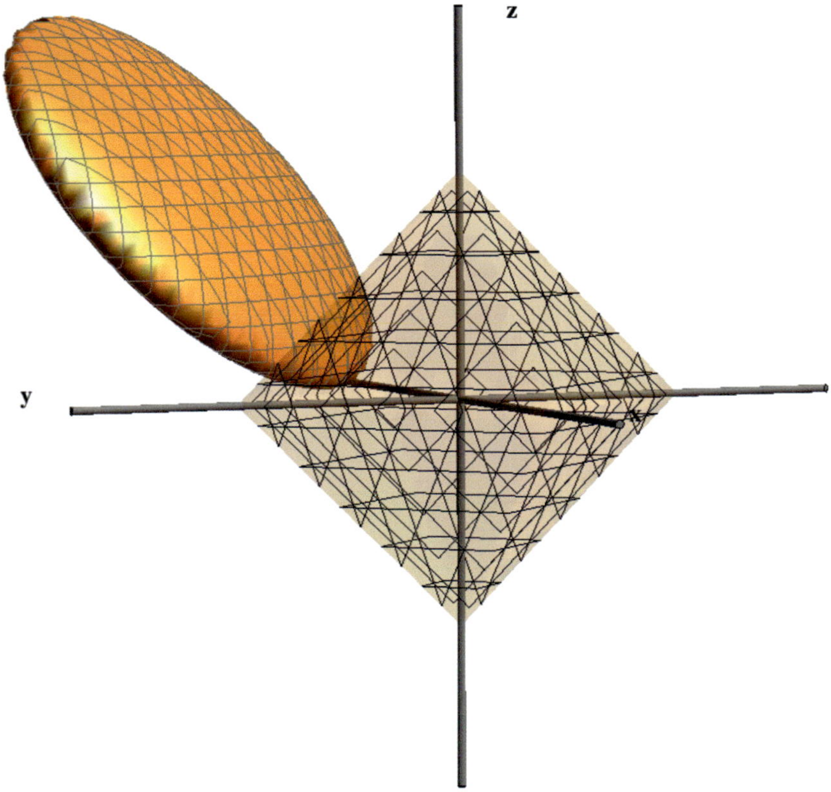

FIGURE 4.7
The lasso solution is the point where the ellipsoid touches the lasso constraint region $|x|+|y|+|z| \leq t$.

property is of interest. As shown by Fan and Li [7], it turns out that the lasso does not possess the oracle property.

To discuss this, we first review the asymptotic theory for the (full model) OLS. For this analysis, we make the following assumptions:

(A1) $\{\varepsilon_i\}_{i=1}^n$ are i.i.d. with mean 0 and variance σ^2.

(A2) $\mathbf{Q}_n = \frac{1}{n}\sum_{i=1}^n \mathbf{x}_i\mathbf{x}_i^T \to \mathbf{Q} > 0$.

(A3) $\frac{1}{n}\max_{1\leq i\leq n}\{\mathbf{x}_i^T\mathbf{x}_i\} \to 0$.

Let $\hat{\boldsymbol{\beta}}_{n,\text{ols}}$ denote the OLS. Under the above conditions, it can be shown that

$$\sqrt{n}\left(\hat{\boldsymbol{\beta}}_{n,\text{ols}} - \boldsymbol{\beta}\right) \overset{\text{d}}{\rightsquigarrow} \text{N}_p\left(\mathbf{0}, \sigma^2\mathbf{Q}^{-1}\right). \tag{4.21}$$

Therefore the asymptotic variance for the OLS is $\sigma^2\mathbf{Q}^{-1}$. Obviously the OLS cannot possess the oracle property except when the true model is the full-model; i.e., $\mathscr{S}_0 = \{1,\ldots,p\}$.

Now consider bridge estimators for $\gamma > 1$

$$\hat{\boldsymbol{\beta}}_{n,\gamma} = \underset{\boldsymbol{\beta}\in\mathbb{R}^p}{\arg\min}\left\{||\mathbf{Y}-\mathbf{D}\boldsymbol{\beta}||^2 + \lambda_n \sum_{k=1}^{p} |\beta_k|^\gamma\right\}, \qquad \gamma > 1.$$

Then $\sqrt{n}$-consistency holds if in addition to the previous conditions (A1)-(A3), we have

$$\lambda_n/\sqrt{n} \to \lambda_0 \geq 0. \tag{4.22}$$

Under these conditions, it can be shown [11]

$$\sqrt{n}\left(\hat{\boldsymbol{\beta}}_{n,\gamma} - \boldsymbol{\beta}\right) \overset{\mathrm{d}}{\rightsquigarrow} \arg\min_{\mathbf{u}}(V(\mathbf{u}))$$

where

$$\begin{aligned} V(\mathbf{u}) &= -2\mathbf{u}^T\mathbf{Z} + \mathbf{u}^T\mathbf{Q}\mathbf{u} + \lambda_0 \sum_{k=1}^{p} u_k \mathrm{sign}(\beta_k)|\beta_k|^{\gamma-1} \\ \mathbf{Z} &\sim \mathrm{N}_p(\mathbf{0}, \sigma^2\mathbf{Q}). \end{aligned}$$

Notice that if $\lambda_0 = 0$ we obtain the OLS limit, $\mathbf{Q}^{-1}\mathbf{Z}$.

Now consider ridge regression, which is the bridge estimator for $\gamma = 2$, which we denote by $\hat{\boldsymbol{\beta}}_{n,\mathrm{ridge}}$. Under (A1)-(A3), and assuming its ridge parameter λ_n satisfies (4.22), it follows that $\hat{\boldsymbol{\beta}}_{n,\mathrm{ridge}} = \left(\mathbf{D}^T\mathbf{D} + \lambda_n\mathbf{I}\right)^{-1}\mathbf{D}^T\mathbf{Y}$ and

$$\sqrt{n}\left(\hat{\boldsymbol{\beta}}_{n,\mathrm{ridge}} - \boldsymbol{\beta}\right) \overset{\mathrm{d}}{\rightsquigarrow} \mathbf{Q}^{-1}(\mathbf{Z} - \lambda_0\boldsymbol{\beta}/2) \overset{d}{=} \mathrm{N}_p(-\lambda_0\mathbf{Q}^{-1}\boldsymbol{\beta}/2, \sigma^2\mathbf{Q}^{-1}). \tag{4.23}$$

Notice (4.23) shows the ridge estimator has a non-zero asymptotic bias if $\lambda_0 > 0$ and yet at the same time it has the same asymptotic variance as the OLS. Thus if the ridge parameter is non-vanishing, this is not flattering for ridge as it suggests it is inferior to OLS. However in finite sample settings, ridge regression may have better MSE performance for an appropriately selected λ_n.

Now finally let us consider the asymptotic behavior of lasso. Denote the lasso estimator by $\hat{\boldsymbol{\beta}}_{n,\mathrm{lasso}}$. Assuming (A1)-(A3), and that the lasso parameter λ_n satisfies (4.22), it can be shown that [11]

$$\sqrt{n}\left(\hat{\boldsymbol{\beta}}_{n,\mathrm{lasso}} - \boldsymbol{\beta}\right) \overset{\mathrm{d}}{\rightsquigarrow} \arg\min_{\mathbf{u}}(V^*(\mathbf{u})) \tag{4.24}$$

where

$$\begin{aligned} V^*(\mathbf{u}) &= -2\mathbf{u}^T\mathbf{Z} + \mathbf{u}^T\mathbf{Q}\mathbf{u} + \lambda_0 \sum_{\beta_k\neq 0} u_k \mathrm{sign}(\beta_k) + \lambda_0 \sum_{\beta_k=0} |u_k| \\ &= -2\mathbf{u}^T\mathbf{Z} + \mathbf{u}^T\mathbf{Q}\mathbf{u} + \lambda_0 \sum_{k=1}^{p_0} u_k \mathrm{sign}(\beta_{0,k}) + \lambda_0 \sum_{k=p_0+1}^{p} |u_k|. \end{aligned}$$

This shows that for the lasso to possess the oracle property, $V^*(\mathbf{u})$ should be minimized at $\mathbf{u}_2 = \mathbf{0}$ where $\mathbf{u} = (\mathbf{u}_1, \mathbf{u}_2)$. However, this only holds if

$$-\frac{\lambda_0}{2}\mathbf{1} \leq \mathbf{Q}_{21}\mathbf{Q}_{11}^{-1}\left(\mathbf{Z}_1 - \lambda_0\mathrm{sign}(\boldsymbol{\beta}_0)/2\right) - \mathbf{Z}_2 \leq \frac{\lambda_0}{2}\mathbf{1} \tag{4.25}$$

It is easy to construct examples where this fails (see Exercise 4.21).

4.9.2 Prediction-loss property

The failure of the lasso to achieve the oracle property is disappointing. So why does the lasso receive such widespread recognition and use? The reason is that the lasso is an excellent predictor. It has the ability to shrink coefficients to zero which allows it to adapt to sparsity and produce good prediction in high dimensional settings.

To demonstrate this last point we can show that the lasso possesses the so-called *prediction-loss* consistency property. To simplify this result, we make the following assumptions. Note that in the following we assume the covariates $\mathbf{X} = (X^{(1)}, \ldots, X^{(p)})^T$ are random. We denote the true $\boldsymbol{\beta}$ as $\boldsymbol{\beta}_0 = (\beta_{01}, \ldots, \beta_{0p})^T$ and the true conditional mean by $\mu(\mathbf{X}) = \mathbb{E}(Y|\mathbf{X})$. The following conditions are assumed.

(B1) $\{\varepsilon_i\}_{i=1}^n$ are i.i.d. $\mathrm{N}(0, \sigma^2)$.
(B2) $|X^{(k)}| \le M < \infty$ for $k = 1, \ldots, p$.

Then the following holds [3].

Theorem 4.3. *Under assumptions (B1) and (B2), let K be any constant such that* $\sum_{k=1}^p |\beta_{0k}| \le K$. *Then the lasso predictor* $\hat{\mu}_{n,K}(\mathbf{X})$ *obtained from the lasso coefficient estimate* $\hat{\boldsymbol{\beta}}_{n,\text{lasso}}$ *with lasso parameter* $\lambda = K$ *satisfies*

$$\mathbb{E}(\hat{\mu}_{n,K}(\mathbf{X}) - \mu(\mathbf{X}))^2 \le 2KM\sigma\sqrt{\frac{2\log(2p)}{n}} + 8K^2M^2\sqrt{\frac{2\log(2p^2)}{n}}.$$

Therefore, the lasso is prediction consistent if p is smaller than $\exp(n)$ assuming that $\lambda = K$ is selected correctly. The condition for K requires at the very least that $K/\sqrt{n} = o(1)$, which is interesting because this is condition (4.22) with the limiting value $\lambda_0 = 0$. By (4.24) this implies

$$\sqrt{n}\left(\hat{\boldsymbol{\beta}}_{n,\text{lasso}} - \boldsymbol{\beta}\right) \overset{\text{d}}{\rightsquigarrow} \mathrm{N}_p(\mathbf{0}, \sigma^2\mathbf{Q}^{-1})$$

which implies that the lasso would have the same asymptotic distribution as the OLS. This suggests that the lasso cannot achieve optimal estimation and prediction-loss consistency simultaneously. Theorem 4.3 is however, still impressive for the lasso. It shows that prediction consistency holds as long as p is not exponentially larger than n. This explains why lasso is so successful in high dimensions: it is a great predictor due to its sparsity property.

4.9.3 Brief discussion of algorithms for fitting lasso

Several algorithms have been proposed for fitting the lasso. Tibshirani [19] proposed quadratic programming with sign constraints. Chen, Donoho and Saunders [4] introduced basis pursuit and used an interior point program based on quadratic programming. More recently, Efron et al. [6] introduced LARS (least angle regression), a similar idea also appeared earlier in the works of Osborne, Presnell and Turlach [14] and Osborne, Presnell and Turlach [15]. A key feature of LARS is that it provides the entire lasso solution path $\{\hat{\boldsymbol{\beta}}_{\text{lasso}}(\lambda) : \lambda \ge 0\}$ using only $O(p^3 + np^2)$ calculations, equivalent to the cost of calculating a single fit using OLS. This makes LARS one of the fastest methods for obtaining the lasso.

For a more detailed discussion on lasso computations, see Chapter 14, where we discuss coordinate descent, another efficient algorithm for obtaining lasso solutions. However, before ending this chapter, we return to the body fat analysis considered earlier to provide an illustration of lasso.

Figure 4.8 shows the lasso solution as a thick red line superimposed on Figure 4.1. Each red point on the line represents the lasso-selected model for that dimension from 1 to 13, obtained using the LARS algorithm. Lasso chooses a high R^2 model in each instance, though not always the one with the highest goodness of fit (especially for model size 3). This figure illustrates how lasso can efficiently assess the model space to identify high-value models through penalization, unlike FPE, which requires evaluating all $2^p - 1$ models. Remember also that FPE faces the problem of identifying a high-value model within a fixed dimension due to crowding which is another issue that is avoided by using lasso.

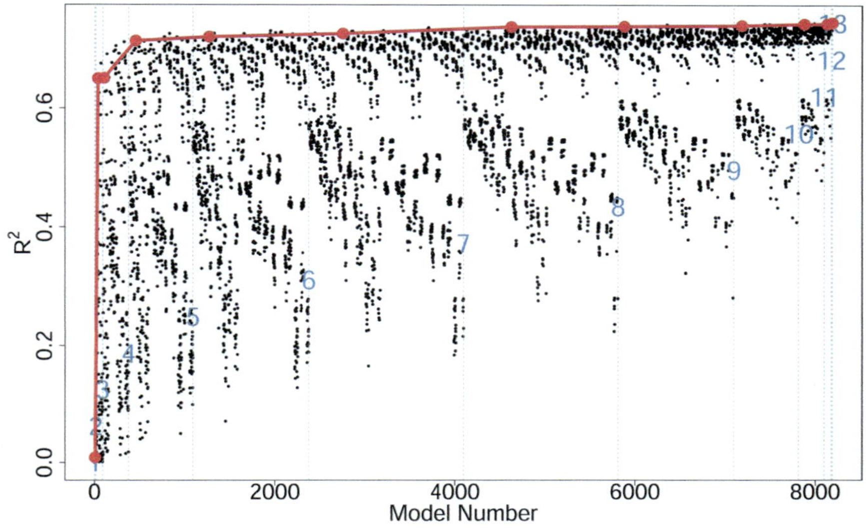

FIGURE 4.8
We have modified Figure 4.1 for the body fat data by superimposing a red line that shows models selected by the lasso for each dimension.

Exercises

Exercise 4.1. Let span$\{\mathbf{v}_1,\ldots,\mathbf{v}_r\}$ = column span$\{\mathbf{D}\}$ where $\{\mathbf{v}_1,\ldots,\mathbf{v}_n\}$ is an orthonormal basis spanning $\mathbb{R}^n$. Prove that $\eta_i = 0$ for $i > r$ where $\eta_i = \boldsymbol{\mu}^T\mathbf{v}_i$.

Exercise 4.2. Let $\{\mathbf{v}_1,\ldots,\mathbf{v}_n\}$ be an orthonormal basis spanning $\mathbb{R}^n$. Prove that $||\boldsymbol{\epsilon}^T\mathbf{v}_i||^2$ are i.i.d. $\sigma^2\chi_1^2$ if $\boldsymbol{\epsilon}$ is $\mathrm{N}(\mathbf{0},\sigma^2\mathbf{I}_n)$.

Exercise 4.3. Suppose that $X_1,\ldots,X_n$ are i.i.d. $N(0,\sigma^2)$ random variables. Using a geometric approach, show that $s_n^2 = \sum_{i=1}^n (X_i - \overline{X})^2/(n-1) \sim \sigma^2\chi_{n-1}^2$ and is independent of $\overline{X} \sim \mathrm{N}(0,\sigma^2/n)$.

Exercise 4.4. Consider fitting a regression model using restricted least squares where we fit the model using the first k columns of an $n \times p$ design matrix. Let $\hat{S}(k)$ be the RSS for model k and define $\hat{\sigma}_n^2(k) = \hat{S}(k)/(n-k)$ to be the k model estimator of the variance. Show using a geometric argument that $\mathbb{E}[\hat{\sigma}_n^2(k)] \geq \mathbb{E}[\hat{\sigma}_n^2(p)]$ for $k = 1,\ldots,p$. Be sure to clearly state all your assumptions.

Exercise 4.5. Let $\mathbf{Z} \in \mathbb{R}^n$ be a multivariate normal vector. Let $\mathscr{A}$ and $\mathscr{B}$ be subspaces of $\mathbb{R}^n$ such that $\mathscr{A} \perp \mathscr{B}$. Prove that $\mathbf{Z}_{\mathscr{A}}$ is independent of $\mathbf{Z}_{\mathscr{B}}$, where $\mathbf{Z}_{\mathscr{A}}$ and $\mathbf{Z}_{\mathscr{B}}$ are the projections of $\mathbf{Z}$ onto $\mathscr{A}$ and $\mathscr{B}$ respectively.

Exercise 4.6. Consider the linear regression model

$$Y_i = \alpha + x_{i1}\beta_1 + \cdots + x_{ip}\beta_p + \varepsilon_i, \qquad i = 1, \ldots, n,$$

where ε_i are i.i.d. $\mathrm{N}(0, \sigma^2)$. Using a geometric approach, derive the distribution for R^2 under the null hypothesis $H_0 : \beta_1 = \cdots = \beta_p = 0$.

Exercise 4.7. Let a plane P pass through points $\mathbf{0}, \mathbf{x}_1, \mathbf{x}_2, \ldots, \mathbf{x}_p \in \mathbb{R}^n$ so that P is defined by

$$P = \left\{ \mathbf{y} : \mathbf{y} = \sum_{k=1}^{p} \beta_k \mathbf{x}_k \right\}.$$

Let $\mathbf{D} = (\mathbf{x}_1, \ldots, \mathbf{x}_p)_{n\times p}$ and denote its rank by $\mathrm{r}(\mathbf{D}) = k$. Therefore P is a k-dimensional subspace of $\mathbb{R}^n$.

1. Suppose that $n \geq p \geq k$. Given $\mathbf{y} \in \mathbb{R}^n$ find the projection of $\mathbf{y}$ on to P.
2. Suppose that $p \geq n$. Given $\mathbf{y} \in \mathbb{R}^n$ find the projection of $\mathbf{y}$ on to P.

Exercise 4.8. Verify (4.7). Show that the terms on the right-hand side can be identified as variance and bias-squared values eq. (4.8).

Exercise 4.9. Show that the asymptotic MSE for the oracle estimator of Chapter 4.4 equals $p_0\sigma^2$.

Exercise 4.10. List all the assumptions used to show that $\sigma^2 C_P / n$ is an unbiased estimator for MSE (assume that σ^2 is known). Briefly discuss the cons of each assumption.

Exercise 4.11. Prove (4.14).

Exercise 4.12. Prove (4.15).

Exercise 4.13. As in Chapter 4.6.1, assume a simple slope-intercept model: $Y_i = \alpha + x_i\beta + \varepsilon_i$, for $i = 1, \ldots, n$, where ε_i are i.i.d. $\mathrm{N}(0, \sigma^2)$ errors and $\overline{x} = 0$ and $\sum_{i=1}^n x_i^2 = n$. Let $\hat{Y}(x_0) = \hat{\alpha} + \hat{\beta} x_0 1_{\{\hat{\delta} = \mathcal{M}_1\}}$ be the predictor for Y using the model selection procedure $\hat{\delta}$ and let $Y(x_0) = \alpha + \beta x_0$ be the true value of Y at x_0.

In this exercise, we will use simulated data to study the risk of a procedure $\hat{\delta}$ where the risk at x_0 is defined as $R(\hat{\delta}, x_0) = \mathbb{E}(Y(x_0) - \hat{Y}(x))^2$.

1. Simulate data using an equally spaced design for x between $[-1, 1]$. Calculate the risk functions for AIC, BIC and hypothesis testing for $x_0 = 0.5$ for sample sizes $n = 25, 100, 200, 1000$ with $\sigma = .5$. Repeat your simulations for different values of β.
2. Display your results as 4 graphs, one for each sample size, where the horizontal axis is the value for β and the vertical axis is the risk.
3. Discuss your results. In particular, how do the various procedures perform as a function of n and β? Which procedure is best?

Exercise 4.14. In this exercise we will prove classical results about the OLS. We assume a regression model as in (4.2) where $\mathbb{E}(\boldsymbol{\epsilon}) = \mathbf{0}$ and $\mathbb{E}(\boldsymbol{\epsilon}\boldsymbol{\epsilon}^T) = \sigma^2 \mathbf{I}_n$.

1. Assuming $\mathbf{D}$ has full rank, give another proof that $\hat{\boldsymbol{\beta}} = (\mathbf{D}^T\mathbf{D})^{-1}\mathbf{D}^T\mathbf{Y}$.
2. Show that $\mathrm{Var}(\hat{\boldsymbol{\beta}}) = \sigma^2(\mathbf{D}^T\mathbf{D})^{-1}$.
3. Let $L_1 = (\hat{\boldsymbol{\beta}} - \boldsymbol{\beta})^T(\hat{\boldsymbol{\beta}} - \boldsymbol{\beta})$ be the distance from $\hat{\boldsymbol{\beta}}$ to $\boldsymbol{\beta}$. Show that

$$\mathbb{E}(L_1^2) = \sigma^2 \mathrm{tr}\left(\mathbf{D}^T\mathbf{D}\right)^{-1}.$$

4. From (2) show that $\mathbb{E}(\hat{\boldsymbol{\beta}}^T\hat{\boldsymbol{\beta}}) = \boldsymbol{\beta}^T\boldsymbol{\beta} + \sigma^2\mathrm{tr}\left(\mathbf{D}^T\mathbf{D}\right)^{-1}$ and if $\boldsymbol{\epsilon}$ is normally distributed $\mathrm{Var}(L_1^2) = 2\sigma^4\mathrm{tr}\left(\mathbf{D}^T\mathbf{D}\right)^{-2}$.

5. Let $\lambda_1 \geq \lambda_2 \geq \lambda_p \geq 0$ be the ordered eigenvalues of $\mathbf{D}^T\mathbf{D}$. Show that

$$\mathbb{E}(L_1^2) = \sigma^2 \sum_{k=1}^{p} (1/\lambda_i).$$

6. If $\boldsymbol{\epsilon}$ is normally distributed show that

$$\text{Var}(L_1^2) = 2\sigma^4 \sum_{k=1}^{p} (1/\lambda_i)^2.$$

Exercise 4.15. (Continuation of Exercise 4.14). The OLS can become highly unstable due to collinearity. In this exercise we show that the ridge estimator (4.19) stabilizes the OLS. For notational convenience, let $\hat{\boldsymbol{\beta}}^*$ denote the ridge estimator $\hat{\boldsymbol{\beta}}_{\text{ridge}}$ and let $k \geq 0$ denote the ridge value.

1. Show that $\hat{\boldsymbol{\beta}}^* = (\mathbf{D}^T\mathbf{D} + k\mathbf{I}_p)^{-1}\mathbf{D}^T\mathbf{Y} := \mathbf{W}\mathbf{D}^T\mathbf{Y}$.
2. Show that $\hat{\boldsymbol{\beta}}^* = \left(\mathbf{I}_p + k(\mathbf{D}^T\mathbf{D})^{-1}\right)^{-1}\hat{\boldsymbol{\beta}} := \mathbf{Z}\hat{\boldsymbol{\beta}}$.
3. Let $\xi_i(\mathbf{W})$ and $\xi_i(\mathbf{Z})$ be the eigenvalues of $\mathbf{W}$ and $\mathbf{Z}$. Then

$$\xi_i(\mathbf{W}) = 1/(\lambda_i + k), \quad \xi_i(\mathbf{Z}) = \lambda_i/(\lambda_i + k).$$

4. $\mathbf{Z} = \mathbf{I}_p - k(\mathbf{D}^T\mathbf{D} + k\mathbf{I}_p)^{-1} = \mathbf{I}_p - k\mathbf{W}$.
5. If $k > 0$, then $\hat{\boldsymbol{\beta}}^*$ is shorter than $\hat{\boldsymbol{\beta}}$:

$$(\hat{\boldsymbol{\beta}}^*)^T(\hat{\boldsymbol{\beta}}^*) < \hat{\boldsymbol{\beta}}^T\hat{\boldsymbol{\beta}}.$$

6. The RSS for $\hat{\boldsymbol{\beta}}^*$ equals

$$||\mathbf{Y} - \mathbf{D}\hat{\boldsymbol{\beta}}^*||^2 = \mathbf{Y}^T\mathbf{Y} - (\hat{\boldsymbol{\beta}}^*)^T\mathbf{D}^T\mathbf{Y} - k(\hat{\boldsymbol{\beta}}^*)^T(\hat{\boldsymbol{\beta}}^*).$$

   Interpret $||\mathbf{Y} - \mathbf{D}\hat{\boldsymbol{\beta}}^*||^2$ in terms of $||\mathbf{Y} - \mathbf{D}\hat{\boldsymbol{\beta}}||^2$.

Exercise 4.16. Consider the model (3.1)-(3.3) described in Chapter 3.3.

1. Show that $\text{Cor}(Y,\mathbf{X})^T = \boldsymbol{\Sigma\beta}/\text{Var}(Y)^{1/2}$.

2. Notice that $\mathbb{E}(Y|\mathbf{X})$ does not depend on X_4 since the coefficient for X_4 equals 0, however, the above calculation shows that $\text{Cor}(Y, X_4)$ is non-zero. What is the theoretical value from your calculation? Explain what this result implies in practice when using linear regression.

3. Simulate data from the regression model described above for $\rho = .9$ and $\sigma^2 = 0.5$ multiple times to obtain an empirical estimate for $\text{Cor}(Y,\mathbf{X})$ and compare it to the theoretical value derived above.

Exercise 4.17. Verify equation (4.20).

Exercise 4.18. Prove that (4.21) holds under the stated conditions.

Exercise 4.19. Prove that (4.23) holds under the stated conditions.

Exercise 4.20. Consider a simple regression problem with one variable. Let

$$\hat{\beta}(\lambda) = \underset{\beta \in \mathbb{R}}{\arg\min}\left\{\sum_{i=1}^{n}(Y_i - x_i\beta)^2 + \lambda|\beta|\right\}$$

be the lasso solution where $\lambda \geq 0$. Let $\hat{\beta}$ be the OLS solution (i.e., $\hat{\beta} = \hat{\beta}(0)$). Reparameterize λ according to $\lambda \leftarrow \lambda/(2\sum_{i=1}^n x_i^2)$. Show that

$$\hat{\beta}(\lambda) = \begin{cases} \hat{\beta} - \lambda & \text{if } \hat{\beta} > 0 \text{ and } \lambda < |\hat{\beta}| \\ \hat{\beta} + \lambda & \text{if } \hat{\beta} < 0 \text{ and } \lambda < |\hat{\beta}| \\ 0 & \lambda \geq |\hat{\beta}|. \end{cases}$$

Note that the above can also be written as

$$\hat{\beta}(\lambda) = \text{sign}(\hat{\beta})(\hat{\beta} - \lambda)_+$$

which shows that the lasso solution is a soft-thresholded function of the OLS.

Exercise 4.21. Using (4.25) construct a scenario that explicitly shows the lasso does not possess the oracle property.

Exercise 4.22. The following example is designed to further illustrate the lack of oracle property for the lasso. Consider the quadratic regression model

$$Y_i = \beta_0 + \beta_1 x_i + \beta_2 x_i^2 + \varepsilon_i, \quad \text{where } i = 1, \ldots, n.$$

Here ε_i are i.i.d. $\text{N}(0, \sigma^2)$ and the covariates X_i are i.i.d. sampled values from a $\text{U}[0,1]$. Standardize the covariates as follows. Let

$$\overline{x}^{(1)} = \frac{1}{n}\sum_{i=1}^n x_i, \quad \overline{x}^{(2)} = \frac{1}{n}\sum_{i=1}^n x_i^2,$$

and

$$\sigma_x^{(1)} = \left(\frac{1}{n}\sum_{i=1}^n (x_i - \overline{x}^{(1)})^2\right)^{1/2}, \quad \sigma_x^{(2)} = \left(\frac{1}{n}\sum_{i=1}^n (x_i^2 - \overline{x}^{(2)})^2\right)^{1/2}.$$

Define the standardized quadratic regression model

$$\begin{aligned} Y_i &= \beta_0 + \beta_1 \left(\frac{x_i - \overline{x}^{(1)}}{\sigma_x^{(1)}}\right) + \beta_2 \left(\frac{x_i - \overline{x}^{(2)}}{\sigma_x^{(2)}}\right) + \varepsilon_i \\ &= \beta_0 + \beta_1 z_{i1} + \beta_2 z_{i2} + \varepsilon_i. \end{aligned}$$

Calculate the lasso estimator for (β_1, β_2) using simulated data where in your simulations set $\beta_2 = 0$. Repeat this experiment independently $B > 1$ times.

1. Use the B simulations to approximate the asymptotic bias and asymptotic variance for the lasso estimated (β_1, β_2) values. Do this for several different lasso parameter values.
2. Using the same simulations as above, calculate the full-model OLS and report its estimated asymptotic variance and bias.
3. Using the B simulations, plot the values for the lasso estimator. Do this for several different lasso parameter values.
4. Using the B simulations, plot the values for the OLS estimator.
5. Comment on the difference between the OLS and lasso. What are your conclusions regarding the oracle property?

Exercise 4.23. The lasso can be expected to be a biased estimator in most situations due to the L_1-regularization it applies to the regression coefficients. This regularization introduces bias in the estimates to achieve variance reduction and promote sparsity. Using correlated features, design a simulation that shows that the lasso produces biased estimates for the β coefficients. Does this result still hold if the features are uncorrelated?

Exercise 4.24. Conduct a simulation to test the performance of bridge estimators from the sparse family $0 < \gamma < 1$. Test different values of γ and compare this to lasso ($\gamma = 1$). What are your conclusions? Test both the correlated and uncorrelated feature settings.

Exercise 4.25. Repeat the previous exercise but now study the effect of increasing the dimension, p.

Bibliography

[1] Akaike H. Statistical predictor identification. *Annals of the Institute of Statistical Mathematics*, 22:203–217, 1970.

[2] Akaike H. Information theory and an extension of the maximum likelihood principle. In *The Second International Symposium on Information Theory*, pp. 267-281. Budapest:Akademia Kiado, 1973.

[3] Chatterjee S. Assumptionless consistency of the lasso. arXiv preprint arXiv:1303.5817, 2013.

[4] Chen S.S., Donoho D.L., and Saunders M A. Atomic decomposition by basis pursuit. *SIAM Journal on Scientific Computing*, 20:33–61, 1998.

[5] Donoho D.L. For most large underdetermined systems of linear equations the minimal ℓ_1-norm solution is also the sparsest solution. *Communications on Pure and Applied Mathematics*, 59(6):797–829, 2006.

[6] Efron B., Hastie T., Johnstone I., and Tibshirani R. Least angle regression (with discussion). *Annals of Statistics*, 32:407–499, 2004.

[7] Fan J. and Li R. Variable selection via nonconcave penalized likelihood and its oracle properties. *Journal of the American statistical Association*, 96:1348–1360, 2001.

[8] Hannan E. J. and Quinn B. G. The determination of the order of an autoregression. *Journal of the Royal Statistical Society*, 41:190–195, 1979.

[9] Hoeting .J.A., Madigan D, Raftery .A.E. and Volinsky C.T. Bayesian model averaging: a tutorial. *Statistical Science*, 1:382–401, 1999.

[10] Hoerl A.E. and Kennard R.W. Ridge regression: Biased estimation for nonorthogonal problems. *Technometrics*, 12:55–67, 1970.

[11] Knight K. and Fu W. Asymptotics for lasso-type estimators. *Annals of Statistics*, (28):1356–1378, 2000.

[12] Mallows C. L. Some comments on C_P. *Technometrics*, 15:661–675, 1973.

[13] Nishii R. Asymptotic properties of criteria for selection of variables in multiple regression. *Annals of Statistics*, 12:758–765, 1984.

[14] Osborne M.R., Presnell B. and Turlach B.A. On the lasso and its dual. *Journal of Computational and Graphical Statistics*, 9(2):319–337, 2000.

[15] Osborne M.R., Presnell B. and Turlach B.A. A new approach to variable selection in least squares problems. *Journal of Numerical Analysis*, 20(3), 389–404, 2000.

[16] Schwarz G. Estimating the dimension of a model. *Annals of Statistics*, 6:461–464, 1978.

[17] Shibata R. Asymptotic mean efficiency of a selection of regression variables. *Annals of the Institute of Statistical Mathematics*, 35:415–423, 1983.

[18] Shibata R. Approximate efficiency of a selection procedure for the number of regression variables. *Biometrika*, 71:43–49, 1984.

[19] Tibshirani R. Regression shrinkage and selection via the lasso. *Journal of the Royal Statistical Society B*, 58:267–288, 1996.

[20] Tibshirani R.J. The lasso problem and uniqueness. Electronic Journal of Statistics, (7) 1456–1490, 2013.

[21] Yang Y. Can the strengths of AIC and BIC be shared? A conflict between model indentification and regression estimation. *Biometrika*, 92:937–950, 2005.

[22] Yang Y. Prediction/estimation with simple linear models: Is it really that simple? *Econometric Theory*, 23(1), 1–36, 2007.

5

Multivariate Regression

5.1 Introduction

Multivariate linear regression extends the concept of simple linear regression, which involves only one dependent variable, to accommodate a q-dimensional response $\mathbf{Y}_i = (Y_{i1}, \ldots, Y_{iq})^T$ that is assumed to be linearly related to covariates $\mathbf{x}_i \in \mathbb{R}^p$ for $i = 1, \ldots, n$. Let $\mathbf{Y}_{n\times q}$ be the response matrix with $\mathbf{Y}_i^T$ as its ith row, and let $\mathbf{D}_{n\times p}$ be the fixed covariate design matrix with $\mathbf{x}_i^T$ as its ith row. The multivariate regression model assumes that

$$\mathbf{Y} = \mathbf{DB} + \mathbf{U} \tag{5.1}$$

where $\mathbf{B}_{p\times q}$ is the unknown regression coefficient matrix and $\mathbf{U}_{n\times q}$ is a measurement error matrix assumed to satisfy

$$\mathbb{E}(\mathbf{U}) = \mathbf{0}_{n\times q}, \quad \mathbb{E}(\mathbf{u}_i\mathbf{u}_i^T) = \boldsymbol{\Sigma}_{q\times q}, \quad \mathbb{E}(\mathbf{u}_i\mathbf{u}_j^T) = \mathbf{0}_{q\times q}, \quad i \neq j,$$

where $\mathbf{u}_i^T$ is the ith row of $\mathbf{U}$. The conditions on $\mathbf{U}$ ensure that

$$\mathbb{E}(\mathbf{Y}|\mathbf{D}) = \mathbf{DB}, \quad \mathrm{Var}(\mathbf{Y}_i|\mathbf{D}) = \boldsymbol{\Sigma}, \quad \mathrm{Cov}(\mathbf{Y}_i, \mathbf{Y}_j|\mathbf{D}) = \mathbf{0}_{q\times q}, \quad i \neq j.$$

Notice that when $q = 1$ we obtain the linear regression model (4.1) of Chapter 4.

An important application of multivariate regression is MANOVA, which generalizes the concept of ANOVA to multiple dependent variables, testing for differences in mean vectors across groups defined by one or more categorical independent variables. This simultaneous testing of multiple outcomes helps in understanding the joint effects of treatments or conditions, reducing the risk of Type-I errors associated with multiple separate tests.

This chapter introduces multivariate linear regression and discusses MANOVA, its various tests, and provides several illustrative examples.

5.2 Least Squares

Similar to the univariate case we define the least squares estimator as the coefficient solution minimizing the residual sums of squares. For the multivariate model (5.1), the least squares estimator is defined as

$$\hat{\mathbf{B}} = \arg\min_{\mathbf{B}\in\mathbb{R}^{p\times q}} \left\{ \mathrm{tr}\left[(\mathbf{Y} - \mathbf{DB})^T (\mathbf{Y} - \mathbf{DB}) \right] \right\}. \tag{5.2}$$

The solution to (5.2) can be determined directly from the univariate case. We present this solution using the same geometric approach as before. For the following, let $\mathbf{Y}_{(j)} \in \mathbb{R}^n$ and $\mathbf{B}_{(j)} \in \mathbb{R}^p$ denote the jth column of $\mathbf{Y}$ and $\mathbf{B}$ respectively.

DOI: 10.1201/9781003493679-5

Theorem 5.1. *Let $\{\boldsymbol{v}_1, \dots, \boldsymbol{v}_n\}$ be an orthonormal basis spanning $\mathbb{R}^n$ such that the first r vectors $\boldsymbol{v}_1, \dots, \boldsymbol{v}_r$ spans the column space of $\boldsymbol{D}$ where $1 \le r \le p$ is the rank of $\boldsymbol{D}$. Let $\hat{\boldsymbol{\mu}}_{(j)} = \sum_{i=1}^{r} Z_{ij}\boldsymbol{v}_i$, where $Z_{ij} = \boldsymbol{Y}_{(j)}^T \boldsymbol{v}_i$. Then, for $1 \le j \le q$,*

(i) $\boldsymbol{D}\hat{\boldsymbol{B}}_{(j)} = \hat{\boldsymbol{\mu}}_{(j)}$.

(ii) If $r = p$, *then* $\hat{\boldsymbol{B}}_{(j)} = (\boldsymbol{D}^T\boldsymbol{D})^{-1}\boldsymbol{D}^T\boldsymbol{Y}_{(j)} = (\boldsymbol{D}^T\boldsymbol{D})^{-1}\boldsymbol{D}^T\hat{\boldsymbol{\mu}}_{(j)}$.

Thus, $\hat{\boldsymbol{B}}_{(j)}$ is the least-squares solution obtained from regressing $\boldsymbol{Y}_{(j)} = \sum_{i=1}^{n} Z_{ij}\boldsymbol{v}_i$ (the jth column of $\boldsymbol{Y}$) onto $\boldsymbol{D}$. Furthermore, when $r = p$,

$$\hat{\boldsymbol{B}} = [\hat{\boldsymbol{B}}_{(1)}, \dots, \hat{\boldsymbol{B}}_{(q)}] = (\boldsymbol{D}^T\boldsymbol{D})^{-1}\boldsymbol{D}^T\boldsymbol{Y} = (\boldsymbol{D}^T\boldsymbol{D})^{-1}\boldsymbol{D}^T[\hat{\boldsymbol{\mu}}_{(1)}, \dots, \hat{\boldsymbol{\mu}}_{(q)}].$$

Proof of Theorem 5.1. By the definition of the trace function

$$\mathrm{tr}\Big[(\mathbf{Y}-\mathbf{DB})^T(\mathbf{Y}-\mathbf{DB})\Big] = \sum_{j=1}^{q}(\mathbf{Y}_{(j)} - \mathbf{DB}_{(j)})^T(\mathbf{Y}_{(j)} - \mathbf{DB}_{(j)}).$$

Therefore, the solution to (5.2) can be determined by minimizing each of the terms in the above sum separately, each of which is a univariate least squares problem with response $\mathbf{Y}_{(j)}$ and design matrix $\mathbf{D}$. The result follows by applying Theorem 4.1 to each term separately. □

Remark 5.1. An equivalent way to formulate the least squares problem is by the optimization

$$\hat{\mathbf{B}} = \arg\min_{\mathbf{B}\in\mathbb{R}^{p\times q}}\Big\{\mathrm{tr}\Big[(\mathbf{Y}-\mathbf{DB})^T\mathbf{\Sigma}^{-1}(\mathbf{Y}-\mathbf{DB})\Big]\Big\}$$

which is similar to (5.2) but uses the inverse covariance matrix of the measurement error to scale the quadratic optimization function. This alternate optimization arises from maximum likelihood estimation (MLE) under the assumption of normality. A consequence is that the MLE for $\mathbf{B}$ and the least-squares solution coincide when the data is assumed to be normal (see Exercise 5.1). □

Remark 5.2. The same calculations as above show that that the MLE for $\mathbf{\Sigma}$ is (Exercise 5.3)

$$\hat{\mathbf{\Sigma}} = \frac{1}{n}\big(\mathbf{Y}-\mathbf{D}\hat{\mathbf{B}}\big)^T\big(\mathbf{Y}-\mathbf{D}\hat{\mathbf{B}}\big).$$

Likewise, one can obtain the restricted MLE for $\mathbf{\Sigma}$ under an imposed constraint to $\mathbf{B}$. These estimators for $\mathbf{\Sigma}$ play an important role in likelihood ratio MANOVA tests for $\mathbf{B}$ to be discussed. □

Remark 5.3. The least squares solution for multivariate linear regression has a geometric interpretation similar to linear regression in Chapter 4. Let $\hat{\mathbf{Y}} = \mathbf{D}\hat{\mathbf{B}}$ be the least squares predictor and $\hat{\mathbf{U}} = \mathbf{Y} - \hat{\mathbf{Y}}$ the residual matrix. By Theorem 5.1, the columns $\hat{\mathbf{Y}}_{(j)} = \hat{\boldsymbol{\mu}}_{(j)}$ for the predictor matrix are orthogonal to the columns $\hat{\mathbf{U}}_{(j)} = \mathbf{Y}_{(j)} - \hat{\mathbf{Y}}_{(j)}$ for the residual matrix, for $1 \le j \le p$. Thus, by orthogonality,

$$\hat{\mathbf{U}}^T\hat{\mathbf{U}} = \mathbf{Y}^T\mathbf{Y} - \hat{\mathbf{Y}}^T\hat{\mathbf{Y}}.$$

The left-hand side is called the residual sum of squares and cross-product (SSCP) matrix. This shows that the total SSCP matrix $\mathbf{Y}^T\mathbf{Y}$ can be decomposed into the fitted SSCP matrix $\hat{\mathbf{Y}}^T\hat{\mathbf{Y}}$ and the residual SSCP matrix $\hat{\mathbf{U}}^T\hat{\mathbf{U}}$.

This orthogonal decomposition is the basis for MANOVA testing. Under normality, $\hat{\mathbf{U}}^T\hat{\mathbf{U}}$ has a Wishart distribution (Appendix 3.G). This can be used to test the null hypothesis that $\mathbf{B} = \mathbf{0}$, since under the null, and assuming normality, $\hat{\mathbf{Y}}^T\hat{\mathbf{Y}}$ has a Wishart distribution (not depending on $\mathbf{B}$) independent of $\hat{\mathbf{U}}^T\hat{\mathbf{U}}$. More general MANOVA hypothesis tests for $\mathbf{B}$ (e.g., using contrasts) can be described using this principle. □

5.3 Likelihood Ratio Tests

Likelihood ratio tests can be used to test specific hypotheses about $\mathbf{B}$. Suppose that $\mathbf{B}$ is partitioned into two matrices $\mathbf{B}_1$ and $\mathbf{B}_2$ of dimension $p_1 \times q$ and $p_2 \times q$ where $p = p_1 + p_2$

$$\mathbf{B} = \begin{pmatrix} \mathbf{B}_1 \\ \mathbf{B}_2 \end{pmatrix}$$

and we wish to test the null hypothesis $H_0 : \mathbf{B}_2 = \mathbf{0}$.

To test the null, partition $\mathbf{D}$ so that $\mathbf{D} = [\mathbf{D}_1, \mathbf{D}_2]$, where $\mathbf{D}_1$ correspond to the first p_1 columns of $\mathbf{D}$, and $\mathbf{D}_2$ the remaining p_2 columns. Let $L(\mathbf{B}, \boldsymbol{\Sigma})$ denote the log-likelihood under the assumption that $\mathbf{u}_i$ are i.i.d. $\mathrm{N}_q(\mathbf{0}, \boldsymbol{\Sigma})$. The likelihood ratio test (LRT) for H_0 is

$$\lambda = \frac{\sup_{(\mathbf{B},\boldsymbol{\Sigma})\in H_0} L(\mathbf{B}, \boldsymbol{\Sigma})}{\sup_{(\mathbf{B},\boldsymbol{\Sigma})} L(\mathbf{B}, \boldsymbol{\Sigma})} = \frac{L(\hat{\mathbf{B}}_1, \hat{\boldsymbol{\Sigma}}_1)}{L(\hat{\mathbf{B}}, \hat{\boldsymbol{\Sigma}})} = \left(\frac{|\hat{\boldsymbol{\Sigma}}|}{|\hat{\boldsymbol{\Sigma}}_1|}\right)^{n/2} \tag{5.3}$$

where

$$\hat{\boldsymbol{\Sigma}}_1 = \frac{1}{n}\left(\mathbf{Y} - \mathbf{D}_1\hat{\mathbf{B}}_1\right)^T\left(\mathbf{Y} - \mathbf{D}_1\hat{\mathbf{B}}_1\right), \quad \hat{\boldsymbol{\Sigma}} = \frac{1}{n}\left(\mathbf{Y} - \mathbf{D}\hat{\mathbf{B}}\right)^T\left(\mathbf{Y} - \mathbf{D}\hat{\mathbf{B}}\right).$$

We can assess the significance of λ using Wilks' lambda (Appendix 3.G), defined as

$$\Lambda = \lambda^{2/n} = \frac{|\hat{\boldsymbol{\Sigma}}|}{|\hat{\boldsymbol{\Sigma}}_1|} \sim \Lambda(q, n-p, p_2).$$

Asymptotically the LRT satisfies

$$-n \log \Lambda \asymp \chi^2_{qp_2}$$

which is another way to determine the level of significance. Bartlett's approximation may also be used

$$-\left[n - p - \frac{1}{2}(q - p_2 + 1)\right] \log \Lambda \asymp \chi^2_{qp_2}$$

and often works better.

5.3.1 General hypotheses

More generally, the LRT can be used to test hypotheses of the form

$$H_0 : \mathbf{CBM} = \mathbf{0} \tag{5.4}$$

where $\mathbf{C}_{g\times p}$ and $\mathbf{M}_{q\times r}$ are of rank $g \le p$ and $r \le q$. Here $\mathbf{C}$ is called the *contrast matrix* and $\mathbf{M}$ is the *M-matrix* (M is for MANOVA). For example, to test $\mathbf{B}_2 = \mathbf{0}_{p_2\times q}$, where $p_1 + p_2 = p$,

$$\mathbf{C} = [\mathbf{0}_{p_2\times p_1}, \mathbf{I}_{p_2\times p_2}] \quad \text{and} \quad \mathbf{M} = \mathbf{I}_{q\times q}.$$

We will assume that $\mathbf{D}$ has full rank $p \le n$ so that $(\mathbf{D}^T\mathbf{D})^{-1}$ exists. The sum of squares and cross-product (SSCP) matrix due to the regression under the null (5.4) is

$$\mathbf{H} = (\mathbf{C}\hat{\mathbf{B}}\mathbf{M})^T\left[\mathbf{C}\left(\mathbf{D}^T\mathbf{D}\right)^{-1}\mathbf{C}^T\right]^{-1}\mathbf{C}\hat{\mathbf{B}}\mathbf{M}.$$

The residual (error) SSCP matrix is

$$\mathbf{E} = (\mathbf{YM})^T\left[\mathbf{I} - \mathbf{D}\left(\mathbf{D}^T\mathbf{D}\right)^{-1}\mathbf{D}^T\right]^{-1}\mathbf{YM}.$$

The LRT for the hypothesis **CBM** = **0** gives Wilks' lambda

$$\Lambda = \lambda^{2/n} = \frac{|\mathbf{E}|}{|\mathbf{H}+\mathbf{E}|} \sim \Lambda(r, n-p, g).$$

There are several related tests based on the **H** and **E** SSCP matrices. Table 5.1 lists some commonly used tests (including Wilks' lambda for completeness).

TABLE 5.1
Commonly used tests based on the ***H*** *and* ***E*** *SSCP matrices.*

Test	Formula
Wilks' lambda	$\frac{\lvert\mathbf{E}\rvert}{\lvert\mathbf{H}+\mathbf{E}\rvert}$
Hotelling-Lawley trace	$\text{tr}(\mathbf{HE}^{-1})$
Pillai trace	Sum of the eigenvalues (or trace) of $\mathbf{H}(\mathbf{H}+\mathbf{E})^{-1}$
Roy's largest root	Largest eigenvalue of $\mathbf{HE}^{-1}$

5.3.2 Dose-response example

Dose-response data frequently appear in the physical sciences and are often analyzed by using linear regression models. Here we consider a multivariate dose-response example taken from the book of Srivastava [1].

In this experiment, twenty-five tanks, each containing fish of a certain species, received varying dosages of copper (mg/liter). The average weight of the fish was measured. Additionally, the proportions (p_{ij}, $j = 1,\ldots,5$) of fish that died after 8, 14, 24, 36, and 48 hours were recorded. The response variables p_{ij} were then transformed using an arcsine transformation to stabilize their variance

$$Y_{ij} = \arcsin\sqrt{p_{ij}}, \quad i = 1,\ldots,25, \; j = 1,\ldots,5.$$

Figure 5.1 shows the transformed responses for each time point, plotted against log-dosage and average weight. The plot reveals a strong relationship between log-dosage and fish survival, but there appears to be no significant relationship between survival and average weight.

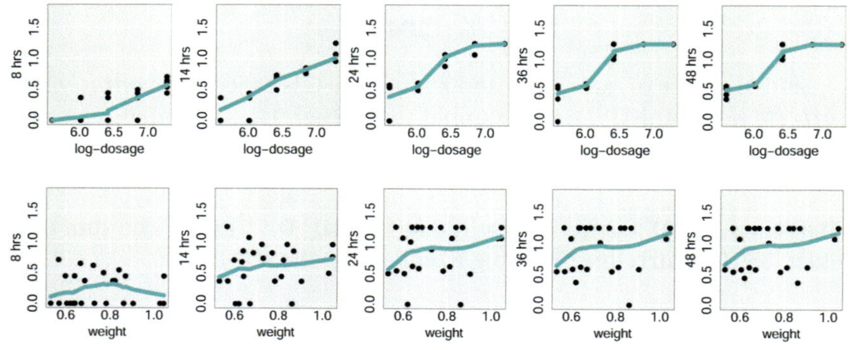

FIGURE 5.1
Proportion (arcsin *transformed) of species dying as a function of log-dosage and weight, stratified by time.*

There are two variables, log-dosage and weight, and including an overall intercept, we have three parameters in total. Hence $n = 25$, $p = 3$ and $q = 5$. The response vector for tank i is $\mathbf{Y}_i = (Y_{i1}, \ldots, Y_{i5})^T$ and the design matrix and coefficient matrix are

$$\mathbf{D} = \begin{bmatrix} 1 & d_1 & w_1 \\ \vdots & \vdots & \vdots \\ 1 & d_{25} & w_{25} \end{bmatrix}, \quad \mathbf{B} = \begin{bmatrix} \beta_{01} & \beta_{02} & \beta_{03} & \beta_{04} & \beta_{05} \\ \beta_{11} & \beta_{12} & \beta_{13} & \beta_{14} & \beta_{15} \\ \beta_{2,1} & \beta_{2,2} & \beta_{2,3} & \beta_{2,4} & \beta_{2,5} \end{bmatrix}$$

where d_i and w_i are the log-dosage of the treatment and the average species weight in tank i. The first row of **B** are coefficient parameters for the intercept term, the second row are parameters for log-dosage, and the third row are parameters for weight.

Before testing for **B**, we first assess the multivariate normality of the data by examining the normality of the residuals using the methods described in Chapter 3.6. The results, shown in Figure 5.2, indicate that the data is reasonably normally distributed, although two outliers are present. These outliers can be seen in the extreme right tail of the first panel and in the upper right corner of the Q-Q plot in the second panel. Consequently, we removed these two outliers for our subsequent analyses, resulting in a sample size of $n = 23$.

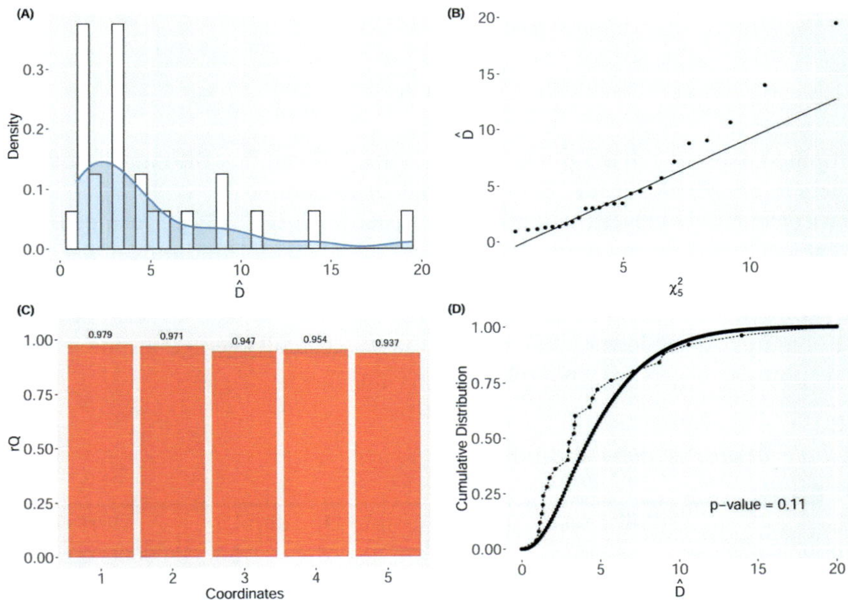

FIGURE 5.2
Assessing multivariate normality for residuals from dose-response data: (A) Density plot for Mahalanobis distance squared, $\hat{D}$; (B) Q-Q plot for $\hat{D}$; (C) rQ correlations; (D) Estimated distribution and theoretical distribution for $\hat{D}$.

We begin by testing whether there is a dosage effect. This corresponds to the hypothesis $H_0: \beta_{1j} = 0,\ j = 1,\ldots,5$. To calculate the LRT we must determine the MLE under the null, which corresponds to regressing **Y** on the design matrix $\mathbf{D}_1$ with coefficient matrix $\mathbf{B}_1$

$$\mathbf{D}_1 = \begin{bmatrix} 1 & w_1 \\ \vdots & \vdots \\ 1 & w_{25} \end{bmatrix}, \quad \mathbf{B}_1 = \begin{bmatrix} \beta_{01} & \beta_{02} & \beta_{03} & \beta_{04} & \beta_{05} \\ \beta_{2,1} & \beta_{2,2} & \beta_{2,3} & \beta_{2,4} & \beta_{2,5} \end{bmatrix}.$$

Direct calculations give

$$\hat{\mathbf{B}} = \begin{bmatrix} -1.97 & -2.59 & -2.32 & -2.23 & -2.19 \\ 0.39 & 0.52 & 0.50 & 0.48 & 0.47 \\ -0.42 & -0.18 & 0.06 & 0.13 & 0.13 \end{bmatrix}$$

$$\hat{\mathbf{B}}_1 = \begin{bmatrix} -2.10 & -2.65 & -2.30 & -2.19 & -2.15 \\ 0.36 & 0.51 & 0.50 & 0.49 & 0.48 \end{bmatrix}$$

yielding

$$\Lambda = \lambda^{2/n} = \frac{|(\mathbf{Y}-\mathbf{D}\hat{\mathbf{B}})^T(\mathbf{Y}-\mathbf{D}\hat{\mathbf{B}})|}{|(\mathbf{Y}-\mathbf{D}\hat{\mathbf{B}}_1)^T(\mathbf{Y}-\mathbf{D}\hat{\mathbf{B}}_1)|} = \frac{6.081 \times 10^{-6}}{5.929 \times 10^{-5}} = 0.10256.$$

Under normality, this has a $\Lambda(q, n-p, p_2)$ distribution $(n = 23, q = 5, p = 3, p_2 = 1)$. Using Bartlett's approximation, we obtain

$$-\left[n - p - \frac{1}{2}(q - p_2 + 1)\right] \log\Lambda = -17.5 \times \log(0.10256) = 39.85$$

which compared to a χ^2_5 yields a p-value of 1.6×10^{-7}. Under the assumption of normality we can determine the significance using an F-statistic (Appendix 3.G)

$$\frac{n-p-q+1}{q} \times \frac{1-\Lambda}{\Lambda} \sim F_{q,n-p-q+1}.$$

This yields

$$\frac{16}{5} \times \frac{1-.10256}{.10256} = 28.00125$$

which compared to an $F_{5,16}$ has a p-value of 2.24×10^{-7}. Given that the data is close to normality (with outliers removed), the latter calculation may be preferable.

While we calculated the LRT manually, this calculation is typically performed using statistical software. Table 5.2 illustrates a typical analysis conducted with standard software. The first line entry indicates a strong effect of log-dosage, which aligns with our manual analysis. The second line entry shows that weight has no significant effect. This finding is consistent with the results depicted in Figure 5.1: the lower panel reveals that the overall average response increases with time, but responses are generally not related to weight.

TABLE 5.2
MANOVA table for dose-response example.

	Wilks' test	approx F	num Df	den Df	Pr(>F)
log-dosage	0.10256	28.001	5	16	0.0000002243
weight	0.80543	0.773	5	16	0.583

We can test specific hypotheses for **B** using (5.4). For example, from Figure 5.2, it appears that the slope for log-dosage for time points $j = 3,4,5$ are almost identical. We can test this using the null hypothesis $\beta_{13} = \beta_{14} = \beta_{15}$, which can be written in the form (5.4) with

$$\mathbf{C} = [0,1,0] \text{ and } \mathbf{M} = \begin{bmatrix} 0 & 0 \\ 0 & 0 \\ -1 & 0 \\ 1 & -1 \\ 0 & 1 \end{bmatrix}, \quad g = \text{rank}(\mathbf{C}) = 1, r = \text{rank}(\mathbf{M}) = 2.$$

This corresponds to testing $\beta_{13} = \beta_{14}$ and $\beta_{14} = \beta_{15}$, which is equivalent to testing $\beta_{13} = \beta_{14} = \beta_{15}$. The results (Table 5.3) indicate no evidence of a departure from the null, thus suggesting that log-dosage effects are equivalent for time points $j = 3,4,5$.

TABLE 5.3
Results from testing the null $\beta_{13} = \beta_{14} = \beta_{15}$.

	Wilks' test	approx F	num Df	den Df	Pr(>F)
$\beta_{13} = \beta_{14} = \beta_{15}$	0.92332	0.788914	2	19	0.46867

5.4 MANOVA

It is well known that analysis of variance (ANOVA) can be reformulated as a univariate linear regression model. Similarly, multivariate analysis of variance (MANOVA) can be described as a special case of the multivariate regression model (5.1), thus allowing tests for treatment effects and other inferential questions of interest to be run using the same LRT approach discussed in Chapter 5.3.

5.4.1 One-way MANOVA

For simplicity we first focus on the one-way MANOVA model but extensions to two-way models and higher order models follow straightforwardly and will be discussed next. The one-way MANOVA model is written as

$$\mathbf{Y}_{ij} = \boldsymbol{\mu} + \boldsymbol{\tau}_i + \mathbf{u}_{ij}, \quad i = 1, \ldots, G, \ j = 1, \ldots, n_i \tag{5.5}$$

where $\mathbf{Y}_{ij} \in \mathbb{R}^q$ is the q-dimensional response for treatment i in replicate experiment j, $\boldsymbol{\mu} \in \mathbb{R}^q$ is the overall mean and $\boldsymbol{\tau}_i \in \mathbb{R}^q$ is the treatment effect for i. We assume $\mathbf{u}_{ij} \overset{\text{iid}}{\sim} \mathrm{N}_q(\mathbf{0}, \boldsymbol{\Sigma})$. By stacking $\mathbf{Y}_{ij}$ and defining $\mathbf{D}$ and $\mathbf{B}$ appropriately, (5.5) can be written as a multivariate linear model (5.1).

Without additional constraints, the parameters in model (5.5) are unidentified. This is because we could add a term, say $\boldsymbol{\tau}_0$, to $\boldsymbol{\mu}$ yielding a new mean $\boldsymbol{\mu} + \boldsymbol{\tau}_0$. Then by subtracting $\boldsymbol{\tau}_0$ from $\boldsymbol{\tau}_i$, we obtain a different set of treatment parameters $\boldsymbol{\tau}_i - \boldsymbol{\tau}_0$ but without changing the model. There are many ways to enforce identification of (5.5). One is the "sum-to-zero" constraint:

$$\mathbf{Y}_{ij} = \boldsymbol{\mu} + \boldsymbol{\tau}_i + \mathbf{u}_{ij}, \quad \sum_{i=1}^{G} n_i \boldsymbol{\tau}_i = \mathbf{0}.$$

This ensures identification holds because

$$\mathbb{E}(\mathbf{Y}_{..}) = \frac{1}{N}\mathbb{E}\left(\sum_{i=1}^{G}\sum_{j=1}^{n_i} \mathbf{Y}_{ij}\right) = \boldsymbol{\mu} + \frac{1}{N}\sum_{i=1}^{G}\sum_{j=1}^{n_i} \boldsymbol{\tau}_i = \boldsymbol{\mu} + \frac{1}{N}\sum_{i=1}^{G} n_i \boldsymbol{\tau}_i$$

where $N = \sum_i n_i$. Therefore in order for $\boldsymbol{\mu}$ to represent the overall mean, the second term on the right must be zero, which is satisfied by the sum-to-zero condition.

Another parameterization used for identification is to set one of the treatment effects to zero. Typically the first treatment effect is chosen, i.e. $\boldsymbol{\tau}_1 = \mathbf{0}$. This is treated as the baseline condition. Then (5.5) is written as

$$\mathbf{Y}_{ij} = \boldsymbol{\mu} + \sum_{k=2}^{G} \boldsymbol{\tau}_k \mathbf{1}_{\{i=k\}} + \mathbf{u}_{ij}, \quad i = 1, \ldots, G, \ j = 1, \ldots, n_i. \tag{5.6}$$

In this case, $\boldsymbol{\tau}_k$ for $k \geq 2$ represents the treatment effect k relative to the baseline treatment effect.

The null hypothesis of no treatment effect can be written as

$$H_0 : \boldsymbol{\tau}_1 = \cdots = \boldsymbol{\tau}_G = \mathbf{0}.$$

The one-way MANOVA for testing H_0 uses the following decomposition of the total SSCP matrix

$$\underbrace{\sum_{i=1}^{G}\sum_{j=1}^{n_i}\left(\mathbf{Y}_{ij}-\overline{\mathbf{Y}}\right)\left(\mathbf{Y}_{ij}-\overline{\mathbf{Y}}\right)^T}_{\text{total SSCP}} \tag{5.7}$$

$$= \underbrace{\sum_{i=1}^{G} n_i\left(\overline{\mathbf{Y}}_i-\overline{\mathbf{Y}}\right)\left(\overline{\mathbf{Y}}_i-\overline{\mathbf{Y}}\right)^T}_{\text{SSCP Treatment}} + \underbrace{\sum_{i=1}^{G}\sum_{j=1}^{n_i}\left(\mathbf{Y}_{ij}-\overline{\mathbf{Y}}_i\right)\left(\mathbf{Y}_{ij}-\overline{\mathbf{Y}}_i\right)^T}_{\text{SSCP error}}.$$

The first term on the right is the SSCP for the treatment which equals the between sum of squares and corresponds to $\mathbf{H}$ which under the assumption of normality is distributed as Wishart$_q(\mathbf{\Sigma}, g-1)$. The second term is the sum of squares for the error and corresponds to $\mathbf{E}$ and is distributed as Wishart$_q(\mathbf{\Sigma}, N-g)$. Because $\mathbf{H}$ and $\mathbf{E}$ are independent, as well as cancelling a common factor $|\mathbf{\Sigma}|$ in the numerator and denominator, we arrive at the following test statistic for H_0:

$$\frac{|\mathbf{E}|}{|\mathbf{E}+\mathbf{H}|} \sim \Lambda(q, N-g, g-1).$$

From this we obtain Table 5.4.

TABLE 5.4
One-way MANOVA table. The test statistic $|\mathbf{E}|/|\mathbf{E}+\mathbf{H}|$ has distribution $\Lambda(q, N-g, g-1)$.

Source	SSCP	Df	Test
Treatment	$\mathbf{H}$	$g-1$	$\lvert\mathbf{E}\rvert/\lvert\mathbf{E}+\mathbf{H}\rvert$
Error	$\mathbf{E}$	$N-g$	
Total	$\mathbf{E}+\mathbf{H}$	$N-1$	

Example 5.1. The following data consists of treadmill exercise pulmonary function measurements recorded on 25 men and 25 women. For each individual, four variables were measured on oxygen consumption: resting volume (L/min), resting volume (mL/kg/min), maximum oxygen (L/min), maximum oxygen (ml/kg/min). We use a subscript of 1 to denote women and 2 to denote men. We wish to test

$$H_0 : \boldsymbol{\tau}_1 = \boldsymbol{\tau}_2 = \mathbf{0}.$$

The sample means for the two groups are

$$\overline{\mathbf{Y}}_1 = \begin{bmatrix} 0.31 \\ 5.18 \\ 2.32 \\ 38.15 \end{bmatrix}, \quad \overline{\mathbf{Y}}_2 = \begin{bmatrix} 0.40 \\ 5.33 \\ 3.69 \\ 49.42 \end{bmatrix}.$$

The sample variance-covariance matrices are

$$\mathbf{\Sigma}_1 = \begin{bmatrix} 0.01 & 0.15 & 0.00 & 0.03 \\ 0.15 & 2.78 & -0.04 & 1.28 \\ 0.00 & -0.04 & 0.12 & 1.10 \\ 0.03 & 1.28 & 1.10 & 23.26 \end{bmatrix}$$

$$\mathbf{\Sigma}_2 = \begin{bmatrix} 0.01 & 0.07 & 0.03 & 0.15 \\ 0.07 & 1.14 & 0.15 & 3.43 \\ 0.03 & 0.15 & 0.46 & 3.31 \\ 0.15 & 3.43 & 3.31 & 55.25 \end{bmatrix}.$$

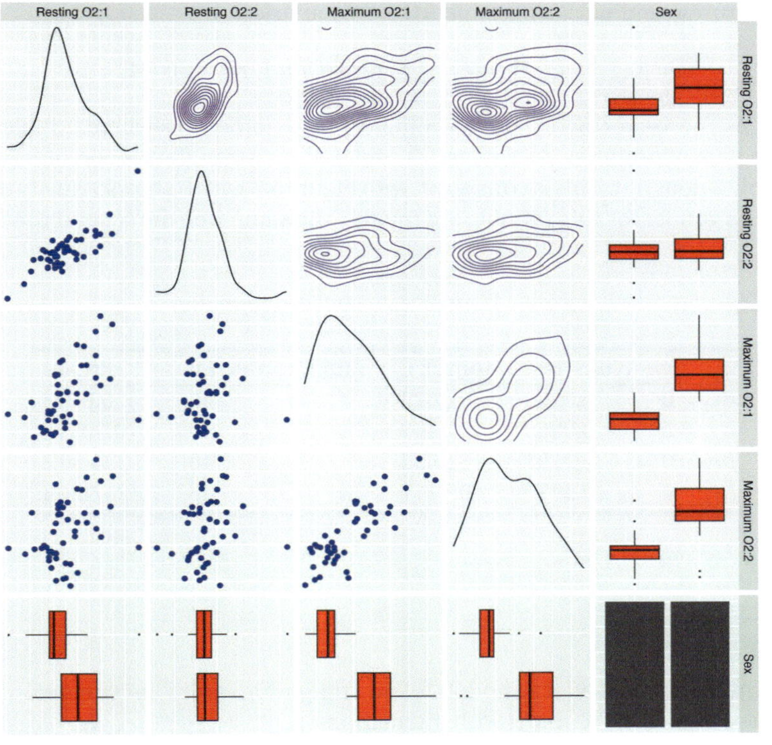

FIGURE 5.3
Exploratory data plot for oxygen consumption data.

The values are quite similar except for the fourth attribute. This suggests some evidence of a departure from the equality of means and variances. Figure 5.3 displays the data and reveals a bimodal pattern for this attribute, specifically for maximum oxygen measurement 2. This bimodal distribution seems to be influenced by a sex effect.

TABLE 5.5
One-way MANOVA for gender effect from oxygen consumption data.

	test	approx F	num Df	den Df	Pr(>F)
Wilks	0.3324	22.587	4	45	2.7805e-10

The results from fitting a one-way MANOVA are given in Table 5.5. There is strong evidence that the null hypothesis is false and we can conclude that there is a gender effect.

The previous result included an intercept term and sets the first treatment level to zero as in (5.6). Hence

$$\mathbf{B} = \begin{bmatrix} \beta_{01} & \beta_{02} & \beta_{03} & \beta_{04} \\ \beta_{11} & \beta_{12} & \beta_{13} & \beta_{14} \end{bmatrix}$$

where β_{0j} is the intercept term for variable j and β_{1j} is the treatment effect for men for variable j (women are the baseline effect in this parameterization). We would like to determine which of the four oxygen consumption variables differ between men and women. Thus we would like to test each of the following hypotheses

$$H_{01}: \beta_{11} = 0 \quad H_{02}: \beta_{12} = 0 \quad H_{03}: \beta_{13} = 0 \quad H_{04}: \beta_{14} = 0.$$

Using (5.4) the corresponding contrast and MANOVA matrices are

$$\mathbf{C} = [0, 1] \text{ and } \mathbf{H}_j = \mathbf{e}_j, \quad g = \text{rank}(\mathbf{C}) = 1, r = \text{rank}(\mathbf{M}) = 1$$

where $\mathbf{e}_j \in \mathbb{R}^4$ is the vector with 1 in coordinate j and zeroes elsewhere. Table 5.6 lists the results for the four hypotheses. Variables 1,3 and 4 show strong evidence of a gender effect, whereas variable 2 does not. This is consistent with Figure 5.3 which shows visually that the mean difference across gender for this variable are roughly the same (see the second row).

TABLE 5.6
Testing gender effect for each variable from oxygen consumption data.

	Wilks' test	approx F	num Df	den Df	Pr(>F)
$\beta_{11} = 0$	0.8223	10.3673	1	48	0.0023
$\beta_{12} = 0$	0.9969	0.1448	1	48	0.70518
$\beta_{13} = 0$	0.3701	81.693	1	48	6.226e-12
$\beta_{14} = 0$	0.5429	40.4117	1	48	7.1257e-08

□

5.4.2 Multiway MANOVA

Extensions to multiway fixed effects models follow straightforwardly. For concreteness, below we specify the two-way MANOVA model with interactions

$$\mathbf{Y}_{ijk} = \boldsymbol{\mu} + \boldsymbol{\tau}_i + \boldsymbol{\beta}_j + \boldsymbol{\gamma}_{ij} + \mathbf{u}_{ijk}, \quad i = 1, \ldots, I, \ j = 1, \ldots, J, \ k = 1, \ldots, K. \tag{5.8}$$

As in the one-way MANOVA model, (5.8) is unidentified without parameter restrictions. A sufficient condition to ensure identification is

$$\sum_{i=1}^{I} \boldsymbol{\tau}_i = \sum_{j=1}^{J} \boldsymbol{\beta}_j = \sum_{i=1}^{I} \boldsymbol{\gamma}_{ij} = \sum_{j=1}^{J} \boldsymbol{\gamma}_{ij} = \mathbf{0}.$$

The total SSCP decomposes into SSCP matrices for the two treatments (factors), their interactions and an SSCP matrix for the error (residual). From this we obtain the following two-way MANOVA table with interactions (Table 5.7).

Example 5.2. For this example, we use CDC cancer statistics from the state of Florida. Crude rates were obtained for all cancers (per 100,000) from 1999 through 2009 for men and women, stratified by age group (stratified into 14 groups) and by race (African American and Caucasian). We tested whether rates varied by age, gender and race. Thus, the outcome was the $p = 11$ dimensional vector of cancer rates over the 11 year period 1999-2009. A three-way MANOVA model with interactions was applied. Due to the large number of levels for age, only an interaction between race and gender could be fit. The results are listed in Table 5.8.

There is strong evidence of an age effect and a marginal effect due to gender. The age effect is not surprising, as cancer rates are expected to increase with age. This is confirmed by Figure 5.4, which shows crude cancer rates over an 11-year span, stratified by age group. Cancer rates are significantly higher, by thousands of times, for individuals over 60 years old. Additionally, it is noteworthy that cancer rates have been systematically decreasing over time for older individuals.

□

TABLE 5.7
Two-way MANOVA with interactions table.

Source	SSCP	Df	Test
Factor 1	$\mathbf{H}_1 = JK\sum_i(\overline{\mathbf{Y}}_{i.} - \overline{\mathbf{Y}})(\overline{\mathbf{Y}}_{i.} - \overline{\mathbf{Y}})^T$	I^*	$\lvert\mathbf{E}\rvert/\lvert\mathbf{E}+\mathbf{H}_1\rvert$
Factor 2	$\mathbf{H}_2 = IK\sum_j(\overline{\mathbf{Y}}_{.j} - \overline{\mathbf{Y}})(\overline{\mathbf{Y}}_{.j} - \overline{\mathbf{Y}})^T$	J^*	$\lvert\mathbf{E}\rvert/\lvert\mathbf{E}+\mathbf{H}_2\rvert$
Interaction	$\mathbf{H}_{\text{intr}} = K\sum_{ij}(\overline{\mathbf{Y}}_{ij} - \overline{\mathbf{Y}}^*_{ij})(\overline{\mathbf{Y}}_{ij} - \overline{\mathbf{Y}}^*_{ij})^T$	I^*J^*	$\lvert\mathbf{E}\rvert/\lvert\mathbf{E}+\mathbf{H}_{\text{intr}}\rvert$
Error	$\mathbf{E} = \sum_{ijk}(\overline{\mathbf{Y}}_{ijk} - \overline{\mathbf{Y}}_{ij})(\overline{\mathbf{Y}}_{ijk} - \overline{\mathbf{Y}}_{ij})^T$	IJK^*	
Total	$\mathbf{E} + \mathbf{H}_1 + \mathbf{H}_2 + \mathbf{H}_{\text{intr}}$	$IJK - 1$	

$I^* = (I-1)$ $\quad J^* = (J-1)$ $\quad K^* = (K-1)$

$\overline{\mathbf{Y}}_{i.} = \sum_{jk} Y_{ijk}/(JK)$ $\quad \overline{\mathbf{Y}}_{.j} = \sum_{ik} Y_{ijk}/(IK)$

$\overline{\mathbf{Y}}_{ij} = \sum_k Y_{ijk}/K$ $\quad \overline{\mathbf{Y}}^*_{ij} = \overline{\mathbf{Y}}_{i.} + \overline{\mathbf{Y}}_{.j} - \overline{\mathbf{Y}}$

TABLE 5.8
Testing age, gender and race effects for cancer rates in Florida.

	Wilks' test	approx F	num Df	den Df	Pr(>F)
Age	0.0011	1.770	13	143	0.00008943
Sex	0.4656	2.295	1	143	0.04676
Race	0.5981	1.344	1	143	0.26635
Race×Sex	0.5629	1.553	1	143	0.18273

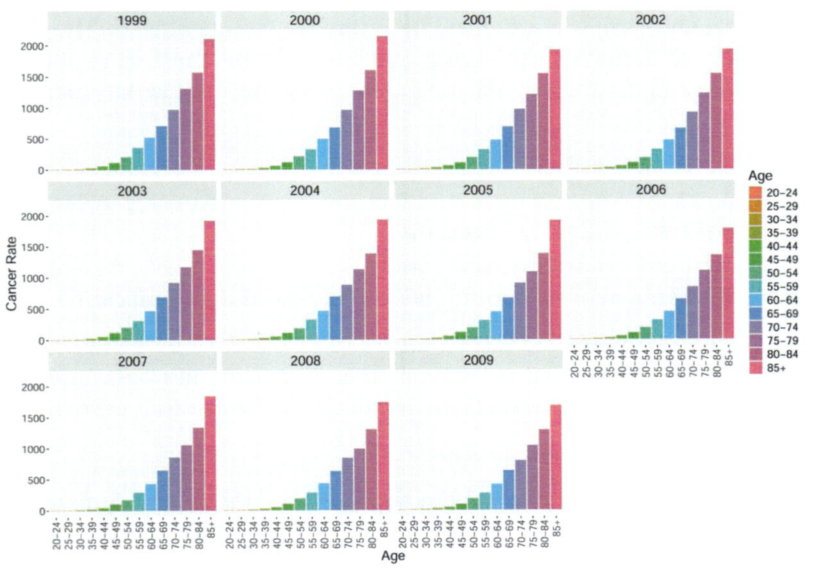

FIGURE 5.4
Florida cancer rates from 1999-2009 stratified by age group.

Exercises

Exercise 5.1. In this exercise we will show that the MLE and least squares estimator coincide under the assumption of normality.

1. Show that the log-likelihood for (5.1) assuming $\mathbf{U}$ is normal can be described by:

$$l(\mathbf{B},\mathbf{\Sigma}) = -\frac{n}{2}\log|2\pi\mathbf{\Sigma}| - \frac{1}{2}\text{tr}\left[\mathbf{\Sigma}^{-1}(\mathbf{Y}-\mathbf{DB})^T(\mathbf{Y}-\mathbf{DB})\right].$$

2. Let $\hat{\mathbf{Y}} = \mathbf{D}\hat{\mathbf{B}}$ and $\hat{\mathbf{U}} = \mathbf{Y}-\mathbf{D}\hat{\mathbf{B}}$ where $\hat{\mathbf{B}}$ is the least squares estimator. Therefore $\mathbf{Y}-\mathbf{DB} = \hat{\mathbf{U}} + \mathbf{D}(\hat{\mathbf{B}}-\mathbf{B})$. Use this to show that the log-likelihood can be rewritten as

$$l(\mathbf{B},\mathbf{\Sigma}) = -\frac{n}{2}\log|2\pi\mathbf{\Sigma}| - \frac{1}{2}\text{tr}\left\{\mathbf{\Sigma}^{-1}\left[\hat{\mathbf{U}}^T\hat{\mathbf{U}} + (\hat{\mathbf{B}}-\mathbf{B})^T\mathbf{D}^T\mathbf{D}(\hat{\mathbf{B}}-\mathbf{B})\right]\right\}.$$

Therefore, show that the MLE and the least squares estimator coincide when the data is normal.

Exercise 5.2. Prove that for any matrix $\mathbf{A}_{p\times p} > 0$

$$\psi(\mathbf{\Sigma}) = |\mathbf{\Sigma}|^{-n/2}\exp\left(-\frac{1}{2}\text{tr}\left[\mathbf{\Sigma}^{-1}\mathbf{A}\right]\right)$$

is maximized over $\mathbf{\Sigma} > 0$ by $\mathbf{\Sigma} = n^{-1}\mathbf{A}$.

Exercise 5.3. Using Exercise 5.1 and Exercise 5.2 determine the MLE for $\mathbf{\Sigma}$ when the data is normal. For completeness, state the MLE for $(\mathbf{B},\mathbf{\Sigma})$.

Exercise 5.4. Verify (5.3) using Exercise 5.3.

Exercise 5.5. Consider the `wine` dataset from the UCI Machine Learning Repository, which contains chemical analysis of wines derived from three different cultivars. The dataset includes measurements for 13 different constituents found in each of the three types of wines. Perform a MANOVA to test whether there are significant differences in the multivariate mean vectors across the three cultivars.

1. Test for normality.
2. Using one-way MANOVA, formulate and test the hypothesis using Wilks' lambda, Pillai's trace, Hotelling's trace, and Roy's largest root.
3. Interpret the results in the context of the study.
4. Perform post-hoc analyses to determine which cultivars differ significantly.

Exercise 5.6. Write the one-way MANOVA model of Exercise 5.5 as a multivariate regression model (5.1). Using the design matrix $\mathbf{D}$ from this model, fit the data directly using $\mathbf{D}$ as in Chapter 5.3. Confirm that your results match those from the previous exercise.

Exercise 5.7. Verify the SSCP decomposition (5.7).

Exercise 5.8. Present a geometric argument showing why the $\mathbf{H}$ and $\mathbf{E}$ matrices in (5.7) are independent.

Exercise 5.9. Find an interesting data set and test for normality. Removing any outliers, so that the remaining data can be assumed to be multivariate, standardize the features. Then from the correlation matrix, identify a set of variables that appear highly correlated with other variables. Use that set of variables as the multivariate response values in a multivariate normal regression. Test whether the model is statistically significant.

Exercise 5.10. Simulate a synthetic data set similar to Chapter 5.3.2 and test for a linear and quadratic dosage effect.

Exercise 5.11. Analyze the `o2cons` dataset from the R-package `MANOVA.RM` which contains oxygen consumption measurements of leukocytes at various times points under a treatment effect.

1. Test for a linear effect in the response in time.
2. Use MANOVA to test the overall multivariate effects of group, treatment and time.
3. Assess the adequacy of the models and discuss the biological implications of your findings.

Exercise 5.12. Consider the `sleepstudy` dataset from the `lme4` package in R, which contains reaction times on a series of tests for subjects under different sleep deprivation conditions measured at different times.

1. Conduct a MANOVA to examine the interaction between the sleep deprivation and days on the reaction times.
2. Interpret the multivariate test results and follow up with univariate analyses if needed.

Exercise 5.13. Consider the `College Scorecard` dataset from the U.S. Department of Education, which contains information about higher education institutions, including average SAT scores, graduation rates, and average family income of students. Investigate the interaction between two categorical predictors (e.g., public/private and region) on multiple outcome variables (e.g., average SAT scores, graduation rates, average family income) using multivariate regression analysis. Include appropriate main effects in the model as well. Interpret the practical implications of your findings.

Exercise 5.14. Download a data involving rates similar to Example 5.2. Using contrasts test whether rates are changing over time.

Exercise 5.15. Setting the first level of each factor and their interaction to zero, describe an alternate parameterization for (5.8) ensuring model identification. Provide a proof of why your parameterization works.

Bibliography

[1] Srivastava M.S. *Methods of Multivariate Statistics.* John Wiley & Sons, 2002.

6

Discriminant Analysis and Classification

6.1 Introduction

Discriminant analysis is a statistical technique used for classifying observations into one of $J \geq 2$ predefined distinct populations, $\Pi_1, \ldots, \Pi_J$. Using a training data set $\mathbf{X}_1, \ldots, \mathbf{X}_n$ of observations with known population labels, the goal is to find a linear combination of features that characterizes or separates the data appropriately. Then, given a feature $\mathbf{X} \in \mathbb{R}^p$, the discriminant rule assigns $\mathbf{X}$ to the most appropriate population. There are several types of discriminant analysis, but the most commonly used are Linear Discriminant Analysis (LDA) and Quadratic Discriminant Analysis (QDA), which will be covered here. We will also discuss the more general area of classification analysis, providing some illustrations such as the nearest neighbor classifier. Additionally, we will state an important result about optimality in classification related to the Bayes rule.

6.2 Maximum Likelihood Approach

The maximum likelihood (ML) discriminant rule assigns each $\mathbf{x} \in \mathbb{R}^p$ to the density with the largest value. If $f_j(\mathbf{x})$ is the density for population j, then $\mathbf{X} = \mathbf{x}$ is assigned to population j^* if

$$f_{j^*}(\mathbf{x}) \geq f_j(\mathbf{x}), \quad \text{for } j = 1, \ldots, J.$$

For example, if the densities are assumed to be $\mathrm{N}_p(\boldsymbol{\mu}_j, \boldsymbol{\Sigma})$ where $\boldsymbol{\Sigma} > 0$, then $\mathbf{X}$ is assigned to j^* if

$$j^* = \underset{j=1,\ldots,J}{\arg\min} \left\{ (\mathbf{X} - \boldsymbol{\mu}_j)^T \boldsymbol{\Sigma}^{-1} (\mathbf{X} - \boldsymbol{\mu}_j) \right\}. \tag{6.1}$$

Applying (6.1) requires knowing the population parameters $\boldsymbol{\mu}_j$ and $\boldsymbol{\Sigma}$. Because these values are generally unknown, a sample discriminant rule is used. Thus, one assigns $\mathbf{X}$ to j^* if

$$j^* = \underset{j=1,\ldots,J}{\arg\min} \left\{ (\mathbf{X} - \hat{\boldsymbol{\mu}}_j)^T \hat{\boldsymbol{\Sigma}}^{-1} (\mathbf{X} - \hat{\boldsymbol{\mu}}_j) \right\},$$

where $\hat{\boldsymbol{\mu}}_j = \overline{\mathbf{X}}_j$ and $\hat{\boldsymbol{\Sigma}}$ is the unbiased estimator for $\boldsymbol{\Sigma}$ defined by

$$\hat{\boldsymbol{\Sigma}} = \frac{1}{n-J} \sum_{j=1}^{J} n_j \hat{\boldsymbol{\Sigma}}_j,$$

DOI: 10.1201/9781003493679-6

where $\hat{\boldsymbol{\Sigma}}_j$ is the sample covariance matrix for population j with sample size n_j

$$\hat{\boldsymbol{\Sigma}}_j = \frac{1}{n_j} \sum_{i \in \Pi_j} (\mathbf{X}_i - \overline{\mathbf{X}}_j)(\mathbf{X}_i - \overline{\mathbf{X}}_j)^T.$$

6.3 Bayes Discrimination Rule

In certain settings it may be reasonable to assign a prior probability of being a member of population Π_j which is denoted by $\pi_j = \mathbb{P}\{\mathbf{X} \in \Pi_j\}$. To incorporate this information we adopt a Bayesian approach where the model is of the form

$$\mathbf{X}|\zeta \sim f_\zeta(\mathbf{x}), \quad \mathbb{P}\{\zeta = j\} = \pi_j.$$

Here ζ is a variable that identifies the population for $\mathbf{X}$. Notice that this model is similar to the multivariate normal mixture model described previously in Chapter 3.3.6. To form the discriminant procedure in this framework, we use the Bayes discriminant rule, a procedure which has certain optimality properties.

Definition 6.1. Let $\pi_1,\ldots,\pi_J$ be prior probabilities for $\Pi_1,\ldots,\Pi_J$. The Bayes discriminant rule assigns $\mathbf{X} = \mathbf{x}$ to the population j^* with maximum posterior probability, i.e.

$$\begin{aligned} j^* &= \underset{j=1,\ldots,J}{\arg\max}\left\{\mathbb{P}\{\zeta = j|\mathbf{X} = \mathbf{x}\}\right\} \\ &= \underset{j=1,\ldots,J}{\arg\max}\left\{\frac{\mathbb{P}\{\mathbf{X} = \mathbf{x}|\zeta = j\}\pi_j}{\sum_k \mathbb{P}\{\mathbf{X} = \mathbf{x}|\zeta = k\}\pi_k}\right\} \\ &= \underset{j=1,\ldots,J}{\arg\max}\left\{\pi_j f_j(\mathbf{x})\right\}. \end{aligned}$$

□

Let $\delta : \mathbb{R}^p \to \{1,\ldots,J\}$ be an arbitrary discriminant rule which assigns $\mathbf{x}$ to one of the J populations. We can code δ as the collection of 0-1 functions $\{\phi_1,\ldots,\phi_J\}$ such that $\phi_j(\mathbf{x}) = 1$ if and only if $\mathbf{x}$ is classified to Π_j. For example, for the Bayes discriminant rule, denoted by δ^*, its functions are $\{\phi_1^*,\ldots,\phi_J^*\}$ where

$$\phi_j^*(\mathbf{x}) = 1_{\{\pi_j f_j(\mathbf{x}) = \pi_{j^*} f_{j^*}(\mathbf{x})\}}, \quad j = 1,\ldots,J.$$

For any rule δ, the probability that $\mathbf{X}$ is assigned to $\Pi_{j'}$, when in fact $\mathbf{X}$ belongs to Π_j, is

$$p_{j'j} = \int \phi_{j'}(\mathbf{x}) f_j(\mathbf{x})\, d\mathbf{x}.$$

Specifically, if $j' = j$, then p_{jj} represents the probability of correct classification into Π_j. The following theorem demonstrates that the Bayes discriminant rule is optimal in terms of minimizing misclassification under the given prior.

Theorem 6.1. *If $\pi_1,\ldots,\pi_J$ are prior probabilities for $\Pi_1,\ldots,\Pi_J$, then there is no discriminant rule δ that has larger posterior probability of correct classification than the Bayes discriminant rule under the prior.*

Let $\{p^*_{jj}\}$ be the probability of correct classification under the Bayes discriminant rule δ^*. Likewise, let $\{p_{jj}\}$ be the probability of correct classification under an arbitrary decision rule. Then, Theorem 6.1 asserts that

$$\sum_{j=1}^{J} \pi_j p_{jj} = \sum_{j=1}^{J} \int \phi_j(\mathbf{x})\pi_j f_j(\mathbf{x})\, d\mathbf{x} \le \sum_{j=1}^{J} \pi_j p^*_{jj} = \sum_{j=1}^{J} \int \phi^*_j(\mathbf{x})\pi_j f_j(\mathbf{x})\, d\mathbf{x}.$$

Proof. For any $\{\phi_1,\ldots,\phi_J\}$ where $0 \le \phi_j \le 1$ and $\sum_{j=1}^{J} \phi_j = 1$

$$\begin{aligned}
\sum_{j=1}^{J} \pi_j p_{jj} &= \sum_{j=1}^{J} \pi_j \int \phi_j(\mathbf{x}) f_j(\mathbf{x})\, d\mathbf{x} \\
&= \int \sum_{j=1}^{J} \left[\phi_j(\mathbf{x})\pi_j f_j(\mathbf{x})\right] d\mathbf{x} \\
&\le \int \sum_{j=1}^{J} \left[\phi_j(\mathbf{x})\pi_{j^*} f_{j^*}(\mathbf{x})\right] d\mathbf{x} \\
&= \int \pi_{j^*} f_{j^*}(\mathbf{x})\, d\mathbf{x} \\
&= \int \sum_{j=1}^{J} \left[\phi^*_j(\mathbf{x})\pi_j f_j(\mathbf{x})\right] d\mathbf{x} \\
&= \sum_{j=1}^{J} \pi_j p^*_{jj}.
\end{aligned}$$

□

We can generalize these results by assigning a loss to a misclassification, turning this into a decision theory problem. Let $L(j', j)$ be the *loss function* for allocating an observation $\mathbf{x}$ to $\Pi_{j'}$ when $\mathbf{x}$ comes from Π_j. It is assumed that $L(j', j) > 0$ for each $j' \neq j$ and $L(j, j) = 0$ for $j = 1,\ldots,J$. Let δ be a decision rule described as above in terms of the 0-1 functions $\{\phi_1,\ldots,\phi_J\}$. The *risk function* for δ for Π_j is

$$R(\delta, j) = \sum_{j'=1}^{J} L(j', j) \int \phi_{j'}(\mathbf{x}) f_j(\mathbf{x})\, d\mathbf{x} = \sum_{j'=1}^{J} L(j', j) p_{j'j}.$$

The risk, $R(\delta, j)$, equals the expected loss for δ given that $\mathbf{x}$ comes from Π_j. For example, if $L(j', j) = 1$ for $j' \neq j$, then

$$R(\delta, j) = 1 - p_{jj}$$

equals the misclassification probability for Π_j.

Using the loss function we obtain a general formulation of the Bayes risk.

Definition 6.2. If π_j are prior probabilities for Π_j, define the cost-weighted Bayes risk for δ as

$$r(\delta, (\pi_j)_1^J) = \sum_{j=1}^{J} \pi_j R(\delta, j).$$

In other words, the cost-weighted risk $r(\delta, (\pi_j)_1^J)$ equals the posterior expected loss for δ. □

Define the cost-weighted Bayes rule δ^* as the rule that assigns $\mathbf{x}$ to the population Π_{j^*} with minimal cost-weighted loss

$$\sum_{j \neq j^*} L(j^*, j)\pi_j f_j(\mathbf{x}) = \min_{j'} \left\{ \sum_{j \neq j'} L(j', j)\pi_j f_j(\mathbf{x}) \right\}.$$

Theorem 6.2. *If $\pi_1, \ldots, \pi_J$ are prior probabilities for $\Pi_1, \ldots, \Pi_J$, then there is no decision rule δ with cost-weighted risk $r(\delta, (\pi_j)_1^J)$ smaller than the cost-weighted risk for the Bayes rule, $r(\delta^*, (\pi_j)_1^J)$.*

6.4 Linear Discrimination

A limitation of ML estimation (MLE) is that it requires specifying the underlying density. Fisher proposed the nonparametric method called *linear discriminant analysis* (LDA), which avoids the need to specify the population distribution. LDA determines a linear combination $Y_i = \mathbf{a}^T \mathbf{X}_i$, where $\mathbf{a} \in \mathbb{R}^p$ is selected to maximize the ratio of within-group to between-group sums of squares of $Y_1, \ldots, Y_n$. The within-group sums of squares is $\mathbf{a}^T \mathbf{W} \mathbf{a}$, where

$$\mathbf{W} = \sum_{i=1}^{n} (\mathbf{X}_i - \overline{\mathbf{X}})(\mathbf{X}_i - \overline{\mathbf{X}})^T.$$

The between groups sums of squares is $\mathbf{a}^T \mathbf{B} \mathbf{a}$, where

$$\mathbf{B} = \sum_{j=1}^{J} n_j (\overline{\mathbf{X}}_j - \overline{\mathbf{X}})(\overline{\mathbf{X}}_j - \overline{\mathbf{X}})^T.$$

LDA chooses the vector $\hat{\mathbf{a}}$ which maximizes the ratio of these two values:

$$\hat{\mathbf{a}} = \arg\max_{\mathbf{a} \in \mathbb{R}^p} \frac{\mathbf{a}^T \mathbf{B} \mathbf{a}}{\mathbf{a}^T \mathbf{W} \mathbf{a}}.$$

The transformation $Y = \hat{\mathbf{a}}^T \mathbf{X}$ is called the linear discriminant function, and it can be shown that $\hat{\mathbf{a}}$ equals the eigenvector of $\mathbf{W}^{-1}\mathbf{B}$ with the largest eigenvalue. Once the linear discriminant function is calculated, $\mathbf{X}$ is allocated to the population j^* with sample mean $\hat{\mathbf{a}}^T \overline{\mathbf{X}}_{j^*}$ closest to $\hat{\mathbf{a}}^T \mathbf{X}$. Thus, $\mathbf{X}$ is assigned to j^* if

$$|\hat{\mathbf{a}}^T \mathbf{X} - \hat{\mathbf{a}}^T \overline{\mathbf{X}}_{j^*}| \le |\hat{\mathbf{a}}^T \mathbf{X} - \hat{\mathbf{a}}^T \overline{\mathbf{X}}_j|, \quad j \ne j^*.$$

The discriminant function for populations $\Pi_{j'}$ and Π_j is therefore

$$\mathbb{Q}_{j'j}(\mathbf{X}) = |\hat{\mathbf{a}}^T (\mathbf{X} - \overline{\mathbf{X}}_{j'})| - |\hat{\mathbf{a}}^T (\mathbf{X} - \overline{\mathbf{X}}_j)|$$

with decision boundary $\{\mathbf{x} : \mathbb{Q}_{j'j}(\mathbf{x}) = 0\}$. In the special case when $J = 2$ it can be shown that $\hat{\mathbf{a}} = \mathbf{W}^{-1}\mathbf{d}$ where $\mathbf{d} = \overline{\mathbf{X}}_1 - \overline{\mathbf{X}}_2$, so that $\mathbf{X}$ is allocated to Π_1 if

$$\mathbf{d}^T \mathbf{W}^{-1} \left(\mathbf{X} - \overline{\mathbf{X}}\right) > 0, \quad \text{where } \overline{\mathbf{X}} = (\overline{\mathbf{X}}_1 + \overline{\mathbf{X}}_2)/2.$$

We illustrate LDA using Fisher's famous Iris data set. Figure 6.1 displays scatter plots of length and width measurements for the three iris species, with linear discriminant functions shown as straight lines. The three lines represent discriminant functions for *Setosa* versus *Versicolor*, *Setosa* versus *Virginica*, and *Versicolor* versus *Virginica*. The diagonal panels of the scatter plot show density estimates for each length and width measurement. The last two diagonal panels for petal length and width indicate a bimodal shape because *Setosa* has much smaller values than the other two species (see bottom off-diagonal panels). It appears that *Setosa* is easy to separate, while the other two species require more effort. In an actual picture of the iris (Figure 6.2), we observe that *Setosa* looks strikingly different.

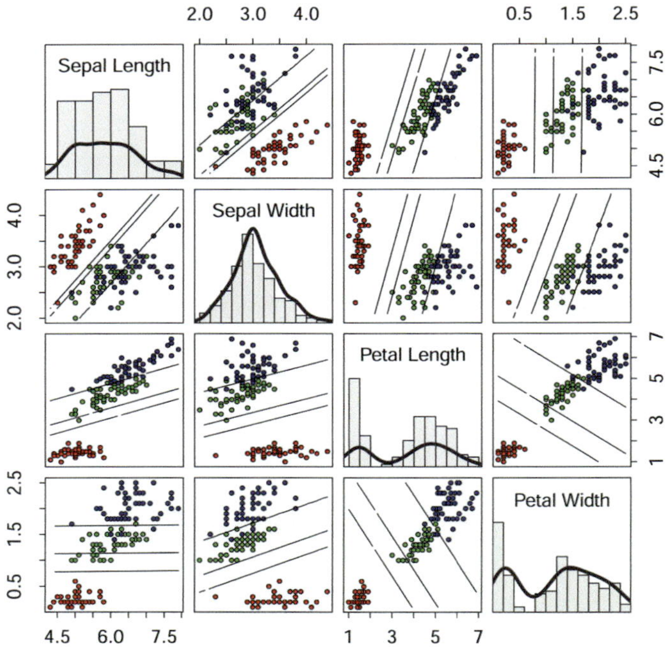

FIGURE 6.1
Linear discriminant analysis (LDA) for Fisher's Iris dataset (Setosa •, Versicolor •, Virginica •). Straight lines in each panel are LDA functions for Setosa versus Versicolor, Setosa versus Virginica, and Versicolor versus Virginica.

FIGURE 6.2
Iris Setosa, Versicolor, Virginica species.

6.5 Quadratic Discrimination

The MLE approach for multivariate normal data assumes equal covariance matrices across populations which yields linear discriminant functions. If covariance matrices are unequal the resulting discriminant function is quadratic in **X** and the resulting method is called *quadratic discriminant analysis* (QDA).

More precisely, assume $\Pi_j \sim \mathrm{N}_p(\boldsymbol{\mu}_j, \boldsymbol{\Sigma}_j)$ where $\boldsymbol{\Sigma}_j > 0$. Then **X** is assigned to j^* if

$$j^* = \underset{j=1,\ldots,J}{\arg\min}\left\{\log|\boldsymbol{\Sigma}_j| + (\mathbf{X}-\boldsymbol{\mu}_j)^T\boldsymbol{\Sigma}_j^{-1}(\mathbf{X}-\boldsymbol{\mu}_j)\right\}.$$

Observe that when $\boldsymbol{\Sigma}_1 = \cdots = \boldsymbol{\Sigma}_J = \boldsymbol{\Sigma}$ this is equivalent to

$$j^* = \underset{j=1,\ldots,J}{\arg\max}\left\{2\boldsymbol{\mu}_j^T\boldsymbol{\Sigma}^{-1}\mathbf{X} + \boldsymbol{\mu}_j^T\boldsymbol{\Sigma}^{-1}\boldsymbol{\mu}_j\right\}.$$

Then the discriminant function for populations $\Pi_{j'}$ and Π_j is

$$\mathbb{Q}_{j'j}(\mathbf{x}) = 2(\boldsymbol{\mu}_{j'}-\boldsymbol{\mu}_j)^T\boldsymbol{\Sigma}^{-1}\mathbf{x} + \boldsymbol{\mu}_{j'}^T\boldsymbol{\Sigma}^{-1}\boldsymbol{\mu}_{j'} - \boldsymbol{\mu}_j^T\boldsymbol{\Sigma}^{-1}\boldsymbol{\mu}_j$$

which is linear in **x**. However, when the covariances are unequal, the discriminant function is

$$\mathbb{Q}_{j'j}(\mathbf{x}) = \log\frac{|\boldsymbol{\Sigma}_{j'}|}{|\boldsymbol{\Sigma}_j|} + (\mathbf{x}-\boldsymbol{\mu}_{j'})^T\boldsymbol{\Sigma}_{j'}^{-1}(\mathbf{x}-\boldsymbol{\mu}_{j'}) - (\mathbf{x}-\boldsymbol{\mu}_j)^T\boldsymbol{\Sigma}_j^{-1}(\mathbf{x}-\boldsymbol{\mu}_j)$$

which is quadratic in **x**.

Figure 6.3 displays the data (n = 500) and the quadratic decision boundary from a two-dimensional synthetic experiment with two classes. We generated 250 data points for each class, where data points for each class were sampled from bivariate normal distributions such that the two classes overlap. This simulation demonstrates how QDA improves upon LDA in more challenging classification problems with overlapping classes.

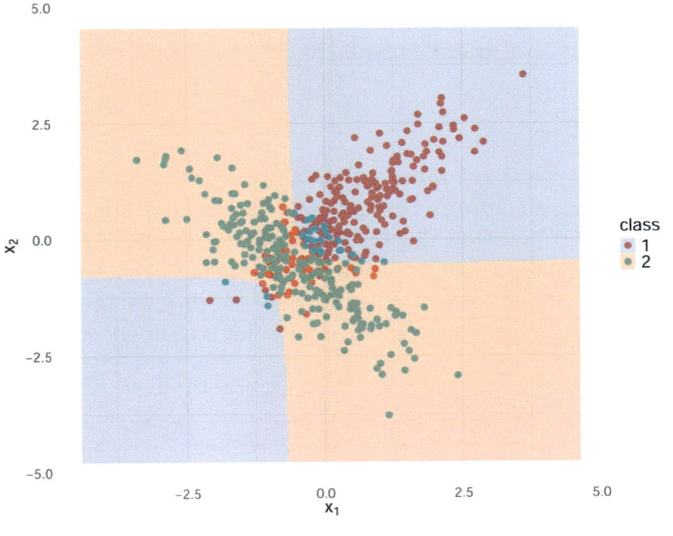

FIGURE 6.3
Quadratic discriminant analysis (QDA) for synthetic two-dimensional data.

A test set of size $n' = 5000$ was used to determine the misclassification error rate. The use of a quadratic boundary works well here and the results are reasonably good (Table 6.1). QDA achieves an overall misclassification error of approximately 14%.

TABLE 6.1
Confusion matrix for QDA on test data.

	1	2
1	2020	216
2	480	2284

6.6 Classification Analysis and Bayes Rule

Classification analysis refers to a broad range of techniques and methods used to assign labels to observations based on their feature values. Discriminant analysis is the multivariate task of *separating* distinct sets of objects. The discriminant rule is trained on learning data and used to assign a new test data point $\mathbf{X}$ to the population deemed most appropriate. As we saw, it typically uses statistical properties of the data, such as means and covariances, to model the differences between classes. As such, discriminant analysis is a specialized technique within the larger field of classification analysis.

Classification analysis is an example of *supervised learning*, which involves a response outcome value used for training. In this context, two or more groups or populations are known *a priori*, which are the response values, and new observations are *classified* into one of these known populations using a trained classifier.

Mathematically, the classification problem is described as follows. The learning data is denoted by $(\mathbf{X}_1, Y_1), \ldots, (\mathbf{X}_n, Y_n) \in \mathscr{X} \times \mathscr{Y}$, where $Y \in \mathscr{Y} = \{C_1, \ldots, C_J\}$ is the observed response recording the class membership. For convenience of notation, we will code the class labels as $\mathscr{Y} = \{1, \ldots, J\}$. When $J = 2$, this is called the *two-class problem*, and the response is coded as binary values, $Y \in \{0, 1\}$. The goal in classification analysis is to use the learning data to assign new $\mathbf{X}$ to the correct class while minimizing error related to the conditional probability $\mathbb{P}\{Y = y|\mathbf{X} = \mathbf{x}\}$.

6.6.1 Bayes classifier

The Bayes rule was shown to have minimal posterior misclassification probability among all decision rules under a specified prior (Theorem 6.1). We can consider the same problem from a non-Bayesian perspective to see how this fits into the larger problem of classification.

We start by considering the two-class supervised problem, $J = 2$, where the response is $Y \in \{0, 1\}$. Let $(\mathbf{X}, Y)$ be a generic data point and denote the conditional probability $Y = j$ given $\mathbf{X} = \mathbf{x}$ by $f_{j|\mathbf{X}}(\mathbf{x}) = \mathbb{P}\{Y = j|\mathbf{X} = \mathbf{x}\}$ for $j = 0, 1$. The Bayes classifier is defined as

$$\delta^*(\mathbf{x}) = 1_{\{f_{1|\mathbf{X}}(\mathbf{x}) \geq 0.5\}}. \tag{6.2}$$

The following result shows that (6.2) is the optimal classifier.

Theorem 6.3. *For any classifier* $\delta : \mathscr{X} \to \{0, 1\}$,

$$\mathbb{P}\{\delta^*(\mathbf{X}) \neq Y\} \leq \mathbb{P}\{\delta(\mathbf{X}) \neq Y\}.$$

The Bayes risk (Bayes error rate) is the misclassification error for the Bayes classifier and equals

$$\mathbb{P}\{\delta^*(\mathbf{X}) \neq Y\} = \mathbb{E}\left[\min\left\{f_{0|\mathbf{X}}(\mathbf{X}), f_{1|\mathbf{X}}(\mathbf{X})\right\}\right]. \tag{6.3}$$

Thus, the Bayes error rate (6.3) is the lowest misclassification error achievable by a classifier.

According to Theorem 6.3, the optimal rule is based on whether $f_{1|\mathbf{x}}(\mathbf{x}) \geq 0.5$, which is equivalent to testing

$$\mathbb{P}\{Y=1|\mathbf{X}=\mathbf{x}\} \geq \mathbb{P}\{Y=0|\mathbf{X}=\mathbf{x}\}.$$

Applying Bayes' theorem, we have

$$\frac{\mathbb{P}\{Y=1, \mathbf{X}=\mathbf{x}\}}{f_{\mathbf{X}}(\mathbf{x})} \geq \frac{\mathbb{P}\{Y=0, \mathbf{X}=\mathbf{x}\}}{f_{\mathbf{X}}(\mathbf{x})}.$$

Cancelling the common value $f_{\mathbf{X}}$ (the density of $\mathbf{X}$) and appealing to Bayes' theorem once again, yields

$$\mathbb{P}\{Y=1\} f_{\mathbf{X}|1}(\mathbf{x}) \geq \mathbb{P}\{Y=0\} f_{\mathbf{X}|0}(\mathbf{x})$$

where $f_{\mathbf{X}|j}(\mathbf{x})$ is the conditional density of $\mathbf{X}$ given $Y=j$. Therefore, this is equivalent to the Bayes rule of Theorem 6.1 if the prior probabilities π_j are chosen to match the true population probabilities $\mathbb{P}\{Y=j\}$.

6.6.2 Cost-weighted Bayes classifier

More generally, consider the J classification problem where $J \geq 2$. According to Theorem 6.2, for a prior chosen such that $\pi_j = \mathbb{P}\{Y=j\}$, the (cost-weighted) Bayes classifier δ^* is the rule that assigns $\mathbf{x}$ to the class with minimal cost-weighted loss

$$\min_{1\leq j'\leq J}\left\{\sum_{j\neq j'} L(j',j)\pi_j f_{\mathbf{X}|j}(\mathbf{x})\right\} = \min_{1\leq j'\leq J}\left\{\sum_{j\neq j'} L(j',j) f_{j|\mathbf{X}}(\mathbf{x})\right\} \tag{6.4}$$

where $f_{j|\mathbf{X}}(\mathbf{x}) = \mathbb{P}\{Y=j|\mathbf{X}=\mathbf{x}\}$ and $L(j',j)$ is the loss when $\mathbf{x}$ is assigned to j' instead of the correct j classification and $L(j,j)=0$.

The cost-weighted Bayes classifier has the smallest cost-weighted Bayes risk, where the cost-weighted risk for a decision rule $\delta : \mathscr{X} \to \{1,\ldots,J\}$ with loss function L and prior $\pi_j = \mathbb{P}\{Y=j\}$ equals

$$\begin{aligned}
r(\delta, L) &= \sum_{j=1}^{J} \pi_j R(\delta, j) \\
&= \sum_{j=1}^{J}\left[\pi_j \sum_{j'=1}^{J} L(j',j) \int 1_{\{\delta(\mathbf{x})=j'\}} f_{\mathbf{X}|j}(\mathbf{x})\, d\mathbf{x}\right] \\
&= \sum_{j=1}^{J}\left[\sum_{j'=1}^{J} L(j',j)\mathbb{E}\left[1_{\{\delta(\mathbf{X})=j', Y=j\}}\right]\right].
\end{aligned} \tag{6.5}$$

Simple manipulation of (6.4) shows that if $L(j',j) = L_j$ for all $j' \neq j$, the cost-weighted Bayes classifier simplifies to

$$\delta^*(\mathbf{x}) = \arg\max_{1\leq j\leq J}\left\{L_j f_{j|\mathbf{X}}(\mathbf{x})\right\}. \tag{6.6}$$

If $L_j = L$ for all j, this further simplifies to

$$\delta^*(\mathbf{x}) = \arg\max_{1\leq j\leq J}\left\{f_{j|\mathbf{X}}(\mathbf{x})\right\}$$

which is a generalization of the two-class Bayes classifier since the cost-weighted risk is

$$r(\delta, L) = L\sum_{j=1}^{J}\left[\sum_{j'\neq j}\mathbb{E}\left[1_{\{\delta(\mathbf{X})=j', Y=j\}}\right]\right]$$

which equals the misclassification error when $L = 1/2$.

The following result summarizes the properties of the cost-weighted Bayes classifier.

Theorem 6.4. *Let $L_j > 0$ be the loss for misclassifying label $j = 1, \ldots, J$ and let*

$$r(\delta, \{L_1, \ldots, L_J\}) = \sum_{j=1}^{J}\sum_{j'\neq j} L_j\mathbb{E}\left[1_{\{\delta(\mathbf{X})=j', Y=j\}}\right]$$

be the cost-weighted risk for a classifier $\delta : \mathscr{X} \to \{1, 2, \ldots, J\}$. Then

$$r(\delta, \{L_1, \ldots, L_J\}) \geq r(\delta_{WB}, \{L_1, \ldots, L_J\})$$

where δ_{WB} is the cost-weighted Bayes classifier defined by

$$\delta_{WB}(\boldsymbol{x}) = \arg\max_{1\leq j\leq J}\left\{L_j f_{j|\mathbf{X}}(\boldsymbol{x})\right\}.$$

Notice in particular if $L_j = L$ are the same for all j, then the cost-weighted Bayes classifier is the majority voting rule. Also, for the $J = 2$ classification problem, if L_0 and L_1 are the losses for class labels $j = 0$ and $j = 1$, then

$$\delta_{WB}(\boldsymbol{x}) = 1\left\{f_{1|\mathbf{X}}(\boldsymbol{x}) \geq \frac{L_0}{L_0 + L_1}\right\}. \tag{6.7}$$

For example, if $L_0 = L_1$ then (6.7) is the Bayes classifier (6.2).

Remark 6.1. We will return to the cost-weighted Bayes classifier in Chapter 16.12 when we consider class imbalanced problems. It will be discussed that for the $J = 2$ classification problem that the cost-weighted Bayes classifier (6.7) is an instance of what is called a *quantile classifier* where the classification decision is based on a threshold value $0 < q < 1$ (which in this instance is $q = L_0/(L_0 + L_1)$). This key fact paves the way to an optimal solution to the class imbalanced problem. □

Remark 6.2. The Bayes classifier (6.6) obtained for the loss function $L(j', j) = L_j$, $j' \neq j$, has another interpretation called *one-versus-one classification* [3] which is a technique for recasting the multiclassification problem into a series of simpler two-class problems.

Using the identity

$$\arg\max_{1\leq j\leq J}\alpha_j = \arg\max_{1\leq j\leq J}\sum_{j'=1}^{J} 1_{\{\alpha_j > \alpha_{j'}\}}$$

where $\alpha_1, \ldots, \alpha_J$ are any set of distinct real numbers, (6.6) can be alternatively described by

$$\begin{aligned}\delta^*(\mathbf{x}) &= \arg\max_{1\leq j\leq J}\sum_{j'=1}^{J} 1_{\left\{L_j f_{j|\mathbf{X}}(\mathbf{x}) > L_{j'} f_{j'|\mathbf{X}}(\mathbf{x})\right\}} \\ &= \arg\max_{1\leq j\leq J}\sum_{j'\neq j} 1\left\{\frac{L_j f_{j|\mathbf{X}}(\mathbf{x})}{f_{j|\mathbf{X}}(\mathbf{x}) + f_{j'|\mathbf{X}}(\mathbf{x})} > \frac{L_{j'} f_{j'|\mathbf{X}}(\mathbf{x})}{f_{j|\mathbf{X}}(\mathbf{x}) + f_{j'|\mathbf{X}}(\mathbf{x})}\right\}.\end{aligned}$$

Each entry in the sum of the last line is equivalent to the Bayes classifier for the two-class problem with class labels $\{j', j\}$ (where the other labels and their data are removed). The sum counts the number of times $Y = j$ is the predicted class label for $\mathbf{x}$ in each of the $J - 1$ two-class

FIGURE 6.4
Four-class problem where one-versus-one (OvO) may be more efficient.

problems of j versus j'. This shows that the Bayes multiclassifier can be obtained by running $J(J-1)/2$ one-versus-one classification analyses. Then, $\mathbf{x}$ is classified to the class label with the most wins. For example, if $\hat{f}^{(jj')}_{y|\mathbf{X}}(\mathbf{x})$ is the estimated conditional probability for $Y = y$ for the two-class problem of j versus j', we obtain the one-versus-one (OvO) classifier

$$\hat{\delta}(\mathbf{x}) = \arg\max_{1\leq j\leq J} \sum_{j'\neq j} 1_{\left\{L_j \hat{f}^{(jj')}_{j|\mathbf{X}}(\mathbf{x}) > L_{j'} \hat{f}^{(jj')}_{j'|\mathbf{X}}(\mathbf{x})\right\}}.$$

This equivalent formulation has a useful advantage. It may be difficult in practice to devise an effective multiclassifier for all J class labels since the classifier has to work well over all j. However, since it can sometimes be more practical to devise effective two-class predictors, this approach makes it possible to develop an effective multiclassifier using one-versus-one. An illustration is given in Figure 6.4. □

6.7 Nearest Neighbor Classification

To provide an example of a classifier, we consider nearest neighbor classifiers. Nearest neighbors are effective and computationally efficient procedures used in various supervised problems. Given $\mathbf{x} \in \mathbb{R}^p$, the kth nearest neighbor classifier ($k \geq 1$) assigns the majority class label of the k closest neighbors to $\mathbf{x}$, where closeness is defined by an underlying metric. Formally, let $\mathbf{X}_{(1)}(\mathbf{x}), \ldots, \mathbf{X}_{(n)}(\mathbf{x})$ be the ordering of $\mathbf{X}_1, \ldots, \mathbf{X}_n$ according to distance from $\mathbf{x}$. Let $Y_{(1)}(\mathbf{x}), \ldots, Y_{(n)}(\mathbf{x})$ be the Y values for the ordered $\mathbf{X}_i$. The kth nearest neighbor classifier for $\mathbf{x}$ is defined as

$$\delta(\mathbf{x}) = \arg\max_{j=1,\ldots,J} \left\{ \sum_{l=1}^{k} 1_{\{Y_{(l)}(\mathbf{x})=j\}} \right\}.$$

For example, the nearest neighbor classifier is simply $Y_{(1)}(\mathbf{x})$, the Y value of the closest neighbor.

Figure 6.5 illustrates the flexibility of the decision boundary for nearest neighbors using the synthetic "spiral" data set from the `mlbench` package [4]. There are two classes within two

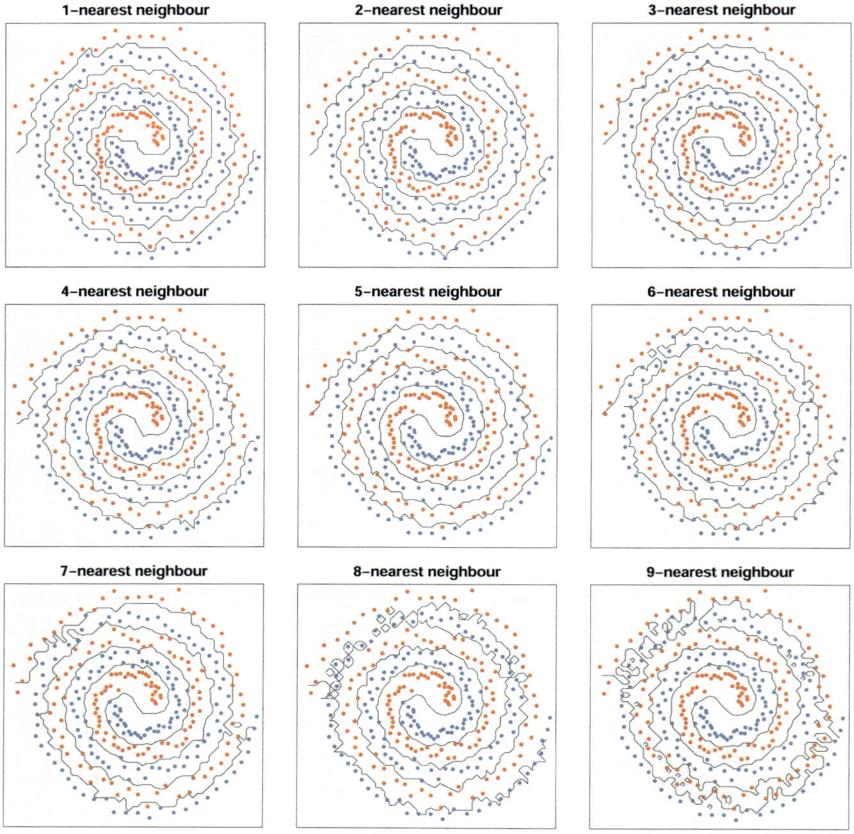

FIGURE 6.5
k-nearest neighbors, $k = 1, \ldots, 9$, for the "spirals" two-class problem.

entangled spirals in $\mathbb{R}^2$. Blue and orange points represent the observed data, with the color indicating the class label. The black lines trace the nearest neighbor decision boundary, with small blue and orange points identifying class assignments. In this example, increasing the number of neighbors k fragments the decision boundary and reduces accuracy, but the results remain good even for moderately large k. The spiral data is challenging for linear or quadratic decision boundaries.

6.7.1 Consistency of nearest neighbors

For two-class problems, the asymptotic misclassification error for the $k = 1$ nearest neighbors equals

$$e_{\text{NN}} = \mathbb{E}\left[2 f_{0|\mathbf{x}}(\mathbf{X}) f_{1|\mathbf{x}}(\mathbf{X})\right].$$

By Theorem 6.3, the misclassification error for the Bayes decision rule (which is the optimal rule) equals $e^* = \mathbb{E}\left[\min\{f_{0|\mathbf{x}}(\mathbf{X}), f_{1|\mathbf{x}}(\mathbf{X})\}\right]$. It can be shown that

$$e^* \leq e_{\text{NN}} \leq 2e^*(1-e^*) \leq 2e^*. \tag{6.8}$$

This is called the Cover-Hart theorem [1]. In general, Devroye et al. [2] showed

$$\begin{aligned} e^* &\leq \cdots \leq e_{(2k+1)\text{NN}} \leq e_{(2k)\text{NN}} = e_{(2k-1)\text{NN}} \\ &\leq \cdots \leq e_{3\text{NN}} \leq e_{2\text{NN}} = e_{\text{NN}} \leq 2e^* \end{aligned}$$

and note that $e_{(2k-1)\text{NN}} = e_{2k\text{NN}}$ for all $k \geq 1$.

The above result is interesting because it suggests that k should increase with n if we want to achieve optimal risk performance using nearest neighbors. On the other hand, the analysis for the spiral data set (Figure 6.5) suggests that the value for k should be relatively small to successfully recovery the true boundary.

Therefore what value should we use? In practice, in a finite sample data setting, the value for k can be selected using cross-validation, a topic we will cover in Chapter 7.9. For the spiral data set this will show that performance is best when k is small (see Exercise 6.14). This is not a contradiction to the asymptotic results given above since those results apply to pointwise risk consistency whereas cross-validation is designed to achieve good prediction error performance. This point will be discussed in Chapter 7.9.

Exercises

Exercise 6.1. Prove that (6.1) is the ML discriminant rule assuming that population Π_k is distributed as $\text{N}_p(\boldsymbol{\mu}_k, \boldsymbol{\Sigma})$.

Exercise 6.2. Prove that Fisher's linear discriminant function is the first eigenvector of $\mathbf{W}^{-1}\mathbf{B}$.

Exercise 6.3. Show that boundary hyperplane for the ML discriminant rule between $\text{N}_p(\boldsymbol{\mu}_1, \boldsymbol{\Sigma})$ and $\text{N}_p(\boldsymbol{\mu}_2, \boldsymbol{\Sigma})$, $\boldsymbol{\Sigma} > 0$, is orthogonal to $\boldsymbol{\mu}_1 - \boldsymbol{\mu}_2$ if and only if $\boldsymbol{\mu}_1 - \boldsymbol{\mu}_2$ is an eigenvector of $\boldsymbol{\Sigma}$.

Exercise 6.4. In this exercise, we consider the ML discriminant rule for two populations with the *same* mean but *different* covariance matrices. The populations are $\text{N}_p(\boldsymbol{\mu}, \boldsymbol{\Sigma}_1)$ and $\text{N}_p(\boldsymbol{\mu}, \boldsymbol{\Sigma}_2)$ such that

$$\boldsymbol{\Sigma}_j = \sigma_j^2\left[(1-\rho)I + \rho\mathbf{1}\mathbf{1}^T\right], \quad j = 1, 2,$$

where $-1 < \rho < 1$ is the correlation and $\mathbf{1}$ is the p-dimensional vector of 1's.

(a) Show that ML discriminant function is proportional to

$$z_1 - \rho\left[1 + (p-1)\rho\right]^{-1} z_2 + C$$

where $z_1 = \sum_{i=1}^{p} x_i^2$ and $z_2 = (\sum_{i=1}^{p} x_i)^2$ for some constant C.

(b) Under what situation will this result in equal misclassification probability for the two populations?

Exercise 6.5. Prove Theorem 6.2

Exercise 6.6. Let π_j be the prior probability that $\mathbf{x} = (x_1, x_2)^T$ comes from population Π_j for $j = 1,2$, where $\pi_1 + \pi_2 = 1$. The density for population Π_j is bivariate Cauchy,

$$f_j(\mathbf{x}) = (2\pi)^{-1} k_j (k_j^2 + x_1^2 + x_2^2)^{-3/2}, \quad \mathbf{x} \in \mathbb{R}^2.$$

The cost of misclassifying $\mathbf{x}$ to Π_j is $c_j > 0$ for $j = 1,2$. Find the decision rule that minimizes the posterior expected cost of misclassifying $\mathbf{x}$.

Exercise 6.7. Prove Theorem 6.3.

Exercise 6.8. A bank wants to segment its customer base to better tailor its marketing strategies for promoting term deposits. The bank has collected data from previous marketing campaigns, including customer demographics, account information, and details about the marketing campaigns. The goal is to use discriminant analysis to classify new customers into predefined segments based on their likelihood to subscribe to a term deposit. You can download the `Bank Marketing` dataset from the UCI Machine Learning Repository for this analysis.

1. Preprocess the data by dealing with missing values, encode categorical variables as necessary, and standardize the continuous variables if required.
2. Perform LDA to build a model that classifies customers into predefined segments based on their likelihood to subscribe to a term deposit.
3. Evaluate the performance of the model using appropriate metrics (e.g., confusion matrix, classification accuracy, etc.).
4. Interpret the results of the LDA, including the discriminant functions. Provide insights on which variables are most influential in distinguishing between the customer segments.

Exercise 6.9. Credit card fraud is a significant problem in the financial sector, causing substantial financial losses each year. Using discriminant analysis, we aim to classify transactions as either fraudulent or non-fraudulent based on various features derived from the transaction data. We will use the `Credit Card Fraud Detection Dataset` from Kaggle for this analysis. The dataset contains $n = 284{,}807$ transactions with $p = 30$ features.

1. Preprocess the data by dealing with missing values, encode categorical variables as necessary, and standardize the continuous variables if required.
2. Perform Linear Discriminant Analysis (LDA) to build a model that classifies transactions as fraudulent or non-fraudulent.
3. Evaluate the performance of the model using appropriate metrics such as confusion matrix, classification accuracy, precision, recall, F1-score, and AUC-ROC.
4. Interpret the results of the LDA, including the discriminant function and the importance of each predictor variable.
5. Use the LDA model to classify a set of new transactions into fraudulent or non-fraudulent categories. Discuss potential strategies for implementing this model in a real-time fraud detection system.

Exercise 6.10. Install the `mlbench` package and simulate data from the "ringnorm" data. Run a QDA analysis. Determine what the Bayes classifier is for this example and what the theoretical Bayes error rate is. How does your QDA compare?

Exercise 6.11. Assuming a loss $L(j', j) = L_j$ for $j' \neq j$, prove that the cost-weighted Bayes classifier can be written as (6.6).

Exercise 6.12. Select a classification data with a suitably large number of labels $J > 2$. Run a supervised multiclassifier analysis. Then for comparison apply one-versus-one (OvO) classification to the same data. Compare the results and discuss your findings.

Exercise 6.13. The wine industry uses various chemical and sensory tests to determine the quality of wine. Using a dataset containing chemical properties of wine samples, we aim to classify the quality of wine into different categories using OvO classification. A recommended dataset for this exercise is the `Wine` dataset available from the UCI Machine Learning Repository. This data consists of $n = 4898$ wine samples where the quality of the wine is measured on an ordinal categorical scale from 0 to 10.

1. Train an OvO classifier using a suitable algorithm (e.g., support vector machines, logistic regression, etc.; readers can also try methods discussed in later chapters like Chapter 16).
2. Evaluate the performance of the OvO classifier and interpret the results.

Exercise 6.14. Simulate training and testing data from the spirals data set. Fit the training data using k nearest neighbors for $k = 1, \ldots, 20$, and use the testing data to determine the misclassification error rate. Plot the test set misclassification error rate against k. Comment on your findings.

Exercise 6.15. Consider a query point $\mathbf{X}_0$ in a p-dimensional unit hypercube $[0,1]^p$. The training set consists of n points uniformly distributed in $[0,1]^p$. We are interested in the distance D_k from $\mathbf{X}_0$ to its k-th nearest neighbor as $p \to \infty$.

1. Consider a p-dimensional ball of radius r centered at $\mathbf{X}_0$. The volume $V_p(r)$ of this ball is given by

$$V_p(r) = \frac{\pi^{p/2}}{\Gamma(p/2+1)} r^p.$$

   Therefore, the probability that all n points fall outside the ball is $(1 - V_p(r))^n$. Therefore, the CDF of the distance D to the nearest neighbor is $F_D(r) = 1 - (1 - V_p(r))^n$. Using Stirling's approximation show that $V_p(r) \to 0$ as $p \to \infty$. What can you conclude about D from this?
2. The distance to the k-th nearest neighbor, D_k, is the k-th order statistic of the distribution of D. The CDF of D_k is given by

$$F_{D_k}(r) = \sum_{i=k}^{n} \binom{n}{i} F_D(r)^i (1 - F_D(r))^{n-i}.$$

   Using the first part, what can you conclude about D_k?

Exercise 6.16. The previous exercise shows that for large p, the distance D_k from $\mathbf{X}_0$ to its k-th nearest neighbor is likely to be very large. In this exercise, your task is to try to determine what the theoretical mean of D_k is by using simulations.

1. Generate n points uniformly distributed in a p-dimensional unit hypercube. Define a query point $\mathbf{X}_0$ randomly chosen within the unit hypercube and calculate the Euclidean distances from the query point to all other points. Find the k-th nearest neighbor distance.

2. Repeat the process multiple times to obtain an empirical distribution of the k-th nearest neighbor distances.
3. What is your conclusion regarding $\mathbb{E}(D_k)$? Vary n, p and k in making your final assessment.

Exercise 6.17. In this exercise, you will empirically test the Cover-Hart theorem by comparing the error rates of the nearest neighbor (1-NN) classifier and the Bayes classifier on a synthetic dataset, demonstrating that the bound (6.8) holds with high probability.

1. Create a binary classification problem where the true decision boundary is known. Since the true distribution is known, calculate the Bayes error rate analytically.
2. Simulate data from your classification problem and determine the 1-NN classifier.
3. Repeat the process multiple times to observe the proportion of times the 1-NN error rate is within the bound. Vary the sample size n and state your final assessment keeping n in mind.

Bibliography

[1] Cover T.M. and Hart P.E. Nearest neighbor pattern classification. *IEEE Transactions on Information Theory,* 13(1):21–27, 1967.

[2] Devroye, L., Györfi, L., and Lugosi, G. *A Probabilistic Theory of Pattern Recognition.* Springer, 1996.

[3] Friedman J.H., Another approach to polychotomous classification. *Technical Report, Statistics Department, Stanford University,* 1996.

[4] Leisch F. and Dimitriadou E. mlbench: Machine Learning Benchmark Problems. R package version 2.1-3.1. `https://cran.r-project.org/package=mlbench`, 2023.

7

Generalization Error

7.1 Introduction

Chapter 4 studied Ordinary Least Squares (OLS) and evaluated goodness-of-fit methods for measuring performance on training data. We found that methods like R^2 tend to select larger, overfitted models with high mean-squared error (MSE). Since MSE is related to prediction error, it might seem that choosing models based on training error would lead to inferior performance. But is this really true? Is there a substantial difference between training error and prediction error, and can we study these differences ourselves?

This chapter aims to explore this question in detail, focusing on the principle of empirical risk minimization, which is a general technique for training learning procedures by reducing training error. This study will lead us to the statistical theory of learning, where we will find that consistent learning is achievable using empirical risk minimization under certain regularity conditions pertaining to the model space size, and thus represents, in certain ways, the generalization of what we learned previously in regards to regularization in Chapter 4.3. We will provide a general overview of these concepts and state some important theoretical results in this area.

Another objective is to provide a general definition of generalization error (prediction error) that extends beyond the linear models we previously explored. We will also discuss methods for empirically estimating this error, such as cross-validation and the leave-one-out bootstrap. Additionally, we will delve into why cross-validation is effective in the context of linear regression, linking these concepts back to penalization methods like AIC and BIC discussed earlier. It will be demonstrated that leave-one-out cross-validation is reliable for minimizing prediction error. However, when the focus is on model selection, a more rigorous form of cross-validation is required.

7.2 Definition of Generalization Error

We start with a discussion of generalization error, providing a general definition and indicating several examples of its use. The following general setup will be considered. Let $\mathscr{L} = \{(\mathbf{X}_i, Y_i)\}_1^n$ be the learning data, where $(\mathbf{X}_i, Y_i)$ are i.i.d. from a common distribution $\mathbb{P}$. Here, $\mathbf{X}_i \in \mathbb{R}^p$ are p-dimensional covariates, and $Y_i \in \mathscr{Y}$ is the response value (which may or may not be continuous). We write $(\mathbf{X}, Y) \in \mathbb{R}^p \times \mathscr{Y}$ for a generic data point from $\mathbb{P}$, where $(\mathbf{X}, Y)$ is assumed to be independent of $\mathscr{L}$.

Let $\hat{F}(\mathbf{X}) := \hat{F}(\mathbf{X}, \mathscr{L})$ be an estimator for the unknown target function $F(\mathbf{X})$. Previously, we introduced the concept of *conditional prediction error* for the linear regression model. More generally, we can define the conditional prediction error for an estimator $\hat{F}$ as

$$\mathrm{err}(\hat{F}) = \mathbb{E}_{\mathbf{X},Y}\left\{L(Y, \hat{F}(\mathbf{X}, \mathscr{L}))|\mathscr{L}\right\} = \mathbb{E}_{\mathbf{X},Y}\left\{L(Y, \hat{F}(\mathbf{X}))\right\},$$

DOI: 10.1201/9781003493679-7

where $L(Y,\hat{F}) \geq 0$ is a prespecified function, called the *loss function,* that measures discrepancy between Y and the predicted value $\hat{F}$.

Recall from Chapter 4 that the *generalization error,* or unconditional prediction error, is obtained by integrating the conditional prediction error over $\mathscr{L}$. Thus, we have the following general definition for the generalization error,

$$\mathrm{Err}(\hat{F}) = \mathbb{E}_{\mathscr{L}}\mathbb{E}_{\mathbf{X},Y}\left\{L(Y,\hat{F}(\mathbf{X}))\right\}. \tag{7.1}$$

Below we list several examples that fall within the above framework.

1. For regression, the target function is typically $F(\mathbf{X}) = \mathbb{E}(Y|\mathbf{X})$ and $L(Y,\hat{F}) = (Y-\hat{F})^2$ is called the squared-error loss, or *the* L_2*-loss function.*
2. Quantile regression is an example where the goal is slightly different than (1). The target function of interest is a prespecified quantile for the conditional cumulative distribution function $F(y|\mathbf{X}) = \mathbb{P}\{Y \leq y|\mathbf{X}\}$. If $0 < \alpha < 1$ is the quantile of interest, the target is $F(\mathbf{X}) = \inf\{y : F(y|\mathbf{X}) \geq \alpha\}$. The loss used is
$$L(Y,\hat{F}) = \phi_\alpha(Y-\hat{F})$$
where $\phi_\alpha(u) = u(\alpha - 1_{\{u<0\}})$ is the so-called *check loss-function* (or tick-loss).

   The quantile loss function is also referred to as the *pinball loss.* The formula for the pinball loss for a given quantile α can be rewritten as
$$L(Y,\hat{F}) = \begin{cases} \alpha(Y-\hat{F}) & \text{if } Y \geq \hat{F} \\ (\alpha-1)(Y-\hat{F}) & \text{if } Y < \hat{F}. \end{cases}$$
   Notice that when written in terms of the check-loss function, this is equivalent to
$$\phi_\alpha(u) = \begin{cases} \alpha U & \text{if } U \geq 0 \\ (\alpha-1)U & \text{if } U < 0 \end{cases}$$
   where $U = Y - \hat{F}$ is the residual. Therefore, the quantile loss function is the check-loss function for the residual evaluated at the quantile value α.
3. In quantile regression, we might be interested in estimating the entire distribution function $F(y|\mathbf{X}) = \mathbb{P}\{Y \leq y|\mathbf{X}\}$ rather than focusing on a specific quantile value. In this case, the *continuous ranked probability score* (CRPS) loss can be used, defined as
$$\begin{aligned} L(Y,\hat{F}) &= \int_{-\infty}^{\infty} \left(\hat{F}(u) - 1_{\{Y\leq u\}}\right)^2 du \\ &= \mathbb{E}_U|U-Y| - \frac{1}{2}\mathbb{E}_{U,U'}|U-U'| \end{aligned}$$
   where in the second line, U, U' are i.i.d. copies under the distribution $\mathbb{P}$ for Y. The first line defining CRPS shows it is a type of integrated L_2-loss function (closely related to the Brier score discussed below). The identity in the second line shows that the loss is related to the absolute error and is measured on the original scale of Y.
4. In classification with a binary response $Y \in \{0,1\}$, the unknown target function is $F(\mathbf{X}) = \mathbb{P}\{Y = 1|\mathbf{X}\}$. Often the goal in such analyses is accurate classification and for this purpose performance is measured using 0-1 loss or *misclassification loss* which corresponds to
$$L(Y,\hat{F}) = 1_{\{Y\neq\hat{\delta}\}}$$

where $\hat{\delta} := \hat{\delta}(\hat{F})$ is the Bayes classifier $\hat{\delta}(\hat{F}) = 1_{\{\hat{F} \geq 1/2\}}$. As discussed in Chapter 6.6 the Bayes decision rule is the optimal classifier under 0-1 loss.

5. Other examples used in binary classification are losses $L(Y, \hat{F})$ of the form $\phi(Y\hat{F})$ where $\phi : \mathbb{R} \to \mathbb{R}^+$ is a convex function and where Y is recoded as $\{-1, +1\}$. Examples include: (i) *logit loss*, $\phi(u) = \log_2(1 + \exp(-u))$; (ii) *exponential loss*, $\phi(u) = \exp(-u)$; and (iii) *hinge loss*, $\phi(u) = \max(1 - u, 0)$.
6. Zero-one loss is also used in multiclassification for responses $Y \in \{1, \ldots, J\}$ with $J \geq 2$ labels. Let $\hat{F} = (\hat{F}_1, \ldots, \hat{F}_J)^T$ where $\hat{F}_j := \hat{F}_j(\mathbf{X})$ is the estimator for the conditional probability $F_j(\mathbf{X}) = \mathbb{P}\{Y = j|\mathbf{X}\}$. Then $L(Y, \hat{F}) = 1_{\{Y \neq \hat{\delta}\}}$, where $\hat{\delta} := \hat{\delta}(\hat{F})$ is the Bayes multiclassifier (Chapter 6.6)

$$\hat{\delta}(\hat{F}) = \arg\max_{1 \leq j \leq J} \left\{ \hat{F}_j \right\}.$$

7. Another loss used in classification problems, $J \geq 2$, is the *Brier score* [2] defined by

$$L(Y, \hat{F}) = \frac{1}{J} \sum_{j=1}^{J} \left(1_{\{Y=j\}} - \hat{F}_j \right)^2. \tag{7.2}$$

This differs from 0-1 loss due to the use of L_2-error. The Brier score is used when the goal is to estimate the conditional j-probability, F_j.

8. A deficiency of (7.2) is that the reference value for random guessing depends on the number of class labels. For example, if $J = 2$, a classifier that randomly guesses the class label will have a Brier score of $1/4$. But as the number of classes J increases, the Brier score under random guessing converges to 1. A *normalized Brier score* resolves this issue and is defined as

$$L(Y, \hat{F}) = \frac{J}{J-1} \sum_{j=1}^{J} \left(1_{\{Y=j\}} - \hat{F}_j \right)^2. \tag{7.3}$$

Under (7.3), the random classifier has normalized Brier score equal to 1, independently of J.

9. The *cross-entropy loss* has similarties to the Brier score as it also measures the difference between the true class labels and the conditional j-probabilities. The cross-entropy loss for multiclassification is defined by

$$L(Y, \hat{F}) = -\sum_{j=1}^{J} 1_{\{Y=j\}} \log \hat{F}_j.$$

When $J = 2$, the cross-entropy loss is called *log-loss* because of its connection to the negative log-likelihood for logistic regression.

Table 7.1 summarizes some of the different loss functions used in regression and classification problems and lists the function that minimizes the loss. In each of these cases the function that minimizes the generalization error is the target function. Therefore, these are all examples of what are called *proper scoring rules* [7, 8]. A proper scoring rule is defined to be a loss function that satisfies

$$\mathbb{E}_{(\mathbf{X},Y)\sim F}\left\{L(Y, F(\mathbf{X}))\right\} \leq \mathbb{E}_{(\mathbf{X},Y)\sim F}\left\{L(Y, F^*(\mathbf{X}))\right\}, \quad \text{for all } F^*.$$

Thus, a scoring rule, or loss function, is called proper if the true distribution as the forecast distribution is an optimal strategy in expectation.

TABLE 7.1
Different loss functions used in statistical problems and the target function that minimizes the loss.

Statistical Problem	Type of Loss	Loss Function $L(Y,\hat{F})$	Optimal Function $\arg\min\{L(Y,F)\}$
Regression	L_2-loss	$(Y-\hat{F})^2$	$\mathbb{E}(Y\|\mathbf{X})$ Conditional mean
Quantile Regression	Quantile loss (Pinball loss)	$\phi_\alpha(Y-\hat{F})$ $\phi_\alpha(u)=u(\alpha-1_{\{u<0\}})$	$\inf\{y:\mathbb{P}\{Y\le y\|\mathbf{X}\}\ge\alpha\}$ α-quantile
Quantile Regression	CRPS	$\int_{-\infty}^{\infty}\left(\hat{F}(u)-1_{\{Y\le u\}}\right)^2 du$	$\mathbb{P}\{Y\le y\|\mathbf{X}\}$ Conditional CDF
Classification $J\ge2$	Zero-one	$1_{\{Y\neq\hat{\delta}(\hat{F})\}}{}^{\dagger}$	$\sum_{j=1}^{J}\sum_{j'\neq j}\mathbb{E}\left[1_{\{\delta(\mathbf{X})=j',Y=j\}}\right]^{\ddagger}$ Bayes Multiclass Error
Classification $J\ge2$	Brier score	$\frac{1}{J}\sum_{j=1}^{J}\left(1_{\{Y=j\}}-\hat{F}_j\right)^2$	$\mathbb{P}\{Y=j\|\mathbf{X}\}$ Conditional j-probability
Classification $J\ge2$	Cross-entropy	$-\sum_{j=1}^{J}1_{\{Y=j\}}\log\hat{F}_j$	$\mathbb{P}\{Y=j\|\mathbf{X}\}$ Conditional j-probability

$^{\dagger}\hat{\delta}(\hat{F})=\arg\max_{1\le j\le J}\{\hat{F}_j\}$

$^{\ddagger}\delta(\mathbf{X})=\arg\max_{1\le j\le J}\{F_j(\mathbf{X})\},\quad F_j(\mathbf{X})=\mathbb{P}\{Y=j|\mathbf{X}\}$

7.3 Generalization Error for Regression

In this section we take a closer look at the generalization error for regression and consider how its value can be decomposed. It is assumed that the nonparametric regression model satisfies

$$\mathbb{E}[Y|\mathbf{X}]=F(\mathbf{X}),\quad \mathbb{E}[(Y-F(\mathbf{X}))^2|\mathbf{X}]=\sigma^2. \tag{7.4}$$

Therefore, under L_2-loss, the generalization error for an estimator $\hat{F}$ of F is

$$\mathrm{Err}(\hat{F})=\mathbb{E}_{\mathscr{L}}\mathbb{E}_{\mathbf{X},Y}\left[Y-\hat{F}(\mathbf{X})\right]^2.$$

Recall that in Chapter 4 we showed that this can be further decomposed as

$$\mathrm{Err}(\hat{F})=\sigma^2+\mathbb{E}_{\mathscr{L},\mathbf{X}}\left[\hat{F}(\mathbf{X})-F(\mathbf{X})\right]^2.$$

Thus, the generalization error is described by the underlying noise σ^2 and the MSE. Furthermore, it was shown that the MSE itself can be described into terms for the bias-squared and variance of the estimator. Thus, in total, the generalization error is an expression of three quantities.

The previous result for the MSE relies on what is called the fixed **X** case, where features are pre-determined and conditioned on. Here, we provide a more general description where we consider **X** to be random. This will prove useful for later parts in the book.

We introduce the following definitions for the conditional bias and variance (conditional on **X**) for this purpose:

$$\begin{aligned}\text{Bias}\{\hat{F}\,|\,\mathbf{X}\} &= \mathbb{E}_{\mathscr{L}}\{\hat{F}(\mathbf{X})\} - F(\mathbf{X})\\ \text{Var}\{\hat{F}\,|\,\mathbf{X}\} &= \mathbb{E}_{\mathscr{L}}\left\{\left(\hat{F}(\mathbf{X}) - \mathbb{E}_{\mathscr{L}}[\hat{F}(\mathbf{X})]\right)^2\right\}.\end{aligned}$$

Observe that in the above definitions we used the independence between $\mathscr{L}$ and **X** to write $\mathbb{E}_{\mathscr{L}}$ in place of $\mathbb{E}_{\mathscr{L}|\mathbf{X}}$. Using this notation, it is now easy to check that

$$\begin{aligned}\text{Err}(\hat{F}) &= \sigma^2 + \mathbb{E}_{\mathscr{L},\mathbf{X}}\left[\hat{F}(\mathbf{X}) - \mathbb{E}_{\mathscr{L}}[\hat{F}(\mathbf{X})] + \mathbb{E}_{\mathscr{L}}[\hat{F}(\mathbf{X})] - F(\mathbf{X})\right]^2\\ &= \sigma^2 + \underbrace{\mathbb{E}_{\mathbf{X}}\left\{\text{Bias}\{\hat{F}\,|\,\mathbf{X}\}^2\right\}}_{\text{integrated bias-squared}} + \underbrace{\mathbb{E}_{\mathbf{X}}\left\{\text{Var}\{\hat{F}\,|\,\mathbf{X}\}\right\}}_{\text{integrated variance}}. \qquad (7.5)\end{aligned}$$

We can summarize this as follows.

The generalization error (i.e. prediction error) for nonparametric regression under L_2-loss equals the overall variance plus the MSE (also called mean squared prediction error). The MSE equals the integrated bias-squared plus the integrated variance. Thus, the prediction error is comprised of three terms: (i) the overall variance; (ii) the integrated bias-squared; (iii) the integrated variance.

For future reference, we note that the key step in the above is the MSE decomposition

$$\text{MSE} = \mathbb{E}\left[\hat{F}(\mathbf{X}) - F(\mathbf{X})\right]^2 = \mathbb{E}_{\mathbf{X}}\left\{\text{Bias}\{\hat{F}\,|\,\mathbf{X}\}^2 + \text{Var}\{\hat{F}\,|\,\mathbf{X}\}\right\}.$$

Readers might also find it useful to compare the above right-hand side to the MSE decomposition for the linear model given in (4.8). Chapter 7.10.1 returns to the linear model for further discussion about the generalization error in this case.

7.4 Generalization Error for Classification

Here we take a closer look at generalization error in the setting of binary classification, $Y \in \{0,1\}$. By our previous discussion, the generalization error for a classifier $\hat{\delta}$ under 0-1 loss is

$$\text{Err}(\hat{F}) = \mathbb{E}\left[1_{\{Y \neq \hat{\delta}(\mathbf{X})\}}\right] = \mathbb{P}\{Y \neq \hat{\delta}(\mathbf{X})\}$$

where $\hat{\delta}(\mathbf{X})$ is the classifier for the estimator $\hat{F}(\mathbf{X})$ of $F(\mathbf{X}) = \mathbb{P}\{Y = 1|\mathbf{X}\}$.

The use of 0-1 loss presupposes that classification is the goal, but sometimes we might be more interested in estimating the conditional probability. For example, classifying a tumor as benign or malignant is important, but knowing the probability that the tumor is malignant can provide more information to the surgical oncologist treating a patient.

As was discussed, loss functions like the Brier score and cross-entropy can be used for this purpose. Unlike 0-1 loss, which is used when the goal is classification, these metrics are specifically designed with the purpose of estimating the J-class probabilities. It is important to be aware of these differences from both a subject matter and statistical perspective. Interestingly, estimating the regression function F is harder than classifying a case. In fact, successful classification follows as a corollary to successful regression.

Consider the following result, which examines estimating F under L_2-loss.

Theorem 7.1. *[Theorem 1.1, 9]. Let $\hat{F}_n : \mathbb{R}^p \to [0,1]$ be an estimator for F. The classifier for $\hat{F}_n$ is defined as $\hat{\delta}_n(\boldsymbol{x}) = 1_{\{\hat{F}_n(\boldsymbol{x}) \geq 0.5\}}$ and let $\delta(\boldsymbol{x})$ denote the Bayes classifier for F. Then,*

$$\mathbb{P}\{\hat{\delta}_n(\mathbf{X}) \neq Y\} \leq \mathbb{P}\{\delta(\mathbf{X}) \neq Y\} + 2\left(\int_{\mathbb{R}^p} \left(\hat{F}_n(\boldsymbol{x}) - F(\boldsymbol{x})\right)^2 \mathbb{P}(d\boldsymbol{x})\right)^{1/2}.$$

Theorem 7.1 shows that if $\hat{F}_n$ is weakly consistent for F, then the classifier for $\hat{F}$ is automatically Bayes risk consistent. This makes sense intuitively because optimal classification of Y only requires determining whether F is greater than 1/2, whereas consistently estimating F requires $\hat{F}(\mathbf{x})$ to be close to $F(\mathbf{x})$ for almost all $\mathbf{x} \in \mathbb{R}^p$, which is much more difficult. Figure 7.1 illustrates this point.

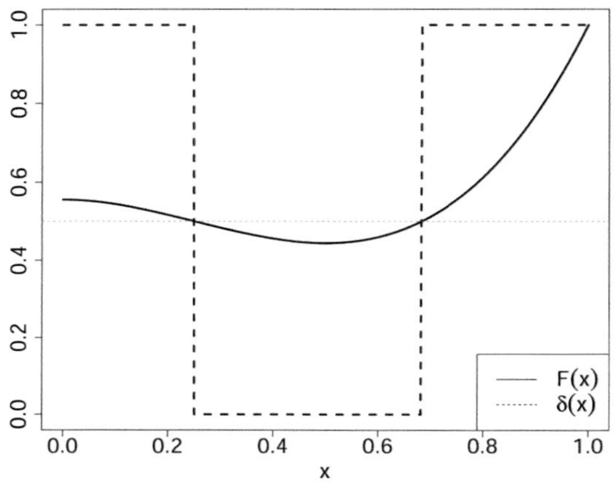

FIGURE 7.1
Regression is harder than classification. The thick line is $F(x) = \mathbb{P}\{Y = 1|X = x\}$ where $x \in [0,1]$. Dashed line is the Bayes classifier $\delta(x)$. It is harder to estimate F than δ because δ only requires knowing whether $F \geq 1/2$ (dashed blue line) and not the actual value of F.

7.4.1 Decomposition of the Brier score

As mentioned, the squared value appearing in the Brier score loss (7.2) is reminiscent of the L_2-loss used in regression. Although these two are not equivalent, we can still use an L_2-loss decomposition like (7.5) to understand the Brier score. For the two-class problem, the loss can be written as

$$L(Y, \hat{F}(\mathbf{X})) = \frac{1}{2}\sum_{j=0}^{1}\left(1_{\{Y=j\}} - \hat{F}_j(\mathbf{X})\right)^2$$

where $\hat{F}_1(\mathbf{X}) = 1 - \hat{F}_0(\mathbf{X}) = \mathbb{P}\{Y = 1|\mathbf{X}\}$. Let us consider the generalization error and its decomposition for the two-class problem under this loss.

In order to proceed, we will have to use a different decomposition than (7.5). This is because (7.5) relies on assumption (7.4) which stipulates an equal variance model, which obviously does not hold for the two-class problem. Therefore, for this decomposition, we can rely only on the model assumption $F(\mathbf{X}) = \mathbb{P}\{Y = 1|\mathbf{X}\} = \mathbb{E}[Y|\mathbf{X}]$.

We have

$$\begin{aligned}
\mathrm{Err}(\hat{F}) &= \mathbb{E}\left[Y - F(\mathbf{X}) + F(\mathbf{X}) - \hat{F}(\mathbf{X})\right]^2 \\
&= \mathbb{E}\left[(Y - F(\mathbf{X}))^2\right] + \mathbb{E}\left[\hat{F}(\mathbf{X}) - F(\mathbf{X})\right]^2 \\
&= \mathbb{E}_{\mathbf{X}}\left[(Y - \mathbb{E}(Y|\mathbf{X}))^2|\mathbf{X}\right] + \mathbb{E}\left[\hat{F}(\mathbf{X}) - \mathbb{E}(Y|\mathbf{X})\right]^2 \\
&= \mathbb{E}_{\mathbf{X}}\left[\mathrm{Var}(Y|\mathbf{X})\right] + \mathbb{E}\left[\hat{F}(\mathbf{X}) - \mathbb{E}(Y|\mathbf{X})\right]^2 \\
&= \mathrm{Var}(Y) - \mathrm{Var}\left(\mathbb{E}[Y|\mathbf{X}]\right) + \mathbb{E}\left[\hat{F}(\mathbf{X}) - \mathbb{E}(Y|\mathbf{X})\right]^2
\end{aligned}$$

where the last line holds from

$$\mathrm{Var}(Y) = \mathbb{E}_{\mathbf{X}}\left[\mathrm{Var}(Y|\mathbf{X})\right] + \mathrm{Var}\left(\mathbb{E}[Y|\mathbf{X}]\right).$$

As before, there are three terms in the decomposition, however, the values are different than (7.5). The first term measures the overall variance of the outcome and is called the *uncertainty*. Since it is a constant, the choice of $\hat{F}$ does not affect it. The second term is also a constant value. This term is called the *resolution* and measures the variability of the posterior mean of the data. It quantifies the informativeness of the learning problem. The third term is the integrated MSE. This is the value we would like to minimize in order to reduce generalization error. When working with the Brier score, this term is often referred to as the *reliability*.

7.5 Training Error

Now we turn to another goal of this chapter, which is to consider the issue of using training error for performance assessment. The training error, also called the resubstitution estimate, is defined as

$$\mathrm{err}_{\mathrm{naive}}(\hat{F}) = \frac{1}{n}\sum_{i=1}^{n} L(Y_i, \hat{F}(\mathbf{X}_i)).$$

This estimator is optimistically biased because it approximates the integral $\mathbb{E}_{(\mathbf{X},Y)}$ using the learning data, which was previously used to construct $\hat{F}$. For example, if we had additional data, we could construct a much better estimator by using an additional i.i.d. sample from $\mathbb{P}$ to approximate $\mathbb{E}_{(\mathbf{X},Y)}$. Denote this sample by $(\mathbf{X}_1^*, Y_1^*), \ldots, (\mathbf{X}_m^*, Y_m^*)$ and define the "iid" estimator as

$$\mathrm{err}_{\mathrm{iid}}(\hat{F}) = \frac{1}{m}\sum_{i=1}^{m} L(Y_i^*, \hat{F}(\mathbf{X}_i^*)).$$

This estimator works because, conditional on $\mathscr{L}$,

$$\mathrm{err}_{\mathrm{iid}}(\hat{F}) \xrightarrow{a.s.} \mathbb{E}_{\mathbf{X},Y} L(Y, \hat{F}(\mathbf{X})), \quad \text{as } m \to \infty$$

due to the law of large numbers (assuming $\mathscr{L}$ is fixed).

The optimism of performance might seem to preclude using training error for effective learning. Surprisingly, the training error is closer to the prediction error than we might think. In fact, we can quantify exactly how far apart these two values are.

To show this, let us simplify things by considering the classification problem, $Y \in \{0, 1\}$, where $F(\mathbf{X}) = \mathbb{P}\{Y = 1 \mid \mathbf{X}\}$ is the target under 0-1 loss. Let $\hat{\delta}$ be the Bayes classifier derived from $\hat{F}$, and, for the moment, we assume that $\hat{F}$ is a pre-chosen function not depending on the learning data. Then we have,

$$\mathrm{err}_{\mathrm{naive}}(\hat{F}) = \frac{1}{n}\sum_{i=1}^{n} 1_{\{Y_i \neq \hat{\delta}(\mathbf{X}_i)\}} = \frac{1}{n}\sum_{i=1}^{n} Z_i, \quad \text{where } Z_i = 1_{\{Y_i \neq \hat{\delta}(\mathbf{X}_i)\}}.$$

The generalization error is

$$\text{Err}(\hat{F}) = \mathbb{P}\{Y \neq \hat{\delta}(\mathbf{X})\} = \mathbb{E}Z, \quad \text{where } Z = 1_{\{Y\neq\hat{\delta}(\mathbf{X})\}}.$$

Because $Z, Z_1, \ldots, Z_n$ are i.i.d., by Hoeffding's inequality (Chapter 2.18), noting $a_i = 0 < Z_i < 1 = b_i$, we have

$$\mathbb{P}\left\{\left|\sum_{i=1}^{n}(Z_i - \mathbb{E}Z)\right| \geq t\right\} \leq 2\exp\left(-\frac{2t^2}{n}\right), \quad t > 0.$$

Dividing by n inside the probability, deduce for any $t > 0$

$$\mathbb{P}\left\{|\text{err}_{\text{naive}}(\hat{F}) - \text{Err}(\hat{F})| \geq t\right\} \leq 2\exp(-2t^2 n). \tag{7.6}$$

In particular for any $0 < \delta \leq 1$, by setting $2\exp(-2t^2 n) = \delta$ and solving for t, we have with probability at least $1-\delta$,

$$|\text{err}_{\text{naive}}(\hat{F}) - \text{Err}(\hat{F})| \leq \sqrt{\frac{\log(2/\delta)}{2n}},$$

or equivalently

$$\text{err}_{\text{naive}}(\hat{F}) - \sqrt{\frac{\log(2/\delta)}{2n}} \leq \text{Err}(\hat{F}) \leq \text{err}_{\text{naive}}(\hat{F}) + \sqrt{\frac{\log(2/\delta)}{2n}}.$$

Therefore, this shows that the training error can be used to estimate the generalization error. However, the accuracy of this approximation depends heavily on the sample size and how tight we want the bound to be in probability. Also, remember that this applies to a single $\hat{F}$. But suppose we have a collection of pre-chosen estimators $\hat{F}_1, \ldots, \hat{F}_N$ (also *not* depending on the learning data) and the goal is to select the best one using the training data. Then, do we arrive at the same bound?

The issue is that if we select the estimator with the smallest training error, then inequality (7.6) used to derive the bound no longer applies. This is because (7.6) holds only if $\hat{F}$ is *fixed in advance*. Let $\hat{F}^*$ be the estimator with the smallest training error. Then, an appropriate inequality for $\hat{F}^*$ is

$$\mathbb{P}\left\{\left|\text{err}_{\text{naive}}(\hat{F}^*) - \text{Err}(\hat{F})\right| \geq t\right\} \leq 2N\exp(-2t^2 n). \tag{7.7}$$

The above inequality is referred to as a *uniform inequality* because it holds over a collection of events.

> *The training error can be used to approximate the generalization error, but this approximation depends on the sample size, n, the tightness of the bound in probability, and how many estimators, N, we have.*

This shows that the more estimators we have, the less likely it is for the training error to be informative. Using the same argument as before, with probability at least $1-\delta$,

$$|\text{err}_{\text{naive}}(\hat{F}^*) - \text{Err}(\hat{F})| \leq \sqrt{\frac{\log(N) + \log(2/\delta)}{2n}}.$$

Because this depends on N, this suggests that too many estimators will degrade the effectiveness of training error for learning. The goal in learning theory is to consider how to make N so large that it can include a large class of estimators, including those obtained from the data, and for which the bound above can hold.

7.6 An Overview of the Theory of Learning

The last section gives just a small taste of what is encapsulated in the so-called statistical theory of learning. In this section we cover some of this material in more detail. The goal will be to provide a general overview of concepts and to state some of the more important theoretical results. For readers interested in a more comprehensive treatment of statistical learning we recommend consulting the work of Vapnik and Chervonenkis. A good starting point is the overview paper by Vapnik [16] which contains many useful references.

The general framework for studying learning is as follows. As before, there is a loss function $L(y, F(\mathbf{x}, \theta))$ used for performance evaluation where y is the observed response and $\mathbf{x}$ is the observed input. Here $F(\mathbf{x}, \theta)$ is the target function which is indexed by a parameter $\theta \in \Theta$. The goal is to minimize the *population risk functional*

$$R(\theta) = \int L(y, F(\mathbf{x}, \theta))\, d\mathbb{P}(\mathbf{x}, y) = \mathbb{E}[L(Y, F(\mathbf{X}, \theta))]$$

over $\theta \in \Theta$ where $\mathbb{P}$ is the distribution for $(\mathbf{X}, Y)$.

The above assumes a well defined loss function, but we can further generalize and abstract the learning problem without this requirement. Let $\mathbf{z} = (\mathbf{x}, y) \in \mathbb{R}^d$ and $Q(\mathbf{z}, \theta) = L(y, F(\mathbf{x}, \theta))$. Then, the general goal is to minimize

$$R(\theta) = \int Q(\mathbf{z}, \theta)\, d\mathbb{P}(\mathbf{z}) = \mathbb{E}[Q(\mathbf{Z}, \theta)]. \tag{7.8}$$

For example, for classification under zero-one loss, $Q(\mathbf{z}, \theta) = 1_{\{y \neq \delta(\mathbf{x}, \theta)\}}$ where $\delta(\mathbf{x}, \theta) = 1_{\{F(\mathbf{x}, \theta) \geq 1/2\}}$ is a zero-one classifier from a space of classifiers indexed by $\theta \in \Theta$. An example of a problem without a formal loss is the maximum likelihood estimator (MLE). Suppose the goal is to maximize the density $f(\mathbf{Z}|\theta) = f(\mathbf{X}, Y|\theta)$ for θ over some parameter space, Θ. Setting $Q(\mathbf{Z}, \theta) = -\log f(\mathbf{Z}|\theta)$, the population MLE minimizes the population negative log-likelihood,

$$R(\theta) = -\int \log f(\mathbf{z}|\theta)\, d\mathbb{P}(\mathbf{z}). \tag{7.9}$$

Remark 7.1. The previous example illustrates a useful feature of statistical learning: it allows for misspecification. Thus, the distribution $\mathbb{P}$ for the data in (7.9) may not necessarily be the same as the assumed model value $f(\cdot|\theta)$. This is because the goal is to find the best fitting density from some class $\mathscr{F}_\Theta = \{f(\cdot|\theta) : \theta \in \Theta\}$ where it is possible that $\mathscr{F}_\Theta$ can be misspecified by excluding the true data generating $\mathbb{P}$. This is common in machine learning, where the objective is not necessarily to recover the true model, which may not even exist, but to find the best fitting procedure from a sufficiently large space. □

While the optimization task (7.8) represents the theoretical goal in statistical learning, we have to realize that we cannot actually achieve this for real data since $\mathbb{P}$ will generally be unknown. Instead, we minimize the *empirical risk functional*

$$R_n(\theta) = \frac{1}{n} \sum_{i=1}^{n} Q(\mathbf{Z}_i, \theta) \tag{7.10}$$

where $\mathbf{Z}_1, \ldots, \mathbf{Z}_n \in \mathbb{R}^d$ are the i.i.d. learning data. For example, in place of minimizing the theoretical objective function (7.9), we minimize

$$R_n(\theta) = -\frac{1}{n} \sum_{i=1}^{n} \log f(\mathbf{Z}_i|\theta),$$

which is the negative log-likelihood function and yields the MLE.

This principle of replacing the population risk functional (7.8) with the data-based functional (7.10) is called the *empirical risk minimization principle* (ERMP). Thus, essentially the idea behind ERMP is to replace $\mathbb{P}$ with the empirical measure $\mathbb{P}_n$ (Chapter 2.13):

$$R(\theta) = \int Q(\mathbf{z},\theta)\, d\mathbb{P}(\mathbf{z}) = \mathbb{E}\left[Q(\mathbf{Z},\theta)\right]$$
$$\overset{\text{replaced by}}{\longrightarrow} R_n(\theta) = \int Q(\mathbf{z},\theta)\, d\mathbb{P}_n(\mathbf{z}) = \mathbb{E}_n\left[Q(\mathbf{Z},\theta)\right].$$

7.6.1 ERMP consistency

For the ERMP to be consistent, the empirical risk minimizer of (7.10) should converge to the minimizer of (7.8). Let $\hat{\theta}_n$ be the estimator which minimizes (7.10). Formally, we need the following to hold:

$$R_n(\hat{\theta}_n) \xrightarrow{\text{p}} R(\theta_0), \quad \text{where } \hat{\theta}_n = \arg\min_{\theta\in\Theta} R_n(\theta), \ \ \theta_0 = \arg\min_{\theta\in\Theta} R(\theta).$$

A sufficient and necessary condition for this is that the empirical risk functional converges uniformly to the population risk functional

$$\lim_{n\to\infty} \mathbb{P}\left\{\sup_{\theta\in\Theta}(R(\theta) - R_n(\theta)) > \varepsilon\right\} = 0, \quad \text{for all } \varepsilon > 0. \tag{7.11}$$

This one-sided convergence result implies that even the worst case scenario will converge. Define the worst solution $\hat{\theta}_n^*$ as

$$\hat{\theta}_n^* = \arg\max_{\theta\in\Theta}(R(\theta) - R_n(\theta)).$$

Then (7.11) asserts that $R(\hat{\theta}_n^*) = R_n(\hat{\theta}_n^*) + o_p(1)$.

Let us pause for a moment and summarize how estimation for a target function $F(\cdot,\theta)$ proceeds in the general approach to learning. Here, the empirical risk represents the training error and the estimator $\hat{\theta}_n$ is the estimator that minimizes the training error over a collection of models $\theta \in \Theta$:

$$\hat{\theta}_n = \arg\min_{\theta\in\Theta}\left\{\frac{1}{n}\sum_{i=1}^{n} L(Y_i, F(\mathbf{X}_i,\theta))\right\}.$$

The ERMP is based on the idea that if (7.11) holds, then the training error for the optimized estimator $F(\cdot,\hat{\theta}_n)$ converges to the population error of the optimal population estimator $F(\cdot,\hat{\theta}_0)$:

$$\frac{1}{n}\sum_{i=1}^{n} L(Y_i, F(\mathbf{X}_i,\hat{\theta}_n)) \xrightarrow{\text{p}} \int L(y, F(\mathbf{x},\theta_0))\, d\mathbb{P}(\mathbf{x}, y).$$

Therefore, optimization over the training error can be used to obtain estimators $F(\cdot,\hat{\theta}_n)$ with the property that their risk converges to the optimal population risk.

7.6.2 VC dimension

The ERMP approach is impressive since it allows θ to be selected from a space Θ of our choosing. However, in Chapter 7.5 we saw that we have to be careful not to allow Θ to be too large, since it is possible to over-train on the data. This is where a critical feature of the theory comes into play. It will be shown that even though Θ can be very general, there is a constraint to how large Θ is allowed to be.

The key to all of this is (7.11). A remarkable result is that there exists a simple condition that is both sufficient and necessary for (7.11) which holds for *any probability measure* $\mathbb{P}$. This condition depends on the *VC-dimension* which is a combinatorial quantity that measures the complexity of the space of learners. If this dimension is finite then (7.11) holds and the ERMP is consistent. Thus the VC-dimension is the value that tells us how large Θ is allowed to be if we want to avoid over-training.

> *The empirical risk minimization principle (ERMP) constructs an estimator from a class of models Θ by minimization of the empirical risk functional. The ERMP is consistent for Θ for any $\mathbb{P}$ if and only if the VC-dimension is finite.*

To define the VC-dimension we need to introduce the *growth function* which is a function that measures the ability of Θ to separate the learning sample. As this is a very general concept, we will simplify things and consider the classification problem under zero-one loss, in which case $Q(\mathbf{z},\theta) = 1_{\{y \neq \delta(\mathbf{x},\theta)\}}$ are zero-one functions.

When evaluated over the sample, each $Q(\mathbf{z},\theta)$ yields a binary vector of length n consisting of zero-one values

$$q(\theta, n) = \left(1_{\{y_1 \neq \delta(\mathbf{x}_1,\theta)\}}, \ldots, 1_{\{y_n \neq \delta(\mathbf{x}_n,\theta)\}}\right)^T \in \{0,1\}^n, \quad \theta \in \Theta.$$

We can think of the vector $q(\theta, n)$ as a way of encoding how a learner θ separates the space, with zeroes going into one group, and ones going into the other group. Let $G_z(\Theta, \mathbf{z}_1, \ldots, \mathbf{z}_n)$ be the total number of different ways Θ is able to separate the sample in this way. The growth function $G(\Theta, n)$ is defined as

$$G(\Theta, n) = \log\left\{ \sup_{\mathbf{z}_1,\ldots,\mathbf{z}_n \in \mathbb{R}^d} G_z(\Theta, \mathbf{z}_1, \ldots, \mathbf{z}_n) \right\}.$$

It can be shown that (7.11) holds for every $\mathbb{P}$ if and only if

$$\lim_{n \to \infty} \frac{G(\Theta, n)}{n} = 0. \tag{7.12}$$

Therefore, in order for the ERMP to be consistent for Θ, the growth function must be sublinear in n. This is a remarkable fact that the ability to learn using Θ can be quantified by knowing just this one fact. What is also impressive, is that the growth function possesses the unique property that it must either be linear or it can be bounded by a logarithmic function. More precisely, the growth function must either be $G(\Theta, n) = n \log 2$ or it can be bounded by

$$G(\Theta, n) < V\left(\log \frac{n}{V} + 1\right)$$

for some finite integer V where V is the smallest integer satisfying $G(\Theta, V) = V \log 2$. In the first case we say that the VC-dimension is infinite and in the second case we say that the VC-dimension is finite, where the value V denotes the VC-dimension. Consequently, from (7.12), we can conclude that the ERMP is consistent for Θ for every $\mathbb{P}$ if and only if the VC-dimension is finite.

7.7 Shatter Coefficient, Uniform Convergence of Empirical Measure

At an abstract level, consistency of the ERMP for Θ comes down to the question of uniform convergence of the empirical measure. If the underlying space of learners is not too large, then uniform convergence will be possible, and empirical risk minimization using Θ will be successful.

But quantifying the size of a space of learners Θ is not always straightforward and it will be easier to think about learning in terms of sets. Thus, given a class of sets $\mathscr{A}$ over $\mathbb{R}^d$, we are interested in the conditions for $\mathscr{A}$ to ensure convergence of

$$\mathbb{P}\left\{\sup_{A\in\mathscr{A}} |\mathbb{P}_n(A) - \mathbb{P}(A)| > \varepsilon\right\}. \tag{7.13}$$

This might seem strange at first, because how is $\mathscr{A}$ related to Θ, where the former is made up of sets on $\mathbb{R}^d$, and the latter is a parameter space indexing the class of learning procedures? However, if we think back to our last example involving classification, learning came down to quantifying the complexity of Θ by considering the ability to separate points. For empirical risk minimization to work, the number of n-dimensional binary vectors $q(\theta, n)$ generated by Θ was required to be sub-linear on the logarithmic scale for the VC-dimension to be finite. Therefore, by looking at the problem in just the right way, we turned this into a question about separating points in space. As will be seen, this is not much different than thinking about sets.

An important tool is the shatter coefficient which is combinatorial quantity for measuring the complexity of $\mathscr{A}$. This value plays a direct role in bounding (7.13).

Definition 7.1. Let $\zeta_1,\ldots,\zeta_n$ be vectors in $\mathbb{R}^q$ and let $\mathscr{A}$ be a collection of sets over $\mathbb{R}^q$. Define $N(\mathscr{A},\zeta_1,\ldots,\zeta_n)$ to be the number of distinct sets

$$\left\{\{\zeta_1,\ldots,\zeta_n\}\cap A\right\}$$

over $A\in\mathscr{A}$. Then the n-shatter coefficient for $\mathscr{A}$ is

$$s(\mathscr{A}, n) = \max_{\zeta_1,\ldots,\zeta_n\in\mathbb{R}^q} N(\mathscr{A},\zeta_1,\ldots,\zeta_n).$$

□

The shatter coefficient $s(\mathscr{A}, n)$ equals the number of ways that $\mathscr{A}$ can separate n points in $\mathbb{R}^q$. Below we list some useful examples (note we have set $q = p$ similar to the dimension of our feature space).

1. Let $\mathscr{A}$ be the set of all p-dimensional rectangles. Then $s(\mathscr{A}, n) \le n^{2p+1}$ for $n > 4p$ [4, Theorem 13.3 and 13.8].
2. Let $\mathscr{A}$ be the set of all half-spaces on $\mathbb{R}^p$ (i.e. the set of all hyperplanes on $\mathbb{R}^p$). Then $s(\mathscr{A}, n) \le n^p$ [3].
3. Let $\mathscr{A} = \{\sum_{l=1}^p (x_l - a_l)^2 \le b\}$ be the set of all closed balls on $\mathbb{R}^p$ where $\{a_l\}_1^p, b$ are real valued. By Theorem 13.9 of [4] [see also 3, 5, 13], if $\mathcal{V}$ is a finite dimensional vector space, then $\{g(\mathbf{x}) \le 0 : g\in\mathcal{V}\}$ has shatter coefficient $n^V + 1 \le n^{V+1}$ where $V \ge 1$ is the dimension of $\mathcal{V}$ for $n > 1$. Observe that $\sum_{l=1}^p (x_l - a_l)^2 - b$ is a finite dimensional vector space spanned by the intercept and polynomials of degree 2 in p variables. Therefore, $s(\mathscr{A}, n) \le n^{V+1}$ for $V \le p+1$.

Spaces like those listed above are common in machine learning. For example, half-spaces encompass decision boundaries produced by classification trees. Consider Figure 7.2, which shows a classification tree for the Iris data (see Chapter 15.2 for more about trees). Here, the data is restricted to the two species, *Versicolor* and *Setosa*, making it a two-class problem. Only sepal length and width features are used in the tree for clarity. The top figure displays the tree, and the bottom figure shows the tree partition. Notice that tree cuts are always parallel to the coordinate axis, referred to as *axis-aligned* cuts. Such cuts are special instances of half-spaces. For example, the top cut is based on sepal length and separates the data into regions where sepal

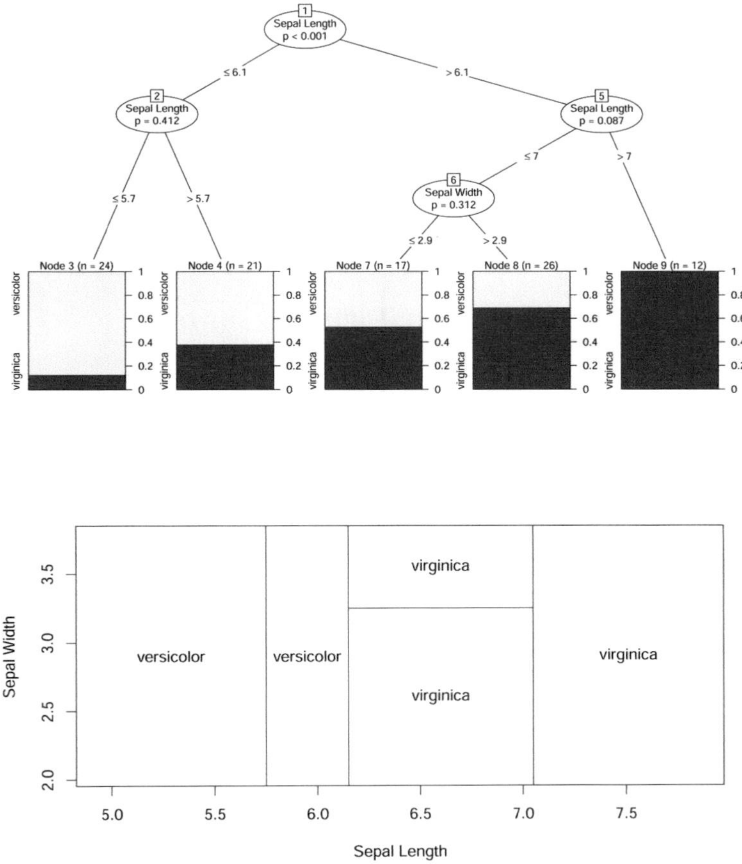

FIGURE 7.2
Classification tree (top) and its decision boundary (bottom) for Iris data (restricted to sepal length and sepal width features and Versicolor and Setosa species). Notice that tree cuts are always parallel to the coordinate axis, which is a special instance of half-space cuts, thus resulting in a partition of rectangles for the feature space.

length is ≤ 6.1, which is a hyperplane cut defined by one variable, and thus a special instance of a half-space.

Observe that in all the shatter coefficients listed above, they have a polynomial rate in n; i.e.

$$\lim_{n\to\infty} \frac{\log s(\mathscr{A}, n)}{n} = 0.$$

This type of behavior is important because it is sufficient to ensure the convergence of (7.13). This is the famous theorem of Vapnik and Chervonenkis.

Theorem 7.2. *(Vapnik and Chervonenkis, [17]) For any probability measure $\mathbb{P}$ and class of sets $\mathscr{A}$, for each $n \geq 1$ and each $\varepsilon > 0$,*

$$\mathbb{P}\left\{\sup_{A\in\mathscr{A}} |\mathbb{P}_n(A) - \mathbb{P}(A)| > \varepsilon\right\} \leq 8s(\mathscr{A}, n)\exp(-n\varepsilon^2/32).$$

Notice how much more general Theorem 7.2 is than (7.7), which was used to bound the generalization error for a fixed number of models. That bound was obtained from the Hoeffding inequality using Bonferroni's inequality. Theorem 7.2 replaces N, the number of models in that bound, with the shatter coefficient $s(\mathscr{A}, n)$. Importantly, observe that the bound remains small when the shatter coefficient is polynomial.

Theorem 7.2 tells us about the asymptotic behavior of the empirical measure. However, as stated it does not seem to directly apply to learners θ in an index set Θ, or to empirical and population risk functional, as in our original motivating problem.

To make this connection, consider again the classification problem under 0-1 loss. Then

$$R_n(\theta) = \frac{1}{n}\sum_{i=1}^{n} L(Y_i, F(\mathbf{X}_i, \theta)) = \frac{1}{n}\sum_{i=1}^{n} 1_{\{Y_i \neq \delta(\mathbf{X}_i,\theta)\}}$$

and

$$R(\theta) = \mathbb{E}[L(Y, F(\mathbf{X}, \theta))] = \mathbb{P}\{Y \neq \delta(\mathbf{X}, \theta)\}.$$

Consider sets of the following form

$$A_\theta = \Big\{\{\delta(\mathbf{x},\theta) = 1\} \times 0\Big\} \bigcup \Big\{\{\delta(\mathbf{x},\theta) = 0\} \times 1\Big\}, \qquad \theta \in \Theta. \tag{7.14}$$

Notice that A_θ can be equivalently described as

$$\begin{aligned} A_\theta &\equiv \Big\{(\mathbf{x}, y) : \delta(\mathbf{x},\theta) = 1, y = 0\Big\} \bigcup \Big\{(\mathbf{x}, y) : \delta(\mathbf{x},\theta) = 0, y = 1\Big\} \\ &= \Big\{(\mathbf{x}, y) : \delta(\mathbf{x},\theta) \neq y\Big\}. \end{aligned}$$

Consequently, $|\mathbb{P}_n(A_\theta) - \mathbb{P}(A_\theta)| > \varepsilon$ if and only if

$$\left| \frac{1}{n}\sum_{i=1}^{n} 1_{\{Y_i \neq \delta(\mathbf{X}_i,\theta)\}} - \mathbb{P}\{Y \neq \delta(\mathbf{X},\theta)\} \right| > \varepsilon,$$

or put another way: $|\mathbb{P}_n(A_\theta) - \mathbb{P}(A_\theta)| > \varepsilon$ if and only if $|R_n(\theta) - R(\theta)| > \varepsilon$. Let $\mathscr{A} = \{A_\theta : \theta \in \Theta\}$ be the set of all sets of the form (7.14) and let $s(\mathscr{A}, n)$ be its shatter coefficient. Then Theorem 7.2 applies and we obtain

$$\begin{aligned} \mathbb{P}\Big\{\sup_{\theta \in \Theta} |R_n(\theta) - R(\theta)| > \varepsilon\Big\} &= \mathbb{P}\Big\{\sup_{A \in \mathscr{A}} |\mathbb{P}_n(A) - \mathbb{P}(A)| > \varepsilon\Big\} \\ &\leq 8 s(\mathscr{A}, n) \exp(-n\varepsilon^2/32). \end{aligned}$$

It should also be clear that $s(\mathscr{A}, n)$ equals the number of ways that Θ can separate n points in space.

7.8 Partitioning Number and Super Greedy Trees

We present an example using classification trees to illustrate how the previous ideas can be extended. Let Π be a partition of $\mathbb{R}^p$. A partition of $\mathbb{R}^p$ of size $K+1$ is a finite collection $\Pi = \{A_1, \ldots, A_{K+1}\}$ of non-intersecting sets whose union equals $\mathbb{R}^p$. Given Π, we can define a classifier

$$\delta(\mathbf{x}, \Pi) = \sum_{A \in \Pi} \theta_A 1_{\{\mathbf{x} \in A\}} \tag{7.15}$$

where $\theta_A \in \{0, 1\}$. Such a classifier is also called a partitioning classifier.

Consider learning using partitioning classifiers of the form (7.15). We want to determine if learning is feasible in this setting using empirical risk minimization. This requires verifying that the empirical risk converges uniformly, which means controlling the underlying dimension of the problem. Previously, we used the shatter coefficient to bound the VC-dimension, but since this space involves partitions, we need a new measure of complexity for partitions.

Let $\mathscr{P}_K$ be a set containing partitions Π of $\mathbb{R}^p$ of size K. The complexity of $\mathscr{P}_K$ is quantified by the *partitioning number* [12] which is a combinatorial value similar to the growth function. Let $\mathbf{x}_1,\dots,\mathbf{x}_n$ be vectors in $\mathbb{R}^p$ and define $\mathfrak{p}_x(\mathscr{P}_K,\{\mathbf{x}_1,\dots,\mathbf{x}_n\})$ to be the total number of distinct partitions

$$\{\Pi \cap \{\mathbf{x}_1,\dots,\mathbf{x}_n\}\}$$

over $\Pi \in \mathscr{P}_K$ where order of appearance of sets is not counted. The partitioning number is

$$\mathfrak{p}(\mathscr{P}_K,n) = \max_{\mathbf{x}_1,\dots,\mathbf{x}_n\in\mathbb{R}^p} \mathfrak{p}_x(\mathscr{P}_K,\{\mathbf{x}_1,\dots,\mathbf{x}_n\})$$

and equals the maximum number of distinct partitions that can be obtained by separating any set of n vectors using $\Pi \in \mathscr{P}_K$.

To study if consistent learning can be achieved, we seek conditions on $\mathfrak{p}(\mathscr{P}_K,n)$ that ensures

$$\sup_{\Pi\in\mathscr{P}_K} \sum_{A\in\Pi} \left|\mathbb{P}_n(A) - \mathbb{P}(A)\right| \xrightarrow{\text{p}} 0. \tag{7.16}$$

It turns out that the partitioning number plays a role in the analysis of (7.16) just as the shatter coefficient does for (7.13). Consider the following result (see Lemma 1 from Lugosi and Nobel [12]).

Theorem 7.3. *For each $n \ge 1$ and each $\varepsilon > 0$*

$$\mathbb{P}\left\{\sup_{\Pi\in\mathscr{P}_K} \sum_{A\in\Pi} \left|\mathbb{P}_n(A) - \mathbb{P}(A)\right| > \varepsilon\right\} \le 2^{K+2}\cdot \mathfrak{p}(\Pi_K,2n)\cdot \exp(-n\varepsilon^2/32).$$

To see how Theorem 7.3 would apply to our example, observe that the empirical risk for the classifier (7.15) is

$$\begin{aligned} R_n(\Pi) &= \frac{1}{n}\sum_{i=1}^{n}\sum_{A\in\Pi} 1_{\{Y_i\neq\theta_A,\,\mathbf{X}_i\in A\}} \\ &= \sum_{A\in\Pi}\left[\frac{1}{n}\sum_{i=1}^{n} 1_{\{Y_i\neq\theta_A,\,\mathbf{X}_i\in A\}}\right] \\ &= \sum_{A\in\Pi}\left[\mathbb{P}_n\left(g(Y,\theta_A)1_{\{\mathbf{X}\in A\}}\right)\right], \end{aligned}$$

where $g(y,\theta) = 1_{\{y\neq\theta\}}$ and $\mathbb{P}_n$ is the empirical measure of $(\mathbf{X}_1,Y_1),\dots,(\mathbf{X}_n,Y_n)$. The population risk for the classifier is

$$R(\Pi) = \sum_{A\in\Pi}\left[\mathbb{P}\left(g(Y,\theta_A)1_{\{\mathbf{X}\in A\}}\right)\right].$$

We have the following crude upper bound:

$$\begin{aligned} |R_n(\Pi) - R(\Pi)| &\le \sum_{A\in\Pi}\left|\mathbb{P}_n\left(g(Y,\theta_A)1_{\{\mathbf{X}\in A\}}\right) - \mathbb{P}\left(g(Y,\theta_A)1_{\{\mathbf{X}\in A\}}\right)\right| \\ &\le \sum_{\theta\in\{0,1\}}\sum_{A\in\Pi}\left|\mathbb{P}_n\left(g(Y,\theta)1_{\{\mathbf{X}\in A\}}\right) - \mathbb{P}\left(g(Y,\theta)1_{\{\mathbf{X}\in A\}}\right)\right| \\ &\le \sum_{\theta\in\{0,1\}}\sup_{\Pi\in\mathscr{P}_K}\sum_{A\in\Pi}\left|\mathbb{P}_n\left(g(Y,\theta)1_{\{\mathbf{X}\in A\}}\right) - \mathbb{P}\left(g(Y,\theta)1_{\{\mathbf{X}\in A\}}\right)\right|. \end{aligned} \tag{7.17}$$

However, comparing this to Theorem 7.3, we can see this is not quite the same. Theorem 7.3 considers convergence for the quantity

$$\sup_{\pi \in \mathscr{P}_K} \sum_{A \in \pi} \left| \mathbb{P}_n(A) - \mathbb{P}(A) \right|$$

whereas we need to bound

$$\sup_{g \in \mathscr{G}} \left\{ \sup_{\pi \in \mathscr{P}_K} \sum_{A \in \pi} \left| \mathbb{P}_n \left(g(Y) 1_{\{\mathbf{X} \in A\}} \right) - \mathbb{P} \left(g(Y) 1_{\{\mathbf{X} \in A\}} \right) \right| \right\}$$

where $\mathscr{G}$ is a space of functions over $\mathscr{Y} = \{0, 1\}$. In our example, $\mathscr{G}$ is the relatively simple space consisting of the two functions

$$g(y, 0) = 1_{\{y \neq 0\}}, \quad g(y, 1) = 1_{\{y \neq 1\}}.$$

To deal with this, we will need a slightly more general result. Consider the following, which follows as a special case of Lemma 1 from Ishwaran and Lee [10].

Lemma 7.4. *Let $\mathscr{G}$ be a bounded class of real-valued functions on $\mathscr{Y} \subseteq \mathbb{R}$ which are either indicator functions or functions like $g(y) = \alpha y$ for $|\alpha| \leq C$ for some finite constant, C. Then if $\mathfrak{p}(\mathscr{P}_K, n)$ is polynomial in n and $K := K(n)$ satisfies $K n^{\gamma - 1} \log n \to 0$ as $n \to \infty$ for some $0 < \gamma < 1$,*

$$\sup_{g \in \mathscr{G}} \left\{ \sup_{\pi \in \mathscr{P}_K} \sum_{A \in \pi} \left| \mathbb{P}_n \left(g(Y) 1_{\{\mathbf{X} \in A\}} \right) - \mathbb{P} \left(g(Y) 1_{\{\mathbf{X} \in A\}} \right) \right| \right\} = o_p(n^{-\gamma/2}).$$

Combining (7.17) with Lemma 7.4, we observe that if the partitioning number for $\mathscr{P}_K$ is polynomial, and the size K of partitions in $\mathscr{P}_K$ is sufficiently smaller than n, then the empirical risk converges uniformly.

7.8.1 Super greedy tree illustration

As an illustration of a partitioning classifier, consider the following theoretical tree construction. This tree induction uses rectangular splits, unlike typical trees that use axis-aligned splits, as in our previous example using the Iris data. This more general construction is an example of a *super greedy tree*, which will be discussed in Chapter 15.11 in more detail.

The classifier is built as follows. The algorithm is initialized at step $k = 0$ by setting $\pi_0 = \mathbb{R}^p$ and assigning to π_0 the classifier that classifies all values as $\theta(\mathbb{R}^p) = 1_{\{\mathbb{P}\{Y=1\} \geq 1/2\}}$. This classifier represents the optimal partitioning classifier for π_0 because it minimizes the classification error for the set $\mathbb{R}^p$. To see this, let

$$\delta(\mathbf{x}, \theta, A) = \theta 1_{\{\mathbf{x} \in A\}}$$

be the elementary classifier that classifies all data values in a set A as $\theta \in \{0, 1\}$. Then

$$\delta^*(\mathbf{x}, A) = \delta(\mathbf{x}, \theta(A), A), \quad \text{where } \theta(A) = 1_{\{\mathbb{P}\{Y=1|\mathbf{X} \in A\} \geq 1/2\}}$$

is the *elementary Bayes classifier* that classifies all data values in a set A as $\theta(A)$. By Theorem 6.3,

$$\mathbb{P}\{Y \neq \theta(A) \,|\, \mathbf{X} \in A\} \leq \mathbb{P}\{Y \neq \theta \,|\, \mathbf{X} \in A\}, \quad \theta \in \{0, 1\}.$$

Thus, the elementary Bayes classifier is the optimal classifier for a set. In particular, $\delta^*(\mathbf{x}, \mathbb{R}^p) = \theta(\mathbb{R}^p)$.

The partition π_0 is then split at step $k = 1$ along a rectangle A yielding the updated partition $\pi_1 = \{A, A^c\}$ where $A^c = \mathbb{R}^p \setminus A$ is the complement of A. The rectangle used for splitting is chosen from the set of all possible rectangles so as to maximize the decrease in population risk

$$\mathbb{E}\left[1_{\{Y \neq \delta^*(\mathbf{X}, \mathbb{R}^p)\}}\right] - \left[\mathbb{E}\left[1_{\{Y \neq \delta^*(\mathbf{X}, A)\}} 1_{\{\mathbf{X} \in A\}|}\right] + \mathbb{E}\left[1_{\{Y \neq \delta^*(\mathbf{X}, A^c)\}} 1_{\{\mathbf{X} \in A^c\}|}\right]\right]. \quad (7.18)$$

The updated classifier is $\delta^*(\mathbf{x}, A) + \delta^*(\mathbf{x}, A^c)$.

So to summarize so far: For step $k = 0$, the partition is trivially set to $\Pi_0 = \mathbb{R}^p$ and the initialized tree classifier is defined by

$$\delta(\mathbf{x}, \Pi_0) = \theta(\mathbb{R}^p).$$

Then for step $k = 1$, an optimal A is found by maximizing (7.18), and Π_0 is refined to $\Pi_1 = \{A, A^c\}$ and the tree classifier is updated to

$$\delta(\mathbf{x}, \Pi_1) = \theta(A)1_{\{\mathbf{x}\in A\}} + \theta(A^c)1_{\{\mathbf{x}\in A^c\}}.$$

See Figure 7.3 (a) for illustration of a $k = 1$ rectangle cut.

Next, for step $k = 2$, the algorithm seeks the rectangle that splits A to maximize decrease in population risk over A. The same is done for A^c (see Figure 7.3 (b) and (c) for illustration). Then the region A or A^c which yields the largest decrease in empirical risk is split. This process is repeated: at each step k all current regions of the space are considered, and the region yielding the largest decrease in empirical risk using a rectangle cut is split. At the end of step k, the current partition Π_{k-1} is refined to Π_k, and the partitioning classifier is updated from $\delta(\mathbf{x}, \Pi_{k-1})$ to $\delta(\mathbf{x}, \Pi_k)$. The tree construction is run for $k = 1, 2, \ldots K$ steps, yielding at the end of the procedure, a partition over $\mathbb{R}^p$ of size $K + 1$, Π_K, and its partitioning classifier, $\delta(\mathbf{x}, \Pi_K)$, which is an example of a super greedy tree.

This tree differs from a typical tree in two ways: (1) it uses greedy cuts, and (2) it uses geometric objects (rectangles in this example) for cuts. These differences are important and can significantly improve performance. Chapter 15.11 will show that the complexity of the super greedy tree, as measured by the partitioning number, is bounded by the shatter coefficient. Since the shatter coefficient for interesting geometric objects can be much larger than for half-spaces used by standard trees, this results in a larger space of partitions to reduce empirical risk, explaining why a super greedy tree can perform better.

7.9 Cross-Validation

Generalization error (prediction error) is the gold standard for assessing a predictor's performance, but since it requires knowing the true data-generating mechanism $\mathbb{P}$, which is generally unknown, we must use empirical techniques to estimate it. In this section, we discuss cross-validation, one of the most commonly used methods for estimating generalization error.

Formally, let $s \subset \{1, \ldots, n\}$ be a random set with distribution $\mathbb{P}_s$ and let $\mathscr{L}_{(s)}$ be the learning set restricted to s and $\mathscr{L}_{(-s)}$ be the learning set restricted to $\{1, \ldots, n\} \setminus s$. Then $\mathscr{L}_{(s)}$ denotes the test data and $\mathscr{L}_{(-s)}$ the training data corresponding to s. The cross-validation estimate for the generalization error (7.1) is defined as

$$\mathrm{Err}_{\mathrm{CV}}(\hat{F}) = \mathbb{E}_s \left[\frac{1}{|s|} \sum_{i \in s} L\left(Y_i, \hat{F}\left(\mathbf{X}_i, \mathscr{L}_{(-s)}\right)\right) \right].$$

Observe that $\hat{F}$ is constructed using $\mathscr{L}_{(-s)}$ while the performance of the predictor is estimated using $\mathscr{L}_{(s)}$.

Different distributions for s give rise to different cross-validation estimates. The deleting-d multifold cross-validation estimator [19] assumes a uniform distribution on all subsets of size d. The training data is of size $n - d$ and the test data of size d. For regression under L_2-loss, the

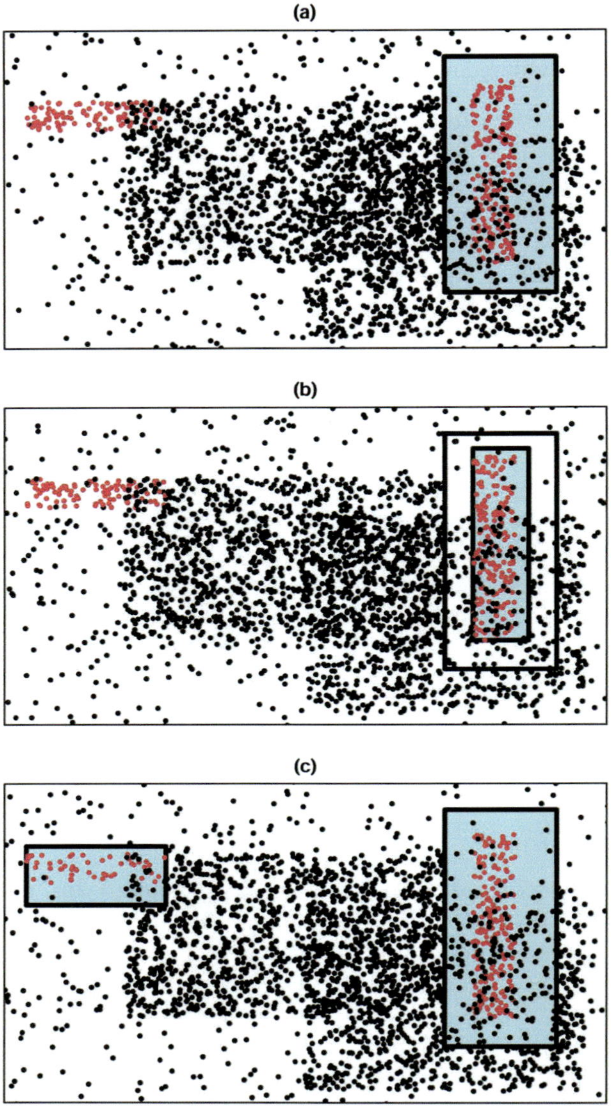

FIGURE 7.3
Rectangle splits for two-class data set using a super greedy tree (class labels indicated by black and red points): (a) $k = 1$ split cuts the space using a blue rectangle (top right); (b) $k = 2$ split cuts the rectangle from the $k = 1$ split into two regions where the blue region is the cut and remaining region is the complement; (c) $k = 2$ split cuts the complementary region from the $k = 1$ split using the blue rectangle in the top left.

estimator becomes

$$\text{MCV}_d = \left[d\binom{n}{d} \right]^{-1} \sum_{s} \sum_{i \in s} \left(Y_i - \hat{F}(\mathbf{X}_i, \mathcal{L}_{(-s)}) \right)^2 \tag{7.19}$$

where the summation over s is over all subsets of size d.

Expression (7.19) is also called M-fold CV (cross-validation) when $M = n/d$. Due to computational cost, it is often not possible to run M-fold CV by summing over all subsets of size d. In place of this, the sum is taken over a fixed set of M mutually exclusive sets each of size d. Given $s_1, \ldots, s_M \subseteq \{1, \ldots, n\}$ of mutually exclusive sets of size d such that $\bigcup_{j=1}^{M} s_j = \{1, \ldots, n\}$, the M-fold CV estimate is [1]

$$\text{MCV}_M = \frac{1}{n} \sum_{j=1}^{M} \sum_{i \in s_j} \left(Y_i - \hat{F}(\mathbf{X}_i, \mathcal{L}_{(-s_j)}) \right)^2 .$$

The special case when $M = n$ and $d = 1$ corresponds to leave-one-out cross-validation, often simply abbreviated as CV or loo-CV. In this case the training data is of size $n - 1$ and the test data set has one observation. The CV estimator is [14, 15]

$$\text{CV} = \frac{1}{n} \sum_{i=1}^{n} \left(Y_i - \hat{F}(\mathbf{X}_i, \mathcal{L}_{(-i)}) \right)^2 . \tag{7.20}$$

Another variation of M-fold validation makes use of repeated sampling. Rather than working with a pre-chosen partition $s_1, \ldots, s_M$ of $\{1, \ldots, n\}$ we randomly sample $B > 1$ sets $s_1^*, \ldots, s_B^*$ each of size d. This is sometimes called the repeated learning-testing (RLT) method. The RLT estimator is

$$\text{RLT}_M = \frac{1}{Bd} \sum_{b=1}^{B} \sum_{i \in s_b^*} \left(Y_i - \hat{F}(\mathbf{X}_i, \mathcal{L}_{(-s_b^*)}) \right)^2 .$$

7.10 Cross-Validation for the Linear Model

To gain insight and better understand cross-validation, we return to the linear model. We consider L_2-loss and study how cross-validation can be used to determine the optimal model using restricted least squares. Assume $\mathbf{Y} = \mathbf{D}\boldsymbol{\beta} + \boldsymbol{\epsilon}$, where $\boldsymbol{\beta} \in \mathbb{R}^p$ is the regression parameter vector, $\mathbf{Y} = (Y_1, \ldots, Y_n)^T$, $\boldsymbol{\epsilon} = (\varepsilon_1, \ldots, \varepsilon_n)^T$, and $\varepsilon_1, \ldots, \varepsilon_n$ are i.i.d. with mean zero and variance $\sigma^2 > 0$. $\mathbf{D}$ is the $n \times p$ design matrix with ith row entry $(x_{i1}, \ldots, x_{ip})$. For simplicity, $\mathbf{D}$ is taken to be fixed.

We assume that $\boldsymbol{\beta}$ is ordered such that its first k_0 coefficients are non-zero. The value k_0 is the true dimension of the model but is unknown. Denote by $\mathcal{M}_k$ the regression model involving only the *first* k coefficients of $\boldsymbol{\beta}$, and $\mathbf{D}_k$ the corresponding design matrix, for $k = 1, \ldots, p$. The restricted least squares estimator for $\mathcal{M}_k$ is

$$\hat{\boldsymbol{\beta}}_k = (\mathbf{D}_k^T \mathbf{D}_k)^{-1} \mathbf{D}_k^T \mathbf{Y}.$$

Consider using cross-validation to determine k_0. The optimal value for k_0 is found by choosing the k having the smallest cross-validation value. Let s denote a subset of $\{1, \ldots, n\}$ and assuming all inverses exists, define:

$$\begin{aligned}
\mathbf{Y}_s &= (Y_i, i \in s)^T \\
\mathbf{D}_{sk} &= (x_{ij}), \quad i \in s,\ j = 1,\ldots,k \\
\mathbf{D}_k &= (x_{ij}), \quad i = 1,\ldots,n,\ j = 1,\ldots,k \\
\mathbf{H}_{sk} &= \mathbf{D}_{sk}\mathbf{A}_k^{-1}\mathbf{D}_{sk}^T \\
\mathbf{H}_k &= \mathbf{D}_k\mathbf{A}_k^{-1}\mathbf{D}_k^T \\
\mathbf{A}_k &= \mathbf{D}_k^T\mathbf{D}_k.
\end{aligned}$$

Observe that $\hat{\boldsymbol{\beta}}_k = \mathbf{A}_k^{-1}\mathbf{D}_k^T\mathbf{Y}$ and the predicted value under $\hat{\boldsymbol{\beta}}_k$ is $\hat{\mathbf{Y}}_k = \mathbf{D}_k\hat{\boldsymbol{\beta}}_k = \mathbf{H}_k\mathbf{Y}$. The residual vector is $\hat{\mathbf{r}}_k = \mathbf{Y} - \hat{\mathbf{Y}}_k = (\mathbf{I} - \mathbf{H}_k)\mathbf{Y} = \mathbf{H}_k^{\perp}\mathbf{Y}$.

Consider selecting k_0 using the deleting-d multifold CV estimator (7.19)

$$\mathrm{MCV}_d(k) = \left[d\binom{n}{d}\right]^{-1} \sum_s \left\|\mathbf{Y}_s - \mathbf{D}_{sk}\hat{\boldsymbol{\beta}}_{(-s),k}\right\|^2, \tag{7.21}$$

where $\hat{\boldsymbol{\beta}}_{(-s),k}$ is the restricted least squares estimator under $\mathscr{M}_k$ calculated using only cases in $\mathscr{L}_{(-s)}$. The summation runs over all possible subsets s of size d.

The special case of CV (loo-CV) corresponds to $s = i$ and $d = 1$. Then, (7.21) becomes

$$\mathrm{CV}(k) = n^{-1}\sum_{i=1}^{n}\left(Y_i - \mathbf{D}_{ik}\hat{\boldsymbol{\beta}}_{(-i),k}\right)^2,$$

where $\mathbf{D}_{ik} = (x_{i1},\ldots,x_{ik})$.

Expression (7.21) can be expanded using the residual sum-of-squares, $\hat{S}_k = \hat{\mathbf{r}}_k^T\hat{\mathbf{r}}_k$. It can be shown that (see Exercise 7.10)

$$\mathrm{MCV}_d(k) = n^{-1}\hat{S}(k) + \left[d\binom{n}{d}\right]^{-1} \sum_s \hat{\mathbf{r}}_{sk}^T\mathbf{W}_{sk}\hat{\mathbf{r}}_{sk}, \tag{7.22}$$

where $\mathbf{W}_{sk} = (\mathbf{I} - \mathbf{H}_{sk})^{-2} - \mathbf{I}$ and $\hat{\mathbf{r}}_{sk} = \mathbf{Y}_s - \mathbf{D}_{sk}\hat{\boldsymbol{\beta}}_k$. Expansion (7.22) is interesting because it shows $\mathrm{MCV}_d(k)$ is composed of two pieces: The first being the mean residual sum-of-squares; the second being a weighted average of residuals. So from this we can see that (7.22) is a FPE criterion where the second term represents the penalty.

For loo-CV, (7.22) implies

$$\mathrm{CV}(k) = n^{-1}\hat{S}(k) + n^{-1}\sum_{i=1}^{n}\frac{H_{ik}(2 - H_{ik})}{(1 - H_{ik})^2}\hat{r}_{ik}^2. \tag{7.23}$$

This is a useful computational formula. For each k, the value $\mathrm{CV}(k)$ is entirely determined from $\mathbf{H}_k$ since $\hat{S}(k) = \sum_{i=1}^n \hat{r}_{ik}^2$, $\hat{r}_{ik} = ((\mathbf{I} - \mathbf{H}_k)\mathbf{Y})_i$ and $H_{ik} = \mathbf{e}_i^T\mathbf{H}_k\mathbf{e}_i$ where $\mathbf{e}_i$ is the ith Euclidean vector. Only $\mathbf{H}_k$ has to be calculated for model k, in comparison to brute-force cross-validation, which requires refitting the least squares estimator n times (to remove each i), for each k.

The second term of $\mathrm{CV}(k)$ is the penalty term and equals a weighted average of the squared residuals $\hat{r}_{ik}^2$, with weights calculated by combining H_{ik} in a specific way. It is useful to compare how these weights differ from another closely related metric, the generalized cross-validation (GCV) criteria [18]

$$\begin{aligned}
\mathrm{GCV}(k) &= \frac{1}{n}\|(\mathbf{I} - \mathbf{H}_k)\mathbf{Y}\|^2 \Big/ \left[\frac{1}{n}\mathrm{tr}(\mathbf{I} - \mathbf{H}_k)\right]^2 \\
&= n^{-1}\hat{S}(k)\Big/\left(1 - n^{-1}k^*\right)^2, \quad \text{where } k^* = \mathrm{tr}(\mathbf{H}_k).
\end{aligned}$$

We expand this expression using the following identity. For any $0 < h_k < 1$,

$$\frac{1}{(1-h_k)^2} = 1 + \frac{h_k(2-h_k)}{(1-h_k)^2}.$$

Therefore, setting $h_k = n^{-1}k^*$,

$$\mathrm{GCV}(k) = n^{-1}\hat{S}(k) + \frac{h_k(2-h_k)}{n(1-h_k)^2}\sum_{i=1}^{n}\hat{r}_{ik}^2$$

which is nearly the same as (7.23) expect that H_{ik} is replaced with h_k for each i.

7.10.1 Connection to prediction error

Now we look at how cross-validation estimates prediction error, We begin by obtaining a closed form expression for prediction error.

Consider new independent data $\mathbf{Y}^{\mathrm{new}} = \mathbf{D}\boldsymbol{\beta} + \boldsymbol{\epsilon}^{\mathrm{new}}$, where $\boldsymbol{\epsilon}^{\mathrm{new}} = (\epsilon_1^{\mathrm{new}}, \ldots, \epsilon_n^{\mathrm{new}})^T$ are i.i.d. measurements with the same distribution as $\boldsymbol{\epsilon}$. The conditional prediction error for $\hat{\mathbf{Y}}_k = \mathbf{H}_k\mathbf{Y}$ is

$$\mathrm{err}(k) = n^{-1}\mathbb{E}_{\mathrm{new}}\|\mathbf{Y}^{\mathrm{new}} - \hat{\mathbf{Y}}_k\|^2.$$

Let $\boldsymbol{\mu} = \mathbf{D}\boldsymbol{\beta}$ be the true mean. Then

$$\begin{aligned}
\mathrm{err}(k) &= n^{-1}\mathbb{E}_{\mathrm{new}}\|(\boldsymbol{\mu} + \boldsymbol{\epsilon}^{\mathrm{new}}) - \mathbf{H}_k(\boldsymbol{\mu} + \boldsymbol{\epsilon})\|^2 \\
&= n^{-1}\mathbb{E}_{\mathrm{new}}\|\boldsymbol{\epsilon}^{\mathrm{new}} + \mathbf{H}_k^{\perp}\boldsymbol{\mu} - \mathbf{H}_k\boldsymbol{\epsilon}\|^2 \\
&= n^{-1}\mathbb{E}_{\mathrm{new}}\|\boldsymbol{\epsilon}^{\mathrm{new}}\|^2 + n^{-1}\|\mathbf{H}_k^{\perp}\boldsymbol{\mu} - \mathbf{H}_k\boldsymbol{\epsilon}\|^2 \\
&= \sigma^2 + n^{-1}\boldsymbol{\mu}^T\mathbf{H}_k^{\perp}\boldsymbol{\mu} + n^{-1}\boldsymbol{\epsilon}^T\mathbf{H}_k\boldsymbol{\epsilon}.
\end{aligned}$$

Taking expectations over $\boldsymbol{\epsilon}$, and using $\mathbb{E}(\boldsymbol{\epsilon}^T\mathbf{H}_k\boldsymbol{\epsilon}) = k^*\sigma^2$, we obtain the prediction error

$$\mathrm{Err}(k) = \sigma^2 + n^{-1}\boldsymbol{\mu}^T\mathbf{H}_k^{\perp}\boldsymbol{\mu} + n^{-1}k^*\sigma^2.$$

The first term, σ^2, is the noise in the model; the second term, $n^{-1}\boldsymbol{\mu}^T\mathbf{H}_k^{\perp}\boldsymbol{\mu}$, is the bias; the final term, $n^{-1}k^*\sigma^2$, is the penalty. Readers might find it interesting to compare these last two terms to (4.8).

To compare this to cross-validation, we need to work out the mean of $n^{-1}\hat{S}_k$. Remembering that $\hat{S}_k = \hat{\mathbf{r}}_k^T\hat{\mathbf{r}}_k = \mathbf{Y}^T\mathbf{H}_k^{\perp}\mathbf{Y}$,

$$\begin{aligned}
n^{-1}\mathbb{E}(\hat{S}(k)) &= n^{-1}\mathbb{E}\left[(\boldsymbol{\mu} + \boldsymbol{\epsilon})^T\mathbf{H}_k^{\perp}(\boldsymbol{\mu} + \boldsymbol{\epsilon})\right] \\
&= n^{-1}\boldsymbol{\mu}^T\mathbf{H}_k^{\perp}\boldsymbol{\mu} + n^{-1}\mathbb{E}\left(\boldsymbol{\epsilon}^T\mathbf{H}_k^{\perp}\boldsymbol{\epsilon}\right) \\
&= \sigma^2 + n^{-1}\boldsymbol{\mu}^T\mathbf{H}_k^{\perp}\boldsymbol{\mu} - n^{-1}k^*\sigma^2.
\end{aligned}$$

The last line holds from $\mathbb{E}(\boldsymbol{\epsilon}^T\mathbf{H}_k^{\perp}\boldsymbol{\epsilon}) = \mathbb{E}(\boldsymbol{\epsilon}^T\boldsymbol{\epsilon}) - \mathbb{E}(\boldsymbol{\epsilon}^T\mathbf{H}_k\boldsymbol{\epsilon}) = n\sigma^2 - k^*\sigma^2$.

Therefore taking the mean of (7.23), we have

$$\mathbb{E}(\mathrm{CV}(k)) = \sigma^2 + n^{-1}\boldsymbol{\mu}^T\mathbf{H}_k^{\perp}\boldsymbol{\mu} - n^{-1}k^*\sigma^2 + n^{-1}\sum_{i=1}^{n}\frac{H_{ik}(2-H_{ik})}{(1-H_{ik})^2}\mathbb{E}(\hat{r}_{ik}^2).$$

Therefore, $\mathbb{E}(\mathrm{CV}(k))$ is the same as prediction error, but with a different penalty:

$$\mathrm{Err}(k)\ \text{penalty} = n^{-1}k^*\sigma^2 \tag{7.24}$$

$$\mathrm{CV}(k)\ \text{penalty} = -n^{-1}k^*\sigma^2 + n^{-1}\sum_{i=1}^{n}\frac{H_{ik}(2-H_{ik})}{(1-H_{ik})^2}\mathbb{E}(\hat{r}_{ik}^2). \tag{7.25}$$

We expect the difference to be asymptotically negligible.

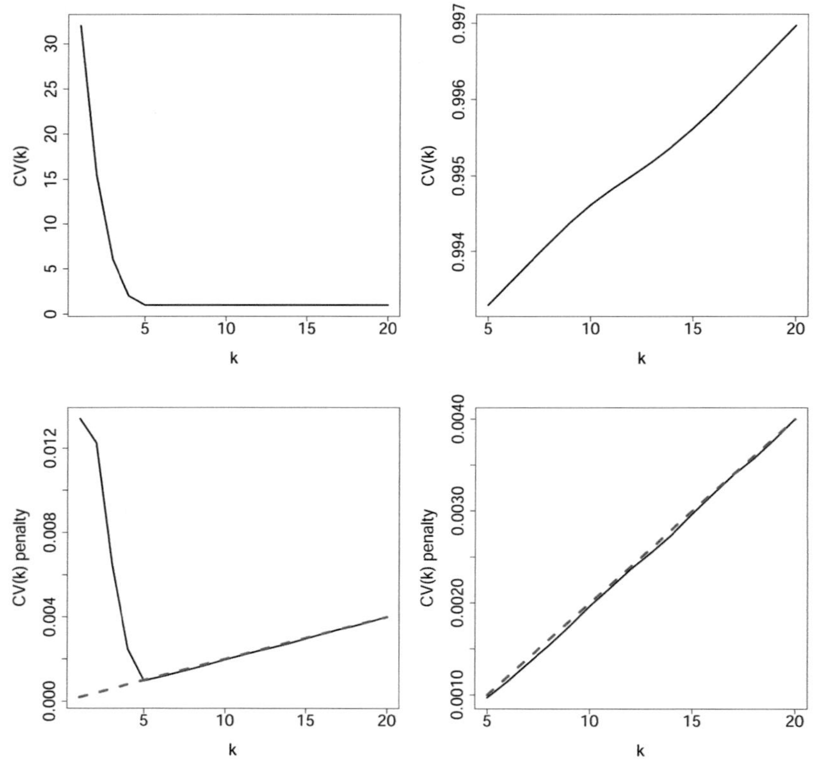

FIGURE 7.4
Top panels: CV(k) values for models $k = 1,\ldots,p$ from linear model simulation; right-panel is zoom in for $k \geq k_0$ (i.e. true model and overfit models). Bottom panels: solid black line shows CV(k) penalty (7.25) for different models k; red dashed line is Err(k) (prediction error) penalty (7.24).

To provide some empirical evidence of this, the following simulation experiment was conducted. The model was $Y_i = \sum_{k=1}^{p} x_{ik}\beta_k + \varepsilon_i$ for $i = 1,\ldots,n$, where x_{ik} are i.i.d. N(0,1) and ε_i are i.i.d. N$(0,\sigma^2)$ with $\sigma^2 = 1$ and $n = 5000$. The signal variables are $\beta_k = (6-k)$ for $k = 1,\ldots,k_0 = 5$ and the non-signal variables are $\beta_k = 0$ for $k = k_0+1,\ldots,p = 20$.

The top left panel of Figure 7.4 displays CV(k) for $k = 1,\ldots,p$. A classic elbow shape can be seen with the elbow joint occurring at $k = k_0 = 5$, exactly as we would hope for. However, CV(k) values for $k \geq k_0$ all appear to be the same and suggests that CV(k) cannot distinguish the true model $k = k_0$ from overfit models $k > k_0$. The right panel focuses on $k \geq k_0$ helping to see that in fact CV(k) does increase with increasing k. However, the effect is small (compare the ranges on the vertical axes) suggesting that CV(k) can overfit. On the other hand, if CV(k) overfits than resulting prediction error is unlikely to be much higher than for the true model.

This last point can be confirmed by looking at the bottom panels which displays the penalties used by the two methods. The thick black line is the CV(k) penalty (7.25), whereas the dashed red line is the Err(k) penalty (7.24). The two lines are nearly identical over the true model $k = k_0$ and overfit models $k > k_0$, thus demonstrating the point we wanted to make. It is also interesting that the penalty for CV(k) over the underfit models $k < k_0$ is substantially larger than Err(k). Therefore, we can conclude that CV(k) will tend not to underfit, but is likely to "overfit", although in the latter case, this does not affect the generalization error performance.

7.10.2 Consistency

The asymptotic distribution for $\text{MCV}_d(k)$ can be described in detail. This will show from the point of view of model selection performance, that consistent model selection requires the size of the test fold d to be nearly equal to the sample size. More formally, it will be required for $d := d(n)$ that

$$\lim_{n\to\infty} \frac{d}{n} = \lambda > 0. \tag{7.26}$$

Remember that since $\text{CV}(k)$ corresponds to $d = 1$, this shows that leaving one observation out for testing is going in the opposite direction of what is needed. Instead, we need to hold out almost all the observations for testing if we want consistent model selection.

Under condition (7.26), and certain additional mild regularity conditions, Zhang [19] established the following result:

$$\text{MCV}_d(k) = \begin{cases} \sigma^2 - n^{-1}\boldsymbol{\epsilon}^T\mathbf{H}_k\boldsymbol{\epsilon} + \dfrac{2-\lambda}{1-\lambda}\cdot\dfrac{k\sigma^2}{n} + o_p(1), & \text{if } k \geq k_0 \\ \sigma^2 + n^{-1}\boldsymbol{\mu}^T\mathbf{H}_k^{\perp}\boldsymbol{\mu} + o_p(1), & k < k_0. \end{cases} \tag{7.27}$$

The second line for $k < k_0$ addresses underfit models, where one or more signal variables have been excluded. The first line for $k \geq k_0$ includes the true model and all overfit models. Since we expect for $k \geq k_0$

$$n^{-1}\boldsymbol{\epsilon}^T\mathbf{H}_k\boldsymbol{\epsilon} = n^{-1}\mathbb{E}(\boldsymbol{\epsilon}^T\mathbf{H}_k\boldsymbol{\epsilon}) + o_p(1) = n^{-1}k\sigma^2 + o_p(1)$$

and for $k < k_0$,

$$n^{-1}\boldsymbol{\mu}^T\mathbf{H}_k^{\perp}\boldsymbol{\mu} = b_k + o_p(1)$$

for some positive bias term $b_k > 0$, this shows that models $k \geq k_0$ will have smaller $\text{MCV}_d(k)$ than $k < k_0$ and therefore will be preferentially selected. At the same time, for $k \geq k_0$, the term

$$\frac{2-\lambda}{1-\lambda}\cdot\frac{k\sigma^2}{n}$$

will discourage larger overfit models. Thus for $k \geq k_0$, $\text{MCV}_d(k)$ is expected to be

$$\sigma^2 + \frac{k\sigma^2}{n}\left(\frac{2-\lambda}{1-\lambda} - 1\right) + o_p(1),$$

while for $k < k_0$, we expect $\sigma^2 + b_k + o_p(1)$. Asymptotically the smallest value will occur when $k = k_0$; thus leading to a consistent procedure.

7.11 Summarizing what we have learned about cross-validation

Looking back at what we just learned, things might seem murky. We defined d to equal the size of the test set. We showed $\text{CV}(k)$, which corresponds to $d = 1$, is a good estimator for prediction error. But then we saw $\text{MCV}_d(k)$ is preferred for model selection when d is very large. How do we reconcile these apparently different messages?

Actually, these results are very similar to the behavior of AIC and BIC for risk estimation and model selection performance previously discussed in Chapter 4.6. We found that due to its heavy penalty, BIC was model selection consistent if the space of models is finite and if a true model exists. This is similar to (7.27). Increasing d, which increases the test size, acts like a heavier penalty. Then $\mathrm{MCV}_d(k)$ becomes a FPE criterion asymptotically equal to [19]

$$n^{-1}\left(\hat{S}(k) + \frac{2-\lambda}{(1-\lambda)}k\sigma^2\right), \quad 0 < \lambda := \lim_{n\to\infty}\frac{d}{n} < 1.$$

The penalty increases as λ becomes closer to 1, or equivalently the closer d becomes to n. This leads to a pointwise consistent procedure for model selection.

But what about CV(k)? We showed CV(k) accurately estimates the prediction error Err(k) for overfit models. However, because the difference in Err(k) may be small between overfit models, CV(k) may have a hard time choosing the best one. This means is that if the goal is to achieve low prediction error, then minimizing CV(k) achieves that goal. But this does not mean CV(k) chooses the true model each time, since it can easily overfit. This was confirmed in the theoretical simulation example of CV(k) in Figure 7.4. This same phenomenon is observed with AIC, where BIC's heavy penalty, while good for pointwise consistency, is bad for uniform consistency, whereas AIC, which tends towards larger models, achieves good prediction.

Therefore, we can summarize our results by saying that model selection and prediction error are dichotomous goals and achieving them requires different sets of conditions.

Weaker penalties (smaller test fold sizes) are better for prediction error, while heavier penalties (larger test fold sizes) yield pointwise model selection consistency. Thus:

1. *CV and AIC perform well when the goal is achieving good prediction performance.*
2. *MCV and BIC perform well when the goal is to select a true model from a finite class.*

7.12 Out-of-Bag Estimate

A valuable technique for estimating generalization error is out-of-bag (OOB) estimation, which employs bootstrap resampling. This method offers a smoothed alternative to leave-one-out cross-validation (CV) and reduces variance. CV can occasionally exhibit high variance due to the significant correlation between training folds, resulting in overly similar trained models and affecting the accuracy of generalization error estimation.

To explain the OOB estimator, we first need to define the bootstrap. The bootstrap is an independent and identically distributed sample from the empirical measure, which essentially means it is a draw from the original dataset using sampling with replacement.

Definition 7.2. Let $\mathbb{P}_n$ be the empirical measure for $\mathbf{X}_1,\dots,\mathbf{X}_n$. Draw n values $\mathbf{X}_1^*,\dots,\mathbf{X}_n^*$ independently from $\mathbb{P}_n$. Then $\mathbf{X}_1^*,\dots,\mathbf{X}_n^*$ is called an (Efron) bootstrap sample. □

Sampling with replacement results in a bootstrap sample that on average uses approximately 63.2% of the original data. This bootstrap data is referred to as *in-bag* data in machine learning. Since only 63.2% of the data is used by the bootstrap this means there is approximately 36.8% that remains untouched. This data serves the role of test data and is called *out-of-bag* (see Figure 7.5 for illustration).

The OOB estimator treats the OOB data as the test data. Let $\hat{F}^*$ be the predictor $\hat{F}$ derived on the basis of a bootstrap draw from $\mathscr{L}$. Define indicator functions for identifying OOB data

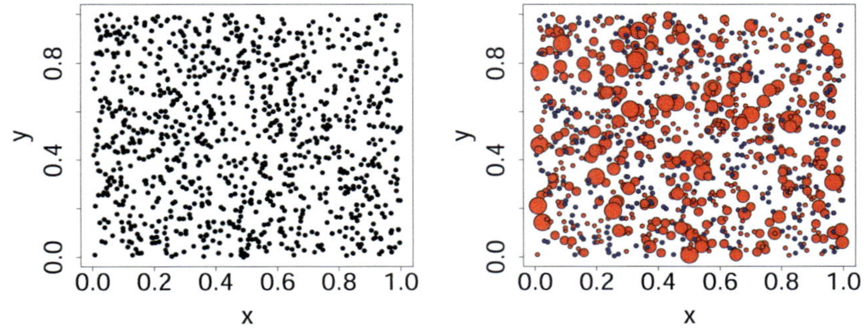

FIGURE 7.5
Illustration of a bootstrap sample. Left-hand side are 1000 points sampled uniformly from $[0,1] \times [0,1]$. *Right-hand side depicts the bootstrap sampled points (red) with the relative frequency of their selection being displayed by the size of a point. Blue points indicate out-of-bootstrap points; refered to as the out-of-bag data.*

points:

$$I_i = \begin{cases} 1 & \text{if } i \text{ is OOB} \\ 0 & \text{otherwise.} \end{cases}$$

The OOB estimator for the generalization error (7.1) is defined as

$$\text{Err}_{\text{oob}} = \mathbb{E}^* \left\{ \frac{1}{n} \sum_{i=1}^{n} I_i L(Y_i, \hat{F}^*(\mathbf{X}_i)) \right\} \tag{7.28}$$

where $\mathbb{E}^*$ denotes the expectation under the bootstrap distribution $\mathbb{P}_n$. Thus, (7.28) averages the loss over the OOB data over all bootstrap draws. For this reason, (7.28) is called the *OOB error*.

(7.28) can be written in another way by resampling the data with a single data point removed. For this reason, the estimator is also referred to as the *leave-one-out bootstrap* estimator [6]. Let $\mathbb{P}_n^{(i)}$ be the empirical distribution function for the data with $(\mathbf{X}_i, Y_i)$ removed. Because a bootstrap sample from $\mathbb{P}_n$ with $I_i = 1$ has the same distribution as a bootstrap sample from $\mathbb{P}_n^{(i)}$, we have

$$\text{Err}_{\text{oob}} = \frac{1}{n} \sum_{i=1}^{n} \mathbb{E}^* \left\{ I_i L(Y_i, \hat{F}^*(\mathbf{X}_i)) \right\} = \frac{1}{n} \sum_{i=1}^{n} \mathbb{E}^{*(i)} \left\{ L(Y_i, \hat{F}^{*(i)}(\mathbf{X}_i)) \right\}, \tag{7.29}$$

where $\hat{F}^{*(i)}$ denotes the predictor $\hat{F}$ calculated from a bootstrap sample of size n drawn from $\mathbb{P}_n^{(i)}$.

Identity (7.29) suggests that the OOB estimator is nearly the same as loo-CV (7.20). However, it differs in certain important ways. Instead of calculating the loss from the estimator $\hat{F}^{(i)}$ with i removed as in loo-CV,

$$\text{CV} = \sum_{i=1}^{n} L(Y_i, \hat{F}^{(i)}(\mathbf{X}_i)),$$

the OOB estimator averages the loss of bootstrap estimators $\hat{F}^{*(i)}$ over all bootstraps that exclude i. This averaging process over a much larger collection of fitted models reduces variance and is one of the reasons it is preferred to loo-CV [6].

In practice, (7.28) is approximated by drawing a large number of bootstrap samples $B \geq 1$. Let $\hat{F}_b^*$ be $\hat{F}$ obtained using the bth bootstrap draw from $\mathbb{P}_n$. Define

$$Q_{ib} = L(Y_i, \hat{F}_b^*(\mathbf{X}_i)).$$

Define indicator functions for identifying OOB data points for each bootstrap sample

$$I_{ib} = \begin{cases} 1 & \text{if } i \text{ is OOB for the } b\text{th draw} \\ 0 & \text{otherwise.} \end{cases}$$

Then, (7.28) is approximated by

$$\widehat{\text{Err}}_{\text{oob}} = \frac{1}{B}\sum_{b=1}^{B}\left(\frac{1}{\sum_{j=1}^{n} I_{jb}}\sum_{i=1}^{n} I_{ib}Q_{ib}\right). \tag{7.30}$$

This treats the OOB data as test data to obtain a test set error, which is then averaged over all the bootstrap draws. We can summarize the estimator as the following three steps:

1. Draw a bootstrap sample and use this to calculate $\hat{F}_b^*$.
2. Estimate the generalization error $\widehat{\text{Err}}_{\text{oob},b}$ for $\hat{F}_b^*$ using the OOB data.
3. Repeat B times and average the values to obtain the OOB estimator, $\widehat{\text{Err}}_{\text{oob}} = \sum_{b=1}^{B} \widehat{\text{Err}}_{\text{oob},b}/B$.

The proof that (7.30) approximates (7.28) uses the fact that 0.368 of the bootstrap data is OOB on average. We have

$$\begin{aligned}
\widehat{\text{Err}}_{\text{oob}} &= \frac{1}{B}\sum_{b=1}^{B}\sum_{i=1}^{n}\frac{I_{ib}Q_{ib}}{\sum_{j=1}^{n} I_{jb}} \\
&\asymp \frac{1}{0.368nB}\sum_{b=1}^{B}\sum_{i=1}^{n} I_{ib}Q_{ib}, \quad (\text{as } n \to \infty) \\
&= \frac{1}{0.368nB}\sum_{i=1}^{n}\sum_{b=1}^{B} I_{ib}Q_{ib} \\
&\asymp \frac{1}{n}\sum_{i=1}^{n}\sum_{b=1}^{B}\frac{I_{ib}Q_{ib}}{\sum_{b=1}^{B} I_{ib}} \quad (\text{as } B \to \infty).
\end{aligned}$$

This approximates the right-hand side of (7.29) and therefore (7.28).

Exercises

Exercise 7.1. Give 5 examples not already discussed in the book of learning problems describing the target function of interest and the loss function used for assessing performance. For each of your examples give a real-world illustration.

Exercise 7.2. Present a real example using data demonstrating how over-using training error leads to overfitting as in (7.7).

Exercise 7.3. Prove that each of the loss functions in Table 7.1 is a proper scoring rule.

Exercise 7.4. Consider a linear regression model of the form:

$$Y = \beta_0 + \beta_1 X_1 + \beta_2 X_2 + \boldsymbol{\epsilon}$$

where $\mathbf{X} = (X_1, X_2)^T$ is bivariate normal with mean and covariance

$$\boldsymbol{\mu} = \begin{pmatrix} 3 \\ -1 \\ 2 \end{pmatrix}, \quad \boldsymbol{\Sigma} = \begin{pmatrix} 1 & .25 \\ .25 & 1 \end{pmatrix}$$

and $\mathbf{X}$ is independent of $\boldsymbol{\epsilon} \sim \text{N}(0, \sigma^2)$.

1. Derive analytically the expression for the generalization error under L_2-loss, expressing the generalization error in terms of the variance and bias-squared.
2. Simulate data from the regression model (choosing values for the β's and σ^2). Fit the linear regression model on the training set and calculate the prediction error on the test set as an estimate of the generalization error. Repeat many times.
3. Compare the empirical generalization error estimator to the theoretical value.

Exercise 7.5. Consider a two-class classification problem where the class label Y is sampled from class 0 with probability $0 < \pi < 1$ and class 1 with probability $1 - \pi$, where class 0 is $\mathbf{X} \sim \mathrm{N}_2(\boldsymbol{\mu}_1, \boldsymbol{\Sigma}_1)$ and class 1 is $\mathbf{X} \sim \mathrm{N}_2(\boldsymbol{\mu}_2, \boldsymbol{\Sigma}_2)$. A linear classifier $\hat{\delta} : \mathbb{R}^2 \to \{0, 1\}$ is used to classify the observations with misclassification error $\mathbb{P}\{\mathbf{Y} \neq \hat{\delta}(\mathbf{X})\}$.

1. Derive analytically the expression for the misclassification error.
2. Simulate data from the true model (assign values for $\pi, \boldsymbol{\mu}_1, \boldsymbol{\mu}_2, \boldsymbol{\Sigma}_1, \boldsymbol{\Sigma}_2$). Split the dataset into training and test sets. Fit a linear classifier (LDA) on the training set and calculate the misclassification error on the test set as an estimate of the generalization error. Repeat this process many times.
3. Compare the empirical generalization error with the theoretical value.
4. How do your results change if π becomes very close to 1?

Exercise 7.6. A data set has $n = 600$ values. To properly test the final hypothesis, you set aside $n' = 200$ cases which are not used in the training. You define $N = 1000$ different pre-chosen predictors $\hat{F}_1, \ldots, \hat{F}_N$ and select the final model $\hat{F}$ from these. The goal is to estimate $\mathrm{Err}(\hat{F})$. You use two estimators: the training error $\mathrm{err}_{\mathrm{naive}}(\hat{F})$ and the test set error $\mathrm{err}_{\mathrm{iid}}(\hat{F})$; the latter being calculated using the $n' = 200$ test cases.

1. Using a 5% error tolerance, which estimate has a larger confidence interval?
2. Should you set aside a larger test sample size n'?

Exercise 7.7. Write an R-function to implement M-fold cross-validation for linear regression. Apply your function to the body fat data set from Chapter 4 using $M = 5, 10$. Now try CV (i.e. loo-CV with $M = n$). Comment on the results.

Exercise 7.8. Using the R-function from the previous example, devise simulation experiments to test the difference between M-fold cross-validation and loo-CV. Summarize your findings in terms of bias and MSE performance.

Exercise 7.9. The bottom panel of Figure 7.4 compares the penalty of $\mathrm{CV}(k)$ to the penalty of $\mathrm{Err}(k)$. Write R-code to reproduce this example with the following changes:

1. Calculate and plot the theoretical value for $\mathrm{Err}(k)$ against $k = 1, \ldots, p$. Superimpose the values of $\mathrm{CV}(k)$ for $k = 1, \ldots, p$ on the same figure. Describe your findings.
2. Repeat the previous analysis but this time modifying the simulation using correlated $\mathbf{X}$ variables. Compare your findings to (1).

Exercise 7.10. In this exercise we will prove (7.22). The proof makes use of the Sherman-Morrison-Woodbury (SMW) formula for matrices (Chapter 2.7), which is stated as follows. For any matrices $\mathbf{A}_{p\times p}$, $\mathbf{B}_{p\times m}$, $\mathbf{C}_{m\times m}$ and $\mathbf{D}_{m\times p}$, and assuming the necessary inverses exist:

$$(\mathbf{A} + \mathbf{BCD})^{-1} = \mathbf{A}^{-1} - \mathbf{A}^{-1}\mathbf{B}(\mathbf{C}^{-1} + \mathbf{DA}^{-1}\mathbf{B})^{-1}\mathbf{DA}^{-1}.$$

1. Define $\mathbf{D}_{(-s),k}$ to be $\mathbf{D}_k$ with only those rows $i \notin s$. Notice that

$$\mathbf{D}_{(-s),k}^T \mathbf{D}_{(-s),k} = \mathbf{D}_k^T \mathbf{D}_k - \mathbf{D}_{sk}^T \mathbf{D}_{sk}.$$

   Using this and SMW show that

$$(\mathbf{D}_{(-s),k}^T \mathbf{D}_{(-s),k})^{-1} = \mathbf{A}_k^{-1} + \mathbf{A}_k^{-1}\mathbf{D}_{sk}^T(\mathbf{I} - \mathbf{H}_{sk})^{-1}\mathbf{D}_{sk}\mathbf{A}_k^{-1}.$$

2. Let $\mathbf{Y}_{(-s)}$ be $\mathbf{Y}$ with only those coordinates $i \notin s$. Show using (1) and

$$\hat{\boldsymbol{\beta}}_{(-s),k} = (\mathbf{D}_{(-s),k}^T \mathbf{D}_{(-s),k})^{-1} \mathbf{D}_{(-s),k}^T \mathbf{Y}_{(-s)}$$

that

$$\mathbf{D}_{sk}\hat{\boldsymbol{\beta}}_{(-s),k} = \mathbf{D}_{sk}\hat{\boldsymbol{\beta}}_k - \mathbf{H}_{sk}(\mathbf{I}-\mathbf{H}_{sk})^{-1}\left[\mathbf{Y}_s - \mathbf{D}_{sk}\hat{\boldsymbol{\beta}}_k\right]$$

and therefore that

$$\begin{aligned} \mathbf{Y}_s - \mathbf{D}_{sk}\hat{\boldsymbol{\beta}}_{(-s),k} &= \left[\mathbf{I} + \mathbf{H}_{sk}(\mathbf{I}-\mathbf{H}_{sk})^{-1}\right]\left[\mathbf{Y}_s - \mathbf{D}_{sk}\hat{\boldsymbol{\beta}}_k\right] \\ &= \left[(\mathbf{I}-\mathbf{H}_{sk})^{-1}\right]\left[\mathbf{Y}_s - \mathbf{D}_{sk}\hat{\boldsymbol{\beta}}_k\right]. \end{aligned}$$

3. Now use (2) to prove that

$$\text{MCV}_d(k) = \left[d\binom{n}{d}\right]^{-1} \sum_s \left\|(\mathbf{I}-\mathbf{H}_{sk})^{-1}\hat{\mathbf{r}}_{sk}\right\|^2.$$

4. To complete the proof of (7.22), use the following combinatorial fact. Notice that when summing $\hat{\mathbf{r}}_{sk}^T\hat{\mathbf{r}}_{sk}$ over s, each coordinate of $\hat{\mathbf{r}}_k$ is counted $\binom{n-1}{d-1}$ times. Hence show that

$$\left[d\binom{n}{d}\right]^{-1} \sum_s \hat{\mathbf{r}}_{sk}^T\hat{\mathbf{r}}_{sk} = n^{-1}\hat{S}(k)$$

and from this complete the proof of (7.22).

Exercise 7.11. Here you will evaluate the performance of k-NN (k-nearest neighbor) classification using cross-validation and determine the optimal value of k for the classifier.

1. Use a publicly available dataset or a dataset from a public repository such as the UCI Machine Learning Repository.
2. Use loo-CV to evaluate the performance of k-NN for different values of k.
3. Determine the value of k that yields the best CV performance.

Exercise 7.12. This example will have you evaluate the performance of k-NN using leave-one-out bootstrap and determine the optimal value of k for the classifier on a complex dataset.

1. Find a suitable real-world dataset for classification.
2. Use leave-one-out bootstrap to estimate the error rate for different values of k. Determine the value of k that yields the best out-of-sample performance.
3. Compare your results to loo-CV. Comment on your findings.

Exercise 7.13. Here you will evaluate how well the loo-CV estimator approximates the true generalization error for the k-NN classifier.

1. Consider a two-class problem where the data is generated from two distinct bivariate normal distributions. Let the classes be $Y = 0$ and $Y = 1$, where class $j = 0, 1$ is $\text{N}(\boldsymbol{\mu}_j, \boldsymbol{\Sigma})$ and where $\boldsymbol{\mu}_0 = (0,0)^T$, $\boldsymbol{\mu}_1 = (d, 0)$ and $\boldsymbol{\Sigma} = I$. Determine the true Bayes error rate.
2. Decide on a value for d which controls the separation in the two classes. Simulate n samples from each class. Implement k-NN classifier for a range of k values, For each k, compute the loo-CV error rate.
3. Repeat the simulation many times. Give a plot that shows the estimated error rates for the different values of k along with the Bayes error rate. What can you infer about the reliability of loo-CV for estimating the generalization error?

Exercise 7.14. Let $(\mathbf{X}, Y) \in \mathbb{R}^p \times \{0,1\}$ have distribution $\mathbb{P}$. Let $\Pi = \{A_1, \ldots, A_{K+1}\}$ be a partition of size $K+1$ for $\mathbb{R}^p$ and define the partitioning classifier for Π as

$$\delta(\mathbf{x}, \Pi) = \sum_{k=1}^{K+1} \theta_{A_k} 1_{\{\mathbf{x} \in A_k\}}, \quad \text{where } \theta_{A_k} = 1_{\{\mathbb{P}\{Y=1|\mathbf{X} \in A_k\} \geq 1/2\}}.$$

Prove that

$$\mathbb{P}\{Y \neq \delta(\mathbf{x}, \Pi)\} \leq \mathbb{P}\{Y \neq \delta(\mathbf{x}, \{\theta_k\}_1^{K+1}, \Pi)\}$$

for any set of partitioning values $0 \leq \{\theta_k\}_1^{K+1} \leq 1$.

Exercise 7.15. OOB error can be used as a simple yet effective technique for variable selection [11]. This method calculates the variable importance (VIMP) for each variable in a regression model by comparing the difference in OOB prediction error. First, the full model prediction error is calculated. Then, for each variable, the coefficient is set to zero, and the OOB error for the resulting "marginal" model is calculated. This procedure is repeated for each bootstrap sample $b = 1, \ldots, B$. The VIMP for a variable is defined as the averaged difference between the OOB error with the coefficient set to zero and the full model OOB error.

1. Write an *R* function to calculate OOB variable importance and apply it to a regression data set of your choosing. Comment on the results.
2. Compare the results to fitting least squares and using p-values to select variables.

Bibliography

[1] Breiman L., Friedman J.H., Olshen R.A., and Stone C.J. *Classification and Regression Trees*, Belmont, California, 1984.

[2] Brier G.W. Verification of forecasts expressed in terms of probability. *Monthly Weather Review*, 78(1), 1–3, 1950.

[3] Cover, T. M. Geometrical and statistical properties of systems of linear inequalities with applications in pattern recognition. *IEEE Transactions on Electronic Computers*, 3:326–334, 1965.

[4] Devroye, L., Györfi, L., and Lugosi, G. *A Probabilistic Theory of Pattern Recognition*. Springer, 1996.

[5] Dudley, R. M. Central limit theorems for empirical measures. *The Annals of Probability*, 6 (6):899–929, 1978.

[6] Efron B. and Tibshirani R. Improvements on cross-validation: the 632+ bootstrap method. *Journal of the American Statistical Association*, 92(438), 548–560, 1997.

[7] Gneiting T. and Raftery A.E. Strictly proper scoring rules, prediction, and estimation. *Journal of the American Statistical Association*, 102(477):359–378, 2007.

[8] Gneiting T. and Katzfuss M. Probabilistic forecasting. *Annual Review of Statistics and Its Application*, 1: 125–151, 2014.

[9] Gyorfi L., Kohler M., Krzyzak A., and Walk H. *A Distribution-Free Theory of Nonparametric Regression*, Springer, 2002.

[10] Ishwaran H. and Lee D.K.K. Super greedy trees: some asymptotic results. *Technical Report, University of Miami*, 2024.

[11] Lu M. and Ishwaran H. A prediction-based alternative to P values in regression models. *Journal of Thoracic and Cardiovascular Surgery*, 155(3):1130–1136, 2018.

[12] Lugosi G. and Nobel A. Consistency of data-driven histogram methods for density estimation and classification. *The Annals of Statistics*, 24(2), 687–706, 1996.

[13] Steele, J. M. *Combinatorial Entropy and Uniform Limit Laws.* PhD Thesis, Department of Mathematics, Stanford University, 1975.

[14] Stone M. Cross-validatory choice and assessment of statistical predictions. *Journal of the Royal Statistical Society: Series B*, 36(2), 111–133, 1974.

[15] Stone M. An asymptotic equivalence of choice of model by cross-validation and Akaike's criterion. *Journal of the Royal Statistical Society: Series B*, 39(1), 44–47, 1977.

[16] Vapnik V.N. An overview of statistical learning theory. *IEEE Transactions on Neural Networks*, 10(5), 988–999, 1999.

[17] Vapnik V.N. and Chervonenkis A.Y. On the uniform convergence of relative frequencies of events to their probabilities. *Theory of Probability and its Applications*, 16(2), 264–281, 1971.

[18] Golub G.H. and Heath M. and Wahba G. Generalized cross-validation as a method for choosing a good ridge parameter. *Technometrics*, 21:215–223, 1979.

[19] Zhang P. Model selection via multifold cross validation. *Annals of Statistics*, 21:299–313, 1993.

8

Principal Component Analysis

8.1 Introduction

Principal Component Analysis (PCA) is a multivariate unsupervised dimension reduction technique. It transforms a set of possibly correlated variables into a smaller number of uncorrelated variables called principal components. These principal components are linear combinations of the original variables, chosen to capture the maximum variance in the data.

PCA is useful and popular for several reasons:

1. *Dimensionality Reduction*: PCA reduces the number of variables while retaining most of the original data's variability. This simplification makes it easier to visualize and analyze high-dimensional data.
2. *Decorrelation*: By transforming correlated variables into uncorrelated principal components, PCA simplifies the structure of the data, which can be beneficial for subsequent analyses and modeling.
3. *Noise Reduction*: By focusing on the principal components that capture the most variance, PCA can help filter out noise and less important details in the data, leading to more robust models.
4. *Data Compression*: PCA enables efficient data compression by reducing the storage and computational requirements while preserving essential information.
5. *Feature Extraction*: PCA helps in identifying the most informative features in the data, which can be useful for tasks like clustering, classification, and regression.
6. *Computational Scalability*: PCA is a relatively light procedure in terms of computational costs. For example, it will be shown it can be calculated using singular value decomposition which is a highly efficient numerical procedure used for matrix factorization.

8.2 Spectral Decomposition Theorem

The Spectral Decomposition Theorem (SDT) plays an important role in PCA.

Theorem 8.1. *(Spectral Decomposition) Any square symmetric matrix $\mathbf{A}_{p\times p}$ can be written as $\mathbf{A} = \mathbf{P}\boldsymbol{\Lambda}\mathbf{P}^T$ where $\boldsymbol{\Lambda}_{p\times p} = diag\{\lambda_j\}_{j=1}^{p}$ are the real-valued eigenvalues of $\mathbf{A}$ and $\mathbf{P}_{p\times p}$ is the orthonormal matrix ($\mathbf{P}^T\mathbf{P} = \mathbf{P}\mathbf{P}^T = \mathbf{I}$) comprised of the unit eigenvectors $\mathbf{e}_1, \ldots, \mathbf{e}_p$ of $\mathbf{A}$. More precisely,*

$$\mathbf{A} = [\mathbf{e}_1, \ldots, \mathbf{e}_p]_{p\times p} \begin{bmatrix} \lambda_1 & & \\ & \ddots & \\ & & \lambda_p \end{bmatrix}_{p\times p} \begin{bmatrix} \mathbf{e}_1^T \\ \vdots \\ \mathbf{e}_p^T \end{bmatrix}_{p\times p} = \sum_{j=1}^{p} \lambda_j \mathbf{e}_j \mathbf{e}_j^T.$$

DOI: 10.1201/9781003493679-8

Important properties that follow from Theorem 8.1 are as follows.

1. We can assume without loss of generality that $\lambda_1 \geq \lambda_2 \geq \cdots \geq \lambda_p$.
2. If $\mathbf{A} > 0$ is positive definite, $|\mathbf{A}| > 0$ and $\lambda_1 \geq \lambda_2 \geq \cdots \geq \lambda_p > 0$.
3. If $\mathbf{A} \geq 0$ is positive semidefinite, $|\mathbf{A}| \geq 0$ and $\lambda_1 \geq \lambda_2 \geq \cdots \geq \lambda_p \geq 0$.
4. If $\mathbf{A} > 0$, then $\mathbf{A}^{-1} = \mathbf{P}\boldsymbol{\Lambda}^{-1}\mathbf{P}^T = \sum_{j=1}^{p} \lambda_j^{-1}\mathbf{e}_j\mathbf{e}_j^T$.
5. $\mathbf{A}^{1/2} = \mathbf{P}\boldsymbol{\Lambda}^{1/2}\mathbf{P}^T = \sum_{j=1}^{p} \lambda_j^{1/2}\mathbf{e}_j\mathbf{e}_j^T$.
6. If $\mathbf{A} > 0$, then $\mathbf{A}^{-1/2} = \mathbf{P}\boldsymbol{\Lambda}^{-1/2}\mathbf{P}^T = \sum_{j=1}^{p} \lambda_j^{-1/2}\mathbf{e}_j\mathbf{e}_j^T$.

As an illustration of Theorem 8.1, suppose that $\mathbf{A}_{p\times p}$ is an equicorrelation matrix with parameter $-1 < \rho < 1$. Then $\mathbf{A} = (1-\rho)\mathbf{I} + \rho\mathbf{J}$ where $\mathbf{J} = \mathbf{1}\mathbf{1}^T$ and $\mathbf{1} = (1,\ldots,1)^T$ (see Chapter 2.7). Define $\mathbf{e}_p = \mathbf{1}/\sqrt{p}$ and let $\mathbf{e}_1,\ldots,\mathbf{e}_{p-1}$ be orthonormal vectors orthogonal to $\mathbf{e}_p$. Then $\{\mathbf{e}_1,\ldots,\mathbf{e}_p\}$ is an orthonormal basis for $\mathbb{R}^p$. It follows therefore that $\mathbf{I} = \sum_{j=1}^{p} \mathbf{e}_j\mathbf{e}_j^T$ and

$$
\begin{aligned}
(1-\rho)\mathbf{I} + \rho\mathbf{J} &= (1-\rho)\sum_{j=1}^{p} \mathbf{e}_j\mathbf{e}_j^T + p\rho\,\mathbf{e}_p\mathbf{e}_p^T \\
&= (1-\rho)\sum_{j=1}^{p-1} \mathbf{e}_j\mathbf{e}_j^T + \left[1+(p-1)\rho\right]\mathbf{e}_p\mathbf{e}_p^T. \qquad (8.1)
\end{aligned}
$$

Then, (8.1) is the spectral representation for $\mathbf{A}$. To check this, observe that since $\mathbf{e}_1,\ldots,\mathbf{e}_p$ are unit eigenvectors for $\mathbf{I}$ and $\mathbf{J}$ they are also unit eigenvectors for $\mathbf{A}$. Also, their eigenvalues are $\lambda_j = (1-\rho)$ for $j = 1,\ldots,p-1$ and $\lambda_p = 1+(p-1)\rho$.

Also as a side note, observe that from property 4 above, assuming that $-1/(p-1) < \rho < -1$ so that $\mathbf{A}$ is invertible, we have

$$
\begin{aligned}
\left[(1-\rho)\mathbf{I} + \rho\mathbf{J}\right]^{-1} &= (1-\rho)^{-1}\sum_{j=1}^{p-1} \mathbf{e}_j\mathbf{e}_j^T + \left[1+(p-1)\rho\right]^{-1}\mathbf{e}_p\mathbf{e}_p^T \\
&= \frac{1}{1-\rho}\left[\mathbf{I} - \frac{p\rho}{1+(p-1)\rho}\mathbf{e}_p\mathbf{e}_p^T\right] \\
&= \frac{1}{1-\rho}\left[\mathbf{I} - \frac{\rho}{1+(p-1)\rho}\mathbf{J}\right]
\end{aligned}
$$

which is another way to derive the inverse of an equicorrelation matrix; see Chapter 2.7.

8.3 Principal Component Transformation

The principal component transformation is the orthogonal transformation that maps the random vector $\mathbf{X} \in \mathbb{R}^p$ to a new vector $\mathbf{Y} \in \mathbb{R}^p$ such that $\mathbf{Y}$ has coordinates that are uncorrelated with one another. In the following, let $\boldsymbol{\Sigma} = \mathbf{P}\boldsymbol{\Lambda}\mathbf{P}^T$ be the spectral decomposition of the covariance matrix $\boldsymbol{\Sigma} = \text{Cov}(\mathbf{X})$. The PC transformation is obtained directly from the spectral decomposition of $\boldsymbol{\Sigma}$ and equals $\mathbf{P}$.

Definition 8.1. If $\mathbf{X}$ is a p-dimensional random vector with mean $\boldsymbol{\mu}$ and variance $\boldsymbol{\Sigma}$, the principal component transformation of $\mathbf{X}$ is

$$
\mathbf{X} \mapsto \mathbf{Y} = (Y_1,\ldots,Y_p)^T = \mathbf{P}^T(\mathbf{X}-\boldsymbol{\mu}),
$$

where $\mathbf{P} = [\mathbf{e}_1, \ldots, \mathbf{e}_p]_{p \times p}$ is the orthonormal matrix comprised of the eigenvectors of $\boldsymbol{\Sigma}$ and $\boldsymbol{\Lambda} = \text{diag}\{\lambda_j\}_{j=1}^p$ are the eigenvalues of $\boldsymbol{\Sigma}$. Notice we can write $Y_j = \mathbf{e}_j^T(\mathbf{X} - \boldsymbol{\mu})$. We call $Y_j \in \mathbb{R}$ the jth *principal component score* and $\mathbf{e}_j \in \mathbb{R}^p$ the jth *principal component loading.* □

8.3.1 Properties

Call $\ell^T\mathbf{X}$ a standardized linear combination (SLC) of a p-dimensional random vector $\mathbf{X}$ if $\ell^T\ell = 1$. The following result, which holds as a consequence of the principal component transformation, shows that the principal components scores are orthogonal, ensuring no correlation between them, and each captures the maximum possible variance in the data over any SLC uncorrelated with the previous scores.

These properties are summarized as follows.

1. $\mathbb{E}(\mathbf{Y}) = \mathbf{0}$.
2. $\text{Var}(\mathbf{Y}) = \boldsymbol{\Lambda}$. Therefore, $\text{Var}(Y_1) = \lambda_1 \geq \cdots \geq \text{Var}(Y_p) = \lambda_p \geq 0$.
3. $\text{Var}(Y_1) \geq \text{Var}(\ell^T\mathbf{X})$ for any SLC, ℓ.
4. $\text{Var}(Y_j) \geq \text{Var}(\ell^T\mathbf{X})$ for any SLC, ℓ, such that $\ell^T\mathbf{X}$ is uncorrelated with $Y_1, \ldots, Y_{j-1}$.
5. $\sum_{j=1}^p \text{Var}(X_j) = \sum_{j=1}^p \lambda_j = \sum_{j=1}^p \text{Var}(Y_j)$.
6. $|\boldsymbol{\Sigma}| = \prod_{j=1}^p \lambda_j = \prod_{j=1}^p \text{Var}(Y_j)$.

Thus, the first PC explains the most variance of all possible SLCs, equal to the first eigenvalue, the second PC explains the second highest variance, equal to the second eigenvalue, and so forth. The total variance of the new scores is equal to the total variance of the original data, thus the PC solution satisfies a conservation of variance. Effectively, these results show that the PC solution has the property that it distributes the available total variance as efficiently as possible, concentrating most of it into as few orthogonal scores as possible.

Example 8.1. Let $\mathbf{X} = (X_1, X_2)^T$ where $\boldsymbol{\mu} = \mathbf{0}$ and

$$\boldsymbol{\Sigma} = \begin{bmatrix} 1 & \rho \\ \rho & 1 \end{bmatrix}, \quad \text{where } 0 < \rho < 1.$$

Since $\boldsymbol{\mu} = \mathbf{0}$, the principal component transformation is $\mathbf{Y} = (Y_1, Y_2)^T = \mathbf{P}^T\mathbf{X}$ and since $\boldsymbol{\Sigma}$ is an equicorrelation matrix, it follows from (8.1) that $\lambda_1 = 1 + \rho$, $\lambda_2 = 1 - \rho$, $\mathbf{e}_1 = (1,1)^T/\sqrt{2}$, $\mathbf{e}_2 = (1,-1)^T/\sqrt{2}$ and

$$\begin{aligned} \boldsymbol{\Sigma} &= (1+\rho)\,\mathbf{e}_1\mathbf{e}_1^T + (1-\rho)\,\mathbf{e}_2\mathbf{e}_2^T \\ &= \frac{1+\rho}{2}\begin{bmatrix} 1 & 1 \\ 1 & 1 \end{bmatrix} + \frac{1-\rho}{2}\begin{bmatrix} 1 & -1 \\ -1 & 1 \end{bmatrix} \\ &= \begin{bmatrix} 1 & \rho \\ \rho & 1 \end{bmatrix}. \end{aligned}$$

Therefore the principal components are

$$\begin{aligned} Y_1 &= \mathbf{e}_1^T\mathbf{X} = \frac{1}{\sqrt{2}}(X_1 + X_2) \\ Y_2 &= \mathbf{e}_2^T\mathbf{X} = \frac{1}{\sqrt{2}}(X_1 - X_2), \end{aligned}$$

with variances $1+\rho$ and $1-\rho$. We see that Y_1 represents an average, while Y_2 is a contrast. Also observe that

$$\lambda_1 + \lambda_2 = (1+\rho) + (1-\rho) = 2 = \sum_{j=1}^{2} V(X_j).$$

Notice that when $\rho = 1$ the total variance is entirely explained by the first PC. □

8.4 Sample Principal Component Analysis

To use PCA for data analysis, we replace population parameters with sample estimators. We assume the data is $\mathbf{X}_1, \ldots, \mathbf{X}_n$ with unknown mean $\boldsymbol{\mu}$ and unknown variance $\boldsymbol{\Sigma}$. We estimate $\boldsymbol{\mu}$ by $\overline{\mathbf{X}}$ and $\boldsymbol{\Sigma}$ by $\mathbf{S}$. Let

$$\mathbf{S} = \hat{\mathbf{P}}\hat{\boldsymbol{\Lambda}}\hat{\mathbf{P}}^T = \sum_{j=1}^{p} \hat{\lambda}_j \hat{\mathbf{e}}_j \hat{\mathbf{e}}_j^T \tag{8.2}$$

be the spectral decomposition of $\mathbf{S}$ where $\hat{\mathbf{P}} = [\hat{\mathbf{e}}_1, \ldots, \hat{\mathbf{e}}_p]$ is the orthogonal matrix comprised of the eigenvectors of $\mathbf{S}$ and $\hat{\boldsymbol{\Lambda}} = \text{diag}\{\hat{\lambda}_j\}_{j=1}^{p}$ are the eigenvalues of $\mathbf{S}$.

Definition 8.2. The sample principal component transformation is

$$\mathbf{X}_i \mapsto \mathbf{Y}_i = \hat{\mathbf{P}}^T(\mathbf{X}_i - \overline{\mathbf{X}}), \quad i = 1, \ldots, n.$$

Let $\hat{Y}_{ij} = \hat{\mathbf{e}}_j^T(\mathbf{X}_i - \overline{\mathbf{X}})$ and $\hat{\mathbf{Y}}_i = (\hat{Y}_{i1}, \ldots, \hat{Y}_{ip})^T$. We call $\hat{Y}_{1j}, \ldots, \hat{Y}_{nj}$ the jth principal component scores sample and $\hat{\mathbf{Y}}_1, \ldots, \hat{\mathbf{Y}}_n$ the principal component sample. □

8.4.1 Properties

1. The sample mean of $\hat{\mathbf{Y}}_1, \ldots, \hat{\mathbf{Y}}_n$ is $\mathbf{0}$.
2. The sample variance-covariance of $\hat{\mathbf{Y}}_1, \ldots, \hat{\mathbf{Y}}_n$ is $\hat{\boldsymbol{\Lambda}}$.
3. The total sample variance is $\sum_{j=1}^{p} \hat{\lambda}_j$.
4. The sum of the first k eigenvalues divided by the sum of all eigenvalues, $\sum_{j=1}^{k} \hat{\lambda}_j / \sum_{j=1}^{p} \hat{\lambda}_j$, represents the proportion of total variance explained by the first k sample principal components.

8.4.2 PCA is not scale invariant

An important limitation of PCA is that it is not scale invariant. Thus, depending on the how $\mathbf{X}$ is measured (e.g., pounds versus kilograms, or inches versus centimeters), we can obtain substantially different principal components. When variables are measured on different scales, the preferred way to analyze PCA is to use the sample correlation matrix $\mathbf{R}$ in place of the sample variance $\mathbf{S}$. If $\mathbf{R} = (R_{jk})_{p\times p}$, then $\mathbf{R}$ is obtained from $\mathbf{S} = (S_{jk})_{p\times p}$ via

$$R_{j,k} = \frac{S_{jk}}{\sqrt{S_{jj}S_{kk}}}.$$

Equivalently, the data can be transformed so that each variable has unit sample variance. This amounts to the transformation

$$\mathbf{X}_i \mapsto \mathbf{X}_i^* = \begin{pmatrix} X_{i1}/\sqrt{S_{11}} \\ \vdots \\ X_{ip}/\sqrt{S_{pp}} \end{pmatrix}, \quad i = 1, \ldots, n.$$

8.5 Determining the Number of Components to Keep

In PCA we reduce the original p-dimensional data into univariate linear components, the PC scores, which is a form of dimension reduction. Additionally, we should only retain the top PC scores which are informative, which is another form of data reduction. But how many components should we keep?

One method for determining this is the *scree plot*. The percentage of variance explained by the remaining PC components is plotted against the number of components:

$$\left(j, \sum_{k=j+1}^{p} \lambda_k \Big/ \sum_{k=1}^{p} \lambda_k \right), \quad j = 1, \ldots, p-1.$$

The scree plot indicates where large eigenvalues cease and smaller eigenvalues start: the point where an elbow forms on the graph indicates the number of components to use. Under normality one can test specific hypotheses about the percentage of variance explained, but it is more common for the number of components to be selected by informally "eye-balling" the scree plot.

Another method is the PC rank-trace plot [4, Chapter 7.2]. In this case, a plot is constructed with values

$$\left((1 - j/p)^{1/2}, \sum_{k=j+1}^{p} \lambda_k^2 \Big/ \sum_{k=1}^{p} \lambda_k^2 \right), \quad j = 1, \ldots, p-1.$$

The horizontal coordinate (first coordinate) equals the estimated difference between the reduced rank and full rank regression, while the vertical coordinate equals the estimated reduction in the residual matrix using a full rank model versus a reduced rank model. Similar to the scree plot, the number of components are determined by finding the elbow in the graph.

8.6 Singular Value Decomposition

The Singular Value Decomposition (SVD) is a powerful and widely used matrix factorization technique in linear algebra. It decomposes a given matrix into three simpler matrices, revealing the intrinsic structure of the original matrix. SVD is especially useful in various applications such as signal processing, machine learning, and genomics.

Theorem 8.2. *(Singular Value Decomposition) Any matrix* $\boldsymbol{A}_{n\times p}$ *where* $n \geq p$ *can be written as* $\boldsymbol{A} = \boldsymbol{UDV}^T$ *where* $\boldsymbol{U}_{n\times p}$ *is an orthonormal matrix* ($\boldsymbol{U}^T\boldsymbol{U} = \boldsymbol{I}_p$), $\boldsymbol{V}_{p\times p}$ *is an orthogonal matrix* ($\boldsymbol{V}^T\boldsymbol{V} = \boldsymbol{VV}^T = \boldsymbol{I}_p$), *and* $\boldsymbol{D}_{p\times p} = diag\{d_j\}_{j=1}^{p}$ *is a diagonal matrix with non-negative entries.*

The preferred way to calculate PCA is by SVD applied to the centered and possibly scaled data matrix, and not by a spectral decomposition of the covariance or correlation matrix. Working with the original (transformed) matrix and not a $p \times p$ sum of squares matrix, improves numerical accuracy and makes it possible to use PCA for high dimensions.

A numerically stable PCA can be achieved by the following. First, we center the data by subtracting the sample mean

$$\mathbf{X}_i \mapsto \mathbf{X}_i^* = \mathbf{X}_i - \overline{\mathbf{X}}, \quad i = 1, \ldots, n.$$

The sample covariance for the transformed data $\mathbf{S}^*$ equals the sample covariance matrix $\mathbf{S}$ for the original data. We have

$$\begin{aligned}
\mathbf{S}^* &= \frac{1}{n}\sum_{i=1}^{n}(\mathbf{X}_i^* - \overline{\mathbf{X}}^*)(\mathbf{X}_i^* - \overline{\mathbf{X}}^*)^T \\
&= \frac{1}{n}\sum_{i=1}^{n}\mathbf{X}_i^*\mathbf{X}_i^{*T} \\
&= \frac{1}{n}\left(\mathbf{U}\mathbf{D}\mathbf{V}^T\right)^T\left(\mathbf{U}\mathbf{D}\mathbf{V}^T\right) \\
&= \frac{1}{n}\mathbf{V}\mathbf{D}^2\mathbf{V}^T
\end{aligned}$$

where $\mathbf{A} = \mathbf{U}\mathbf{D}\mathbf{V}^T$ is the SVD for the centered data matrix

$$\mathbf{A} = \begin{pmatrix} \mathbf{X}_1^{*T} \\ \vdots \\ \mathbf{X}_n^{*T} \end{pmatrix}_{n \times p}. \tag{8.3}$$

Notice that

$$\mathbf{S}^* = \frac{1}{n}\mathbf{V}\mathbf{D}^2\mathbf{V}^T = \sum_{j=1}^{p}\frac{d_j^2}{n}\mathbf{v}_j\mathbf{v}_j^T,$$

where $\mathbf{V} = [\mathbf{v}_1, \ldots, \mathbf{v}_p]_{p \times p}$. Comparing this with (8.2), and keeping in mind that $\mathbf{S}^* = \mathbf{S}$, we have that $\hat{\mathbf{P}} = \mathbf{V}$ and $\hat{\mathbf{\Lambda}} = \mathbf{D}^2/n$ and therefore the sample principal component transformation is

$$\mathbf{X}_i \mapsto \mathbf{Y}_i = \mathbf{V}^T\mathbf{X}_i^* = \mathbf{V}^T(\mathbf{X}_i - \overline{\mathbf{X}}), \quad i = 1, \ldots, n$$

and the jth principal component scores are $\hat{Y}_{ij} = \mathbf{v}_j\mathbf{X}_i^* = \mathbf{v}_j(\mathbf{X}_i - \overline{\mathbf{X}})$. Therefore, the PC transformation of the original problem can be calculated from the SVD of the centered data matrix, $\mathbf{A}$. In this case, the PC loadings equal the column vectors of $\mathbf{V}$. Algorithm 8.1 summarizes this procedure.

8.7 More Singular Value Decomposition

SVD has many useful applications beyond PCA. Here we collect together some of these results. Throughout let $\mathbf{A}$ be an $n \times p$ matrix with rank $r \le \min(n, p)$. Assume that $\mathbf{A}$ is centered so that its overall mean is zero. Write its SVD as $\mathbf{A} = \mathbf{U}\mathbf{D}\mathbf{V}^T$ where without loss of generality the singular values are assumed ordered so that

$$d_1 \ge d_2 \ge \cdots \ge d_r > 0 \quad \text{and} \quad d_{r+1} = \cdots = d_p = 0.$$

Notice that $d_j > 0$ for $j = 1, \ldots, p$ if $\mathbf{A}$ has full column rank $r = p$ when $p < n$. Observe also that we can write $\mathbf{A}$ in summation form

$$\mathbf{A} = \sum_{k=1}^{r} d_k\mathbf{u}_k\mathbf{v}_k^T$$

where $\mathbf{u}_k$ and $\mathbf{v}_k$ are the kth column vectors from $\mathbf{U}_{n \times p}$ and $\mathbf{V}_{p \times p}$ respectively.

Algorithm 8.1 *PCA from SVD*

1: For PCA using the covariance matrix, let $\mathbf{A}$ be the centered data matrix defined by (8.3) using centered values $\mathbf{X}_i^* = \mathbf{X}_i - \overline{\mathbf{X}}$. For PCA using the correlation matrix, define $\mathbf{A}$ by using centered *and* standardized data values

$$\mathbf{X}_i^* = \begin{pmatrix} (X_{i1} - \overline{X}_1)/\sqrt{S_{11}} \\ \vdots \\ (X_{ip} - \overline{X}_p)/\sqrt{S_{pp}} \end{pmatrix}, \qquad i = 1, \dots, n.$$

2: Let $\mathbf{UDV}^T$ be the SVD for $\mathbf{A}$. Then the jth principal component scores are

$$\hat{Y}_{ij} = \mathbf{v}_j \mathbf{X}_i^*, \quad i = 1, \dots, n$$

where $\mathbf{v}_j$ is the jth column of $\mathbf{V}$. Equivalently, the jth principal component scores equal the jth column from the matrix $\mathbf{AV}$.

3: The variance explained by the jth principal component equals d_j^2/n where $\{d_j\}_1^p$ are the diagonal values of $\mathbf{D}$. The proportion of total variance explained by the jth principal component equals $d_j^2 / \sum_{k=1}^p d_k^2$

8.7.1 Connection to PCA and PC regression

We start by pointing out some further aspects of PCA due to SVD. Let $\mathbf{A}$ denote the centered data matrix (8.3) used for the PC analysis where $\overline{\mathbf{X}} = \mathbf{0}$. Thus the data matrix is

$$\mathbf{A} = \begin{pmatrix} \mathbf{X}_1^T \\ \vdots \\ \mathbf{X}_n^T \end{pmatrix} = \mathbf{UDV}^T.$$

Recall that we can use the SVD to extract values for PCA. For example, we discussed that the PC loading values are obtained from the columns of $\mathbf{V}$. However, $\mathbf{U}$ also contains useful information. The columns of $\mathbf{U}$ contain the scaled principal component scores. This is because the PC scores are

$$\mathbf{Y}_i = (Y_{i1}, \dots, Y_{ip})^T = (\mathbf{v}_1^T \mathbf{X}_i, \dots, \mathbf{v}_p^T \mathbf{X}_i)^T = \mathbf{V}^T \mathbf{X}_i, \quad i = 1, \dots, n.$$

Therefore

$$\begin{pmatrix} \mathbf{Y}_1^T \\ \vdots \\ \mathbf{Y}_n^T \end{pmatrix} = \begin{pmatrix} \mathbf{X}_1^T \\ \vdots \\ \mathbf{X}_n^T \end{pmatrix} [\mathbf{v}_1, \dots, \mathbf{v}_p] = \mathbf{AV}.$$

Hence the PC scores equal $\mathbf{AV}$. But also notice that

$$\mathbf{AV} = (\mathbf{UDV}^T)\mathbf{V} = \mathbf{UD} \tag{8.4}$$

Therefore, the columns of $\mathbf{U}$ contain the PC scores up to a scalar transformation (the scale multiples being the diagonal entries of $\mathbf{D}$).

Another interesting property of $\mathbf{U}$ is that it equals the orthogonally rotated data space. This is because by (8.4), $\mathbf{U} = \mathbf{AVD}^{-1}$ and because $\mathbf{AV}$ is the rotation of $\mathbf{A}$ by the orthogonal marix $\mathbf{V}$, it follows that $\mathbf{U}$ (up to a scalar transformation) equals the rotated space. We can also show that the span of **U** equals the span of **A**. Observe that for any $\mathbf{z} \in \mathbb{R}^p$, we can write $\mathbf{z} = \sum_{j=1}^{p} \alpha_j \mathbf{v}_j$. Hence

$$\begin{aligned}\mathbf{Az} &= \mathbf{A}\left(\sum_{j=1}^{p} \alpha_j \mathbf{v}_j\right) \\ &= \mathbf{UDV}^T\left(\sum_{j=1}^{p} \alpha_j \mathbf{v}_j\right) \\ &= \sum_{j=1}^{p} \alpha_j d_j \mathbf{u}_j.\end{aligned}$$

Related to PCA is the method of principle component regression. In PC regression the design matrix **A** in the regression model is replaced by the orthogonal matrix **U** from its SVD. Then the least squares estimator is calculated in the transformed orthogonal space. Consider the regression model

$$\mathbf{Y}_{n\times 1} = \mathbf{A}_{n\times p}\boldsymbol{\beta}_{p\times 1} + \boldsymbol{\epsilon}_{n\times 1}$$

where $n \geq p$ and we can assume **Y** and **A** are centered with mean values of zero. Replacing **A** with $\mathbf{A} = \mathbf{UDV}^T$, we can rewrite the model as

$$\mathbf{Y} = \mathbf{UDV}^T\boldsymbol{\beta} + \boldsymbol{\epsilon} = \mathbf{W}\boldsymbol{\theta} + \boldsymbol{\epsilon}$$

where $\mathbf{W}_{n\times p} = \mathbf{U}$ is the new design matrix and $\boldsymbol{\theta} = \mathbf{DV}^T\boldsymbol{\beta} \in \mathbb{R}^p$ is the reparameterized regression coefficient vector. By (8.4), **W** equals the PC scores up to a scale transformation. Also $\mathbf{W}^T\mathbf{W} = \mathbf{I}$, hence the least squares estimator for $\boldsymbol{\theta}$ is

$$\hat{\boldsymbol{\theta}} = (\mathbf{W}^T\mathbf{W})^{-1}\mathbf{W}^T\mathbf{Y} = \mathbf{W}^T\mathbf{Y}.$$

The least squares estimator $\hat{\boldsymbol{\beta}}$ for $\boldsymbol{\beta}$ is obtained from

$$\hat{\boldsymbol{\theta}} = \mathbf{DV}^T\hat{\boldsymbol{\beta}}.$$

For example if **D** has all positive entries, then $\hat{\boldsymbol{\beta}} = \mathbf{VD}^{-1}\hat{\boldsymbol{\theta}}$.

8.7.2 Rank optimality

Next we discuss the so-called rank optimality property of SVD. For any rank $r' < r$ matrix $\hat{\mathbf{A}}$ [1]

$$||\mathbf{A} - \hat{\mathbf{A}}||_F^2 \geq \left\| \sum_{j=r'+1}^{r} d_j \mathbf{u}_j \mathbf{v}_j^T \right\|_F^2 = \sum_{j=r'+1}^{r} d_j^2$$

where $||\cdot||_F$ is the Frobenius norm. The right-hand side becomes an equality when $\hat{\mathbf{A}}$ equals the r'-rank SVD of **A**

$$\hat{\mathbf{A}}^{(r')} = \sum_{j=1}^{r'} d_j \mathbf{u}_j \mathbf{v}_j^T.$$

Thus, for any rank $r' < r$ matrix $\hat{\mathbf{A}}$

$$||\mathbf{A} - \hat{\mathbf{A}}||_F^2 \geq ||\mathbf{A} - \hat{\mathbf{A}}^{(r')}||_F^2 = \sum_{j=r'+1}^{r} d_j^2.$$

This shows that the SVD has a rank-optimality property. Among low-rank matrices $r' < r$, the matrix closest to **A** is the SVD decomposition constructed using the first r' singular values. Note that we have the following monotonicity property for the rank-reduced SVD's of **A**:

$$0 \leq ||\hat{\mathbf{A}}^{(1)}||_F^2 \leq ||\hat{\mathbf{A}}^{(2)}||_F^2 \leq \cdots \leq ||\hat{\mathbf{A}}^{(r)}||_F^2 = ||\mathbf{A}||_F^2, \quad ||\hat{\mathbf{A}}^{(r')}||_F^2 = \sum_{j=1}^{r'} d_j^2,\ 1 \leq r' \leq r.$$

This suggests that in tasks involving approximating **A**, we should consider approximations constructed from the singular values.

8.7.3 Sparse PCA

In high dimensions it sometimes occurs that the PC loadings consist of many non-zero values of roughly the same size, making them hard to interpret. One solution is to use penalization to encourage interpretable sparse PC loadings [7]. The following theorem which makes use of rank optimality is instrumental for this.

Theorem 8.3. *For any matrices* $\hat{\boldsymbol{U}}_{n\times p}$ *and* $\hat{\mathbf{V}}_{p\times p}$ *such that* $\hat{\boldsymbol{U}}^T\hat{\boldsymbol{U}} = \boldsymbol{I}_p$ *and* $\hat{\mathbf{V}}^T\hat{\mathbf{V}} = \hat{\boldsymbol{V}}\hat{\boldsymbol{V}}^T = \boldsymbol{I}_p$ *and any diagonal matrix* $\hat{\mathbf{D}}_{p\times p}$ *with entries* $\{\hat{d}_j\}_1^p$, *we have*

$$||\boldsymbol{A} - \hat{\boldsymbol{U}}\hat{\boldsymbol{D}}\hat{\boldsymbol{V}}^T||_F^2 = ||\boldsymbol{A}||_F^2 - 2\sum_{j=1}^p \hat{d}_j\hat{\boldsymbol{u}}_j^T\boldsymbol{A}\hat{\boldsymbol{v}}_j + \sum_{j=1}^p \hat{d}_j^2$$

where $\{\hat{\boldsymbol{u}}_j\}_1^p$ *and* $\{\hat{\boldsymbol{v}}_j\}_1^p$ *are the column vectors for* $\hat{\boldsymbol{U}}$ *and* $\hat{\mathbf{V}}$.

For example, applying Theorem 8.3 to the SVD for **A**, we obtain

$$||\mathbf{A}||_F^2 = 2\sum_{j=1}^r d_j\mathbf{u}_j^T\mathbf{A}\mathbf{v}_j - \sum_{j=1}^r d_j^2 = \sum_{j=1}^r d_j^2.$$

Therefore, the SVD of **A** is the minimizer of $||\mathbf{A} - \hat{\mathbf{U}}\hat{\mathbf{D}}\hat{\mathbf{V}}^T||_F^2$ over all $\hat{\mathbf{U}}\hat{\mathbf{D}}\hat{\mathbf{V}}^T$ decompositions of **A**. However, since this is generally not a sparse solution, we reframe the optimization problem as

$$||\mathbf{A} - \hat{\mathbf{U}}\hat{\mathbf{D}}\hat{\mathbf{V}}^T||_F^2 + P(\hat{\mathbf{U}}, \hat{\mathbf{V}})$$

where P is some suitably chosen penalty.

However, trying to solve this directly is difficult. Instead, we can break up the optimization into simple simpler 1-rank problems that are iterated until convergence.

Setting $\hat{d}_j = 0$ for $j > 1$, Theorem 8.3 shows that for any $d > 0$ and any $\mathbf{u} \in \mathbb{R}^n$ and $\mathbf{v} \in \mathbb{R}^p$ where $||\mathbf{u}||_2 = 1$ and $||\mathbf{v}||_2 = 1$,

$$||\mathbf{A} - d\mathbf{u}\mathbf{v}^T||_F^2 = ||\mathbf{A}||_F^2 - 2d\mathbf{u}^T\mathbf{A}\mathbf{v} + d^2.$$

Hence the solution to

$$\operatorname*{arg\,min}_{d>0,\mathbf{u},\mathbf{v}} \left\{\frac{1}{2}||\mathbf{A} - d\mathbf{u}\mathbf{v}^T||_F^2\right\} \quad \text{subject to } ||\mathbf{u}||_2 = 1,\ ||\mathbf{v}||_2 = 1,\ P_1(\mathbf{u}) \le c_1,\ P_2(\mathbf{v}) \le c_2,$$

where P_1 and P_2 are penalties, is equivalent to

$$\operatorname*{arg\,max}_{\mathbf{u},\mathbf{v}} \{\mathbf{u}^T\mathbf{A}\mathbf{v}\} \quad \text{subject to } ||\mathbf{u}||_2 = 1,\ ||\mathbf{v}||_2 = 1,\ P_1(\mathbf{u}) \le c_1,\ P_2(\mathbf{v}) \le c_2.$$

(The optimized value d is $\hat{d} = \hat{\mathbf{u}}^T\mathbf{A}\hat{\mathbf{v}}$).

It turns out that directly optimizing the above jointly over $(\mathbf{u}, \mathbf{v})$ is still too challenging. The solution is to solve **u** by holding **v** fixed and then solving **v** by holding **u** fixed and repeat this iteratively. Each of these steps is convex and can be quickly solved. For example, holding **v** fixed, the convex step for **u** is

$$\operatorname*{arg\,max}_{\mathbf{u}} \{\mathbf{u}^T\mathbf{A}\mathbf{v}\} \quad \text{subject to } ||\mathbf{u}||_2 = 1,\ P_1(\mathbf{u}) \le c_1.$$

The steps for **u** and **v** are iterated until convergence. Once the solution $(\hat{d}, \hat{\mathbf{u}}, \hat{\mathbf{v}})$ is obtained, **A** is updated according to

$$\mathbf{A} \leftarrow \mathbf{A} - \hat{d}\hat{\mathbf{u}}\hat{\mathbf{v}}^T$$

and a new $(\mathbf{u}, \mathbf{v})$ iterative cycle is applied to the updated matrix. This entire process is repeated for a total of r steps.

8.7.4 Generalized inverse

Recall from Chapter 2.8 that the generalized inverse $\mathbf{A}^+$ of a matrix $\mathbf{A}$ is the best approximate solution to the inversion problem $\mathbf{AB} = \mathbf{I}$. Therefore, the generalized inverse is very valuable, but how can we calculate its value in practice? It turns out that a solution can be obtained easily from the SVD.

Theorem 8.4. *The generalized inverse for $\boldsymbol{A}_{n\times p}$ given that $n \ge p$ is*

$$\boldsymbol{A}^+_{p\times n} = \boldsymbol{V}\boldsymbol{D}^+\boldsymbol{U}^T$$

where $\boldsymbol{D}^+$ is the generalized inverse of $\boldsymbol{D}$ defined as the $p \times p$ diagonal matrix with entries $1/d_j$ for $j = 1,\dots,r$ and zero for coordinates $j = r+1,\dots,p$. If $p > n$ then transpose $\boldsymbol{A}$ so that the above result applies and transpose the resulting inverse. Thus, $\boldsymbol{A}^+ = ((\boldsymbol{A}^T)^+)^T$ when $p > n$.

The theorem can be proved by directly verifying the properties of a generalized inverse (see Exercise 8.8):

$$\mathbf{AA^+A} = \mathbf{A},\ \ \mathbf{A^+AA^+} = \mathbf{A}^+,\ \ (\mathbf{AA}^+)^T = \mathbf{AA}^+.$$

8.7.5 Ellipsoids

We can describe ellipsoids geometrically by using the SVD. Let $\mathbf{A}_{p\times p}$ be positive definite (notice that $n = p$). Let P be defined by

$$P = \{\mathbf{x} : \mathbf{x}^T\mathbf{A}^{-1}\mathbf{x} = c^2\}.$$

Then P is a p-dimensional ellipsoid centered at the origin. From the SVD of $\mathbf{A}$ we have the following result

Theorem 8.5. *Suppose that $d_1 > d_2 > \cdots > d_p > 0$ are distinct. Then the jth principal axes for P is $\boldsymbol{v}_j$ with length $cd_j^{1/2}$.*

The theorem shows that the principal axes for P are the eigenvectors of $\mathbf{A}$ with length proportional to their eigenvalues.

8.8 PCA Examples

8.8.1 Frog audio syllables

Our first example examines a dataset commonly used to illustrate multiclassification, but which is also well suited for PCA. The data consists of processed audio signals recorded from frogs, each identified by their family, genus, and species. There are 4 families, 8 genera, and 10 species in the data, representing the multiclass labels used for multiclass analysis. Additionally, the audio information for each frog was processed by separating it into "syllables." Mel-frequency cepstral coefficients (MFCCs) were used to describe these syllables. For each extracted syllable, 22 normalized MFCCs were calculated. These values represent the **X**-features for the data (p = 22) used for the PCA. The number of syllables per frog varied from 1 to 458, resulting in varying numbers of records per specimen. In total, there were 60 frogs and $n = 7195$ records.

Figure 8.1 displays some of the data, including the phenotypes (multiclass labels) for better understanding. The scatter plots (lower diagonal) and bivariate contour plots (upper diagonal) indicate heterogeneity in the features. Notably, the features exhibit strong correlations with each other, which is promising for PCA (see also Figure 8.2 for correlations of all $p = 22$ features). Due

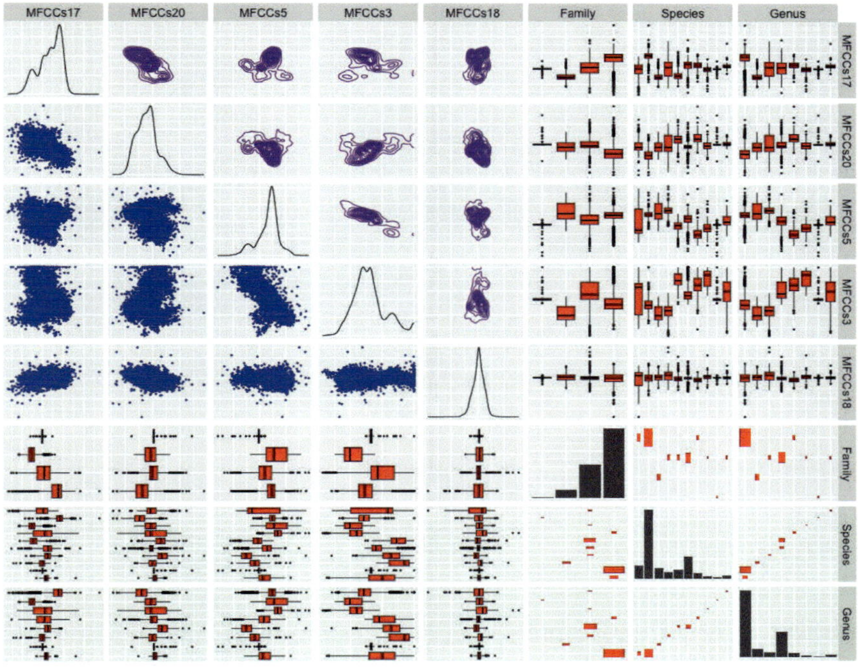

FIGURE 8.1
Graphical visualization of the frog data set.

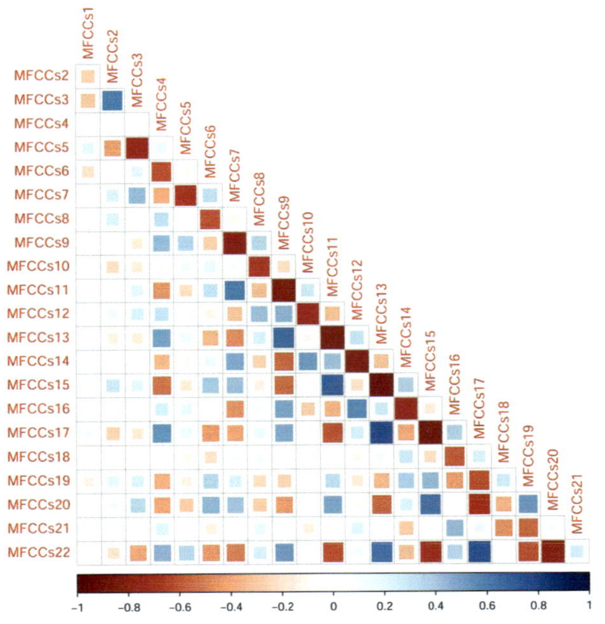

FIGURE 8.2
Correlations for features from frog data set.

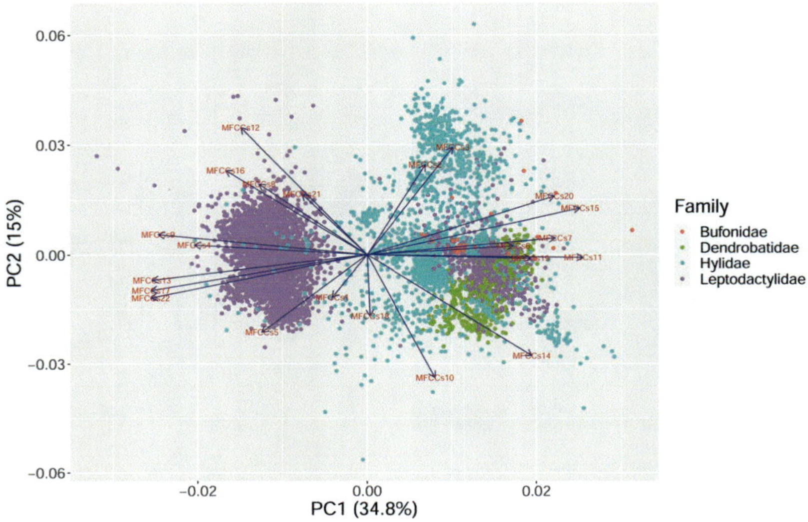

FIGURE 8.3
Biplot from PCA of frog data set.

to the variability in the features, confirmed by investigating the standard deviations of all $p = 22$ variables, PCA used the correlation matrix by scaling variables to a variance of 1.

Let $\mathbf{UDV}^T$ be the SVD of the centered and scaled data matrix, where

$$\mathbf{U} = [\mathbf{u}_1, \mathbf{u}_2, \ldots, \mathbf{u}_p]_{n\times p}, \quad \mathbf{D} = \text{diag}\{d_j\}_{j=1}^p, \quad \mathbf{V}^T = [\mathbf{v}_1^T, \ldots, \mathbf{v}_p^T]_{p\times p}.$$

Figure 8.3 displays the results of the PCA solution by showing bivariate points and arrows

$$\mathbf{g}_i = (u_{i1}, u_{i2})^T,\ i = 1, \ldots, n \quad \text{and} \quad \mathbf{h}_j = (d_1 v_{j1}, d_2 v_{j2})^T,\ j = 1, \ldots, p,$$

which is a type of figure referred to as a biplot [2]. Specifically, arrows $\mathbf{h}_j$ are the jth values from the first two eigenvectors (loadings) scaled by the first two eigenvalues, while points $\mathbf{g}_i$ are the ith row of the first two columns of $\mathbf{U}$ and equal the scaled first two principal component scores (recall (8.4)).

The theory behind the biplot is that the bivariate points $\{\mathbf{g}_i\}_1^n$ display distances, so points further apart are better separated. The bivariate vectors $\{\mathbf{h}_j\}_1^p$ display correlations, with arrows that are tangential identifying clusters that are uncorrelated with each other (see Chapter 8.8.2 for more details about biplots). For example, the points for *Leptodactylidae* are distant from *Hylidae*, indicating that these two families, which make up the bulk of the data (61% and 30%, respectively), are well separated. Additionally, the arrows pointing left toward *Leptodactylidae* are nearly tangential to those pointing upper right toward *Hylidae*, suggesting these two sub-clusters are nearly uncorrelated.

Figure 8.4 shows the percentage and cumulative percentage of variance explained for the PC components. Overall, 10 components explain approximately 90% of variance and probably are enough to summarize the data properly. While this might seem like an informal way to determine useful PC scores, this is often the way PCA is performed in practice.

However, we can use a more systematic method to determine the number of components by leveraging the outcome information available in this data. Recall that the family, species, and genus subtype were recorded for each frog, providing phenotype information for a more structured analysis. The idea is to run a classification analysis using the PC scores for the **X**-features and the phenotype information for the class labels. By increasing the number of PC

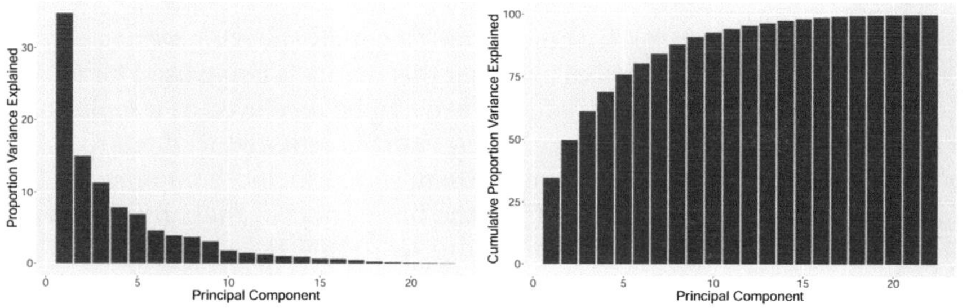

FIGURE 8.4
Scree plot of variance explained by PC components from frog data set.

scores, we can assess how informative the PC values are. This approach is a generalization of the PC regression discussed in Chapter 8.7.1.

We will use a random forest classifier for this purpose (Chapter 16). Performance of the random forests classifier was determined by misclassification error which was estimated by cross-validation. The classification analysis used family for the class label outcome. Thus there are four possible labels for the outcome value.

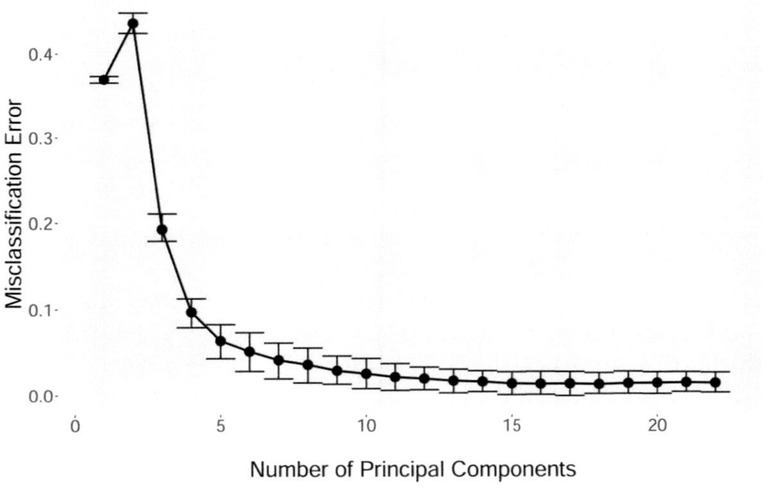

FIGURE 8.5
Cross-validated misclassification error for random forest classifier built using PC scores as ***X****-features. The number of PC scores used as features are increased sequentially from* $j = 1, 2, \ldots, p$.

Figure 8.5 displays the misclassification error for the random forest classifiers as the number of PC scores used for features increases for $j = 1, \ldots, p$. Superimposed are 95% confidence regions (see Chapter 17.3). From the figure, we observe that the misclassification error drops very quickly, indicating that about 10 features are likely sufficient to effectively predict the outcome. This finding aligns with the results from the scree plot. However, an interesting observation is the increase in error when moving from the first to the second PC score. This is unexpected, as we would typically expect the error to decrease, especially for the top PC scores. We will revisit this point in our discussion of PC regression.

8.8.2 Biplot

In this section we discuss the biplot in more detail. Generally speaking we can consider a biplot to be a tool for visualizing data matrices (whether this matrix is being used for PCA or not). It is based on the principle that any matrix of rank two can be displayed as a biplot of a vector for each row and each column. For matrices of higher rank, a rank-reduced matrix is used. Biplots were originally named by Gabriel [2], who used them with principal components.

The rationale for the method follows from the basic concept that any matrix $\mathbf{A}_{n\times p}$ of rank $r \le p$ can be factorized as $\mathbf{A} = \mathbf{G}\mathbf{H}^T$ of dimensions $\mathbf{G}_{n\times r'}$ and $\mathbf{H}_{p\times r'}$ where $2 \le r' \le r$. The goal of the biplot is to plot the rows of $\mathbf{G}$ against the rows of $\mathbf{H}$. The solution $\mathbf{A} = \mathbf{G}\mathbf{H}^T$ is not unique so a strategy for resolving this has to be used. As was discussed briefly in our previous analysis, in PCA, the biplot shows the first two scaled PC loadings simultaneously alongside the first two scaled PC scores.

To understand why this is used for PCA, suppose as before that $\mathbf{A}$ is the centered data matrix (8.3) and let $\mathbf{A} = \mathbf{U}\mathbf{D}\mathbf{V}^T$ be its SVD. Then we replace the full r rank SVD with the $r' = 2$ rank reduced SVD

$$\begin{aligned}\mathbf{A}^{(r')} &= [\mathbf{u}_1, \mathbf{u}_2]\begin{bmatrix} d_1 & 0 \\ 0 & d_2 \end{bmatrix}[\mathbf{v}_1, \mathbf{v}_2]^T \\ &= [\mathbf{u}_1, \mathbf{u}_2]\begin{bmatrix} d_1\mathbf{v}_1^T \\ d_2\mathbf{v}_2^T \end{bmatrix} \\ &:= \mathbf{G}\mathbf{H}^T.\end{aligned}$$

In PCA, the biplot uses $\mathbf{G}$ equal to the first two columns of $\mathbf{U}$ and $\mathbf{H}$ equal to the first two columns of $\mathbf{VD}$. Using $n\mathbf{S}^{(r')} = (\mathbf{A}^{(r')})^T\mathbf{A}^{(r')}$ and $\mathbf{A}^{(r')} = \mathbf{G}\mathbf{H}^T$, we have [2]

$$\begin{aligned} n\mathbf{S}^{(r')} &= \mathbf{H}\mathbf{H}^T \quad \text{(covariance) } s_{jk} \sim \mathbf{h}_j^T\mathbf{h}_k \\ n^{-1}\mathbf{A}^{(r')}\left(\mathbf{S}^{(r')}\right)^{-1}(\mathbf{A}^{(r')})^T &= \mathbf{G}\mathbf{G}^T \quad \text{(distance) } d_{ii'} \sim ||\mathbf{g}_i - \mathbf{g}_{i'}||^2 \end{aligned}$$

Therefore, $\mathbf{g}_i = (u_{1,i}, u_{2,i})$, $i = 1,\ldots,n$, measures Mahalanobis distance squared, while $\mathbf{h}_j = (d_1 v_{1,j}, d_2 v_{2,j})$, $j = 1,\ldots,p$ assesses correlation.

Thus, this shows that the biplot lends itself to understanding the data from the two perspectives of distance and correlation simultaneously. We use a simulation from Gabriel [2] to illustrate this further. Let $Z_{i,j}$ be i.i.d. N(0,1) for $i = 1,\ldots,30$, $j = 1,\ldots,4$. The features $\mathbf{X}_i = (X_{i1},\ldots,X_{i4})^T$ are simulated according to:

i	$X_{i,1}$	$X_{i,2}$	$X_{i,3}$	$X_{i,4}$
1-10	$Z_{i,1}+10$	$Z_{i,1}+Z_{i,2}$	$-Z_{i,1}-Z_{i,2}-Z_{i,3}$	$Z_{i,1}+\cdots+Z_{i,4}$
11-20	$Z_{i,1}$	$Z_{i,1}+Z_{i,2}$	$-Z_{i,1}-Z_{i,2}-Z_{i,3}$	$Z_{i,1}+\cdots+Z_{i,4}$
21-30	$Z_{i,1}$	$Z_{i,1}+Z_{i,2}$	$-Z_{i,1}-Z_{i,2}-Z_{i,3}$	$Z_{i,1}+\cdots+Z_{i,4}+10$

Therefore, the data is made up of three distinct groups. In group 1 ($i = 1,\ldots,10$), the first coordinate is inflated, in group 2 ($i = 11,\ldots,20$), no coordinates are inflated, while in group 3 ($i = 21,\ldots,30$), coordinate 4 is inflated. In all three groups, coordinates 1 and 4 should be the least similar, while coordinates 2 and 3 should be negatively correlated.

Due to the way $\mathbf{X}$ is simulated, the variance for the first and last coordinates will be much larger than the other two coordinates. Therefore, we used the correlation matrix for PCA. For illustration, we list $\mathbf{D}$ and $\mathbf{V}$ from the SVD of the data matrix and the first two columns of $\mathbf{VD}$

(which as discussed above equal the **h** arrows of the biplot):

$$d_1 = 8.22, \quad d_2 = 5.96, \quad d_3 = 3.07, \quad d_4 = 1.88$$

$$\mathbf{V} = \begin{bmatrix} -0.13 & 0.85 & -0.51 & -0.09 \\ 0.60 & 0.22 & 0.34 & -0.69 \\ -0.59 & -0.29 & -0.22 & -0.72 \\ 0.52 & -0.38 & -0.76 & -0.04 \end{bmatrix}, \quad \mathbf{VD}_{4\times 2} = \begin{bmatrix} -1.04 & 5.04 \\ 4.95 & 1.31 \\ -4.86 & -1.76 \\ 4.27 & -2.29 \end{bmatrix}.$$

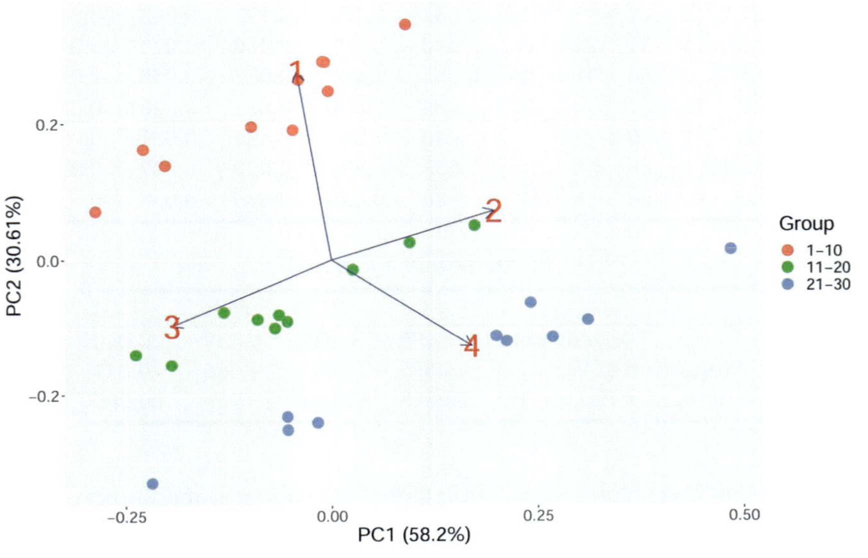

FIGURE 8.6
Biplot of Gabriel's data.

The biplot is displayed in Figure 8.6. Notice how the scaled PC scores, the $\mathbf{g}_i = (u_{1,i}, u_{2,i})$ points, are well separated into the three underlying groups (identified by color). The arrows $\mathbf{h}_j = (d_1 v_{1,j}, d_2 v_{2,j})$, shown in blue, also correctly identify the relationship between the coordinates of the data (the upper horizontal axis represents the first coordinate of $\mathbf{h}_j$, while the right vertical axis represents the second coordinate of $\mathbf{h}_j$). Arrows 2 and 3 point in opposite directions, suggesting that coordinates 2 and 3 are negatively correlated, whereas arrows 1 and 4 appear tangential to one another, indicating a weak relationship between coordinates 1 and 4.

8.8.3 A cautionary tale about PC regression

In PC regression (PCR), the PC scores obtained from PCA are used as features in a regression model (Chapter 8.7.1). Since the PC scores have the property that they explain variance in the original data, we might expect that the top PC scores will always be more informative than lower PC scores in the PCR analysis.

However, this is not always the case, as the next example shows. The example is taken from the paper of Hadi and Ling [3] who used it to illustrate a cautionary tale of PCR. Their data listed in Table 8.1 has 4 explanatory variables $X_1, \ldots, X_4$ and one response value Y. The PC scores for the data are listed alongside the values of $\mathbf{X}$ in the table. Underneath the table we list the variance explained by each component.

TABLE 8.1
Data used for PC regression example.

Y	X_1	X_2	X_3	X_4	$PC1$	$PC2$	$PC3$	$PC4$	Y^*
78.5	7	26	6	60	-1.467	-1.903	-0.530	0.039	0.077
74.3	1	29	15	52	-2.136	-0.238	-0.290	-0.030	-0.060
104.3	11	56	8	20	1.130	-0.184	-0.011	-0.094	-0.187
87.6	11	31	8	47	-0.660	-1.577	0.179	-0.033	-0.066
95.9	7	52	6	33	0.359	-0.484	-0.740	0.019	0.038
109.2	11	55	9	22	0.967	-0.170	0.086	-0.012	-0.024
102.7	3	71	17	6	0.931	2.135	-0.173	0.008	0.017
72.5	1	31	22	44	-2.232	0.692	0.460	0.023	0.045
93.1	2	54	18	22	-0.352	1.432	-0.032	-0.045	-0.090
115.9	21	47	4	26	1.663	-1.828	0.851	0.020	0.040
83.8	1	40	23	34	-1.640	1.295	0.494	0.031	0.063
113.3	11	66	9	12	1.693	0.392	-0.020	0.037	0.074
109.4	10	68	8	12	1.746	0.438	-0.275	0.037	0.074

PC analysis of ***X*** *features.*

	$PC1$	$PC2$	$PC3$	$PC4$
Eigenvalues	1.495	1.255	0.432	0.040
Proportion of Variance	55.893	39.402	4.665	0.041
Cumulative Proportion	55.893	95.294	99.959	100.000

Therefore, the first 2 PC scores explain over 95% of the variance, and the first 3 scores explain over 99% of the variance. Thus, we would expect the first 2, or maybe the first 3 components, to be enough for effective PC regression.

In a surprising twist, we find something very different. Table 8.2 lists the p-values for each of the PC scores used in the PC regression of Y. Both component 1 and component 3 are highly significant, but components 2 and 4 are not. Table 8.2 also lists the results from a regression of Y on the original **X** variables (after standardization). Only the first coordinate X_1 is marginally significant. Therefore, in hindsight, the results from the PC regression might not be surprising since loadings that focus on X_1 appear to be better for predicting Y, but there is no guarantee that loadings focusing on X_1 reduce variance.

The message here is that there is a difference between unsupervised and supervised regression. In PCR, we perform supervised regression, but PCA creates combinations of **X** features in an unsupervised manner without knowledge of Y. Therefore, we should not always expect that the importance of a PC score in regression will be related to its percent variance explained.

To further illustrate this point, we run a PC regression using the variable Y^* listed in Table 8.2 as our outcome. Regressing Y^* against the PC scores, we find an even more interesting result: the fourth PC component is the only significant score, while the other PC scores are uninformative.

8.8.4 Breast cancer (p bigger than n)

Our next example considers a gene expression data set from a breast cancer study [6]. There are $n = 286$ patients in this study. A total of 22,000 gene expressions were collected for each individual which after processing left a total of $p = 17,816$ genes, which are the data used for the PC analysis.

TABLE 8.2
PC regression results.

Regression of Y (centered) on **X** *(all standardized)*

	Estimate	Std. Error	t-statistic	p-value
X_1	9.124	4.130	2.209	0.055
X_2	7.939	10.619	0.748	0.474
X_3	0.653	4.558	0.143	0.889
X_4	-2.411	11.189	-0.215	0.834

Regression of Y (centered) on PC scores (all standardized)

	Estimate	Std. Error	t-statistic	p-value
*PC*1	9.8831	0.4452	22.20	0.0000
*PC*2	0.1250	0.5303	0.24	0.8189
*PC*3	4.5548	1.5411	2.96	0.0161
*PC*4	5.8375	16.5208	0.35	0.7320

Regression of Y^* *on PC scores (all standardized)*

	Estimate	Std. Error	t-value	Pr(>|t|)
*PC*1	0.0001	0.0001	1.56	0.1524
*PC*2	0.0000	0.0001	0.23	0.8215
*PC*3	0.0003	0.0002	1.25	0.2416
*PC*4	1.9983	0.0022	907.78	0.0000

Notice that p is much larger than n in this example. Therefore, we have to be careful with the PC analysis. The issue is that the SVD decomposition stated in Theorem 8.2 only applies when $n \geq p$. Therefore we have to "trick" the SVD into thinking that this is the case even though we have a p bigger than n scenario. We do this by working with the transposed data matrix (this was the same trick used to obtain the generalized inverse for $p > n$ in Theorem 8.4). In our example since the expression data has high variability, we work with the correlation matrix instead of the covariance matrix. Therefore, in our example, the data matrix **A** is created by stacking the centered and standardized gene expression values.

We take the transpose of **A** and obtain the SVD of the transpose denoted by $\mathbf{UDV}^T$ and then we transpose this value back. By taking the transpose, the loadings for the PCA are the column vectors of the **U** matrix and not the **V** matrix as previously. The PC scores are obtained from the columns of **AU** and not **AV**. A summary of the SVD procedure for $p > n$ is given in Algorithm 8.2. Notice that the effective dimension of the problem is n since this is the limiting (smaller) dimension at play. This is why the number of PC scores is n (equal to the number of columns of **U**).

In the p bigger than n problem, the effective dimension is n. For PCA this means that the number of PC scores is n and not p as in standard PCA.

Figure 8.7 displays the first two PC scores from the analysis. Superimposed on the figure is patient clinical information that was also available with this data. This information is displayed using symbols with different sizes and colors. Yellow and red points distinguish ER– and ER+ status (estrogen receptor status). We observe that the two PC's do a good job of delineating

Algorithm 8.2 *PCA from SVD when $p > n$*

1: For PCA using the covariance matrix, let $\mathbf{A}$ be the data matrix defined by (8.3) by stacking the centered $\mathbf{X}$ features (for PCA based on the correlation matrix, we use centered *and* standardized $\mathbf{X}$ features).
2: Let $\mathbf{UDV}^T$ be the SVD for $\mathbf{A}^T$ where $\mathbf{U}$ has dimension $p \times n$ and $\mathbf{D}$ is a $n \times n$ diagonal matrix. Then the jth loading is the jth column of $\mathbf{U}$ and the jth principal component score equals the jth column of the matrix $\mathbf{AU}$.
3: The proportion of total variance explained by the jth principal component equals $d_j^2 / \sum_{k=1}^n d_k^2$ for $j = 1, \ldots, n$.

ER status. Within the ER+ subgroup, the molecular subtypes are mostly LumA and normal tumors. LumA tumors typically have low proliferation gene expression scores and do very well. In comparison, most ER– tumors are basal, errb2 (HER2) and have high proliferation scores and do more poorly. Furthermore, both aggressive tumors (larger symbols) and non-aggressive tumors (smaller symbols) appear within the ER– subgroup. Unfortunately, PCA is unable to delineate the ER– subgroup along these lines. The ER– subtype is difficult to study and is a challenge to current work in breast cancer.

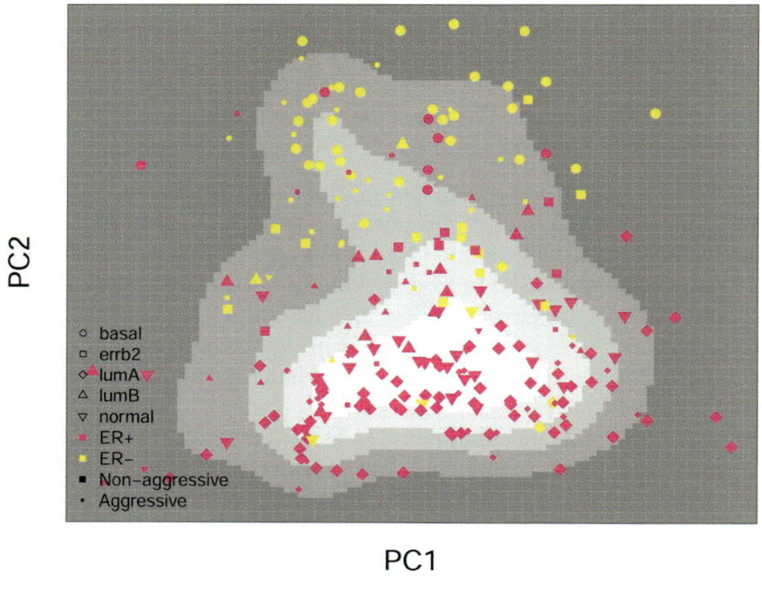

FIGURE 8.7
First two principal component scores for breast cancer microarray data, stratified by molecular subtype, ER+/ER– status and aggressiveness of tumor.

8.8.5 Facial recognition: eigenfaces

Automated algorithms for facial recognition are used widely in video surveillance, homeland security, smart cards, passport license registration, and numerous other applications. Computerized systems called *eigenfaces* are a tool based on PCA that operates directly on the data matrix of an image. Each face starts as a picture image which is then associated with a $r \times c$ data matrix where each (i, j) entry records the intensity value of the original image at pixel location (i, j). Typically intensity values are converted to an 8-bit grayscale which is a number between 0-255, with 0 representing black and 255 representing white.

Figure 8.8 displays the images of a set of faces taken between April 1992 and 1994 at the Olivetti Research Laboratory in Cambridge, UK. The size of each image is 92×112 using 8-bit grey levels. There are 10 different images of 40 distinct subjects. For certain subjects, the images were captured at different times, with variations in lighting, facial expressions (e.g., open or closed eyes, smiling or not smiling), and facial details such as wearing glasses or not. All the images are taken against a dark homogeneous background and the subjects are in up-right, frontal position with tolerance for some side movement.

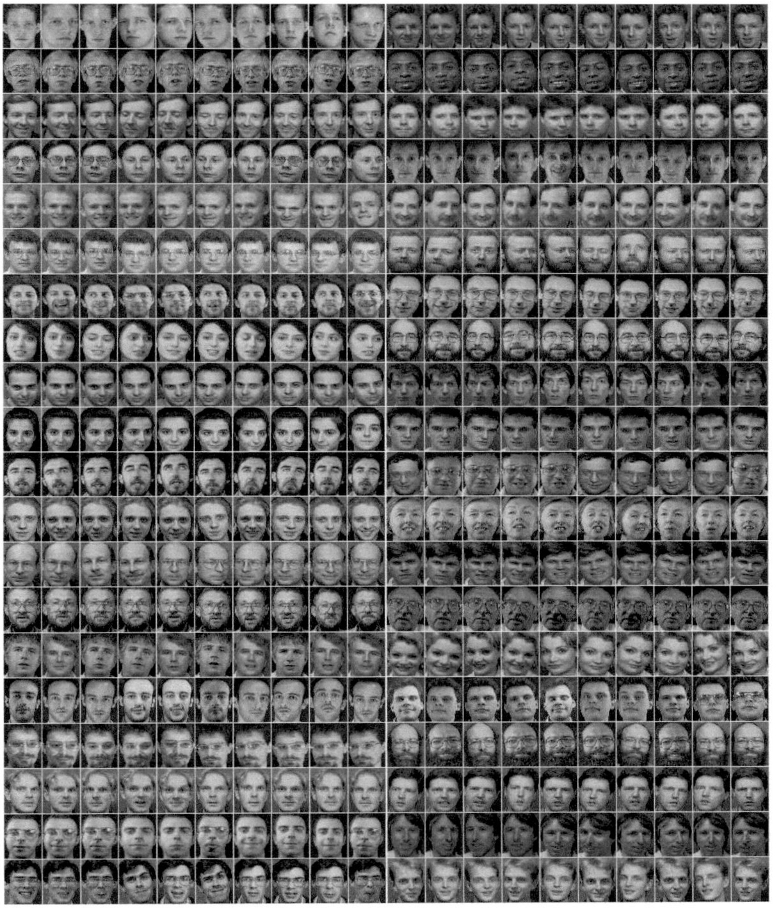

FIGURE 8.8
Facial images of 40 distinct subjects under 10 different conditions. Data from the Olivetti Research Laboratory in Cambridge, UK.

To construct an eigenface the matrix of an image is stacked by converting it into a long vector $\mathbf{X}$ of length $r \times c$. For each individual this yields the collection of vectors $\mathbf{X}_1, \ldots, \mathbf{X}_n$ where $n = 10$ and $\mathbf{X}_i \in \mathbb{R}^p$ where $p = r \times c = 10304$. Since $p > n$ in this problem, we used Algorithm 8.2 to obtain the PC solution. The covariance matrix was used for the solution; therefore, the data matrix was formed by stacking centered values $\mathbf{X}_i - \overline{\mathbf{X}}$.

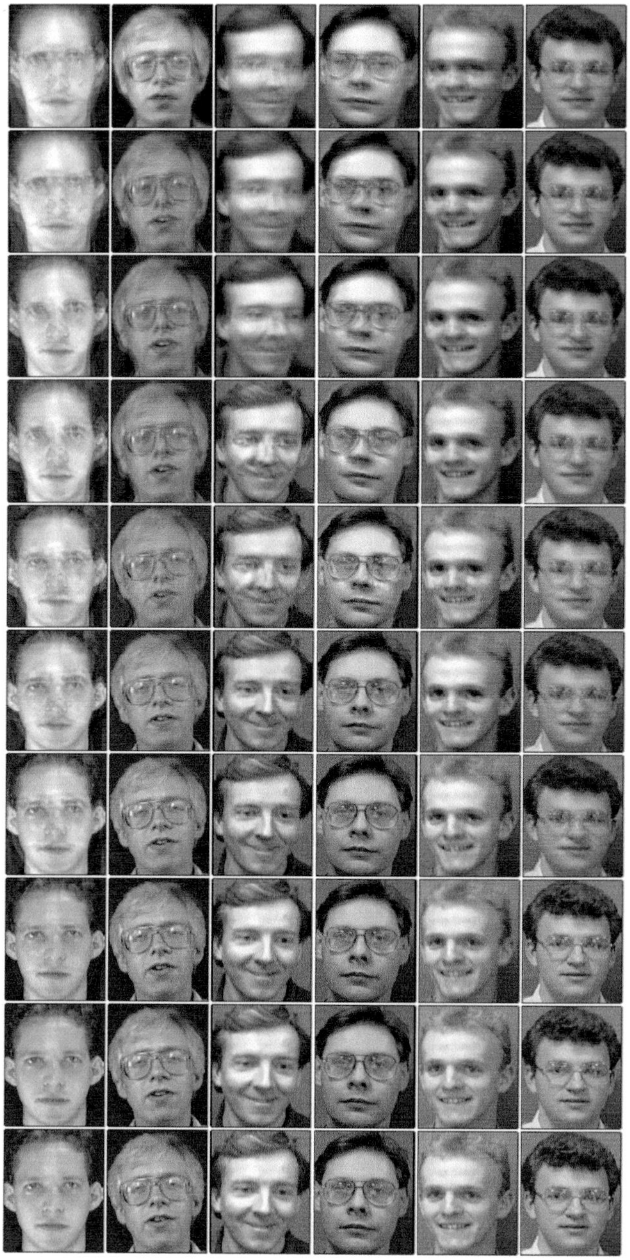

FIGURE 8.9
Eigenface reconstruction for first image for individuals 1 through 6 (top left hand side of Figure 8.8). Going from top to bottom, eigenface reconstruction incrementally adds prinicipal components 1 through 10.

Let $\mathbf{U} = [\mathbf{u}_1, \ldots, \mathbf{u}_n]_{p\times n}$ be the eigenvectors from the PCA. Define the rank-q reduced eigenface, $1 \le q \le n$, for i to be

$$\hat{\mathbf{X}}_i^{(q)} = \overline{\mathbf{X}} + \hat{\mathbf{Q}}^{(q)}(\mathbf{X}_i - \overline{\mathbf{X}}), \quad \text{where } \hat{\mathbf{Q}}^{(q)} = \sum_{i=1}^{q} \mathbf{u}_i \mathbf{u}_i^T. \tag{8.5}$$

We can explain the motivation for this estimator by considering the eigenface for the population vector $\mathbf{X}$. Let $\boldsymbol{\mu}$ and $\boldsymbol{\Sigma}$ be the population mean and covariance for $\mathbf{X}$. We seek to approximate $\mathbf{X}$ using an estimator

$$\hat{\mathbf{X}} = \boldsymbol{\mu} + \mathbf{Q}(\mathbf{X} - \boldsymbol{\mu}).$$

Notice that $\mathbb{E}(\hat{\mathbf{X}}) = \boldsymbol{\mu}$ which shows that the estimator matches the first moment. The estimator also matches the second moment. To establish this fact, we need to show

$$\text{Var}(\hat{\mathbf{X}}) = \mathbf{Q}\boldsymbol{\Sigma}\mathbf{Q}^T = \boldsymbol{\Sigma}.$$

Let $\boldsymbol{\Sigma} = \mathbf{P}\boldsymbol{\Lambda}\mathbf{P}^T$ be the spectral decomposition of $\boldsymbol{\Sigma}$ where $\mathbf{P} = [\mathbf{e}_1, \ldots, \mathbf{e}_n]$. Therefore since $\mathbf{Q} = \mathbf{P}\mathbf{P}^T$, we have

$$\mathbf{Q}\boldsymbol{\Sigma}\mathbf{Q}^T = \mathbf{P}\mathbf{P}^T\left(\mathbf{P}\boldsymbol{\Lambda}\mathbf{P}^T\right)\mathbf{P}\mathbf{P}^T = \mathbf{P}\boldsymbol{\Lambda}\mathbf{P}^T = \boldsymbol{\Sigma}.$$

Therefore, this provides an explanation for $\hat{\mathbf{X}}$. For rank reduction, we can approximate $\mathbf{Q}$ using the first q eigenvectors. This reduces dimensionality and yields an accurate approximation if the eigenvalues for the remaining $n - q$ eigenvectors are small. The rank-q reduced eigenface for $\mathbf{X}$ is

$$\hat{\mathbf{X}}^{(q)} = \boldsymbol{\mu} + \mathbf{Q}^{(q)}(\mathbf{X} - \boldsymbol{\mu}), \quad \text{where } \mathbf{Q}^{(q)} = \sum_{i=1}^{q} \mathbf{e}_i \mathbf{e}_i^T.$$

The estimator (8.5) is simply the sample analog of the population estimator.

Figure 8.9 shows the eigenface of the first face for each of the first six individuals from Figure 8.8 (first six rows, column 1, top left). From top to bottom, the eigenfaces are of rank $q = 1, \ldots, 10$. The images rapidly denoise as more eigenvectors are incorporated into the eigenface estimator. With 10 eigenvectors, all six faces become distinct and clear. Note that with a larger training sample, eigenface estimators provide a highly compressed solution space. If the training data consists of n images of dimension $r \times c$, the rank-reduced eigenface solution will often approximate the data accurately using $q \ll n$ matrices $\{\mathbf{u}_1\mathbf{u}_1^T, \ldots, \mathbf{u}_q\mathbf{u}_q^T\}$.

Remark 8.1. We observe (8.5) can also be interpreted as a least squares solution. Assume $\overline{\mathbf{X}} = \mathbf{0}$. The rank-$q$ reduced eigenface can be written as

$$\hat{\mathbf{X}}_i^{(q)} = \hat{\mathbf{Q}}^{(q)}\mathbf{X}_i = \sum_{i=1}^{q} \mathbf{u}_i(\mathbf{u}_i^T\mathbf{X}_i).$$

The right-hand side is the projection of $\mathbf{X}_i$ onto the span of $\{\mathbf{u}_1, \ldots, \mathbf{u}_q\}$. Therefore $\hat{\mathbf{X}}_i^{(q)}$ is the least squares solution from projecting $\mathbf{X}_i$ onto the design matrix $\mathbf{U}^{(q)} = [\mathbf{u}_1, \ldots, \mathbf{u}_q]$. □

Exercises

Exercise 8.1. Prove the properties of the PC transformation given in Chapter 8.3.1.

Exercise 8.2. Find a data set suitable for PCA and run PCA. Discuss your results.

Exercise 8.3. Here you will investigate the robustness of PCA to noise in the data.

1. Load the `Wine` dataset from the UCI repository.
2. Split the data into training and testing sets.
3. Add Gaussian noise with different variance levels to the training data.
4. Apply PCA to the noisy training data and project it onto a lower-dimensional space.
5. Train a classifier (e.g., SVM or logistic regression) on the lower-dimensional representation.
6. Evaluate the classifier's performance on the original, unnoisy testing set.
7. Analyze how the presence of noise affects the PCA components and the classifiers performance.

Exercise 8.4. Show for a square $p \times p$ symmetric matrix $\mathbf{A}$, that its singular value decomposition $\mathbf{A} = \mathbf{U}\mathbf{D}\mathbf{V}^T$ coincides with its spectral decomposition $\mathbf{A} = \mathbf{P}\boldsymbol{\Lambda}\mathbf{P}^T$.

Exercise 8.5. Let $\boldsymbol{\mu} = (\mu_1, \ldots, \mu_p)^T$ where $-1 < \mu_j < 1$ such that $\sum_{j=1}^{p} \mu_j = 0$. Define $\mathbf{D}$ to be the $p \times p$ diagonal matrix with entries $\mu_1^2, \ldots, \mu_p^2$ and define $\mathbf{A} = \mathbf{I} + \mathbf{D} - \boldsymbol{\mu}\boldsymbol{\mu}^T$. Under what conditions is $\mathbf{A}$ invertible?

Exercise 8.6. Choose a large data set $\mathbf{A}$. Center $\mathbf{A}$ so that its mean is zero. Compare $\mathbf{A}$ to the low-rank approximation $\sum_{k=1}^{r'} d_k \mathbf{u}_k \mathbf{v}_k^T$ for $r' = 1, 2, \ldots, r$ where r is the rank of $\mathbf{A}$. For example, your comparison of these values could use something like the metric:

$$||\mathbf{A} - \sum_{k=1}^{r'} d_k \mathbf{u}_k \mathbf{v}_k^T||_F^2.$$

However, be creative and consider other ways to assess differences. Summarize your findings.

Exercise 8.7. Prove Theorem 8.3

Exercise 8.8. Prove Theorem 8.4.

Exercise 8.9. What does $||\mathbf{A}\mathbf{A}^+ - \mathbf{I}||_F^2$ equal? Prove your answer.

Exercise 8.10. Let $\mathbf{A}_{p \times p}$ be a symmetric matrix of rank $r \le p$. Show there exists an orthogonal transformation such that for $\mathbf{x}$ restricted to the column span($\mathbf{A}$), $\mathbf{x}^T \mathbf{A}^+ \mathbf{x}$ can be written as

$$\mathbf{x}^T \mathbf{A}^+ \mathbf{x} = \sum_{k=1}^{r} u_k^2 / \lambda_k$$

where $\lambda_1 > \lambda_2 > \cdots > \lambda_r > 0$ are the ordered non-zero eigenvalues of $\mathbf{A}$.

Exercise 8.11. In this excercise, you will develop a PCA implementation from scratch without using any high-level libraries.

1. Load a high-dimensional dataset (e.g., `MNIST` dataset with $p = 784$ features).
2. Standardize the dataset.
3. Compute the covariance matrix of the standardized data.
4. Calculate the eigenvalues and eigenvectors of the covariance matrix.
5. Sort the eigenvalues and select the top k eigenvectors to form the transformation matrix.
6. Project the original data onto the new k-dimensional subspace.
7. Compare the results to PCA software implementations like `prcomp` in R.

Exercise 8.12. Here we will use PCA to visualize high-dimensional data in a lower-dimensional space.

1. Load the `CIFAR-10` dataset, which consists of $n = 60,000$, 32x32 color images in 10 classes.
2. Preprocess the data by flattening the images into a high-dimensional feature space ($p = 3072$ features per image).

3. Standardize the dataset.
4. Apply PCA to reduce the dimensionality of the data to 2 or 3 dimensions.
5. Create a scatter plot to visualize the data points in the reduced-dimensional space, color-coded by class.
6. Discuss how well the PCA projection separates the different classes and the limitations of PCA for this task.

Exercise 8.13. Find a data set suitable for regression where the number of features p is larger than the sample size n.

1. Describe how to reparametrize the model for PC regression (PCR) and describe how to implement PCR in this setting in order to obtain useful answers (keep in mind that $p > n$).
2. Implement your PCR algorithm on your data. Discuss your results in the context of the original parameterization.

Exercise 8.14. Recall that sparse PCA is a variation of PCA that incorporates sparsity constraints to produce sparse loadings. This is particularly useful in high-dimensional settings where interpretability is key, such as in gene expression data analysis, where the goal is to identify a small number of genes (features) that are most important for explaining the variance in the data. Unlike traditional PCA, which uses all variables to construct the principal components, sparse PCA imposes a penalty on the loadings, similar to how lasso and elastic net impose penalties in regression (see Chapter 4.8). This penalty enforces sparsity, allowing only a subset of the original variables to have non-zero loadings. In this exercise, you will gain familiarity with this technique using a high-dimensional gene expression dataset.

1. Load the `Leukemia` dataset, which contains gene expression measurements for two types of leukemia: acute lymphoblastic leukemia (ALL) and acute myeloid leukemia (AML).
2. Preprocess the data by standardizing the gene expression measurements.
3. Apply standard PCA to the dataset and visualize the first two principal components.
4. Apply sparse PCA using the `PMA` R-package to the same dataset. Set appropriate sparsity constraints to obtain sparse loadings.
5. Visualize the first two sparse principal components and identify the top contributing genes.
6. Compare the variance explained by the standard PCA and sparse PCA.
7. Discuss the biological relevance of the top contributing genes identified by sparse PCA.

Exercise 8.15. Using Theorem 8.4 write a program to calculate the generalized inverse $\mathbf{A}^+$. Illustrate the program on an interesting example and directly verify that the three properties of a generalized inverse hold in your example.

Exercise 8.16. Prove Theorem 8.5.

Exercise 8.17. Obtain the first 10 facial images of the first individual from Figure 8.8 (search for the data online). Convert each image to a vector of length $p = 92 \times 112$ where each entry is the intensity value for one of the pixels from the original image encoded on an 8-bit gray scale.

1. Determine the rank-q reduced eigenface $\hat{\mathbf{X}}_i^{(q)}$ for $q = 1,\ldots,10$ for each facial image $i = 1,\ldots,10$.
2. Determine the average eigenface $n^{-1}\sum_{i=1}^{n}\hat{\mathbf{X}}_i^{(q)}$ for $q = 1,\ldots,10$. What does the average eigenface look like? Why?

Exercise 8.18. Kernel Principal Component Analysis (Kernel PCA) is an extension of PCA that allows for the extraction of principal components in a high-dimensional feature space induced by a kernel function [5]. While traditional PCA is limited to capturing linear relationships in the data, kernel PCA can capture nonlinear relationships by mapping the original data into a higher-dimensional space using a kernel function and then performing PCA in this transformed space. Common kernel functions include the radial basis function (RBF) kernel, polynomial kernel, and sigmoid kernel. In this exercise you will learn to apply kernel PCA to extract nonlinear features from data and compare it with linear PCA.

1. Load the `Digits` dataset from the `scikit-learn` package or a similar dataset in R.
2. Split the data into training and testing sets.
3. Apply linear PCA to reduce the dimensionality of the data to 2 or 3 dimensions.
4. Visualize the data points in the reduced-dimensional space using a scatter plot.
5. Apply kernel PCA with a radial basis function (RBF) kernel to the same dataset.
6. Visualize the data points in the reduced-dimensional space using a scatter plot.
7. Compare the linear PCA and kernel PCA projections in terms of data separability and clustering quality.
8. Discuss the advantages and limitations of using kernel PCA for nonlinear feature extraction.

Bibliography

[1] Eckart C. and Young G. The approximation of one matrix by another of lower rank. *Psychometrika,* 1:211–218, 1936.

[2] Gabriel K.R. The biplot graphic display of matrices with application to principal component analysis. *Biometrika,* 58(3):453–467, 1971.

[3] Hadi A.S. and Ling R.F., 1998. Some cautionary notes on the use of principal components regression. *The American Statistician,* 52(1),15–19, 1998.

[4] Izenman A.J. *Modern Multivariate Statistical Techniques.* Springer: New York, 2008.

[5] Schölkopf B., Smola A. and Müller K.-R. Nonlinear component analysis as a kernel eigenvalue problem, *Neural computation,* vol. 10, no. 5, pp. 1299–1319, 1998.

[6] Wang Y., Klijn J.G., Zhang Y., Sieuwerts A.M., Look M.P., Yang F., Talantov D., Timmermans M., Meijer-van Gelder M.E., Yu J., Jatkoe T., Berns E.M., Atkins D. and Foekens J.A. Gene-expression profiles to predict distant metastasis of lymph-node-negative primary breast cancer. *Lancet,* 365(9460):671–679, 2005.

[7] Witten D.M. and Tibshirani R. and Hastie T. A penalized matrix decomposition, with applications to sparse principal components and canonical correlation analysis. *Biostatistics,* 10(3):515–534, 2009.

9

Canonical Correlation Analysis

9.1 Introduction

Canonical Correlation Analysis (CCA) is an unsupervised multivariate statistical technique used to identify and quantify the relationships between two sets of variables. It finds linear combinations of each set of variables that are maximally correlated with each other. Formally, CCA seeks to find pairs of canonical variables $(\mathbf{a}^T\mathbf{X}, \mathbf{b}^T\mathbf{Y})$, where $\mathbf{X} = (X_1, \ldots, X_p)^T$ and $\mathbf{Y} = (Y_1, \ldots, Y_q)^T$ are two sets of vectors, and $\mathbf{a}$ and $\mathbf{b}$ are the weight vectors that maximize the correlation between the canonical variables.

CCA has many useful properties:

1. CCA provides a way to explore and quantify the relationships between two sets of variables, making it valuable in fields where understanding these multivariate relationships is crucial, such as biology, psychology, and economics.
2. CCA can integrate and analyze data from different sources or types, such as combining gene expression data with clinical data, thereby providing a comprehensive view of the relationships between the datasets.
3. By focusing on the canonical variables that capture the most significant correlations, CCA reduces the complexity of the data, making it easier to interpret and analyze.
4. CCA helps in identifying the most informative features from each set of variables that are related to the other set, which can be useful for predictive modeling and other analytical tasks.

9.2 CCA Optimization Problem

Let $\mathbf{X} \in \mathbb{R}^p$ and $\mathbf{Y} \in \mathbb{R}^q$ be the random vectors of interest with means $\boldsymbol{\mu}$ and $\boldsymbol{\nu}$ and covariance matrices

$$\begin{aligned}
\mathrm{Var}(\mathbf{X},\mathbf{X}) &= \mathbb{E}(\mathbf{X}-\boldsymbol{\mu})(\mathbf{X}-\boldsymbol{\mu})^T = \boldsymbol{\Sigma}_{11} \quad (p \times p)\\
\mathrm{Var}(\mathbf{Y},\mathbf{Y}) &= \mathbb{E}(\mathbf{Y}-\boldsymbol{\nu})(\mathbf{Y}-\boldsymbol{\nu})^T = \boldsymbol{\Sigma}_{22} \quad (q \times q)\\
\mathrm{Cov}(\mathbf{X},\mathbf{Y}) &= \mathbb{E}(\mathbf{X}-\boldsymbol{\mu})(\mathbf{Y}-\boldsymbol{\nu})^T = \boldsymbol{\Sigma}_{12} \quad (p \times q)\\
\mathrm{Cov}(\mathbf{Y},\mathbf{X}) &= \mathbb{E}(\mathbf{Y}-\boldsymbol{\nu})(\mathbf{X}-\boldsymbol{\mu})^T = \boldsymbol{\Sigma}_{21} \quad (q \times p).
\end{aligned}$$

The CCA optimization problem aims to find linear combinations $\mathbf{a}^T\mathbf{X}$ and $\mathbf{b}^T\mathbf{Y}$, where $\mathbf{a} \in \mathbb{R}^p$ and $\mathbf{b} \in \mathbb{R}^q$, achieve maximal correlation. The correlation is defined by

$$\rho(\mathbf{a},\mathbf{b}) = \frac{\mathbf{a}^T\boldsymbol{\Sigma}_{12}\mathbf{b}}{\sqrt{\mathbf{a}^T\boldsymbol{\Sigma}_{11}\mathbf{a}}\sqrt{\mathbf{b}^T\boldsymbol{\Sigma}_{22}\mathbf{b}}}.$$

The vectors $\mathbf{a}$ and $\mathbf{b}$ are called the canonical correlation vectors.

DOI: 10.1201/9781003493679-9

However, because this problem is invariant under scaling of $\mathbf{a}$ and $\mathbf{b}$, we must impose a constraint to obtain a unique solution. We assume the variance of $\mathbf{a}^T\mathbf{X}$ and $\mathbf{b}^T\mathbf{Y}$ is unity:

$$\mathrm{Var}(\mathbf{a}^T\mathbf{X}) = \mathbf{a}^T\boldsymbol{\Sigma}_{11}\mathbf{a} = 1, \quad \mathrm{Var}(\mathbf{b}^T\mathbf{Y}) = \mathbf{b}^T\boldsymbol{\Sigma}_{22}\mathbf{b} = 1.$$

The optimization problem then becomes

$$\max_{\mathbf{a},\mathbf{b}} \mathbf{a}^T\boldsymbol{\Sigma}_{12}\mathbf{b} \quad \text{subject to} \quad \mathbf{a}^T\boldsymbol{\Sigma}_{11}\mathbf{a} = \mathbf{b}^T\boldsymbol{\Sigma}_{22}\mathbf{b} = 1.$$

9.3 Canonical Correlation Solution

To define the solution, let $\mathbb{U} = \mathbf{K}\mathbf{K}^T$ and $\mathbb{V} = \mathbf{K}^T\mathbf{K}$, where

$$\mathbf{K} = \boldsymbol{\Sigma}_{11}^{-1/2}\boldsymbol{\Sigma}_{12}\boldsymbol{\Sigma}_{22}^{-1/2}.$$

Let $(\mathbf{u}_j)_{j=1}^p$ and $(\mathbf{v}_k)_{k=1}^q$ denote the standardized eigenvectors for $\mathbb{U}$ and $\mathbb{V}$. It follows that $\mathbb{U}$ and $\mathbb{V}$ have the same non-zero eigenvalues $\lambda_1 \geq \ldots \geq \lambda_r > 0$, where $r = \mathrm{r}(\boldsymbol{\Sigma}_{12}) \leq \min(p,q)$. Furthermore, $(\mathbf{u}_j)_{j=1}^p$ and $(\mathbf{v}_k)_{k=1}^q$ are the column vectors of $\mathbf{U}$ and $\mathbf{V}$ from the singular value decomposition (SVD) of $\mathbf{K} = \mathbf{U}\mathbf{D}\mathbf{V}^T$ and $(\lambda_l^{1/2})_{l=1}^r$ are the diagonal elements of $\mathbf{D}$.

With this notation, we state the canonical correlation solution.

Theorem 9.1. *For $l = 1,\ldots,r$:*

1. $\boldsymbol{a}_l = \boldsymbol{\Sigma}_{11}^{-1/2}\boldsymbol{u}_l$ *and* $\boldsymbol{b}_l = \boldsymbol{\Sigma}_{22}^{-1/2}\boldsymbol{v}_l$ *are the lth canonical correlation vectors for $\boldsymbol{X}$ and $\boldsymbol{Y}$. Note that* $\boldsymbol{a}_l^T\boldsymbol{\Sigma}_{11}\boldsymbol{a}_l = \boldsymbol{b}_l^T\boldsymbol{\Sigma}_{22}\boldsymbol{b}_l = 1$.
2. *The random variables* $U_l = \boldsymbol{a}_l^T\boldsymbol{X}$ *and* $V_l = \boldsymbol{b}_l^T\boldsymbol{Y}$ *are called the lth canonical correlation variables. Note that* $Cov(U_l, U_j) = Cov(V_l, V_j) = 0$ *if* $l \neq j$.
3. $\rho_l = \lambda_l^{1/2}$ *is called the lth canonical correlation.*

9.3.1 Sample canonical correlation analysis

CCA is applied by replacing population parameters with sample estimators. Given a sample $(\mathbf{X}_1,\mathbf{Y}_1),\ldots,(\mathbf{X}_n,\mathbf{Y}_n)$, the sample estimates for $\boldsymbol{\Sigma}_{11},\boldsymbol{\Sigma}_{22},\boldsymbol{\Sigma}_{12},\boldsymbol{\Sigma}_{21}$ are

$$\begin{aligned}
\mathbf{S}_{11} &= \frac{1}{n}\sum_{i=1}^n \mathbf{X}_i\mathbf{X}_i^T - \overline{\mathbf{X}}\overline{\mathbf{X}}^T, \qquad \mathbf{S}_{22} = \frac{1}{n}\sum_{i=1}^n \mathbf{Y}_i\mathbf{Y}_i^T - \overline{\mathbf{Y}}\overline{\mathbf{Y}}^T \\
\mathbf{S}_{12} &= \frac{1}{n}\sum_{i=1}^n \mathbf{X}_i\mathbf{Y}_i^T - \overline{\mathbf{X}}\overline{\mathbf{Y}}^T, \qquad \mathbf{S}_{21} = \frac{1}{n}\sum_{i=1}^n \mathbf{Y}_i\mathbf{X}_i^T - \overline{\mathbf{Y}}\overline{\mathbf{X}}^T.
\end{aligned}$$

9.4 Simple Algorithm for Regularized CCA

CCA can become numerically unstable because it requires inverting the square-root covariance matrices $\mathbf{S}_{11}^{1/2}$ and $\mathbf{S}_{22}^{1/2}$. These matrices can easily become singular or near-singular, especially in higher-dimensional problems. To address this, we present Algorithm 9.1 for performing regularized CCA. This procedure uses a "ridge-type" penalty to regularize the matrices,

Algorithm 9.1 *Regularized CCA*

1: Input $(\mathbf{X}_1, \mathbf{Y}_1), \ldots, (\mathbf{X}_n, \mathbf{Y}_n)$ and ridge parameters $\kappa_x > 0$ and $\kappa_y > 0$

2: Standardize $\mathbf{X}$ and $\mathbf{Y}$ to have zero mean and variance equal to one

$$\mathbf{X}_i \leftarrow \begin{pmatrix} (X_{i1} - \overline{X}_1)/\sqrt{S_1^x} \\ \vdots \\ (X_{ip} - \overline{X}_p)/\sqrt{S_p^x} \end{pmatrix}, \mathbf{Y}_i \leftarrow \begin{pmatrix} (Y_{i1} - \overline{Y}_1)/\sqrt{S_1^y} \\ \vdots \\ (Y_{iq} - \overline{Y}_q)/\sqrt{S_q^y} \end{pmatrix}, \quad i = 1, \ldots, n.$$

3: Calculate the sample variance-covariance matrices $\mathbf{S}_{11}$ and $\mathbf{S}_{22}$ and $\mathbf{S}_{12}$

4: Ridge stabilize the diagonal values for the variance matrices

$$\mathbf{S}_{11} \leftarrow \mathbf{S}_{11} + \kappa_x \mathbf{I}_p, \quad \mathbf{S}_{22} \leftarrow \mathbf{S}_{22} + \kappa_y \mathbf{I}_q$$

5: Let $\mathbf{S}_{11} = \mathbf{U}_{11}\mathbf{D}_{11}\mathbf{V}_{11}^T$ and $\mathbf{S}_{22} = \mathbf{U}_{22}\mathbf{D}_{22}\mathbf{V}_{22}^T$ be the SVD for $\mathbf{S}_{11}$ and $\mathbf{S}_{22}$ and from this calculate

$$\mathbf{S}_{11}^{-1/2} = \mathbf{U}_{11}(\mathbf{D}_{11})^{-1/2}\mathbf{V}_{11}^T, \quad \mathbf{S}_{22}^{-1/2} = \mathbf{U}_{22}(\mathbf{D}_{22})^{-1/2}\mathbf{V}_{22}^T$$

6: Calculate $\mathbf{K} = \mathbf{S}_{11}^{-1/2}\mathbf{S}_{12}\mathbf{S}_{22}^{-1/2}$ and obtain the SVD for $\mathbf{K} = \mathbf{U}\mathbf{D}\mathbf{V}^T$

7: Return canonical correlation vectors $\mathbf{a}_l = \mathbf{S}_{11}^{-1/2}\mathbf{u}_l$ and $\mathbf{b}_l = \mathbf{S}_{122}^{-1/2}\mathbf{v}_l$ and canonical correlations $\rho_l = d_{ll}$ for $l = 1, \ldots, r$

improving numerical stability, and being simple to use in practice (an example is given later). Recall that ridge estimation was discussed in Chapter 4.8.

Line 4 of the algorithm modifies $\mathbf{S}_{11}$ and $\mathbf{S}_{22}$ by adding user-specified ridge parameters, $\kappa_x > 0$ and $\kappa_y > 0$. To see why this stabilizes the inverse, consider $\mathbf{S}_{11}$ (a similar argument applies to $\mathbf{S}_{22}$). Let

$$\mathbf{S}_{11} = \mathbf{P}\mathbf{\Lambda}\mathbf{P}^T = \sum_{k=1}^{p} \lambda_k \mathbf{e}_k \mathbf{e}_k^T$$

be the spectral decomposition for $\mathbf{S}_{11}$, where $\mathbf{e}_k$ is the kth unit eigenvector with eigenvalue $\lambda_k \geq 0$. Then

$$\mathbf{S}_{11} + \kappa_x \mathbf{I}_p = \sum_{k=1}^{p} (\lambda_k + \kappa_x)\mathbf{e}_k \mathbf{e}_k^T.$$

Notice that by introducing $\kappa_x > 0$, this ensures that the inverse is well defined. In particular, we have

$$\left(\mathbf{S}_{11} + \kappa_x \mathbf{I}_p\right)^{-1/2} = \sum_{k=1}^{p} (\lambda_k + \kappa_x)^{-1/2} \mathbf{e}_k \mathbf{e}_k^T.$$

We refer to this type of stabilization as a ridge-type regularization.

Line 5 of Algorithm 9.1 uses the SVD decomposition of the modified $\mathbf{S}_{11}$ and $\mathbf{S}_{22}$ to determine $\mathbf{S}_{11}^{-1/2}$ and $\mathbf{S}_{22}^{-1/2}$. Line 7 extracts the eigenvectors $\mathbf{u}_l$ and $\mathbf{v}_l$ for $\mathbb{U}$ and $\mathbb{V}$ using the SVD of $\mathbf{K}$ from line 6, transforming these to canonical vectors $\mathbf{a}_l = \mathbf{S}_{11}^{-1/2}\mathbf{u}_l$ and $\mathbf{b}_l = \mathbf{S}_{22}^{-1/2}\mathbf{v}_l$.

9.5 Nutrigenomic Study

To illustrate CCA, and compare this with regularized CCA, we reanalyze data from the nutrigenomic study of Martin et al. [1]. This study examined the effect of diet on hepatic lipid expression and gene expression in wild type and PPARα mice genotypes. In total, there were $n = 40$ mice subjected to 5 different diets. Expression of 21 hepatic fatty acids (lipids) was

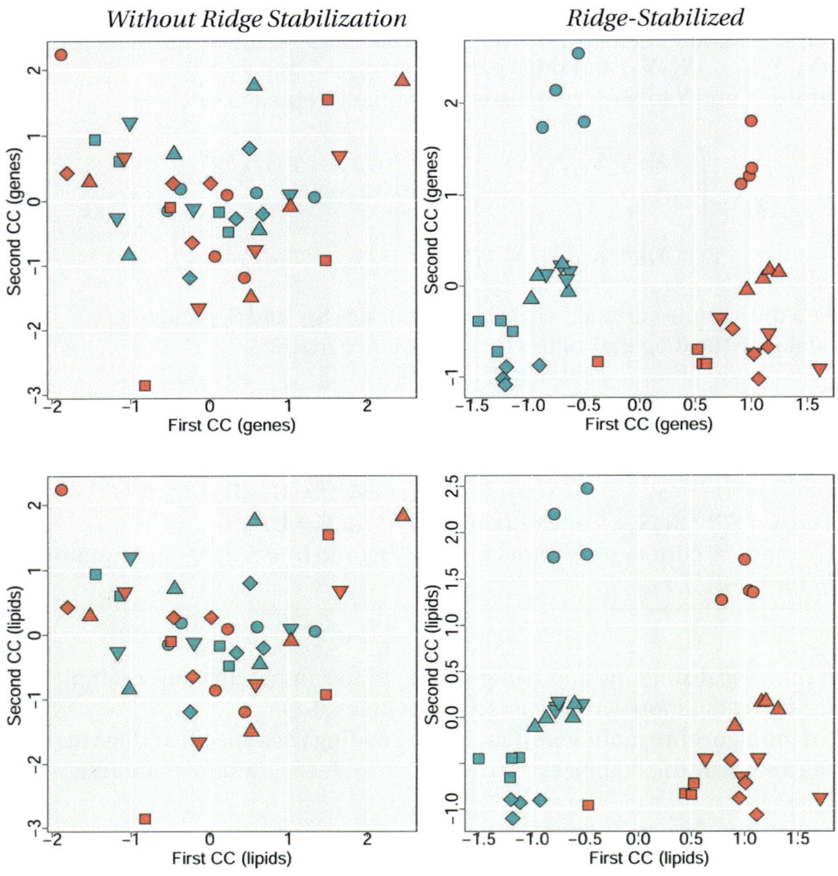

FIGURE 9.1
Nutrigenomic analysis. First and second canonical correlation variables for genes and lipids (red points are PPARα mouse genotypes, symbols indicate type of diet). Left side is unregularized. Right side is regularized using ridge values $\kappa_x = \kappa_y = 0.2$. Notice that for both procedures, the plots of (U_1, U_2) (top) and (V_1, V_2) (bottom) are almost identical, This is a reflection of a high correlation between the first two canonical variables.

monitored, along with gene expression. Although thousands of gene expressions were collected, we restrict our analysis to the 120 most interesting genes.

For the analysis, we use the gene expressions for the **X** variables ($p = 120$) and lipid expressions for **Y** ($q = 21$). Ridge values $\kappa_x = \kappa_y = 0.2$ were used. Figure 9.1 displays the first two canonical variables (U_1, U_2) and (V_1, V_2) on the top and bottom, respectively. To help understand the results, we include the phenotype and diet experimental conditions. Red points identify the PPARα mouse genotype, and blue points are the wild type mouse genotype. The type of symbol identifies the type of diet. Figures on the left are the non-regularized CCA, while the right figures are regularized CCA using ridge stabilization. We immediately see the benefit of ridge stabilization, which is able to delineate mouse genotype and the effect of diet, while the non-regularized procedure is unable to find anything.

To gain more insight into the data, we use a supervised multivariate random forest analysis (random forests and multivariate forests are discussed in detail in Chapter 16.5). An important feature of random forests is that it provides an internal measure of variable importance, called permutation importance, or VIMP, which can be used for variable selection (Chapter 17.2).

FIGURE 9.2
Permutation variable importance (VIMP) using multivariate random forests (Chapter 16.5) for each of the $q = 21$ responses from the nutrigenomic study. Red signifies influential variables; blue are noisy variables.

The multivariate random forest was constructed by using lipid expressions **Y** for the $q = 21$ dimensional response vector. The features were gene expressions **X**, and we also included mouse genotype and type of diet as additional covariates. Thus, there was a total of $p+2 = 122$ features. The VIMP for each variable for each of the responses from the random forests analysis is shown in Figure 9.2. To improve visual clarity, the figure displays only a subset of the most meaningful features.

The most influential variable across the responses is diet, followed by genotype. These two variables are strong predictors of lipid expression. Recall that the regularized CCA also identified diet and genotype as important factors, further confirming their significance. The VIMP plot shows that certain genes are also predictive. For example, PMDCI has a high VIMP for the C20.1n.9 lipid response, and CYP3A11 is another notable gene. This indicates that gene expression adds value to the prediction model beyond phenotype and diet.

Exercises

Exercise 9.1. Show that for an arbitrary fixed $\mathbf{b}$,

$$\max_{\mathbf{a}}\left\{\frac{(\mathbf{a}^T\boldsymbol{\Sigma}_{12}\mathbf{b})^2}{(\mathbf{a}^T\boldsymbol{\Sigma}_{11}\mathbf{a})(\mathbf{b}^T\boldsymbol{\Sigma}_{22}\mathbf{b})}\right\}=\frac{\mathbf{b}^T\boldsymbol{\Sigma}_{21}\boldsymbol{\Sigma}_{11}^{-1}\boldsymbol{\Sigma}_{12}\mathbf{b}}{\mathbf{b}^T\boldsymbol{\Sigma}_{22}\mathbf{b}}$$

and that the maximum of this quantity over $\mathbf{b}$, i.e.

$$\max_{\mathbf{b}}\left\{\frac{\mathbf{b}^T\boldsymbol{\Sigma}_{21}\boldsymbol{\Sigma}_{11}^{-1}\boldsymbol{\Sigma}_{12}\mathbf{b}}{\mathbf{b}^T\boldsymbol{\Sigma}_{22}\mathbf{b}}\right\},$$

equals the largest eigenvalue of $\boldsymbol{\Sigma}_{11}^{-1}\boldsymbol{\Sigma}_{12}\boldsymbol{\Sigma}_{22}^{-1}\boldsymbol{\Sigma}_{21}$.

Exercise 9.2. Let $\mathbf{K}=\boldsymbol{\Sigma}_{11}^{-1/2}\boldsymbol{\Sigma}_{12}\boldsymbol{\Sigma}_{22}^{-1/2}$. Define

$$\mathbb{U}=\mathbf{K}\mathbf{K}^T,\quad \mathbb{V}=\mathbf{K}^T\mathbf{K}$$

and

$$\mathbb{U}^*=\boldsymbol{\Sigma}_{11}^{-1/2}\mathbb{U}\boldsymbol{\Sigma}_{11}^{1/2},\quad \mathbb{V}^*=\boldsymbol{\Sigma}_{22}^{-1/2}\mathbb{V}\boldsymbol{\Sigma}_{22}^{1/2}.$$

Show that $\mathbb{U},\mathbb{V},\mathbb{U}^*,\mathbb{V}^*$ all have the same non-zero eigenvalues and that these eigenvalues are real and strictly positive.

Exercise 9.3. Using the notation of the previous exercise, show that

$$\mathbb{U}=\mathbf{I}-\boldsymbol{\Sigma}_{11}^{-1}\boldsymbol{\Sigma}_{11.2},\quad \text{where } \boldsymbol{\Sigma}_{11.2}=\boldsymbol{\Sigma}_{11}-\boldsymbol{\Sigma}_{12}\boldsymbol{\Sigma}_{22}^{-1}\boldsymbol{\Sigma}_{21}.$$

Hence, the first canonical correlation vector for $\mathbf{X}$ is given by the eigenvector corresponding to the smallest eigenvalue of $\boldsymbol{\Sigma}_{11.2}$.

Exercise 9.4. Show that the squared canonical correlations are the roots of the equation

$$|\boldsymbol{\Sigma}_{12}\boldsymbol{\Sigma}_{22}^{-1}\boldsymbol{\Sigma}_{21}-\lambda\boldsymbol{\Sigma}_{11}|=0$$

and that the canonical correlation vectors for $\mathbf{X}$ satisfy

$$\boldsymbol{\Sigma}_{12}\boldsymbol{\Sigma}_{22}^{-1}\boldsymbol{\Sigma}_{21}\mathbf{a}_i=\lambda_i\boldsymbol{\Sigma}_{11}\mathbf{a}_i,\quad i=1,\dots,p.$$

Exercise 9.5. Let $\mathbf{A}=\sum_{j=1}^p\lambda_j\mathbf{e}_j\mathbf{e}_j^T$ be the spectral decomposition of a square symmetric matrix $\mathbf{A}_{p\times p}$. Let $\mathbf{A}(\lambda)=\mathbf{A}+\lambda\mathbf{I}_p$ where $\lambda>0$ (we call $\mathbf{A}$ the ridge stabilized matrix for $\mathbf{A}$). Prove that

1. $\mathbf{A}(\lambda)=\sum_{j=1}^p(\lambda_j+\lambda)\mathbf{e}_j\mathbf{e}_j^T$.

2. If $\mathbf{A} > 0$ show that $\mathbf{A}^{-1}(\lambda) = \sum_{j=1}^{p} (\lambda_j + \lambda)^{-1} \mathbf{e}_j \mathbf{e}_j^T$.
3. Suppose that $\mathbf{A}$ is singular such that $\lambda_1 > \cdots > \lambda_q > 0$ and $\lambda_{q+1} = \cdots = \lambda_p = 0$. Determine $\mathbf{A}^{-1}(\lambda)$.

Exercise 9.6. In this example we will analyze the relationship between environmental factors and socioeconomic indicators using CCA.

1. Use the `World Bank` open data platform to download datasets for environmental factors (e.g., air quality, water quality) and socioeconomic indicators (e.g., GDP, education levels).
2. Preprocess the data by aligning the data by country and year. Standardize the features in each dataset.
3. Apply CCA to identify the canonical variables that maximize the correlation between environmental factors and socioeconomic indicators.
4. Interpret the canonical variables and discuss the implications for policy and decision-making.
5. Create visualizations such as canonical correlation plots and geographic maps to illustrate the findings.

Exercise 9.7. Here we will implement sparse CCA to handle high-dimensional datasets where the number of features exceeds the number of samples, and interpret the results.

1. Use the `TCGA` (The Cancer Genome Atlas) dataset, available through the `TCGAbiolinks` R package.
2. Select a subset of gene expression data and corresponding clinical data for a specific cancer type. Standardize the features in each dataset.
3. Apply sparse CCA to identify the most relevant features in the gene expression and clinical datasets.
4. Interpret the canonical variables and discuss the biological significance of the selected features.
5. Validate the findings using external datasets or literature.

Exercise 9.8. In this exercise we will explore multi-view learning by using CCA to integrate and analyze data from different modalities (e.g., text and images) to improve classification performance.

1. Use the `Flickr30k` dataset, which contains images and their textual descriptions. This dataset can be accessed online.
2. In this step you need to extract image features and text features from the data and then standardize them. In practice, you would extract these features using a pre-trained convolution neural network (like VGG16 or ResNet) for images and a text embedding model (like Word2Vec or BERT) for textual descriptions. You will need to read up on this as this can be a little bit tricky.
3. The data is very large, therefore you should restrict attention to a subset of the images. You should also consider using PCA to reduce the dimensionality of the image and text features prior to CCA.
4. Apply regularized CCA to identify canonical variables that maximize the correlation between image and text features.

5. Use the canonical variables as input to a classifier of your choice to improve classification performance compared to using raw features alone.
6. Compare the classification performance using appropriate metrics.

Bibliography

[1] Martin P.G.P. et al. Novel aspects of PPARα-mediated regulation of lipid and xenobiotic metabolism revealed through a nutrigenomic study. *Hepatology*, 45(3): 767–777, 2007.

10

Newton's Method

10.1 Introduction

Newton's method (also known as the Newton-Raphson method) is an iterative numerical method for finding roots (the zeroes) of a real-valued function $f : \mathbb{R} \to \mathbb{R}$ in one variable. The goal is to find values $z \in \mathbb{R}$ such that $f(z) = 0$. Newton's method accomplishes this by using the derivative f' of f, making it a gradient-based numerical optimization method. The method starts with an initial guess $z_0 \in \mathbb{R}$ and then iteratively updates each previous step by

$$z_m = z_{m-1} - \frac{f(z_{m-1})}{f'(z_{m-1})}. \tag{10.1}$$

Provided the function is reasonably well-behaved and that z_0 is sufficiently close to the true root, the method is guaranteed to converge.

The Newton-Raphson method can be explained as follows. To find the root z such that $f(z) = 0$, we calculate the slope of $f(z)$ at the current value z_{m-1}

$$f'(z_{m-1}) = \lim_{\delta \to 0} \left[\frac{f(z_{m-1} + \delta) - f(z_{m-1})}{z_{m-1} + \delta - z_{m-1}} \right].$$

If the distance $z_m - z_{m-1}$ is small, then $z_m = z_{m-1} + \delta$ for some small δ so that

$$f'(z_{m-1}) \asymp \frac{f(z_m) - f(z_{m-1})}{z_m - z_{m-1}}.$$

If z_m is the required solution $f(z_m) = 0$, then with some rearrangement we obtain (10.1). Geometrically, each update z_m is the intersection with the z-axis of a line tangent to f at z_{m-1}.

Figure 10.1 illustrates Newton's method applied to the function

$$f(z) = 5z^3 - 7z^2 - 40z + 100.$$

This function is challenging, with one unique root at $z = -3.2$ and a valley near $z = 3$ that can trap the algorithm. The top figures display the steps taken by Newton's method using an initial value $z_0 = 6$. The first panel shows the first step, where the tangent at z_0 intersects the z-axis at $z = 4.3$. The next two figures show the progress after 5 and 20 steps. At some point (the fourth step), the algorithm takes a large step to the left, luckily escaping the valley. However, a larger starting value causes the algorithm to get stuck, oscillating between values to the left and right of the valley. Thus, a poor starting value can cause Newton's method to fail to converge in this example. The bottom panels use an initial value of $z_0 = -4$. The three panels show the first, second, and third steps of the algorithm. The algorithm converges rapidly to the root $z = -3.2$ thanks to the good starting value. This demonstrates a general property of Newton's method: it converges rapidly if properly initialized.

DOI: 10.1201/9781003493679-10

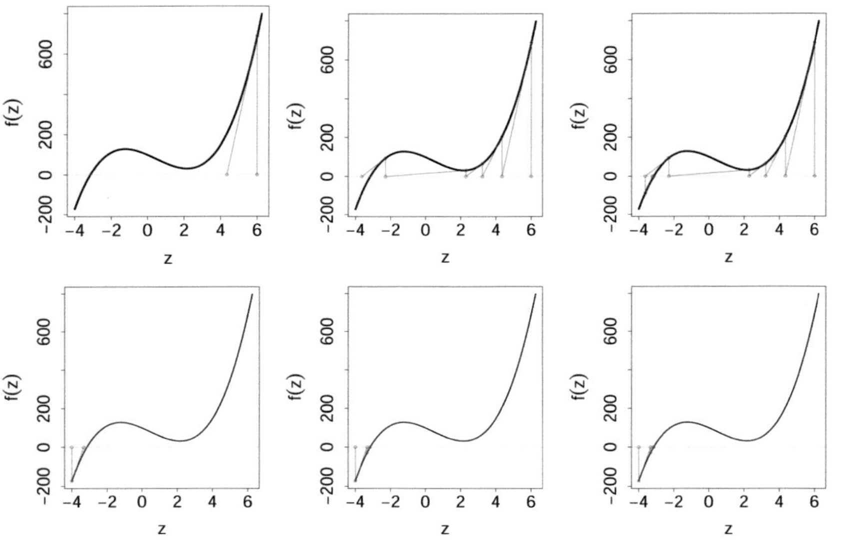

FIGURE 10.1
Illustration of Newton's method applied to $f(z) = 5z^3 - 7z^2 - 40z + 100$ *using different starting values* z_0 *(top figures use* $z_0 = 6$ *while the bottom figures use* $z_0 = -4$*).*

10.2 Multivariable Maximization

In statistical problems, we often need to find the maximum (or minimum) of a multivariate function. The extension of Newton's method to multivariable functions can be used for this purpose. Let f be a multivariate real-valued function $f : \mathbb{R}^p \to \mathbb{R}$ with gradient (see Chapter 2.9)

$$\nabla f(\mathbf{z}) = \frac{\partial f(\mathbf{z})}{\partial \mathbf{z}} = \begin{bmatrix} \dfrac{\partial f(\mathbf{z})}{\partial z_1} \\ \vdots \\ \dfrac{\partial f(\mathbf{z})}{\partial z_p} \end{bmatrix}_{p\times 1},$$

where the partial derivative along j is defined by

$$\frac{\partial f(\mathbf{z})}{\partial z_j} = \lim_{\delta \to 0} \frac{f(z_1, \ldots, z_{j-1}, z_j + \delta, z_{j+1}, \ldots, z_p) - f(\mathbf{z})}{\delta}.$$

To find the maximum of f we seek the value $\mathbf{z}^* \in \mathbb{R}^p$ such that

$$f(\mathbf{z}^*) \geq f(\mathbf{z}), \quad \text{for all } \mathbf{z} \in \mathbb{R}^p.$$

To find this maxima we solve for the root of its gradient. Thus the task is to find the solution $\mathbf{z}^*$ such that

$$\nabla f(\mathbf{z}^*) = \mathbf{0}.$$

For this we will need the Hessian matrix of f defined as the matrix of second-order derivatives of f (see Chapter 2.9)

$$\mathbf{H}(\mathbf{z}) = \begin{pmatrix} \frac{\partial^2 f(\mathbf{z})}{\partial z_1^2} & \frac{\partial^2 f(\mathbf{z})}{\partial z_1 \partial z_2} & \cdots & \frac{\partial^2 f(\mathbf{z})}{\partial z_1 \partial z_p} \\ \frac{\partial^2 f(\mathbf{z})}{\partial z_2 \partial z_1} & \frac{\partial^2 f(\mathbf{z})}{\partial z_2^2} & \cdots & \frac{\partial^2 f(\mathbf{z})}{\partial z_2 \partial z_p} \\ \vdots & \vdots & \ddots & \vdots \\ \frac{\partial^2 f(\mathbf{z})}{\partial z_p \partial z_1} & \frac{\partial^2 f(\mathbf{z})}{\partial z_p \partial z_2} & \cdots & \frac{\partial^2 f(\mathbf{z})}{\partial z_p^2} \end{pmatrix}_{p\times p}.$$

We write $\mathbf{H} = (f_{ij})$ for convenience. If all f_{ij} are continuous, then $\mathbf{H}(\mathbf{z})$ is symmetric; i.e. $f_{ij} = f_{ji}$. Let $\mathbf{g}(\mathbf{z}) = \nabla f(\mathbf{z})$ denote the gradient of f. Assuming that $\mathbf{g}$ is differentiable, we have (Chapter 2.9)

$$\mathbf{g}(\mathbf{z}^*) = \mathbf{g}(\mathbf{z}) + \mathbf{J}(\mathbf{z}^* - \mathbf{z}) + \mathbf{r}(\mathbf{z}^*, \mathbf{z}), \tag{10.2}$$

where $\mathbf{r}(\mathbf{z}^*, \mathbf{z}) = o(||\mathbf{z}^* - \mathbf{z}||)$ and $\mathbf{J} := \mathbf{J}(\mathbf{z})$ is the linear operator with jth column equal to

$$\frac{\partial \mathbf{g}}{\partial z_j} = \left(\frac{\partial^2 f}{\partial z_j \partial z_1}, \ldots, \frac{\partial^2 f}{\partial z_j \partial z_p}\right)^T,$$

which under assumed symmetry, $f_{ij} = f_{ji}$, implies $\mathbf{J} = \mathbf{H}$. Discarding the remainder term $\mathbf{r}$ this suggests that if $\mathbf{z}^*$ is a stationary point so that $g(\mathbf{z}^*) = 0$, then $\mathbf{z}^*$ is determined from (10.2) by

$$\mathbf{z}^* = \mathbf{z} - \mathbf{H}(\mathbf{z})^{-1}\mathbf{g}(\mathbf{z}).$$

This is the basis of the multivariate Newton method. Specifically, given an initial estimate $\mathbf{z}_0 \in \mathbb{R}^p$, one generates a sequence of values $\mathbf{z}_1, \mathbf{z}_2, \ldots \in \mathbb{R}^p$, where

$$\mathbf{z}_m = \mathbf{z}_{m-1} - \mathbf{H}(\mathbf{z}_{m-1})^{-1}\mathbf{g}(\mathbf{z}_{m-1}). \tag{10.3}$$

10.3 Convergence

We observe that multivariate Newton-Raphson uses both the gradient and the Hessian, which is why the method is able to converge so rapidly. Here, we will use a Taylor's expansion to show that Newton's method actually possesses superlinear convergence.

Suppose $\mathbf{z}_m \to \mathbf{z}^*$ where $\mathbf{z}^*$ is such that $\nabla f(\mathbf{z}^*) = \mathbf{0}$. Therefore, we are assuming the algorithm is initialized correctly and converges. We have

$$\mathbf{0} = \mathbf{g}(\mathbf{z}^*) = \mathbf{g}(\mathbf{z}_m) + \mathbf{H}((\mathbf{z}_m)(\mathbf{z}^* - \mathbf{z}_m) + o\left(||\mathbf{z}_m - \mathbf{z}^*||\right).$$

Multiplying throughout by $\mathbf{H}(\mathbf{z}_m)^{-1}$, we have

$$\mathbf{z}_m - \mathbf{z}^* - \mathbf{H}(z_m)^{-1}\mathbf{g}(\mathbf{z}_m) = o\left(||\mathbf{z}_m - \mathbf{z}^*||\right).$$

The $m+1$ Newton step is $\mathbf{z}_{m+1} = \mathbf{z}_m - \mathbf{H}(\mathbf{z}_m)^{-1}\mathbf{g}(\mathbf{z}_m)$ which implies

$$\mathbf{z}_{m+1} - \mathbf{z}^* = o\left(||\mathbf{z}_m - \mathbf{z}^*||\right).$$

Therefore for $\mathbf{z}_m \neq \mathbf{z}^*$

$$\lim_{m\to\infty} \frac{||\mathbf{z}_{m+1} - \mathbf{z}^*||}{||\mathbf{z}_m - \mathbf{z}^*||} = \lim_{m\to\infty} \frac{o\left(||\mathbf{z}_m - \mathbf{z}^*||\right)}{||\mathbf{z}_m - \mathbf{z}^*||} = 0.$$

This implies a superlinear convergence rate.

10.4 Numerical Issues

We mention numerical issues that can arise in practical settings.

1. The inverse of the Hessian matrix may fail to exist. For example, in regions of the space where f is linear the Hessian is zero, in which case the inverse does not exist. Non-invertibility is not necessarily a reflection of smoothness.
2. It assumes that $\nabla f(\mathbf{z}^*) = \mathbf{0}$ corresponds to a maxima. But without further conditions imposed on f, there may be multiple local maxima, or $\mathbf{z}^*$ may be a saddle point (neither a local minima or local maxima).
3. A local maxima is a point $\mathbf{z}^*$ such that

$$f(\mathbf{z}^*) \geq f(\mathbf{z}), \quad \text{for all } \mathbf{z} \in \mathbb{D} \subset \mathbb{R}^p.$$

Generally, convergence of a numerical algorithm to a global maxima is unlikely, unless the function is known to be everywhere concave.

4. There is no guarantee that $f(\mathbf{z}_{m+1}) \geq f(\mathbf{z}_m)$. Thus, Newton's method is different from algorithms that seek to always increase the value of the function. Chapter 11 discusses steepest descent, which is an example of the latter.

We can summarize the key points in the following theorem.

Theorem 10.1. *Let $f: \mathbb{R}^p \to \mathbb{R}$ be a real valued function.*

1. *If $-\boldsymbol{H}(\mathbf{z}) > 0$ for all $\mathbf{z}$, then $\mathbf{z}^*$ is a global maxima.*
2. *If $f(\mathbf{z})$ has a relative extremum at $\mathbf{z}^*$ and its partial derivatives $f_i(\mathbf{z})$ exist at $\mathbf{z}^*$, then $\nabla f(\mathbf{z}^*) = \mathbf{0}$.*
3. *Suppose that $\nabla f(\mathbf{z}^*) = \mathbf{0}$ and all second order derivatives f_{ij} are continuous. Let D_k be the determinant of the Hessian matrix at $\mathbf{z}^*$ formed by the first k rows and the first k columns, $1 \leq k \leq p$,*

$$D_k = \begin{vmatrix} f_{11}(\mathbf{z}^*) & f_{12}(\mathbf{z}^*) & \cdots & f_{1k}(\mathbf{z}^*) \\ f_{21}(\mathbf{z}^*) & f_{22}(\mathbf{z}^*) & \cdots & f_{2k}(\mathbf{z}^*) \\ \vdots & \vdots & \ddots & \vdots \\ f_{k1}(\mathbf{z}^*) & f_{k2}(\mathbf{z}^*) & \cdots & f_{kk}(\mathbf{z}^*) \end{vmatrix}.$$

Then D_k is called the kth leading principal minor. It follows that:

   (a) *If $D_k > 0$ for $1 \leq k \leq p$, then f has a relative minimum at $\mathbf{z}^*$.*
   (b) *If $(-1)^k D_k > 0$ for $1 \leq k \leq p$, then f has a relative maxiumum at $\mathbf{z}^*$.*
   (c) *If for some $1 \leq 2k \leq p$ we have $D_{2k} < 0$, then $\mathbf{z}^*$ is a saddle point.*

10.5 Logistic Regression

Logistic regression is a special example of generalized linear models (GLM) used for regression classification. The logistic regression framework assumes independent binary responses $Y_1, \ldots, Y_n \in \{0, 1\}$ with covariates $\mathbf{x}_1, \ldots, \mathbf{x}_p \in \mathbb{R}^p$, where it is assumed that

$$\mathbb{P}\{Y_i = 1 | \mathbf{x}_i\} = \pi_i = \frac{1}{1 + \exp(-F_i)}.$$

The predictor $F_i = \alpha + \mathbf{x}_i^T \boldsymbol{\beta}$ is assumed to be linear in the covariates where α is the intercept and $\boldsymbol{\beta} \in \mathbb{R}^p$ are the regression parameters.

We use the maximum likelihood estimator (MLE) to estimate $(\alpha, \boldsymbol{\beta})^T$. The likelihood for the observed data $(y_i, \mathbf{x}_i)_{i=1}^n$ is

$$\begin{aligned} L &= \prod_{i=1}^{n} \pi_i^{y_i} (1-\pi_i)^{1-y_i} \\ &= \exp\left[\sum_{i=1}^{n} y_i \log\left(\frac{\pi_i}{1-\pi_i}\right) + \sum_{i=1}^{n} \log(1-\pi_i) \right]. \end{aligned}$$

The relationship of π_i to the linear predictor F_i gives $1 - \pi_i = (1 + \exp(F_i))^{-1} = \exp(-F_i)(1 + \exp(-F_i))^{-1}$ which implies that

$$\log\left(\frac{\pi_i}{1-\pi_i}\right) = \log\left(\frac{(1+\exp(-F_i))^{-1}}{\exp(-F_i)(1+\exp(-F_i))^{-1}}\right) = F_i.$$

The function $\log(p/(1-p))$ is called the logit transformation. It transforms the probability to a predictor that is linear in the covariates. Using the above, the MLE is obtained by maximizing the log-likelihood.

$$\begin{aligned} \log L &= \sum_{i=1}^{n} y_i \log\left(\frac{\pi_i}{1-\pi_i}\right) + \sum_{i=1}^{n} \log(1-\pi_i) \\ &= \sum_{i=1}^{n} \Big[y_i F_i - \log(1+\exp(F_i)) \Big]. \end{aligned}$$

We will use Newton-Raphson to obtain the MLE. This amounts to finding the maxima of $f(\mathbf{z})$ where $f(\mathbf{z}) = \log L$ and $\mathbf{z} = (\alpha, \boldsymbol{\beta})^T \in \mathbb{R}^{p+1}$.

Therefore, we want to maximize

$$f(\mathbf{z}) = \sum_{i=1}^{n} \Big[y_i F_i - \log(1+\exp(F_i)) \Big], \quad \mathbf{z} = (\alpha, \boldsymbol{\beta})^T \in \mathbb{R}^{p+1}.$$

Let $\mathbf{y} = (y_1, \ldots, y_n)^T$ and $\boldsymbol{\pi} = (\pi_1, \ldots, \pi_n)$ where $\pi_i = (1 + \exp(-F_i))^{-1}$. Using straightforward differentiation (Chapter 2.9), and recalling that $F_i = \alpha + \mathbf{x}_i^T \boldsymbol{\beta}$, it can be verified that

$$\nabla f(\mathbf{z}) = \mathbf{D}^T (\mathbf{y} - \boldsymbol{\pi}), \tag{10.4}$$

where

$$\mathbf{D} = \begin{pmatrix} (1, \mathbf{x}_1)^T \\ \vdots \\ (1, \mathbf{x}_n)^T \end{pmatrix}_{n \times (p+1)}.$$

Differentiation again shows that $\partial \pi_i / \partial F_i = w_i$, where $w_i = \pi_i(1-\pi_i)$. Let $\mathbf{F} = (F_1, \ldots, F_n)^T$. Because $\mathbf{H}(\mathbf{z}) = \nabla^2 f(\mathbf{z})$, deduce from the above that

$$\begin{aligned} \mathbf{H}(\mathbf{z}) &= \frac{\partial \mathbf{D}^T (\mathbf{y} - \boldsymbol{\pi})}{\partial \mathbf{z}} \\ &= \frac{\partial \mathbf{D}^T (\mathbf{y} - \boldsymbol{\pi})}{\partial \boldsymbol{\pi}} \times \frac{\partial \boldsymbol{\pi}}{\partial \mathbf{F}} \times \frac{\partial \mathbf{F}}{\partial \mathbf{z}} \\ &= -\mathbf{D}^T \mathbf{W} \mathbf{D}, \end{aligned}$$

where $\mathbf{W}_{n \times n} = \mathrm{diag}\{w_i\}_{i=1}^n$.

Example 10.1. We use a simple intercept-slope model for illustration of the gradient and Hessian. There is one covariate x_i and $F_i = \alpha + x_i\beta$ satisfies a straight line relationship. The gradient is

$$\nabla f(\alpha,\beta) = \begin{pmatrix} \sum_{i=1}^n (y_i - \pi_i) \\ \sum_{i=1}^n (y_i - \pi_i)x_i \end{pmatrix}$$

and the Hessian is

$$\begin{aligned} \mathbf{H}(\alpha,\beta) &= -\begin{pmatrix} 1 & \cdots & 1 \\ x_1 & \cdots & x_n \end{pmatrix} \begin{pmatrix} w_1 & \cdots & 0 \\ & \ddots & \\ 0 & \cdots & w_n \end{pmatrix} \begin{pmatrix} 1 & x_1 \\ \vdots & \vdots \\ 1 & x_n \end{pmatrix} \\ &= -\begin{pmatrix} \sum_{i=1}^n w_i & \sum_{i=1}^n w_i x_i \\ \sum_{i=1}^n w_i x_i & \sum_{i=1}^n w_i x_i^2 \end{pmatrix}. \end{aligned}$$

A sufficient condition for a maxima is that the negative Hessian matrix is positive definite. Therefore, it suffices to verify that $-\mathbf{H}(\alpha,\beta)$ has a positive determinant. The determinant is

$$D = \left(\sum_{i=1}^n w_i\right)\left(\sum_{i=1}^n w_i x_i^2\right) - \left(\sum_{i=1}^n w_i x_i\right)^2.$$

Recall the Cauchy-Schwarz inequality (Chapter 2.17). For any real values $(u_i)_{i=1}^n$ and $(v_i)_{i=1}^n$

$$\left(\sum_{i=1}^n u_i v_i\right)^2 \le \left(\sum_{i=1}^n u_i^2\right)\left(\sum_{i=1}^n v_i^2\right)$$

with equality occuring if and only if $u_i = cv_i$ for some constant c. Setting $u_i = w_i^{1/2}$ and $v_i = w_i^{1/2}x_i$ we have

$$\left(\sum_{i=1}^n w_i x_i\right)^2 \le \left(\sum_{i=1}^n w_i\right)\left(\sum_{i=1}^n w_i x_i^2\right),$$

which implies that $D \ge 0$. Furthermore $D = 0$ if and only if $x_i = c$ for some constant c. □

Returning to our calculation, setting $\tilde{\mathbf{y}} = (\mathbf{y} - \boldsymbol{\pi})$, the Newton-Raphson update is

$$\mathbf{z}_m = \mathbf{z}_{m-1} + \left(\mathbf{D}^T\mathbf{W}_{m-1}\mathbf{D}\right)^{-1}\mathbf{D}^T\tilde{\mathbf{y}}_{m-1},$$

where we have used a subscript of $m-1$ to indicate dependence on the current estimate $\mathbf{z}_{m-1}$. The update is a solution to a system of linear equations. This can be seen by multiplying throughout by $\mathbf{D}^T\mathbf{W}_{m-1}\mathbf{D}$, yielding

$$\begin{aligned} (\mathbf{D}^T\mathbf{W}_{m-1}\mathbf{D})\,\mathbf{z}_m &= (\mathbf{D}^T\mathbf{W}_{m-1}\mathbf{D})\,\mathbf{z}_{m-1} + \mathbf{D}^T\tilde{\mathbf{y}}_{m-1} \\ &= \mathbf{D}^T\left[\mathbf{W}_{m-1}\mathbf{F}_{m-1} + \tilde{\mathbf{y}}_{m-1}\right] \\ &= \mathbf{D}^T\mathbf{b}_{m-1}, \end{aligned} \tag{10.5}$$

where $\mathbf{b}_{m-1} = \mathbf{W}_{m-1}\mathbf{F}_{m-1} + \tilde{\mathbf{y}}_{m-1}$.

We can recognize (10.5) as a solution to a weighted least squares problem with design matrix $\mathbf{D}$, response values $\mathbf{c}_{m-1} = \mathbf{D}^T\mathbf{b}_{m-1}$, and weights $\mathbf{W}_{m-1}$. Rewrite (10.5) as

$$\mathbf{B}_{m-1}^T\mathbf{B}_{m-1}\mathbf{z}_m = \mathbf{c}_{m-1}, \tag{10.6}$$

where $\mathbf{B}_{m-1} = \mathbf{W}_{m-1}^{1/2}\mathbf{D}$. This can be efficiently solved by using the QR decomposition (or by using the SVD of $\mathbf{B}_{m-1}$).

10.5.1 Correlation and its effect on the MLE

The MLE performs well in large sample size settings when the design matrix has full rank (i.e. the covariates are reasonably uncorrelated). However, its performance can break down outside of this. Here we demonstrate that the log-likelihood becomes elongated due to correlation which degrades performance of the MLE.

For our illustration, we simulated $n = 200$ values from the logistic regression model

$$F_i = \beta_1 x_{1,i} + \beta_2 x_{i,2}, \qquad i = 1, \ldots, n,$$

where $\beta_1 = -1$ and $\beta_2 = 1$. Notice there are two covariates ($p = 2$) and no intercept term. The $x_{1,i}$ is drawn from a standard normal and

$$x_{2,i} = x_{1,i} + \sigma Z_i$$

where Z_i is standard normal. Therefore, the two covariates are correlated. The value $\sigma > 0$ measures the strength of the correlation. As σ decreases, the correlation increases. Similar to the slope-intercept model previously considered, the Cauchy-Schwarz inequality shows that the negative Hessian is positive definite regardless of the value of $\sigma > 0$. However, even though a positive definite negative Hessian guarantees a concave surface, the correlation plays a significant role in shaping the likelihood surface.

Values of σ ranging from $\sigma = 2.5$ and $\sigma = 0.25$ were used. Figure 10.2 plots the log-likelihood surface and the Newton-Raphson solution (path in red). As σ decreases (going from left to right, top to bottom), the log-likelihood becomes elongated. This clearly affects the accuracy of the MLE. Table 10.1 records the GLM results for each model. The p-value is near zero in all cases showing that the MLE correctly determines that the two features are relevant, however, the estimated values for the regression parameters degrades noticeably with increasing correlation. For the smallest value $\sigma = .25$ corresponding to the highest correlation, the estimated values are approximately twice the size of the true value $(\beta_1, \beta_2)^T = (-1, 1)^T$.

10.6 Fisher Scoring

Fisher's method of scoring is often used in place of Newton-Raphson for maximum likelihood estimation (in the case of logistic regression, the two methods coincide). Let $Y_1, \ldots, Y_n$ be i.i.d. with density $f(\cdot|\boldsymbol{\theta})$ where $\boldsymbol{\theta} \in \mathbb{R}^p$. The log-likelihood is

$$l(\boldsymbol{\theta}) = \sum_{i=1} \log f(y_i|\boldsymbol{\theta}).$$

To determine the MLE for $\boldsymbol{\theta}$, we find the root of the score function:

$$\mathbf{S}(\boldsymbol{\theta}) = \frac{\partial l(\boldsymbol{\theta})}{\partial \boldsymbol{\theta}} = \mathbf{0}.$$

A Taylor expansion around an initial value $\boldsymbol{\theta}_*$ yields (Theorem 2.11)

$$\mathbf{S}(\boldsymbol{\theta}) = \mathbf{S}(\boldsymbol{\theta}_*) + \mathbf{H}(\boldsymbol{\theta}_*)(\boldsymbol{\theta} - \boldsymbol{\theta}_*) + o(||\boldsymbol{\theta} - \boldsymbol{\theta}_*||),$$

where $\mathbf{H}(\boldsymbol{\theta}) = \nabla \mathbf{S}(\boldsymbol{\theta})$ is the derivative of the score at $\boldsymbol{\theta}$ equal to the Hessian of the log-likelihood. If $\mathbf{S}(\boldsymbol{\theta}) = \mathbf{0}$, rearranging terms and discarding the remainder, yields

$$\boldsymbol{\theta} = \boldsymbol{\theta}_* - \mathbf{H}(\boldsymbol{\theta}_*)^{-1}\mathbf{S}(\boldsymbol{\theta}_*)$$

which can be seen to be a Newton-Raphson update.

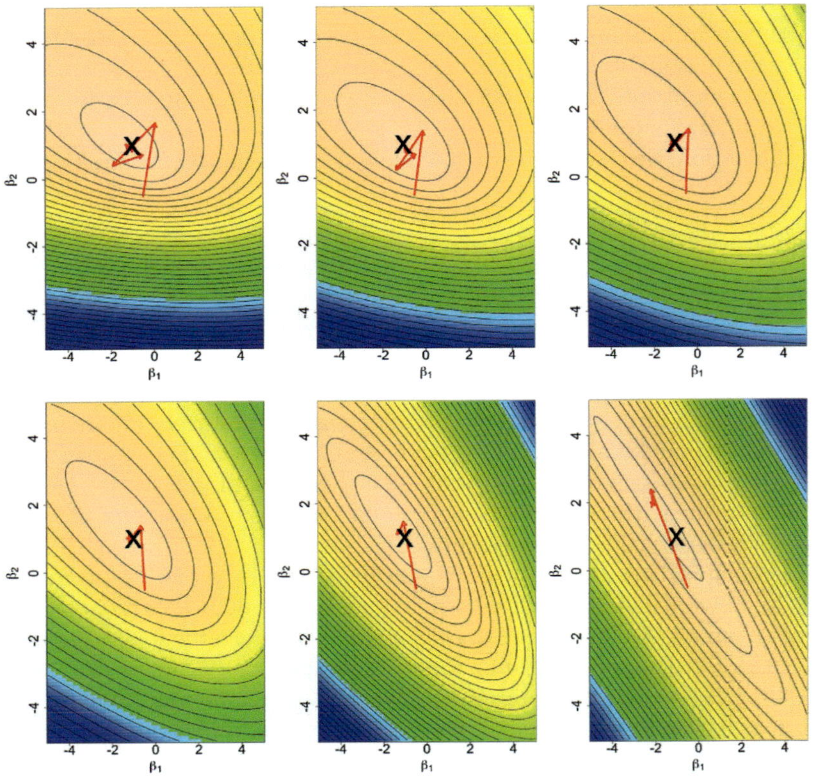

FIGURE 10.2
Log-likelihood surfaces from logistic regression where the amount of correlation between two variables is increased going from left to right and top to bottom. As the correlation increases, the log-likelihood becomes elongated, which affects the MLE (Newton-Raphson solution indicated by red arrows). True coefficient value is indicated by a cross.

Fisher's method modifies Newton-Raphson by replacing $\mathbf{H}(\boldsymbol{\theta})$ with $-n\mathbf{I}(\boldsymbol{\theta})$, where $\mathbf{I}(\boldsymbol{\theta})$ is the Fisher information for $\boldsymbol{\theta}$ from a sample size of $n = 1$ (Chapter 3.4) defined by

$$\begin{aligned}\mathbf{I}(\boldsymbol{\theta}) &= \mathbb{E}\left[\left(\frac{\partial \log f(Y|\boldsymbol{\theta})}{\partial \boldsymbol{\theta}}\right)\left(\frac{\partial \log f(Y|\boldsymbol{\theta})}{\partial \boldsymbol{\theta}}\right)^T\right] \\ &= \mathbb{E}\begin{bmatrix} l_1 l_1 & \cdots & l_1 l_p \\ \vdots & & \vdots \\ l_p l_1 & \cdots & l_p l_p \end{bmatrix}, \quad \text{where } l_j = \partial l(\boldsymbol{\theta})/\partial \theta_j.\end{aligned}$$

Fisher's update is

$$\boldsymbol{\theta} = \boldsymbol{\theta}_* + n^{-1}\mathbf{I}(\boldsymbol{\theta}_*)^{-1}\mathbf{S}(\boldsymbol{\theta}_*).$$

Under standard regularity conditions, $\mathbb{E}(l_j l_k) = \mathbb{E}(l_{jk})$, where $l_{jk} = \partial^2 l(\boldsymbol{\theta})/\partial\theta_j \partial\theta_k$, which leads to the following important identity

$$\mathbf{I}(\boldsymbol{\theta}) = -\mathbb{E}\left[\frac{\partial^2 \log f(Y|\boldsymbol{\theta})}{\partial \boldsymbol{\theta}\, \partial \boldsymbol{\theta}^T}\right] = -\mathbb{E}(\mathbf{H}(\boldsymbol{\theta}))/n.$$

TABLE 10.1
Logistic regression (GLM) for $p = 2$ dimensional regression model with increasing correlation between features.

		$\hat{\beta}$	Std. Error	Z-statistic	p-value
$\sigma = 2.5$	x_1	-1.27	0.27	-4.72	0.00
	x_2	1.02	0.14	7.06	0.00
$\sigma = 2.05$	x_1	-0.95	0.23	-4.03	0.00
	x_2	0.87	0.13	6.66	0.00
$\sigma = 1.60$	x_1	-1.16	0.26	-4.54	0.00
	x_2	1.10	0.17	6.63	0.00
$\sigma = 1.15$	x_1	-1.22	0.26	-4.62	0.00
	x_2	1.13	0.19	5.98	0.00
$\sigma = .70$	x_1	-1.35	0.31	-4.33	0.00
	x_2	1.22	0.26	4.71	0.00
$\sigma = .25$	x_1	-2.14	0.66	-3.23	0.00
	x_2	1.98	0.64	3.10	0.00

Therefore, the Fisher update can be described by

$$\boldsymbol{\theta} = \boldsymbol{\theta}_* - \mathbb{E}(\mathbf{H}(\boldsymbol{\theta}_*))^{-1}\mathbf{S}(\boldsymbol{\theta}_*). \tag{10.7}$$

Thus, the Hessian used in Newton-Raphson is replaced by the expected value in Fisher scoring. This is advantageous because it can simplify calculations. Also, using expectations acts like a smoothing operation and improves numerical performance.

Chapter 3.4 described the asymptotic theory for unbiased estimating equations which included the asymptotic theory for the MLE as a special case. Under standard regularity conditions, denoting $\hat{\boldsymbol{\theta}}$ for the MLE and $\boldsymbol{\theta}_0$ for the true parameter, we have shown

$$\sqrt{n}(\hat{\boldsymbol{\theta}} - \boldsymbol{\theta}_0) \overset{d}{\rightsquigarrow} N_p\left(\mathbf{0}, \mathbf{I}(\boldsymbol{\theta}_0)^{-1}\right).$$

Therefore, as a by-product, the Fisher scoring algorithm provides an estimate for the MLE's asymptotic variance. This value is estimated by $n^{-1}\mathbf{I}(\boldsymbol{\theta}_m)^{-1}$ where $\boldsymbol{\theta}_m$ is the m step Fisher solution.

By the same reasoning, we can use Newton-Raphson to estimate the asymptotic variance of the MLE. By (10.7), Newton-Raphson and Fisher scoring coincide when $\mathbf{H}(\boldsymbol{\theta}) = \mathbb{E}(\mathbf{H}(\boldsymbol{\theta}))$ (which is the case for logistic regression; see Exercise 10.4). Thus, in this case, if $\boldsymbol{\theta}_m$ is the Newton-Raphson m step solution, the variance of $\hat{\boldsymbol{\theta}}$ can be estimated by

$$-\mathbf{H}(\boldsymbol{\theta}_m)^{-1} = -\mathbb{E}(\mathbf{H}(\boldsymbol{\theta}_m))^{-1} = n^{-1}\mathbf{I}(\boldsymbol{\theta}_m)^{-1},$$

the negative inverse of the Hessian matrix at $\boldsymbol{\theta}_m$. As an illustration, the columns "Std. Error" reported in Table 10.1 are the square root of the diagonal values of $-\mathbf{H}(\boldsymbol{\theta}_m)^{-1}$.

Exercises

Exercise 10.1. Prove that $z_m = z_{m-1} - f(z_{m-1})/f'(z_{m-1})$, the Newton update (10.1) for a real valued function is the intersection with the z-axis of a line tangent to f at z_{m-1}.

Exercise 10.2. Find the roots of the function $f(z) = z^3 - 2z + 2$. Starting with the initial guess $z_0 = 1$ and searching within the region $(-1, 2)$ yields numerical issues. Explain why.

Exercise 10.3. Prove that (10.4) holds. Hint: let $L(y_i, F_i)$ denote the ith term in the sum for $f(\mathbf{z})$ and use the fact that

$$\frac{\partial L(y_i, F_i)}{\partial \mathbf{z}} = \frac{\partial L(y_i, F_i)}{\partial F_i} \times \frac{\partial F_i}{\partial \mathbf{z}}.$$

Exercise 10.4. Show that Fisher's method of scoring is equivalent to Newton-Raphson for logistic regression.

Exercise 10.5. Assume a logistic regression model with p variables such that

$$\log\left(\frac{\pi_i}{1-\pi_i}\right) = \beta_0 + \sum_{k=1}^{p} \beta_k x_{i,k}, \qquad i = 1, \ldots, n.$$

Determine the Hessian matrix. Under what conditions is the Hessian matrix positive definite?

Exercise 10.6. Here we apply Newton's method to a non-convex function and analyze its convergence properties.

1. Consider the function $f(x, y) = x^4 + y^4 - 4x^2 - 4y^2$.
2. Implement Newton's method to find the critical points of this function.
3. Analyze the Hessian matrix at each iteration to determine whether the method converges to a local minimum, maximum, or saddle point.
4. Visualize the convergence path on a contour plot of the function.
5. Compare the results with those obtained using gradient descent (Chapter 11) and discuss the differences.

Exercise 10.7. Here we solve a constrained optimization problem using Newton's method with a penalty approach.

1. Consider the function $f(x, y) = x^2 + y^2$ subject to the constraint $x + y = 1$.
2. Formulate the penalized objective function incorporating the constraint.
3. Implement Newton's method to minimize the penalized objective function.
4. Experiment with different penalty parameters and analyze their impact on the solution.
5. Compare the results with those obtained using the method of Lagrange multipliers.

Exercise 10.8. What would happen if we used Newton's method to fit least squares for a linear regression model? Calculate the gradient and Hessian from the optimization problem and use this in your explanation.

Exercise 10.9. Consider the real-valued function

$$f(\mathbf{x}) = \mathbf{x}^T \mathbf{A} \left(\text{diag}(\mathbf{x}\mathbf{x}^T)\mathbf{x}\right) - \mathbf{w}^T \mathbf{x},$$

where $\mathbf{A}$ is a fixed positive definite matrix, $\mathbf{w}$ is a fixed vector and $\text{diag}(\mathbf{x}\mathbf{x}^T)$ is a diagonal matrix where each diagonal element is the sum of the squares of the elements of $\mathbf{x}$.

1. Implement Newton's method to minimize this function. Feel free to use any implications that you might find helpful.
2. Compare the performance and convergence rate of Newton's method with steepest descent (Chapter 11). Discuss your results.

11

Steepest Descent

11.1 Introduction

In contrast to Newton's method (Chapter 10), there are algorithms where each step is guaranteed to move the solution to a lower function evaluation. The strategy in this case for minimizing a function can be described as follows:

1. Given the current point $\mathbf{z}_0$, choose a direction $\mathbf{z}$ to move in.
2. Find a point $\mathbf{z}^* = \mathbf{z}_0 + \lambda\mathbf{z}$ such that $f(\mathbf{z}^*) \leq f(\mathbf{z}_0)$.

The steepest descent method, also known as the gradient descent method, is an iterative optimization algorithm used to find the minimum of a differentiable function that follows this principle. It is one of the simplest and most widely used methods for optimization problems. It is based on the following key observation.

A real-valued, differentiable function $f : \mathbb{R}^n \to \mathbb{R}$ decreases fastest from a point $\boldsymbol{z}$ along $-\nabla f(\boldsymbol{z})$, the negative gradient of f at $\boldsymbol{z}$.

This key fact can be deduced from the following.

Theorem 11.1. *Among all directions $\boldsymbol{b} \in \mathbb{R}^n$ with unit norm $\|\boldsymbol{b}\| = 1$, the direction*

$$\boldsymbol{b}^* = -\frac{\nabla f(\boldsymbol{z})}{\|\nabla f(\boldsymbol{z})\|}$$

has the minimal inner product with $\nabla f(\boldsymbol{z})$ (assuming that $\nabla f(\boldsymbol{z}) \neq \boldsymbol{0}$). Hence the direction of maximal decrease of the function at the point $\boldsymbol{z}$ is along $-\nabla f(\boldsymbol{z})$.

Proof. The result is an immediate consequence of the following inequality:

$$\nabla f(\mathbf{z})^T\mathbf{b} \geq -\|\nabla f(\mathbf{z})\|\,\|\mathbf{b}\| = \nabla f(\mathbf{z})^T\left(\frac{-\nabla f(\mathbf{z})}{\|\nabla f(\mathbf{z})\|}\right) = \nabla f(\mathbf{z})^T\mathbf{b}^*.$$

Note that the inequality implicitly uses $\nabla f(\mathbf{z}) \neq \mathbf{0}$. □

11.2 Steepest Descent Algorithm

The basic idea of steepest descent is to move iteratively from an initial point $\mathbf{z}$ in the direction of the negative gradient $-\nabla f(\mathbf{z})$ of the function at the current point. By Theorem 11.1, this is the

DOI: 10.1201/9781003493679-11

direction of steepest descent because it is the direction in which the function decreases most rapidly.

To determine how far we should move along this direction, let $\mathbf{g} = -\nabla f(\mathbf{z})$ and define $G(\rho) = f(\mathbf{z} + \rho\mathbf{g})$. Then G has a negative derivative at $\rho = 0$ (Exercise 11.1). Thus, there exists a $\rho^* > 0$, such that $G(\rho^*) < G(0)$. Hence, the update point $\mathbf{z}^*$ defined by

$$\mathbf{z}^* = \mathbf{z} - \rho^* \nabla f(\mathbf{z})$$

yields $f(\mathbf{z}^*) < f(\mathbf{z})$. Applying this principle iteratively we obtain a sequence $\mathbf{z}_0, \mathbf{z}_1, \ldots, \mathbf{z}_M$, where

$$\mathbf{z}_m = \mathbf{z}_{m-1} - \rho_m \nabla f(\mathbf{z}_{m-1}), \quad m = 1, \ldots, M,$$

where if $\{\rho_m\}_1^M$ are appropriately selected as above, we have

$$f(\mathbf{z}_0) \geq f(\mathbf{z}_1) \geq \cdots \geq f(\mathbf{z}_M).$$

Algorithm 11.1 describes this procedure.

Algorithm 11.1 *Steepest Descent*

1: Initialize $\mathbf{z}_0$; set $m = 0$.
2: **while** $\nabla f(\mathbf{z}_{m-1}) \neq \mathbf{0}$ **do**
3: Update $m \leftarrow m + 1$.
4: Set $\mathbf{g}_m = -\nabla f(\mathbf{z}_{m-1})$.
5: Find ρ_m by a line-search: $\rho_m = \underset{0 \leq \rho \leq \infty}{\arg\min} f(\mathbf{z}_{m-1} + \rho\mathbf{g}_m)$.
6: Let $\mathbf{z}_m = \mathbf{z}_{m-1} + \rho_m \mathbf{g}_m$.
7: **end while**

The following result due to Goldstein [4] states conditions that are needed for steepest descent to converge.

Theorem 11.2. *If $f: \mathbb{R}^n \to \mathbb{R}$ is a real-valued function with continuous first partial derivatives, $\mathbf{z}_0$ is such that $S = \{\mathbf{z}: f(\mathbf{z}) \leq f(\mathbf{z}_0)\}$ is bounded, then $\{\mathbf{z}_m\}_{m=0}^{\infty}$ has a subsequence converging to $\mathbf{z}_\infty \in S$, where $\mathbf{z}_\infty$ is a stationary point satisfying $\nabla f(\mathbf{z}_\infty) = \mathbf{0}$. Furthermore, $f(\mathbf{z}_m) \downarrow f(\mathbf{z}_\infty)$.*

The step size ρ_m used in the algorithm can be chosen in various ways. Line 5 which uses an exact line-search for the minimum can be replaced by a simpler update $\mathbf{z}_m = \mathbf{z}_{m-1} + \rho_m^* \mathbf{g}_m$ using any $0 < \rho_m^* \leq \hat{\rho}_m$, where $\hat{\rho}_m$ is the smallest $\rho > 0$ such that

$$f(\mathbf{z}_{m-1} + \rho\mathbf{g}_m) \leq f(\mathbf{z}_{m-1}).$$

For example, ρ_m^* can be chosen by setting the directional derivative to zero:

$$f'(\mathbf{z}_{m-1} + \rho_m^* \mathbf{g}_m) = 0.$$

Goldstein [4] called this *modified steepest descent* (see his remark (ii) where he credits the method to Curry [2]). Often, when steepest descent is used, it is actually the modified version that is being applied [1, 3, 5–7].

Remark 11.1 (Orthogonal behavior of steepest descent). Modified steepest descent finds a local minimum along the negative gradient vector at each step of the iteration. This choice forces the updated gradient to be orthogonal to the previous direction. This follows because the directional derivative at step m is

$$\frac{\partial}{\partial \rho} f(\mathbf{z}_{m-1} + \rho\mathbf{g}_m) = \nabla f(\mathbf{z}_m)^T \frac{\partial}{\partial \rho}\left\{\mathbf{z}_{m-1} + \rho\mathbf{g}_m\right\} = \nabla f(\mathbf{z}_m)^T \mathbf{g}_m.$$

Thus, the solution $\rho = \rho_m^*$ forces $\mathbf{g}_m$ to be orthogonal to $\nabla f(\mathbf{z}_m)$.

This orthogonal property is a weakness of the method. If the curvature of the target function varies widely along different dimensions, then because each descent direction must be orthogonal to the previous direction, it may take many small zig-zag steps to reach the function minimum rather than taking a more efficient direct path. □

11.3 Illustration

Figure 11.2 provides an illustration of modified steepest descent. The figure plots the contours of the function

$$f(\mathbf{z}) = -5 + 2z_1 + z_1^2 + 3\sin(z_2)$$

on the $\mathbf{z} = (z_1, z_2)$ plane. Notice that f has multiple local minima located within the circular contours (Figure 11.1). However, the found minima depends strongly on the starting location. The four descent paths, defined by the different initial positions as marked, trace the steps of modified steepest descent to its local minima.

At each step, due to the orthogonality property, the gradient vector is orthogonal to the function contour lines as well as the previous gradient direction. For example, the gradient at the initial point of path 1 points almost directly toward the minimum, so path 1 takes one large step, then two additional small steps to reach its minima.

Table 11.1 lists the modified steepest descent parameter values ρ_m^* at each step along the specified path. Notice that each of the three steps in path 1 have roughly the same sized parameter value, even though the actual vector lengths are dramatically different. The size of ρ_m^* is relative to the current gradient. In the remaining three paths, the parameter size of the second step increases to move the path to the local minimum. This is because there is an unstable

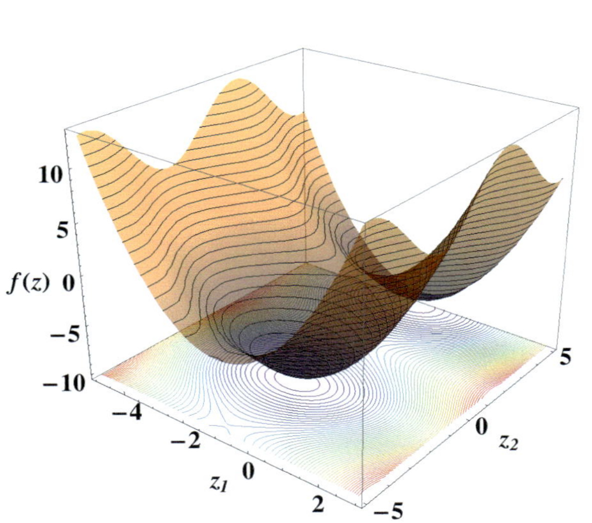

FIGURE 11.1
Three-dimensional plot of $f(\mathbf{z}) = -5 + 2z_1 + z_1^2 + 3\sin(z_2)$.

TABLE 11.1
Modified steepest descent parameter estimates for paths of Figure 11.2.

path	1	2	3	4
ρ_1^*	0.494	0.629	0.507	0.504
ρ_2^*	0.343	1.302	1.488	1.313
ρ_3^*	0.482	0.479	0.499	0.499
ρ_4^*		0.343	0.334	0.334
ρ_5^*		0.479		

stationary point, or *saddle point*, in the center, with a near zero gradient. Path 2 is interesting. It starts closer to a local minimum than any other path, but it requires more iterations to converge (as shown in Table 11.1). The slower convergence occurs because the gradient direction at the initial point is not directed toward a minima, but across a parabolic contour of the function along the z_1-axis. This initial gradient direction forces the path to repeatedly step beyond a more direct path toward the minimum.

11.4 Fixed Step Size

A simpler choice for step size, which avoids solving the directional derivative, is to set the descent parameter to a fixed value ρ, chosen small enough to ensure that $\rho < \rho_m^*$ for all m. Figure 11.3 demonstrates this method using the same four initial descent path locations from Figure 11.2. We used a fixed descent parameter of $\rho = 0.2$ (which is smaller than any ρ_m^* from Table 11.1).

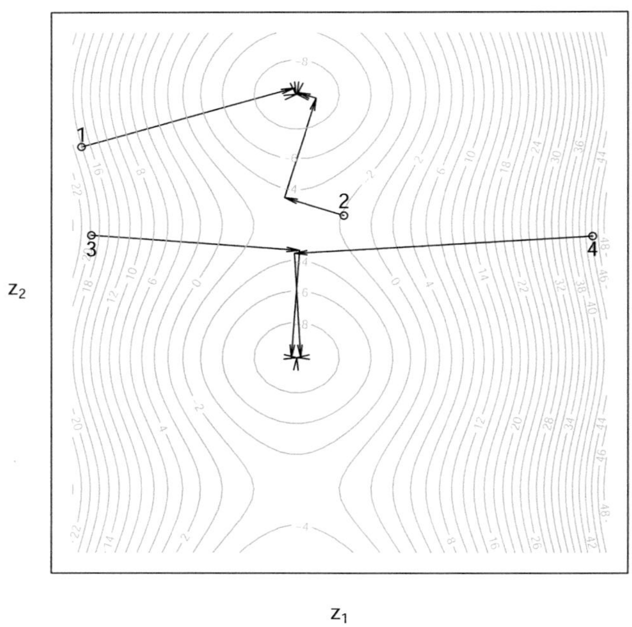

FIGURE 11.2
Modified steepest descent solution paths. Function contours of $f(\mathbf{z}) = -5 + 2z_1 + z_1^2 + 3\sin(z_2)$.

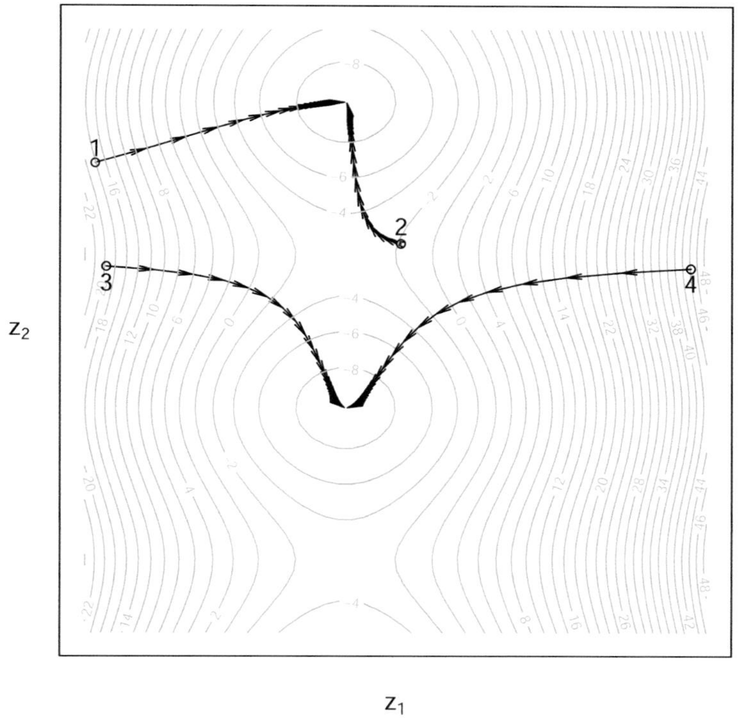

FIGURE 11.3
Steepest descent solution paths, fixed descent parameter $\rho = 0.2$. Function contours and descent initial points as in Figure 11.2.

Observe how each path proceeds in a more direct manner to the minima. Though this requires more iterations, the shorter steps adjust the path direction as it progresses toward a minimum.

Table 11.2 compares the number of iterations for modified steepest descent (indicated as ρ_m^*) to two different fixed-step size descent solutions. Note that the difficulty with the convergence of path 2 no longer exists with the fixed descent parameters. In fact, when $\rho = 0.1$, path 2 converges faster than the other paths.

The point that is relevant here is that a smaller fixed parameter value will take a proportionally higher number of iterations to converge. This is not desirable when trying to minimize a deterministic function. However, in the case of gradient boosting to be discussed in the next chapter, this type of behavior is called *slow-learning* and will be advantageous because of its regularization effect on the statistical problem.

TABLE 11.2
Number of iterations to convergence for modified steepest descent (ρ_m^*) and fixed descent parameter values of ρ.

path	1	2	3	4
ρ_m^*	3	5	4	4
$\rho = 0.2$	10	10	10	11
$\rho = 0.1$	21	20	22	23

Exercises

Exercise 11.1. Let $\mathbf{g} = -\nabla f(\mathbf{z})$ be the gradient for a differentiable real-valued function $f : \mathbb{R}^p \to \mathbb{R}$. Define $G(\rho) = f(\mathbf{z}+\rho\mathbf{g})$. Show that G has a negative derivative at $\rho = 0$ and prove that this implies there exists a $\rho^* > 0$ such that $G(\rho^*) < G(0)$.

Exercise 11.2. Explore the function $f(\mathbf{z}) = -z_1 z_2 z_3(1 - z_1 - z_2 - z_3)$ using steepest descent. Try different starting values. Summarize your findings.

Exercise 11.3. Find the minimum of $f(\mathbf{z}) = (z_1 - 2)^2 + (z_2 - 1)^2 + (z_3 - 3)^2$ using steepest descent.

Exercise 11.4. In this example, we implement steepest descent to find the minimum of a non-convex function and compare it with other optimization methods.

1. Consider the function $f(x, y) = x^2 + y^2 + 10\sin(x) + 10\sin(y)$.
2. Implement the steepest descent algorithm to find the minimum of this function.
3. Initialize the algorithm from different starting points to observe its behavior.
4. Compare the results with those obtained using other optimization methods such as Newton's method (Chapter 10) and the BFGS algorithm.
5. Visualize the convergence path on the function's surface plot.

Exercise 11.5. This exercise applies steepest descent to solve a constrained optimization problem with penalization.

1. Consider the function $f(x, y) = x^2 + y^2$ subject to the constraint $x + y = 1$.
2. Use a penalty method to incorporate the constraint into the objective function.
3. Implement the steepest descent algorithm to minimize the penalized objective function.
4. Experiment with different penalty parameters to observe their impact on the solution.
5. Compare the solution with the one obtained using a method specifically designed for constrained optimization (e.g., Lagrange multipliers).

Exercise 11.6. Here we will implement modified steepest descent to optimize the parameters of a logistic regression model.

1. Load a binary classification dataset.
2. Define the logistic regression loss function.
3. Implement modified steepest descent to minimize the logistic loss function. You should decide what type of line search parameter you want to use in order to get the best performance.
4. Initialize the parameters randomly and iterate until convergence.
5. Evaluate the test set performance of your model.
6. Compare your results to GLM (generalized linear models).

Exercise 11.7. This exercise applies steepest descent to optimize a portfolio of assets by minimizing the risk subject to a return constraint.

1. Load historical stock prices for a set of assets (for example Yahoo or Google Finance are good resources for this type of data).
2. Compute the expected returns and covariance matrix of the assets. To convert stock prices to asset returns, you can use the daily closing prices. Asset returns can be calculated using the formula $(Y_t - Y_{t-1})/Y_{t-1}$ where Y_t is the closing price at day t.
3. Define the objective function as the portfolio variance (risk) and include a return constraint.

4. Implement the steepest descent algorithm to minimize the portfolio variance subject to the return constraint.
5. Compare the optimized portfolio with the results obtained using quadratic programming.

Exercise 11.8. Here you will use steepest descent to reconstruct a blurred image by minimizing the mean squared error.

1. Load a grayscale image and artificially blur it using a known kernel (e.g., Gaussian blur).
2. Define the objective function as the mean squared error between the blurred image and the reconstructed image.
3. Implement the steepest descent algorithm to minimize the objective function and reconstruct the original image.
4. Experiment with different initialization strategies and step sizes.
5. Compare the reconstructed image with the original and blurred images.

Bibliography

[1] Burden R. L. and Faires J. *Numerical Analysis.* PWS-Kent Publishing Co., Boston, fifth edition, 1993.

[2] Curry H.B. The method of steepest descent for non-linear minimization problems. *Quarterly of Applied Mathematics*, 2:258, 1944.

[3] Dennis J. and Schnabel R. *Numerical Methods for Unconstrained Optimization and Nonlinear Equations.* Society for Industrial Mathematics, 1996.

[4] Goldstein A. A. Cauchy's method of minimization. *Nuerishce Mathematik*, 4:146–150, 1962.

[5] Golub G. H. and Van Loan C. F. *Matrix Computations. Third Edition.* The Johns Hopkins University Press, 1996.

[6] Kincaid D. and Cheney W. *Numerical Analysis: Mathematics of Scientific Computing.* Brooks/Cole Publishing Company, Pacific Grove, CA, 1991.

[7] Press W. H., Teukolsky S. A., Vetterling W. T., and Flannery B. P. *Numerical Recipes: The Art of Scientific Computing.* Cambridge University Press, New York, NY, USA, third edition, 2007.

12

Gradient Boosting

12.1 Introduction

Advances in machine learning using boosting developed rapidly in the mid-90s. Freund and Schapire [4] introduced *Adaboost* for classification, a precursor to the widely used *gradient boosting* method developed by Friedman [5]. A remarkable feature of Adaboost was that it required little supervision and seemed immune to overfitting. It often outperformed traditional methods, especially in challenging scenarios where standard model assumptions failed. However, the mathematical explanation for Adaboost's success was initially unclear. The algorithm iteratively re-weighted observations, fitting a simple learner (such as a tree) in each iteration with higher weights given to data points that were difficult to classify.

Also of relevance is the method of matching pursuit [9], which was developed in the signal processing literature for the purpose of reconstructing a signal function, f. Matching pursuit seeks to recover f by using an iterative sequence of low-dimensional additive solutions. This is a theoretical construction without involving data. Gradient boosting, on the other hand, works with real data and produces an ensemble of weighted learners by iterative gradient descent. Adaboost, as already described, reweights observations iteratively. Therefore, these methods appear to be very different from one another, without any apparent connection. However, they are more closely related than one might think. In this chapter, we discuss how these methods are related while developing theories and applications to aid in understanding boosting.

12.2 What is Boosting?

Boosting is a machine learning procedure for supervised learning. To formally explain boosting, consider the following setting. Let $\{(\mathbf{X}_i, Y_i)\}_1^n$ denote the i.i.d. learning data, and let $(\mathbf{X}, Y) \in \mathscr{X} \times \mathscr{Y}$ denote a generic feature-response pair. The feature space is $\mathscr{X} \subseteq \mathbb{R}^p$. The value Y can be continuous (regression) with $\mathscr{Y} = \mathbb{R}$, or it can take integer values (classification) with $\mathscr{Y} = \{C_1, \ldots, C_J\}$ for $J \geq 2$. The goal in boosting is to estimate an unknown function $F : \mathbb{R}^p \to \mathbb{R}$ of $\mathbf{X} = (X_1, \ldots, X_p) \in \mathbb{R}^p$ by minimizing a specified loss function $L(Y, F)$. This framework is similar to that outlined in Chapter 7, although we use slightly different notation and terminology to align with the boosting literature.

Boosting approximates F by using a large collection of functions, $\mathscr{H}$, defined on $\mathscr{X}$. If $\mathscr{H}$ is rich enough, then it is reasonable to expect F to be well approximated by an additive expansion of the form

$$F(\mathbf{x}) = \sum_{k=1}^{K} \beta_k h_k(\mathbf{x}). \tag{12.1}$$

DOI: 10.1201/9781003493679-12

Here $\beta_k \in \mathbb{R}$ are real-valued coefficients used to form linear combinations of $h_k \in \mathscr{H}$. The functions $h_k : \mathscr{X} \to \mathbb{R}$ are referred to as *base learners* and are chosen to be relatively simple so they can be rapidly fit to data. For this reason, they are sometimes called *weak base learners*.

To determine an appropriate additive expansion for F (12.1), boosting employs empirical risk minimization (Chapter 7.6). The theoretical goal is to minimize the population risk functional

$$R(F) = \mathbb{E}L(Y, F(\mathbf{X})).$$

Assuming F can be expressed as an additive expansion (12.1), the optimization problem can be described as finding F_0 such that

$$F_0 = \underset{F\in\overline{\text{lin}(\mathscr{H})}}{\arg\min}\{R(F)\} = \underset{F\in\overline{\text{lin}(\mathscr{H})}}{\arg\min}\{\mathbb{E}\left[L(Y, F(\mathbf{X}))\right]\}$$

where $\text{lin}(\mathscr{H})$ is the space of all finite linear combinations of elements of $\mathscr{H}$ and $\overline{\text{lin}(\mathscr{H})}$ is its closure. We will write $\mathscr{F}$ to denote the space F belongs to, thus $F \in \mathscr{F}$ where it is assumed that $\overline{\text{lin}(\mathscr{H})} \subseteq \mathscr{F}$. Also for ease of notation, we write $\mathscr{D} = \overline{\text{lin}(\mathscr{H})}$ for the closure of the span of the base learners. We call $\mathscr{D}$ the vocabulary of the dictionary $\mathscr{H}$.

Because the true distribution is unknown, the population risk functional $R(F)$ is replaced with the empirical functional $R_n(F)$ and F is estimated using

$$\hat{F}_n = \underset{F\in\mathscr{D}}{\arg\min}\{R_n(F)\} = \underset{F\in\mathscr{D}}{\arg\min}\left\{\frac{1}{n}\sum_{i=1}^{n} L(Y_i, F(\mathbf{X}_i))\right\}.$$

However, due to the size of the dictionary, direct empirical risk minimization may be ineffective, or even computationally infeasible, in many settings. We can see why this might be the case by parameterizing the weak learners by $h(\mathbf{x}) := h_\theta(\mathbf{x})$ where $\theta \in \Theta$ for some parameter set Θ. Then

$$\{\hat{\beta}_k, \hat{\theta}_k\}_{k=1}^{K} = \underset{\{\beta_k,\theta_k\}_{k=1}^{K}}{\arg\min} \sum_{i=1}^{n} L\left(Y_i, \sum_{k=1}^{K} \beta_k h_{\theta_k}(\mathbf{x}_i)\right). \tag{12.2}$$

It is not hard to see that direct optimization can become challenging for high-dimensional Θ. However, an even more concerning problem is that (12.2) is unregularized and therefore may easily result in overfitting.

To regularize the boosting optimization, one approach is to use stagewise descent algorithms with a capped number of iterations. Friedman's gradient boosting algorithm [5] to be discussed uses this principle. Capping iterations and similar techniques restrict the model space $\mathscr{D}$ to $\mathscr{D}_n$, improving generalization error. Because of this, in many settings, the optimization problem used by boosting is better described as:

$$\hat{F}_n = \underset{F\in\mathscr{D}_n}{\arg\min}\left\{\frac{1}{n}\sum_{i=1}^{n} L(Y_i, F(\mathbf{X}_i))\right\}.$$

12.3 Matching Pursuit

In matching pursuit [9], the parameter of interest is a prechosen function $f \in \mathscr{F}$, representing the signal, and the goal is to reconstruct f using an additive expansion. The reconstruction is obtained from a sequence of low-dimensional optimization problems, which can even be allowed to be infinite. For each step $m = 1, 2, \ldots$, matching pursuit solves

$$\{\rho_m, h_m\} = \underset{\rho\in\mathbb{R}, h\in\mathscr{H}}{\arg\min}\ L\big(f, f_{m-1} + \rho h\big) \tag{12.3}$$

which is used to update the m-th step estimator

$$f_m = f_{m-1} + \rho_m h_m.$$

By induction, this gives $f_m = f_0 + \sum_{i=1}^{m} \rho_i h_i$.

Algorithm 12.1 presents a formal description of the method which we refer to as *theoretical matching pursuit*. The reconstruction is in terms of linear combinations of base learners from a space $\mathscr{H} \subseteq \mathscr{F}$. No data is involved, but rather the algorithm is a proof by construction that an additive solution for f can be achieved by the principle of matching pursuit. For this reason, we refer to it as "theoretical".

For generality, we will assume that our space of functions $\mathscr{F}$ is a Hilbert space. A Hilbert space is a vector space that is complete under its inner product (a space is complete if every Cauchy sequence in the space has a limit that is also in the space). The inner product is denoted by (f_1, f_2) for $f_1, f_2 \in \mathscr{F}$. The base learners used for the construction are taken from the set $\mathscr{H} = \{h_\theta : \theta \in \Theta\} \subseteq \mathscr{F}$ where h_θ is such that $||h_\theta|| = 1$ and $||\cdot||$ is the norm for $\mathscr{F}$. As before, $\mathscr{D} = \overline{\text{lin}(\mathscr{H})}$ is the vocabulary equal to the closed linear span of $\mathscr{H}$.

Algorithm 12.1 *Theoretical Matching Pursuit*

1: Signal $f \in \mathscr{F}$.
2: Initialize $m = 0$, $r_0 = f$.
3: **while** $||r_m|| > 0$ **do**
4:     Update $m \leftarrow m + 1$.
5:     Find $h_m \in \mathscr{H}$ with α-maximal inner product (r_{m-1}, h_m).
6:     Let $\rho_m = (r_{m-1}, h_m)$.
7:     Update $r_m = r_{m-1} - \rho_m h_m$.
8: **end while**

Algorithm 12.1 builds the additive expansion of f by repeated residual fitting. In particular, r_m of Line 7 is a residual function obtained from projecting the previous r_{m-1} onto $\mathscr{H}$. As illustration, the residual r_1 for $m = 1$ after projecting $r_0 = f$ onto a learner $h \in \mathscr{H}$ is defined by

$$f = (f, h)h + r_1.$$

Because r_1 is orthogonal to h, we must have

$$||f||^2 = |(f, h)|^2 \cdot ||h||^2 + ||r_1||^2 = |(f, h)|^2 + ||r_1||^2.$$

Observe that if $||r_1||^2$ is small then we have discovered a base learner that approximates f reasonably well.

However a single base learner is unlikely to yield a zero residual. Therefore, the algorithm is applied iteratively by repeated residual fitting. Line 2 initializes the first residual, $r_0 = f$: the goal is to drive the residual r_m to zero as $m \to \infty$ to obtain the additive expansion for f.

At step m, Line 5 finds the base learner h_m with α-maximal inner product with the current residual r_{m-1}. Specifically, we wish to find h_m satisfying

$$|(r_{m-1}, h_m)| \geq \alpha \sup_{h \in \mathscr{H}} |(r_{m-1}, h)| \tag{12.4}$$

where $0 < \alpha \leq 1$ is a prespecified shrinkage factor. An important feature of matching pursuit is that α can be less than 1, thus h_m does not need to be the closest base learner to the current residual, instead it suffices to find a learner that is within the preset α level. This is a useful feature that allows flexibility in matching and provides a way to control the greediness of the algorithm.

It will be convenient to rewrite Line 7 as

$$r_{m-1} = (r_{m-1}, h_m)h_m + r_m \tag{12.5}$$

and since r_m is orthogonal to h_m, we must have

$$||r_{m-1}||^2 = |(r_{m-1}, h_m)|^2 + ||r_m||^2. \tag{12.6}$$

Writing f as

$$f = \sum_{n=0}^{m-1} (r_n - r_{n+1}) + r_m,$$

and $||f||^2$ as

$$||f||^2 = \sum_{n=0}^{m-1} \left(||r_n||^2 - ||r_{n+1}||^2\right) + ||r_m||^2,$$

we have by (12.5) that f can be decomposed as

$$f = \sum_{n=1}^{m} (r_{n-1}, h_n)h_n + r_m \tag{12.7}$$

and by (12.6), that

$$||f||^2 = \sum_{n=1}^{m} |(r_{n-1}, h_n)|^2 + ||r_m||^2.$$

Thus, the original f is decomposed into a sum of base learners, each of which is chosen to match the current residual. Notice that even though the base learners are very abstract in matching pursuit, the decomposition has an orthogonal description reminiscent of least squares in linear models.

The proof is completed if $||r_m|| \to 0$, since by induction

$$f_m = \sum_{n=1}^{m} (r_{n-1}, h_n)h_n,$$

which implies by (12.7) that $f - f_m = r_m$ which yields the desired reconstruction if the residual converges to zero. However, there is a subtle point related to the size of the underlying base learner space which can affect this. The following result states what the limit is in general, and is due to Mallat and Zhang [9]. For this result, we use the fact that a non-zero orthogonal projection operator P on a Hilbert space $\mathcal{H}$ is a linear map $P : \mathcal{H} \to \mathcal{H}$ that satisfies

$$P^2 = P, \qquad ||P|| = 1, \qquad (Ph, h') = (h', Ph) \text{ for all } h, h' \in \mathcal{H}.$$

Theorem 12.1. *Let $P_{\mathcal{D}}$ be the orthogonal projection operator on $\mathcal{D}$ and let $P_{\mathcal{D}^\perp}$ be the orthogonal projection operator on the orthogonal complement of $\mathcal{D}$. The residual r_m defined in (12.5) satisfies*

$$\lim_{m\to\infty} ||r_m - P_{\mathcal{D}^\perp} f|| = 0.$$

Hence

$$P_{\mathcal{D}} f = \sum_{n=1}^{\infty} (r_{n-1}, h_n)h_n$$

and

$$||P_{\mathcal{D}} f||^2 = \sum_{n=1}^{\infty} |(r_{n-1}, h_n)|^2.$$

Theorem 12.1 shows that the theoretical matching pursuit algorithm recovers the components of f that belong to the span of $\mathscr{H}$ but it cannot recover components outside of this space. This makes sense because the best we can hope to do in terms of recovering f is ultimately limited by the dictionary. This concept is not unique to matching pursuit, but also applies to gradient boosting, and other boosting procedures.

Matching pursuit can only recover the true signal up to the closed linear span of the base learners. Thus the vocabulary should be rich enough for accurate signal recovery, but at the same base learners should also be simple and elementary enough to be easily computed in applied settings.

Notice that in the case where the base learner space is complete, $\mathscr{D} = \mathscr{F}$, we have $P_{\mathscr{D}^{\perp}} f = 0$ and $P_{\mathscr{D}} f = f$. In other words,

$$f = \sum_{n=1}^{\infty} (r_{n-1}, h_n) h_n.$$

Thus f is recovered by the underlying dictionary.

12.4 Tight Bounds for the Weak Greedy Algorithm

The weak greedy algorithm (WGA) [10, 11] is similar to matching pursuit and provides a theoretical construction for expressing a function f in terms of unit-norm basis functions. Several useful bounds have been developed for WGA which make it worthwhile to study even though in many ways it is nearly identical to matching pursuit. In this section we will consider one such result.

We consider a finite-dimensional subspace $\mathscr{D}$ equal to the span of unit-norm functions $h_1, h_2, \ldots, h_d \in \mathscr{H}$, where $\mathscr{H}$ is a Hilbert space. WGA approximates a function $f \in \mathscr{D}$ using the $\mathscr{D}$-basis functions. The construction utilizes an iterative sequence of remainder functions:

$$r_0 f = f, \quad r_m f = r_{m-1} f - \nu_m \langle r_{m-1} f, h_{\delta_m} \rangle h_{\delta_m}, \quad m = 1, 2, \ldots, \tag{12.8}$$

where $0 < \nu_m \leq 1$ are prechosen learning parameters and δ_m is the index selector that selects a basis learner h_{δ_m} satisfying

$$\langle r_{m-1} f, h_{\delta_m} \rangle \geq \alpha \cdot \sup_{1 \leq j \leq d} \langle r_{m-1} f, h_j \rangle \quad \text{for some fixed } 0 < \alpha \leq 1. \tag{12.9}$$

From this, it is easily verified that

$$f = \sum_{j=0}^{m-1} \nu_{j+1} \langle r_j f, h_{\delta_{j+1}} \rangle h_{\delta_{j+1}} + r_m f,$$

where the remainder function r_m satisfies

$$||r_m f||^2 = ||r_{m-1} f||^2 - \nu_m (2 - \nu_m) |\langle r_{m-1} f, h_{\delta_m} \rangle|^2,$$

and note that $0 < \nu_m (2 - \nu_m) \leq 1$. By Theorem 5.1 of [11] (with some slight modifications), if f can be expressed as an element of $\mathscr{D}$, such that

$$f(\mathbf{x}) = \sum_{k=1}^{d} \beta_k h_k(\mathbf{x}), \quad \sum_{k=1}^{d} |\beta_k| \leq B < \infty,$$

then

$$\|r_m f\|^2 \leq B(m, \alpha, B, \{v_j\}_1^m) = B^2 \left(1 + \alpha^2 \sum_{j=1}^{m} v_j^2 (2 - v_j)^2\right)^{-\alpha/(2+\alpha)}. \quad (12.10)$$

(12.10) bounds the remainder function in terms of the WGA parameters. This provides insight into how these values could be selected to improve the reconstruction of f. For example, $\alpha = 1$ gives a 1/3 cubic rate of convergence in m, while $\alpha = 1/2$ gives a 1/5 rate. Therefore, increasing α increases the rate that the remainder converges to zero, with $\alpha = 1$ producing the fastest rate. This makes sense since in both WGA and matching pursuit, α controls the accuracy of the base learner in matching the current residual (compare (12.9) to (12.4)).

(12.10) also shows that v_j strongly influences the rate. Setting $0 < v_j = v < 1$, the rate is of order $(mv^2(2-v)^2)^{-\alpha/(2+\alpha)}$, which becomes faster with increasing v. Since v scales the amount of projection onto the remainder function (see (12.8)), we can interpret v as a slow-learning parameter. However, a larger v (greedier) is not always better. The same type of v learning parameter is used in gradient boosting, where smaller values induce a regularized procedure similar to L_1-penalization, which can be advantageous.

Remark 12.1. (12.10) can be used to prove consistency for gradient boosting [1]. To apply (12.10), the learning parameters are fixed at a common value $v_j = v$. The dictionary $\mathscr{H}$ is taken to be the $L_2(\mathbb{P})$ space $\{f : \int f(\mathbf{x})^2 d\mathbb{P}(\mathbf{x}) < \infty\}$ with inner product $\langle f, g\rangle = \int f(\mathbf{x})g(\mathbf{x})d\mathbb{P}(\mathbf{x})$ and norm $\|f\|^2 = \langle f, f\rangle = \int f(\mathbf{x})^2 d\mathbb{P}(\mathbf{x})$. Here $\mathbb{P}$ is the measure for a data point $(\mathbf{X}, Y)$. Then we wish to show that the population (real-world) gradient boosting algorithm matches the theoretical WGA asymptotically, which then makes it possible to use the bound (12.10) to prove consistency for boosting. □

12.5 Adaboost

Adaboost developed by Freund and Schapire [4] can be considered to be the grandfather of modern boosting and a precursor to gradient boosting. This procedure builds a classifier according to

$$F_M(\mathbf{x}) = \sum_{m=1}^{M} c_m h_m(\mathbf{x}).$$

Thus, Adaboost adopts the boosting view of estimating F by additively combining weak base learners. However, unlike boosting, the algorithm does not appear to directly minimize empirical risk.

Algorithm 12.2 (reproduced from Friedman, Hastie and Tibshirani [6]) provides a description of the procedure. The key idea is to iteratively reweight observations by increasing weights on misclassified observations. This forces the procedure to concentrate more effort on correctly classifying difficult cases. We use the notation $(\mathbf{x}_1, y_1), \ldots, (\mathbf{x}_n, y_n)$ to denote the observed learning data where y_i is coded as $\{-1, +1\}$.

Figure 12.1 illustrates how AdaBoost uses weights to construct a classifier. Data ($n = 500$, $p = 2$) was simulated from the circle in the square synthetic experiment from the `mlbench` R-package [8]. Stumpy trees with a maximum depth of 3 were used as weak learners. The figure shows how weights $\{w_i\}_1^n$ change with the number of iterations, m. As seen, weights quickly shift from a uniform distribution to small and large values, with large values near the classification boundary (where the circle touches the square, which are the hardest points to classify). The last panel displays test set prediction error (test set size $n = 10000$). The error rate drops substantially with only a few trees, rapidly classifying easy cases away from the boundary.

Algorithm 12.2 *Discrete AdaBoost*

Initialize weights $w_i = 1/n$.
for $m = 1, \dots, M$ **do**
  Fit the weak learner $h_m(\mathbf{x}) \in \{-1, 1\}$ to the data using weights $\{w_i\}_1^n$.
  Let $\text{err}_m = \sum_{i=1}^n w_i 1_{\{y_i \neq h_m(\mathbf{x}_i)\}}$.
  Let $c_m = \log((1 - \text{err}_m)/\text{err}_m)$.
  Normalize and update weights: $w_i \leftarrow w_i \exp\left(c_m 1_{\{y_i \neq h_m(\mathbf{x}_i)\}}\right)$.
end for
The M step classifier is $\text{sign}\left(\sum_{m=1}^M c_m h_m(\mathbf{x})\right)$.

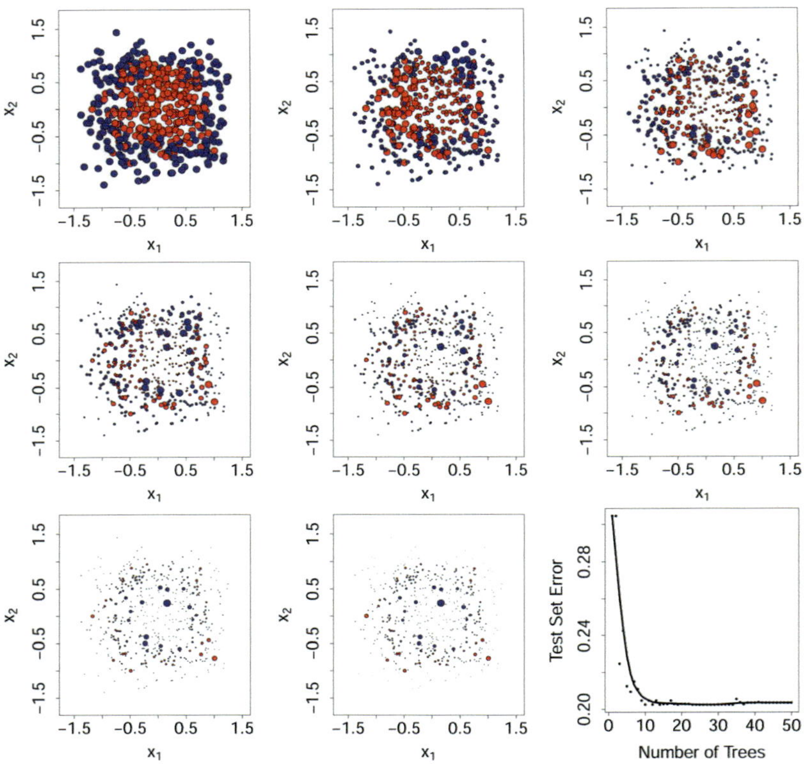

FIGURE 12.1
AdaBoost weights $\{w_i\}_1^n$ for the circle in a square two-class synthetic experiment on $\mathbb{R}^2$ (see the R-package `mlbench` *for details). Values are shown for iterations $m = 1, 2, 3, 4, 5, 6, 10, 50$ using Adaboost with stumpy trees of depth 3. Size of weights are proportional to the size of points: values near the boundary of the circle with the square are hardest to classify and receive the largest weights as m increases. The bottom right plot shows test set prediction error as a function of m (number of trees).*

AdaBoost then focuses on improving classification near the boundary, with a slower effect on reducing prediction error.

Considering Algorithm 12.2, there is no reason to believe that AdaBoost is an empirical risk reduction strategy, or for that matter a residual fitting procedure like matching pursuit. Surprisingly, however, Friedman, Hastie and Tibshirani [6] showed that AdaBoost is an additive *stagewise procedure* for minimizing the exponential loss function $L(y,F)=\exp(-yF)$. A stagewise procedure repeatedly refits residuals.

Thus, AdaBoost is more related to matching pursuit than it first appears. As a stagewise procedure, it opens the door to other loss functions and problems, such as regression, using a similar strategy. This is the idea behind gradient boosting, discussed next.

12.6 Generic Gradient Boosting

As mentioned, a major theoretical advancement in understanding boosting was made by Friedman, Hastie and Tibshirani [6], who showed that AdaBoost could be seen as an additive stagewise procedure, similar to matching pursuit. The gradient boosting method developed by Friedman [5] generalizes this idea, allowing for arbitrary loss functions applicable to almost any setting. It unites the principles of matching pursuit and gradient descent under one framework.

Because matching pursuit is a theoretical algorithm unrelated to data, the challenge in developing gradient boosting was adapting these concepts to data settings. Let $(\mathbf{x}_1,y_1),\ldots,(\mathbf{x}_n,y_n)$ denote the learning data. One of the key ideas of Friedman [5] is to replace the matching pursuit optimization (12.3) with the empirical analog:

$$(\rho_m,h_m)=\underset{\rho\in\mathbb{R},h\in\mathscr{H}}{\arg\min}\sum_{i=1}^{n}L\big(y_i,F_{m-1}(\mathbf{x}_i)+\rho h(\mathbf{x}_i)\big). \tag{12.11}$$

A second innovation is to solve the matching pursuit step using a two-step optimization. The reason for recasting the matching pursuit step as two separate optimizations is that it makes the problem easier to implement in data settings. In the first step, the optimized base learner h_m is obtained by using the gradient of the loss function, after which, in the second step, the scale parameter ρ_m is obtained by a simple one-dimensional line-search. The estimator for F is then updated by

$$F_m(\mathbf{x})=F_{m-1}(\mathbf{x})+\rho_m h_m(\mathbf{x}). \tag{12.12}$$

Regarding the first step of obtaining the optimized learner, define the ith negative gradient evaluated at the current estimator by

$$g_m(\mathbf{x}_i)=-\left[\frac{\partial L(y_i,F(\mathbf{x}_i))}{\partial F(\mathbf{x}_i)}\right]_{F_{m-1}(\mathbf{x}_i)}=-L'(y_i,F_{m-1}(\mathbf{x}_i)),\quad i=1,\ldots,n.$$

The n-dimensional negative gradient vector is $\mathbf{g}_m=(g_m(\mathbf{x}_1),\ldots,g_m(\mathbf{x}_n))^T$. Recall in theoretical matching pursuit that the optimized base learner was obtained by finding the maximal α-inner product

$$|(r_{m-1},h_m)|\geq\alpha\sup_{h\in\mathscr{H}}|(r_{m-1},h)|$$

where $0<\alpha\leq 1$ is a prespecified shrinkage factor. As discussed, allowing α to be less than 1 permits flexibility in identifying a suitable base learner. This same principle is used by gradient boosting. Here the residual r_m is defined by

$$r_m=F_m-F_{m-1}.$$

To decrease r_m as fast as possible, we know by the steepest descent principle (Chapter 11) we should move along the negative gradient. This is the strategy used by gradient boosting.

However, keep in mind that the negative gradient is not necessarily a member of $\mathscr{H}$, the class of weak learners being boosted. Therefore instead of selecting the negative gradient, gradient boosting selects the base learner from $\mathscr{H}$ closest to the gradient. Typically this is done using least squares and the optimal base learner is defined by

$$h_m = \arg\min_{h\in\mathscr{H}} \sum_{i=1}^{n} \left[g_m(\mathbf{x}_i) - h(\mathbf{x}_i)\right]^2. \tag{12.13}$$

Thus, similar to theoretical matching pursuit, gradient boosting uses the strategy that the best match to the residual may not be necessary, instead it suffices to use an adequate match.

Remark 12.2. The learner search (12.13) can be written in a slightly different way to allow for a scaling coefficient in the optimization. Thus, (12.13) could be replaced with

$$(h_m, \beta_m) = \arg\min_{h\in\mathscr{H},\beta\in\mathbb{R}} \sum_{i=1}^{n} \left[g_m(\mathbf{x}_i) - \beta h(\mathbf{x}_i)\right]^2. \tag{12.14}$$

However, there is no loss of generality in using (12.13), as the scaling coefficient β could always be absorbed into the base learner h_m. The choice of which formulation is more convenient will depend on the base learner used. □

Following the determination of h_m, a line-search is used to calculate ρ_m

$$\rho_m = \arg\min_{\rho\in\mathbb{R}} \sum_{i=1}^{n} L\big(y_i, F_{m-1}(\mathbf{x}_i) + \rho h_m(\mathbf{x}_i)\big). \tag{12.15}$$

This gives (ρ_m, h_m) and completes the optimization of (12.11). The estimator is then updated using (12.12).

While this completes the steps needed for gradient boosting, an additional modification is used in practice. For greater control over learning, the following update is used in place of (12.12):

$$F_m(\mathbf{x}) = F_{m-1}(\mathbf{x}) + \nu\rho_m h_m(\mathbf{x}).$$

The difference here is the inclusion of the parameter $0 < \nu \le 1$ which has the effect of scaling the contribution of (ρ_m, h_m) and therefore acts as a slow-learning parameter. This is similar to the ν_j parameters used by WGA, where recall from the bound (12.10) that decreasing ν_j induces a less greedy procedure with slower convergence. Chapter 13 discusses the effect of ν in further detail. Algorithm 12.3 provides a complete description of the algorithm that includes the slow-learning adjustment. We will refer to this algorithm as (generic) gradient boosting.

12.6.1 Connection to steepest descent

The motivation for using the gradient boosting in the matching step comes from the steepest descent principle. Recall that the goal of boosting is to minimize the empirical risk functional

$$R_n(F) = \frac{1}{n}\sum_{i=1}^{n} L(y_i, F(\mathbf{x}_i)).$$

Applying a Taylor expansion to $R_n(F)$ for $F = F_m$ around $\nu_m = 0$, where

$$F_m(\cdot) = F_{m-1}(\cdot) + \nu_m h_m(\cdot), \quad \nu_m = \nu\rho_m,$$

Algorithm 12.3 *Generic Gradient Boosting*

Initialize $F_0(\mathbf{x}) = \arg\min_{\rho\in\mathbb{R}} \sum_{i=1}^{n} L\big(y_i, \rho\big)$.
for $m = 1, \ldots, M$ **do**
  $g_m(\mathbf{x}_i) = -L'(y_i, F_{m-1}(\mathbf{x}_i))$.
  $h_m = \arg\min_{h\in\mathcal{H}} \sum_{i=1}^{n} [g_m(\mathbf{x}_i) - h(\mathbf{x}_i)]^2$.
  $\rho_m = \arg\min_{\rho\in\mathbb{R}} \sum_{i=1}^{n} L\big(y_i, F_{m-1}(\mathbf{x}_i) + \rho h_m(\mathbf{x}_i)\big)$.
  Update $F_m(\mathbf{x}) = F_{m-1}(\mathbf{x}) + \nu\rho_m h_m(\mathbf{x})$.
end for

is the m step updated learner, we have

$$\begin{aligned} R_n(F_m) &= \frac{1}{n}\sum_{i=1}^{n} L(y_i, F_m(\mathbf{x}_i)) \\ &\asymp \frac{1}{n}\sum_{i=1}^{n} L(y_i, F_{m-1}(\mathbf{x}_i)) - \frac{\nu_m}{n}\sum_{i=1}^{n} g_m(\mathbf{x}_i)h_m(\mathbf{x}_i). \end{aligned}$$

Therefore, in order to decrease $R_n(F_m)$ the fastest, we should choose $h_m(\mathbf{x}_i) = g_m(\mathbf{x}_i)$, which is the steepest descent principle.

To make the connection to the steepest descent even more transparent, let us simplify things and assume the loss function can be expressed as $L(y, F) = \phi(y - F)$ for some function ϕ. Then (12.15) reduces to

$$\rho_m = \arg\min_{\rho\in\mathbb{R}} \sum_{i=1}^{n} \phi\big(y_i - F_{m-1}(\mathbf{x}_i) - \rho h_m(\mathbf{x}_i)\big),$$

which by (12.13) should closely approximate

$$\arg\min_{\rho\in\mathbb{R}} \sum_{i=1}^{n} \phi\big(\underbrace{y_i - F_{m-1}(\mathbf{x}_i)}_{z_{i,m-1}} - \rho g_m(\mathbf{x}_i)\big).$$

Updating $F_m(\mathbf{x}_i)$ yields an approximate updated $z_{i,m}$ as follows:

$$z_{i,m} = y_i - F_m(\mathbf{x}_i) \approx y_i - F_{m-1}(\mathbf{x}_i) - \nu\rho_m g_m(\mathbf{x}_i) = z_{i,m-1} - \nu\rho_m g_m(\mathbf{x}_i).$$

This m step update to z is nearly the same as that used by steepest descent (Algorithm 11.1), which can be seen by considering Algorithm 12.4 which is an alternate description of Algorithm 12.3 more aligned with steepest descent. Here, the function $f(\mathbf{z}_m)$ being minimized is the scaled empirical risk, where $\mathbf{z}_m = (y_1 - F_m(\mathbf{x}_1), \ldots, y_n - F_m(\mathbf{x}_n))^T$ is the residual vector. Similar to steepest descent each step minimizes the optimization function by moving along the negative gradient. The difference is that the negative gradient in boosting is approximated by weak learners, whereas steepest descent directly uses the gradient. Thus, gradient boosting applies the steepest descent principle with the caveat that the approximation must be within the allowed vocabulary.

Gradient boosting selects the base learner closest to the negative gradient. Thus, gradient boosting follows the principle of steepest descent but where directional moves are constrained to base learners from the underlying dictionary.

Algorithm 12.4 *Generic Gradient Boosting (connection to steepest descent)*

1: Initialize $z_{i,0} = y_i - F_0(\mathbf{x}_i)$.
2: **for** $m = 1, \ldots, M$ **do**
3: Let $g_{i,m} = -\phi'(z_{i,m-1})$ where $\phi(y - F) = L(y, F)$.
4: The two-step optimization is:

$$\begin{aligned} f(\{z_{i,m-1} - \rho g_{i,m}\}_1^n) &= \sum_{i=1}^n \phi(z_{i,m-1} - \rho g_{i,m}) \\ h_m(\mathbf{x}_i) &= \arg\min_{h \in \mathcal{H}} \sum_{i=1}^n [g_{i,m} - h(\mathbf{x}_i)]^2 \\ \rho_m &= \arg\min_{\rho \in \mathbb{R}} f(\{z_{i,m-1} - \rho h_m(\mathbf{x}_i)\}_1^n) \\ &\approx \arg\min_{\rho \in \mathbb{R}} f(\{z_{i,m-1} - \rho g_{i,m}\}_1^n). \end{aligned}$$

5: Update $z_{i,m} = z_{i,m-1} - \nu \rho_m h_m(\mathbf{x}_i) \approx z_{i,m-1} - \nu \rho_m g_{i,m}$.
6: **end for**

12.7 Gradient Boosting for Linear Regression

In this section we consider gradient boosting for linear regression. The linear regression model is $\mathbb{E}[Y|\mathbf{X} = \mathbf{x}] = \mathbf{x}^T \boldsymbol{\beta}$ where $\boldsymbol{\beta} = (\beta_1, \ldots, \beta_p)^T \in \mathbb{R}^p$ is the regression parameter vector. Using L_2-loss we boost

$$F(\mathbf{x}_i) = \mathbf{x}_i^T \boldsymbol{\beta}, \quad \mathbf{x}_i = (x_{i1}, \ldots, x_{ip})^T, \quad i = 1, \ldots, n,$$

by using linear learners. Since we are taking F to be a linear regression model, this means that the base learners $h_k(\mathbf{x})$ are the coordinates of $\mathbf{x}$; i.e. $h_k(\mathbf{x}) = x_k$ for $k = 1, \ldots, p$ and $\mathcal{H} = \{h_1, \ldots, h_p\}$. Thus, the assumption of an additive expansion for F corresponds to

$$F(\mathbf{x}) = \sum_{k=1}^p \beta_k h_k(\mathbf{x}) = \sum_{k=1}^p \beta_k x_k.$$

Although the focus of boosting is to boost model F, we will see that an advantage of parametric gradient boosting is that this naturally yields an estimate for the boosted regression parameter $\boldsymbol{\beta}$ as a by-product.

Remark 12.3. Notice that the joint optimization problem (12.2) in this setting corresponds to

$$\{\hat{\beta}_k\}_1^p = \arg\min_{\{\beta_k\}_1^p} \sum_{i=1}^n L\left(y_i, \sum_{k=1}^p \beta_k h_k(\mathbf{x}_i)\right),$$

where the loss function is $L(y, F) = \frac{1}{2}(y - F)^2$ (the factor of 1/2 is added for convenience to simplify the gradient calculations). In this framework, the full model least squares estimator has the smallest residual sum of squares, and therefore the smallest empirical risk:

$$\hat{\boldsymbol{\beta}}_{\text{OLS}} = \arg\min_{\boldsymbol{\beta} \in \mathbb{R}^p} \sum_{i=1}^n \left(y_i - \sum_{k=1}^p \beta_k x_{ik}\right)^2.$$

This highlights that joint optimization can lead to overfitting due to a lack of regularization. □

Now let's turn to the details of how gradient boosting applies in this setting. Interestingly, matching pursuit optimization and the two-stage gradient boosting optimization step are identical due to the use of L_2-loss. This is a unique aspect of linear regression that generally does not hold in other settings.

To see this, observe that matching pursuit (12.11) solves

$$(\rho_m, k_m) = \underset{\rho \in \mathbb{R}, 1 \le k \le p}{\arg\min} \sum_{i=1}^{n} \Big(y_i - \big[F_{m-1}(\mathbf{x}_i) + \rho x_{ik}\big]\Big)^2. \tag{12.16}$$

Using the two-stage optimization, first notice that the negative gradient is

$$g_m(\mathbf{x}_i) = y_i - F_{m-1}(\mathbf{x}_i)$$

which is the residual r_m from the mth step. This simplification allows us to combine the line-search (12.15)

$$\rho_m = \underset{\rho \in \mathbb{R}}{\arg\min} \sum_{i=1}^{n} \Big(y_i - \big[F_{m-1}(\mathbf{x}_i) + \rho x_{ik_m}\big]\Big)^2$$

and the learner-search (12.14)

$$(k_m, \beta_m) = \underset{1 \le k \le p, \beta \in \mathbb{R}}{\arg\min} \sum_{i=1}^{n} \Big[g_m(\mathbf{x}_i) - \beta x_{ik}\Big]^2$$

into a single step because

$$\Big(y_i - \Big[F_{m-1}(\mathbf{x}_i) + \rho x_{ik}\Big]\Big)^2 = \Big[g_m(\mathbf{x}_i) - \rho x_{ik}\Big]^2.$$

Thus, in the special case of L_2-loss, (12.16) is equivalent to solving (12.13) and (12.15). This leads to Algorithm 12.5, which we call LS Boost.

Algorithm 12.5 *LS Boost*

1: $F_0(\mathbf{x}) = \sum_i y_i / n$.
2: **for** $m = 1$ to M **do**
3: $\quad (\rho_m, k_m) = \underset{\rho \in \mathbb{R}, 1 \le k \le p}{\arg\min} \sum_{i=1}^{n} \big[y_i - F_{m-1}(\mathbf{x}_i) - \rho x_{ik}\big]^2$.
4: $\quad F_m(\mathbf{x}_i) = F_{m-1}(\mathbf{x}_i) + \nu \rho_m x_{ik_m}$.
5: **end for**

Remark 12.4. The interpretation of the gradient as a residual makes it clear that LS Boost is a stagewise algorithm that repeatedly refits the residual similar to matching pursuit. □

Remark 12.5. Because F and the regression coefficients are related due to linearity, Algorithm 12.5 can be modified to provide regression coefficient estimates. In particular, we have $F_M(\mathbf{x}_i) = F_0(\mathbf{x}_i) + \sum_{k=1}^{p} \hat{\beta}_k x_{ik}$, where for a given coordinate $1 \le k \le p$, the boosted kth coefficient estimate equals the sum of $\nu \rho_m$ values where the active direction for step m is the coordinate direction k. Thus, the boosted coefficient estimate for coordinate k is $\hat{\beta}_k = \nu \sum_{m=1}^{M} \rho_m 1_{\{k_m = k\}}$. □

12.8 Gradient Boosting for Standarized Linear Regression: L_2Boost

Standardizing the data simplifies the interpretation of gradient boosting in linear regression, making its similarity to matching pursuit even clearer. This leads to the so-called L_2Boost procedure outlined below.

Let $\mathbf{x}_{(k)} = (x_{i1},\ldots,x_{in})^T$ be the n-dimensional vector for attribute k. We will assume the features are standardized with a squared-length of one:

$$(\mathbf{x}_{(k)},\mathbf{x}_{(k)}) = ||\mathbf{x}_{(k)}||^2 = \sum_{i=1}^{p} x_{ik}^2 = 1, \quad \text{for } k = 1,\ldots,p. \tag{12.17}$$

The notation $(\mathbf{u},\mathbf{v}) = \sum_{i=1}^{n} u_i v_i$ is used for the inner product for two vectors $\mathbf{u},\mathbf{v} \in \mathbb{R}^n$ and $||\cdot||$ for the norm.

Assuming that the data has been standardized according to (12.17), least squares yields

$$\arg\min_{\rho\in\mathbb{R}} \sum_{i=1}^{n} \left[g_m(\mathbf{x}_i) - \rho x_{ik}\right]^2 = ||\mathbf{g}_m - (\mathbf{x}_{(k)}\mathbf{x}_{(k)}^T)\mathbf{g}_m||^2.$$

From this, we obtain the solution to the matching pursuit step:

$$\rho_m = (\mathbf{g}_m,\mathbf{x}_{(k_m)}), \qquad k_m = \arg\max_{1\le k\le p} |(\mathbf{g}_m,\mathbf{x}_{(k)})|.$$

This follows as a consequence of the following result.

Theorem 12.2. $\arg\min_{1\le k\le p} ||\mathbf{g}_m - (\boldsymbol{x}_{(k)}\boldsymbol{x}_{(k)}^T)\mathbf{g}_m||^2 = \arg\max_{1\le k\le p} |(\mathbf{g}_m,\boldsymbol{x}_{(k)})|.$

Proof. With a slight abuse of notation, let $\rho_k = (\mathbf{g}_m,\mathbf{x}_{(k)})$. Then $(\mathbf{x}_{(k)}\mathbf{x}_{(k)}^T)\mathbf{g}_m$ can be written as $\rho_k\mathbf{x}_{(k)}$ for $1 \le k \le p$. Hence, for any two coordinates $j \ne l$:

$$||\mathbf{g}_m - (\mathbf{x}_{(l)}\mathbf{x}_{(l)}^T)\mathbf{g}_m||^2 \le ||\mathbf{g}_m - (\mathbf{x}_{(j)}\mathbf{x}_{(j)}^T)\mathbf{g}_m||^2$$

if and only if

$$||\mathbf{g}_m - \rho_l\mathbf{x}_{(l)}||^2 \le ||\mathbf{g}_m - \rho_j\mathbf{x}_{(j)}||^2. \tag{12.18}$$

Expanding and collecting terms, (12.18) becomes

$$2\left(\rho_j(\mathbf{g}_m,\mathbf{x}_{(j)}) - \rho_l(\mathbf{g}_m,\mathbf{x}_{(l)})\right) \le \rho_j^2(\mathbf{x}_{(j)},\mathbf{x}_{(j)}) - \rho_l^2(\mathbf{x}_{(l)},\mathbf{x}_{(l)}) = \rho_j^2 - \rho_l^2,$$

where the right-hand side uses assumption (12.17). Now rewriting the left-hand side in terms of ρ_j and ρ_l, we have

$$2(\rho_j^2 - \rho_l^2) \le \rho_j^2 - \rho_l^2.$$

Thus for (12.18) to hold, we must have $\rho_j^2 \le \rho_l^2$, or eqivalently $|(\mathbf{g}_m,\mathbf{x}_{(j)})| \le |(\mathbf{g}_m,\mathbf{x}_{(l)})|$. □

Theorem 12.2 shows that the favorable direction k_m equals the coordinate k with the largest absolute value of $(\mathbf{g}_m,\mathbf{x}_{(k)})$; the latter being proportional to the *gradient-correlation* for k defined as

$$\text{corr}(\mathbf{g}_m,\mathbf{x}_{(k)}) = \frac{(\mathbf{g}_m,\mathbf{x}_{(k)})}{||\mathbf{g}_m||\cdot||\mathbf{x}_{(k)}||} = \frac{(\mathbf{g}_m,\mathbf{x}_{(k)})}{||\mathbf{g}_m||}.$$

Therefore, we can conclude that in this setting, gradient boosting is a stagewise procedure of repeated residual refitting by maximal correlation, which is a type of inner product, and thus similar to the α-maximal inner product used by matching pursuit (12.4). Also, recall that matching pursuit uses stagewise residual refitting with unit-norm basis functions. Therefore, we can see that by standardizing the features, we have made LS Boost more closely aligned with matching pursuit.

Gradient boosting in linear regression with standardized coefficients repeatedly refits the residual, seeking each time to move in the direction of the coordinate with the largest absolute gradient-correlation.

In addition to scaling the features, it is common to center the response to a zero mean. This eliminates the problem of including an intercept in the model. Because Y is centered, the features should also be centered. Let $\mathbf{1}$ denote the n-dimensional vector $(1,\ldots,1)^T$. Then, the data is standardized according to:

$$(\mathbf{y},\mathbf{1}) = 0, \quad (\mathbf{x}_{(k)},\mathbf{1}) = 0, \quad (\mathbf{x}_{(k)},\mathbf{x}_{(k)}) = 1, \quad k = 1,\ldots,p. \tag{12.19}$$

The following algorithm, Algorithm 12.6, which will be referred to as L_2Boost, highlights the connection to stagewise refitting and recasts LS Boost in terms of the gradient-correlation. Algorithm 12.7 is an alternate version of L_2Boost which recasts the algorithm in terms of the β coefficients.

Algorithm 12.6 *L_2Boost for standardized data (12.19)*

Initialize $F_{0,i} = 0$ for $i = 1,\ldots,n$ and let $\mathbf{F}_0 = (F_0(\mathbf{x}_1),\ldots,F_0(\mathbf{x}_n))^T$.
for $m = 1$ to M **do**
  $k_m = \arg\max_{1\le k\le p} |(\mathbf{g}_m,\mathbf{x}_{(k)})|$ for $\mathbf{g}_m = \mathbf{y} - \mathbf{F}_{m-1}$.
  $\mathbf{F}_m = \mathbf{F}_{m-1} + \nu\rho_m\mathbf{x}_{(k_m)}$, where $\rho_m = (\mathbf{g}_m,\mathbf{x}_{(k_m)})$.
end for

Algorithm 12.7 *L_2Boost (alternate description in terms of β coefficients)*

Initialize $\hat{\beta}_j = 0$ for $j = 1,\ldots,p$
for $m = 1$ to M **do**
  Find k most correlated with the current residual

$$k = \arg\max_{1\le j\le p}\left\{\left|\operatorname{corr}\left(y_i - \sum_{l=1}^{p} x_{il}\hat{\beta}_l,\, x_{ij}\right)\right|\right\}.$$

  Update $\hat{\beta}_k \leftarrow \hat{\beta}_k + \nu\rho$ where $\rho = \sum_{i=1}^{n} x_{ik}\left(y_i - \sum_{j=1}^{p} x_{ij}\hat{\beta}_j\right)$.
end for

12.8.1 Diabetes data

To illustrate L_2Boost, we analyze the diabetes data from Efron et al. [2]. The data consists of $n =$ 442 patients in which the response of interest, y, is a quantitative measure of disease progression for a patient. In total there are 64 variables, including 10 baseline measurements for each patient, 45 interactions, and 9 quadratic terms.

Three different parameters were used: $\nu = 0.05, 0.1, 0.25$. Ten-fold cross-validation (CV) estimated the prediction error. The rounded CV error was 2972, 3003, and 2952, optimized at

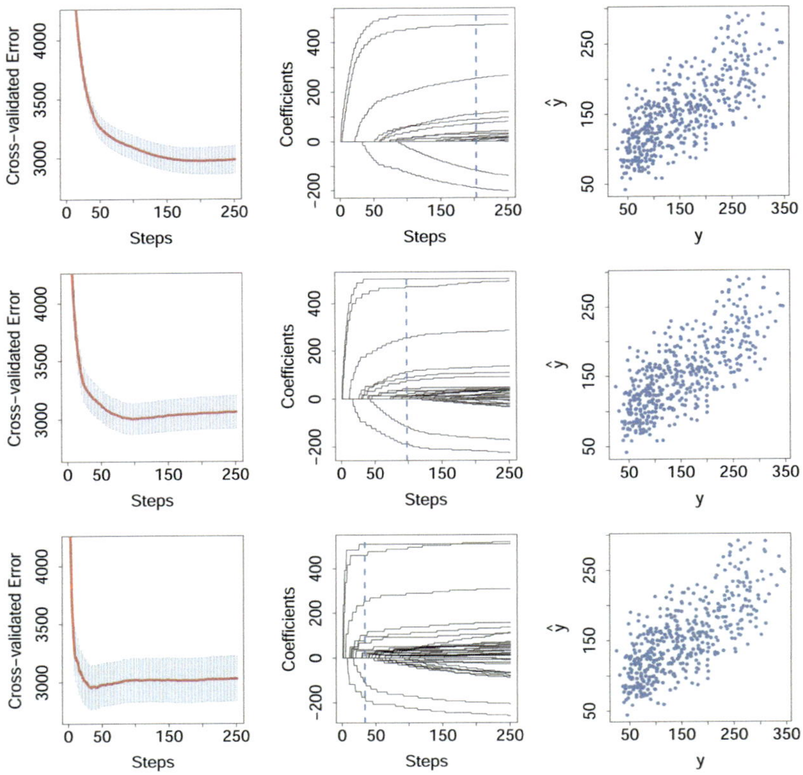

FIGURE 12.2
Gradient boosting for the diabetes data with regularization parameters $\nu = 0.05, 0.1, 0.25$ *(top, middle, bottom).*

203, 98, and 34 steps for the three ν values, respectively. Although the error rate differences are small, we favor $\nu = 0.1$ because it produces the best CV curve, as seen in the first column panels of Figure 12.2. The smallest value, $\nu = 0.05$, takes too long to reach its minimum, and the largest value, $\nu = 0.25$, has a sharp minimum, suggesting it is less stable. The rightmost column panel shows predicted values $\hat{y}$ against observed y, with all three analyses yielding nearly identical results. Thus, the sensitivity seen in the CV curves does not affect prediction performance but does affect estimated regression coefficients, as shown in the middle column panels. These panels display the *coefficient solution path*, where each line shows the estimated coefficient for a feature in terms of the number of boosting steps.

Remark 12.6. We have seen this apparent dichotomy between estimation and prediction before. For example, in our discussion of AIC versus BIC in Chapter 4.6, and the difference between the deleting-d multifold CV estimator and leave-one-out CV in Chapter 7.9. The same principle is at play here. This is because CV is being minimized to tune M, which leads to similar prediction error performance for the different ν. But we cannot expect that estimation under the tuned M will be similar, because achieving optimal model selection performance and estimation requires a different type of cross-validation strategy (Chapter 7.9). □

12.9 Consistency of L_2Boost

We have compared gradient boosting to methods like steepest descent and matching pursuit, which are proven to work, giving us some confidence in gradient boosting. However, these methods are theoretical and not applied to data, so the theoretical evidence for boosting's effectiveness is indirect.

A more direct way to understand if boosting works is to study its asymptotic properties. In this section, we review a study by Bühlmann [1], who proved consistency for L_2Boost in high-dimensional settings (in that work the algorithm was referred to as L_2Boosting with componentwise linear least squares, but for simplicity, we will continue to refer to the method as L_2Boost).

Bühlmann [1] considered the model

$$\begin{aligned} Y_i &= F_n(\mathbf{X}_i) + \varepsilon_i, \quad i = 1, \ldots, n \\ F_n(\mathbf{x}) &= \sum_{k=1}^{p_n} \beta_{k,n} x_k, \quad \mathbf{x} \in \mathbb{R}^{p_n}. \end{aligned}$$

Notice that in this linear model, the number of variables p_n is allowed to vary with the sample size, n.

The following assumptions were used:

(A1) The dimension of the feature space satisfies $p_n = O(\exp(Cn^{1-\zeta}))$ for some $0 < \zeta < 1$, $0 < C < \infty$.

(A2) $\sup_n \sum_{k=1}^{p_n} |\beta_{k,n}| < \infty$.

(A3) $\mathbf{X}_i$ are i.i.d. and bounded and $\mathbb{E}(X_j^2) = 1$ for $j = 1, \ldots, p_n$.

(A4) ε_i are i.i.d. and $\mathbb{E}(\varepsilon_1) = 0$, $\mathbb{E}(\varepsilon_1^s) < \infty$ where $s > 4/\zeta$.

Bühlmann [1] proved the following remarkable result.

Theorem 12.3. *Let $F_{M,n}$ be the M-step L_2Boost predictor constructed from the learning data. Under conditions (A1)-(A4) there exists a subsequence $M_n \to \infty$ such that*

$$\mathbb{E}_{\mathbf{X}} |F_{M,n} - F_n(\mathbf{X})|^2 = o_p(1)$$

where $\mathbf{X} = (X_1, \ldots, X_{p_n})^T$ is a new predictor variable with the same distribution as the learning data covariates.

Theorem 12.3 shows that L_2Boost is mean-squared-error consistent when the number of parameters p_n increases with n. Remarkably, the dimension p_n can be nearly as large as $\exp(n)$. The proof details are in Bühlmann [1], but roughly, the proof shows that L_2Boost can be viewed as a semi-population version of WGA. This allows the WGA bound (12.10) to be used to bound the mean-squared error.

12.10 Tuning Parameters and Connection of L_2Boost to the Lasso

L_2Boost has two parameters, M and ν, that are used to regulate its greediness. Therefore, it is helpful to understand their theoretical properties.

1. Regarding M, Theorem 12.3 shows that controlling the number of boosting iterations can achieve MSE consistency in high-dimensional settings. Therefore, capping the boosting steps is an effective way to regularize prediction error.
2. The parameter ν also significantly affects L_2Boost's behavior. When ν is small, L_2Boost mimics the behavior of the lasso. Efron et al. [2] discussed this phenomenon, showing that the solution path for forward stagewise regression procedures like L_2Boost approximates the lasso and converges as $\nu \to 0$ (see also Ehrlinger and Ishwaran [3]). This is useful because the lasso can shrink coefficients exactly to zero, so setting ν to small values allows L_2Boost to achieve the same property.
3. However, using L_2Boost as an alternative to the lasso is discouraged. This approximation is highly inefficient due to its slow convergence and cycling behavior (see Chapter 13). If a small ν is desired, it is more practical to directly calculate the lasso.

Regarding the second point, to show the similarity between L_2Boost and lasso, Figure 12.3 displays the standardized coefficient paths for the two procedures using the diabetes data.

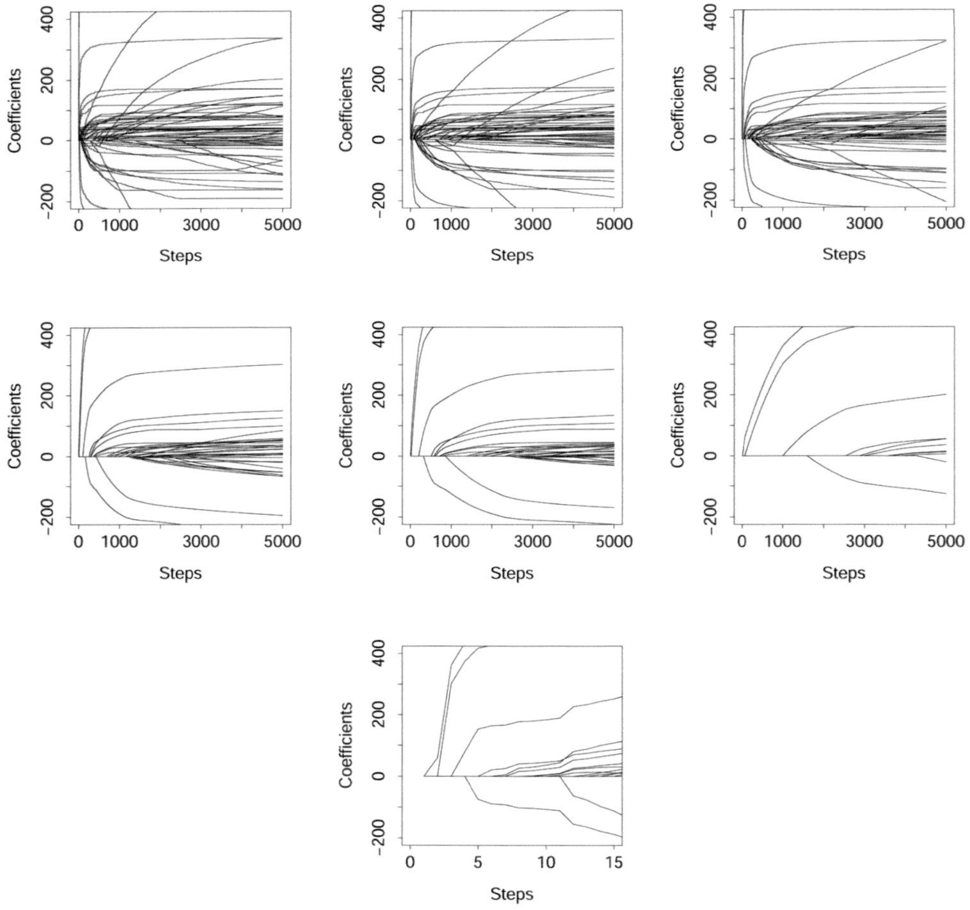

FIGURE 12.3
Bottom figure is the lasso coefficient path for the diabetes data. First six panels are gradient boosting with regularization parameters $\nu = .2, .1, .05, .01, .005, .001$, respectively.

The first six panels show boosting with increasingly smaller v. The bottom panel shows lasso calculated using the `lars` R-package [2] (see Chapter 14). Notice how the coefficient paths for gradient boosting resemble lasso as v becomes smaller.

In summary, parameters M and v in L_2Boost are designed to be tuned for achieving accurate prediction error. Capping M leads to MSE consistency, while small values for v promote sparsity similar to lasso and will encourage the lasso prediction-loss consistency property (see Theorem 4.3).

12.11 Gradient Boosting for Logistic Regression

The flexibility of gradient boosting makes it easy to use in many data settings. We describe here how it can be used for logistic regression by boosting weak linear learners. The log-likelihood will be taken to be

$$\log\left\{\prod_{i=1}^{n} p(\mathbf{x}_i)^{y_i}(1-p(\mathbf{x}_i))^{1-y_i}\right\}$$

where $p(\mathbf{x}_i) = \mathbb{P}(Y = 1|\mathbf{X} = \mathbf{x}_i)$ and $Y \in \{0, 1\}$ is the binary outcome.

By definition, the loss function is the negative of the log-likelihood,

$$L(y_i, F(\mathbf{x}_i)) = -y_i \log p(\mathbf{x}_i) - (1-y_i)\log(1-p(\mathbf{x}_i)).$$

We assume that

$$p(\mathbf{x}_i) = \frac{1}{1+\exp(-F(\mathbf{x}_i))}, \quad \text{where } F(\mathbf{x}_i) = \alpha + \sum_{k=1}^{p} \beta_k x_{ik},$$

which implies that

$$L(y_i, F(\mathbf{x}_i)) = -y_i F(\mathbf{x}_i) - \log(1-p(\mathbf{x}_i)).$$

Therefore, it follows that the ith negative gradient evaluated at the current estimator is

$$g_m(\mathbf{x}_i) = -\left[\frac{\partial L(y_i, F(\mathbf{x}_i))}{\partial F(\mathbf{x}_i)}\right]_{F_{m-1}(\mathbf{x}_i)} = y_i - p_m(\mathbf{x}_i),$$

where $p_m(\mathbf{x}_i) = (1+\exp(-F_{m-1}(\mathbf{x}_i)))^{-1}$.

Thus, the two-step optimization (12.13) and (12.15) can be written as

$$h_m = \arg\min_{h\in\mathscr{H}} \sum_{i=1}^{n} \left[y_i - p_m(\mathbf{x}_i) - h(\mathbf{x}_i)\right]^2,$$

and

$$\rho_m = \arg\min_{\rho\in\mathbb{R}} \sum_{i=1}^{n} \phi\big(y_i, F_{m-1}(\mathbf{x}_i) + \rho h_m(\mathbf{x}_i)\big)$$

where $\phi(y, F) = -yF + \log(1+\exp(F))$.

No closed-form solution exists for the line-search but it can be readily approximated. Due to linearity, we use parameteric weak learners $h = x_k$ similar to linear regression for this step. Also because F includes an intercept we always include an intercept when we boost F. Therefore, the line-search is a two-dimensional optimization over the intercept and regression coefficient for the active coordinate direction $k = k_m$:

$$(\hat{\alpha}_m, \hat{\beta}_m)^T = \arg\min_{(\alpha,\beta)\in\mathbb{R}^2} \sum_{i=1}^{n} \phi(y_i, F_{m-1}(\mathbf{x}_i) + \alpha + \beta x_{ik_m}). \tag{12.20}$$

This is identical to fitting GLM (generalized linear models) for a slope-intercept model, which is one method that could be used to solve (12.20). However, a more efficient solution is to use a one-step Newton approximation (see Chapter 12.12).

Determining the most favorable direction k_m is simplified when the features are standardized as in (12.19). Then similar to linear regression, k_m equals the largest absolute gradient correlation. This follows from an argument similar to Theorem 12.2:

$$\begin{aligned} k_m &= \underset{1\le j\le p,\alpha,\beta}{\arg\min} \|\mathbf{g}_m - \alpha\mathbf{1} - \beta\mathbf{x}_{(j)}\|^2 \\ &= \underset{1\le j\le p}{\arg\min} \|(\mathbf{g}_m - \overline{\mathbf{g}}_m\mathbf{1}) - (\mathbf{x}_{(j)}\mathbf{x}_{(j)}^T)\mathbf{g}_m\|^2 \\ &= \underset{1\le j\le p}{\arg\min} \|\mathbf{g}_m - (\mathbf{x}_{(j)}\mathbf{x}_{(j)})\mathbf{g}_m\|^2, \qquad (\text{from } (\mathbf{x}_{(j)}, \mathbf{1}) = 0) \\ &= \underset{1\le j\le p}{\arg\max} |(\mathbf{g}_m, \mathbf{x}_{(j)})|, \qquad (\text{from } (\mathbf{x}_{(j)}, \mathbf{x}_{(j)}) = 1). \end{aligned}$$

The mth update for F becomes

$$F_m(\mathbf{x}_i) = F_{m-1}(\mathbf{x}_i) + \hat{\alpha}_m + \nu\hat{\beta}_m x_{ik_m}.$$

Remark 12.7. Observe that ν is only applied to the active coordinate k_m and not the intercept. The intercept is an offset term that is always included, and since it is an ancillary parameter, we aim to eliminate its effect quickly, hence no regularization is imposed on it. Linear regression addresses this by centering Y, but this strategy doesn't apply to logistic regression due to the nonlinearity of F. By including the intercept, we have defined a weak learner space slightly larger than linear regression. Here, $\mathscr{H}$ equals the set of translates of the coordinate functions, $\mathscr{H} = \cup_{k=1}^{p}\{h : h(\mathbf{x}) = \alpha + x_k, \alpha \in \mathbb{R}\}$. □

12.11.1 Prostate cancer

We provide an illustration using a prostate cancer dataset. The outcome is pathologic presence or absence of prostate cancer in prostate cancer patients. The attributes include an assortment of clinical measurements, including PSA (protein specific antigen) level, tumor volume size and Gleason score information. The data contain $n = 158$ patients and a total of $p = 20$ features. The results from running standard logistic regression (GLM) on the data is provided in Table 12.1.

Observe the large coefficient value for the variable "tumor frac" which is an attribute measuring tumor size. The large value indicates numerical instability (in this example, all covariates have been standardized to allow for comparison across coefficient estimates).

We compare these results to logistic gradient boosting. The coefficient solution path is shown in Figure 12.4. Tuning parameters were set to $M = 500$ and $\nu = 0.1$. Around step $m = 400$, we observe a rapid increase in the coefficient estimates for the variable tumor frac, reflecting the same instability seen in GLM. However, cross-validation for tuning the boosting steps shows that the optimal number of iterations is well below this value. We used 5 and 10-fold validation and leave-one-out CV (loo-CV). All three methods indicate an optimal step size of approximately $m = 200$. The loo bootstrap (OOB validation) was also used, selecting a sparser model with $m = 100$. The optimal m for all four methods is superimposed on Figure 12.4 using thin dashed lines. In all cases, the coefficient estimates at these optimized values are stable. Thus, we can use regularization to avoid the numerical problems seen with GLM.

TABLE 12.1
Logistic regression (GLM) results from prostate cancer data. Observe the large coefficient value for the attribute tumor frac which is a sign of numerical instability.

	$\hat{\beta}$	Std. Error	Z-statistic	p-value
Intercept	-12.84	5.11	-2.51	0.01
age	0.04	0.04	0.92	0.36
blood	0.01	0.17	0.05	0.96
capsular	-0.32	0.30	-1.08	0.28
cen	0.56	0.40	1.40	0.16
depth	1.84	1.53	1.21	0.23
fat	0.39	0.30	1.32	0.19
nvbtu56	0.01	0.16	0.04	0.97
prgleason	0.71	0.42	1.71	0.09
psa	1.40	0.52	2.69	0.01
rptco49	0.14	0.59	0.23	0.82
rptf51	0.12	0.20	0.60	0.55
secgleason	1.21	0.36	3.33	0.00
surgery	-0.10	0.28	-0.37	0.71
svtumor 58	0.26	0.15	1.81	0.07
tumor ap23p	0.04	0.14	0.27	0.79
tumor frac	17.78	16.79	1.06	0.29
tumor orcc24p	-0.08	0.09	-0.91	0.36
tumor sm31	0.19	0.12	1.62	0.11
tumor tr22p	-0.01	0.10	-0.14	0.89
tumor volume	0.25	1.03	0.25	0.80

12.12 Gradient Boosting with Standardized Linear Learners for General Regression Models

Using standardized linear learners, we found that gradient boosting in logistic and linear regression is quite similar. Both methods involve repeatedly refitting residuals using matching pursuit, solved by maximal correlation. This approach is not limited to these settings. As long as the empirical risk can be described by a linear predictor F, this method can be applied to any regression model. In this section, we summarize this procedure and provide practical details for running the algorithm.

The learning data is denoted by $\mathscr{L} = \{(\mathbf{x}_1, y_1), \ldots, (\mathbf{x}_n, y_n)\}$ where $\mathbf{x}_i$ are the features and y_i are the response values. The data is standardized as in (12.19). We assume that the empirical risk can be written as

$$R_n(F) = \frac{1}{n}\sum_{i=1}^{n} L(y_i, F(\mathbf{x}_i)),$$

where $F(\mathbf{x}) = \alpha + \sum_{k=1}^{p} \beta_k x_k$ is a linear predictor for a regression model. For example in logistic regression, $L(y, F(\mathbf{x})) = -yF(\mathbf{x}) + \log(1 + \exp(F(\mathbf{x})))$ where F is the linear predictor on the logit scale for the probability $\mathbb{P}\{Y = 1|\mathbf{X} = \mathbf{x}\}$.

We boost F by using a linear learner equal to coordinate k of the feature vector translated by the intercept term

$$F(\mathbf{x}) \leftarrow F(\mathbf{x}) + \alpha + \nu x_k, \tag{12.21}$$

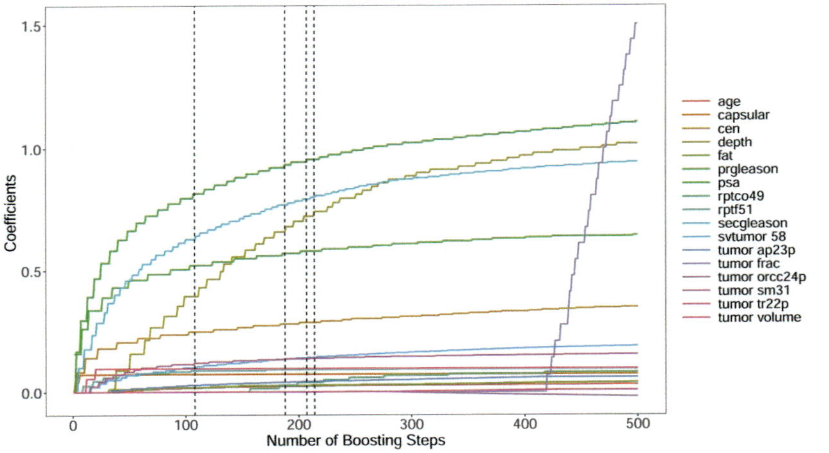

FIGURE 12.4
Coefficient solution path for prostate data using gradient boosting.

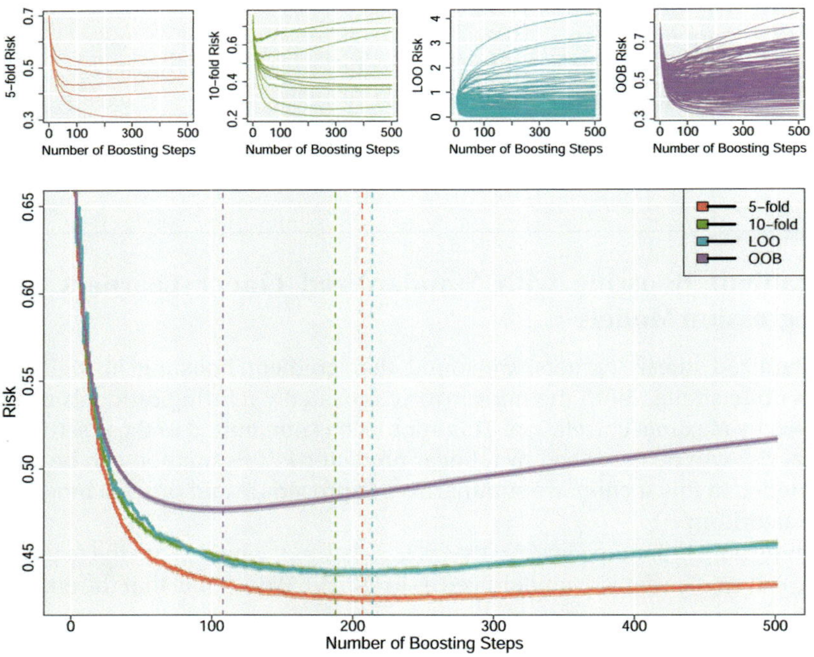

FIGURE 12.5
Tuning the number of boosting steps for the prostate data using K-fold cross-validation, LOO (leave-one-out CV) and OOB (out-of-bag) validation.

where by Theorem 12.2, because the data is standardized, the most favorable direction k equals the coordinate with maximal gradient-correlation

$$k = \arg\max_{1\le j\le p} |(\mathbf{g}, \mathbf{x}_{(j)})|, \quad \text{where } \mathbf{g} = -\nabla R_n(\mathbf{F}) := -\frac{\partial R_n(F)}{\partial (F(\mathbf{x}_1),\dots,F(\mathbf{x}_n))^T}.$$

As previously discussed, the ν-regularization in (12.21) is not applied to the intercept which functions as an ancillary parameter and therefore is not regularized.

The general procedure is summarized in Algorithm 12.8.

Algorithm 12.8 *Gradient Boosting with Standardized Linear Learners*

1: Intialize $F_0(\mathbf{x}_i) = \arg\min_{\rho\in\mathbb{R}} \sum_{i=1}^n L(y_i, \rho)$ for $i = 1,\dots,n$.
2: **for** $m = 1$ to M **do**
3: Determine the most favorable direction using maximal gradient-ocorrelation

$$k_m = \arg\max_{1\le k\le p} \left|\sum_{i=1}^n g_m(\mathbf{x}_i)\mathbf{x}_{ik}\right|, \quad \text{where } g_m(\mathbf{x}_i) = -L'(y_i, F_{m-1}(\mathbf{x}_i)).$$

4: Solve the two-dimensional line-search

$$(\hat\alpha_m, \hat\beta_m)^T = \arg\min_{(\alpha,\beta)\in\mathbb{R}^2} \sum_{i=1}^n L(y_i, F_{m-1}(\mathbf{x}_i) + \alpha + \beta x_{ik_m})$$

5: Update: $F_m(\mathbf{x}_i) = F_{m-1}(\mathbf{x}_i) + \hat\alpha_m + \nu\hat\beta_m x_{ik_m}$.
6: **end for**

12.12.1 Some practical details

1. The algorithm returns the estimator $F_M(\mathbf{x}_i)$ for $i = 1,\dots,n$ for the linear predictor $F(\mathbf{x}_i) = \alpha + \sum_{k=1}^p \beta_k x_{ik}$. Written in terms of estimated β coefficients, the boosted predictor is $F_M(\mathbf{x}_i) = \hat\alpha + \sum_{k=1}^p \hat\beta_k x_{ik}$ where

$$\hat\alpha = \sum_{m=1}^M \hat\alpha_m, \quad \hat\beta_k = \nu \sum_{m=1}^M \hat\beta_m 1_{\{k_m=k\}}.$$

   This assumes the data is standardized, so to convert the results to the original scale, the coefficients should be appropriately rescaled.

2. Tuning M can be accomplished by K-fold cross-validation. Let $\mathscr{L}_{(s)}$ denote the s test fold of size n_s for $s = 1,\dots,K$, $\cup_{s=1}^K \mathscr{L}_{(s)} = \mathscr{L}$, and let $F_{m,s}$ be the boosted predictor for step $m = 1,\dots,m$ constructed from the training fold $\mathscr{L} \setminus \mathscr{L}_{(s)}$. Then the cross-validated risk for $\mathscr{L}_{(s)}$ is

$$R_{m,s} = \frac{1}{K}\sum_{s=1}^K \left[\frac{1}{n_s}\sum_{i\in\mathscr{L}_s} L(y_i, F_{m,s}(\mathbf{x}_i))\right].$$

   Calculate $R_{m,s}$ for each fold and take the average, $\overline{R}_m = \sum_{s=1}^K R_{m,s}/K$. The optimized step size is determined by

$$\hat m = \arg\min_{1\le m\le M}\{\overline{R}_m\}.$$

3. The two-dimensional line-search (Line 4) can be efficiently determined by a one-step Newton approximation (Chapter 10). Define

$$f_m(\alpha,\beta) = \sum_{i=1}^{n} L(y_i, F_{m-1}(\mathbf{x}_i) + \alpha + \beta x_{ik_m}).$$

For notational convenience let $z_{im} = F_{m-1}(\mathbf{x}_i) + \alpha + \beta x_{ik_m}$. Then the gradient for f_m is

$$\nabla f_m(\alpha,\beta) = \mathbf{D}_m^T \mathbf{g}_{fm}$$

where

$$\mathbf{D}_m = \begin{pmatrix} 1 & x_{1k_m} \\ \vdots & \vdots \\ 1 & x_{nk_m} \end{pmatrix}, \quad \mathbf{g}_{fm} = \begin{pmatrix} L'(y_1, z_{1m}) \\ \vdots \\ L'(y_n, z_{nm}) \end{pmatrix}.$$

Assuming L is twice-differentiable, the Hessian $\mathbf{H}_m(\alpha,\beta) = \nabla^2 f_m(\alpha,\beta)$ is (Chapter 2.9)

$$\mathbf{H}_m(\alpha,\beta) = \mathbf{D}_m^T \mathbf{G}_{fm} \mathbf{D}_m, \quad \text{where } \mathbf{G}_{fm} = \begin{pmatrix} L''(y_1, z_{1m}) & & \\ & \ddots & \\ & & L''(y_n, z_{nm}) \end{pmatrix}.$$

The one-step approximation using a $(0,0)^T$ starting value is (10.3)

$$(\hat{\alpha}_m, \hat{\beta}_m)^T = -(\mathbf{D}_m^T \mathbf{G}_{fm} \mathbf{D}_m)^{-1} \mathbf{D}_m^T \mathbf{g}_{fm}.$$

For example, for logistic regression, we have by Chapter 10.5

$$\nabla f_m(\alpha,\beta) = \mathbf{D}_m^T(\mathbf{y} - \boldsymbol{\pi}_m), \quad \text{where } \boldsymbol{\pi}_m = \begin{pmatrix} (1+\exp(-z_{1m}))^{-1} \\ \vdots \\ (1+\exp(-z_{nm}))^{-1} \end{pmatrix}$$

and

$$\mathbf{H}_m(\alpha,\beta) = -\mathbf{D}_m^T \mathbf{W}_m \mathbf{D}_m, \quad \text{where } \mathbf{W}_m = \text{diag}\{\pi_{im}(1-\pi_{im})\}_{i=1}^n.$$

Therefore the one-step approximation is

$$(\hat{\alpha}_m, \hat{\beta}_m)^T = (\mathbf{D}_m^T \mathbf{W}_m \mathbf{D}_m)^{-1} \left[\mathbf{D}_m^T (\mathbf{y} - \boldsymbol{\pi}_m)\right]$$

which is the solution to a weighted least squares problem.

Exercises

Exercise 12.1. Choose an interesting real-valued function $f : \mathbb{R} \to \mathbb{R}$. Decide on a space of base learners $\mathscr{H}$ and apply the theoretical matching pursuit algorithm (Algorithm 12.1) to recover f using a finite number of steps m.

1. How accurate is the approximation? Display graphically the accuracy of the approximation as a function of the number of steps.
2. Provide a graphical representation of the values $\{(r_{n-1}, h_n)h_n\}_1^m$ in your decomposition $f_m = \sum_{n=1}^m (r_{n-1}, h_n)h_n$.

Exercise 12.2. The goal of this exercise is to use the matching pursuit algorithm to compress an image by representing it as a sparse combination of basis functions and then reconstructing the image from this sparse representation. The following steps should help complete this excercise:

1. Create a suitable set of basis functions for the sparse representation (e.g., Discrete Cosine Transform, DCT, can be used).
2. Implement the matching pursuit algorithm to approximate the image using a sparse combination of basis functions.
3. Reconstruct the image from the sparse representation and evaluate the compression performance.

Exercise 12.3. Let $f : \mathbb{R}^d \to \mathbb{R}$ be a real-valued multivariable function. Let $h_1^*(\mathbf{x}) = x_1, h_2^*(\mathbf{x}) = x_2, \dots, h_d^*(\mathbf{x}) = x_d$ be the coordinate functions and let $h_{d+1} = 1$ be the constant function. Let $h_1, h_2, \dots, h_{d+1}$ be the normalized functions (normalized to have length one) and let $\mathscr{D}$ be the finite-dimensional space spanned by $h_1, \dots, h_{d+1}$. Use the WGA to approximate f using $\mathscr{D}$.

1. Use the bound (12.10) to give an explicit expression for the accuracy WGA approximates f. (Hint: express f as two terms of the projection of f onto $\mathscr{D}$ and its orthogonal complement, $\mathscr{D}^\perp$).
2. Does this approximation go to zero as $m \to \infty$?
3. What conditions for f are needed for the approximation to go to zero?

Exercise 12.4. Compare the Adaboost algorithm to gradient boosting on several different data sets. Summarize your findings.

Exercise 12.5. Using the `ozone` dataset from the `mlbench` R-package:

1. Analyze the data using gradient boosting and lasso. Tune the parameters in each case using cross-validation. Describe your findings.
2. Fit the data using least-squares and compare the results.

Exercise 12.6. Here you will use gradient boosting to predict housing prices in California. Additionally, analyze the estimated coefficients from the parametric learners to understand the relationships between housing prices and various features.

1. Load the `California Housing` dataset, which contains information about houses in California, such as median income, median house age, average rooms, and location.
2. Handle missing values and scale the features.
3. Implement gradient boosting with parametric learners of your choice to predict housing prices.
4. Evaluate the predictive performance of your boosted model.
5. Analyze the estimated coefficients from the boosted model to understand the relationships between the features and housing prices.

Exercise 12.7. If we set $\nu = 1$ in Algorithm 12.6 for L_2Boost what happens? According to the algorithm, if k is the most favorable direction at step m then the updated predictor is

$$F_m(\mathbf{x}) = F_{m-1}(\mathbf{x}) + \rho_m x_k$$

where $\rho_m = \mathbf{X}_{(k)}^T \mathbf{g}_m$. In other words, the updated predictor equals the stagewise predictor from regressing coordinate k on to the residual, which is a very greedy operation. This suggests the algorithm will converge in $M = p$ steps after fitting all the variables. Is this true? Prove your answer.

Exercise 12.8. Write your own gradient boosting algorithm for logistic regression using parametric learners.

1. Run the algorithm on a data set of your choosing and compare the results to GLM.
2. Repeat your analysis of part 1 but where the boosting algorithm is modified so that the design matrix includes an intercept term which is treated just like the other covariates (dimension is $p+1$). Thus the line-search (12.20) is modified to

$$\hat{\beta}_m = \arg\min_{\beta \in \mathbb{R}} \sum_{i=1}^{n} \phi(y_i, F_{m-1}(\mathbf{x}_i) + \beta x_{ik_m})$$

and the mth update for F is modified to

$$F_m(\mathbf{x}_i) = F_{m-1}(\mathbf{x}_i) + \nu \hat{\beta}_m x_{ik_m}.$$

where $k_m \in \{1, \ldots, p+1\}$ and $k_m = 1$ occurs when the intercept term is selected as the most promising direction. Comment on your analysis and compare it to part 1.

Exercise 12.9. The goal of this exercise is to use gradient boosting with parametric learners to predict customer churn based on various customer features.

1. Find a dataset containing information about customers and whether they have churned. One suggestion is the `Telco Customer Churn` dataset.
2. Split the dataset into training and testing sets.
3. Implement gradient boosting using parametric learners to predict customer churn.
4. Train the model on the training set and evaluate its performance on the testing set using metrics such as accuracy, precision, recall, and F1-score.

Exercise 12.10. The objective here is to develop a gradient boosting model using parametric learners for multiclass classification. This will be applied to the problem of classifying handwritten digits.

1. We will use the `MNIST` dataset which consists of 28×28 pixel grayscale images of handwritten digits (0-9) where each image is additionally provided with the true class labeled digit.
2. Flatten the 28×28 images into a 784-dimensional vector. Normalize the pixel values to range between 0 and 1. Divide the dataset into training and testing sets.
3. We will use the softmax function for multiclass classification. The softmax function converts raw class scores into probabilities

$$\mathbb{P}\{Y_i = j | \mathbf{X} = \mathbf{x}_i\} = \frac{\exp(\mathbf{x}_i^T \boldsymbol{\beta}_j)}{\sum_{l=1}^{J} \exp(\mathbf{x}_i^T \beta_l)}, \qquad j = 1, \ldots, J.$$

Let $F = \{F_j\}_{j=1}^{J}$ be the target function where $F_j(\mathbf{x}) = \mathbb{P}\{Y = j | \mathbf{X} = \mathbf{x}\}$. Use the cross-entropy loss (Chapter 7) as the loss function which measures the difference between the true labels and the conditional j-probabilities,

$$L(Y, F) = -\sum_{j=1}^{j} 1_{\{Y=j\}} \log F_j.$$

Thus the empirical loss is

$$\text{Empirical loss} = -\frac{1}{n} \sum_{i=1}^{n} \sum_{j=1}^{j} 1_{\{y_i=j\}} \log \hat{F}_j(\mathbf{x}_i)$$

4. Train the gradient boosting algorithm using cross-validation to tune the hyperparameters.
5. Test the model and report the accuracy of your procedure using suitable metrics. Show some instances of misclassified instances to understand the limitations of the procedure and ways for potential improvements.

Bibliography

[1] Bühlmann P. Boosting for high-dimensional linear models. *Annals of Statistics*, 34:559–583, 2006.

[2] Efron B., Hastie T., Johnstone I., and Tibshirani R. Least angle regression (with discussion). *Annals of Statistics*, 32:407–499, 2004.

[3] Ehrlinger J. and Ishwaran H. Characterizing L_2Boosting. *Annals of Statistics*, 40, 1074-1101, 2012.

[4] Freund Y. and Schapire R.E. Experiments with a new boosting algorithm. *In Proceedings of the 13th International Conference on Machine Learning*, 148-156, 1996.

[5] Friedman J.H. Greedy function approximation: A gradient boosting machine. *Annals of Statistics*, 29:1189–1232, 2001.

[6] Friedman J.H, Hastie T. and Tibshirani R. Additive logistic regression: A statistical view of boosting (with discussion). *Annals of Statistics*, 28:337–407, 2000.

[7] Hastie T., Taylor J., Tibshirani R. and Walther G. Forward stagewise regression and the monontone lasso. *Electronic Journal of Statistics*, 1:1–29, 2007.

[8] Leisch F. and Dimitriadou E. mlbench: Machine Learning Benchmark Problems. R package version 2.1-3.1. `https://cran.r-project.org/package=mlbench`, 2023.

[9] Mallat S. and Zhang Z. Matching pursuits with time-frequency dictionaries. *IEEE Transactions on Signal Processing*, 41:3397–3415, 1993.

[10] Devore R.A. and Temlyakov V.N. Some remarks on greedy algorithms. *Advances in Computational Mathematics*, 5:173–187, 1996.

[11] Temlyakov V.N. Weak greedy algorithms. *Advances in Computational Mathematics*, 12:213–227, 2000.

13

Detailed Analysis of L_2Boost

13.1 Introduction

In this chapter, we examine the impact of the boosting parameter $0 < \nu \leq 1$ on the L_2Boost algorithm (also known as L_2Boosting with componentwise linear least squares) introduced in Chapter 12.8. We focus on the number of steps until favorability. During L_2Boost's descent along a fixed coordinate direction, a new coordinate becomes favorable when its basis is closest to the current gradient. But how many iterations does it take to reach favorability? We present a closed-form expression for this quantity and use it to describe L_2Boost's solution path and to characterize some of the algorithm's properties. One notable property we study is active set cycling, where the algorithm cycles between the same coordinates when ν is small. This also shows how correlation affects L_2Boost's variable selection ability. Much of this content is summarized from Ehrlinger and Ishwaran [1]. Interested readers should refer to this source for more details.

13.2 The Fixed Direction Case

For our analysis, we use the same notation as Chapter 12.8. We remind readers that the features are standardized to a norm of one and are centered to a mean of zero. The response is also centered. See (12.19).

We introduce the following additional notation for describing L_2Boost's solution path. Let $\{l_1, \ldots, l_{M^*}\}$ be the $M^* \leq M$ non-duplicated values in order of appearance of the selected coordinate directions $\mathscr{B}_M = \{k_1, \ldots, k_M\}$. We refer to these ordered, non-duplicated values as *critical directions* of the path. For example, if $\mathscr{B}_M = \{5,5,5,3,5,1,4,4,5\}$, the critical directions are the unique values $\{5,3,5,1,4,5\}$ and $M^* = 6$. To formally describe the solution path we introduce the following nomenclature.

Definition 13.1. The descent length along a critical direction l_r is denoted by L_r. The critical point S_r is the step number at which the descent along l_r ends. Thus, L_2Boost begins by descending along direction l_1 (the first critical direction) for L_1 steps, after which it switches to a descent along direction l_2 (the second critical direction) for L_2 steps. The last descent is along l_{M^*} (the final critical direction) for L_{M^*} steps. We refer to $(l_r, L_r, S_r)_{r=1}^{M^*}$ as the solution path of L_2Boost. See Figure 13.1 for a simple illustration. □

A key observation is that L_2Boost's behavior along a given descent is deterministic except for its descent length L_r (number of steps). If we could determine the descent length, then an exact description of the solution path becomes possible as L_2Boost is essentially a collection of fixed paths.

DOI: 10.1201/9781003493679-13

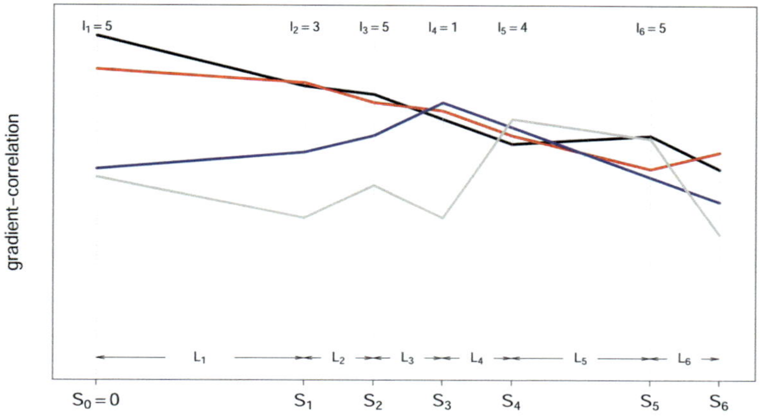

FIGURE 13.1
Solution path for L_2Boost where $\mathscr{B}_M = \{5,5,5,3,5,1,4,4,5\}$. The $M^* = 6$ critical directions are $(l_r)_1^6 = (5,3,5,1,4,5)$ with critical descent step lengths $(L_r)_1^6 = (3,1,1,1,2,1)$ and critical points $(S_r)_1^6 = (3,4,5,6,8,9)$.

Algorithm 13.1 *L_2Boost (Fixed Direction, k)*

$\mathbf{F}_{k,0} = \mathbf{F}_{m'}$
for $m = 1$ to M' **do**
  $\mathbf{F}_{k,m} = \mathbf{F}_{k,m-1} + \nu\rho_{k,m}\mathbf{X}_{(k)}$, where $\rho_{k,m} = \mathbf{X}_{(k)}^T(\mathbf{y} - \mathbf{F}_{k,m-1})$
end for

Imagine that we are at step m' of the algorithm and in the following step a new critical direction k is selected. We study the descent along k for the next $m = 1,\ldots,M'$ steps (see Algorithm 13.1). Thus, for the mth step of the descent along k the predictor is

$$\mathbf{F}_{k,m} = \mathbf{F}_{k,m-1} + \nu\rho_{k,m}\mathbf{X}_{(k)}, \quad \text{where} \quad \rho_{k,m} = \mathbf{X}_{(k)}^T(\mathbf{y} - \mathbf{F}_{k,m-1}).$$

The following result states a closed form solution for the m-step predictor.

Theorem 13.1. *$\boldsymbol{F}_{k,m} = \boldsymbol{F}_{k,0} + \nu_m\rho_{k,1}\boldsymbol{X}_{(k)}$, where $\nu_m = 1-(1-\nu)^m$ and $\rho_{k,1} = \boldsymbol{X}_{(k)}^T(\boldsymbol{y} - \boldsymbol{F}_{k,0})$.*

Theorem 13.1 shows that taking a single step with learning parameter ν_m yields the same limit as taking m steps with the smaller learning parameter ν. The result also sheds insight into how ν slows the descent relative to stagewise regression. Notice that the m-step predictor can be written as

$$\mathbf{F}_{k,m} = \underbrace{\mathbf{F}_{k,0} + \rho_{k,1}\mathbf{X}_{(k)}}_{\text{stagewise}} - \underbrace{(1-\nu)^m\rho_{k,1}\mathbf{X}_{(k)}}_{\text{slow-learning}}.$$

The first term on the right is the predictor from a greedy stagewise step, while the second term represents the effect of slow-learning. This latter term is what slows the descent relative to a greedy step. When $m \to \infty$ this term vanishes and we end up with stagewise fitting, $\nu = 1$.

13.3 Directional Change in the Descent

Theorem 13.1 shows we can take one large boosting step in place of many small steps, but it does not indicate how many steps are required before a new favorable direction arises. By knowing this the entire k-descent could be described by a single step.

To determine the descent length, suppose that a total of m steps are taken. At step $m+1$ the algorithm must decide whether to continue along k or to select a new direction j. To determine when to switch directions, we introduce the following definition.

Definition 13.2. A direction j is said to be more favorable than k at step $m+1$ if $|\rho_{k,m}| \geq |\rho_{j,m}|$ and $|\rho_{k,m+1}| < |\rho_{j,m+1}|$. Thus, if j is more favorable at $m+1$, the descent switches to j for step $m+1$. □

To determine when j becomes more favorable, it will be useful to have a closed form expression for $\rho_{k,m+1}$ and $\rho_{j,m+1}$. By Theorem 13.1,

$$\begin{aligned} \rho_{j,m+1} &= \mathbf{X}_{(j)}^T(\mathbf{y}-\mathbf{F}_{k,m}) \\ &= \mathbf{X}_{(j)}^T\left[(\mathbf{y}-\mathbf{F}_{k,0}) - \nu_m \rho_{k,1}\mathbf{X}_{(k)}\right] \\ &= \rho_{j,1} - \nu_m \rho_{k,1} R_{j,k}, \end{aligned}$$

where $R_{j,k} = \mathbf{X}_{(j)}^T\mathbf{X}_{(k)}$. Setting $j = k$ yields $\rho_{k,m+1} = (1-\nu)^m \rho_{k,1}$. Therefore, $|\rho_{k,m+1}| < |\rho_{j,m+1}|$ if and only if

$$(1-\nu)^{2m}\rho_{k,1}^2 < \left(\rho_{j,1} - \nu_m \rho_{k,1} R_{j,k}\right)^2.$$

Dividing throughout by $\rho_{k,1}$, with a little bit of rearrangement, this becomes

$$(1-\nu)^{2m} < \left[(1-\nu)^m R_{j,k} + \left(d_{j,k} - R_{j,k}\right)\right]^2,$$

where $d_{j,k} = \rho_{j,1}/\rho_{k,1}$. Notice importantly that $|d_{j,k}| \leq 1$ because k is the direction with maximal gradient-correlation at the start of the descent. It is also useful to keep in mind that $R_{j,k}$ is the sample correlation of $\mathbf{X}_{(j)}$ and $\mathbf{X}_{(k)}$ due to the unit norm-constraints imposed on the columns of $\mathbf{X}$; thus, $|R_{j,k}| \leq 1$. The following result states the number of steps taken along k before j becomes more favorable.

Theorem 13.2. *The number of steps $m_{j,k}$ taken along k so that j becomes more favorable than k at $m_{j,k}+1$ is the largest integer m such that*

$$(1-\nu)^{m-1} \geq \frac{|d_{j,k} - R_{j,k}|}{1 - R_{j,k}\text{sign}(d_{j,k} - R_{j,k})}. \tag{13.1}$$

It follows that for $0 < \nu < 1$

$$m_{j,k} = \text{floor}\left[1 + \frac{\log|d_{j,k} - R_{j,k}| - \log\left(1 - R_{j,k}\text{sign}(d_{j,k} - R_{j,k})\right)}{\log(1-\nu)}\right], \tag{13.2}$$

where floor(z) *is the largest integer less than or equal to z.*

Remark 13.1. In particular, notice that $m_{j,k} = \infty$ when $d_{j,k} = R_{j,k}$ (adopting the standard convention that sign$(0) = 0$ and assuming that $\nu < 1$). We call $d_{j,k} = R_{j,k}$ the *repressible condition.* Repressibility plays a key role in L_2Boost's behavior in correlated settings as shown shortly. □

Remark 13.2. We also obtain $m_{j,k} = 1$ when $\nu = 1$ from (13.1) which corresponds to greedy stagewise fitting. Because this makes the $\nu = 1$ case uninteresting, we hereafter assume that $0 < \nu < 1$. □

13.4 Defining the Solution Path

Theorem 13.2 shows that the problem of determining the next variable to enter the solution path can be recast as finding the direction requiring the fewest number of steps, $m_{j,k}$, to favorability. We show that when combined with Theorem 13.1, this characterizes L_2Boost's solution path.

As before, assume that k corresponds to the first critical direction of the path; i.e., $l_1 = k$. By Theorem 13.2, L_2Boost descends along k for a total of $S_1 = L_1$ steps, where $L_1 = m_{l_2,k}$ and l_2 is the coordinate requiring the smallest number of steps to become more favorable than k. By Theorem 13.1, the predictor at step S_1 is

$$\mathbf{F}_{S_1} = \mathbf{F}_0 + \nu L_1 \rho_{l_1}^{(1)} \mathbf{X}_{(l_1)}, \quad \text{where } \rho_{l_1}^{(1)} = \mathbf{X}_{(l_1)}^T (\mathbf{y} - \mathbf{F}_0).$$

Applying Theorem 13.1 once again, but now using a descent along l_2 initialized at $\mathbf{F}_{S_1}$, and continuing this argument recursively, as well as using the representation for the number of steps from Theorem 13.2, yields the following recursive description of L_2Boost's solution path.

Theorem 13.3. $\boldsymbol{F}_{S_r} = \boldsymbol{F}_{S_{r-1}} + \nu L_r \rho_{l_r}^{(r)} \boldsymbol{X}_{(l_r)}$, *where* $\{(l_i, L_i, S_i, \rho_{l_i}^{(i)})\}_1^{M^*}$ *are determined recursively from*

$$l_1 = \arg\max_{1 \le j \le p} |\boldsymbol{X}_{(j)}^T (\boldsymbol{y} - \boldsymbol{F}_0)|, \quad l_{r+1} = \arg\max_{j \ne l_r} |\rho_j^{(r+1)}|,$$

$$M_j^{(r)} = \text{floor}\left[1 + \frac{\log|D_j^{(r)} - R_{j,l_r}| - \log\left(1 - R_{j,l_r}\,\text{sign}(D_j^{(r)} - R_{j,l_r})\right)}{\log(1-\nu)}\right],$$

$$L_r = M_{l_{r+1}}^{(r)}, \quad S_r = S_{r-1} + L_r, \quad S_0 = 0,$$

$$D_j^{(r)} = \frac{\rho_j^{(r)}}{\rho_{l_r}^{(r)}}$$

$$\rho_j^{(r+1)} = \boldsymbol{X}_{(j)}^T (\boldsymbol{y} - \boldsymbol{F}_{S_r}) = \rho_j^{(r)} - \nu L_r \rho_{l_r}^{(r)} R_{j,l_r}.$$

Remark 13.3. A technical issue arises in Theorem 13.3 when $M_j^{(r)}$ is not unique. Non-uniqueness can occur due to rounding which is caused by the floor function used in the definition of $m_{j,k}$. This is why line 1 selects the next critical value, l_{r+1}, by maximizing the absolute gradient-correlation $|\rho_j^{(r+1)}|$ and not by minimizing the step number $M_j^{(r)}$. This definition for l_{r+1} is equivalent to the two-step solution: $l_{r+1} \leftarrow \arg\max_{j \in l_{r+1}} |\rho_j^{(r+1)}|$, where $l_{r+1} = \arg\min_{j \ne l_r} \{M_j^{(r)}\}$. □

13.5 Cycling Behavior

It has been observed that decreasing the regularization parameter slows the convergence of stagewise descent algorithms. We show that L_2Boost exhibits this phenomenon, building on our discussion from Chapter 12.10. We do so by describing this property as an active set cycling phenomenon. Using results from the earlier fixed descent analysis, we show in the case of an active set of two variables that L_2Boost systematically switches (cycles) between its two variables when v is small. For an arbitrarily small v this forces the absolute gradient-correlations for the active set variables to be nearly equal. This point of equality represents a singularity point that triggers a near-perpetual deterministic cycle between the variables, ending only when a new variable enters the active set with nearly the same absolute gradient-correlation.

Our insight will come from looking at Theorem 13.2 in more depth. As before, assume the algorithm has been initialized so that k is the first critical step. Previously the descent along k was described in terms of steps, but this can be equivalently expressed in units of the "step size" taken. Define

$$v_{j,k} = v_{m_{j,k}} = 1-(1-v)^{m_{j,k}}.$$

Recall that Theorem 13.1 showed that a single step along k with v replaced with $v_{j,k}$ yields the same limit as $m_{j,k}$ steps using v. We call $v_{j,k}$ the step size taken along k. Because j becomes more favorable than k at $m_{j,k}+1$, the gradient following a step size of $v_{j,k}$ along k satisfies

$$\left|\mathbf{X}_{(j)}^T\left(\mathbf{y}-\mathbf{F}_0-v_{j,k}\rho_{k,1}\mathbf{X}_{(k)}\right)\right| < \left|\mathbf{X}_{(k)}^T\left(\mathbf{y}-\mathbf{F}_0-v_{j,k}\rho_{k,1}\mathbf{X}_{(k)}\right)\right|. \tag{13.3}$$

This applies to all coordinates $j \neq k$, and in particular holds for the second critical direction, l_2, which rephrased in terms of step size, is the smallest $v_{j,k}$ value:

$$l_2 = \arg\min_{j\neq k}\{v_{j,k}\}.$$

Although the inequality (13.3) is strict, it becomes arbitrarily close to equality by shrinking v. With a little bit of rearranging, (13.1) implies that

$$\hat{v}_j < v_{j,k}, \quad \text{where } \hat{v}_j = 1-\frac{|d_{j,k}-R_{j,k}|}{1-R_{j,k}\text{sign}(d_{j,k}-R_{j,k})}.$$

We will show $\hat{v}_j$ is the step size making the absolute gradient-correlation between j and k equal:

$$\left|\mathbf{X}_{(j)}^T\left(\mathbf{y}-\mathbf{F}_0-\hat{v}_j\rho_{k,1}\mathbf{X}_{(k)}\right)\right| = \left|\mathbf{X}_{(k)}^T\left(\mathbf{y}-\mathbf{F}_0-\hat{v}_j\rho_{k,1}\mathbf{X}_{(k)}\right)\right|. \tag{13.4}$$

The next Theorem shows that $v_{l_2,k}$ converges to the smallest $\hat{v}_j$ satisfying (13.4); thus, (13.3) becomes an equality in the limit. For convenience, we define $v_{j,k}^- = v_{m_{j,k}-1}$.

Theorem 13.4. *Let* $\hat{\rho}_j = \mathbf{X}_{(j)}^T(\mathbf{y}-\mathbf{F}_0-\hat{v}_j\rho_{k,1}\mathbf{X}_{(k)})$. *Then* $|\hat{\rho}_j| = |\hat{\rho}_k|$. *Furthermore, if* $l^* = \arg\min_{j\neq k}\{\hat{v}_j\}$ *and* $\hat{v} = \hat{v}_{l^*}$, *then* $v_{l_2,k}^- \leq \hat{v} < v_{l_2,k}$ *and* $v_{l_2,k} \to \hat{v}$ *as* $v \to 0$.

Therefore, for arbitrarily small v, $v_{l_2,k} \asymp \hat{v}$ and k and l_2 will have near-equal absolute gradient-correlations. This latter property triggers two-cycling. To see why, let us assume for the moment that the active set variables have equal absolute gradient-correlations. Then by a direct application of Theorem 13.2, one can show that the number of steps taken along l_2 before k becomes more favorable is $m=1$. Thus, following the descent along k, the algorithm switches to l_2, but then immediately switches back to k. If v is small enough, this process is repeated, setting off a two-cycling pattern.

The next result is a formal statement of these arguments. Define

$$d_{j,k}^{(m)} = \frac{\rho_{j,m}}{\rho_{k,m}}, \qquad \text{where } \rho_{l,m} = \mathbf{X}_{(l)}^T(\mathbf{y}-\mathbf{F}_{m-1}).$$

For notational convenience, set $j = l_2$ and $m = m_{j,k}$. For technical reasons, we shall assume $d_{j,k}^{(m)} \neq R_{j,k}$. Recall Remark 13.1 showed that $d_{j,k}^{(m)} = R_{j,k}$, the repressible condition, yields an infinite number of steps to favorability. Thus, for k to even be eligible for favorability we must have $d_{j,k}^{(m)} \neq R_{j,k}$.

Theorem 13.5. *If the first two critical directions are* (k, j) *and* $v_{j,k} = \hat{v}_j$, *then* k *is favored over* j *for the next step after* j *if* $d^{(m)}_{j,k} \neq R_{j,k}$.

Theorem 13.5 assumes that $v_{j,k} = \hat{v}_j$. While this only holds in the limit, for arbitrarily small v the two values should be nearly equal, and thus the assumption is reasonable. Notice also that Theorem 13.5 only shows that k is more favorable than j, and not that the algorithm switches to k. However, we can see that this must be the case. For arbitrarily small v, k's gradient-correlation should be near to j's and by definition j has maximal absolute gradient-correlation along the second descent.

Indeed, the following result shows that the absolute gradient-correlations for k and j can be made arbitrarily close for small enough v for any step $r \geq 1$ following the descent along k. The result also shows that the sign of the gradient-correlation is preserved when v is arbitrarily small.

Theorem 13.6. $\rho_{j,m+r}/\rho_{k,m+r} \to \text{sign}(\hat{\rho}_j)/\text{sign}(\hat{\rho}_k)$ *as* $v \to 0$ *for each* $r \geq 1$.

Combining Theorems 13.5 and 13.6, we see that if v is small enough, the first three critical directions of the path must be (k, j, k) with critical points $(m, m+1, m+2)$. And once the descent switches back to k, it is clear from the same argument that the next critical direction, l_4, will be j, and so forth.

13.6 Illustration of Two-Cycling

We present a numerical example demonstrating two-cycling. For our example we simulated data according to a linear regression model with $n = 100$, and $p = 40$. The first 10 coordinates of $\boldsymbol{\beta}$ were set to 5, with the remaining coordinates set to 0. The design matrix $\mathbf{X}$ was simulated by drawing its entries independently from a standard normal distribution. Measurement error was simulated from a standard normal.

Figure 13.2 plots the standardized gradient-correlations in terms of the number of steps. We define the standardized gradient-correlation by

$$\rho^*_{k,m} = \frac{\mathbf{X}^T_{(k)}(\mathbf{y} - \mathbf{F}_{m-1})}{||(\mathbf{y} - \mathbf{F}_{m-1})||}.$$

The plots show the behavior of each coordinate within an active set descent. The rug marks show each step m for clarity, and the dashed vertical lines indicate the step $m_{j,k}$ where the next step adds a new critical direction to the solution path. The top left panel shows the complete descent along the first four active variables. The remaining panels detail the coordinate behavior as the active set increases from one to four coordinates.

The top right panel shows a repeated selection of the l_1 direction shown in black. The last step along l_1 occurs at $m_{j,k}$ marked with the vertical dashed line, where the next step is along the l_2 direction shown in red. This point marks the beginning of the two-cycling behavior, which continues in the lower left panel. At each step, the algorithm systematically switches direction between the l_1 and l_2 directions, until an additional direction becomes more favorable. The cycling pattern is $\{l_1, l_2, l_1, l_2, \ldots\}$. The lower right panel demonstrates three-cycling behavior ending with the fourth coordinate entering the active set. It is instructive to note that the order of selection within three-cycling is non-deterministic. Non-deterministic cycling patterns are typical behavior of higher-order cycling.

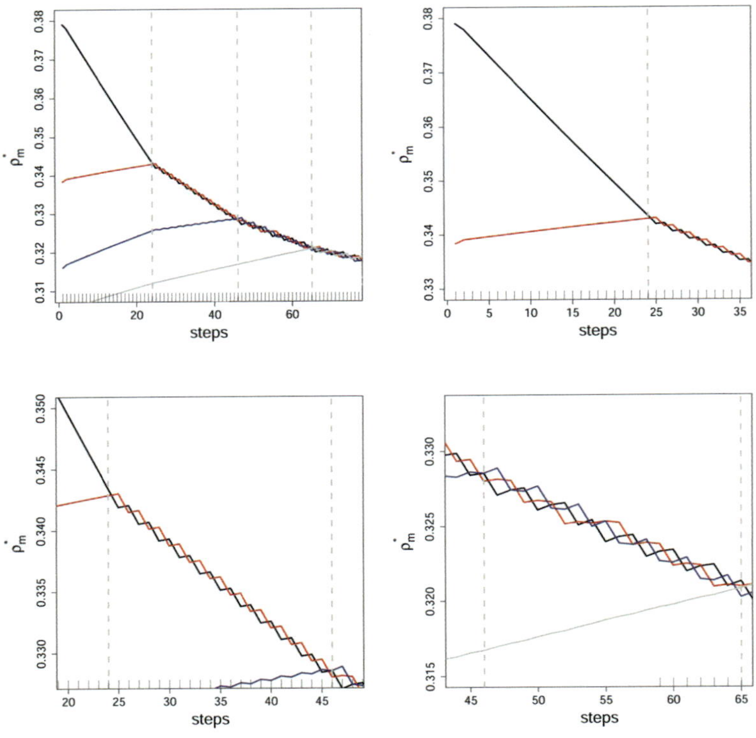

FIGURE 13.2
L_2Boost standardized gradient-correlation path for ν = 0.005. Top left panel details the path through the first three active variables, the remaining panels detail each active variable descent.

13.7 Repressibility Affects Variable Selection in Correlated Settings

Now we turn our attention to the issue of correlation. We have shown that a new direction j becomes more favorable than the current direction k at step $m_{j,k}+1$ where $m_{j,k}$ is the smallest integer value satisfying

$$1 - \frac{|d_{j,k} - R_{j,k}|}{1 - R_{j,k}\text{sign}(d_{j,k} - R_{j,k})} < 1 - (1-\nu)^{m_{j,k}}. \tag{13.5}$$

Using our previous notation, let $\hat{\nu}_j$ and $\nu_{j,k}$ denote the left and right-hand sides of the above inequality, respectively. Importantly, this relationship holds regardless of the size of the active set.

It is clear that large values of $m_{j,k}$ are designed to hinder non-informative variables from entering the solution path. If j requires a large number of steps to become favorable, it is non-informative relative to the current gradient and therefore unattractive as a candidate. Surprisingly, however, such an interpretation does not always apply in correlated problems. There are situations where j is informative, but $m_{j,k}$ can be artificially large due to correlation.

To see why, suppose that j is informative with a relatively large value of $d_{j,k}$. Now, if j and k are correlated, so much so that $R_{j,k} \asymp d_{j,k}$, then $|d_{j,k} - R_{j,k}| \asymp 0$. Hence, $m_{j,k} \asymp \infty$ and $\nu_{j,k} \asymp 1$. The infinitely large number of steps (equivalently, a large step size of 1) precludes j from being selected. Thus, even though j is promising with a relatively large gradient-correlation, it may not be selected because of its large correlation with k.

The problem is that j becomes an unlikely candidate for selection when $d_{j,k}$ is close to $R_{j,k}$. In fact, (13.5) shows that $m_{j,k} = \infty$ when $d_{j,k} = R_{j,k}$ so that j can never become more favorable than k when the two values are equal. We refer to the condition $d_{j,k} = R_{j,k}$ as the *repressible condition.* Repressibility plays an important role in correlated settings. We distinguish between two types of repressibility: weak and strong repressibility. Weak repressibility occurs in the trivial case when $|R_{j,k}| = 1$. Weak repressibility implies that $|d_{j,k}| = |R_{j,k}| = 1$. Hence the gradient-correlation for j and k are equal in absolute value and j and k are perfectly correlated. This trivial case simply reflects a numerical issue arising from the redundancy of the j and k columns of the $\mathbf{X}$ design matrix. The stronger notion of repressibility, which we refer to as strong repressibility, is required to address the non-trivial case $|R_{j,k}| \neq 1$ in which j is repressed without being perfectly correlated with k. The following definition summarizes these ideas.

Definition 13.3. We say j has the strong repressible condition if $d_{j,k} = R_{j,k}$ and $|R_{j,k}| < 1$. We say that j is (strongly) repressed by k when this happens. On the other hand, j has the weakly repressible condition if $d_{j,k} = R_{j,k}$ and $|R_{j,k}| = 1$. □

13.8 Elastic Net Boosting

The tendency of the lasso to select only a handful of variables from among a group of correlated variables was noted in Zou and Hastie [2]. To address this, Zou and Hastie [2] described a different penalization than lasso. Rather than relying only on L_1-penalization, they included an additional L_2-regularization parameter designed to encourage a ridge-type grouping effect, and termed the resulting estimator *the elastic net.* Specifically, for a fixed $\lambda > 0$ (the ridge parameter) and a fixed $\lambda_0 > 0$ (the lasso parameter), the elastic net was defined as

$$\hat{\boldsymbol{\beta}}_{\text{enet}} = (1+\lambda) \operatorname*{argmin}_{\boldsymbol{\beta}\in\mathbb{R}^p} \left\{ ||\mathbf{y} - \mathbf{D}\boldsymbol{\beta}||^2 + \lambda \sum_{k=1}^{p} \beta_k^2 + \lambda_0 \sum_{k=1}^{p} |\beta_k| \right\}. \tag{13.6}$$

To calculate the elastic net, (13.6) can be recast as a lasso optimization problem by replacing the original data with augmented values. Replace $\mathbf{y}$ $(n \times 1)$ and $\mathbf{D}$ $(n \times p)$ with augmented values $\mathbf{y}^*$ and $\mathbf{D}^*$ defined as follows:

$$\mathbf{y}^* = \begin{bmatrix} \mathbf{y} \\ 0 \\ \vdots \\ 0 \end{bmatrix}_{(n+p)\times 1}, \quad \mathbf{D}^* = \frac{1}{\sqrt{1+\lambda}} \begin{bmatrix} \mathbf{D} \\ \sqrt{\lambda}\mathbf{I} \end{bmatrix}_{(n+p)\times p} := \left[\mathbf{X}^*_{(1)}, \ldots, \mathbf{X}^*_{(p)}\right]. \tag{13.7}$$

Reparameterize $\boldsymbol{\beta}$ as $\boldsymbol{\beta}^* = \boldsymbol{\beta}\sqrt{1+\lambda}$. It follows that (13.6) can be expressed as

$$\hat{\boldsymbol{\beta}}_{\text{enet}} = \sqrt{1+\lambda} \operatorname*{argmin}_{\boldsymbol{\beta}\in\mathbb{R}^p} \left\{ ||\mathbf{y}^* - \mathbf{D}^*\boldsymbol{\beta}||^2 + \frac{\lambda_0}{\sqrt{1+\lambda}} \sum_{k=1}^{p} |\beta_k| \right\},$$

which is an L_1-optimization problem that can be solved using the lasso.

One explanation for why the elastic net is so successful in correlated problems is due to its decorrelation property. Let $R^*_{j,k} = \mathbf{X}^{*T}_{(j)}\mathbf{X}^*_{(k)}$. Because the data is standardized such that $\mathbf{X}^T_{(j)}\mathbf{X}_{(j)} = \mathbf{X}^T_{(k)}\mathbf{X}_{(k)} = 1$, we have

$$R^*_{j,k} = \begin{cases} \dfrac{\mathbf{X}^T_{(j)}\mathbf{X}_{(k)}}{1+\lambda} = \dfrac{R_{j,k}}{1+\lambda} & \text{if } j \neq k \\[2ex] \dfrac{\mathbf{X}^T_{(j)}\mathbf{X}_{(j)}+\lambda}{1+\lambda} = 1 & \text{if } j = k. \end{cases}$$

One can see that λ is a decorrelation parameter, with larger values reducing the correlation between coordinates. Zou and Hastie [2] stated this effect promotes a *grouping property* for the elastic net that overcomes the lasso's inability to select groups of correlated variables.

Decorrelation is an important ingredient for elastic net's success. However, we will argue that in addition to its role in decorrelation, λ has a surprising connection to repressibility that further explains λ's role in regularizing the elastic net.

13.8.1 How λ regularizes the solution path

To address repressibility we introduce a generalized L_2Boost procedure termed elasticBoost. The elasticBoost algorithm is a modification of L_2Boost applied to the augmented problem: see Algorithm 13.2. For arbitrarily small ν, the solution path for elasticBoost approximates the elastic net (similar to how L_2Boost approximates the lasso). We study elasticBoost to present an explanation of how λ imposes L_2-regularization, thus giving further insight into why elastic net works.

Algorithm 13.2 *elasticBoost*

1: Augment the data (13.7). Set $F^*_{0,i} = 0$ for $i = 1, \ldots, n+p$.
2: Run L_2Boost for M iterations using the augmented data.
3: Let $F^*_{M,i}$ denote the M-step predictor (discard $F^*_{M,i}$ for $i > n$). Let $\beta^*_{M,k}$ denote the M-step coefficient estimate for β_k.
4: Rescale the regression estimates: $\beta_{M,k} = \sqrt{1+\lambda}\,\beta^*_{M,k}$.

We consider how λ affects $m^*_{j,k}$, the number of steps to favorability (defined as in (13.2) but with $\mathbf{y}$ and $\mathbf{X}$ replaced by their augmented values $\mathbf{y}^*$ and $\mathbf{X}^*$). At initialization, the gradient-correlation is

$$\rho^*_{j,1} = \mathbf{X}^{*T}_{(j)}(\mathbf{y}^* - \hat{\mathbf{F}}^*_0) = \frac{1}{\sqrt{1+\lambda}}\mathbf{X}^T_{(j)}\mathbf{y} - \frac{1}{\sqrt{1+\lambda}}\left(\sum_{i=1}^n x_{i,j}F^*_{0,i} + \sqrt{\lambda}F^*_{0,n+j}\right).$$

We consider the special case when $F^*_{0,i} = 0$, corresponding to the first descent of the algorithm. Then

$$\rho^*_{j,1} = \frac{1}{\sqrt{1+\lambda}}\mathbf{X}^T_{(j)}\mathbf{y} = \frac{1}{\sqrt{1+\lambda}}\rho_{j,1}.$$

Therefore, $d^*_{j,k} = \rho_{j,1}/\rho_{k,1} = d_{j,k}$, and hence

$$m^*_{j,k} = \text{floor}\left[1 + \frac{\log|d_{j,k} - R^*_{j,k}| - \log\left(1 - R^*_{j,k}\text{sign}(d_{j,k} - R^*_{j,k})\right)}{\log(1-\nu)}\right].$$

This equals the number of steps in the original (non-augmented) problem but where $\mathbf{X}$ is replaced with variables decorrelated by a factor of $\sqrt{1+\lambda}$. For large λ this addresses the problem discussed earlier. Recall that $m_{j,k}$ can become inflated due to the near equality of $d_{j,k}$ with $R_{j,k}$. However, $R^*_{j,k} = R_{j,k}/\sqrt{1+\lambda}$ shrinks to zero with increasing λ, which stops $m^*_{j,k}$ from becoming inflated.

This provides one explanation for λ's role in regularization, at least for the case when λ is large. But we now suggest another theory that applies to both small and large λ. We argue that regularization is imposed not just by decorrelation, but through a combination of decorrelation and reversal of repressibility.

To show this, let us suppose that near-repressibility holds. We assume therefore that $R_{j,k} = d_{j,k}(1+\delta)$ for some small $|\delta| < 1$, where δ can be either positive or negative. Then,

$$\begin{aligned}
&\log|d_{j,k} - R^*_{j,k}| - \log\Big(1 - R^*_{j,k}\mathrm{sign}(d_{j,k} - R^*_{j,k})\Big) \\
&= \underbrace{\left[\log|d_{j,k}| + \log\left|1 - \frac{1+\delta}{\sqrt{1+\lambda}}\right|\right]}_{\text{Repressibility Effect}} \\
&\quad \underbrace{- \log\left(1 - \frac{R_{j,k}}{\sqrt{1+\lambda}}\mathrm{sign}\left(R_{j,k}\left[\frac{1}{1+\delta} - \frac{1}{\sqrt{1+\lambda}}\right]\right)\right)}_{\text{Decorrelation Effect}}.
\end{aligned} \tag{13.8}$$

The first term on the right captures the effect of repressibility. When δ is small, λ plays a crucial role in controlling its size. If $\lambda = 0$, the expression reduces to $\log|d_{j,k}| + \log|\delta|$ which converges to $-\infty$ as $|\delta| \to 0$; thus precluding j from being selected (keep in mind that (13.8) is divided by $\log(1-\nu)$, which is negative; thus $m^*_{j,k} \to \infty$). On the other hand, any $\lambda > 0$, even a relatively small value, ensures that the expression remains small even for arbitrarily small δ, thus reversing the effect of repressibility.

The second term on the right of (13.8) is related to decorrelation. Let us look at how this term varies with λ. If $1+\lambda > (1+\delta)^2$ (which holds if λ is large enough when $\delta > 0$, or for all $\lambda > 0$ if $\delta < 0$), the term reduces to

$$-\log\left(1 - \frac{R_{j,k}}{\sqrt{1+\lambda}}\mathrm{sign}\left(R_{j,k}\right)\right),$$

which remains bounded when $\lambda > 0$ if $R_{j,k} \to 1$. In contrast, if $1+\lambda < (1+\delta)^2$, the term reduces to

$$-\log\left(1 + \frac{R_{j,k}}{\sqrt{1+\lambda}}\mathrm{sign}\left(R_{j,k}\right)\right),$$

which remains bounded if $R_{j,k} \to 1$ and shrinks in absolute size as λ increases.

Taken together, these arguments show λ imposes L_2-regularization through a combination of decorrelation and reversal of repressibility which applies even when λ is relatively small.

These arguments apply to the first descent. For the general case, readers should consult [1].

13.9 Comparing elasticBoost to L_2Boost in the Presence of Correlation

To compare elasticBoost to L_2Boost we simulated data from a linear model with dimension $n = 1000$, $p = 40$. The first 15 coordinates of β were set to 3; all other coordinates were equal to 0. The features were simulated so that coordinates were correlated in 3 blocks of size 5 for the first 15 coordinates. The remaining 25 coordinates were uncorrelated. The measurement error was sampled independently from a normal with a standard deviation $\sigma = 15$ (see Ehrlinger and Ishwaran [1] for details).

To approximate lasso a small regularization parameter $\nu = .05$ was used. The total allowed steps for both procedures was $M = 500$ where tuning for M used 10-fold cross-validation. For elasticBoost since it has an additional regularization parameter λ, a double cross-validation loop was used to tune (λ, M). Using the optimally tuned M (and λ) we obtained the coefficient estimates for both procedures. The experiment was run 1000 times.

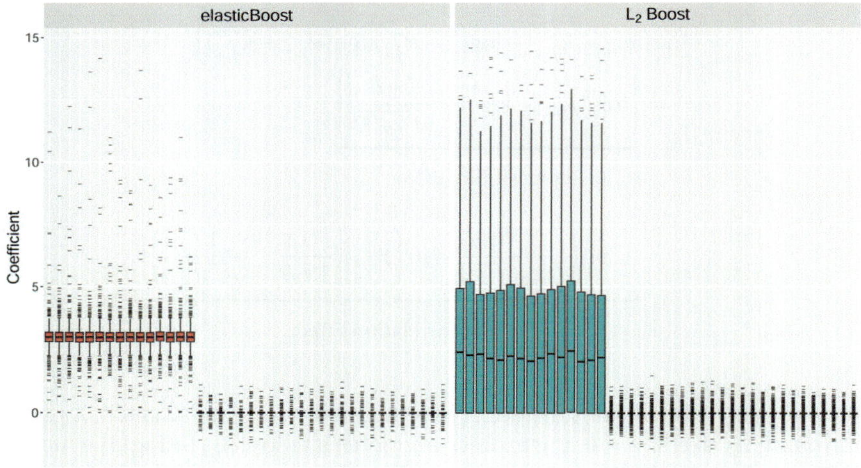

FIGURE 13.3
Coefficient estimates from 1000 runs of elasticBoost and L_2Boost using correlated feature simulation. The first 15 coefficients have a true value of 3 which is correctly identified by elasticBoost. However, due to the blocked correlation structure of the features, L_2Boost does not obtain the correct signal value of 3 and spreads its values over a wide range from 0 to 15 over the first 15 signal coordinates.

The coefficient estimates are displayed in Figure 13.1. For elasticBoost, the estimates are correctly centered at the true value of 3 for the signal variables (first 15 features) and 0 for the remaining noise variables (last 25 coordinates). However, for L_2Boost, due to the blocked correlation structure of the features, coefficient estimates for the signal variables are distributed across a range of values from 0 to 15. This is because lasso (which is being approximated here by L_2Boost) will select only a handful of variables from within a correlated block and spread the true signal value over those variables. For example, if lasso selects three variables from the first block it will assign the value of 5 to the 3 variables. This property of the lasso has been noted in Zou and Hastie [2].

Exercises

Exercise 13.1. Investigate how different values of the boosting parameter $0 < \nu < 1$ influence the behavior of L_2Boost.

1. Generate a synthetic dataset with $n = 1000$ samples and $p = 20$ features. Ensure that the features have varying degrees of correlation.

2. Run L_2Boost with different values of v keeping the number of steps fixed (e.g., 500 iterations).
3. Track and record which coordinates are selected at each iteration. Create a plot showing the selection frequency of each coordinate over the boosting iterations for different values of v.

Exercise 13.2. Apply L_2Boost to a real dataset. Display the standardized gradient-correlations as in Figure 13.2 for the first 100–200 steps.

Exercise 13.3. Write an R-program to implement elasticBoost. Simulate a regression model to compare elasticBoost and L_2Boost for variable selection. Vary n, p and feature correlation in your simulation. Summarize your findings.

Exercise 13.4. Explain how to extend L_2Boost to bivariate regression where the outcome is two dimensional, $\mathbf{Y} = (Y_1, Y_2)^T$. Write an R-program to implement the bivariate L_2Boost procedure. Explain why active cycling will still be present and demonstrate this using real or synthetic data by varying the regularization parameter $0 < v < 1$ going from large to small values.

Bibliography

[1] Ehrlinger J. and Ishwaran H. Characterizing L_2Boosting. *Annals of Statistics*, 40, 1074-1101, 2012.

[2] Zou H. and Hastie T. Regularization and variable selection via the elastic net. *Journal of the Royal Statistical Society Series B*, 67(2):301–320, 2005.

14

Coordinate Descent

14.1 Introduction

We have explored several algorithms for minimizing a real-valued function $f(\mathbf{z})$ over $\mathbf{z} \in \mathbb{R}^p$. However, in some situations, we need to minimize f with constraints on $\mathbf{z}$. One important example is the L_1-penalization for least squares, known as lasso, which we discussed earlier and will be the focus of this chapter. Here, we consider a general method called coordinate descent (also known as block coordinate descent). This method is a powerful technique for minimizing functions under constraints.

We consider the setting where the goal is to minimize

$$f(\mathbf{z}_1,\dots,\mathbf{z}_K) = f_0(\mathbf{z}_1,\dots,\mathbf{z}_K) + \sum_{k=1}^{K} f_k(\mathbf{z}_k), \tag{14.1}$$

where $\mathbf{z}_k \in \mathbb{R}^{p_k}$ are block-coordinates of $\mathbf{z}$ and $f_k : \mathbb{R}^{p_k} \to \mathbb{R}$ are real-valued functions representing the restriction functions. The dimension of $\mathbf{z}$ is p where $p = p_1 + \cdots + p_K$; i.e. the vector $\mathbf{z}$ is broken into K blocks with each block $\mathbf{z}_k$ of dimension p_k and where $\mathbf{z}_k$ is constrained by the restriction function f_k.

The coordinate descent algorithm cycles through each block of coordinates, one at a time, performing a low-dimensional minimization. Specifically, the algorithm works as follows.

1. Initialize the algorithm at some value $\mathbf{z}_0$ (typically chosen to be an interior point in the effective domain of f).
2. Let $\tilde{\mathbf{z}} = (\tilde{\mathbf{z}}_1,\dots,\tilde{\mathbf{z}}_K)^T$ be the current value. Update $\tilde{\mathbf{z}}_k$ one at a time, $k = 1,\dots,K$, according to

$$\tilde{\mathbf{z}}_k \leftarrow \arg\min_{\mathbf{z}_k \in \mathbb{R}^{p_k}} f(\tilde{\mathbf{z}}_1,\dots,\tilde{\mathbf{z}}_{k-1},\mathbf{z}_k,\tilde{\mathbf{z}}_{k+1},\dots,\tilde{\mathbf{z}}_K).$$

There are various versions of this algorithm. For instance, it is not necessary to cycle through all K blocks in each iteration. Instead, it is sufficient if there is a constant $T \geq K$ such that every block $k = 1,\dots,K$ is selected at least once between iteration r and iteration $(r+T)$, for all iterations r. Under certain technical conditions, it can be shown that this approach will converge to a stationary point. For a comprehensive theoretical analysis and additional references, see Tseng [5].

14.2 Coordinate Descent for Regression with L_1-Penalization: Fitting the Lasso

Let's delve into how coordinate descent is applied to linear regression with L_1-penalization, known as lasso. Much of this explanation is based on the excellent paper by Friedman, Hastie

DOI: 10.1201/9781003493679-14

and Tibshirani [2]. Consider $Y_i \in \mathbb{R}$ as the response and $\mathbf{X}_i \in \mathbb{R}^p$ as the covariate in a linear regression model where $\mathbb{E}(Y_i|\mathbf{X}_i = \mathbf{x}_i) = \beta_0 + \mathbf{x}_i^T \boldsymbol{\beta}$ and $\boldsymbol{\beta} = (\beta_1, \ldots, \beta_p)^T$ represents the regression coefficients. We have n observation pairs $(\mathbf{x}_i, y_i)$ for $i = 1, \ldots, n$. Denote $\mathbf{x}_{(j)}$ as the n-dimensional vector containing the values of the jth covariate and $\mathbf{y} = (y_1, \ldots, y_n)^T$ as the response vector. The inner product of two vectors $\mathbf{u}, \mathbf{v} \in \mathbb{R}^n$ is given by $(\mathbf{u}, \mathbf{v}) = \sum_{i=1}^n u_i v_i$.

Prior to running the algorithm the data is first standardized so that

$$\frac{1}{n}\sum_{i=1}^{n} x_{ij} = \frac{1}{n}(\mathbf{x}_{(j)}, \mathbf{1}) = 0, \quad \frac{1}{n}\sum_{i=1}^{n} x_{ij}^2 = \frac{1}{n}(\mathbf{x}_{(j)}, \mathbf{x}_{(j)}) = 1, \quad j = 1, \ldots, p.$$

This is achieved by transforming the data by $x_{ij} \leftarrow (x_{ij} - \overline{x}^{(j)})/s_j$ where $\overline{x}^{(j)}$ and s_j^2 are the sample mean and sample variance (using n and not $n-1$) of the original data.

The goal is to find the solution to the lasso optimization problem

$$\underset{(\beta_0, \boldsymbol{\beta}) \in \mathbb{R}^{p+1}}{\arg\min}\ R_n(\beta_0, \boldsymbol{\beta}) := \underset{(\beta_0, \boldsymbol{\beta}) \in \mathbb{R}^{p+1}}{\arg\min} \left[\frac{1}{2n} \sum_{i=1}^{n} (y_i - \beta_0 - \mathbf{x}_i^T \boldsymbol{\beta})^2 + \lambda \|\boldsymbol{\beta}\|_1 \right] \tag{14.2}$$

where $\lambda > 0$ is the lasso shrinkage parameter. This is clearly a special case of f described by (14.1). Ignoring the intercept β_0 for the moment (which will be estimated prior to running the algorithm), $f_0(\mathbf{z}) = \sum_{i=1}^n (y_i - \beta_0 - \mathbf{x}_i^T \boldsymbol{\beta})^2/(2n)$ where $\mathbf{z} = \boldsymbol{\beta}$, $\mathbf{z}_k = \beta_k$ and $f_k(\mathbf{z}_k) = \lambda|\beta_k|$. Thus the lasso solution can be obtained by coordinate descent, to be described next.

14.2.1 Coordinate descent algorithm for lasso

Consider one cycle of the coordinate descent algorithm. Let $\tilde{\boldsymbol{\beta}} \in \mathbb{R}^p$ be the current coefficient vector estimate and let $\tilde{\beta}_j \in \mathbb{R}$ for $j = 1, \ldots, p$ be the j coordinate of $\tilde{\beta}$. The estimate for the intercept is denoted by $\tilde{\beta}_0 \in \mathbb{R}$. As mentioned earlier, this value is set at the start of the algorithm and is never changed. It is easy to see that $\tilde{\beta}_0 = \overline{y}$ is an appropriate estimator.

For a given coordinate j, the update for $\tilde{\beta}_j$ is obtained by calculating the gradient of $R_n(\tilde{\beta}_0, \boldsymbol{\beta})$ at β_j. It follows that the update for $\tilde{\beta}_j$ (see Exercise 14.1) is

$$\tilde{\beta}_j \leftarrow S\left(\frac{1}{n} \sum_{i=1}^{n} x_{ij}(y_i - y_i^{(j)}), \lambda \right) \tag{14.3}$$

where $y_i^{(j)} = \tilde{\beta}_0 + \sum_{l \neq j} x_{il} \tilde{\beta}_l$ is the fitted value obtained by excluding the contribution from covariate j and and $S(z, \lambda)$ is the soft-thresholding operator equal to

$$\operatorname{sign}(z)(|z| - \lambda)_+ = \begin{cases} z - \lambda & \text{if } z > 0 \text{ and } \lambda < |z| \\ z + \lambda & \text{if } z < 0 \text{ and } \lambda < |z| \\ 0 & \text{if } \lambda \geq |z|. \end{cases}$$

Define r_i to be the current residual

$$r_i = y_i - \tilde{\beta}_0 - \sum_{j=1}^{p} x_{ij} \tilde{\beta}_j = y_i - \tilde{\beta}_0 - \sum_{k: |\tilde{\beta}_k| > 0} x_{ik} \tilde{\beta}_k.$$

Then with a little bit of manipulation, it is easily seen that

$$\tilde{\beta}_j \leftarrow S\left(\tilde{\beta}_j + \frac{1}{n} \sum_{i=1}^{n} x_{ij} r_i,\ \lambda \right). \tag{14.4}$$

One can see from (14.4) why coordinate descent is computationally efficient. Many coefficients are zero after thresholding with S and they remain that way and are never updated.

14.2.2 Efficient cycling updates

High computational efficiency of the algorithm can be obtained by a few simple steps. First define the gradient for coordinate j to be

$$\tilde{g}_j = \frac{1}{n}(\mathbf{x}_{(j)}, \mathbf{y}) - \frac{1}{n} \sum_{k:|\tilde{\beta}_k|>0} (\mathbf{x}_{(j)}, \mathbf{x}_{(k)})\tilde{\beta}_k.$$

Note that the sum is for indices $k > 1$ and excludes $\tilde{\beta}_0$ (the intercept term) which is $k = 0$. Using the following identity

$$\tilde{\beta}_j + \tilde{g}_j = \tilde{\beta}_j + \frac{1}{n} \sum_{i=1}^{n} x_{ij} r_i$$

we can rewrite (14.4) as

$$\tilde{\beta}_j \leftarrow S\big(\tilde{\beta}_j + \tilde{g}_j,\, \lambda\big). \tag{14.5}$$

Now we use (14.5) for efficient implementation.

1. Prior to starting the algorithm, compute inner products $n^{-1}(\mathbf{x}_{(j)}, \mathbf{y})$ for $j = 1, \ldots, p$. Then these do not need to be recomputed each time in (14.5). This requires $O(pn)$ operations.

2. Maintain a list of coordinates comprising the *active variables*, i.e. a variable enters the active set when their coefficient first becomes non-zero. When a new variable j enters the active set, calculate and store (forever) the scaled inner product of j with all the rest of the features, $n^{-1}(\mathbf{x}_{(j)}, \mathbf{x}_{(k)})$ for $k \neq j$, a total of $O(pn)$ operations. This step of computing and storing the inner products of an active variable should not be used when $p \gg n$ or when p is sufficiently large, say $p > 500$, as this will become inefficient. We discuss this other case later which is treated differently.

3. **Variables can leave the active set!** Although it is rare, an active variables' coefficient can sometimes become zero, and when this occurs it should be removed from the active set (however, we continue to retain inner products that were previously calculated by keeping a master list of variables that were active at some point).

4. (14.5) is calculated rapidly by keeping an updated vector of length p containing the current gradient values

$$\tilde{g}_j = \frac{1}{n}(\mathbf{x}_{(j)}, \mathbf{y}) - \frac{1}{n} \sum_{k:|\tilde{\beta}_k|>0} (\mathbf{x}_{(j)}, \mathbf{x}_{(k)})\tilde{\beta}_k, \quad j = 1, \ldots, p. \tag{14.6}$$

   If one of the coefficients changes, we update **all** p values of (14.6), but this only requires $O(p)$ operations due to step 2. For example, suppose coordinate l changes from $\tilde{\beta}_l$ to $\tilde{\beta}_l^*$, then

$$\tilde{g}_j \leftarrow \tilde{g}_j + \frac{1}{n}(\mathbf{x}_{(j)}, \mathbf{x}_{(l)})\tilde{\beta}_l - \frac{1}{n}(\mathbf{x}_{(j)}, \mathbf{x}_{(l)})\tilde{\beta}_l^* = \tilde{g}_j + \frac{1}{n}(\mathbf{x}_{(j)}, \mathbf{x}_{(l)})(\tilde{\beta}_l - \tilde{\beta}_l^*). \tag{14.7}$$

   With m non-zero terms in the model, using (14.7), a complete cycle costs at most $O(pm)$ operations as long as no new variables enter the active set. The cost is $O(pn)$ for each new variable entered.

5. Further efficiency is obtained by cycling over the active set. After a complete cycle (C_1) through all the variables, we iterate on only the coordinates in the currently active set, doing so until those coordinates converge (to be defined shortly). Following the active set cycle (A) if another complete cycle (C_2) does not change the active set, we are done; otherwise the process is repeated: $\{(C_1, A, C_2)\}$, $\{(C_1, A, C_2)\}$, etc. A maximum number of iterations is allowed for the active cycle in order to ensure the algorithm eventually terminates. Active set

convergence is defined as follows. Convergence is obtained when the maximum change in the objective function

$$\frac{1}{2n}\sum_{i=1}^{n}(y_i-\tilde{\beta}_0-\mathbf{x}_i^T\tilde{\boldsymbol{\beta}})^2+\lambda||\tilde{\boldsymbol{\beta}}||_1$$

after a coefficient update is less than δ times the null deviance $\frac{1}{n}\sum_{i=1}^{n}(y_i-\overline{y})^2$. A default value is $\delta=10^{-7}$. The maximum change is efficiently determined as follows. Suppose that coordinate l changes from $\tilde{\beta}_l$ to $\tilde{\beta}_l^*$, then to check if convergence is achieved, determine if

$$\left|\frac{1}{n}(\tilde{\beta}_l^*-\tilde{\beta}_l)\left[\sum_{k\neq l}\tilde{\beta}_k(\mathbf{x}_{(l)},\mathbf{x}_{(k)})-(\mathbf{x}_{(l)},\mathbf{y})\right]\right.$$

$$\left.+\frac{1}{2}(\tilde{\beta}_l^*)^2-\frac{1}{2}(\tilde{\beta}_l)^2+\lambda\left[|\tilde{\beta}_l^*|-|\tilde{\beta}_l|\right]\right|\leq\frac{\delta}{n}\sum_{i=1}^{n}(y_i-\overline{y})^2.$$

This can be simplified by substituting $\tilde{g}_j$ (see (14.6)); yielding

$$\left|(\tilde{\beta}_l-\tilde{\beta}_l^*)\left[\tilde{\beta}_l+\tilde{g}_l\right]+\frac{1}{2}(\tilde{\beta}_l^*)^2-\frac{1}{2}(\tilde{\beta}_l)^2+\lambda\left[|\tilde{\beta}_l^*|-|\tilde{\beta}_l|\right]\right|\leq\frac{\delta}{n}\sum_{i=1}^{n}(y_i-\overline{y})^2. \tag{14.8}$$

If (14.8) holds for all coefficient updates in the active set cycle, then convergence is achieved for the active set cycle.

14.2.3 Naive updates

When $p \gg n$ or p is very large, it is inefficient to store the inner products and the p-dimensional gradient. Instead a different strategy "naive-updates" is used by directly employing (14.4) and not (14.5). With this update, we only need to calculate the residual directly once when the algorithm starts ($r_i=y_i-\tilde{\beta}_0$) because the residual only changes when a coefficient changes. Thus if $\tilde{\beta}_l$ changes to $\tilde{\beta}_l^*$, we update the residual by

$$r_i\leftarrow r_i+(\tilde{\beta}_l-\tilde{\beta}_l^*)x_{il}$$

and thus the total cost after applying (14.4) is $O(2n)$ after a coefficient change and the total cost is no more than $O(2pn)$ for a complete cycle. Finally, checking convergence in the active cycle uses a direct calculation and not (14.8). Let r_i be the residual before thresholding (14.4) and r_i^* the residual following a coefficient change from $\tilde{\beta}_l$ to $\tilde{\beta}_l^*$. We check if the following holds:

$$\left|\frac{1}{2n}\sum_{i=1}^{n}\left[(r_i^*)^2-r_i^2\right]+\lambda\left[|\tilde{\beta}_l^*|-|\tilde{\beta}_l|\right]\right|\leq\frac{\delta}{n}\sum_{i=1}^{n}(y_i-\overline{y})^2.$$

14.2.4 Warm starts

The algorithm is applied to a decreasing sequence of λ values with each λ value resulting in a parameter estimate $(\hat{\beta}_0(\lambda),\hat{\beta}_1(\lambda),\ldots,\hat{\beta}_p(\lambda))^T$. This sequence of parameter estimates is called the solution path. To run the algorithm, start with a value $\lambda=\lambda_{\max}$ where $\lambda_{\max}$ is selected to be the smallest λ value for which the entire vector remains at zero.

From (14.5) we can see that if all $\tilde{\beta}_j=0$ then they will remain at zero if $n^{-1}(\mathbf{x}_{(j)},\mathbf{y})<\lambda$. Therefore

$$\lambda_{\max}=\frac{1}{n}\max_{1\leq j\leq p}|(\mathbf{x}_{(j)},\mathbf{y})|.$$

Starting from $\lambda_{\max}$, we create a decreasing sequence of values ending at a suitably chosen small value, $\lambda_{\min}$. The algorithm begins with $\lambda_{\max}$, resulting in all zero values. Then, we move to the next λ in the sequence, using the current solution to initialize the algorithm. This process is repeated for each subsequent λ value, always using the most recent solution as the starting point. This technique, known as warm starts, helps achieve more stable solutions.

Setting the λ sequence correctly is crucial. The `glmnet` R-package [2] employs the following method. Starting with $\lambda_{\max}$, define $\lambda_{\min} = \epsilon\lambda_{\max}$ with $\epsilon = 0.0001$. An equally spaced grid of 100 values is then created between the logarithms of these two values. The grid is transformed back to the original scale, ensuring the first value is always $\lambda_{\max}$. To avoid convergence issues caused by very small λ values, only sufficiently large values are retained. The threshold for what constitutes a "large enough" value is determined relative to $\lambda_{\max}$. Specifically, only those λ_j values that meet the following condition are retained:

$$\frac{\lambda_j}{\lambda_{\max}} > \lambda_R = \begin{cases} 10^{-4} & \text{if } n > p \\ 10^{-2} & \text{otherwise.} \end{cases}$$

14.2.5 Transforming coefficients back to the original data

For a given λ, let $\tilde{\beta}_j$ for $j = 1, \ldots, p$ be the final coefficients returned by the algorithm. These are transformed to the original data scale to obtain the intercept $\hat{\beta}_0$ and coefficient values $\hat{\beta}_j$ as follows:

$$\hat{\beta}_0 = \overline{y} - \sum_{j=1}^{p} \tilde{\beta}_j \frac{\overline{x}^{(j)}}{s_j}, \quad \hat{\beta}_j = \frac{\tilde{\beta}_j}{s_j}, \quad j = 1, \ldots p.$$

14.2.6 Cross-validation for selecting λ

Associated with each λ is an estimated vector $(\hat{\beta}_0(\lambda), \hat{\beta}_1(\lambda), \ldots, \hat{\beta}_p(\lambda))^T$ of regression parameters. To determine the optimal λ, and therefore the final output regression parameters, M-fold validation is used as follows.

First, coordinate descent is applied to the full dataset using a sequence of λ values. After this initial run, the data is split into M folds for cross-validation. Using the same λ sequence, coordinate descent is applied to each training fold. The estimated regression parameters from each fold are then used to calculate the test set error for the corresponding test fold. This process results in a vector of length K of error values, denoted as e_k, for each λ_k value where $k = 1, \ldots, K$ in the λ sequence. For each λ_k, we obtain M different e_k values from the M folds, which are then averaged to obtain $\overline{e}_k$.

Let k^* be the index of the λ_k value with the smallest average error, i.e., $\overline{e}_{k^*} \leq \overline{e}_k$ for all k. The final output regression parameters are then taken from the initial coordinate descent run on the full data, evaluated at $\lambda = \lambda_{k^*}$. These parameters are $(\hat{\beta}_0(\lambda_{k^*}), \hat{\beta}_1(\lambda_{k^*}), \ldots, \hat{\beta}_p(\lambda_{k^*}))^T$.

14.3 Comparing LARS to Coordinate Descent for Lasso

Recall that LARS (least angle regression) can compute the entire lasso solution path with the same computational effort as used by least squares for a single full model evaluation. The number of iterations for LARS is set to $P \leq p$, where P can be less than p. In high-dimensional settings, it is beneficial to choose P much smaller than p. The total number of operations required by LARS is $O(P^3 + nP^2)$. With certain modifications, LARS can be used to obtain the lasso solution, as described in Efron et al. [1].

We compare LARS, lasso (derived from modified LARS), and coordinate descent. Since each method defines an iteration differently, comparing them by iteration count is not appropriate. Instead, we compare their performance in terms of the solution path. This path can be parameterized by the L_1-norm, $\sum_{k=1}^{p}|\hat{\beta}_k|$. However, we prefer to use the fraction of the final L_1-norm, a dimensionless quantity ranging from 0 to 1, defined by

$$\begin{aligned}\text{fraction of final } L_1\text{-norm} &= \frac{L_1\text{-norm of the } j \text{ solution}}{L_1\text{-norm of the } P \text{ saturated solution}} \\ &= \frac{\sum_{k=1}^{p}|\hat{\beta}_{kj}|}{\sum_{k=1}^{p}|\hat{\beta}_{kP}|}, \qquad j = 1,\ldots,P.\end{aligned}$$

The numerator is the L_1-norm of the j coefficient solution, while the denominator is the L_1-norm of the final P coefficient solution (i.e. a procedures final saturated solution), where $\hat{\beta}_{kj}$ denotes the estimate for coefficient β_k from iteration j.

We use two data sets for illustration. The Boston Housing data [3], which is a famous benchmark data set used in machine learning ($n = 506$, $p = 13$), and 2008 close-of-day data for the Standard and Poors 500 (S&P) using the DJIA (Dow Jones Industrial Average) for the y-value ($n = 252$, $p = 492$). The S&P data was converted to daily percentage change to remove time series dependence.

The three methods generally perform similarly on the Boston Housing data (Figure 14.1). The top panel shows the coefficient solution paths, which look quite similar across all methods. The vertical dashed lines indicate the optimized L_1-fraction obtained through 10-fold cross-validation. The bottom panel shows the coefficient values for the optimal L_1-fraction, and these values are nearly identical for all three methods.

In contrast, substantial differences are observed for the S&P stock market data (Figure 14.2). The solution path for coordinate descent is notably different. It features no negative coefficient values and appears sparser, especially for lower L_1-fraction values. The solution paths for LARS and lasso are more similar to each other, but some differences are apparent; for instance, LARS has larger negative coefficient values. The cross-validated optimized L_1-fraction value also varies across the three methods, with coordinate descent having the largest value and LARS the smallest. This results in different optimized coefficient solutions (panel D). Overall, coordinate descent produces the sparsest solution, even though it has the largest optimized L_1-fraction.

Which method provides the closest solution to the lasso? Coordinate descent directly addresses the L_1-optimization problem, aiming to solve it exactly. In contrast, the lasso solution derived from modified LARS relies on certain technical conditions that may or may not hold, especially with highly correlated features like those in the S&P data. These correlations might impact the validity of those conditions. Additionally, the coordinate descent solution tends to be less smooth compared to the piecewise linear path property of LARS-lasso. However, in our examples, the granularity of coordinate descent is reduced by using a very fine λ sequence of length 1000.

To provide a more definitive answer we can use a theoretical property of the lasso to directly test the accuracy of each procedure. Using the Karush-Kuhn-Tucker (KKT) conditions, Tibshirani [4] characterized the property of the lasso in terms of two quantities. The first is the equicorrelation set. Let $\hat{\boldsymbol{\beta}}$ denote a procedure's solution for a λ value. Define the equicorrelation set

$$\mathscr{E} = \left\{ j : \left| \mathbf{x}_{(j)}^T (\mathbf{y} - \mathbf{D}^T \hat{\boldsymbol{\beta}}) \right| = \rho \right\}.$$

This is the set of coordinates with equal absolute correlation with the residual (with absolute correlation equal to some $\rho := \rho(\lambda)$). Here $\mathbf{D} = [\mathbf{x}_{(1)}, \ldots, \mathbf{x}_{(p)}]$ denotes the $n \times p$ design matrix. The second quantity is the equicorrelation sign vector defined as

$$\mathbf{s} = \text{sign}\left(\mathbf{D}_{(\mathscr{E})}^T (\mathbf{y} - \mathbf{D}^T \hat{\boldsymbol{\beta}}) \right) \in \{-1, +1\}^p$$

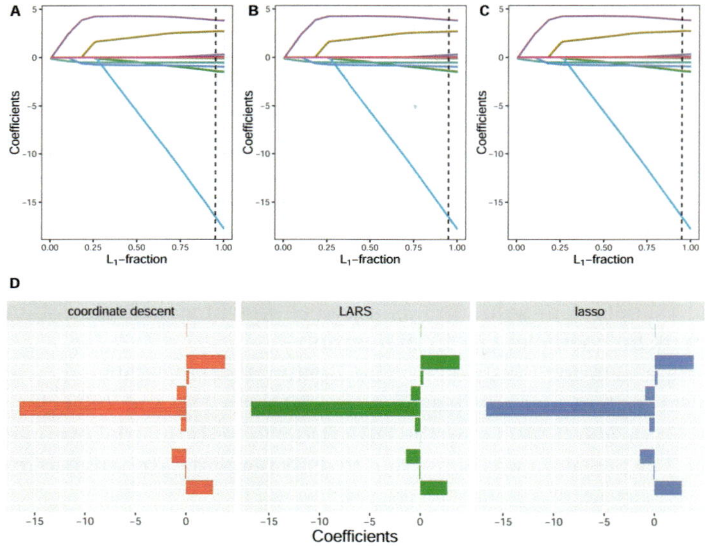

FIGURE 14.1
LARS, lasso and coordinate descent applied to Boston Housing data. Top panel displays coefficient solution path in terms of the fraction of final L_1-norm for: (A) coordinate descent; (B) LARS; (C) lasso. Bottom panel (D) displays the coefficient solution at optimized regularization parameter from 10-fold validation (shown by dashed vertical lines in top panel figures).

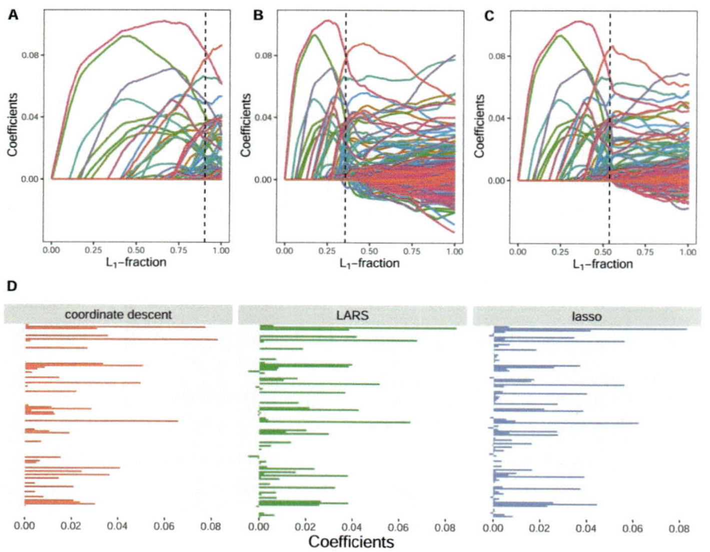

FIGURE 14.2
S&P stock market data. Results are displayed similar to Boston Housing of Figure 14.1.

where $\mathbf{D}_{(\mathscr{E})}$ equals the columns of $\mathbf{D}$ containing the coordinates of $\mathscr{E}$. Then under the full rank condition, $\text{rank}(\mathbf{D}_{\mathscr{E}}) = |\mathscr{E}|$, a true lasso solution $\hat{\boldsymbol{\beta}}^*$ for λ is the unique solution with the property that its coordinates are zero over $\{1,\ldots,p\} \setminus \mathscr{E}$ (i.e. $\hat{\boldsymbol{\beta}}^*_{-(\mathscr{E})} = \mathbf{0}$) and over $\mathscr{E}$ it equals (see Lemma 2 of Tibshirani [4])

$$\hat{\boldsymbol{\beta}}^*_{(\mathscr{E})} = (\mathbf{D}^T_{(\mathscr{E})}\mathbf{D}_{(\mathscr{E})})^{-1}\left(\mathbf{D}^T_{(\mathscr{E})}\mathbf{y} - \rho\mathbf{s}\right). \tag{14.9}$$

Tibshirani [4] observed that the full rank condition is an assumption that is also used by LARS which is needed to obtain the lasso solution.

To test the accuracy of a procedure we compare its estimated lasso value to its theoretical value (14.9). We define the discrepancy as

$$\frac{1}{|\mathscr{E}|}\left\|\hat{\boldsymbol{\beta}}_{(\mathscr{E})} - \hat{\boldsymbol{\beta}}^*_{(\mathscr{E})}\right\|_2^2.$$

Figure 14.3 shows the discrepancy values for LARS-lasso and coordinate descent on the S&P data. For LARS, the discrepancy is zero across the entire range of L_1-fraction values. While the discrepancy for coordinate descent is not zero, it is very small. However, this small difference appears to have a significant impact, as the solution path for coordinate descent was noticeably different from that of LARS. One possible explanation is that when determining $\mathscr{E}$ for coordinate descent, the residual correlation was not identical, and some tolerance was allowed when forming this set. This could help explain the large differences observed in the solution paths. Overall, we can conclude that, in this example, LARS-lasso provides a more accurate lasso solution.

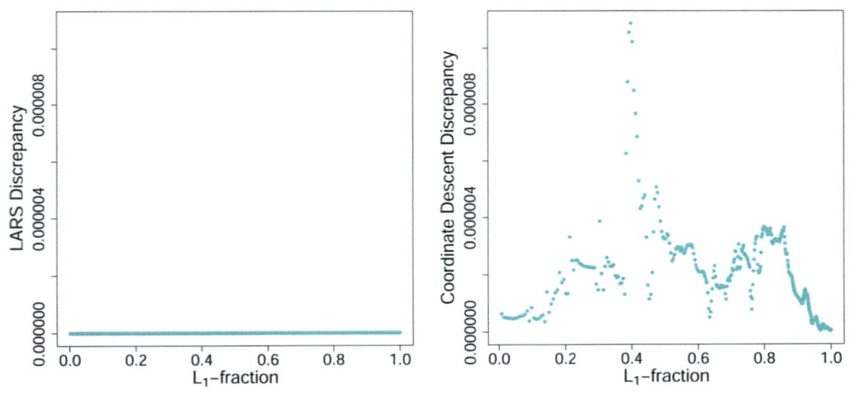

FIGURE 14.3
Discrepancy between lasso estimated value and theoretical value using LARS-lasso (left figure) and coordinate descent (right figure) applied to S&P data.

14.4 Active Cycles When Fitting Lasso

We mentioned that the soft-thresholding operator (14.4) will cause many coefficients to become zero and generally stay that way. This makes coordinate descent highly efficient. One way to visualize this is by looking at the active cycle behavior. For a given λ, coordinate descent iterates between complete updates, followed by active cycling, followed by another

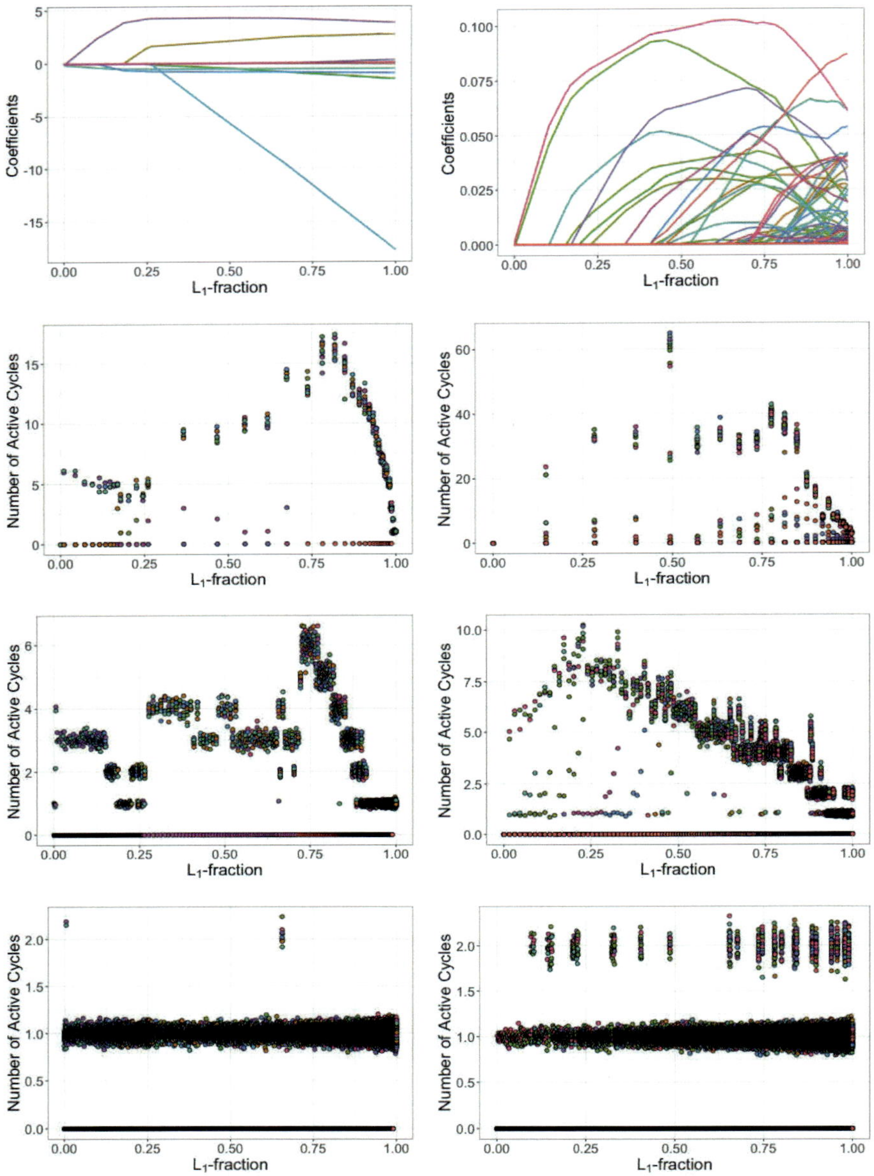

FIGURE 14.4
Number of active cycles using coordinate descent for Boston Housing (left panels) and S&P (right panels) data. Top row shows solution paths from Figure 14.1 and Figure 14.2. Remaining row panels display the number of active cycles for a coordinate where rows were constructed from a λ sequence with 50, 500 and 5000 values respectively. For visual clarity the number of active cycles has been jittered.

complete update, and this process repeats until convergence. Thus, we observe a pattern $\{(C_1, A, C_2)\}, \{(C_1, A, C_2)\}, \ldots$ for a fixed λ.

The active cycle only cycles over coordinates of relevance. Therefore we can look at the number of active cycles for a coordinate to visualize the efficiency of coordinate descent. Figure 14.4 displays this for the Boston Housing and S&P data sets. For reference, the first row of the figure displays the solution paths from the previous analysis. The remaining rows show the number of active cycles for a coordinate where rows were constructed from a λ sequence with 50, 500 and 5000 values respectively. We draw the following conclusions.

1. The number of active cycles for a coordinate decreases as the λ sequence becomes more refined. With 5000 values, a maximum of 2 cycles are observed for any coordinate and most of the active cycles are 1.
2. The number of active cycles is related to warm starts (Chapter 14.2.4). Each coordinate descent run is primed by the previous solution, which leaves little room for the solution to change when using a refined λ sequence.
3. We observe that many coefficients stay at the value of 0 and never change regardless of how λ is set. This demonstrates coordinate descent's efficiency for finding a sparse lasso solution.

Exercises

Exercise 14.1. Prove (14.3). Hint: Exercise 4.20 may be helpful.

Exercise 14.2. Using the naive-update approach (which is much easier to code), write your own coordinate descent algorithm for obtaining the lasso from (14.2).

(a) Compare your algorithm to the R-software package `glmnet` to confirm that your code is working correctly.
(b) Run your procedure on an interesting data set. Compare the results to LARS.

Exercise 14.3. Compare the discrepancy between the theoretical lasso solution to coordinate descent and LARS-lasso as in Figure 14.3 using a dataset of your choosing.

Exercise 14.4. Compare coordinate descent to LARS-lasso using simulated data to study the effects of correlation and increasing dimension p. Do the results between the methods change as these parameters are altered?

Exercise 14.5. We will apply coordinate descent lasso regression to the Iowa housing data previously considered in Figure 17.6 of Chapter 17.3. Recall this is a regression problem where the goal is to predict the sales price of a home. The data is available from the `randomForestSRC` R-package.

1. Impute and preprocess the data as previously described. Fit a lasso regression model using coordinate descent where the λ regularization parameter is tuned using cross-validation. Evaluate the prediction performance of the data (use training and testing data, for example).
2. Extract the coefficients identified by the lasso.
3. Identify significant features using permutation VIMP with confidence intervals as in Figure 17.6.
4. Obtain importance values using VarPro (Chapter 17.6).
5. Compare the results of the three methods and summarize your findings.

Exercise 14.6. For this exercise, we use the `NYC Taxi Trip Duration` dataset from Kaggle, which contains detailed records of taxi trips in NYC. The exercise involves predicting the duration of NYC taxi trips using lasso regression. There are $n = 1{,}458{,}644$ records which makes this a challenging dataset for most machine learning methods.

1. Convert the date-time columns to proper date-time formats and use this to calculate trip duration in seconds which is the Y response. Split the data into training and testing sets.
2. To improve prediction accuracy, try creating new features such as pickup hour and pickup weekday. Convert character vectors to numeric. Generate suitable interaction terms to capture relationships between different features.
3. Fit the regression model on the training data using lasso coordinate descent with cross-validation to select the optimal λ.
4. Similarly, fit the model using LARS-lasso algorithm using cross-validation to select the optimal model.
5. Compare the prediction performance of the two methods over the test data. Compare the computational effort of the two methods.
6. Summarize your findings. Emphasize advantages, disadvantages and differences.

Exercise 14.7. Redo the previous analysis for the NYC taxi data using gradient boosting with linear learners (eg., L_2Boost). Do you get similar results? Discuss differences and summarize your conclusions.

Bibliography

[1] Efron B., Hastie T., Johnstone I., and Tibshirani R. Least angle regression (with discussion). *Annals of Statistics*, 32:407–499, 2004.

[2] Friedman J. and Hastie T. and Tibshirani R. Regularization paths for generalized linear models via coordinate descent. *Journal of Statistical Software*, 33(1), 1–22, 2010.

[3] Harrison D. and Rubinfeld D.L. Hedonic prices and the demand for clean air. *Journal of Environmental Economics and Management*, (5), 81–102, 1978.

[4] Tibshirani R.J. The lasso problem and uniqueness. Electronic Journal of Statistics, (7) 1456–1490, 2013.

[5] Tseng P. Convergence of a block coordinate descent method for nondifferentiable minimization. *Journal of Optimization Theory and Applications*, 109(3), 475–494, 2001.

15

Trees

15.1 Introduction

Trees are intuitive, nonparametric tools for exploring and understanding data. The use of tree-methods for classification has quite a rich history, dating back almost 60 years. Much of this early work emanated from the area of social sciences, starting in the late 1960's. Computational algorithms for automatic construction of classification trees began as early as the 1970's. Algorithms, such as the THAID program developed at the Institute for Social Research, the University of Michigan, laid the groundwork for recursive partitioning algorithms, the predominate algorithm used by many contemporary tree procedures, such as Classification and Regression Trees (CART) [3].

15.2 Classification Trees Using CART

Classification trees constructed using the CART methodology employ recursive partitioning (RPART) to classify a feature $\mathbf{X} \in \mathscr{X}$ into one of $J \geq 2$ class labels for a categorical outcome $Y \in \mathscr{Y} = \{C_1, \ldots, C_J\}$. The tree is constructed from a learning sample $(\mathbf{X}_1, Y_1), \ldots, (\mathbf{X}_n, Y_n) \in \mathscr{X} \times \mathscr{Y}$. Starting at the root node (the top node of the tree, comprising the entire training sample), the node is split into two daughter nodes: a left and a right daughter node. Each daughter node is then recursively split, with each split producing new left and right daughter nodes, and these child nodes are further split. Each split is "greedy," meaning it uses the features and response data within a node to determine the optimal split.

CART specifically uses the concept of *tree impurity* to choose the best split (more on this later). The tree continues to split recursively until it can no longer be partitioned due to lack of data or until a stopping criterion is met, resulting in a collection of terminal nodes. At this stage, a final decision is made regarding the classification of the cases.

Three distinct types of nodes arise when constructing a binary recursive tree.

(a) Root node: The top node of the tree comprising all the data.
(b) Internal node: A node that assigns data to a subgroup.
(c) Terminal node: Final decision (outcome).

Figure 15.1 illustrates the use of CART on the University of Wisconsin breast cancer dataset, a well-known benchmark dataset. The data includes 699 patients classified as having benign or malignant breast cancer, with true disease status confirmed through histology. The goal is to

DOI: 10.1201/9781003493679-15

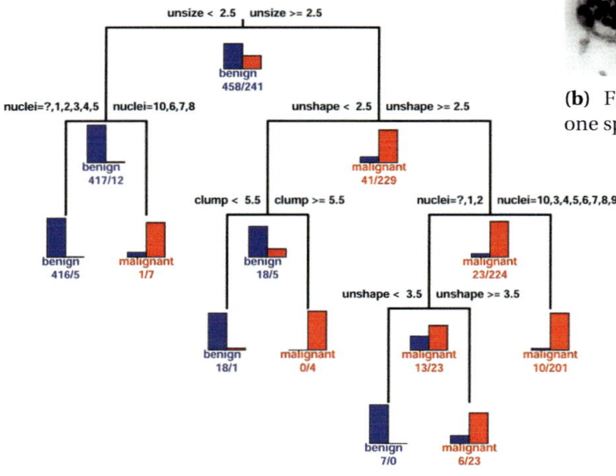

(a) Classification tree using RPART.

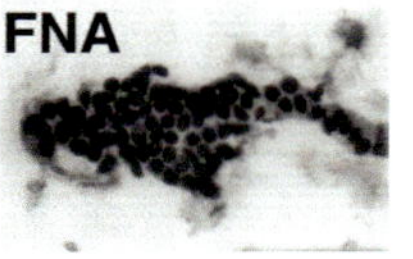

(b) FNA cytology of one specimen.

FIGURE 15.1
Wisconsin Breast Cancer Data: $n = 699$ patients classified with benign or malignant breast cancer, with true disease status confirmed by histology.

accurately predict disease status based on 9 attributes derived from fine needle aspirates (FNA), a minimally invasive biopsy technique used to determine cancer status. FNA helps avoid more invasive procedures, like surgery, which pose additional risks to the patient.

The first split of the tree (at the root node) is based on the variable "unsize," which measures the uniformity of cell size. All patients with a value less than 3 for this variable are assigned to the left node (left daughter node), while those with values of 3 or higher are assigned to the right node (right daughter node). The left and right daughter nodes are then split again (on the variable "unshape" for the right daughter and "clump" for the left daughter), and this process continues recursively until the terminal nodes are reached. Blue and red colors represent the classifier decision for each node (blue indicates benign), and the fraction between 0 and 1 shown in each node represents the relative frequency of malignant cells within the node (note that the percentage shown reflects the percentage of the overall sample size, n). The color of the terminal nodes corresponds to the final classification outcome based on majority voting (i.e. Bayes classification). This estimated classification is the result returned by the tree.

15.3 CART Pruning Process

An important aspect of CART is determining the optimal tree size. To determine a right-sized tree, the tree is initially grown to purity, resulting in an overfitted tree. Then CART uses cost complexity pruning, also known as weakest link pruning, to trim the tree back to a size that results in good performance [3].

The cost complexity measure, $C_\alpha(T)$, for a tree T is defined as:

$$C_\alpha(T) = C(T) + \alpha \cdot |T|,$$

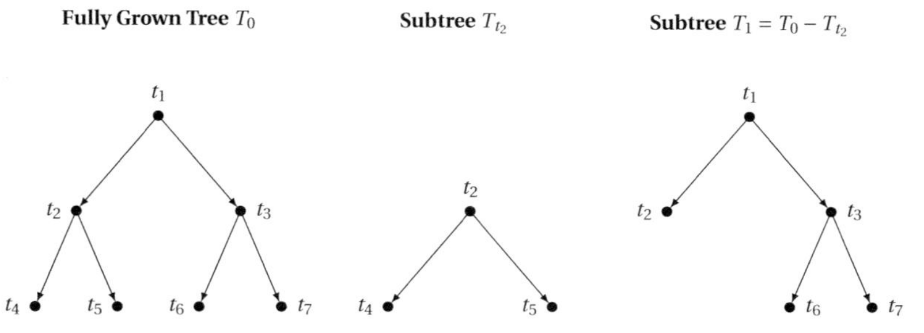

FIGURE 15.2
Pruning a CART tree using cost-complexity.

where $C(T)$ is the total misclassification cost (or some other impurity measure) of the tree T, α is a complexity parameter and $|T|$ is the number of terminal nodes (leaves) in the tree T.

Weakest link pruning starts with a fully grown tree T_0. For an internal node t in T_0, consider the subtree T_t rooted at t (see Figure 15.2 where $t = t_2$). The misclassification cost of T_t is $C(T_t)$, while pruning T_t back to t results in a node with a misclassification cost of $C(t)$. The difference in these values yields the cost complexity of pruning T_t, equaling

$$\Delta C_\alpha(T_t) = C(T_t) - C(t) + \alpha \cdot (|T_t| - 1).$$

Calculate the α value that would make it optimal to prune T_t. This value is

$$\alpha(t) = \frac{C(t) - C(T_t)}{|T_t| - 1}.$$

To determine the optimal pruning, we identify the node t with the smallest $\alpha(t)$. Let this be α_1. Prune this subtree, resulting in a new tree T_1 (in Figure 15.2, $T_1 = T_0 - T_{t_2}$). Repeat the process for the new tree T_1 to find the next α value, α_2, and pruned subtree T_2. Continue iteratively, generating a sequence of trees $T_1 \supset T_2 \supset \cdots \supset T_k \supset \cdots$ and corresponding α values $\alpha_1 \le \alpha_2 \le \cdots \le \alpha_k \le \cdots$ until the tree is pruned to the root, a single tree node.

Each α_i corresponds to a specific pruned subtree in the sequence. By evaluating these subtrees on a validation set or using cross-validation, the optimal α can be selected, leading to the final pruned tree.

Pruning is typically performed when the purpose is to use the tree by itself for statistical inference or prediction. However, in many settings, such as random forests (Chapter 16) and gradient-boosted trees (Chapter 15.14), where the tree is used as a base learner, pruning is typically not done. This is because these methods will already have their own procedures for regularization. In much of the later parts of the book this will be the case.

15.4 Partitioning Classifiers

Recursive binary trees such as CART are examples of a general class of procedures called *partitioning classifiers.* Given a partition $\Pi_n = \{A_1, A_2, \ldots\}$ of $\mathscr{X}$, where we refer to A_k as the cells of $\mathcal{T}_n$, for each $\mathbf{x} \in \mathscr{X}$, the partitioning classifier is defined as

$$\hat{\delta}_n^P(\mathbf{x}) = \arg\max_{1\le j\le J} \sum_{i=1}^n 1_{\{Y_i=C_j\}} 1_{\{\mathbf{X}_i \in \mathcal{T}_n(\mathbf{x})\}} \in \{1,\dots,J\}, \tag{15.1}$$

where $\mathcal{T}_n(\mathbf{x})$ is the cell in $\mathcal{T}_n$ containing $\mathbf{x}$. For simplicity we adopt the notation that class labels C_j are coded as integer values $\{1,\dots,J\}$. Also for $J=2$ it will be convenient to use labels $\{0,1\}$.

Because of the nature of the binary splits used by CART, we can observe that CART functions as a method for partitioning the space, which is why it is considered a partitioning classifier.

Specifically, in Chapter 7.7, we noted that CART's partitioned space consists of hyper-rectangles oriented parallel to the coordinate axes. Given an internal node t in the tree, and assuming continuous features, CART splits t into left and right daughter nodes, t_L and t_R, using a feature X at a real-valued split point $s := s(t)$. This results in left and right daughters of the form

$$t_L = \{\mathbf{X} \in \mathbb{R}^p : X \le s\}, \quad t_R = \{\mathbf{X} \in \mathbb{R}^p : X > s\}.$$

Since the decision is binary, with data going either left or right, this leads to a partition of t in the form $t = t_L \cup t_R$, where the daughters t_L and t_R are disjoint rectangles with sides parallel to the X-coordinate axes. Repeating this process recursively creates a partition $\mathcal{T}_n$ of $\mathscr{X}$, composed of rectangles. This type of partition is a natural consequence of the *axis-aligned cuts* used by CART.

An illustration of this is provided in Figure 15.3. Here CART has been applied to the prostate cancer data considered previously in Chapter 12. To help visualize the partition space, the data has been limited to two variables: prostate-specific antigen (PSA) and tumor volume. In the top figure we have displayed the classification tree, while the bottom panel shows the decision boundary for the tree. Notice how the cuts are always parallel to the horizontal and vertical axes resulting in a partitioning of the feature space into rectangles.

Remark 15.1. The preceding discussions covered splits for continuous features, but CART also handles categorical features. Categorical splits direct data points left and right using complementary pairings of the factor levels of X. For instance, if X is a factor with three labels, $\{A,B,C\}$, there are three complementary pairings (daughters): $\{A\}$ and $\{B,C\}$; $\{B\}$ and $\{C,A\}$; and $\{C\}$ and $\{A,B\}$. In general, for a factor with L distinct labels, there are $2^{L-1}-1$ distinct complementary pairs. □

The partitioning classifier derived from the CART tree induction can be formally defined as follows. Let $\mathbf{x}$ be a given feature and let $\mathcal{T}_n(\mathbf{x})$ be the function identifying the terminal node containing $\mathbf{x}$. Coding Y as $\{0,1\}$, the tree estimates the conditional probability $F(\mathbf{x}) = \mathbb{P}\{Y=1|\mathbf{X}=\mathbf{x}\}$ using

$$\hat{F}_n(\mathbf{x}) = \frac{1}{|\mathcal{T}_n(\mathbf{x})|} \sum_{\mathbf{X}_i \in \mathcal{T}_n(\mathbf{x})} Y_i. \tag{15.2}$$

Classification for $\mathbf{x}$ uses the majority rule which is motivated by the Bayes rule. Recall from Chapter 6.6 that the Bayes classifier is $\delta(\mathbf{x}) = 1_{\{F(\mathbf{x})\ge 0.5\}}$. Thus, the tree classifier is defined as

$$\begin{aligned}\hat{\delta}_n(\mathbf{x}) &= 1_{\left\{\hat{F}_n(\mathbf{x})\ge 0.5\right\}} \\ &= \begin{cases} 0 & \text{if } \sum_{i=1}^n 1_{\{Y_i=1\}} 1_{\{\mathbf{X}_i\in\mathcal{T}_n(\mathbf{x})\}} < \sum_{i=1}^n 1_{\{Y_i=0\}} 1_{\{\mathbf{X}_i\in\mathcal{T}_n(\mathbf{x})\}} \\ 1 & \text{otherwise.}\end{cases}\end{aligned}$$

Equivalently, we have

$$\hat{\delta}_n(\mathbf{x}) = \arg\max_{1\le j\le J} \sum_{i=1}^n 1_{\{Y_i=C_j\}} 1_{\{\mathbf{X}_i\in\mathcal{T}_n(\mathbf{x})\}},$$

which is an example of a partitioning classifier (15.1).

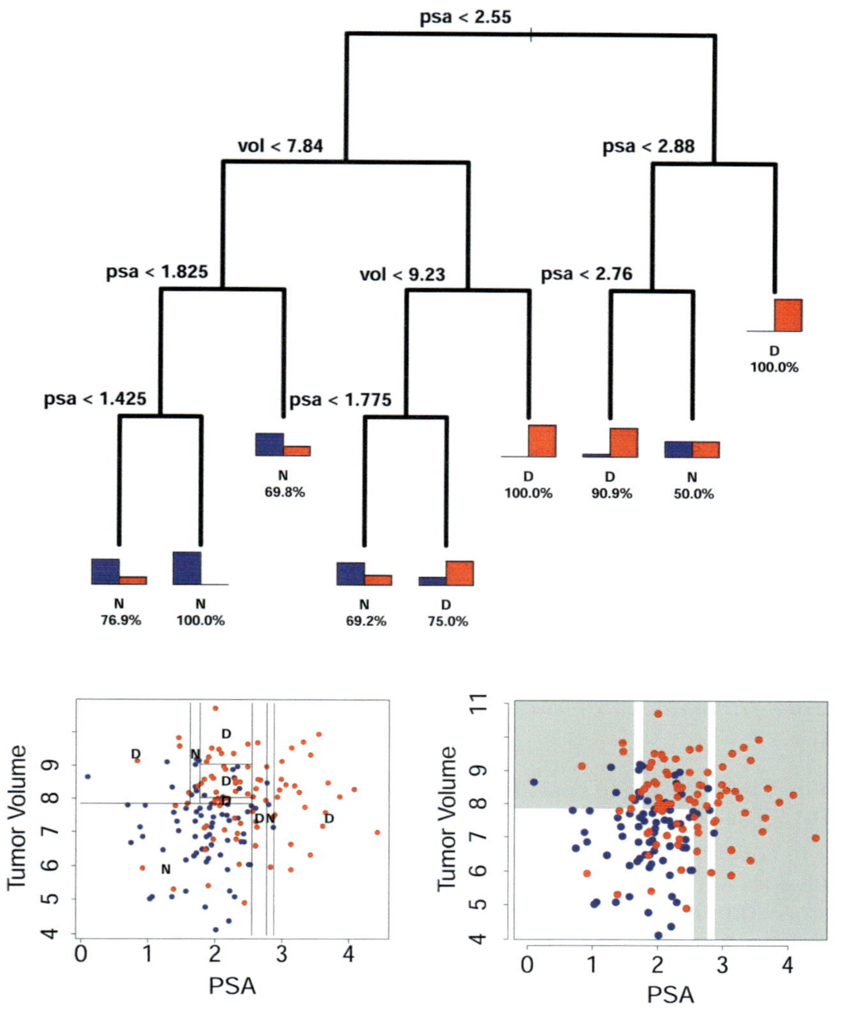

FIGURE 15.3
Classification tree (top) and its decision boundary (bottom) for prostate cancer data. Barplots under terminal nodes of the classification tree indicate the proportion of cases classified as diseased (D) or non-diseased (N), with the predicted class label determined by majority voting. Decision boundary (bottom right) shows using gray and white regions how the tree classifies a new patient using PSA and tumor volume. Blue points are non-diseased patients and red points are diseased patients from the training data.

As an illustration, we can refer back to the classification tree Figure 15.1 for the Wisconsin breast cancer dataset. The terminal nodes at the bottom of the tree are color-coded in blue and red to represent the tree classifier $\hat{\delta}_n(\mathbf{x})$. Malignant cases (in red) are classified when the relative frequency of malignant cases in a node equals or exceeds 1/2. For example, in the leftmost terminal node, the relative frequency of malignant cases is $0.01 = 5/(416+5)$, so the estimated probability $\hat{F}_n(\mathbf{x})$ for $\mathbb{P}\{Y = 1|\mathbf{X} = \mathbf{x}\}$ for a case $\mathbf{x}$ in this node is assigned the value 0.01, and the classification for cases in the node is benign due to majority voting.

15.4.1 Large sample consistency for partitioning classifiers

Partitioning classifiers for the $J = 2$ class problem are Bayes risk consistent under fairly mild conditions, providing a strong rationale for their use. Bayes risk consistency means that the classifier's misclassification rate converges to the Bayes error rate. The Bayes error rate, $\mathbb{P}\{\delta(\mathbf{X}) \neq Y\}$, is the smallest misclassification error achievable under 0-1 loss (Theorem 6.3). This implies that as more data is used, the classifier's performance approaches that of the optimal classifier.

The following result due to Devroye et al. [Theorem 6.1, 10] gives sufficient conditions for this type of consistency. It requires only two conditions. The first is that the number of data points falling in a random cell must increase to infinity. Let

$$N_n(\mathbf{X}) = \sum_{i=1}^{n} 1_{\{\mathbf{X}_i \in \pi_n(\mathbf{X})\}} = |\pi_n(\mathbf{X})|$$

be the random number of data points falling in the same cell as $\mathbf{X}$. A second condition requires that cells should be small enough to detect local changes in $F(\mathbf{x})$. Let

$$\text{diam}(A) = \sup_{\mathbf{x},\mathbf{x}' \in A} ||\mathbf{x} - \mathbf{x}'||$$

denote the diameter of a set $A \subseteq \mathbb{R}^p$. The diameter of a random cell $\pi_n(\mathbf{X})$ containing $\mathbf{X}$ is denoted by $\text{diam}(\pi_n(\mathbf{X}))$.

Theorem 15.1. *Assume $(\mathbf{X}, Y), (\mathbf{X}_1, Y_1), \ldots, (\mathbf{X}_n, Y_n)$ are i.i.d. from a common distribution $\mathbb{P}$ over $\mathbb{R}^p \times \{0,1\}$. For a partioning classifier (15.1), if $\text{diam}(\pi_n(\mathbf{X})) \xrightarrow{\text{p}} 0$ and $N_n(\mathbf{X}) \xrightarrow{\text{p}} \infty$, then $\mathbb{P}\{\hat{\delta}_n^P(\mathbf{X}) \neq Y\} \to \mathbb{P}\{\delta(\mathbf{X}) \neq Y\}$.*

While it may be tempting to conclude from Theorem 15.1 that classification trees like CART are Bayes' risk consistent, we must be cautious when interpreting this result. This is because the partition π_n in Theorem 15.1 is allowed to depend on n and the features, i.e., $\pi_n := \pi_n(\mathbf{X}_1, \ldots, \mathbf{X}_n)$, but the result does not apply to partitions that depend on the response data $Y_1, \ldots, Y_n$. This is the case with procedures like CART, which uses tree impurity for node splitting—a greedy splitting method that uses Y to better separate class labels (see Chapter 15.5). Therefore, its partition depends on both $\mathbf{X}$ and Y, i.e., $\pi_n := \pi_n((\mathbf{X}_1, Y_1), \ldots, (\mathbf{X}_n, Y_n))$.

We still want to establish consistency for a classification tree. One approach is to consider a modified tree procedure that uses random splitting. Instead of using CART's impurity-based split, we randomly select a variable X and split it at a random location within a node t. For example, assume $\mathscr{X} = [0,1]^p$ and $\mathbb{P}$ is uniform over $[0,1]^p$; we then split t along X using a random split-point drawn from a uniform distribution over the range of X values in t. This process is repeated sequentially. At step k, the tree node to be split, t, is randomly chosen from the current set of k internal nodes. The randomly selected t is then split using the described random procedure. Repeating this for $k = 1, \ldots, K$ steps generate a random partition $\pi_{n,K}$ of size $K + 1$.

Theorem 15.2. *(Biau et al. [2]) The random tree construction on $[0,1]^p$ where $\mathbb{P}$ is uniform is Bayes risk consistent if $K \to \infty$ such that $K/n \to 0$ as $n \to \infty$.*

Theorem 15.2 is very promising for tree classification. It shows that even pure random splitting yields a Bayes risk-consistent classifier. However, similar to Theorem 15.1, the theorem does not directly apply to constructions like CART, which uses the response to guide the partition. Nevertheless, since data-dependent partitions typically perform better than random classifiers, we can expect the same large sample consistency to apply to them.

For completeness, we end this section with a brief sketch of the proof of Theorem 15.1. See Section 6.3 from Devroye et al. [10] for more details.

1. It suffices to show that the tree regression estimator (15.2),

$$\hat{F}_n(\mathbf{x}) = \frac{1}{N_n(\mathbf{x})} \sum_{\mathbf{X}_i \in \mathcal{T}_n(\mathbf{x})} Y_i,$$

   converges to the true regression function $F(\mathbf{x}) = \mathbb{P}\{Y = 1|\mathbf{X} = \mathbf{x}\}$. Specifically, we only need to show $\mathbb{E}[|\hat{F}_n(\mathbf{X}) - F(\mathbf{X})|] \to 0$.
2. Let $F_n(\mathbf{x}) = \mathbb{P}\{Y = 1|\mathbf{X} \in \mathcal{T}_n(\mathbf{x})\}$. By the triangle inequality

$$\mathbb{E}[|\hat{F}_n(\mathbf{X}) - F(\mathbf{X})|] \leq \mathbb{E}[|\hat{F}_n(\mathbf{X}) - F_n(\mathbf{X})|] + \mathbb{E}[|F_n(\mathbf{X}) - F(\mathbf{X})|].$$

   Call the terms on the right side, I and II.
3. Noticing that

$$N(\mathbf{x})\hat{F}_n(\mathbf{x}) \sim \text{Binomial}(N(\mathbf{x}), F_n(\mathbf{x}))$$

   the first term I can be bounded by

$$\mathbb{E}[|\hat{F}_n(\mathbf{X}) - F_n(\mathbf{X})|] \leq \frac{1}{2}\mathbb{P}\{N(\mathbf{X}) \leq k\} + \frac{1}{2k^{1/2}}\mathbb{P}\{N(\mathbf{X}) = 0\}$$

   for any $k > 0$ which shows that I converges to zero by the assumption $N_n(\mathbf{X}) \xrightarrow{p} \infty$.
4. For the term II it can be shown this can be bounded using $\mathbb{P}\{\text{diam}(\mathcal{T}_n(\mathbf{X})) > r\}$ for $r > 0$ which converges to zero by assumption.

15.5 CART Impurity Splitting Rules for Classification

A key concept of CART methodology is *tree impurity*. Impurity measures the dissimilarity between observations in a tree. CART chooses the best split for a node by searching over all possible variables and split values, selecting the one that reduces impurity the most. Reducing tree impurity encourages dissimilar cases to be separated. As the number of nodes increases and dissimilar cases are split into daughter nodes, each node becomes more homogeneous and populated by cases with similar outcomes.

CART splits nodes recursively by maximal reduction in impurity. Doing so pushes dissimilar cases apart, thereby creating terminal nodes that are homogeneous in the response outcome. The decrease in impurity for a split s on a tree node t into left and right daughters t_L and t_R equals

$$\hat{\Gamma}(s,t) = \hat{\Gamma}(t) - \left[\hat{P}(t_L)\hat{\Gamma}(t_L) + \hat{P}(t_R)\hat{\Gamma}(t_R)\right] \tag{15.3}$$

where $\hat{\Gamma}(\cdot)$ is the impurity function and $\hat{P}(t_L)$ and $\hat{P}(t_R)$ are the proportion of cases going left and right.

CART splitting rules for classification include the twoing criterion, the entropy criterion and the Gini criterion. The Gini splitting rule is arguably the most popular and is defined as follows. Let t be the tree node to be split. The Gini node impurity for t is defined as

$$\hat{\Gamma}(t) = \sum_{j=1}^{J} \hat{p}_j(t)(1 - \hat{p}_j(t)).$$

where $\hat{p}_j(t)$ is the proportion of class j cases in t. Let s be a proposed split that splits t into left and right daughter nodes t_L and t_R, then the Gini impurity for the daughters are

$$\hat{\Gamma}(t_L) = \sum_{j=1}^{J} \hat{p}_j(t_L)(1-\hat{p}_j(t_L)), \quad \hat{\Gamma}(t_R) = \sum_{j=1}^{J} \hat{p}_j(t_R)(1-\hat{p}_j(t_R)),$$

where $\hat{p}_j(t_L)$ and $\hat{p}_j(t_R)$ are the class j proportions in t_L and t_R, respectively. The decrease in the node impurity (15.3) is

$$\begin{aligned}\hat{\Gamma}(s,t) &= \sum_{j=1}^{J} \hat{p}_j(t)(1-\hat{p}_j(t)) \\ &\quad - \left[\hat{P}(t_L)\sum_{j=1}^{J} \hat{p}_j(t_L)(1-\hat{p}_j(t_L)) + \hat{P}(t_R)\sum_{j=1}^{J} \hat{p}_j(t_R)(1-\hat{p}_j(t_R))\right].\end{aligned}$$

The quantity

$$\hat{\gamma}(s,t) = \hat{P}(t_L)\sum_{j=1}^{J} \hat{p}_j(t_L)(1-\hat{p}_j(t_L)) + \hat{P}(t_R)\sum_{j=1}^{J} \hat{p}_j(t_R)(1-\hat{p}_j(t_R))$$

is referred to as the Gini index or Gini splitting rule. The best split on a variable is the split-point s maximizing the decrease in node impurity, which is equivalent to minimizing the Gini index. In general, the Gini index tries to send all cases with the largest $\hat{p}_j$ to one daughter node, and all other cases to the complementary daughter. Gini splitting favors pure nodes.

Remark 15.2. In $J = 2$ classification with binary response $Y \in \{0,1\}$, the Gini impurity equals the variance of the outcome multiplied by a factor of two. □

The entropy for t is defined as

$$\hat{\Gamma}(t) = -\sum_{j=1}^{J} \hat{p}_j(t)\log\hat{p}_j(t).$$

By (15.3), the best entropy split is the value of s minimizing

$$\hat{\gamma}(s,t) = -\hat{P}(t_L)\sum_{j=1}^{J} \hat{p}_j(t_L)\log\hat{p}_j(t_L) - \hat{P}(t_R)\sum_{j=1}^{J} \hat{p}_j(t)\log\hat{p}_j(t_R).$$

Entropy splitting tries to break up classes into disjoint sets $\mathscr{C}_1, \mathscr{C}_2$ (left/right daughters) such that $\mathscr{C}_1$ minimizes

$$\left|\sum_{j\in\mathscr{C}} \hat{p}_j(t) - \frac{1}{2}\right|$$

over all subsets $\mathscr{C} \subset \{1,\ldots,J\}$.

The twoing criterion [3, pp 104-106] finds the grouping of all J classes into two superclasses that most improve node impurity when considered as a two-class problem. Under Gini impurity, the best twoing split is the value of s maximizing

$$\hat{\gamma}(s,t) = \frac{\hat{P}(t_L)\hat{P}(t_R)}{4}\left[\sum_{j=1}^{J}\left|\hat{p}_j(t_L) - \hat{p}_j(t_R)\right|\right]^2.$$

The superclasses produced by the twoing criterion are

$$\mathscr{C}_1 = \{j : \hat{p}_j(t_L) \geq \hat{p}_j(t_R)\}, \quad \mathscr{C}_2 = \{j : \hat{p}_j(t_L) < \hat{p}_j(t_R)\}.$$

Similar to entropy splitting, the twoing criterion prefers daughters with equal sample sizes.

15.5.1 Mathematical description of impurity

In this section, we provide a mathematical description of impurity and state a key property of the impurity function. This will be used to show that Gini splitting favors pure splits while entropy splitting favors more balanced nodes. Much of this material comes from the paper by Breiman [4] and readers can consult this reference for details.

Assume s splits a node t sending P_L, P_R fraction of the cases left and right, with split-statistic

$$\Gamma(s,t) := \Gamma(P_L, P_R, \mathbf{p}_L, \mathbf{p}_R)$$

where $\mathbf{p}_L, \mathbf{p}_R$ are the left and right J-vector class proportions. Large $\Gamma(s,t)$ define good splits. Equivalently, there exists real-values $0 \le \alpha_j \le 1$ such that

$$\boldsymbol{\alpha} = (\alpha_1,\dots,\alpha_J)^T, \quad \mathbf{p} = (p_1,\dots,p_J)^T$$

$$P_L = \sum_j \alpha_j p_j, \quad P_R = \sum_j (1-\alpha_j) p_j$$

$$p_{jL} = \alpha_j p_j / P_L, \quad p_{jR} = (1-\alpha_j) p_j / P_R$$

and

$$\Gamma(s,t) = f(\boldsymbol{\alpha}, \mathbf{p}),$$

for some function $f : [0,1]^J \to \mathbb{R}$.

Since we wish to maximize $\Gamma(s,t)$, it suffices to determine for a given $\mathbf{p}$ the value of $\boldsymbol{\alpha} \in [0,1]^J$ that maximizes $f(\boldsymbol{\alpha},\mathbf{p})$. We consider functions f built from the class of *impurity functions.* Call $\phi(\mathbf{p})$ an impurity function if

1. It is convex in $\mathbf{p}$.
2. Its maximum occurs when all p_j are equal.
3. Its minimum occurs when one of the $p_j = 1$.

Define the split-statistic for ϕ as

$$\Gamma(s,t) = \phi(\mathbf{p}) - P_L\phi(\mathbf{p}_L) - P_R\phi(\mathbf{p}_R). \tag{15.4}$$

Notice that Gini corresponds to

$$\phi(\mathbf{p}) = \sum_j p_j(1-p_j),$$

while entropy splitting is

$$\phi(\mathbf{p}) = -\sum_j p_j \log(p_j)$$

and that both satisfy the conditions of being an impurity function.

The next result due to Breiman [4] states an important property of (15.4). It shows that the maximum of the split-statistic is achieved when all the α_j are either zero or one. Hence no class is split into both the left and right daughters when using the Gini or entropy splitting and the impurity function has the property that it does not like to separate classes. Note that this result applies to the theoretical split-statistics defined above and therefore only suggests the behavior that we can expect when they are applied in data settings using CART.

Theorem 15.3. *Notice that (15.4) can be written as $f(\boldsymbol{\alpha},\mathbf{p})$, a convex function in $\mathbf{p}$. Its maximum over $\boldsymbol{\alpha} \in [0,1]^J$ is achieved at a vertex.*

Using Theorem 15.3, for Gini, the best vertex minimizes

$$\begin{aligned}&P_L \sum p_{jL}(1-p_{jL}) + P_R \sum p_{jR}(1-p_{jR})\\ &= P_L \sum_{\alpha_j=1} (p_j/P_L)(1-p_j/P_L) + P_R \sum_{\alpha_j=0} (p_j/P_R)(1-p_j/P_R).\end{aligned}$$

Equivalently, maximize $(1/P_L)\sum_{\mathscr{C}_1} p_j^2 + (1/P_R)\sum_{\mathscr{C}_0} p_j^2$ where $\mathscr{C}_0 = \{j : \alpha_j = 0\}$ and $\mathscr{C}_1 = \{j : \alpha_j = 1\}$. The solution is to send all j_0 class cases to the left and all other cases to the right, where $j_0 = \arg\max_j p_j$.

For entropy, the best vertex maximizes

$$\begin{aligned}&P_L \sum p_{jL}\log(p_{jL}) + P_R \sum p_{jR}\log(p_{jR})\\ &= P_L \sum_{\mathscr{C}_1}(p_j/P_L)\log(p_j/P_L) + P_R \sum_{\mathscr{C}_0}(p_j/P_R)\log(p_j/P_R)\\ &= \sum_j p_j \log(p_j) - P_L\log(P_L) - P_R\log(P_R).\end{aligned}$$

Equivalently, maximize $-P_L\log(P_L) - P_R\log(P_R)$. Therefore entropy favors balanced nodes.

15.5.2 Splitting behavior

Here we take a closer look at the splitting behavior of variables using the previously described splitting rules applied to the prostate dataset. A classification tree was grown using each splitting rule, and the splitting behavior for each variable at the root node was recorded. The right panel of Figure 15.4 shows the Gini index for the best split for each variable, with the smallest value observed for tumor volume. The left panel displays the split-statistic for each of the three rules for observed tumor volume values. The optimized split-point is identical ($\hat{s} \approx 8$) for all three rules.

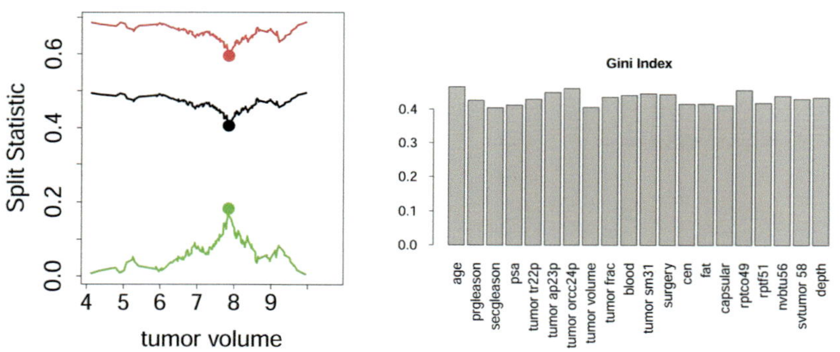

FIGURE 15.4
Left panel: Gini (black), entropy (red), and twoing (green) splitting behavior of tumor volume at the root node for prostate cancer data. Right panel: Gini index for the best split $\hat{s}$ at the root node for each variable. Note that tumor volume has the smallest Gini index.

The splitting behavior of all $p = 20$ variables is shown in Figure 15.5 (solid lines) for each splitting rule. In nearly all cases, the three rules have the same split-point. The dashed lines in Figure 15.5 show the splitting behavior for noisy variables. These are 20 new variables created by permuting each of the original variables. The dashed lines represent the split-statistics for the permuted variables. Notice how the split-point $\hat{s}$ for the randomized variables is often near the edge or located in a region where the split-statistic for the original variable is flat. This behavior

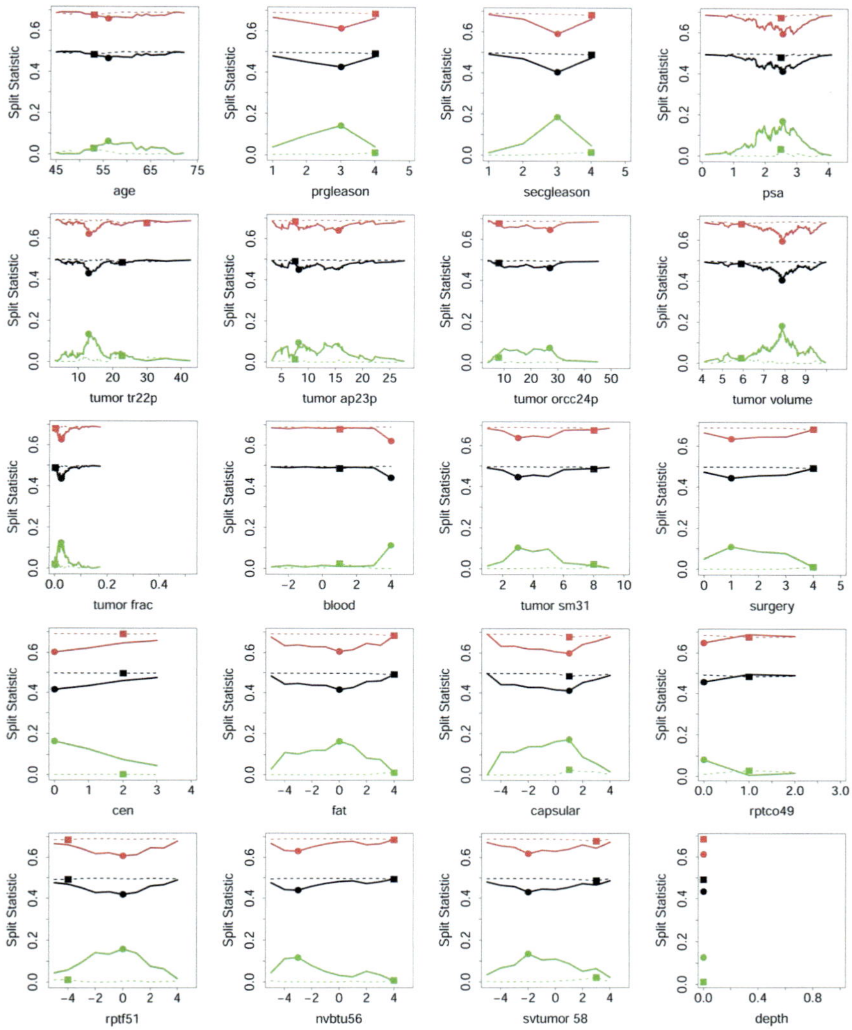

FIGURE 15.5
Solid lines: Gini (black), entropy (red), and twoing (green) splitting behavior for each variable at the root node for prostate cancer data. Dashed lines: splitting behavior of a variable after being randomly permuted.

is called *end-cut preference* (ECP) splitting [3, 22], which favors splits near the edge for noisy variables. As we will see later in the book, this is an extremely useful property, especially in high dimensions.

The prostate data example involves a two-class outcome. To investigate the splitting behavior in a multiclass problem, we used the glass dataset available in the `mlbench` R-package. The goal is to predict the type of glass based on a chemical analysis. There are 9 variables involved in the chemical analysis, and the outcome is one of 6 different types of glass.

As before, we recorded the splits of each variable at the root node. The results are shown in Figure 15.6. Again, we find that the optimized split-point $\hat{s}$ is nearly identical for the three rules. The split-statistic for the optimized split-point for each splitting rule for each variable is given in Figure 15.7. The Gini index and entropy are nearly identical. The minimal split-statistic occurs for Mg (Magnesium). For the twoing rule, the maximum split-statistic also occurs for Mg, so it also splits the root node at this variable. All three rules split Mg at the same value $\hat{s} = 2.68$.

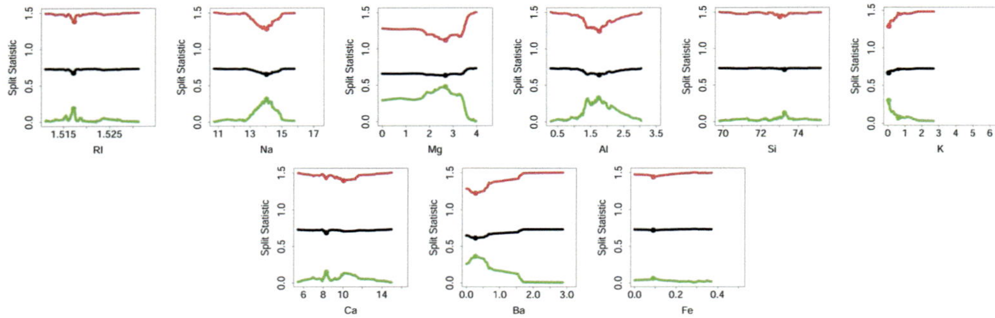

FIGURE 15.6
Gini (black), entropy (red), and twoing (green) splitting behavior for each variable at the root node for glass data.

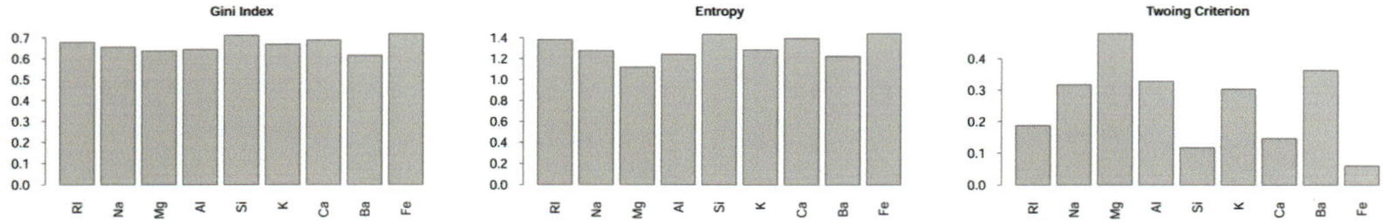

FIGURE 15.7
Gini, entropy, and twoing split-statistic at the optimized split $\hat{s}$ for each variable from glass data.

TABLE 15.1
Relative frequency of glass type for daughter nodes (left and right) when splitting Mg at $\hat{s} = 2.68$. Column entries Gini, entropy, and twoing display the contribution to the split-statistic for each type of glass for each daughter node.

	frequency		Gini		entropy		twoing
glass	left	right	left	right	left	right	
1	0	70	0.00	0.25	0.00	0.36	0.46
2	13	63	0.17	0.24	0.33	0.37	0.20
3	0	17	0.00	0.10	0.00	0.24	0.11
4	13	0	0.17	0.00	0.33	0.00	0.21
5	9	0	0.13	0.00	0.28	0.00	0.15
6	26	3	0.24	0.02	0.36	0.08	0.41

However, when comparing the shapes of the split-statistics for Mg in Figure 15.6, we observe some differences. The entropy and twoing criterion are pronounced at the optimized value $\hat{s}$, in contrast to the Gini index, which is relatively flat across s even at its minimum at $\hat{s}$. The main reason for this is the effect of $\hat{p}_j$ on the statistics. For Gini, the contribution from class label j is $\hat{p}_j(1-\hat{p}_j)$, whereas for entropy it is $-\hat{p}_j \log \hat{p}_j$. The latter can be substantially larger than the Gini when $\hat{p}_j$ is near 0. Therefore, if the split-statistics create daughter nodes with low or high frequencies of certain class labels, the entropy statistic should be larger.

A demonstration of this is provided in Table 15.1, where we compare the contributions to the split-statistics by each glass type label for the split on Mg. Notice how values for entropy are larger than the corresponding entry for Gini whenever the relative frequency of a glass type is either large or small.

15.6 AUC Splitting for Classification

CART splitting rules are centered around impurity. While intuitive, we may prefer a different criterion in some settings. In this section, we develop a classification splitting rule aimed at maximizing the area under the curve (AUC). Chapter 16.12 considers applications of AUC splitting.

AUC refers to the area under the receiver operating characteristic curve (ROC curve). Originally developed for assessing the accuracy of diagnostic tests [15], ROC has become a popular method for evaluating classifier performance. The ROC curve for a test Z plots the true positive rate (sensitivity) against the false positive rate (1−specificity) of the dichotomized variable $Z_\tau = I\{Z > \tau\}$ as $\tau \in \mathbb{R}$ varies over all possible values for Z.

The ROC curve shows all possible trade-offs between sensitivity and specificity for the test. The trace of the curve lies in the square $[0,1] \times [0,1]$; the higher the curve is above the $y = x$ line within this square, the more accurate the test Z is in distinguishing between disease states. The overall accuracy of the test can be assessed by the AUC of the ROC, calculated using the trapezoidal rule. This value estimates the probability that two randomly selected individuals, one diseased and one non-diseased, are correctly ranked by the test [15] (see Theorem 15.4 below).

To see how this applies in practice, let $Y_1, \dots, Y_n$ denote the binary response variables. For consistency with the ROC literature, we assume $Y = 0$ represents the non-diseased group and $Y = 1$ represents the diseased group. Denote by Z the predicted value from the machine learning classifier. Define the ROC function by

$$\begin{aligned}
\mathrm{ROC}(\tau) &= (\text{False Positive Rate}(\tau), \text{True Positive Rate}(\tau)) \\
&= \big(1 - \text{Specificity}(\tau), \text{Sensitivity}(\tau)\big) \\
&= \Big(\mathbb{P}_n\{Z > \tau | Y = 0\}, \mathbb{P}_n\{Z > \tau | Y = 1\}\Big),
\end{aligned}$$

as τ varies over all possible values for $Z_1, \dots, Z_n$. Here $\mathbb{P}_n$ is the empirical measure for $(Z_1, Y_1), \dots, (Z_n, Y_n)$.

Let $\tau_1 < \tau_2 < \cdots < \tau_m$ be the ordered unique Z values and set $\tau_0 = -\infty$. Define the ROC curve as the interpolated curve of $\mathrm{ROC}(\tau_k)$ for $k = 0, \dots, m$. The AUC is defined as the area bounded between this interpolated curve and the horizontal line $\{(x,0) : 0 \le x \le 1\}$. Then

$$\begin{aligned}
\mathrm{AUC} &= \int_{-\infty}^{\infty} \mathbb{P}_n\{Z > \tau | Y = 1\} \mathbb{P}_n\{Z > d\tau | Y = 0\} \\
&= \frac{1}{2} \sum_{k=0}^{m-1} (\alpha_k - \alpha_{k+1})(\beta_k + \beta_{k+1}), \qquad (15.5)
\end{aligned}$$

where $\alpha_k = \mathbb{P}_n\{Z > \tau_k | Y = 0\}$ and $\beta_k = \mathbb{P}_n\{Z > \tau_k | Y = 1\}$.

Figure 15.8 illustrates AUC for a classification analysis from single nucleotide polymorphism (SNP) dataset. The data was obtained from $n = 93$ individuals, 36 of whom had large granular lymphocyte (LGL) leukemia, a rare form of non-Hodgkin's lymphoma. The outcome Y is the presence or absence of LGL. Predictors used in the analysis were allele frequencies from 7 SNPs selected from a 12K SNP array. The predictor Z here is the predicted probability of LGL using a random forest classification analysis (Chapter 16). The AUC in this analysis was 0.598, translating into an error rate of $0.402 = 1 - 0.598$. The ROC curve is shown in the left panel. The right panel displays the values for the predictor Z by LGL status.

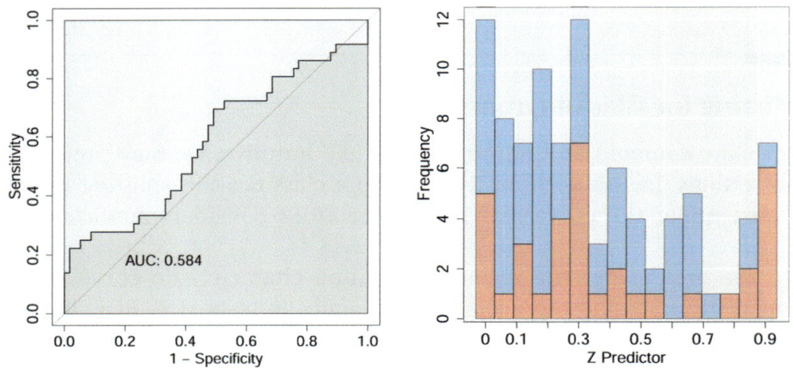

FIGURE 15.8
Illumina 12K SNP classification analysis. The binary outcome Y is the presence or absence of LGL; features are allele frequencies of 7 SNP's from the 12K array believed to be associated with LGL. Left figure is ROC curve where Z is the predicted probability of LGL; Right figure is a histogram of Z by LDL disease status.

15.6.1 AUC measures a test's ability to correctly rank

The AUC calculated by (15.5) equals the Mann-Whitney U-statistic. This is the basis for the AUC's interpretation as a performance metric in machine learning.

Although this result is known [1], for completeness we provide independent proof of this fact. Let $\{Z_{i0}\}_{i=1}^{n_0}$ denote the set of test values having class label $Y = 0$. Also let $\{Z_{j1}\}_{j=1}^{n_1}$ be those test values with $Y = 1$. Note that $n_0 + n_1 = n$. Define

$$S(Z_{i0}, Z_{j1}) = \begin{cases} 0 & \text{if } Z_{i0} > Z_{j1} \\ 1/2 & \text{if } Z_{i0} = Z_{j1} \\ 1 & \text{if } Z_{i0} < Z_{j1}. \end{cases}$$

The Mann-Whitney U-statistic is

$$U = \frac{1}{n_0 n_1} \sum_{i=1}^{n_0} \sum_{j=1}^{n_1} S(Z_{i0}, Z_{j1}). \tag{15.6}$$

By definition (15.6), we can see that U estimates the probability of correctly ranking two individuals randomly selected from the two disease groups:

$$U = \mathbb{P}_n\{Z_{i0} < Z_{j1}\}.$$

The AUC, however, has no such obvious interpretation. Theorem 15.4 given next makes this connection. This shows that the AUC measures the ability of a classifier to correctly separate $Y = 0$ class labels from $Y = 1$ class labels.

Theorem 15.4. *(15.5) equals the Mann-Whitney U-statistic (15.6).*

Proof. By definition, $\alpha_k - \alpha_{k+1} = \mathbb{P}_n\{Z = \tau_{k+1} | Y = 0\}$. Consequently, by (15.5)

$$\begin{aligned}\text{AUC} &= \frac{1}{2}\sum_{k=0}^{m-1} \beta_k \times \mathbb{P}_n\{Z = \tau_{k+1} | Y = 0\} \\ &\quad + \frac{1}{2}\sum_{k=0}^{m-1} \beta_{k+1} \times \mathbb{P}_n\{Z = \tau_{k+1} | Y = 0\}.\end{aligned}$$

The following is a key identity needed for the proof. For each $k = 0, \ldots, m-1$,

$$\begin{aligned}&\beta_k \times \mathbb{P}_n\{Z = \tau_{k+1} | Y = 0\} \\ &= \frac{1}{n_0 n_1} \sum_{Z_{i0}=\tau_{k+1}} \sum_{j=1}^{n_1} I\{Z_{j1} > \tau_k\} \\ &= \frac{1}{n_0 n_1} \sum_{Z_{i0}=\tau_{k+1}} \sum_{j=1}^{n_1} \Big(I\{Z_{j1} > Z_{i0}\} + I\{Z_{j1} = Z_{i0}\} \Big) \\ &= \frac{1}{n_0 n_1} \sum_{Z_{i0}=\tau_{k+1}} \sum_{j=1}^{n_1} \left(S(Z_{i0}, Z_{j1}) + \frac{1}{2} I\{Z_{j1} = Z_{i0}\} \right).\end{aligned}$$

By similar reasoning we have another key identity,

$$\begin{aligned}&\beta_{k+1} \times \mathbb{P}_n\{Z = \tau_{k+1} | Y = 0\} \\ &= \frac{1}{n_0 n_1} \sum_{Z_{i0}=\tau_{k+1}} \sum_{j=1}^{n_1} I\{Z_{j1} > Z_{i0}\} \\ &= \frac{1}{n_0 n_1} \sum_{Z_{i0}=\tau_{k+1}} \sum_{j=1}^{n_1} \left(S(Z_{i0}, Z_{j1}) - \frac{1}{2} I\{Z_{j1} = Z_{i0}\} \right).\end{aligned}$$

Deduce that

$$\begin{aligned}\text{AUC} &= \frac{1}{n_0 n_1} \sum_{k=0}^{m-1} \sum_{Z_{i0}=\tau_{k+1}} \sum_{j=1}^{n_1} S(Z_{i0}, Z_{j1}) \\ &= \frac{1}{n_0 n_1} \sum_{k=1}^{m} \sum_{Z_{i0}=\tau_k} \sum_{j=1}^{n_1} S(Z_{i0}, Z_{j1}).\end{aligned}$$

It is easily seen that the right-hand side is (15.6). □

15.6.2 AUC splitting rule

Here we derive a splitting rule for maximizing AUC. For this we will require a closed form expression for the ROC curve and its area. Without loss of generality consider splitting the root node. Let L and R denote the left and right daughter nodes resulting from this split. We can summarize the frequency of observations in the daughter nodes using the following 2×2 table (Table 15.2).

Following a split to the node let $Z \in [0, 1]$ be the resulting predictor of $Y = 1$. There can be at most two distinct values of Z, one for each daughter node:

$$\tau_1 = \frac{n_{1L}}{n_L} \ (Z \text{ values of left node}), \qquad \tau_2 = \frac{n_{1R}}{n_R} \ (Z \text{ values of right node}),$$

where n_L, n_R are the sample sizes for the left and right daughters ($n_L + n_R = n$; $n_{0L} + n_{1L} = n_L$; $n_{0R} + n_{1R} = n_R$). The ROC curve resulting from the split is therefore

$$\text{ROC}(\tau) = \Big(\mathbb{P}_n\{Z > \tau | Y = 0\}, \mathbb{P}_n\{Z > \tau | Y = 1\} \Big)$$

as τ varies over $\tau \in \{\tau_1, \tau_2\}$.

TABLE 15.2
Frequency of cases in left and right daughter nodes by disease status.

	Left	Right
$Y = 0$	n_{0L}	n_{0R}
$Y = 1$	n_{1L}	n_{1R}

In deriving the AUC it will be helpful to break the calculation into three cases: $\tau_2 > \tau_1$, $\tau_2 < \tau_1$ and $\tau_1 = \tau_2$.

1. Consider first $\tau_2 > \tau_1$. Because the predicted value for the left daughter node is τ_1, it is not possible for $Z > \tau$ for $Z \in L$ when $\tau = \tau_1$. Therefore

$$\text{ROC}(\tau_1) = \Big(\mathbb{P}_n\{Z > \tau_1, Z \in R \mid Y = 0\}, \mathbb{P}_n\{Z > \tau_1, Z \in R \mid Y = 1\}\Big).$$

   Now because all Z in R are larger than τ_1 by assumption, we have

$$\text{ROC}(\tau_1) = \Big(\frac{n_{0R}}{n_{0L} + n_{0R}}, \frac{n_{1R}}{n_{1L} + n_{1R}}\Big). \tag{15.7}$$

   Furthermore, because it is not possible for $Z > \tau_2$ as a result of the assumption τ_2 is the largest observed value, we have $\text{ROC}(\tau_2) = (0,0)$. Therefore the ROC curve contains only three points, $(0,0)$, $\text{ROC}(\tau_1)$ and $(1,1)$.

2. Now consider when $\tau_2 < \tau_1$. When $\tau = \tau_1$, it is not possible for $Z > \tau$ as τ is the largest observed value and therefore $\text{ROC}(\tau_1) = (0,0)$. When $\tau = \tau_2 < \tau_1$ it is not possible for $Z > \tau$ for $Z \in R$; hence

$$\text{ROC}(\tau_2) = \Big(\mathbb{P}_n\{Z > \tau_2, Z \in L \mid Y = 0\}, \mathbb{P}_n\{Z > \tau_2, Z \in L \mid Y = 1\}\Big).$$

   Now because all Z in L are larger than τ_2 because $\tau_1 > \tau_2$, we have

$$\text{ROC}(\tau_2) = \Big(\frac{n_{0L}}{n_{0L} + n_{0R}}, \frac{n_{1L}}{n_{1L} + n_{1R}}\Big). \tag{15.8}$$

   Therefore the ROC curve contains only three points, $(0,0)$, $\text{ROC}(\tau_2)$ and $(1,1)$.

3. Finally, consider the third case $\tau_1 = \tau_2$. Because it is not possible for $Z > \tau$ when $\tau = \tau_1 = \tau_2$, it follows that $R(\tau_1) = R(\tau_2) = (0,0)$ and consequently the ROC curve contains only two points, $(0,0)$ and $(1,1)$.

The AUC can be obtained by revisiting each of the three cases. For (15.7), let

$$(\alpha, \beta) = \Big(\frac{n_{0R}}{n_{0L} + n_{0R}}, \frac{n_{1R}}{n_{1L} + n_{1R}}\Big).$$

Then the AUC equals the area under the three points $(0,0)$, (α, β), $(1,1)$, which equals

$$\begin{aligned}
&\frac{1}{2}\alpha\beta + (1-\alpha)\beta + \frac{1}{2}(1-\alpha)(1-\beta) \\
&\quad = \frac{1}{2}\alpha\beta + \beta - \alpha\beta + \frac{1}{2}(1 - \alpha - \beta + \alpha\beta) \\
&\quad = \frac{1}{2}(1 - \alpha + \beta).
\end{aligned}$$

Now for the second case (15.8), notice that

$$(1-\alpha, 1-\beta) = \left(\frac{n_{0L}}{n_{0L}+n_{0R}}, \frac{n_{1L}}{n_{1L}+n_{1R}}\right).$$

Therefore, the ROC curve is comprised of the three points $(0,0)$, $(1-\alpha, 1-\beta)$, $(1,1)$. The AUC is therefore

$$\begin{aligned}
&\frac{1}{2}(1-\alpha)(1-\beta) + \alpha(1-\beta) + \frac{1}{2}\alpha\beta \\
&\quad = \frac{1}{2}(1-\alpha-\beta+\alpha\beta) + \alpha - \alpha\beta + \frac{1}{2}\alpha\beta \\
&\quad = \frac{1}{2}(1-\beta+\alpha).
\end{aligned}$$

Finally, for the third case, since there are only two points $(0,0)$ and $(1,1)$ on the ROC curve, the AUC equals $1/2$.

By symmetry, the AUC should not depend on the labeling of the terminal nodes, and therefore the AUC is defined as the maximum value of the three cases

$$\max\left(\frac{1}{2}(1-\alpha+\beta), \frac{1}{2}(1-\beta+\alpha), \frac{1}{2}\right) = \max\left(\frac{1}{2}(1-\alpha+\beta), \frac{1}{2}(1-\beta+\alpha)\right),$$

where the right-hand side follows since an AUC of $1/2$ implies $\alpha = \beta$ and because the maximum of $(1-\alpha+\beta)$ and $(1-\beta+\alpha)$ can never be less than 1.

It is now a simple matter to define the AUC splitting rule. For a proposed split into left and right daughter nodes L and R, calculate

$$(\alpha, \beta) = \left(\frac{n_{0R}}{n_{0L}+n_{0R}}, \frac{n_{1R}}{n_{1L}+n_{1R}}\right).$$

The split-statistic is

$$\hat{\gamma}(\alpha, \beta) = \max\left(\frac{1}{2}(1-\alpha+\beta), \frac{1}{2}(1-\beta+\alpha)\right).$$

In order to maximize the AUC, choose the split yielding the maximum split-statistic value $\hat{\gamma}(\alpha, \beta)$.

15.6.3 Extension to the multiclass setting

The AUC splitting rule can be extended to the multiclass setting $Y \in \{1, \ldots, J\}$ where $J > 2$. The interpretation for the diagnostic test Z is as follows. Following a proposed split to a tree node, the diagnostic test Z predicts $Y = j_2$ versus $Y = j_1$ for all possible $j_1 \neq j_2$. To take into account all these comparisons, the AUC is defined as the average over all $j_1 < j_2$ comparisons [14].

We summarize the frequencies resulting from the tree node split as a $J \times 2$ table. Given there are J classes and two nodes, we have $2J$ predicted probabilities (see Table 15.3):

$$\text{left node: } \tau_L^{(1)} = \frac{n_{1L}}{n_L}, \tau_L^{(2)} = \frac{n_{2L}}{n_L}, \cdots, \tau_L^{(J)} = \frac{n_{JL}}{n_L}.$$

$$\text{right node: } \tau_R^{(1)} = \frac{n_{1R}}{n_R}, \tau_R^{(2)} = \frac{n_{2R}}{n_R}, \cdots, \tau_R^{(J)} = \frac{n_{JR}}{n_R}.$$

For each $j_1 < j_2$ comparison, we can treat j_1 as the baseline "no-disease" case and j_2 as the "disease" case. Define

$$(\alpha_{j_1,j_2}, \beta_{j_1,j_2}) = \left(\frac{n_{j_1R}}{n_{j_1}}, \frac{n_{j_2R}}{n_{j_2}}\right).$$

TABLE 15.3
Left: Frequency of cases. Right: Predicted probabilities for left and right daughter nodes for the $J > 2$ multiclass setting.

	Left	Right	
$Y = 1$	n_{1L}	n_{1R}	n_1
$Y = 2$	n_{2L}	n_{2R}	n_2
$\vdots$	$\vdots$	$\vdots$	
$Y = J$	n_{JL}	n_{JR}	n_J
	n_L	n_R	n

	Left	Right
$Y = 1$	$\tau_L^{(1)}$	$\tau_R^{(1)}$
$Y = 2$	$\tau_L^{(2)}$	$\tau_R^{(2)}$
$\vdots$	$\vdots$	$\vdots$
$Y = J$	$\tau_L^{(J)}$	$\tau_R^{(J)}$
	1	1

Define the following statistics

$$U_{j_1,j_2} = \frac{1}{2}\left(1 - \alpha_{j_1,j_2} + \beta_{j_1,j_2}\right), \quad U_{j_2,j_1} = \frac{1}{2}\left(1 - \beta_{j_1,j_2} + \alpha_{j_1,j_2}\right).$$

Using a similar argument as before, the AUC for the comparison between j_1 and j_2 is

$$\text{AUC}_{j_1,j_2} = \max\left(U_{j_1,j_2}, U_{j_2,j_1}\right).$$

The AUC for the proposed split is defined as the average over all $j_1 < j_2$ comparisons. Thus,

$$\begin{aligned} \text{AUC} &= \frac{2}{J(J-1)} \sum_{j_1<j_2} \text{AUC}_{j_1,j_2} \\ &= \frac{2}{J(J-1)} \sum_{j_1<j_2} \max\left(U_{j_1,j_2}, U_{j_2,j_1}\right). \end{aligned}$$

To define the AUC multiclass splitting rule for a proposed split into left and right daughter nodes L and R, calculate the $J \times 2$ frequencies as in Table 15.3. Then calculate U_{j_1,j_2} and U_{j_2,j_1} for all $j_1 < j_2$. The split-statistic is

$$\hat{\gamma} = \frac{2}{J(J-1)} \sum_{1 \le j_1 < j_2 \le J} \max\left(U_{j_1,j_2}, U_{j_2,j_1}\right).$$

To maximize the AUC, choose the split yielding the maximum split-statistic $\hat{\gamma}$.

15.7 Regression Trees Using CART

Regression trees are used for data with continuous outcomes, $Y \in \mathbb{R}$. The induction for regression trees using CART proceeds in nearly the same manner as the classification case. Only the definition of impurity changes.

Let s be a proposed split for a node t sending cases left and right to daughters t_L and t_R. For continuous Y, the impurity of t is defined as the within node sample variance

$$\hat{\Gamma}(t) = \frac{1}{N} \sum_{i \in t} (Y_i - \overline{Y}_t)^2,$$

where $\overline{Y}_t$ is the sample mean for t and N is the sample size of t. The impurity for the daughter nodes are

$$\hat{\Gamma}(t_L) = \frac{1}{N_L}\sum_{i\in t_L}(Y_i - \overline{Y}_{t_L})^2, \quad \hat{\Gamma}(t_R) = \frac{1}{N_R}\sum_{i\in t_R}(Y_i - \overline{Y}_{t_R})^2,$$

where $\overline{Y}_{t_L}$ and $\overline{Y}_{t_R}$ are the sample means for t_L and t_R with sample sizes N_L and N_R.

Maximizing (15.3) is equivalent to minimizing $\hat{P}(t_L)\hat{\Gamma}(t_L) + \hat{P}(t_R)\hat{\Gamma}(t_R)$. Thus we want to minimize

$$\begin{aligned}\hat{\gamma}(s,t) &= \hat{P}(t_L)\hat{\Gamma}(t_L) + \hat{P}(t_R)\hat{\Gamma}(t_R) \\ &= \frac{1}{N}\sum_{i\in t_L}(Y_i - \overline{Y}_{t_L})^2 + \frac{1}{N}\sum_{i\in t_R}(Y_i - \overline{Y}_{t_R})^2. \end{aligned} \tag{15.9}$$

In other words, the optimized split-point is the value minimizing the weighted sample variance. We refer to (15.9) as the weighted variance splitting rule.

In nonparametric regression, the goal is to estimate the conditional mean $F(\mathbf{x}) = \mathbb{E}[Y|\mathbf{X} = \mathbf{x}]$. Using our previous notation, where $\Pi_n(\mathbf{x})$ identifies the terminal node containing $\mathbf{x}$, the tree estimator for F is defined as

$$\hat{F}_n(\mathbf{x}) = \frac{1}{|\Pi_n(\mathbf{x})|}\sum_{\mathbf{X}_i\in\Pi_n(\mathbf{x})} Y_i. \tag{15.10}$$

This estimator is identical to the estimator (15.2) used for the two-class problem where $Y \in \{0,1\}$. This is not surprising since both are trying to estimate a conditional mean. In the two-class problem, we have the identity

$$F(\mathbf{x}) = \mathbb{P}\{Y = 1|\mathbf{X} = \mathbf{x}\} = \mathbb{E}[Y|\mathbf{X} = \mathbf{x}],$$

which explains why the two methods agree.

There is also another interesting connection between the two settings. It turns out that Gini splitting is identical to weighted variance splitting (15.9). To see this, recall that the Gini node impurity for t_L is

$$\sum_{j=1}^{J}\hat{p}_j(t_L)(1-\hat{p}_j(t_L)) = 1 - \sum_{j=1}^{J}\hat{p}_j(t_L)^2 = 1 - \sum_{j=1}^{J}\left(\frac{N_{jL}}{N_L}\right)^2$$

where N_{jL} is the class j frequencies in t_L. Define node impurity for t_R in a similar fashion. Notice that $\sum_j N_{jL} = N_L$, $\sum_j N_{jR} = N_R$ and $N_L + N_R = N$. The Gini index is

$$\begin{aligned}&\frac{N_L}{N}\sum_{j=1}^{J}\hat{p}_j(t_L)(1-\hat{p}_j(t_L)) + \frac{N_R}{N}\sum_{j=1}^{J}\hat{p}_j(t_R)(1-\hat{p}_j(t_R)) \\ &= 1 - \left[\frac{N_L}{N}\sum_{j=1}^{J}\hat{p}_j(t_L)^2 + \frac{N_R}{N}\sum_{j=1}^{J}\hat{p}_j(t_R)^2\right] \\ &= 1 - \left[\frac{N_L}{N}\sum_{j=1}^{J}\left(\frac{N_{jL}}{N_L}\right)^2 + \frac{N_R}{N}\sum_{j=1}^{J}\left(\frac{N_{jR}}{N_R}\right)^2\right] \\ &= 1 - \frac{1}{N}\left[\frac{1}{N_L}\sum_{j=1}^{J}N_{jL}^2 + \frac{1}{N_R}\sum_{j=1}^{J}N_{jR}^2\right]. \end{aligned} \tag{15.11}$$

For $J = 2$, we have the simpler expression (see Remark 15.2)

$$\begin{aligned}&2\frac{N_L}{N}\hat{p}_1(t_L)(1-\hat{p}_1(t_L)) + 2\frac{N_R}{N}\hat{p}_1(t_R)(1-\hat{p}_1(t_R)) \\ &= \frac{2}{N}\left[N_1 - N_L\left(\frac{N_{1L}}{N_L}\right)^2 - N_R\left(\frac{N_{1R}}{N_R}\right)^2\right],\end{aligned}$$

and the goal is to maximize

$$\frac{1}{N_L}N_{1L}^2 + \frac{1}{N_R}N_{1R}^2.$$

Using the identity

$$\sum_{i\in t_L}(Y_i - \overline{Y}_{t_L})^2 + \sum_{i\in t_R}(Y_i - \overline{Y}_{t_R})^2 = \sum_{i=1}^{N} Y_i^2 - \left[N_L\overline{Y}_{t_L}^2 + N_R\overline{Y}_{t_R}^2\right],$$

minimizing (15.9) is equivalent to maximizing

$$N_L\overline{Y}_{t_L}^2 + N_R\overline{Y}_{t_R}^2 = \frac{1}{N_L}\left(\sum_{i\in t_L} Y_i\right)^2 + \frac{1}{N_R}\left(\sum_{i\in t_R} Y_i\right)^2. \tag{15.12}$$

This is identical to Gini splitting since $N_{1L} = \sum_{i\in t_L} Y_i$ and $N_{1R} = \sum_{i\in t_R} Y_i$ when $Y_i \in \{0,1\}$.

As we have also shown the estimators in regression and classification are identical, we have established the following useful result.

Theorem 15.5. *Running a regression tree under weighted variance splitting with binary outcomes $Y \in \{0,1\}$ yields the same tree and conditional mean estimator as running a $J = 2$ classification tree under Gini splitting using the same data.*

15.7.1 Consistency

The regression estimator $\hat{F}_n(\mathbf{x})$ (15.10) for the conditional mean $F(\mathbf{x}) = \mathbb{E}[Y|\mathbf{X} = \mathbf{x}]$ is a partitioning estimator similar to that described for classification (15.2). As this uses continuous Y it could be more correctly called a partitioning histogram estimator.

A *partitioning histogram estimator* for $F : \mathbb{R}^p \to \mathbb{R}$ is an estimator expressible as

$$\hat{F}_n^H(\mathbf{x}) = \frac{1}{|\Pi_n(\mathbf{x})|}\sum_{\mathbf{X}_i\in\Pi_n(\mathbf{x})} Y_i \tag{15.13}$$

for a partition $\Pi_n := \Pi_n(\mathbf{X}_1,\ldots,\mathbf{X}_n)$ of $\mathbb{R}^p$.

Given the strong connection to the partitioning classifier, it should not be surprising that weak consistency for (15.13) holds under the same conditions as Theorem 15.1. The following result due to Gyorfi et al. [Theorem 4.2, 13] makes use of Stone's theorem to be discussed in the next section to prove this.

Theorem 15.6. *Assume $(\mathbf{X}, Y), (\mathbf{X}_1, Y_1), \ldots, (\mathbf{X}_n, Y_n)$ are i.i.d. from a common distribution $\mathbb{P}$ over $\mathbb{R}^p \times \mathbb{R}$ such that $\mathbb{E}(Y^2) < \infty$. If $diam(\Pi_n(\mathbf{X})) \xrightarrow{p} 0$ and $N_n(\mathbf{X}) \xrightarrow{p} \infty$, then*

$$\lim_{n\to\infty} \mathbb{E}\left\{\int_{\mathbb{R}^p} \left(\hat{F}_n^H(\mathbf{x}) - F(\mathbf{x})\right)^2 \mathbb{P}(d\mathbf{x})\right\} = 0.$$

Thus, $\hat{F}_n^H$ is weakly consistent.

The same caveat regarding partitions in Theorem 15.1 applies to Theorem 15.6. The result only applies to partitions that depend on the features. Therefore, it does not apply to the regression estimator (15.10) since its partition depends on both the features and the response.

Putting this aside, Theorem 15.6 has some interesting implications. First, it implies L_2-risk consistency for $\hat{F}_n^H$. The L_2-risk for an estimator $\hat{F}$ is defined as

$$R(\hat{F}) = \mathbb{E}\left[\left(Y - \hat{F}(\mathbf{X})\right)^2\right]$$

which is the generalization error (7.1) under L_2-loss. This can be further decomposed into the underlying variance σ^2 and the mean-squared error (see Chapter 7.3)

$$R(\hat{F}) = \sigma^2 + \mathbb{E}\left[\left(\hat{F}(\mathbf{X}) - F(\mathbf{X})\right)^2\right].$$

Therefore, Theorem 15.6 shows that $\hat{F}_n^H$ is L_2-risk consistent. Also because regression is harder than classification (recall Theorem 7.1), a corollary of Theorem 15.6 is that the estimator is also Bayes risk consistent when $Y \in \{0,1\}$ is a binary response.

15.8 Partitioning Histograms and Classifiers are Locally Weighted Averaged Regression Estimators

We have described regression trees as partitioning histograms and classification trees as partitioning classifiers. More generally, both of these trees can be viewed as weighted average estimators. This connection is summarized below.

1. For a regression tree with partition $\mathcal{T}_n$, (15.10) can be rewritten as

$$\hat{F}_n(\mathbf{x}) = \frac{1}{\sum_{j=1}^n 1_{\{X_j \in \mathcal{T}_n(\mathbf{x})\}}} \sum_{i=1}^n Y_i = \sum_{i=1}^n W_{n,i}(\mathbf{x}) Y_i$$

   where $W_{n,i}(\mathbf{x}) = 1_{\{\mathbf{X}_i \in \mathcal{T}_n(\mathbf{x})\}} / \sum_{j=1}^n 1_{\{\mathbf{X}_j \in \mathcal{T}_n(\mathbf{x})\}}$ are positive weights that sum to 1 for each $\mathbf{x}$. Therefore, this shows $\hat{F}_n$ is a weighted averaged estimator.

2. For classification trees, let $F_j(\mathbf{x}) = \mathbb{P}\{Y = C_j \mid \mathbf{X} = \mathbf{x}\}$ be the conditional probability of the event $Y = C_j$ given $\mathbf{X}$. The tree estimator for this value (c.f. (15.2)) is

$$\hat{F}_{n,j}(\mathbf{x}) = \frac{1}{\sum_{j=1}^n 1_{\{X_j \in \mathcal{T}_n(\mathbf{x})\}}} \sum_{i=1}^n 1_{\{Y_i = C_j\}} = \sum_{i=1}^n W_{n,i}(\mathbf{x}) 1_{\{Y_i = C_j\}}$$

   where $W_{n,i}(\mathbf{x})$ is defined as in the regression case. Therefore, this can also be interpreted as a weighted average estimator. In fact, for $J = 2$ classification, $Y \in \{0,1\}$, the estimators for regression and classification are identical.

This useful fact connects regression and classification trees to the general class of estimators called *local weighted averaging regression functions*. These are functions expressible as

$$\hat{F}_n^W(\mathbf{x}) = \sum_{i=1}^n W_{n,i}(\mathbf{x}) Y_i$$

where $0 \le W_{n,i}(\mathbf{x}) := W_n(\mathbf{x}, \mathbf{X}_1, \ldots, \mathbf{X}_n) \le 1$ are weights summing to one. The weak consistency of such estimators has been established by Stone [24] (this result is in fact what is used to prove Theorem 15.1).

Stone's consistency result relies on three conditions:

1. There exists a $1 \le C < \infty$ such that for each integrable function $F \ge 0$

$$\mathbb{E}\left\{\sum_{i=1}^n W_{n,i}(\mathbf{X}) F(\mathbf{X}_i)\right\} \le C\,\mathbb{E}\{F(\mathbf{X})\}, \quad \text{for all } n \ge 1.$$

2. $\sum_{i=1}^{n} W_{n,i}(\mathbf{X})1_{\{\|\mathbf{X}_i-\mathbf{X}\|>r\}} \xrightarrow{\text{p}} 0$ for all $r > 0$.
3. $\max_{1\le i\le n} W_{n,i}(\mathbf{X}) \xrightarrow{\text{p}} 0$.

The first condition requires the mean estimator of the regression function to be bounded by the population mean. The second condition ensures that the estimator at a point $\mathbf{x}$ is only influenced by nearby points. This is a natural assumption for local weighted estimators, as we usually want the weights to be large for observations $\mathbf{X}_i$ close to $\mathbf{x}$ and small for those far away. Finally, the last condition ensures that the weights become small asymptotically.

Theorem 15.7. *(Stone [24]) Assume $(\boldsymbol{X}, Y), (\boldsymbol{X}_1, Y_1), \ldots, (\boldsymbol{X}_n, Y_n)$ are i.i.d. from a common distribution $\mathbb{P}$ over $\mathbb{R}^p \times \mathbb{R}$ such that $\mathbb{E}(Y^2) < \infty$. If the above three conditions hold, then*

$$\lim_{n\to\infty} \mathbb{E}\left\{\int_{\mathbb{R}^p} \left(\hat{F}_n^W(\boldsymbol{x}) - F(\boldsymbol{x})\right)^2 \mathbb{P}(d\boldsymbol{x})\right\} = 0, \quad \text{where } F(\boldsymbol{x}) = \mathbb{E}[Y|\boldsymbol{X} = \boldsymbol{x}].$$

Thus, the locally weighted averaging estimator is weakly consistent.

15.9 Survival Trees

Survival trees are used for time-to-event data, a common scenario in medical sciences. This type of analysis focuses on understanding survival for the individual based on their variables. Time-to-event can refer to the time to death from a certain disease, time until cancer recurrence, time until the first occurrence of a symptom, or all-cause mortality.

One of the issues in dealing with time-to-event data is the presence of censoring. Generally, this means that the event times for some individuals in a study are not observed exactly and are only known to fall within certain time intervals. Right-censoring is one of the most common types of censoring encountered. This occurs when the event of interest is observed only if it happens before some prespecified time. For example, a patient might be monitored for two weeks without the occurrence of a symptom and then discharged from a hospital. Such a patient is said to be right-censored because the time-to-event is known to exceed two weeks, but the exact event time is unknown. Another example of right-censoring occurs when patients enter a study at different times, and the study is predetermined to end at a certain time. All patients who do not experience an event within the study period are right-censored. A more detailed discussion of survival analysis and censoring is provided in Chapter 19.

Trees can be used to analyze right-censored survival data. We will refer to such trees as survival trees (see Figure 15.9 for an example). As in CART, survival trees are grown by reducing impurity to ensure that terminal nodes are populated by individuals with similar survival behavior. A commonly used splitting rule is the log-rank test. Predicted values for a terminal node in a survival tree are often taken to be the Kaplan-Meier estimator [19] for the survival function.

To make these ideas more concrete, we introduce the following notation. Let $(T_1, \delta_1), \ldots,$ (T_N, δ_N) be the survival times and censoring values for the N individuals in a tree node. An individual i is said to be right-censored at time T_i if $\delta_i = 0$; otherwise, the individual is said to have died at T_i if $\delta_i = 1$. In the case of death, T_i will be referred to as an event time, and the death as an event. We define the log-rank test as follows. Let $t_1 < t_2 < \cdots < t_{N^*}$ be the $N^* \le N$ distinct death times, and let $d_L(t), d_R(t)$ be the number of deaths at time t in the left and right daughter nodes, and let $Y_L(t), Y_R(t)$ be the number of individuals within the left and right daughters who are at risk at time t. Note that the number at risk are those who are alive at time t or who have an

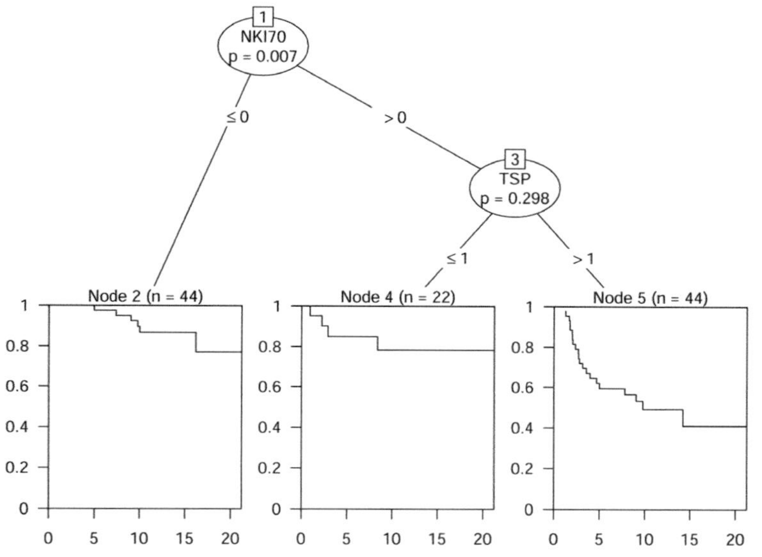

FIGURE 15.9
Survival tree for breast cancer patients where variables NKI70 and TSP are gene signatures. Each terminal node displays the Kaplan-Meier survival curves for patients within that node. For example, the right terminal node (node 5) is the survival function for patients having both the NKI70 and TSP gene signature.

event (death) at t. More precisely, for daughters formed by splitting a variable x at a split-point s,

$$Y_L(t) = \#\{T_i \geq t,\ x_i \leq s\}, \quad Y_R(t) = \#\{T_i \geq t,\ x_i > s\}.$$

Define $Y(t) = Y_L(t) + Y_R(t)$ and $d(t) = d_L(t) + d_R(t)$.

The log-rank splitting rule is

$$\hat{\gamma} = \left| \frac{\sum_{i=1}^{N^*} \left(d_L(t_i) - Y_L(t_i) \dfrac{d(t_i)}{Y(t_i)} \right)}{\sqrt{\sum_{l=1}^{N^*} \dfrac{Y_L(t_i)}{Y(t_i)} \left(1 - \dfrac{Y_L(t_i)}{Y(t_i)}\right) \left(\dfrac{Y(t_i) - d(t_i)}{Y(t_i) - 1} \right) d(t_i)}} \right|.$$

The best split is found by maximizing $\hat{\gamma}$.

The Kaplan-Meier estimator for the survival function (the probability $\mathbb{P}\{T > t\}$ where T is the event time; see Chapter 19) for the daughter nodes is defined as

$$\hat{S}_L(t) = \prod_{t_i \leq t} \left(1 - \frac{d_L(t_i)}{Y_L(t_i)}\right), \quad \hat{S}_R(t) = \prod_{t_i \leq t} \left(1 - \frac{d_R(t_i)}{Y_R(t_i)}\right).$$

All cases in a terminal node receive the same estimated survival function. As illustration, Figure 15.9 displays Kaplan-Meier estimators for the three terminal nodes in a survival tree for breast cancer patients.

15.10 Multivariate Splitting Rules

We describe a class of multivariate splitting rules that can be used to construct multivariate trees. For generality, we will address the scenario where the response values can be either continuous or categorical, or both. We will use the notation $\mathbf{Y} = (Y_1, \ldots, Y_p)^T$ to denote the p-dimensional multivariate outcome. For ease of notation, the first $1 \le l \le q$ coordinates refer to the continuous values and the remaining coordinates $q+1 \le l \le p$ refer to the categorical values (the notation can be modified if all coordinates are continuous or all coordinates are categorical).

We first describe a normalized composite splitting rule [25]. This rule is an additive composition of CART univariate splitting rules. The advantage of the composite rule is computational speed. A disadvantage is that by additively working over the $\mathbf{Y}$ coordinates separately, it does not take into account correlation. To address this, another multivariate splitting rule will be described using Mahalanobis distance that will account for correlation.

15.10.1 Regression

To describe the multivariate composite splitting rule, we begin by considering multivariate regression. First we recall the univariate case. Therefore, let $q = 1$ and let Y be the scalar continuous outcome value. Recall that the CART weighted variance splitting rule (15.9) minimizes

$$\hat{\gamma}_{\text{reg}}(s,t) = \frac{1}{N}\sum_{i \in t_L}(Y_i - \overline{Y}_{t_L})^2 + \frac{1}{N}\sum_{i \in t_R}(Y_i - \overline{Y}_{t_R})^2$$

where s splits a node t into left and right daughters t_L and t_R with sample sizes N, N_L, N_R for t, t_L, t_R, respectively. Values $\overline{Y}_{t_L}, \overline{Y}_{t_R}$ are the sample means for t_L, t_R.

To extend this to the multivariate case $q > 1$, the split-statistic is applied to each coordinate separately. Let $\mathbf{Y}_i = (Y_{i1}, \ldots, Y_{iq})^T$ be the q-dimensional multivariate outcome for case i. The generalization of the univariate split-rule is

$$\hat{\gamma}_{\text{reg},q}(s,t) = \sum_{l=1}^{q}\left\{\sum_{i \in t_L}(Y_{il} - \overline{Y}_{t_{L_l}})^2 + \sum_{i \in t_R}(Y_{il} - \overline{Y}_{t_{R_l}})^2\right\}$$

where $\overline{Y}_{t_{L_l}}$ and $\overline{Y}_{t_{R_l}}$ are the sample means of the l-th response coordinate for left and right daughter nodes. The goal is to minimize $\hat{\gamma}_{\text{reg},q}(s,t)$.

As stated, $\hat{\gamma}_{\text{reg},q}(s,t)$ will only be effective if each outcome coordinate is measured on the same scale, otherwise we could have a coordinate Y_l with very large values whose contribution would dominate the split-rule. This is easily corrected by standardizing each coordinate according to

$$\frac{1}{N}\sum_{i \in t} Y_{il} = 0, \quad \frac{1}{N}\sum_{i \in t} Y_{il}^2 = 1, \quad 1 \le l \le q.$$

The standardization is applied prior to splitting a node. To make this standardization clear, we denote the standardized responses by Y_{il}^*. With some elementary manipulations, it can be verified that minimizing $\hat{\gamma}_{\text{reg},q}(s,j)$, using Y_{il}^* in place of Y_{il}, is equivalent to maximizing (cf. (15.12))

$$\hat{\gamma}^*_{\text{reg},q}(s,t) = \sum_{l=1}^{q}\left\{\frac{1}{N_L}\left(\sum_{i \in t_L} Y_{il}^*\right)^2 + \frac{1}{N_R}\left(\sum_{i \in t_R} Y_{il}^*\right)^2\right\}. \tag{15.14}$$

15.10.2 Classification

For multivariate classification, an averaged standardized Gini splitting rule is used. First consider the univariate case (i.e., the multiclass problem) where the outcome Y_i is a class label from the set $\{1,\dots,J\}$ where $J \geq 2$. By (15.11), the best split is obtained by maximizing

$$\hat{\gamma}^*_{\text{class}}(s,t) = \sum_{j=1}^{J}\left[\frac{1}{N_L}\left(\sum_{i\in t_L} Z_{i(j)}\right)^2 + \frac{1}{N_R}\left(\sum_{i\in t_R} Z_{i(j)}\right)^2\right]$$

where $Z_{i(j)} = 1_{\{Y_i=j\}}$. Now consider multivariate classification where $Y_{il} \in \{1,\dots,J_l\}$ for $q+1 \leq l \leq p$. Notice that the number of class labels can vary over the coordinates. The Gini split-statistic is applied to each coordinate yielding the multivariate Gini splitting rule

$$\hat{\gamma}^*_{\text{class},p-q}(s,t) = \sum_{l=q+1}^{p}\left[\frac{1}{J_l}\sum_{j=1}^{J_l}\left\{\frac{1}{N_L}\left(\sum_{i\in t_L} Z_{il(j)}\right)^2 + \frac{1}{N_R}\left(\sum_{i\in t_R} Z_{il(j)}\right)^2\right\}\right] \quad (15.15)$$

where $Z_{il(j)} = 1_{\{Y_{il}=j\}}$. The normalization $1/J_l$ is included in (15.15) so as to standardize the contribution of the Gini split for each coordinate l.

15.10.3 Multivariate normalized composite splitting rule

Observe that (15.14) and (15.15) are equivalent optimization problems, with optimization over standardized Y^*_{il} for regression and binary values $Z_{il(j)}$ for classification. Given this equivalence, we can combine the two splitting rules to form a composite splitting rule. The mixed outcome splitting rule $\hat{\gamma}_{\text{mix},p}(s,t)$ is a composite standardized split-rule of weighted variance (15.14) and Gini index splitting (15.15); i.e.,

$$\hat{\gamma}^*_{\text{mix},p}(s,t) = \hat{\gamma}^*_{\text{reg},q}(s,t) + \hat{\gamma}^*_{\text{class},p-q}(s,t).$$

The best split is obtained by maximizing $\hat{\gamma}^*_{\text{mix},p}(s,t)$.

15.10.4 Mahalanobis distance

The multivariate regression splitting rule (15.14) is a composite weighted variance rule and as such, since it is additive in the components of the outcomes, it does not take into account correlation between the coordinates that could potentially improve performance.

In order to incorporate this correlation, we introduce a new splitting rule that makes use of the Mahalanobis distance. Let $\mathbf{Z}$ be a random element with mean $\boldsymbol{\mu}_\mathbf{Z}$ and variance-covariance $\boldsymbol{\Sigma}_\mathbf{Z}$. Recall the Mahalanobis distance squared from $\mathbf{Z}$ to its mean value $\boldsymbol{\mu}_\mathbf{Z}$ is defined by

$$D_M(\mathbf{Z}) = (\mathbf{Z}-\boldsymbol{\mu}_\mathbf{Z})^T\boldsymbol{\Sigma}_\mathbf{Z}^{-1}(\mathbf{Z}-\boldsymbol{\mu}_\mathbf{Z}).$$

We cannot directly use this distance for the splitting rule because of the possibility of $\boldsymbol{\Sigma}_\mathbf{Z}$ becoming singular. Such a scenario will happen when growing a tree as the number of observations decreases rapidly which will result in singular covariance matrices for small sample sized nodes. Therefore, we use a modified definition by using the generalized inverse. We use Theorem 8.4 which describes the generalized inverse of a matrix $\mathbf{Q}$ in terms of its SVD (singular value decomposition). We will be interested in symmetric square matrices of the form $\mathbf{Q} = \mathbf{L}^T\mathbf{L}$.

Hereafter, we will assume all coordinates of $\mathbf{Y}$ are continuous: thus $q = p$ and $\mathbf{Y} = (Y_1,\dots,Y_p)^T \in \mathbb{R}^p$. Consider splitting a tree node t. Let $\mathbf{L}^*_t$ be the outcome matrix for t formed by "row-stacking" the centered $\mathbf{Y}$ values in t (we use a $*$ notation to emphasize that the matrix is centered)

$$\mathbf{L}_t^* = \begin{pmatrix} (\mathbf{Y}_1 - \overline{\mathbf{Y}}_t)^T \\ \vdots \\ (\mathbf{Y}_n - \overline{\mathbf{Y}}_t)^T \end{pmatrix}_{n\times p}.$$

Here $\overline{\mathbf{Y}}_t$ is the p-dimensional vector of sample means for $\mathbf{Y}$ in t. If there are n observations in t, then $\mathbf{L}_t^*$ is an $n \times p$ matrix. Because $\mathbf{L}_t^*$ is centered, the sample covariance matrix for the data is $n^{-1}\mathbf{Q}_t^*$ where $\mathbf{Q}_t^* = (\mathbf{L}_t^*)^T\mathbf{L}_t^*$. Apply Theorem 8.4 to $\mathbf{Q}_t^*$ to obtain its generalized inverse $(\mathbf{Q}_t^*)^+$.

Now we define the splitting rule. Suppose that t is split left and right, t_L and t_R. Let $\overline{\mathbf{Y}}_{t_L}$ and $\overline{\mathbf{Y}}_{t_R}$ be the p-dimensional sample mean vectors for $\mathbf{Y}$ in t_L and t_R, respectively. The Mahalanobis multivariate split-statistic is

$$\begin{aligned} &D_{M,t}(t_L, t_R) \\ &= \frac{N_L}{N}\sum_{i\in t_L}(\mathbf{Y}_i - \overline{\mathbf{Y}}_{t_L})^T(\mathbf{Q}_t^*)^+(\mathbf{Y}_i - \overline{\mathbf{Y}}_{t_L}) + \frac{N_R}{N}\sum_{i\in t_R}(\mathbf{Y}_i - \overline{\mathbf{Y}}_{t_R})^T(\mathbf{Q}_t^*)^+(\mathbf{Y}_i - \overline{\mathbf{Y}}_{t_R}). \end{aligned}$$

The best split for t is obtained by minimizing $D_{M,t}(t_L, t_R)$.

15.11 Super Greedy Trees

Even though CART has enjoyed much success, one of its well-known problems is the instability that arises from the nature of its splits. In this section, we will discuss *super greedy trees* (SGTs), which were briefly covered in Chapter 7.8, that can be used to resolve this issue.

Let X denote a generic attribute from the feature vector $\mathbf{X} = (X_1, \ldots, X_p)^T \in \mathbb{R}^p$. For generality, we will refer to a tree node A as a cell and a split on cell A as a cut along the cell. In CART, a univariate cut on X at the value s for cell A results in left and right daughter cells

$$A_L = \{\mathbf{X} \in A : X \le s\}, \quad A_R = \{\mathbf{X} \in A : X > s\}. \tag{15.16}$$

Recall that this cuts A along a univariate coordinate direction parallel to the coordinate axis and is called an axis-aligned cut. The result is a tree partition of hyper-rectangles parallel to the axes, leading to a discontinuous decision boundary with a rough, step-like pattern (see Figure 15.10 (a)). This structure lacks flexibility and requires many splits to be effective, resulting in increased variance and instability.

A two-dimensional classification example demonstrating that the issues with axis-aligned cuts can even lead to an inconsistent classifier is provided in Devroye et al. [10, Chapter 20.8]. An understanding of the issue can be gained by considering the size of CART's partition space. The partition space for axis-aligned cuts is a subset of the space of half-space partitions. A half-space on $\mathbb{R}^p$ is defined by a separating hyperplane:

$$\beta_1 X_1 + \cdots + \beta_p X_p \le \alpha_0.$$

Although this is a fairly rich class, mathematically, the issue with CART is that the partitioning number (Chapter 7.8) for half-space partitions may be too small for effective empirical risk reduction in certain cases.

To address this, the authors suggest an alternative procedure: a greedier tree construction that cuts an entire p-dimensional rectangle rather than a univariate cell, as CART does, thereby creating a more flexible partition space. If T is a rectangle inside A, then A is cut into the

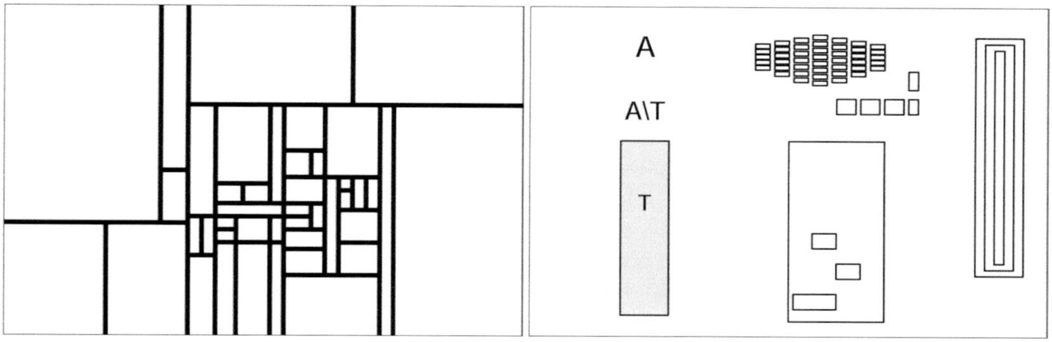

(a) *CART partition.* *Greedy tree cuts.*

FIGURE 15.10
CART versus a greedy tree.

rectangle T and its complement $A \setminus T$ (see Figure 15.10 (b)). The cut with the largest empirical risk reduction is selected. If the number of tree splits K_n satisfies $K_n = o((n/\log n)^{1/3}) \to \infty$, where n is the sample size, then the greedy tree classifier is Bayes risk consistent [10, Theorem 20.9].

Ishwaran [16] described a generalization of this procedure, super greedy trees (SGTs). Previously discussed in Chapter 7.8, the basic principle is to construct a tree by choosing splits from an interesting class of geometric objects. For example, instead of using half-space cuts, we might use rectangular cuts, as in the original greedy tree construction of Devroye et al. [10]. In fact, even more flexible cuts than this are described in Chapter 15.12. Using more flexible cuts results in more efficient risk reduction. To further enhance empirical risk minimization, the SGT procedure even grows the tree sequentially by splitting the best cut first.

15.11.1 SGT induction

We describe the induction steps for building the SGT. The notation used is similar to Chapter 7. Let $Y \in \mathscr{Y}$ denote the scalar response variable and let $\mathbf{X} \in \mathscr{X} \subseteq \mathbb{R}^p$ be the continuous feature vector. We use $F : \mathscr{X} \to \mathbb{R}$ to denote the unknown real-valued target function, $L(Y, F)$ for the loss function and $\mathbb{P}$ for the distribution for $(\mathbf{X}, Y)$. The population risk for F is defined by

$$R(F) = \int L(y, F(\mathbf{x}))\, d\mathbb{P}(\mathbf{x}, y).$$

The empirical risk for F is

$$R_n(F) = \frac{1}{n}\sum_{i=1}^{n} L(Y_i, F(\mathbf{X}_i))$$

where $(\mathbf{X}_1, Y_1), \ldots, (\mathbf{X}_n, Y_n) \in \mathbb{R}^p \times \mathscr{Y}$ are i.i.d. from $\mathbb{P}$. The loss function $L(Y, u)$ is a real-valued differentiable convex function of $u \in \mathbb{R}$ for each Y. Write $L'(Y, u)$ for the partial derivative of L with respect to u.

The SGT construction uses a non-recursive algorithm employing *best split first* (BSF). The tree is initialized at step $k = 0$ by setting the partition $\pi_{n,0} = \mathbb{R}^p$. The algorithm runs sequentially, so that at the start of step $k \geq 1$, there is a partition $\pi_{n,k-1}$ of $\mathscr{X}$ of size k. A cell from $\pi_{n,k-1}$ is cut into two daughters, refining $\pi_{n,k-1}$ to a new partition $\pi_{n,k}$ of size $k+1$. In accordance with BSF, the cell selected to cut the tree is the one that reduces empirical risk the most. After repeating for K steps, the final SGT partition $\pi_{n,K}$ of size $K+1$ is obtained. Associated with each cell A of $\pi_{n,K}$

is a local estimator $\hat{F}_{n,A}$; together, these piecewise estimators form tree estimator for F:

$$\hat{F}_n(\mathbf{x}, \Pi_{n,K}) = \sum_{A \in \Pi_{n,K}} \hat{F}_{n,A} 1_{\{\mathbf{x} \in A\}}, \quad \text{where } \hat{F}_{n,A} \in \mathbb{R}.$$

A key aspect of SGTs is the use of multivariate cuts; however, not all cuts are allowed, and certain rules apply. To cut a cell A, the SGT selects an object from $\mathscr{A}$, which is a predetermined class of geometric objects. This object and its complement become the two daughters of the parent cell A. For convenience, we can call the object selected from $\mathscr{A}$ the "left daughter," denoted by A_L, and the complementary set the "right daughter," denoted by $A_R = A \setminus A_L$. Notice that $A = A_L \cup A_R$. The cut is called a *permissible* SGT cut if it satisfies the following definition.

Definition 15.1. Given a partition Π, an object $T \in \mathscr{A}$ is called permissible if for each $A \in \Pi$, $T \cap A = \emptyset$ or $T \subseteq A$. A permissible SGT cut for a cell $A \in \Pi$ is a cut along a permissible geometric object $T \subset A$ that separates A into new cells T and $A \setminus T$. For example, Figure 15.10 (b) displays a region A and a rectangle $T \subset A$ (in dark gray); thus if $\mathscr{A}$ is the class of rectangles, T is a permissible cut that splits A into T and the complement $A \setminus T$ (light gray region). □

We can now proceed to describe an SGT cut in more detail. Let $\Pi_{n,k-1}$ denote the current partition of size k. To cut $\Pi_{n,k-1}$, the procedure searches for the cell $A \in \Pi_{n,k-1}$ and the permissible cut of A that provides the largest empirical risk reduction. The empirical risk for a partition Π is

$$R_n(\Pi) = \sum_{A \in \Pi} R_n(A), \quad \text{where } R_n(A) = \frac{1}{n} \sum_{i \in A} L(Y_i, \hat{F}_{n,A}).$$

Therefore, to determine the optimal cell for splitting, let

$$\hat{F}_n(\mathbf{x}, \Pi_{n,k-1}) = \sum_{A \in \Pi_{n,k-1}} \hat{F}_{n,A} 1_{\{\mathbf{x} \in A\}}$$

be the current SGT estimator. Consider splitting A using permissible cuts A_L and $A_R = A \setminus A_L$. Then replacing $\Pi_{n,k-1}$ with

$$\Pi_{n,k} = \underbrace{(\Pi_{n,k-1} \setminus A)}_{\text{replace } A} \bigcup \underbrace{A_L}_{\text{with } A_L} \bigcup \underbrace{A_R}_{\text{and } A_R},$$

the piecewise updated estimator is

$$\hat{F}_n(\mathbf{x}, \Pi_{n,k}) = \hat{F}_n(\mathbf{x}, \Pi_{n,k-1}) - \hat{F}_{n,A} 1_{\{\mathbf{x} \in A\}} + \left[\hat{F}_{n,A_L} 1_{\{\mathbf{x} \in A_L\}} + \hat{F}_{n,A_R} 1_{\{\mathbf{x} \in A_R\}}\right].$$

Thus, the task is to find $\{A_L, A_R, \hat{F}_{n,A_L}, \hat{F}_{n,A_R}\}$ that maximizes the empirical risk reduction

$$\begin{aligned}
&R_n(\Pi_{n,k-1}) - R_n(\Pi_{n,k}) \\
&\quad = \frac{1}{n} \sum_{i \in A} L(Y_i, \hat{F}_{n,A}) - \left[\frac{1}{n} \sum_{i \in A} L\Big(Y_i, \hat{F}_{n,A_L} 1_{\{\mathbf{x} \in A_L\}} + \hat{F}_{n,A_R} 1_{\{\mathbf{x} \in A_R\}}\Big)\right] \\
&\quad = \frac{1}{n} \sum_{i \in A} L(Y_i, \hat{F}_{n,A}) - \left[\frac{1}{n} \sum_{i \in A_L} L(Y_i, \hat{F}_{n,A_L}) + \frac{1}{n} \sum_{i \in A_R} L(Y_i, \hat{F}_{n,A_R})\right].
\end{aligned}$$

It suffices to minimize the last two terms in the square brackets, from which it follows by convexity and differentiability of the loss function that $\hat{F}_{n,A_L}$ and $\hat{F}_{n,A_R}$ are the cell estimators defined by

$$\frac{1}{n} \sum_{i \in A_L} L'(Y_i, \hat{F}_{n,A_L}) = 0, \quad \frac{1}{n} \sum_{i \in A_R} L'(Y_i, \hat{F}_{n,A_R}) = 0. \tag{15.17}$$

By induction, the parent cell estimator $\hat{F}_{n,A}$ is defined similarly

$$\frac{1}{n}\sum_{i\in A} L'(Y_i, \hat{F}_{n,A}) = 0. \tag{15.18}$$

Therefore, the optimal cut is the cell with the smallest empirical risk under cell estimators (15.17).

The SGT algorithm is summarized in Algorithm 15.1 where we use the notation

$$\Delta_n(A, A_L, A_R) = \frac{1}{n}\sum_{i\in A} L(Y_i, \hat{F}_{n,A}) - \left[\frac{1}{n}\sum_{i\in A_L} L(Y_i, \hat{F}_{n,A_L}) + \frac{1}{n}\sum_{i\in A_R} L(Y_i, \hat{F}_{n,A_R})\right] \tag{15.19}$$

to define the empirical risk reduction from splitting a cell A into permissible cells A_L, A_R.

Algorithm 15.1 Super Greedy Tree Induction

1: **Input:** Set $\Pi_{n,0} = \mathscr{X}$; set number of cuts K.
2: **for** $k = 1, \ldots, K$ **do**
3: For each cell A in the current partitition $\Pi_{n,k-1}$, calculate $\Delta_n(A, A_L, A_R)$ defined by (15.19) for all permissible cuts A_L, A_R.
4: Let A^* be the cell with permissible cut A_L^*, A_R^* having maximal value, $\Delta_n(A^*, A_L^*, A_R^*) \geq \Delta_n(A, A_L, A_R)$. Then split A^* by A_L^*, A_R^* to obtain the updated partition $\Pi_{n,k} = (\Pi_{n,k-1} \setminus A^*) \bigcup A_L^* \bigcup A_R^*$.
5: **end for**
6: The SGT estimator is $\hat{F}_n(\mathbf{x}, \Pi_{n,K}) = \sum_{A\in\Pi_{n,K}} \hat{F}_{n,A} 1_{\{\mathbf{x}\in A\}}$.

15.11.2 Partioning number for SGTs

SGTs utilize a diverse range of multivariate cuts for tree splitting. Consequently, it is important to compare the size of this tree space to that induced by CART splits and consider the potential for overfitting due to its larger size. In Chapter 7.8, we learned that the complexity of a tree partition space $\mathscr{P}_K$ of size $K+1$ is described by the partitioning number $\mathfrak{p}(\mathscr{P}_K, n)$. This number represents the maximum distinct partitions of any set of n vectors using partitions from $\mathscr{P}_K$. It was shown that if $K < n$ and the partitioning number is polynomial in n, the empirical risk converges uniformly, suggesting that SGTs may be consistent. Since this result depends on bounding the partitioning number, we now show that the partitioning number for an SGT can be bounded by the shatter coefficient defined previously Definition 7.1.

For this proof, let us consider a SGT where $\mathscr{A}$ equals the space of all rectangles on $\mathscr{X}$. Also for simplicity let us take $p = 2$ for now. Then Figure 15.11 displays all the possible partitions of size 2 from the first step ($k = 1$) of the SGT for a specific set of $n = 4$ data points. Here each panel displays the space $\Pi_0 = \mathscr{X}$ being cut into a permissible rectangle (in blue) and its complement (in white), resulting in a partition Π_1 of size 2. Let $\mathscr{P}_1$ be the family of all such partitions on $\mathbb{R}^2$. Figure 15.11 shows there are 16 ways of separating the points using two-dimensional rectangles. In fact, 16 equals the maximum number of ways of separating any 4 points using rectangles.

It can be seen that $\mathfrak{p}(\mathscr{P}_1, 4) \leq 16$. Now consider the next step $k = 2$ in the tree construction where one of the previous rectangles is split. This results in a partition of size $k+1 = 3$. Let $\mathscr{P}_2$ be the family of all partitions of size 3 resulting from the SGT construction following the $k = 2$ step. To calculate the partitioning number for $\mathscr{P}_2$, let us focus on one of the 16 subpanels for the moment and let us call this subpanel A. Then A is comprised of two cells: A rectangle T and its complement $A \setminus T$ with n_T and $n - n_T$ data points, respectively. If the $k = 2$ cut is on T, the shatter coefficient $s(\mathscr{A}, n_T)$ bounds the maximum number of distinct partitions arising

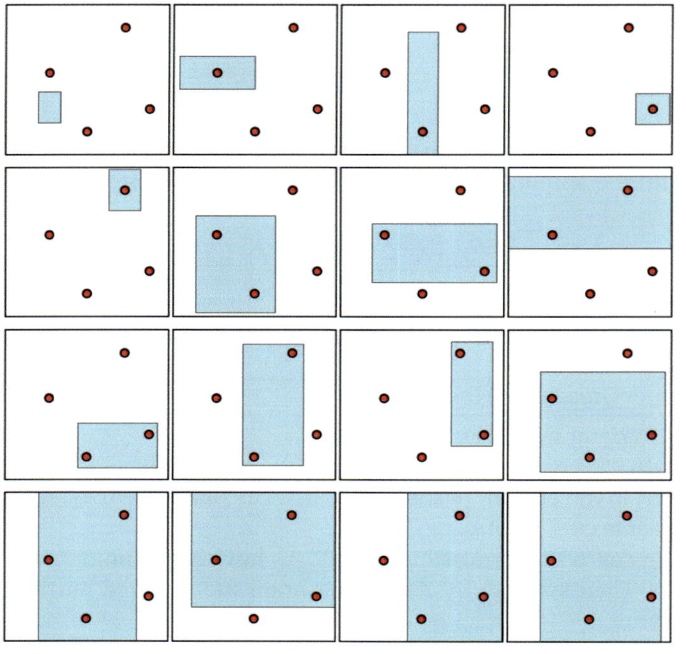

FIGURE 15.11
$2^4 = 16$ *rectangles separating* $n = 4$ *points in* $p = 2$ *dimensions.*

from the rectangle cut. Likewise, if the $k = 2$ cut is on $A \setminus T$, then $s(\mathscr{A}, n - n_T)$ bounds the maximum number of partitions arising from that cut. Thus, the $k = 2$ cut can yield no more than $s(\mathscr{A}, n_T) + s(\mathscr{A}, n - n_T)$ distinct partitions of the n points in this subpanel. Applying this to each of the $s(\mathscr{A}, n) = 16$ subpanels yields the bound

$$\begin{aligned} \mathfrak{p}(\mathscr{P}_2, 4) &\leq s(\mathscr{A}, n) \cdot \{s(\mathscr{A}, n_T) + s(\mathscr{A}, n - n_T)\} \\ &= 16 \cdot \{s(\mathscr{A}, n_T) + s(\mathscr{A}, n - n_T)\}. \end{aligned}$$

This bound can be significantly simplified by the insight that $s(\mathscr{A}, \cdot)$ is superadditive (Theorem 15.8). That is, for $0 \leq m \leq n$,

$$s(\mathscr{A}, m) + s(\mathscr{A}, n - m) \leq s(\mathscr{A}, n). \tag{15.20}$$

Putting (15.20) into the bound above gives $\mathfrak{p}(\mathscr{P}_2, n) \leq s(\mathscr{A}, n)^2 = 16^2$. By induction, $\mathfrak{p}(\mathscr{P}_k, n) \leq s(\mathscr{A}, n)^k$ for $k \geq 1$.

Looking carefully at this argument, it is clear there is nothing special about using rectangles for our class of sets $\mathscr{A}$, or the fact that we used $n = 4$ points, or for that matter that the dimension is $p = 2$. The salient point is that a cut splits a region A into T and its complement $A \setminus T$, and these separate the data in at most $s(\mathscr{A}, n_T) + s(\mathscr{A}, n - n_T)$ ways, which is in turn bounded by (15.20) (and notice if the size of the region to be split is $n^* \leq n$ the bound from (15.20) is $s(\mathscr{A}, n^*)$ and this can be bounded by $s(\mathscr{A}, n^*) \leq s(\mathscr{A}, n)$). Therefore, if superadditivity holds, we have $\mathfrak{p}(\mathscr{P}_k, n) \leq s(\mathscr{A}, n)^k$ for any class of sets $\mathscr{A}$ and any $k \geq 1$. This observation yields the following result.

Theorem 15.8. *For any SGT, if the shatter coefficient is convex in* n *and* $s(\mathscr{A}, 0) = 0$*, then* $\mathfrak{p}(\mathscr{P}_k, n) \leq s(\mathscr{A}, n)^k$.

Proof. We have proven if superadditivity holds, $\mathfrak{p}(\mathscr{P}_k, n) \leq s(\mathscr{A}, n)^k$ for any SGT using k cuts. To complete the proof we show superadditivity holds under the asserted condition that the shatter coefficient $s(\mathscr{A}, n)$ is convex in n and $s(\mathscr{A}, 0) = 0$. This holds because a real-valued function $s^*(u)$ defined on $[0, A]$ satisfies superadditivity $s^*(u_1 + u_2) \geq s^*(u_1) + s^*(u_2)$ for $u_1, u_2, u_1 + u_2 \in [0, A]$ if it is convex on $[0, A]$ and $s^*(0) \leq 0$ [7]. Thus, define $s^*(u) = s(\mathscr{A}, u)$ for $u \in [0, n]$. If s is convex in n, then s^* is convex. Also $s^*(0) = s(\mathscr{A}, 0) = 0$, $s^*(u_1) = s(\mathscr{A}, m)$, $s^*(u_2) = s(\mathscr{A}, n-m)$ and $s^*(u_1+u_2) = s(\mathscr{A}, n)$ for $u_1 = m$, $u_2 = n - m$, $u_1 + u_2 = n \in [0, n]$. □

We can learn several things from Theorem 15.8. First, we can see that in any given step k of the SGT construction, there will be a very large number of cuts available to separate the data points in a cell. Our result shows that this number is bounded by the shatter coefficient. Although this is an upper bound, it suggests that the number of available cuts can increase significantly by increasing the complexity of the geometric object which in turn should lead to greater empirical risk reduction. For example, hyperplane cuts on $\mathbb{R}^p$ as used by CART have a shatter coefficient of $O(n^p)$ (see Chapter 7.7), whereas by Theorem 13.9 of [10], the shatter coefficient using more general cuts of the form $\{g(\mathbf{x}) \leq 0 : g \in \mathscr{V}\}$ is order $O(n^V)$ where $\mathscr{V}$ is a finite-dimensional vector with dimension V. For the lasso implementation of SGTs of Chapter 15.12 (in particular see (15.24)) which makes use of these types of cuts, the shatter coefficient can be as large as $O(n^{p^3})$, which is subtantially larger than CART.

Theorem 15.8 also offers some guidance on the question of overfitting. The VC-theorem (Chapter 7.7) states that consistent learning using empirical risk minimization is only possible if the VC-dimension is finite. To determine whether consistent learning is achievable we can turn to the shatter coefficient as the VC-dimension of a class of sets and the shatter coefficient are related: either the shatter coefficient is subexponential in n and the VC-dimension is finite, or the shatter coefficient is 2^n in which case the VC-dimension is infinite [23]. The same subexponential requirement for the partitioning number has been observed for the consistency of data-dependent partitioning estimators; see [20] and Devroye et al. [10, Chapter 21]. Now according to Theorem 15.8, if the shatter coefficient is polynomial, then the partitioning number is polynomial, therefore if the number of tree splits remains smaller than the sample size, the results from Chapter 7.8 suggest that consistent learning can be achieved using SGTs. In particular, this provides evidence that SGTs will not overfit asymptotically using cuts of the form $\{g(\mathbf{x}) \leq 0 : g \in \mathscr{V}\}$ as used by the SGT lasso implementation to be described in Chapter 15.12.

15.11.3 Regression: CART is a special case of a SGT

Commonly used impurity measures such as within node sum of squares in regression and the Gini index in classification are equivalent to minimizing training error, and thus CART in these special cases amounts to an empirical risk minimization method. Thus, it stands to reason that CART and SGTs should be connected in some way. In this section we describe this connection in more detail. For concreteness, and to aid readers in understanding SGTs, we focus on the regression setting under L_2-loss.

First note that since the loss function is $L(Y, u) = (Y - u)^2$ and its derivative is $L'(Y, u) = 2u$, it follows from convexity that the cell estimators (15.17) and (15.18) are

$$\hat{F}_{n,A_L} = \overline{Y}_{n,A_L} = \frac{\sum_{i \in A_L} Y_i}{|A_L|}, \; \hat{F}_{n,A_R} = \overline{Y}_{n,A_R} = \frac{\sum_{i \in A_R} Y_i}{|A_R|}, \; \hat{F}_{n,A} = \overline{Y}_{n,A} = \frac{\sum_{i \in A} Y_i}{|A|}.$$

Hence, the decrease in empirical risk (15.19) from splitting A into daughters A_L, A_R is

$$\begin{aligned} \Delta_n(A, A_L, A_R) &= R_n(A) - [R_n(A_L) + R_n(A_R)] \qquad (15.21) \\ &= \frac{1}{n} \sum_{i \in A} (Y_i - \overline{Y}_{n,A})^2 - \left[\frac{1}{n} \sum_{i \in A_L} (Y_i - \overline{Y}_{n,A_L})^2 + \frac{1}{n} \sum_{i \in A_R} (Y_i - \overline{Y}^2_{n,A_R} \right]. \end{aligned}$$

Therefore, maximizing $\Delta_n(A)$ is equivalent to minimizing the last two terms in the square brackets, which is the weighted variance splitting rule (15.9).

As a consequence of the convexity of the loss function, the empirical risk cannot increase. However, the following result proves a stronger result showing that (15.21) is strictly positive for impure nodes.

Theorem 15.9. *Tree splits in regression trees using weighted variance splitting are equivalent to splits using empirical risk minimization under L_2-loss. Moreover, any split on a node A into daughters A_L and A_R satisfies $\Delta_n(A, A_L, A_R) \geq 0$. Call the split pure if $\overline{Y}_A = \overline{Y}_{A_L} = \overline{Y}_{A_R}$. Then $\Delta_n(A, A_L, A_R) > 0$ if and only if the split is impure.*

Proof of Theorem 15.9. Let $Z_i = Y_i - \overline{Y}_{n,A}$ for $i \in A$. For notational ease, write $\Delta_n(A)$ for $\Delta_n(A, A_L, A_R)$. Then using $\overline{Z}_A = 0$,

$$
\begin{aligned}
n\Delta_n(A) &= \sum_{i\in A} Z_i^2 - \sum_{i\in A_L} (Z_i - \overline{Z}_{A_L})^2 - \sum_{i\in A_R} (Z_i - \overline{Z}_{A_R})^2 \\
&= \left[\sum_{i\in A} Z_i^2 - \sum_{i\in A_L} Z_i^2 - \sum_{i\in A_R} Z_i^2\right] + |A_L|\overline{Z}_{A_L}^2 + |A_R|\overline{Z}_{A_R}^2 \\
&= |A_L|\overline{Z}_{A_L}^2 + |A_R|\overline{Z}_{A_R}^2.
\end{aligned}
$$

Because the last line is non-negative, $\Delta_n(A) \geq 0$. But not only that, $\Delta_n(A)$ is strictly positive if and only if either $\overline{Z}_{A_L}$ or $\overline{Z}_{A_R}$ is non-zero, which is the purity condition of the theorem. □

Theorem 15.9 states that empirical risk reduction for regression under L_2-loss is equivalent to weighted variance splitting and that risk reduction is strictly positive in the case of impure splits. A sufficient condition for impurity is that Y is distinct in A (which should happen if Y is continuous). Then any split on A (even if it is random) strictly reduces empirical risk. Extensions of this result to classification under proper loss functions are considered in Exercise 15.14, thus showing that this principle is quite general.

Importantly, Theorem 15.9 holds for any type of cut. Thus if splits are selected from $\mathscr{A}$ equal to all axis-aligned cuts, then it follows that a sequentially constructed regression tree using CART is a special case of an SGT under L_2-loss. Thus we have shown the following:

> *Replacing recursive splitting with BSF, a sequentially constructed regression tree using CART with weighted variance splitting is equivalent to an SGT under L_2-loss for $\mathscr{A}$ equal to the space of all axis-aligned cuts. Moreover, empirical risk reduction is always strictly positive for an SGT under L_2-loss whenever the parent node to be split is impure. This holds regardless of the type of cut.*

Algorithm 15.2 states the SGT procedure for regression. Observe in Step 3 where the tree is cut that $\mathscr{A}$ can be any type of geometric cutting space. Setting this to the space of axis-aligned cuts results in a CART implementation.

15.11.4 Consistency for regression

In this section, we establish some asymptotic properties for SGTs. To keep things concrete we continue to focus on regression under L_2-loss. Throughout we assume the following model

$$
Y_i = F(\mathbf{x}_i) + \epsilon_i, \quad i = 1 \ldots, n, \tag{15.22}
$$

where $F(\mathbf{x}) = \mathbb{E}[Y|\mathbf{X} = \mathbf{x}]$ is the unknown regression function. Here $\{\epsilon_i\}_{i=1}^n$ are i.i.d. and independent of $\{\mathbf{X}_i\}_{i=1}^n$ such that $\mathbb{E}(\epsilon_i|\mathbf{X}_i) = 0$ and $\mathbb{E}(\epsilon_i^2|\mathbf{X}_i) = \sigma^2$. The response is bounded such that $|Y| \leq C$ for a fixed $C < \infty$. The following result is due to Ishwaran and Lee [17].

Algorithm 15.2 Super Greedy Tree for Regression under L_2-loss

Input: Set $\Pi_{n,0} = \mathscr{X}$; set number of cuts K.

for $k = 1, \dots, K$ **do**

For each cell $A \in \Pi_{n,k-1}$, calculate

$$\hat{\gamma}_n(A) = \min_{A_L, A_R} \left[\frac{1}{n} \sum_{i \in A_L} (Y_i - \overline{Y}_{n,A_L})^2 + \frac{1}{n} \sum_{i \in A_R} (Y_i - \overline{Y}_{n,A_R})^2 \right]$$

for all permissible cuts A_L, A_R of A. For example if $\mathscr{A}$ is the space of axis-aligned cuts, then the permissible cuts are CART splits (15.16).

Split the tree at the cell A^* with daughters yielding the minimal split-statistic, $\hat{\gamma}_n(A^*) \leq \hat{\gamma}_n(A)$ for $A \in \Pi_{n,k-1}$. Obtain the updated partition $\Pi_{n,k}$ by replacing the cut cell A^* with its daughters.

end for

The SGT estimator is $\hat{F}_n(\mathbf{x}, \Pi_{n,K}) = \sum_{A \in \Pi_{n,K}} \overline{Y}_{n,A} 1_{\{\mathbf{x} \in A\}}$.

Theorem 15.10. *Let* $(\mathbf{X}_1, Y_1), \dots, (\mathbf{X}_n, Y_n) \in \mathbb{R}^p \times \mathscr{Y}$ *be i.i.d. from* $\mathbb{P}$ *where* $\mathscr{Y}$ *is bounded. If* $\mathscr{A}$ *has a polynomial shatter coefficient as in Theorem 15.8 and* $K := K(n) = o(n/\log n)$*, then*

$$\int \left(\hat{F}_n(\boldsymbol{x}, \Pi_{n,K}) - F_n(\boldsymbol{x}, \Pi_{n,K})\right)^2 \mathbb{P}(d\boldsymbol{x}) \xrightarrow{\text{p}} 0, \quad \text{as } n \to \infty,$$

where F_n *is the semipopulation function defined by*

$$F_n(\boldsymbol{x}, \Pi_{n,K}) = \frac{1}{|\Pi_{n,K}(\boldsymbol{x})|} \sum_{i=1}^{n} F(\mathbf{X}_i) 1_{\{\mathbf{X}_i \in \Pi_{n,K}(\boldsymbol{x})\}}.$$

Theorem 15.10 shows that the SGT estimator $\hat{F}_n$ closely matches the semipopulation estimator F_n, which is defined by replacing Y_i with the true regression function $F(\mathbf{X}_i)$. For a cell A, this replaces the SGT estimator for the cell $\hat{F}_{n,A} = \hat{Y}_A$ with the semipopulation estimator $F_{n,A} = \overline{F}_A$, the mean of $F(\mathbf{X}_i)$ for cases in A.

Integrating over $\mathbf{X} \sim \mathbb{P}$, the proof involves controlling the following sum (see Ishwaran and Lee [17] for details):

$$\begin{aligned}
&\sum_{A \in \Pi_{n,K}} \mathbb{P}\left(\left[\hat{F}_n(\mathbf{X}, \Pi_{n,K}) - F_n(\mathbf{X}, \Pi_{n,K})\right]^2 1_{\{\mathbf{X} \in A\}}\right) \\
&= \sum_{A \in \Pi_{n,K}} \mathbb{P}\left(\left[\frac{1}{|A|}\sum_{i=1}^{n} Y_i 1_{\{\mathbf{X}_i \in A\}} - \frac{1}{|A|}\sum_{i=1}^{n} F(X_i) 1_{\{\mathbf{X}_i \in A\}}\right]^2 1_{\{\mathbf{X} \in A\}}\right) \\
&= \sum_{A \in \Pi_{n,K}} \left[\frac{1}{|A|}\sum_{i=1}^{n} [Y_i - F(\mathbf{X}_i)] 1_{\{\mathbf{X}_i \in A\}}\right]^2 \mathbb{P}(A) \\
&\leq 2C \sum_{A \in \Pi_{n,K}} \left|\frac{1}{|A|}\sum_{i=1}^{n} [Y_i - F(\mathbf{X}_i)] 1_{\{\mathbf{X}_i \in A\}}\right| \mathbb{P}(A) \\
&= 2C \sum_{A \in \Pi_{n,K}} \left[\frac{\mathbb{P}(A)}{|A|/n}\right] \left|\mathbb{P}_n\left([Y - F] 1_{\{\mathbf{X} \in A\}}\right)\right| \\
&= 2C \sum_{A \in \Pi_{n,K}} \left[\frac{\mathbb{P}(A)}{|A|/n}\right] \left|\mathbb{P}_n\left([Y - F] 1_{\{\mathbf{X} \in A\}}\right) - \mathbb{P}\left([Y - F] 1_{\{\mathbf{X} \in A\}}\right)\right|.
\end{aligned}$$

The last line holds because $\mathbb{P}\left(\left[Y-F\right]1_{\{\mathbf{X}\in A\}}\right)=0$ for any set A where $\mathbb{P}_n$ denotes the empirical measure. To make the above converge to zero requires uniform convergence of the fraction

$$\frac{\mathbb{P}(A)}{|A|/n}=\frac{\mathbb{P}(A)}{\sum_{i=1}^n 1_{\{X_i\in A\}}/n}\xrightarrow{\text{p}}1$$

and uniform convergence of the empirical difference between Y and F,

$$\left|\mathbb{P}_n\left([Y-F]1_{\{\mathbf{X}\in A\}}\right)-\mathbb{P}\left([Y-F]1_{\{\mathbf{X}\in A\}}\right)\right|\xrightarrow{\text{p}}0.$$

If there are too many cells A in the partition, then for the first term, the number of cases in a given cell, $|A|$, may become too small relative to n for $|A|/n$ to converge uniformly to $\mathbb{P}(A)$, while for the second term, the difference between Y and F may not converge fast enough since each cell has limited data, and we cannot expect the approximation to be accurate enough. To deal with these issues, the number of cuts allowed must be smaller than the sample size, i.e. $K \ll n/\log n$, which explains the first condition required by the theorem. Also, the complexity of the cutting space, $\mathscr{A}$, cannot be too large. This is captured by the second condition of the theorem which requires the shatter coefficient for $\mathscr{A}$ to be polynomial.

Now according to Theorem 15.10, $\hat{F}_n$ closely approximates the semipopulation estimator F_n, thus it is natural to wonder how well F_n approximates the true function F, since after all our ultimate goal is to estimate F. Under the same conditions as Theorem 15.10, Ishwaran and Lee [17] proved the following.

Theorem 15.11. *Assuming the same conditions as Theorem 15.10,*

$$\int\left(F_n(\boldsymbol{x},\Pi_{n,K})-F(\boldsymbol{x})\right)^2\mathbb{P}(d\boldsymbol{x})=R_n(\hat{F}_n(\cdot,\Pi_{n,K}))-R(F)+o_P(1).$$

Combining Theorem 15.10 and Theorem 15.11 shows that the mean-squared error for the SGT is the difference in empirical risk to population risk

$$\int\left(\hat{F}_n(\mathbf{x},\Pi_{n,K})-F(\mathbf{x})\right)^2\mathbb{P}(d\mathbf{x})=R_n(\hat{F}_n(\cdot,\Pi_{n,K}))-R(F)+o_P(1). \tag{15.23}$$

Because the SGT reduces empirical risk at each cut (Theorem 15.9), the empirical risk decreases monotonically for $K=1,\ldots,n-1$

$$\begin{aligned}0=R_n(\hat{F}_n(\cdot,\Pi_{n,n-1}))&\le\cdots\le R_n(\hat{F}_n(\cdot,\Pi_{n,K}))\\&\le\cdots\le R_n(\hat{F}_n(\cdot,\Pi_{n,0}))=\hat{s}_n^2=\frac{1}{n}\sum_{i=1}^n(Y_i-\overline{Y})^2.\end{aligned}$$

The left side bound of 0 is achieved with $K=n-1$ cuts because this partitions the space into n distinct values (splitting to purity). This means that $\hat{F}_n(\mathbf{x})=Y_i$ where i is the single case in $\mathbf{x}$'s terminal node, thus resulting in a training error of 0. The right-hand side bound is because when $K=0$ the tree is stumped and the tree estimator is constant, $\hat{F}_n(\mathbf{x})=\overline{Y}$. Hence, the mean-squared error asymptotically satisfies

$$\begin{aligned}-R(F)&\le\cdots\le R_n(\hat{F}_n(\cdot,\Pi_{n,K}))-R(F)\\&\le R_n(\hat{F}_n(\cdot,\Pi_{n,K-1}))-R(F)\le\cdots\le\hat{s}_n^2-R(F).\end{aligned}$$

As K increases the difference between the training error and the population risk decreases. Eventually, this value will become negative, which is a signal that over-training has happened. The bound on K required by Theorem 15.10 prevents this from happening.

Thus we have established the following general principle:

> *The performance of the SGT is improved by increasing greediness, where greediness is defined as the amount that empirical risk is reduced by a cut. An ideal SGT reduces empirical risk as rapidly as possible using the fewest number of cuts. This applies to all SGTs if the space of cuts $\mathscr{A}$ has a polynomial shatter number and the number of cuts is smaller than n.*

15.12 Using Lasso to Fit Super Greedy Trees

One aspect of SGTs that still needs to be discussed is the calculation of the SGT geometric cut. The computations are straightforward for simple cuts like those used by CART, but we want to consider more complex spaces $\mathscr{A}$ and require an efficient procedure for identifying promising cuts from $\mathscr{A}$. In this section, we address this by defining $\mathscr{A}$ using parametric linear models. As before, we continue to focus on regression. Since the permissible cuts are defined in terms of a linear model, we can use the lasso to efficiently obtain greedy multivariate cuts. Additionally, the lasso is used for the cell estimator, so unlike the previous SGT construction that employs a fixed value estimator for a cell, the lasso SGT uses local regression estimators. This also provides the added benefit of offering a superior terminal node estimator, further enhancing the predictive capabilities of the new procedure.

In this framework [16], the split-value s plays a role in controlling the boundary of the object which is defined by a parametric linear model. We consider linear models of the form (here $\mathbf{x} = (x_1,\ldots,x_p)^T$ denotes the observed covariate vector):

$$\psi(\mathbf{x}) = \beta_0 + \sum_{l=1}^{p} \beta_l x_l + \sum_{l=1}^{p} \beta_{ll} x_l^2 + \sum_{1\le l_1<l_2\le p} \beta_{l_1 l_2} x_{l_1} x_{l_2} + \sum_{1\le l_1\le l_2\le l_3\le p} \beta_{l_1 l_2 l_3} x_{l_1} x_{l_2} x_{l_3} \tag{15.24}$$

and we define a permissible cut using ψ as follows:

Definition 15.2. Given a partition $\mathcal{T}$, a permissible SGT cut for a cell $A \in \mathcal{T}$ is of form $A_L = \{\mathbf{x} : \psi(\mathbf{x}) \le s\}$ where the complementary cell is $A_R = \{\mathbf{x} : \psi(\mathbf{x}) > s\}$ and s is some real-valued split-value. □

Below we highlight several useful and interesting cuts arising from (15.24). It should be mentioned that since ψ is estimated dynamically as the tree is grown, and since the lasso is used to fit (15.24) to avoid over-fitting, cuts can adapt locally to the data; for example, allowing switching from complex cuts to simpler cuts as the need arises. This is important, since as the tree gets deeper, and the sample size in a node decreases, it is expected that splits will naturally tend to simpler elementary cuts such as those used by CART.

15.12.1 CART univariate cuts

CART splits arise under (15.24) when all covariate parameters are zero except for one main effect. Then $\psi(\mathbf{x}) = \beta_0 + \beta_l x_l$ for some $1 \le l \le p$. Splitting according to $\psi(\mathbf{x}) \le s$ amounts to daughters defined by

$$x_l \le \beta_l^{-1}(s - \beta_0) \quad \text{and} \quad x_l > \beta_l^{-1}(s - \beta_0)$$

which are axis-aligned cuts used by CART.

15.12.2 Hyperplane cuts

Hyperplane cuts are obtained when all coefficients are zero except for the main effects: $\psi(\mathbf{x}) = \beta_0 + \sum_{l=1}^{p} \beta_l x_l$. Since A is split according to $\psi(\mathbf{x}) \le s$, taking the intercept from the left to the right side, and setting $\alpha_0 = s - \beta_0$, yields daughters

$$\sum_{l=1}^{p} \beta_l x_l \le \alpha_0 \quad \text{and} \quad \sum_{l=1}^{p} \beta_l x_l \beta_l > \alpha_0.$$

Therefore, the left and right daughters lie on one side or the other of the hyperplane $\sum_{l=1}^{p} \beta_l x_l = \alpha_0$. Notice that varying s has the effect of varying the intercept of the hyperplane.

15.12.3 Elliptical cuts

An elliptical cut can be obtained using only quadratic terms by $\psi(\mathbf{x}) = \beta_0 + \sum_{l=1}^{p} \beta_l x_l^2$ for $\beta_l \ge 0$. Manipulating the intercept as before, splits correspond to $\sum_{l=1}^{p} \beta_l x_l^2 \le (s - \beta_0)$, which is an ellipse for $s > \beta_0$.

15.12.4 Oblique elliptical cuts

The previous example described an ellipse with principal axes parallel to the coordinate axes. For greater generality, rotated ellipses oblique to the coordinate axes are also possible. The contours for such objects can be described by $\{\mathbf{x}^T \mathbf{Q}\mathbf{x} = \alpha_0\}$, for $\mathbf{Q} = (a_{ij})$ a symmetric positive semidefinite matrix, and thus can be accommodated by (15.24). This is because

$$\mathbf{x}^T \mathbf{Q}\mathbf{x} = \sum_{l=1}^{p} a_{ll} x_l^2 + 2 \sum_{1 \le l_1 < l_2 \le p} a_{l_1 l_2} x_{l_1} x_{l_2}$$

and hence oblique elliptical cuts can be obtained using quadratic terms and pairwise interactions via

$$\psi(\mathbf{x}) = \beta_0 + \sum_{l=1}^{p} \beta_{ll} x_l^2 + \sum_{1 \le l_1 < l_2 \le p} \beta_{l_1 l_2} x_{l_1} x_{l_2}.$$

If the ellipse is centered at a nonzero point, then the elliptical contour is $\{(\mathbf{x} - \boldsymbol{\mu})^T \mathbf{Q}(\mathbf{x} - \boldsymbol{\mu}) = \alpha_0\}$ which is slightly more general as it introduces linear terms $\mathbf{x}^T \mathbf{Q}\boldsymbol{\mu}$. In this case, we must add main effect terms $\sum_{l=1}^{p} \beta_l x_l$:

$$\psi(\mathbf{x}) = \beta_0 + \sum_{l=1}^{p} \beta_l x_l + \sum_{l=1}^{p} \beta_{ll} x_l^2 + \sum_{1 \le l_1 < l_2 \le p} \beta_{l_1 l_2} x_{l_1} x_{l_2}.$$

15.12.5 Other interesting cuts

Many other geometric shapes are possible as indicated in Figure 15.12. Subfigure (a) is $x_1^2 - x_2^2 - x_3^2 \le \alpha_0$. Sufigure (b) is $x_1 x_2 x_3 \le \alpha_0$. We observe that another class of cuts can be induced using interval cuts of the form $\alpha_1 < \psi(\mathbf{x}) \le \alpha_2$. For example, Subfigures (c) and (d) use the same model as (b), but using interval cuts. Subfigure (e) is $x_1^2 - x_2^2 - x_3^2 + x_1 x_2 x_3 \le \alpha_0$. Subfigure (f) is the same as (e) but using interval cuts: $\alpha_1 < x_1^2 - x_2^2 - x_3^2 + x_1 x_2 x_3 \le \alpha_2$.

15.12.6 Splitting rule and local regression estimator

This brings us to the second issue which is the local regression estimator used for a tree node. As mentioned, in order to avoid over-fitting, (15.24) is fit by using lasso. For a given cell A, the local

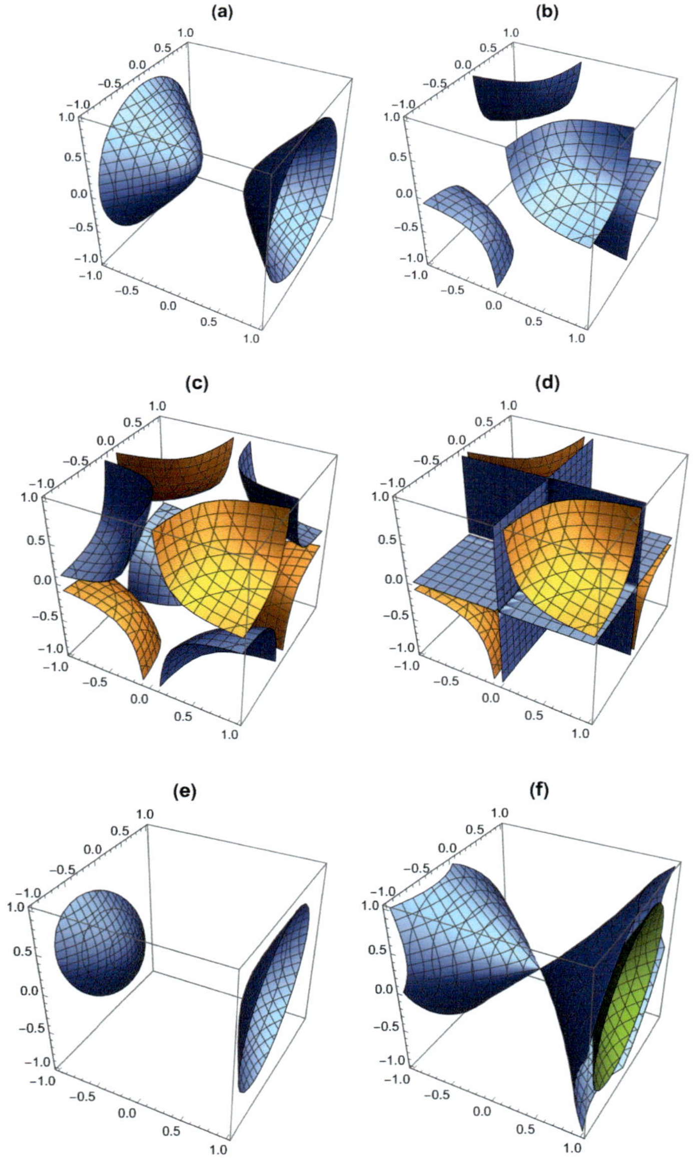

FIGURE 15.12
Different geometric cuts (i.e. tree splits) obtained from the parametric model (15.24).

regression estimator $\hat{\psi}_{n,A}(\mathbf{x})$ is defined to be the lasso fit for (15.24) calculated from the data in A.

The local estimator $\hat{\psi}_{n,A}$ is also used to define the SGT split-statistic. Let $\mathbf{x}_{A,1},\ldots,\mathbf{x}_{A,M}$ be the $M := M(A)$ cases in A and let $\hat{y}_{A,m} = \hat{\psi}_{n,A}(\mathbf{x}_{A,m})$ be the predicted value for $\mathbf{x}_{A,m}$, $m = 1,\ldots,M$. Order the M predicted values

$$\hat{y}_A^{(1)} \le \hat{y}_A^{(2)} \le \cdots \le \hat{y}_A^{(M)}.$$

Calculate the mean-squared-error split-statistic

$$\hat{\gamma}_A(m) = \sum_{i \le m} \left(\hat{y}_A^{(i)} - \frac{1}{m} \sum_{j \le m} \hat{y}_A^{(j)} \right)^2 + \sum_{i > m} \left(\hat{y}_A^{(i)} - \frac{1}{M-m} \sum_{j > m} \hat{y}_A^{(j)} \right)^2, \quad m = 1, \ldots, M-1. \tag{15.25}$$

Let m^* be the value minimizing the split-statistic: $\hat{\delta}_A(m^*) \le \hat{\delta}_A(m)$ for $m = 1, \ldots, M-1$. Then A is split using $\hat{\psi}_{n,A}(x) \le \hat{y}_A^{(m^*)}$. The rationale for (15.25) is that it tries to combine cases together with similar lasso predicted values. An advantage is that once the lasso estimator $\hat{\psi}_{n,A}$ is determined, the split-statistic can be calculated quickly. With some rearrangment, (15.25) can be simplified leading to the following.

Definition 15.3. Define $\hat{S}_A(m) = \sum_{j=1}^{m} \hat{y}_A^{(j)}$. Then minimizing (15.25) is equivalent to maximizing

$$\hat{\gamma}_A^*(m) = \frac{1}{m} \left(\hat{S}_A(m) \right)^2 + \frac{1}{M-m} \left(\hat{S}_A(M) - \hat{S}_A(m) \right)^2, \quad m = 1, \ldots, M. \tag{15.26}$$

We refer to (15.26) as the *SGT split-statistic* for A. If m^* maximizes (15.26), then A is split at $\hat{y}_A^{(m^*)}$ into left and right daughters $A_L = \{\hat{\psi}_{n,A}(x) \le \hat{y}_A^{(m^*)}\}$, $A_R = \{\hat{\psi}_{n,A}(x) > \hat{y}_A^{(m^*)}\}$. □

Definition 15.3 describes an efficient means to calculate promising multivariate cuts. The next step in building the tree is to find the optimal cell to cut. At the start of step $k \ge 1$ we have a partition $\Pi_{n,k-1}$ of size k where the SGT predictor is defined as

$$\hat{F}_n(\mathbf{x}, \Pi_{n,k-1}) = \sum_{A \in \Pi_{n,k-1}} \hat{\psi}_{n,A}(\mathbf{x}) 1_{\{\mathbf{x} \in A\}}.$$

In accordance with BSF, we must split the cell with the largest decrease in empirical risk. Given a cell $A \in \Pi_{n,k-1}$, let A_L and A_R be the left and right daughters selected using Definition 15.3. Then fit the lasso to the data in A_L and A_R to obtain lasso estimators $\hat{\psi}_{n,A_L}$ and $\hat{\psi}_{n,A_R}$, respectively. The SGT predictor for the proposed split is

$$\hat{F}_n(\mathbf{x}, \Pi_{n,k}) = \hat{F}_n(\mathbf{x}, \Pi_{n,k-1}) - \hat{\psi}_{n,A}(\mathbf{x}) 1_{\{\mathbf{x} \in A\}} + \left[\hat{\psi}_{n,A_L}(\mathbf{x}) 1_{\{\mathbf{x} \in A_L\}} + \hat{\psi}_{n,A_R}(\mathbf{x}) 1_{\{\mathbf{x} \in A_R\}} \right]$$

and the decrease in empirical risk is

$$\begin{aligned}
&\Delta_n(A, A_L, A_R) \\
&\quad = R_n(\Pi_{n,k-1}) - R_n(\Pi_{n,k}) \\
&\quad = \sum_{i \in A} (y_i - \hat{\psi}_{n,A}(\mathbf{x}_i))^2 - \sum_{i \in A_L} (y_i - \hat{\psi}_{n,A_L}(\mathbf{x}_i))^2 - \sum_{i \in A_R} (y_i - \hat{\psi}_{n,A_R}(\mathbf{x}_i))^2.
\end{aligned}$$

The SGT splits the cell $A^* \in \Pi_{k-1}$ into new cells A_L^* and A_R^* where A^* is the BSF cell satisfying $\Delta_n(A^*, A_L^*, A_R^*) \ge \Delta_n(A, A_L, A_R)$ for each $A \in \Pi_{k-1}$.

We have discussed several advantages of using lasso, however, one negative aspect is that unlike CART, the empirical risk reduction from a lasso SGT is not guaranteed to be positive. Replacing the fixed value sample average cell estimator with a regression function means that Theorem 15.9 no longer holds in general.

Therefore, in instances where the empirical risk reduction becomes negative, the lasso is replaced with CART. To explain this, the empirical risk reduction from a CART axis-aligned split is

$$\Delta_n^C(A, A_L, A_R) = \sum_{i \in A} (y_i - \overline{y}_A)^2 - \sum_{i \in A_L} (y_i - \overline{y}_{A_L})^2 - \sum_{i \in A_R} (y_i - \overline{y}_{A_R})^2,$$

where $\overline{y}_A, \overline{y}_{A_L}, \overline{y}_{A_R}$ are the sample averages of the responses in A, A_L, A_R. Whenever the risk reduction is negative, cuts are switched to axis-aligned cuts. Lasso estimators are replaced with

CART sample average node estimators and Δ_n is replaced with Δ_n^C for determining empirical risk reduction. Note that the CART sample averaged node estimator can be interpreted as a lasso estimator for the model (15.24) that only includes an intercept term. Finally, in some instances, the lasso can fail due to small sample sizes or for other numerical reasons. For example, for smaller sized nodes it may become impossible to determine the lasso penalization parameter using cross-validation. Therefore in these instances, the same strategy is used to switch to CART.

15.12.7 Illustration of empirical risk reduction

We use a simulation to illustrate SGTs ability to reduce empirical risk. For computations we use the R-package `randomForestSGT` developed for implementing SGTs [18]. The package uses coordinate descent to fit lasso models and 10-fold cross-validation for determining the lasso parameter. Parametric linear models used for splitting by the package are indexed by a parameter `hcut` with values defined as follows:

1. `hcut` = 1: intercept + linear slope terms for each of the p variables.
2. `hcut` = 2: + all quadratic terms.
3. `hcut` = 3: + all pairwise interactions.
4. `hcut` = 4: + all polynomials of degree 3 of two variables.
5. `hcut` = 5: + all polynomials of degree 4 of three variables.
6. `hcut` = 6: + all three-way interactions.

There is also a special case `hcut` = 0 that uses CART axis-aligned splits.

To test the flexibility of the induced multivariate cuts, we used a noiseless simulation

$$Y = \psi(\mathbf{X}) = \theta \sin(\pi X_1 X_2)$$

where covariates $X_1, \ldots, X_p$ were sampled independently from a U(0, 1) distribution. The signal was set to $\theta = 10$ and a sample size $n = 1000$ was drawn where in addition to the two signal variables X_1 and X_2, 1000 noise variables unrelated to the conditional mean were included ($p = 1000 + 2$).

Due to the high dimension the following filtering approach was used to improve performance. Each terminal node A of the tree has an associated lasso fit with estimated β_A coefficients. A shallow SGT using `hcut` = 1 (hyperplane cuts) is calculated and the β_A coefficients are tree averaged to obtain an overall averaged β for each covariate. Features with $\beta = 0$ are removed. Following this, another shallow SGT is fit, but this time under the specified `hcut` value, and the tree averaged β_A coefficients are used to obtain an overall averaged β for each covariate. Features with $\beta = 0$ are removed.

The top left panel of Figure 15.13 displays the regression surface for the simulation function for the two relevant attributes. Following this, moving from left to right and top to bottom, are the predicted regression surfaces for SGTs constructed under different `hcut` settings. All trees were constructed with $K = 40$ splits, and all panels have the same contour values of ψ superimposed. The top middle panel for `hcut` = 0 shows axis-aligned cuts similar to CART. Observe the stepwise boundary, a hallmark of CART axis-aligned splitting. To its right is an SGT using hyperplane cuts. Its performance is noticeably better, but increasing `hcut` improves performance even further.

The last plot on the bottom right gives the empirical risk profile for each procedure in terms of the number of splits. The effect of `hcut` on empirical risk is dramatic. Observe how `hcut` values of 4, 5, and 6 achieve near zero empirical risk after only a few splits (note that their risk profiles are identical in this example). This contrasts with the slow decrease observed for CART (`hcut` = 0), which still has a large empirical risk even after 40 splits.

As a second illustration, we simulated data from the Friedman 1 model [11]

$$Y = \psi(\mathbf{X}) + \epsilon, \quad \psi(\mathbf{X}) = 10\sin(\pi X_1 X_2) + 20(X_3 - 0.5)^2 + 10X_4 + 5X_5,$$

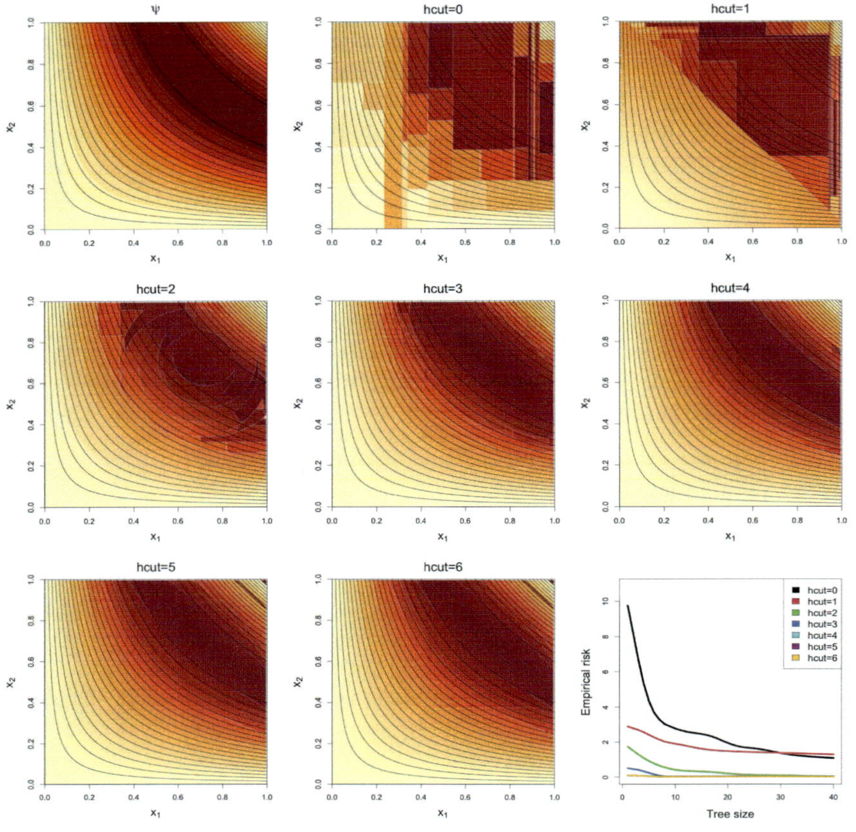

FIGURE 15.13
Nonlinear simulation model with two signal variables and 1000 noise variables. Top left panel displays the true function $\psi(\boldsymbol{x}) = \theta \sin(\pi x_1 x_2)$ as a function of the two relevant attributes. Following this, moving from left to right and top to bottom, are the predicted regression surfaces for SGTs constructed under different `hcut` *settings. All trees use $K = 40$ splits and the same contour values for ψ have been superimposed on all panels. The bottom right figure displays the tree empirical risk profile in terms of number of tree splits (note that* `hcut` $= 4, 5, 6$ *have identical profiles).*

where X_j are i.i.d. U(0,1) for $j = 1, \ldots, p$ and $\epsilon \sim N(0,1)$. The sample size is $n = 500$ and the dimension is set to $p = 10$, therefore there are 5 noisy variables and 5 signal variables.

Empirical risk and OOB (out-of-bag) empirical risk were compared for CART and SGTs. Tree size was set to $K = 50$ and `hcut` $= 3$ was used for SGTs and `hcut` $= 0$ was used for the CART evaluation. Trees were constructed using independent bootstrap samples, where recall that a bootstrap sample leaves an OOB sample of roughly 33% of the original data which can serve as test data for evaluating performance (Chapter 7.12). The results from 100 independent runs are given in Figure 15.14. The top and bottom panels display inbag and OOB empirical risk respectively as a function of the number of tree splits in the form of boxplots where each boxplot gives the summarized values from the 100 runs. The SGTs (right figures) show superior performance with approximately a 5-fold reduction in risk compared with CART (left figures). It is noteworthy that this reduction holds consistently across both the training (inbag) and test set (OOB) empirical risk evaluations.

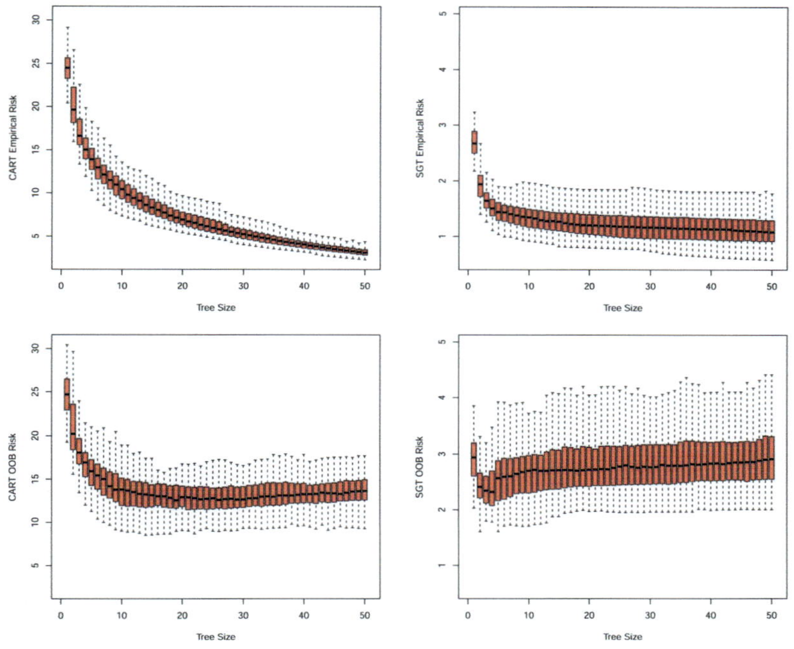

FIGURE 15.14
Comparison of CART (`hcut` = 0) *to SGT* (`hcut` = 3) *using Friedman 1 simulation. Top panel displays bootstrap empirical risk as a function of number of splits where boxplots are the observed empirical risk values from 100 independent runs. Bottom panel displays OOB (out-of-bag) empirical risk. The SGT results (right figures) show superior performance with almost a 5 fold reduction in risk compared with CART (left figures).*

15.13 Bagged Trees

One lesson that emerged from studying SGTs is that the type of cut plays a significant role in performance. More flexible cuts yield richer tree topologies and reduce empirical risk more efficiently. Axis-aligned cuts on the other hand, as used by CART, can be suboptimal in reducing risk due to their inflexibility in some settings.

Another way to view the inefficiency of CART is through the concept of *instability*. Generally speaking a learning method is called unstable if its prediction can change substantially if the learning dataset is perturbed [5, 6]. This is an undesirable property, especially if small perturbations lead to substantial differences, since this leads to excess variance.

We can demonstrate this problem using the prostate data from Figure 15.3. To perturb the data we used bootstrap resampling. 1000 independent bootstrap samples of the data were drawn. For each of these a classification tree was calculated. Each tree was grown to purity so that each terminal node contained one observation. Figure 15.15 shows decision boundaries for four of these trees (bootstrap samples 2, 10, 200 and 1000). Note that tree 1 is the classification tree from Figure 15.3 using the original data without bootstrap. We can clearly observe that the decision spaces for the bootstrapped trees vary considerably, which is a demonstration of tree instability.

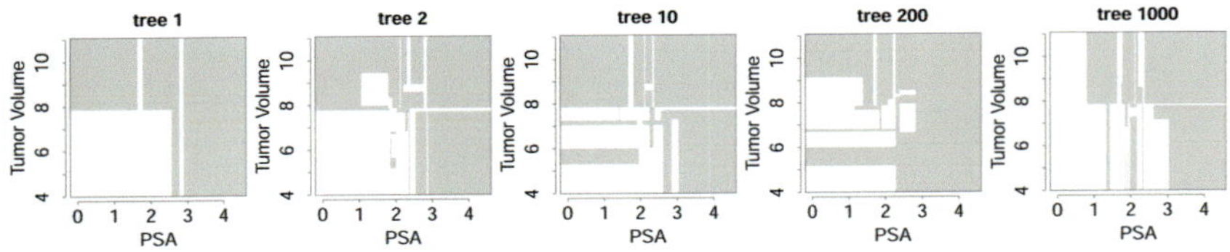

FIGURE 15.15
Decision boundaries from bootstrapped deep trees.

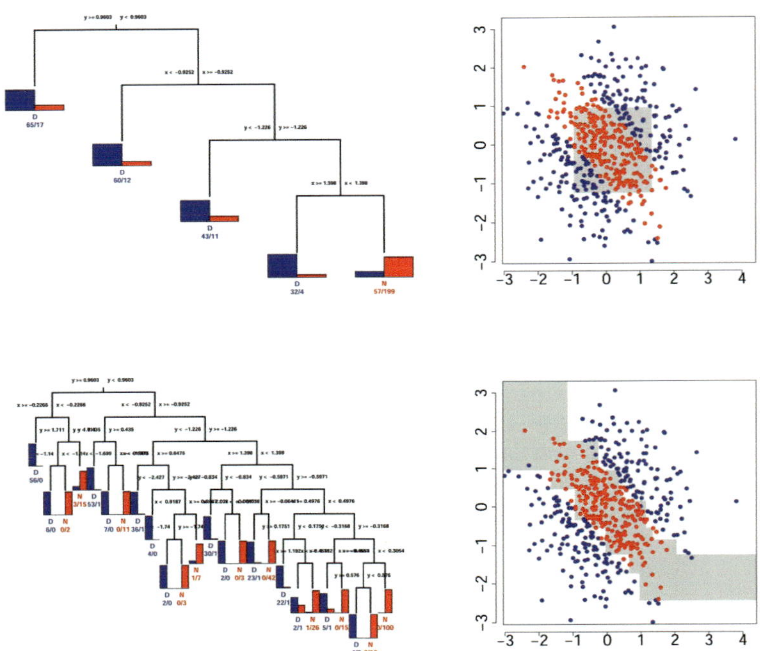

FIGURE 15.16
Oblique decision boundary estimated using a shallow tree decision boundary (top) and deep tree decision boundary (bottom).

Remark 15.3. In our example, we used trees split to purity, but one might argue that pruned trees (Chapter 15.3) could perform better with greater stability. While this argument is reasonable, it is important to remember that deep trees are necessary for capturing complex decision boundaries. An example is shown in Figure 15.16. Here, the data was simulated with a decision boundary that is oblique to the coordinate axes. Notice how a shallow tree results in a less refined decision boundary compared with a deeper tree. □

15.13.1 Bagging procedure

If we base our analysis on a single deep tree, tree instability can certainly be problematic. However, it was interesting to observe that the decision boundaries for some of the trees in our analysis appear to be better than the original tree using the full dataset (i.e., tree number 1). For example, tree 10 identifies problematic misclassified cases within the lower quadrant of the original decision space (see Figure 15.3). Therefore, it seems reasonable that by combining these perturbed boundaries, the resulting classifier may have better overall performance.

This idea is the basis for a powerful method referred to as *bootstrap aggregation*, or simply *bagging*, a general method for reducing the variance of a predictor [5]. The basic premise for bagging is that if the underlying predictor is unstable, then aggregating the predictor over different learning samples can produce a more accurate, and more stable, procedure.

In theory, to bag a predictor, replicates of the predictor are generated over the independent realization of a learning set, $\mathscr{L}$, and averaged. But in practice, because only one learning set is available, bagging is usually applied by bootstrapping the data $\mathscr{L}$ (other resampling procedures can also be used).

As a concrete example, consider bagging a classification tree:

1. Draw a bootstrap sample of the original data.
2. Construct a classification tree using data from Step 1.
3. Repeat Steps 1-2 many times, independently.
4. Use tree majority voting to classify a case.

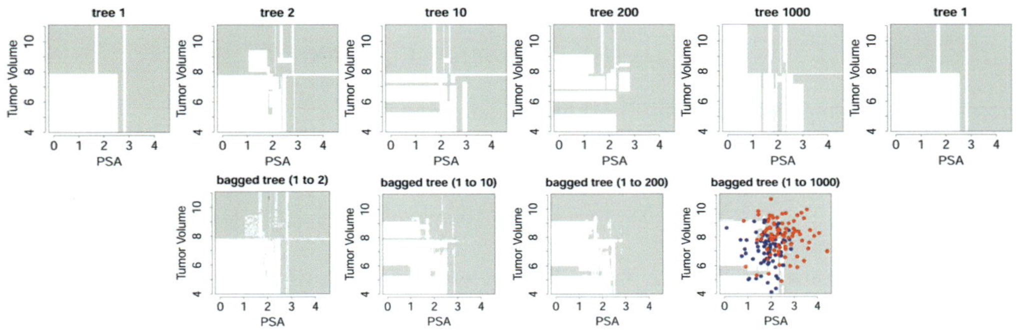

FIGURE 15.17
Top panel shows decision boundary for a bootstrapped tree. Bottom panel shows different aggregated (bagged) decision trees. Bagged trees are more robust to noise (stable) because they utilize information from more than one tree. The most stable bagged tree is the one on extreme right-hand side and shows its decision boundary based on 1000 trees.

Figure 15.17 illustrates this for the prostate data. The bottom panel shows the decision boundary for the bagged classifier as a function of the number of trees (for convenience Figure 15.15 has been reproduced in the top panel). As before, the first plot is the original classifier (tree 1). The second plot is the bagged classifier for tree 1 and the bootstrap tree from the first bootstrap sample, the third plot is the bagged classifier using tree 1 and the first 9 bootstrap trees, and so forth. As the number of trees increases, the bagged classifier becomes more refined. Even with only 10 trees (third plot) the decision boundary is substantially smoother than the original classifier and is better able to classify cases.

Table 1 records the misclassification error for the bagged predictor against the averaged misclassification error for the 1000 bootstrapped trees. The first column is the overall error rate, the second column is the error rate for diseased patients, and the third column is the error rate for non-diseased patients. Error rates were calculated using OOB data (Chapter 7.12). Table 15.4 shows that the bagged classifier is substantially better than any single bootstrapped tree.

TABLE 15.4
Misclassification error (in percentage) for bagged classifier (1000 trees) versus averaged (over 1000 trees) misclassification error using a single tree classifier.

classifier	all	diseased	non-diseased
bagged tree	27.2	28.8	25.9
single tree	34.9	36.7	33.0

15.13.2 Why bagging works

To explain why bagging works, we focus on the regression setting. We assume a nonparametric regression model as in (15.22). We use $\hat{F}(\mathbf{x}) := \hat{F}(\mathbf{x}, \mathscr{L})$ to denote an estimator for the regression function $F(\mathbf{x})$ calculated from the learning data, $\mathscr{L}$. Recall from Chapter 7 that the generalization error (7.1) for $\hat{F}$ is

$$\mathrm{Err}(\hat{F}) = \mathbb{E}_{\mathscr{L}}\mathbb{E}_{\mathbf{X},Y}\left[Y - \hat{F}(\mathbf{X})\right]^2.$$

Define the estimator $\hat{F}_{\text{bag}}(\mathbf{X}) = \mathbb{E}_{\mathscr{L}}[\hat{F}(\mathbf{X})]$. This averages $\hat{F}$ over all realizations of the learning data and equals the theoretical bagged predictor. We have from (7.5)

$$\begin{aligned}\mathrm{Err}(\hat{F}) &= \sigma^2 + \mathbb{E}_{\mathbf{X}}\left\{\mathrm{Bias}\{\hat{F}\,|\,\mathbf{X}\}^2\right\} + \mathbb{E}_{\mathbf{X}}\left\{\mathrm{Var}\{\hat{F}\,|\,\mathbf{X}\}\right\} \\ &= \sigma^2 + \mathbb{E}_{\mathbf{X}}\left[\hat{F}_{\text{bag}}(\mathbf{X}) - F(\mathbf{X})\right]^2 + \mathbb{E}_{\mathscr{L},\mathbf{X}}\left[\hat{F}(\mathbf{X}) - \hat{F}_{\text{bag}}(\mathbf{X})\right]^2\end{aligned}$$

where recall the conditional bias and variance are defined by

$$\begin{aligned}\mathrm{Bias}\{\hat{F}\,|\,\mathbf{X}\} &= \mathbb{E}_{\mathscr{L}}\{\hat{F}(\mathbf{X})\} - F(\mathbf{X}) \\ \mathrm{Var}\{\hat{F}\,|\,\mathbf{X}\} &= \mathbb{E}_{\mathscr{L}}\left\{\left(\hat{F}(\mathbf{X}) - \mathbb{E}_{\mathscr{L}}[\hat{F}(\mathbf{X})]\right)^2\right\}.\end{aligned}$$

This holds for any estimator, and in particular using $\hat{F}_{\text{bag}}$ as our estimator, we obtain

$$\mathrm{Bias}\{\hat{F}_{\text{bag}}\,|\,\mathbf{X}\} = \hat{F}_{\text{bag}} - F(\mathbf{X}), \quad \mathrm{Var}\{\hat{F}_{\text{bag}}\,|\,\mathbf{X}\} = 0,$$

and therefore from the above

$$\begin{aligned}\mathrm{Err}(\hat{F}_{\text{bag}}) &= \sigma^2 + \mathbb{E}_{\mathbf{X}}\left\{\mathrm{Bias}\{\hat{F}_{\text{bag}}\,|\,\mathbf{X}\}^2\right\} \\ &= \sigma^2 + \mathbb{E}_{\mathbf{X}}\left\{\mathrm{Bias}\{\hat{F}\,|\,\mathbf{X}\}^2\right\} \\ &\le \mathrm{Err}(\hat{F}).\end{aligned}$$

Or put another way,

$$\mathrm{Err}(\hat{F}) = \mathrm{Err}(\hat{F}_{\text{bag}}) + \mathbb{E}_{\mathscr{L},\mathbf{X}}\left[\hat{F}(\mathbf{X}) - \hat{F}_{\text{bag}}(\mathbf{X})\right]^2.$$

Hence the prediction error for the theoretical bagged predictor is either reduced, or at the very worst, the same as the original predictor. Not only that, but this difference is entirely due to eliminating the variance from the original predictor. This demonstrates that bagging is a variance reduction technique.

Remark 15.4. A word of caution: The previous argument shows that theoretical bagging guarantees a reduction in variance. However, in practice, we use bootstrap resampling to approximate the theoretical bagged procedure, and in such cases, bagging is no longer guaranteed to reduce variance. Breiman [5] presented examples where bagging actually increased generalization error. In these examples, the underlying procedure was stable, and bagging was unhelpful because the procedure's variance was already controlled.

In general, bagging should be reserved for unstable procedures with low bias and high variance. Bagging does not significantly affect bias; it might slightly increase bias due to the smaller sample size of the bootstrap. However, its dominant effect is in reducing variance. If the variance is high, the net gain is a procedure with improved generalization error. □

15.14 Gradient Boosting Using Trees

Trees are flexible and adaptive nonparametric estimators, and as such they represent ideal base learners for machine learning methods. In Chapter 12 we discussed Friedman's [11] gradient boosting method, a machine learning method for implementing generic boosting. Here we describe how to apply gradient boosting using trees as weak learners. Throughout we use the notation from Chapter 12.

Recall that gradient boosting builds an additive expansion for the target function F in a sequential fashion where at each step a two-step optimization problem is solved. For the mth step, $m = 1, \ldots, M$, one solves the line-search

$$\rho_m = \arg\min_{\rho \in \mathbb{R}} \sum_{i=1}^{n} L\big(y_i, F_{m-1}(\mathbf{x}_i) + \rho h_m(\mathbf{x}_i)\big)$$

where F_{m-1} is the current solution and h_m is the solution to the least squares problem

$$h_m = \arg\min_{h \in \mathscr{H}} \sum_{i=1}^{n} \Big[g_m(\mathbf{x}_i) - h(\mathbf{x}_i)\Big]^2. \tag{15.27}$$

Here $h : \mathscr{X} \to \mathbb{R}$ are base learners from the dictionary space $\mathscr{H}$. The optimized $h_m \in \mathscr{H}$ is the learner closest in L_2-distance to the gradient $\mathbf{g}_m = (g_m(\mathbf{x}_1), \ldots, g_m(\mathbf{x}_n))^T$ defined as the derivative of the loss evaluated at the current value F_{m-1}:

$$g_m(\mathbf{x}_i) = -\left[\frac{\partial L(y_i, F(\mathbf{x}_i))}{\partial F(\mathbf{x}_i)}\right]_{F_{m-1}(\mathbf{x}_i)} = -L'(y_i, F_{m-1}(\mathbf{x}_i)).$$

The mth update for the predictor of the target $F(\mathbf{x})$ is defined as

$$F_m(\mathbf{x}) = F_{m-1}(\mathbf{x}) + \nu \rho_m h_m(\mathbf{x}),$$

where $0 < \nu \leq 1$ is the regularization (learning) parameter.

The underlying principle for identifying the optimized base learner is quite flexible. In each instance we know that the negative gradient g_m represents the ideal solution and the goal is to match g_m as closely as possible to a weak learner from the dictionary $\mathscr{H}$. Solving a least squares problem as in (15.27) is one way to accomplish this. However, for more complex base learners such as trees, we need to extend this idea. Trying to directly optimize least squares for a tree will be difficult. Instead, the idea is to replace the least squares optimization by a tree that is fit directly to data defined using the gradient g_m as the outcome Y and using the $\mathbf{X}$ values for features to split the tree. By using a regression tree with splits defined by mean-squared error, which is an L_2-based splitting mechanism, the idea is that the tree will match the gradient in an L_2 sense. This calculation can be done very rapidly and avoids the problems of a brute force L_2-optimization.

15.14.1 Boosted trees for regression

As a first example, consider tree-based gradient boosting for linear regression. Using our previous notation, the target function is $F(\mathbf{x}) = \mathbb{E}[Y|\mathbf{X} = \mathbf{x}]$. We will boost regression trees using partitions of size K for some small fixed $K > 1$.

In this framework, $\mathscr{H}$ denotes the space of all K-terminal regression trees on $\mathscr{X}$. To make this definition more precise, we write a K-terminal regression tree as

$$h_\theta(\mathbf{x}) = \sum_{k=1}^{K} \theta_k 1_{\{\mathbf{x} \in A_k\}}.$$

Here the tree is parameterized by $\Pi = \{A_1, \dots, A_K\}$, a partition of size K of $\mathscr{X}$, and $\{\theta_k\}_1^K \in \mathbb{R}$, the terminal node predicted values.

The optimized base learner for the mth step is obtained by growing a K-terminal node regression tree using $\{g_m(\mathbf{x}_i)\}_1^n$ as the responses. We can write the optimized base learner as

$$h_m(\mathbf{x}) = \sum_{k=1}^{K} \theta_{mk} 1_{\{\mathbf{x} \in A_{mk}\}}, \quad \text{where } \theta_{mk} = \frac{1}{|A_{mk}|} \sum_{i \in A_{km}} g_m(\mathbf{x}_i)$$

and $\Pi_m = \{A_{m1}, \dots, A_{mK}\}$ is the tree partition.

From this, we obtain the updated estimator

$$F_m(\mathbf{x}) = F_{m-1}(\mathbf{x}) + \nu \rho_m \sum_{k=1}^{K} \theta_{mk} 1_{\{\mathbf{x} \in A_{mk}\}}$$

where the line-search parameter ρ_m is

$$\rho_m = \arg\min_{\rho \in \mathbb{R}} \left\{ \sum_{i=1}^{n} L\left(y_i, F_{m-1}(\mathbf{x}_i) + \rho \sum_{k=1}^{K} \theta_{mk} 1_{\{\mathbf{x}_i \in A_{mk}\}} \right) \right\}.$$

Now, while this leads to a reasonable approach, one problem is that the same line-search parameter ρ is used for each of the terminal nodes. This may not be flexible enough, and a potentially better solution is to use a separate line-search for each of the J nodes.

Thus, combine ρ and θ_{mk} into a single parameter $\gamma_{mk} = \rho\theta_{mk}$, so that the mth update is

$$F_m(\mathbf{x}) = F_{m-1}(\mathbf{x}) + \nu \sum_{k=1}^{K} \gamma_{mk} 1_{\{\mathbf{x} \in A_{mk}\}}.$$

In place of a line-search over a single ρ value, apply a joint optimization over the collection of K parameters $\{\gamma_{mk}\}_1^K$. The single parameter line-search is replaced with

$$\{\gamma_{mk}\}_{k=1}^{K} = \arg\min_{\{\gamma_k\}_1^K} \left\{ \sum_{i=1}^{n} L\left(y_i, F_{m-1}(\mathbf{x}_i) + \sum_{k=1}^{K} \gamma_k 1_{\{\mathbf{x}_i \in A_{mk}\}} \right) \right\}. \tag{15.28}$$

While (15.28) might appear difficult to optimize, remember that the terminal nodes form a partition of the data space. Because of this, (15.28) simplifies to K separate line-searches

$$\gamma_{mk} = \arg\min_{\gamma} \sum_{i \in A_{mk}} L\big(y_i, F_{m-1}(\mathbf{x}_i) + \gamma\big), \quad k = 1, \dots, K.$$

The solution depends on the loss function. Under L_2-loss, since $g_m(\mathbf{x}_i) = y_i - F_{m-1}(\mathbf{x})$, and therefore

$$L(y_i, F_{m-1}(\mathbf{x}_i) + \gamma) = g_m(\mathbf{x}_i) - \gamma,$$

we have

$$\gamma_{mk} = \arg\min_{\gamma} \sum_{i\in A_{mk}} (g_m(\mathbf{x}_i) - \gamma)^2 = \frac{1}{|A_{mk}|} \sum_{i\in A_{km}} g_m(\mathbf{x}_i) = \theta_{mk}.$$

As this is the same as the original terminal node predicted values, this shows separate line-searches are the same as the original single-line-search and in both instances we obtain the same solution

$$F_m(\mathbf{x}) = F_{m-1}(\mathbf{x}) + \nu \sum_{k=1}^{K} \theta_{mk} 1_{\{\mathbf{x}\in A_{mk}\}}.$$

This is a unique feature of L_2-loss and does not hold in general for other loss functions as will be shown in our next example.

15.14.2 Boosted trees for classification

Now we consider the classification problem. We assume the response outcomes are coded as $Y \in \{-1, 1\}$ and use the negative-binomial loss function

$$L(y_i, F(\mathbf{x}_i)) = \log(1 + \exp(-2y_i F(\mathbf{x}_i)))$$

where

$$F(\mathbf{x}_i) = \frac{1}{2}\log\left(\frac{p(\mathbf{x}_i)}{1 - p(\mathbf{x}_i)}\right)$$

and $p(\mathbf{x}_i) = \mathbb{P}\{y_i = 1|\mathbf{x}_i\}$. The negative gradient is

$$g_m(\mathbf{x}_i) = -L'(y_i, F_{m-1}(\mathbf{x}_i)) = \frac{2y_i}{1 + \exp(2y_i F_{m-1}(\mathbf{x}_i))}.$$

As before, a regression tree with K terminal nodes is fit to $g_m(\mathbf{x}_i)$ which are used for the responses. This is followed by separate line-searches to estimate $\{\gamma_{mk}\}_1^K$ for each terminal node. The updated estimator is

$$F_m(\mathbf{x}) = F_{m-1}(\mathbf{x}) + \nu \sum_{k=1}^{K} \gamma_{mk} 1_{\{\mathbf{x}\in A_{mk}\}}$$

where the separate line-search optimizations are

$$\begin{aligned} \gamma_{mk} &= \arg\min_{\gamma} \sum_{i\in A_{mk}} L\big(y_i, F_{m-1}(\mathbf{x}_i) + \gamma\big) \\ &= \arg\min_{\gamma} \sum_{i\in A_{mk}} \log\Big[1 + \exp\big(-2y_i (F_{m-1}(\mathbf{x}_i) + \gamma)\big)\Big]. \end{aligned} \tag{15.29}$$

Unlike linear regression under L_2-loss, there is no closed form solution to (15.29) and it must be numerically solved. A one-step Newton approximation (10.1) yields

$$\gamma_{mk} = \frac{\sum_{i\in A_{mk}} g_m(\mathbf{x}_i)}{\sum_{i\in A_{mk}} |g_m(\mathbf{x}_i)|(2 - |g_m(\mathbf{x}_i)|)}. \tag{15.30}$$

The final predictor obtained from $m = 1, \ldots, M$ iterations is $F_M(\mathbf{x}_i)$ and is converted into a probability estimate $p_M(\mathbf{x}_i) = 1/(1 + \exp(-2y_i F_M(\mathbf{x}_i)))$.

We conclude by emphasizing that the choice of base learner can significantly impact the performance of gradient boosting, and readers should choose carefully. In scenarios where parametric assumptions are violated, boosted trees will typically outperform boosted linear learners. To illustrate this, we conducted an experiment on two-class problems, comparing the performance of boosting with parametric linear learners and tree learners. The linear learners

TABLE 15.5
Gradient boosting using parametric linear learners versus nonparametric tree learners on two-class benchmark data. Last two columns list 10-fold CV estimates of misclassification error.

	n	p	parametric	tree learners
BreastCancer	683	81	0.0439	0.0409
HouseVotes84	232	17	0.0306	0.0350
Ionosphere	351	34	0.1334	0.0627
PimaIndiansDiabetes	768	8	0.2306	0.2332
PimaIndiansDiabetes2	392	8	0.2222	0.2246
Sonar	208	60	0.2686	0.1583

were implemented using the R-package `mboost` [8], while the tree learners were implemented using the R-package `gbm` [12]. Performance was assessed using 10-fold cross-validation, and the average test misclassification error was recorded. The results of the experiment are presented in Table 15.5.

Generally, the two types of boosting agree, but when there is a difference, nonparametric boosting is generally the better procedure, and these differences can be noticeable; for example, in the Ionosphere and Sonar experiments.

Exercises

Exercise 15.1. Implement cost-complexity pruning in CART and evaluate its impact on the model performance.

1. Load the `Adult` dataset from the UCI repository.
2. Split the data into training and testing sets.
3. Train a CART model using the `rpart` package in R.
4. Implement cost-complexity pruning on the trained CART model.
5. Evaluate the pruned and unpruned models on the testing set using accuracy, precision, recall, and F1-score.
6. Visualize the pruned and unpruned trees and discuss the differences.

Exercise 15.2. Find a suitable two-class medical dataset similar to the prostate data. Using the data, grow a shallow classification tree with roughly 4-8 terminal nodes. Calculate the odds ratio in each terminal node. Summarize your results and discuss any high or low-risk patient cohort groups found in your analysis.

Exercise 15.3. Find an interesting classification dataset. Calculate the Gini split statistic for each X variable over all unique values of a variable as was done for the prostate data of Figure 15.5. Comment on your results. Are you able to identify useful variables in your dataset using this method? Explain.

Exercise 15.4. Repeat the previous exercise but this time using a high-dimensional dataset. Comment on the speed of your algorithm. Does this method seem like a practical method for big data settings?

Exercise 15.5. Prove Theorem 15.3.

Exercise 15.6. Use Theorem 15.3 to prove that Gini splitting favors pure splits.

Exercise 15.7. Use Theorem 15.3 to prove that entropy splitting favors balanced nodes.

Exercise 15.8. In this exercise we will investigate AUC-based splitting for classification trees.

1. Load the `Credit Card Fraud Detection` dataset.
2. Split the data into training and testing sets.
3. Implement a custom AUC-based splitting criterion in a decision tree algorithm.
4. Train the custom decision tree on the training set.
5. Evaluate the models performance on the testing set using AUC-ROC, accuracy, precision, recall, and F1-score.
6. Compare the performance of the custom AUC-based tree with a standard Gini impurity-based tree.

Exercise 15.9. The objective of this exercise is to verify that the three conditions of Theorem 15.7 hold for the k-nearest neighbor (k-NN) regression estimator

$$\hat{F}_{n,k}(\mathbf{x}) = \frac{1}{k} \sum_{i \in I_{n,k}(\mathbf{x})} Y_i, \tag{15.31}$$

where $I_{n,k}(\mathbf{x})$ denotes the indices of the k nearest neighbors of $\mathbf{x}$ among the sample points $\{\mathbf{X}_i\}_{i=1}^n$. Therefore, for a suitably selected k, this will show that k-NN is consistent.

1. Show that (15.31) is a locally weighted averaged estimator with weights $W_{n,i}(\mathbf{x}) \in \{0,1\}$ which are 1 if and only if $i \in I_{n,k}(\mathbf{x})$.
2. Verify that the weights for (15.31) satisfy the three conditions under the assumptions that $k \to \infty$ and $k/n \to 0$ as $n \to \infty$.
3. This establishes the consistency of (15.31). Formally state the result you have just proved, including any assumptions made.

Exercise 15.10. The objective here is to train and evaluate a survival tree for time-to-event data.

1. Load the `lung` dataset from the `survival` package in R.
2. Split the data into training and testing sets.
3. Train a survival tree using the `rpart` package with a suitable splitting criterion for survival data.
4. Evaluate the models performance using concordance index and visualize the survival curves for different terminal nodes.
5. Compare the survival trees performance to a Cox regression model [9].

Exercise 15.11. The goal here is to train and evaluate a multivariate regression tree.

1. Load the `mtcars` dataset from base R.
2. Split the data into training and testing sets.
3. Train a multivariate regression tree to predict `mpg`, `hp`, and `wt`.
4. Evaluate the models performance using mean squared error for each target variable.
5. Compare the multivariate trees performance with that of separate univariate trees for each target variable.

Exercise 15.12. Find an interesting multivariate regression dataset with response $(Y_1, \ldots, Y_q)^T$ for $q > 1$. Calculate the Mahalanobis split statistic for each X variable and use this to identify interesting variables. Repeat the exercise but use univariate within node variance splitting for each response coordinate Y_l, $l = 1, \ldots, q$. Compare the X variables identified using this method to the X variables identified using the Mahalanobis split-statistic.

Exercise 15.13. Consider a logistic regression model for $Y \in \{0,1\}$ with $p = 1$ variables such that

$$\mathbb{P}\{Y = 1|X = x\} = \frac{1}{1 + \exp(-\phi(x))}$$

for $\phi(x) = 2x$ with X uniformly distributed on $[0,1]$. Determine the super greedy tree (SGT) of Chapter 15.11 using $k = 1$ splits. Be sure to state clearly the type of splits used and define all your notation.

Exercise 15.14. It was shown in Chapter 15.11.3 that weighted variance splitting reduces empirical risk under L_2-loss for regression trees. In this exercise we will extend this result to classification, $Y \in \{0,1\}$, under various loss functions. Let $\Pi_{n,k-1}$ be the tree classifier partition of size k and let

$$\hat{F}_n(\mathbf{x}, \Pi_{n,k-1}) = \sum_{A \in \Pi_{n,k-1}} \hat{F}_{n,A} 1_{\{\mathbf{x} \in A\}} \in [0,1]$$

be the tree estimator for $\mathbb{P}\{Y = 1|\mathbf{X} = \mathbf{x}\}$. Adopting our previous notation, the empirical risk for $\Pi_{n,k-1}$ is

$$R_n(\Pi_{n,k-1}) = \sum_{A \in \Pi} R_n(A), \quad \text{where } R_n(A) = \frac{1}{n} \sum_{i \in A} L(Y_i, \hat{F}_{n,A}).$$

The cell estimators $\hat{F}_{n,A} \in [0,1]$ are obtained by solving $n^{-1} \sum_{i \in A} L'(Y_i, \hat{F}_{n,A}) = 0$ where the loss function $L(Y, u)$ is assumed to be differentiable and convex in u.

1. Suppose that the tree is updated to $\Pi_{n,k}$ by replacing a cell A with daughters A_L and A_R. The updated tree estimator is

$$\hat{F}_n(\mathbf{x}, \Pi_{n,k}) = \hat{F}_n(\mathbf{x}, \Pi_{n,k-1}) - \hat{F}_{n,A} 1_{\{\mathbf{x} \in A\}} + \left[\hat{F}_{n,A_L} 1_{\{\mathbf{x} \in A_L\}} + \hat{F}_{n,A_R} 1_{\{\mathbf{x} \in A_R\}}\right]$$

   and the empirical risk reduction is

$$R_n(\Pi_{n,k-1}) - R_n(\Pi_{n,k}) \tag{15.32}$$
$$= \frac{1}{n} \sum_{i \in A} L(Y_i, \hat{F}_{n,A}) - \left[\frac{1}{n} \sum_{i \in A_L} L(Y_i, \hat{F}_{n,A_L}) + \frac{1}{n} \sum_{i \in A_R} L(Y_i, \hat{F}_{n,A_R})\right].$$

   Use cross-entropy loss (log-loss) defined in Chapter 7.2. Show that under this loss, the empirical risk reduction (15.32) is always positive except in trivial cases. Using this, propose a cross-entropy splitting rule that can be used for classification with the property that it reduces empirical risk.
2. Repeat part 1 using the Brier loss defined in Chapter 7.2.
3. Both the cross-entropy and Brier loss are examples of proper loss functions, where recall that a proper scoring rule has the property that the true distribution is the optimal strategy in expectation. Using this and parts 1 and 2, state a theoretical result for splitting rules and empirical risk reduction for proper loss functions.

Exercise 15.15. Prove that Theorem 15.10 and Theorem 15.11 implies (15.23).

Exercise 15.16. Why does (15.28) offer a more refined solution?

Exercise 15.17. Show that (15.30) is the one-step Newton solution to (15.29).

Exercise 15.18. Explain why a one-step solution (15.30) is sufficient to approximate (15.29). Demonstrate your answer with a concrete example.

Exercise 15.19. In this example we revisit the Ozone dataset from the `mlbench` package. Fit the data using tree-based boosting using `gbm` and linear boosting using `mboost`. Compare your results and discuss.

Exercise 15.20. This example investigates the effect of overfitting in gradient-boosted trees and methods to mitigate it.

1. Load the `CIFAR-10` dataset.
2. Preprocess the data by resizing the images.
3. Split the data into training and testing sets.
4. Train a gradient boosting tree model using the `xgboost` package.
5. Identify signs of overfitting by analyzing the training and validation loss curves.
6. Experiment with techniques to mitigate overfitting, such as early stopping, regularization, and pruning.
7. Evaluate the effectiveness of these techniques on the testing set using accuracy and confusion matrix.

Exercise 15.21. Illustrate tree instability by using a dataset of your choosing.

Exercise 15.22. In this example you will implement and evaluate bagging for a classification problem.

1. Load the `Breast Cancer` dataset from the `mlbench` package.
2. Split the data into training and testing sets.
3. Implement bagging using the `ipred` package to create an ensemble of classification trees.
4. Evaluate the performance of the bagged model using accuracy, precision, recall, and F1-score.
5. Compare the performance of the bagged model with that of a single decision tree.

Exercise 15.23. Does tree instability become worse with increasing dimension? Use a real example to illustrate your answer.

Exercise 15.24. Write a program to calculate the bagged least squares estimator. Compare results to least squares, lasso and boosting on several different datasets. Measure performance using out-of-sample data. Discuss your findings.

Exercise 15.25. Devise a bagged procedure for doing model selection for linear regression. Compare results for this method to boosting using synthetic data of your own making. Discuss your findings.

Exercise 15.26. Here we will explore the combination of bagging and boosting techniques to improve classification performance.

1. Load the `MNIST` dataset.
2. Preprocess the data by flattening the images.
3. Split the data into training and testing sets.
4. Implement a model that combines bagging and boosting techniques using the `caret` and `gbm` packages.
5. Evaluate the combined models performance.
6. Compare the performance of the combined model with individual bagging and boosting models.

Bibliography

[1] Bamber D. The area above the ordinal dominance graph and the area below the receiver operating characteristic graph. *Journal of Mathematical Psychology*, 12:387–415, 1975.

[2] Biau, G., Devroye, L., and Lugosi, G. Consistency of random forests and other averaging classifiers. *Journal of Machine Learning Research*, 9(Sep):2015–2033, 2008.

[3] Breiman L., Friedman J.H., Olshen R.A., and Stone C.J. *Classification and Regression Trees*, Belmont, California, 1984.

[4] Breiman, L., Some properties of splitting criteria. *Machine Learning*, 24(1), 41–47, 1996.

[5] Breiman L. Bagging predictors. *Machine Learning*, 26:123–140, 1996.

[6] Breiman L. Heuristics of instability and stabilization in model selection. *Annals of Statistics*, 24:2350–2383, 1996.

[7] Bruckner A. Tests for the superadditivity of functions. *Proceedings of the American Mathematical Society*, 13(1):126–130, 1962.

[8] Bühlmann P. and Hothorn T. Boosting Algorithms: Regularization, Prediction and Model Fitting (with Discussion). *Statistical Science*, 22(4), 477–505. 2007

[9] Cox D. R. Regression models and life-tables. *Journal of the Royal Statistical Society. Series B (Methodological)*, 34(2), 187–220, 1972.

[10] Devroye L., Györfi L. and Lugosi G. *A Probabilistic Theory of Pattern Recognition*. Springer, 1996.

[11] Friedman J.H. Greedy function approximation: A gradient boosting machine. *Annals of Statistics*, 29:1189–1232, 2001.

[12] Greenwell B., Boehmke B., Cunningham J. and Developers. *gbm: Generalized Boosted Regression Models*. R package version 2.1.8. `https://CRAN.R-project.org/package=gbm`, 2020.

[13] Gyorfi L., Kohler M., Krzyzak A. and Walk H. *A Distribution-Free Theory of Nonparametric Regression*, Springer, 2002.

[14] D.J. Hand and R.J. Till. A simple generalisation of the area under the ROC curve for multiple class classification problems. *Machine Learning*, 45(2):171–186, 2001.

[15] J. Hanley and B. McNeil. The meaning and use of the area under a receiver operating characteristic (ROC) curve. *Radiology*, 143:29–36, 1982.

[16] Ishwaran H. Super greedy trees *Technical Report, University of Miami*, 2024.

[17] Ishwaran H. and Lee D.K.K. Super greedy trees: some asymptotic results. *Technical Report, University of Miami*, 2024.

[18] Ishwaran H. and Kogalur U.B. randomForestSGT: Random Forest Super Greedy Trees. R package version 1.42 `https://github.com/kogalur/randomForestSGT`, 2024.

[19] Kaplan E.L. and Meier P. Nonparametric estimation from incomplete observations. *Journal of the American Statistical Association*, 53(282), 457–481, 1958.

[20] Lugosi G. and Nobel A. Consistency of data-driven histogram methods for density estimation and classification. *The Annals of Statistics*, 24(2), 687–706, 1996.

[21] Lu M. and Ishwaran H. Variable priority for variable selection in rule-based models. *Technical Report, University of Miami*, 2022.

[22] Morgan J.N. and Messenger R.C. *THAID: a Sequential Search Program for the Analysis of Nominal Scale Dependent Variables*, Survey Research Center, Institute for Social Research, University of Michigan", 1973.

[23] Sauer, N. On the density of families of sets. *Journal of Combinatorial Theory, Series A* 13(1), 145–147, 1972.

[24] Stone C.J. Consistent nonparametric regression. *Annals of Statistics*, 5(4):595–620, 1977.

[25] Tang F. and Ishwaran H. Random forest missing data algorithms. *Statistical Analysis and Data Mining: The ASA Data Science Journal*, 10(6):363–377, 2017.

16

Random Forests

16.1 Introduction

In the 1990s, researchers discovered an interesting phenomenon related to averaging a learner. Surprisingly, the performance of a procedure h could be significantly improved by "tweaking" it over different runs on the data and then aggregating the resulting runs. Modifying h was relatively straightforward, such as by perturbing the data or varying h's tuning parameters. This sparked considerable interest and led to the development of the research area known as ensemble learning. Formally, ensembles are defined as predictors created by aggregating a base learner $h(\cdot,\Theta)$ with respect to a random parameter Θ. Typically, this aggregation involves averaging the predictions. One example is bagging [3], discussed in Chapter 15.13, where Θ is a random element encoding bootstrap instructions. A very popular ensemble method is random forests [4], developed by Leo Breiman, which is closely related to bagging and will be the focus of this chapter.

16.2 Breiman's Random Forests

In random forests, the base learner is a random tree grown under a specific set of instructions which is then averaged. Breiman defined a random forest classifier as follows.

Definition 16.1. A random forest classifier consists of a collection of tree-structured classifiers $\{h(\mathbf{x},\Theta_b), b = 1,2,\ldots\}$ where the $\{\Theta_b\}$ are independent identically distributed random vectors and each tree casts a unit vote for the most popular class at input $\mathbf{x}$. □

According to Breiman, the idea for random forests was shaped by certain earlier papers on ensemble learning. This included the random subspace method [16], random split selection [10], perfect random tree ensembles [9], and shape recognition trees [1]. Because each of these methods can also be considered as a type of random forest, we will use the term "random forests", abbreviated hereafter as RF, to mean the random forest method described in Breiman [4]. An example is Forest-RI (RF random input selection) used for classification which has been reproduced in Algorithm 16.1.

Algorithm 16.1 *Forest-RI*

1: Select at random, at each node, a small group of input variables to split on.
2: Grow the tree using CART methodology to maximum size without pruning.
3: Grow a collection of tree classifiers as in Steps 1 and 2 independently to obtain the random forest classifier.

DOI: 10.1201/9781003493679-16

Step 1 of the algorithm describes the i.i.d. instructions $\{\Theta_b\}$ used to construct the tree. As stated, trees are randomly grown by selecting a different random subset of variables when splitting a tree node. However, Breiman [4] found that performance was significantly improved by using bagging in tandem with random feature selection. Therefore in practice, $\{\Theta_b\}$ is used for two types of randomization: bootstrapping the data and random feature selection.

Generally speaking, splitting to purity in Step 2 is not always desirable. Current research in RF suggests that the optimal choice should depend on the sample size n and number of features p. We know, for example, that large sample consistency for RF requires terminal node sizes to increase with n under certain conditions [7]. Empirically it has been observed that smaller nodes, sometimes even pure nodes, are more effective when the feature space is large, or when the sample size is small [20]. We will come back to this point later.

Incorporating bagging with random feature selection and allowing for a user selected node size leads to a general RF procedure which is described in Algorithm 16.2. In line 2, random feature selection is used for node splitting. To split a node, a subset of $1 \le \text{mtry} \le p$ variables are randomly selected, and the node is cut using the optimal split chosen from these candidate variables. The tree is split until all nodes reach a lower limit of $\text{nodesize} \ge 1$ unique cases. The two tuning parameters mtry and nodesize were coined by Brieman in his Fortran code for random forests and we adhere to this nomenclature throughout. Usually CART (Chapter 15.2, 15.7) is used for the RF tree induction, but this is strictly not necessary.

Algorithm 16.2 *Random Forests*

1: Draw a bootstrap sample of the original data.
2: Grow a tree using data from Step 1. When growing the tree, at each each node in the tree, determine the optimal split for the node using $\text{mtry} \le p$ randomly selected variables. Grow the tree so that each terminal node contains no fewer than $\text{nodesize} \ge 1$ cases.
3: Repeat Steps 1-2, $B > 1$ times independently.
4: Combine the B trees to form the ensemble predictor.

Algorithm 16.2 applies to many data settings, including regression and classification. For classification, RF can be described as follows. Denote the randomized classifier learner by $h(\mathbf{x}, \Pi_b) := h(\mathbf{x}, \Pi_b, \mathscr{L})$, for tree $b = 1, \ldots, B$, where $\Pi_b := \Pi_b(\Theta_b)$ is the randomized partition obtained under instructions Θ_b for tree b and $\mathscr{L}$ is the learning data. For a given input $\mathbf{x}$, $h(\mathbf{x}, \Pi_b)$ returns a value between 0 and 1, which is an estimate for the class probability $\mathbb{P}\{Y = 1|\mathbf{X} = \mathbf{x}\}$. The forest ensemble class probability estimate is obtained by averaging:

$$\hat{F}_{\text{ens}}(\mathbf{x}) = \frac{1}{B} \sum_{b=1}^{B} h(\mathbf{x}, \Pi_b) \in [0, 1].$$

The ensemble forest classifier is defined by

$$\hat{\delta}_{\text{ens}}(\mathbf{x}) = \begin{cases} 1 & \text{if } \hat{F}_{\text{ens}}(\mathbf{x}) = \frac{1}{B}\sum_{b=1}^{B} h(\mathbf{x}, \Pi_b) \ge \frac{1}{2} \\ 0 & \text{otherwise.} \end{cases}$$

Notice this forest is different from Definition 16.1, which uses tree learner classifiers returning binary values. Here the tree learner returns a continuous value between 0 and 1, which is an estimate for $F(\mathbf{x}) = \mathbb{P}\{Y = 1|\mathbf{X} = \mathbf{x}\}$. This gives us a direct estimator for the conditional probability, which is useful in many data applications. Also, when the focus is classification, the ensemble probability $\hat{F}_{\text{ens}}$ can be conveniently converted to a classification value using whatever rule is most appropriate for the problem. Typically, this conversion is by the Bayes rule (as used here), but in some cases, such as class imbalanced data (Chapter 16.12), a different rule is needed.

In the case of regression where the response outcome is continuous, $Y \in \mathbb{R}$, the RF regression tree $h(\mathbf{x}, \Pi_b)$ returns a continuous value which estimates the conditional mean $F(\mathbf{x}) = \mathbb{E}[Y|\mathbf{X} = \mathbf{x}]$. The RF predictor is the averaged randomized function

$$\hat{F}_{\text{ens}}(\mathbf{x}) = \frac{1}{B}\sum_{b=1}^{B} h(\mathbf{x}, \Pi_b) \in \mathbb{R}$$

which is the ensemble estimate for $F(\mathbf{x})$.

16.2.1 Illustration of effectiveness of ensemble learning

RF has proven to be an excellent predictor, performing especially well in challenging problems, such as p bigger than n scenarios and settings with complex interactions between variables.

For illustration, we present results from a large empirical experiment using RF regression (for ease we abbreviate this as RF-R). Details regarding the data can be found in Chapter 17.5.2. Forests were computed using the `randomForestSRC` package [17] which will be used for all RF calculations. Each forest used 500 random regression trees grown using CART. The default settings were used for each dataset. For regression this is $(\text{nodesize}, \text{mtry}, B) = (5, p/3, 500)$.

Performance was measured by standardized prediction (generalization) error, where standardization was by the sample variance of the response outcome Y (see Remark 16.1 for details):

$$\frac{\widehat{\text{Err}}(\hat{F})}{\sum_{i=1}^{n}(Y_i - \overline{Y})^2/n}. \tag{16.1}$$

The value $\widehat{\text{Err}}(\hat{F})$ refers to the estimated prediction error for the estimator $\hat{F}$.

Figure 16.1 displays the results. The red points show the standardized error for RF (estimated using out-of-bag (OOB) data; see Chapter 16.7). The boxplots display the OOB error for each tree in a forest. Notice that the forest error (shown in red) is lower than most, if not all, of its trees. This pattern is consistent across all datasets, and in some cases, the forest error rate drops well below the benchmark value of one (see Remark 16.1), even though many of its trees have values well above one. While certain trees may occasionally perform better than RF, these instances are rare and likely spurious. Therefore, selecting a single tree with a low error would be a poor strategy and perform poorly on test data. The advantage of using an ensemble like RF is that it minimizes high variability due to spurious predictions of a single learner by aggregating the predictions over a class of base learners. This is a variance reduction phenomenon that will be discussed next.

Remark 16.1. A mathematical explanation for (16.1) is as follows. Let $\mathscr{L} = \{(\mathbf{X}_i, Y_i) \in \mathscr{X} \times \mathbb{R}, i = 1, \ldots, n\}$ be the i.i.d. learning data. Assume $Y_i = F(\mathbf{X}_i) + \varepsilon_i$ where $\{\varepsilon_i\}$ are i.i.d. independent of $\{\mathbf{X}_i\}$ and $\mathbb{E}[Y_i|\mathbf{X}_i] = F(\mathbf{X}_i)$, $\text{Var}(Y_i|\mathbf{X}_i) = \sigma^2$. Then

$$\begin{aligned} Y_i - \overline{Y} &= F(\mathbf{X}_i) + \varepsilon_i - \frac{1}{n}\sum_{j=1}^{n}\left(F(\mathbf{X}_j) + \varepsilon_j\right) \\ &= \varepsilon_i\left(1 - \frac{1}{n}\right) - \frac{1}{n}\sum_{j \neq i}\varepsilon_j + F(\mathbf{X}_i) - \frac{1}{n}\sum_{j=1}^{n} F(\mathbf{X}_j) \\ &= U_i + F(\mathbf{X}_i) - \overline{F}, \end{aligned}$$

where

$$U_i = \varepsilon_i\left(1 - \frac{1}{n}\right) - \frac{1}{n}\sum_{j \neq i}\varepsilon_j, \quad \overline{F} = \frac{1}{n}\sum_{j=1}^{n} F(\mathbf{X}_j).$$

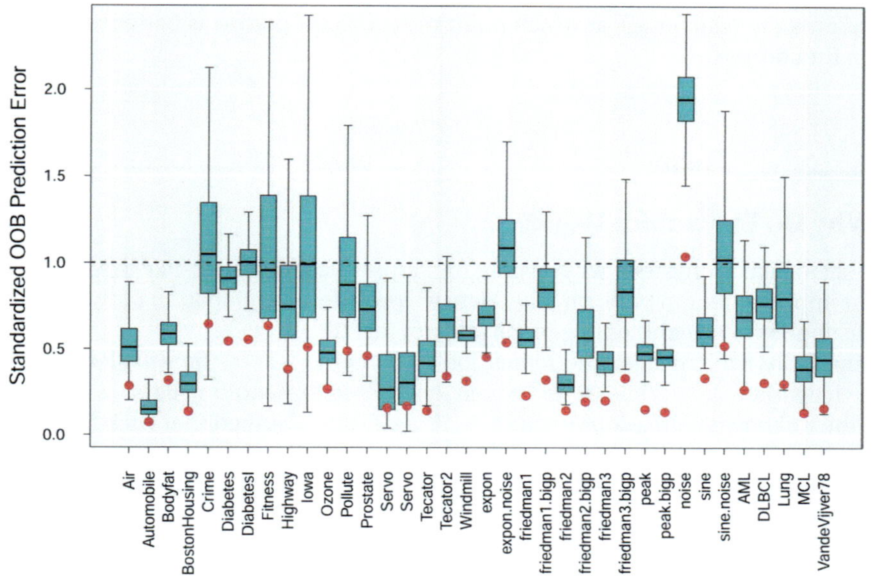

FIGURE 16.1
Performance on benchmark datasets using standardized out-of-bag (OOB) prediction error. Red points record performance for RF-R. Boxplots record the prediction performance of each tree in the forest.

Note that U_i are identically distributed with $\mathbb{E}(U_i) = 0$. Therefore the expected value of the sample variance is

$$\begin{aligned}\frac{1}{n}\mathbb{E}\left[\sum_{i=1}^{n}(Y_i - \overline{Y})^2\right] &= \frac{1}{n}\mathbb{E}\left[\sum_{i=1}^{n} U_i^2 + 2\sum_{i=1}^{n} U_i\left(F(\mathbf{X}_i) - \overline{F}\right) + \sum_{i=1}^{n}\left(F(\mathbf{X}_i) - \overline{F}\right)^2\right] \\ &= \sigma^2\left[\left(1 - \frac{1}{n}\right)^2 + \frac{n-1}{n^2}\right] + \frac{1}{n}\mathbb{E}\left[\sum_{i=1}^{n}\left(F(\mathbf{X}_i) - \overline{F}\right)^2\right] \\ &= \sigma^2\left(1 - \frac{1}{n}\right) + \sigma_X^2.\end{aligned}$$

Because $\hat{\text{Err}}(\hat{F})$ is an estimate of prediction error, by (4.16), we have

$$\frac{\hat{\text{Err}}(\hat{F})}{\sum_{i=1}^{n}(Y_i - \overline{Y})^2/n} \approx \frac{\sigma^2 + \mathbb{E}\left[\hat{F}(\mathbf{X}) - F(\mathbf{X})\right]^2}{\sigma^2(1 - 1/n) + \sigma_X^2}. \tag{16.2}$$

The value $\sigma^2 + \sigma_X^2$ equals the overall variance in the model. Thus, for a large enough n, the denominator in (16.2) approximates the overall variance. The numerator, on the other hand, equals the measurement error plus the MSE (mean-squared error). Comparing it to the denominator leads to an interpretable quantity for comparing results across datasets. Specifically, (16.2) is less than one if and only if

$$\begin{aligned}\mathbb{E}\left[\hat{F}(\mathbf{X}) - F(\mathbf{X})\right]^2 &< -\sigma^2/n + \sigma_X^2 \\ &\le \sigma_X^2 \\ &= \frac{1}{n}\mathbb{E}\left[\sum_{i=1}^{n}\left(F(\mathbf{X}_i) - \overline{F}\right)^2\right] = \mathbb{E}\left[\overline{F} - F(\mathbf{X})\right]^2.\end{aligned}$$

This represents the benchmark at which point the predictor performs better than the averaged true mean function. □

16.3 Why Do Ensembles Work?

It might seem like a mystery why averaging simple base learners like trees leads to such significant improvement in performance. Here we provide an explanation for why this happens. This result applies to ensembles in general and not just RF.

For simplicity, we consider the regression case. Let $h_1, \ldots, h_B$ be a collection of learners where each learner $h_b : \mathscr{X} \to \mathbb{R}$ is trained on the same learning data $\mathscr{L}$. The goal is to estimate the unknown regression function $F : \mathscr{X} \to \mathbb{R}$. It is assumed that each learner is trained separately from one another. Because the learners are trained on the same data, they cannot be independent; however, we can reasonably assume they share a common distribution. This assumption holds for RF, for instance.

Define the ensemble estimator as the averaged value

$$\hat{F}_{\text{ens}}(\mathbf{x}) = \frac{1}{B} \sum_{b=1}^{B} h_b(\mathbf{x}).$$

Let $(\mathbf{X}, \mathbf{Y})$ be an independent test data point with the same distribution as the learning data. Recall from Chapter 7.3 (using the same assumptions as before)

$$\text{Err}(\hat{F}) = \mathbb{E}_{\mathscr{L}} \mathbb{E}_{\mathbf{X},Y} \left[Y - \hat{F}(\mathbf{X}) \right]^2$$

equals the generalization error for an estimator $\hat{F}$ which can be decomposed as

$$\text{Err}(\hat{F}) = \sigma^2 + \mathbb{E}_{\mathbf{X}} \left\{ \text{Bias}\{\hat{F} \,|\, \mathbf{X}\}^2 + \text{Var}\{\hat{F} \,|\, \mathbf{X}\} \right\}$$

where the two terms inside the expectation are the conditional bias and variance for $\hat{F}$ defined by

$$\begin{aligned} \text{Bias}\{\hat{F} \,|\, \mathbf{X}\} &= \mathbb{E}_{\mathscr{L}} \left\{ \hat{F}(\mathbf{X}) \right\} - F(\mathbf{X}) \\ \text{Var}\{\hat{F} \,|\, \mathbf{X}\} &= \mathbb{E}_{\mathscr{L}} \left\{ \left(\hat{F}(\mathbf{X}) - \mathbb{E}_{\mathscr{L}}[\hat{F}(\mathbf{X})] \right)^2 \right\}. \end{aligned}$$

It can be shown that the following holds [37].

Theorem 16.1. *If $h, h_1, \ldots, h_B$ are identically distributed learners constructed from the same learning data, then*

$$\textit{Err}(\hat{F}_{ens}) = \sigma^2 + \mathbb{E}_{\mathbf{X}} \left\{ \textit{Bias}\{h \,|\, \mathbf{X}\}^2 + \frac{1}{B} \textit{Var}\{h \,|\, \mathbf{X}\} + \left(1 - \frac{1}{B} \right) \textit{Cov}(h_1, h_2) \right\}.$$

To understand Theorem 16.1 keep in mind that the number of learners, B, is at the discretion of the user and can be very large (in practice this decision will obviously depend on computational costs, but we will not worry about this here). So, assuming a large enough collection of learners, we can expect the generalization error to closely approximate the limiting value

$$\lim_{B \to \infty} \text{Err}(\hat{F}_{\text{ens}}) = \sigma^2 + \mathbb{E}_{\mathbf{X}} \left\{ \text{Bias}\{h | \mathbf{X}\}^2 + \text{Cov}(h_1, h_2) \right\}.$$

Notice that the variance converges to zero leaving only the bias and covariance, the latter equaling the average covariance between any two learners. Therefore, in order to achieve the smallest possible generalization error (this being σ^2) we should choose base learners with near zero bias while maintaining low covariance.

However, this balancing act is hard to achieve. As bias decreases, learners naturally become more complex, which has the counter effect of increasing covariance. We can see why this is the case for RF. The base learners are randomized trees, so reducing bias seems only possible if trees become less randomized. For example, deep trees without feature selection and bootstrapping would be expected to have a substantially smaller bias than a randomized tree, which uses less data (due to bootstrapping) and is less greedy (since it does not use every variable when splitting). But doing this increases the correlation between trees, as trees become more similar to one another without randomization. Thus, we can summarize the result of the theorem by saying that RF works because it is a variance reduction technique, but its final performance depends on the interplay between bias and covariance of the tree learners.

16.4 The Random Forests Estimator

Now we describe the RF estimator in more detail. We showed in Chapter 15.8 that regression and classification trees could be described as weighted averaged estimators. This interpretation also holds for RF because it is a tree averaged estimator.

For regression with continuous $Y \in \mathbb{R}$, or $J = 2$ classification with binary $Y \in \{0,1\}$, we can describe the bth tree learner with partition $\mathcal{T}_b$ in terms of tree weights similar to (1)

$$h(\mathbf{x}, \mathcal{T}_b) = \sum_{i=1}^{n} W_{i,b}(\mathbf{x}) Y_i, \qquad W_{i,b}(\mathbf{x}) = \frac{1_{\{\mathbf{X}_i \in \mathcal{T}_b(\mathbf{x})\}}}{\sum_{j=1}^{n} 1_{\{\mathbf{X}_j \in \mathcal{T}_b(\mathbf{x})\}}}$$

where $\mathcal{T}_b(\mathbf{x})$ is the function which identifies the cell that $\mathbf{x}$ belongs to for tree b. Then the RF ensemble obtained by averaging can be written as

$$\hat{F}_{\text{ens}}(\mathbf{x}) = \frac{1}{B} \sum_{b=1}^{B} h(\mathbf{x}, \mathcal{T}_b) = \sum_{i=1}^{n} W_i(\mathbf{x}) Y_i, \tag{16.3}$$

where

$$W_i(\mathbf{x}) = \frac{1}{B} \sum_{b=1}^{B} W_{i,b}(\mathbf{x}) \tag{16.4}$$

are the *forest weights*. Because tree weights are convex, $0 \le W_{i,b}(\mathbf{x}) \le 1$, $\sum_{i=1}^{n} W_{i,b}(\mathbf{x}) = 1$, it follows that the forest weights are also convex: $0 \le W_i(\mathbf{x}) \le 1$,

$$\sum_{i=1}^{n} W_i(\mathbf{x}) = \frac{1}{B} \sum_{i=1}^{n} \sum_{b=1}^{B} W_{i,b}(\mathbf{x}) = \frac{1}{B} \sum_{b=1}^{B} \sum_{i=1}^{n} W_{i,b}(\mathbf{x}) = \frac{1}{B} \sum_{b=1}^{B} 1 = 1.$$

More broadly, we can define a general RF predictor. Let $\{(\mathbf{X}_i, Y_i)\}_1^n$ be i.i.d. over $\mathscr{X} \times \mathscr{Y}$. Consider a class of target functions $\{F_\zeta : \zeta \in \mathcal{Z}\}$ indexed by some parameter ζ. Assume that the tree learner for estimating $F_\zeta(\mathbf{x})$ is

$$h_\zeta(\mathbf{x}, \mathcal{T}_b) = \sum_{i=1}^{n} W_{i,b}(\mathbf{x}) g_\zeta(Y_i)$$

for some function $g_\zeta : \mathscr{Y} \to \mathbb{R}$. Then the ensemble estimator for $F_\zeta(\mathbf{x})$ is

$$\hat{F}_{\text{ens},\zeta}(\mathbf{x}) = \frac{1}{B}\sum_{b=1}^{B} h_\zeta(\mathbf{x}, \mathcal{I}_b) = \sum_{i=1}^{n} W_i(\mathbf{x}) g_\zeta(Y_i) \tag{16.5}$$

where $W_1(\mathbf{x}), \ldots, W_n(\mathbf{x})$ are forest weights as in (16.4).

Typically the function g_ζ used is chosen so as to estimate a conditional mean. In other words, if $F_\zeta(\mathbf{x}) = \mathbb{E}[g_\zeta(Y)|\mathbf{X} = \mathbf{x}]$ is a function we wish to estimate, then (16.5) is the estimator we would use to estimate F_ζ. This is because this is an unbiased estimator. Assuming that $Y_1, \ldots, Y_n$ are identically distributed given $W_1(\mathbf{X}), \ldots, W_n(\mathbf{X})$, we have the following unbiased property

$$\begin{aligned}
\mathbb{E}\left[\hat{F}_{\text{ens},\zeta}(\mathbf{X})\right] &= \mathbb{E}\left\{\sum_{i=1}^{n} \mathbb{E}\left[W_i(\mathbf{X}) g_\zeta(Y_i)\Big|\mathbf{X}, W_1(\mathbf{X}_1), \ldots, W_n(\mathbf{X}_n)\right]\right\} \\
&= \mathbb{E}\left\{\sum_{i=1}^{n} W_i(\mathbf{X})\mathbb{E}\left[g_\zeta(Y_i)\Big|\mathbf{X}, W_1(\mathbf{X}_1), \ldots, W_n(\mathbf{X}_n)\right]\right\} \\
&= \mathbb{E}\left\{\mathbb{E}\left[g_\zeta(Y_1)\Big|W_1(\mathbf{X}_1), \ldots, W_n(\mathbf{X}_n)\right]\sum_{i=1}^{n} W_i(\mathbf{X})\right\} \\
&= \mathbb{E}\left\{\mathbb{E}\left[g_\zeta(Y_1)\Big|W_1(\mathbf{X}_1), \ldots, W_n(\mathbf{X}_n)\right]\right\} \\
&= \mathbb{E}\left[g_\zeta(Y_1)\right] = \mathbb{E}\left[g_\zeta(Y)\right] = \mathbb{E}\left\{\mathbb{E}[g_\zeta(Y)|\mathbf{X} = \mathbf{x}]\right\} = \mathbb{E}[F_\zeta(\mathbf{X})].
\end{aligned}$$

Some examples of a general RF predictor are given below.

1. For classification with $Y \in \{1, \ldots, J\}$, let the target function of interest be

$$F_\zeta(\mathbf{x}) = \mathbb{P}\{Y = \zeta|\mathbf{X} = \mathbf{x}\}, \qquad 1 \le \zeta \le J.$$

Then $g_\zeta(y) = 1_{\{y=\zeta\}}$ and the ensemble estimator for $F_j(\mathbf{x})$ for $\zeta = j$ is

$$\hat{F}_{\text{ens},j}(\mathbf{x}) = \sum_{i=1}^{n} W_i(\mathbf{x}) 1_{\{Y_i = j\}}, \qquad 1 \le j \le J.$$

This can be turned into a classifier using the Bayes rule

$$\hat{\delta}_{\text{ens},j}(\mathbf{x}) = \underset{1\le j\le J}{\arg\max}\left\{\hat{F}_{\text{ens},j}(\mathbf{x})\right\}.$$

2. Another example is quantile regression. Let the target function be $F_\zeta(\mathbf{x}) = \mathbb{P}\{Y \le \zeta|\mathbf{X} = \mathbf{x}\}$, for $\zeta \in \mathbb{R}$. Then $g_\zeta(y) = 1_{\{y \le \zeta\}}$ and the ensemble estimator for $F_y(\mathbf{x})$ where $\zeta = y$ is

$$\hat{F}_{\text{ens},y}(\mathbf{x}) = \sum_{i=1}^{n} W_i(\mathbf{x}) 1_{\{Y_i \le y\}}. \tag{16.6}$$

Chapter 16.13 discusses this type of forest which is refered to as a *quantile regression forest.*

16.5 Multivariate Random Forests

We can extend RF using the same principles to define multivariate random forests (mRF) for settings with multivariate responses $\mathbf{Y} = (Y_1, \ldots, Y_q)^T$. The extended mRF allows for response vectors that are continuous, categorical, or a mixture of outcome types.

Two changes are made to the original RF procedure in this extension. (1) Tree nodes are cut using a multivariate tree-splitting rule (Chapter 15.10). Similar to a univariate cut, this cuts a node using a one-dimensional feature X, the only difference being that the split-statistic is calculated using all q coordinates of $\mathbf{Y}$. For example, composite split-statistics which aggregate the contributions from each coordinate can be used. (2) In place of a single target function, we have q target functions and a q-dimensional tree predictor.

We can summarize the algorithm for mRF as follows.

1. Draw a bootstrap sample and grow a multivariate tree using random feature selection where tree nodes are split using a multivariate tree splitting rule.
2. Calculate the multivariate tree predictor.
3. Repeat and average to obtain the mRF predictor.

A mathematical description is as follows. Let $h_j(\mathbf{x}, \Pi_b)$ be the tree predictor for coordinate $j = 1, \ldots, q$ where Π_b is the tree partition for the randomized multivariate tree $b = 1, \ldots, B$. The mRF ensemble is

$$(\hat{F}_{\text{ens,1}}(\mathbf{x}), \ldots, \hat{F}_{\text{ens,q}}(\mathbf{x}))^T = \frac{1}{B}\sum_{b=1}^{B} \left(h_1(\mathbf{x}, \Pi_b), \ldots, h_q(\mathbf{x}, \Pi_b)\right)^T.$$

We can describe the mRF ensemble using forest weights similar to univariate RF. Suppose that $F_j(\mathbf{x}) = \mathbb{E}[g_j(Y)|\mathbf{X} = \mathbf{x}]$ is the target function for coordinate $j = 1, \ldots, q$. Then

$$(\hat{F}_{\text{ens,1}}(\mathbf{x}), \ldots, \hat{F}_{\text{ens,q}}(\mathbf{x}))^T = \sum_{i=1}^{n} W_i(\mathbf{x}) \left(g_1(\mathbf{x}), \ldots, g_q(\mathbf{x})\right)^T.$$

The mRF weights are obtained in exactly the same way as univariate RF.

As an example, consider mRF regression for the case where the coordinates are continuous. Then $F_j = \mathbb{E}[Y_j|\mathbf{X} = \mathbf{x}]$ and $g_j = Y_j$ and

$$(\hat{F}_{\text{ens,1}}(\mathbf{x}), \ldots, \hat{F}_{\text{ens,q}}(\mathbf{x}))^T = \sum_{i=1}^{n} W_i(\mathbf{x}) \left(Y_1, \ldots, Y_q\right)^T. \tag{16.7}$$

As another example, consider mRF for multivariate classification for binary $Y_j \in \{0, 1\}$. Then $F_j = \mathbb{P}[Y_j = 1|\mathbf{X} = \mathbf{x}]$ and $g_j = 1_{\{Y_j = 1\}} = Y_j$ and the mRF ensemble is the same as (16.7).

16.6 In-Bag Ensemble

There is a subtle issue related to the bootstrap that leads to a different type of ensemble called the *in-bag ensemble*. Recall the RF tree is constructed from a bootstrap sample draw. Therefore, the induced RF tree does not use the entire sample but only a fraction of the data (about 0.632 of the original learning sample). But what about the predicted value for the terminal node? What data should be used for determining its value? Although we could run the full sample down the tree, the standard practice is to populate the terminal nodes using the same in-bag data used for the tree induction. Since this only uses the bootstrap sample, the resulting predictor is referred to as *in-bag*. Notice that the predictor we are talking about is different from the predictors used by (16.5), which use the entire data for terminal node prediction.

We can define the RF in-bag ensemble as follows. Let integer values $n_{i,b} \geq 0$ record the bootstrap frequency of case i for the bth tree. For latter use we will also need values recording OOB membership, which we can define by using $n_{i,b}$. Use $I_{i,b} \in \{0,1\}$ to indicate OOB membership for case i:

$$I_{i,b} = \begin{cases} 1 & \text{if } n_{i,b} = 0 \\ 0 & \text{otherwise.} \end{cases}$$

Thus, i is OOB for tree b if $I_{i,b} = 1$, in which case $n_{i,b} = 0$, while i is in-bag if $I_{i,b} = 0$, in which case $n_{i,b} \geq 1$. As before, let $\mathcal{R}_b(\mathbf{x})$ be the function which identifies the cell that $\mathbf{x}$ belongs to. Define the in-bag tree weights for tree b by

$$W^*_{i,b}(\mathbf{x}) = \frac{n_{i,b} 1_{\{\mathbf{X}_i \in \mathcal{R}_b(\mathbf{x})\}}}{\sum_{j=1}^n n_{j,b} 1_{\{\mathbf{X}_j \in \mathcal{R}_b(\mathbf{x})\}}}.$$

We use the notation "$*$" to indicate the dependence on the bootstrap. Notice that a case i's contribution to the terminal node predictor is adjusted by its in-bag frequency $n_{i,b}$ (and if i is OOB, the contribution is $n_{i,b} = 0$).

The in-bag tree estimator is the weighted averaged estimator

$$\begin{aligned} h^*_\zeta(\mathbf{x}, \mathcal{R}_b) &= \sum_{i=1}^n W^*_{i,b}(\mathbf{x}) g_\zeta(Y_i) \\ &= \frac{1}{\sum_{j=1}^n n_{j,b} 1_{\{\mathbf{X}_j \in \mathcal{R}_b(\mathbf{x})\}}} \sum_{i=1}^n n_{i,b} 1_{\{\mathbf{X}_i \in \mathcal{R}_b(\mathbf{x})\}} g_\zeta(Y_i). \end{aligned}$$

Averaging, we obtain the *in-bag ensemble estimator in-bag ensemble*

$$\hat{F}^*_{\text{ens},\zeta}(\mathbf{x}) = \frac{1}{B} \sum_{b=1}^B h^*_\zeta(\mathbf{x}, \mathcal{R}_b) = \sum_{i=1}^n W^*_i(\mathbf{x}) g_\zeta(Y_i), \quad \text{where } W^*_i(\mathbf{x}) = \frac{1}{B} \sum_{b=1}^B W^*_{i,b}(\mathbf{x}).$$

This modifies (16.5) by replacing the forest weights (16.4) with in-bag weights.

16.7 Out-of-Bag Ensemble

The in-bag ensemble is primarily intended for prediction on test data and generally should not be used for inference for the training data. Otherwise, this can yield biased results due to over-training. For inference we instead turn to the *OOB predictor* constructed from OOB data. Because the construction of a RF tree excludes this data, it can be used to define an OOB ensemble, which is a type of cross-validated predictor. This makes it possible to achieve unbiased inference and also to estimate the generalization error.

The OOB ensemble predicted value for case i is obtained by using only trees over which i is OOB. We first illustrate how this works for regression. The OOB predictor for i is defined as

$$\hat{Y}^{(i)} = \sum_{j=1}^n W_j^{(i)} Y_j,$$

where $W_j^{(i)}$ is the forest weight for case j over trees where i is OOB,

$$W_j^{(i)} = \frac{1}{\sum_{b=1}^B I_{i,b}} \sum_{b=1}^B W_{j,b}^{(i)}, \quad \text{where } W_{j,b}^{(i)} = I_{i,b} W^*_{j,b}(\mathbf{X}_i). \tag{16.8}$$

Observe that $W_{j,b}^{(i)}$ equals the in-bag tree weight with the only difference being that it is zero unless i is OOB. Thus, it is easily verified by using the convexity of the in-bag weights that the OOB weights $\{W_j^{(i)}\}_1^n$ are also convex (and thus properly defined):

$$\begin{aligned}\sum_{j=1}^{n} W_j^{(i)} &= \frac{1}{\sum_{b=1}^{B} I_{i,b}} \sum_{j=1}^{n}\sum_{b=1}^{B} W_{j,b}^{(i)} \\ &= \frac{1}{\sum_{b=1}^{B} I_{i,b}} \sum_{b=1}^{B} I_{i,b} \sum_{j=1}^{n} W_{j,b}^{*}(\mathbf{X}_i) \\ &= \frac{1}{\sum_{b=1}^{B} I_{i,b}} \sum_{b=1}^{B} I_{i,b} = 1.\end{aligned}$$

Therefore, $0 \le W_j^{(i)} \le 1$ and $\sum_{j=1}^{n} W_j^{(i)} = 1$.

Let $\hat{F}_{\text{ens}}^{**}$ denote the OOB ensemble $\{\hat{Y}^{(i)}\}_1^n$. The OOB estimate of the prediction error for the in-bag estimator $\hat{F}_{\text{ens}}^{*}$ is defined as

$$\widehat{\text{Err}}(\hat{F}_{\text{ens}}^{*}) = \frac{1}{n}\sum_{i=1}^{n}\left(Y_i - \hat{Y}^{(i)}\right)^2 = R_n(\hat{F}_{\text{ens}}^{**})$$

where R_n is the L_2-loss empirical risk. This is a valid estimate since $\hat{Y}^{(i)}$ is essentially a leave-one-out cross-validated value. Thus the OOB ensemble provides an estimate for the prediction error for the in-bag ensemble, called the OOB error:

$$\widehat{\text{Err}}(\hat{F}_{\text{ens}}^{*}) = \frac{1}{n}\sum_{i=1}^{n}\left(Y_i - \hat{Y}^{(i)}\right)^2 \approx \mathbb{E}_{\mathscr{L}}\mathbb{E}_{\mathbf{X},Y}\left[Y - \hat{F}_{\text{ens}}^{*}(\mathbf{X})\right]^2.$$

This estimator was used for standardized prediction error in Figure 16.1 for our benchmark experiment.

This describes regression. More generally, the OOB ensemble can be defined as follows. Let $h_\zeta^*(\cdot, \mathcal{T}_b)$ be the bth in-bag tree learner. Keeping in mind that the OOB ensemble is determined by averaging in-bag tree estimators over OOB trees, the OOB ensemble estimator for case i for the bth tree is

$$\hat{F}_{\zeta,b}^{(i)} = I_{i,b} h_\zeta^*(\mathbf{X}_i, \mathcal{T}_b) = I_{i,b}\sum_{j=1}^{n} W_{j,b}^{*}(\mathbf{X}_i) g_\zeta(Y_j).$$

Tree-averaging, we obtain the OOB ensemble

$$\hat{F}_\zeta^{(i)} = \frac{\sum_{b=1}^{B}\sum_{j=1}^{n} I_{i,b} W_{j,b}^{*}(\mathbf{X}_i) g_\zeta(Y_j)}{\sum_{b=1}^{B} I_{i,b}} = \sum_{j=1}^{n} W_j^{(i)} g_\zeta(Y_j),$$

where $W_j^{(i)}$ are forest weights defined by (16.8). Equivalently, we can describe the ensemble in terms of the in-bag tree learner

$$\hat{F}_\zeta^{(i)} = \frac{\sum_{b=1}^{B} I_{i,b} h_\zeta^*(\mathbf{X}_i, \mathcal{T}_b)}{\sum_{b=1}^{B} I_{i,b}} = \frac{1}{|\mathcal{O}_i|}\sum_{b\in\mathcal{O}_i} h_\zeta^*(\mathbf{X}_i, \mathcal{T}_b), \tag{16.9}$$

where $\mathcal{O}_i$ are the indices of the OOB trees for i.

Using the notation $\hat{F}_{\text{ens},\zeta}^{**}$ to denote the OOB ensemble $\{\hat{F}_\zeta^{(i)}\}_1^n$, if $L(Y, F)$ is the loss function and R_n the empirical risk, then the OOB estimate for the generalization error (OOB error) for the in-bag ensemble $\hat{F}_{\text{ens},\zeta}^{*}$ is

$$\widehat{\text{Err}}(\hat{F}_{\text{ens},\zeta}^{*}) = \frac{1}{n}\sum_{i=1}^{n} L\left(Y_i, \hat{F}_\zeta^{(i)}\right) = R_n(\hat{F}_{\text{ens},\zeta}^{**}). \tag{16.10}$$

The OOB ensemble is an important aspect of a RF analysis; some of its most important properties are highlighted below.

The OOB ensemble estimator for a case $i = 1,\dots,n$ is constructed using i's OOB trees. This yields a cross-validated estimator that can be used for unbiased inference on the training data and for estimating the generalization error. Importantly, all of this comes with no additional computational effort beyond growing the training forest.

16.8 Illustrating Difference Between In-Bag and OOB Estimation

We use a simulated classification dataset to illustrate the difference between the in-bag and OOB ensemble. The simulation model was logistic regression with a regression function

$$F(\mathbf{x}) = \mathbb{P}\{Y = 1|\mathbf{X} = \mathbf{x}\} = \frac{1}{1 + \exp(-\phi(\mathbf{x}))}$$

where

$$\phi(\mathbf{x}) = \beta_1 x_1 + \beta_2 x_2 + \beta_3 x_1 x_2, \quad \text{where } \beta_1 = 1, \beta_2 = 2, \beta_3 = 2.$$

The response Y was simulated using $n = 100 \times 100$ equally spaced $\mathbf{x}$ values on $[-3,3] \times [-3,3]$.

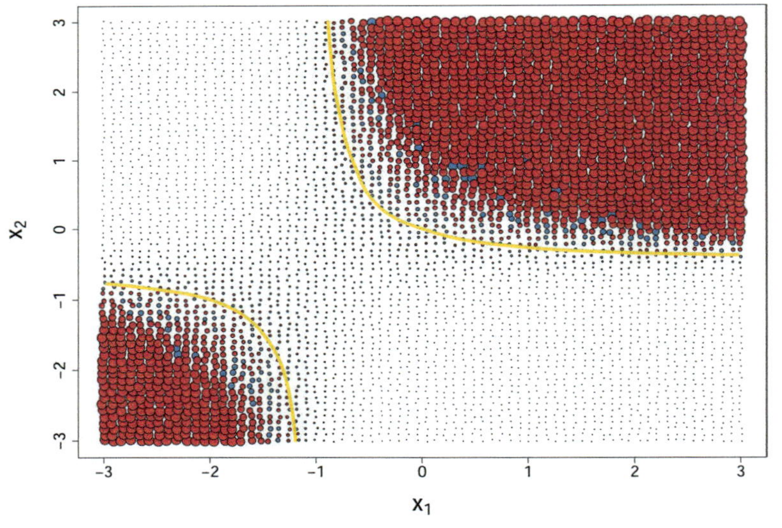

FIGURE 16.2
Logistic regression simulation with $\boldsymbol{x} = (x_1, x_2)^T$ selected from equally spaced points on $[-3,3] \times [-3,3]$. Blue and red points indicate $Y = 0$ and $Y = 1$ responses with size of point scaled to the true regression function $F(\boldsymbol{x})$.

Figure 16.2 displays the data. The data points are colored according to the class label Y, with the symbol size proportional to $F(\mathbf{x})$. The yellow line represents the Bayes rule boundary $\{\mathbf{x} : F(\mathbf{x}) = 1/2\}$. Most of the $Y = 1$ points (shown in red) are located in the bottom left and top right

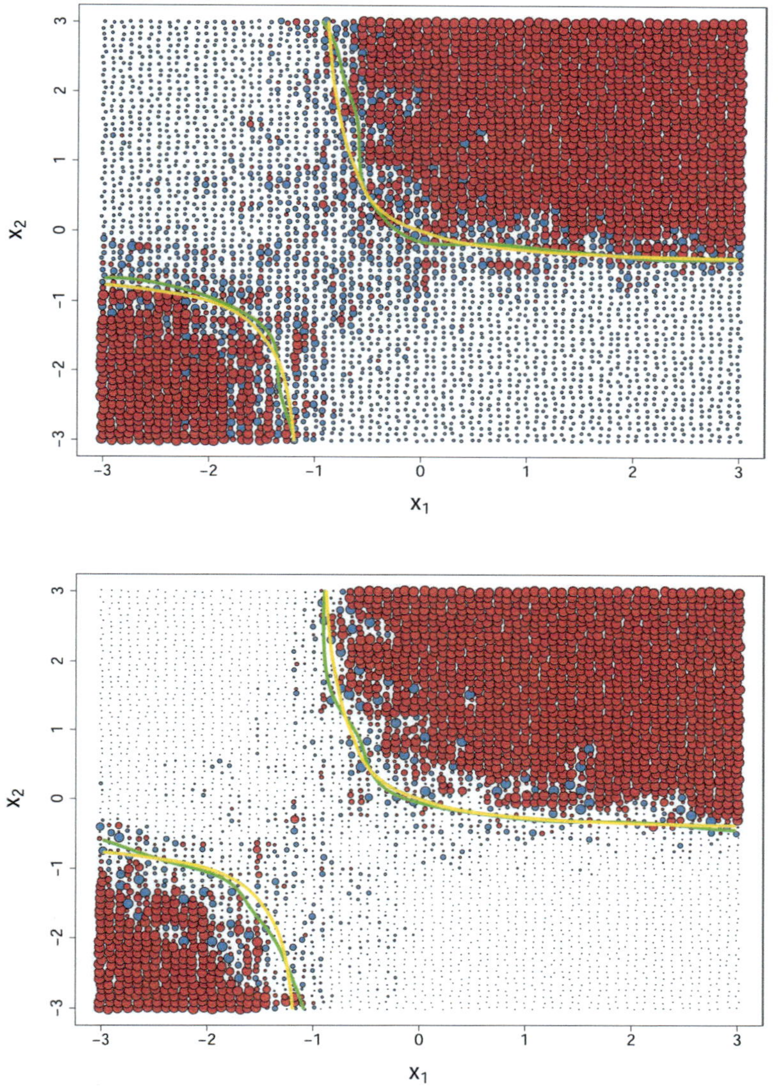

FIGURE 16.3

In-bag ensemble (top) and OOB ensemble (bottom) estimators for $F(\boldsymbol{x})$ from logistic regression simulation using pure trees. Yellow line is true Bayes rule, thick green line is boundary $\{\boldsymbol{x}: \hat{F}(\boldsymbol{x}) = 1/2\}$ defined by the ensemble estimator. The size of a point $\boldsymbol{x}$ has been scaled to the size of $\hat{F}(\boldsymbol{x})$.

quadrants, coinciding with the Bayes rule, while the majority of $Y = 0$ points (shown in blue) are in the opposite quadrants (top left and bottom right). Both class labels are observed near the Bayes decision boundary.

A RF with 500 classification trees was constructed. Trees were grown to purity (nodesize = 1). The in-bag and OOB ensemble estimators for F are given in Figure 16.3. The thick green lines are the estimated Bayes decision boundary $\{\mathbf{x}: \hat{F}(\mathbf{x}) = 1/2\}$ where $\hat{F}$ is the ensemble estimator for F. Both ensembles have lines that track the true boundary (superimposed in yellow). To study estimated values of F, the size of points in the figure has been scaled to $\hat{F}$. Values for the in-bag estimator are exaggerated, especially for $Y = 1$ cases near $(-1,0)^T$, which have been over-inflated to ensure they are classified correctly. This makes it appear the ensemble is performing better than the Bayes rule since values in this region are classified by the Bayes rule to be $Y = 0$.

The in-bag estimator $\hat{F}^*_{\text{ens}}$ is over-trained. In fact, it classifies all cases correctly and achieves a training error of 0, thus it has an empirical risk of zero under 0-1 loss:

$$R_n(\hat{F}^*_{\text{ens}}) = \frac{1}{n}\sum_{i=1}^{n} L(y_i, \hat{F}^*_{\text{ens}}(\mathbf{x}_i)) = 0.$$

The theory we have learned suggests this estimator will perform poorly over test data. But why does this happen, since RF is known to perform so well? One explanation is that we used pure classification trees with a nodesize = 1. This means that $K \approx n$ splits were used for the tree construction, but this is probably too high. Almost all the theory we have seen for trees indicates $K \ll n$ for tree consistency.

Another way to see that the in-bag estimator is overfitting is to compare it to the OOB ensemble, $\hat{F}^{**}_{\text{ens}}$. Since this is cross-validated it should provide unbiased estimation for F over the training data. Figure 16.3 confirms this by showing that $\hat{F}^{**}_{\text{ens}}$ obtains more accurate estimated probabilities (compare this to Figure 16.2).

This demonstrates the earlier statement that the in-bag estimator should not be used for inference over the training data. This is because if it is over-trained, the predicted values will be severely optimistic. But how well does $\hat{F}^*_{\text{ens}}$ actually perform over test data? Recall that a feature of the OOB ensemble is that its empirical risk can be used to estimate the generalization error of the in-bag estimator, so we can use this to assess its performance. In this example, the misclassification error using the OOB ensemble was 0.123, i.e.,

$$R_n(\hat{F}^{**}_{\text{ens}}) = \frac{1}{n}\sum_{i=1}^{n} L(y_i, \hat{F}^{(i)}) = 0.123 = \widehat{\text{Err}}(\hat{F}^*_{\text{ens}}),$$

where $\hat{F}^{(i)}$ is the OOB estimator for case i defined by (16.9): $\hat{F}^{(i)} = \sum_{j=1}^{n} W_j^{(i)} y_j$. This is fairly high when compared to the Bayes error rate in this problem, which equals 0.107 (see Exercise 16.5).

This confirms that trees grown to purity overfit in this example. Therefore, the question is what is right sized RF tree for this data? Since the size of a RF tree is controlled by the value of nodesize, one way to determine the right size is to optimize nodesize.

To determine the optimal nodesize we ran the previous simulation varying nodesize from 1 to 100. A test set was used to estimate the generalization error. This was repeated 100 times and the results averaged. Figure 16.4 plots the test set estimated error against nodesize. The optimal nodesize is 43 with a test set error of 0.108 which nearly equals the Bayes error rate. Figure 16.5 displays the predicted probabilities for the tuned in-bag estimator $\hat{F}^*_{\text{ens}}$ with nodesize of 43. The results are significantly improved. Also (not shown) the empirical risk for the tuned $\hat{F}^*_{\text{ens}}$ is 0.106 and for $\hat{F}^{**}_{\text{ens}}$ it is 0.112, thus showing over-training is significantly reduced.

Figure 16.4 also records the OOB estimated generalization error. It is impressive how close this value is to the test set estimated value. In practice we often only have one dataset without the luxury of a hold out dataset. This provides evidence that the OOB error can be used reliably for estimating the generalization error.

16.9 OOB Residuals for Sampling Conditional Values

The following is another example illustrating the usefulness of the OOB estimator. Consider a regression setting where $Y_i = F(\mathbf{X}_i) + \varepsilon_i$, for $i = 1, \ldots, n$, and assume an equal variance model

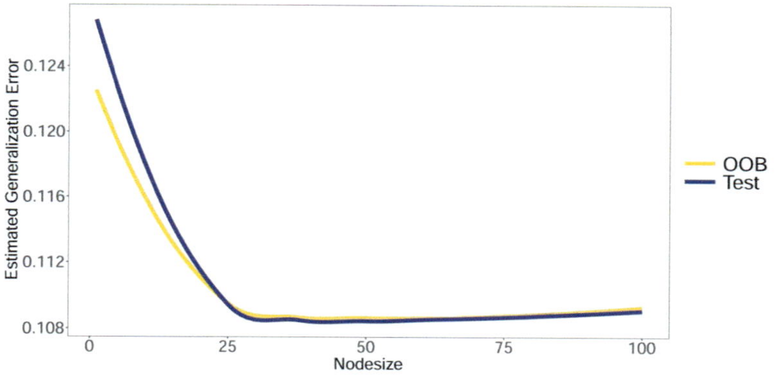

FIGURE 16.4
Test set estimated generalization error (blue line) as a function of RF tree nodesize for logistic regression simulation from Figure 16.2. Also superimposed on the figure is the estimated generalization error from the OOB ensemble (yellow line). Notice how this accurately tracks the test set error.

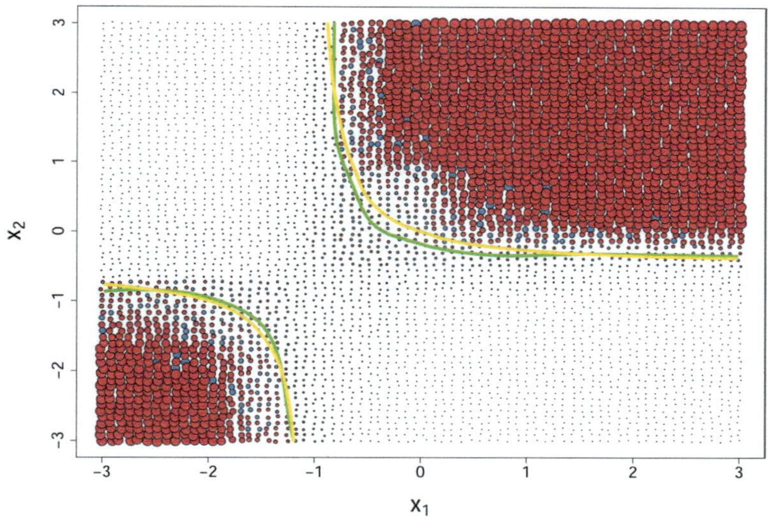

FIGURE 16.5
In-bag ensemble estimator using the optimized nodesize value of 43.

where $\varepsilon_1, \ldots, \varepsilon_n$ are i.i.d. with mean 0 and variance σ^2 independently of $\mathbf{X}_1, \ldots, \mathbf{X}_n$ which are i.i.d. We write the conditional model for $Y_i|\mathbf{X}_i$ as

$$Y_i|\mathbf{X}_i \overset{\text{iid}}{\sim} F_Y(\cdot|\mathbf{X}_i). \tag{16.11}$$

Suppose that we want to draw conditional values of Y_i from (16.11). How can we do this?

Here is an OOB method for doing this. Let $\hat{Y}^{(i)}$ be the OOB estimator for $F(\mathbf{X}_i)$ using RF regression. Define the OOB residual

$$\tilde{\varepsilon}_i = Y_i - \hat{Y}^{(i)}. \tag{16.12}$$

It is relatively easy to see that $\tilde{\varepsilon}_1,\ldots,\tilde{\varepsilon}_n$ are exchangeable random variables. Also Y_i and $\hat{Y}^{(i)}$ are independent, since

$$\hat{Y}^{(i)} := \hat{Y}^{(i)}((\mathbf{X}_1,Y_1),\ldots,(\mathbf{X}_{i-1},Y_i),(\mathbf{X}_{i+1},Y_{i+1}),\ldots,(\mathbf{X}_n,Y_n))$$

is calculated from trees that exclude $(\mathbf{X}_i, Y_i)$ from their construction. Therefore, it is reasonable to use $\tilde{\varepsilon}_1,\ldots,\tilde{\varepsilon}_n$ to approximate $\varepsilon_1,\ldots,\varepsilon_n$ and we can write (approximately), where $\stackrel{d}{=}$ is equal in distribution,

$$Y_i - F(\mathbf{X}_i) \stackrel{d}{=} \tilde{\varepsilon}_i.$$

To sample a value from (16.11), we draw a value from the distribution of $\tilde{\varepsilon}$, call this $\tilde{\varepsilon}_i^*$, obtaining

$$Y_i^* - F(\mathbf{X}_i) = \tilde{\varepsilon}_i^*.$$

Pivot around $F(\mathbf{X}_i)$, and substituting $\hat{Y}^{(i)}$ for $F(\mathbf{X}_i)$, we obtain the desired draw from (16.11)

$$Y_i^* = \hat{Y}^{(i)} + \tilde{\varepsilon}_i^*.$$

The procedure is summarized as follows. Denote $\mathbb{P}_{n,\tilde{\varepsilon}}$ for the empirical measure for $\tilde{\varepsilon}_1,\ldots,\tilde{\varepsilon}_n$.

1. Run a RF regression and obtain OOB residuals (16.12).
2. Draw values $\tilde{\varepsilon}_1^*,\ldots,\tilde{\varepsilon}_n^*$ independently from $\mathbb{P}_{n,\tilde{\varepsilon}}$.
3. Obtain draws for the conditional response (16.11) using $Y_i^* = \hat{Y}^{(i)} + \tilde{\varepsilon}_i^*$ for $i = 1,\ldots,n$.

We give an illustration using the Boston Housing data [15]. To make this a little more interesting we used bivariate RF regression (Chapter 16.5) using median house price and nitric oxide concentration for the bivariate outcomes.

1000 sampled values for the bivariate outcomes were obtained using the above algorithm. See Figure 16.6. The top left panel is the contour plot for the sampled values for all data points. The plot to its right is conditioned on the variable lower status to include only data in the first 20 percentiles of lower status, after which, going from left to right in the middle and bottom panels, lower status is conditioned into percentile groups 20-40, 40-60, 60-80 and 80-100. We observe an increasing trend to the upward left in the bivariate outcomes as lower status increases. This means median price decreases and nitric oxide concentration increases as lower status increases. We also observe that several of the contour plots have a multimodal behavior. This is not unexpected because the Boston Housing data is very challenging and contains a heterogeneous population.

16.10 Random Forests Distance

Trees are very good at capturing the distance between observations, which is important for clustering, grouping, and other multivariate tasks. Because trees are nonparametric, they measure dissimilarity between data points without the limitations of an ad-hoc metric, a common issue with conventional distance-based methods.

Proximity is a metric used by RF to define the closeness between data points. The proximity of $\mathbf{x} \in \mathscr{X}$ to a data point $\mathbf{x}_i$ in the training data equals the tree-averaged number of times $\mathbf{x}$ and $\mathbf{x}_i$ share the same terminal node and is defined by

$$D_{\text{prox}}(\mathbf{x},\mathbf{x}_i) = \frac{1}{B}\sum_{b=1}^{B} 1_{\{\mathbf{x}\in\mathcal{T}_b(\mathbf{x}_i)\}}.$$

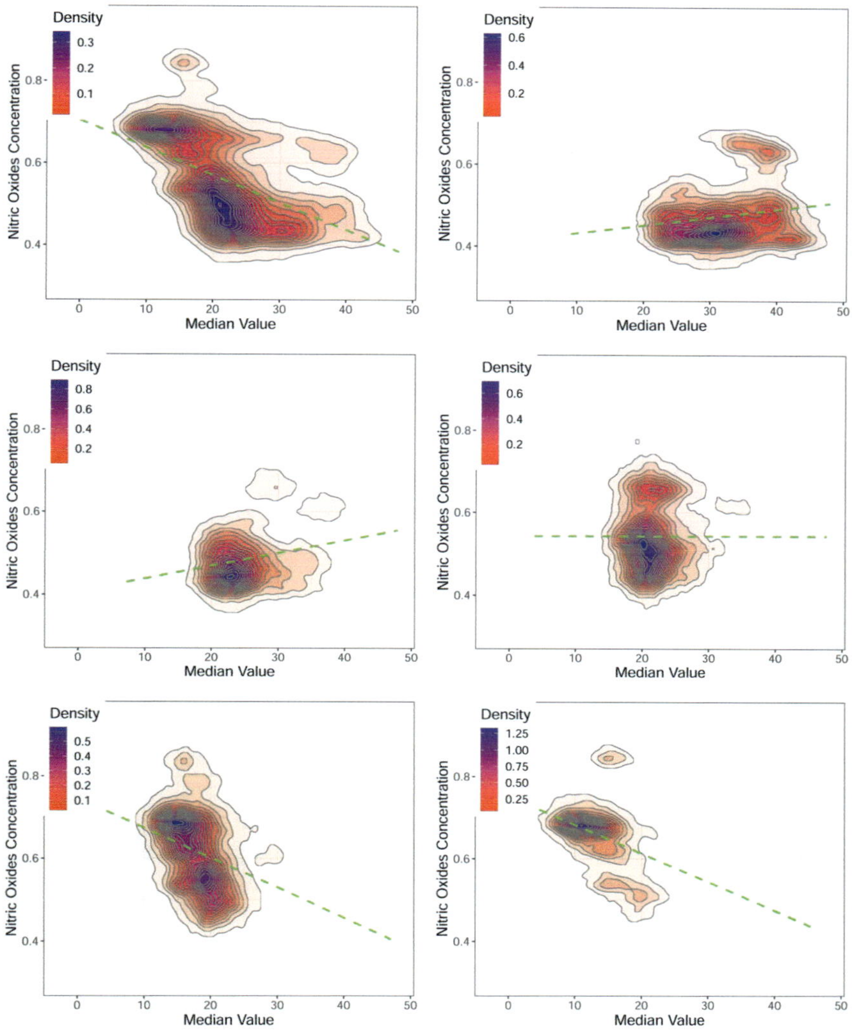

FIGURE 16.6
Bivariate conditional draws for Boston Housing data where the bivariate response is median house price and nitric oxide concentration. The top left panel shows the contour plot from 1000 draws. Remaining panels are contour plots conditional on the attribute lower status, which has been conditioned into 4 groups defined by 20% quantile increments. As plots go from top to bottom, lower status goes from low to high values.

Proximity has certain limitations, which we will describe. To address these, we describe a new random forest distance definition. Similar to proximity, the goal is to measure the closeness of observations using trees. However, unlike proximity, this measure does not use terminal node membership to determine distance. Instead, it determines closeness by counting the split distance from shared ancestors. The issue with proximity is that if two observations split far down the tree versus close to the root node, both scenarios are counted as having a proximity of zero, even though the first scenario involves data points that are closer in the sense of the tree topology.

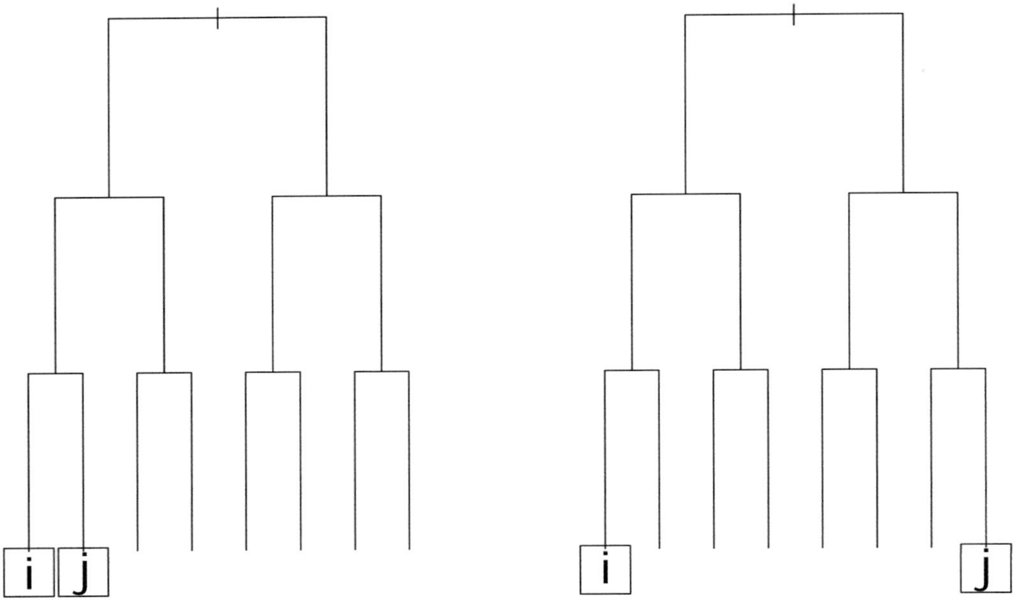

FIGURE 16.7
Two scenarios for comparing random forest proximity and distance.

For each pair of data points $\mathbf{x}_i$ and $\mathbf{x}_j$ in a tree, define $S(\mathbf{x}_i,\mathbf{x}_j)$ to equal the minimum number of splits on the path from the terminal node containing $\mathbf{x}_i$ to the terminal node containing $\mathbf{x}_j$ such that the path includes at least one common ancestor node of $\mathbf{x}_i$ and $\mathbf{x}_j$. Similarly, define $R(\mathbf{x}_i,\mathbf{x}_j)$ as the minimum number of splits on the path from the terminal node containing $\mathbf{x}_i$ to the terminal node containing $\mathbf{x}_j$ such that the path includes the root node. We define the distance between $\mathbf{x}_i$ and $\mathbf{x}_j$ as:

$$D_F(\mathbf{x}_i,\mathbf{x}_j) = \frac{S(\mathbf{x}_i,\mathbf{x}_j)}{R(\mathbf{x}_i,\mathbf{x}_j)}.$$

As a simple illustration, take the tree in Figure 16.7. In the first case (left figure), there is one split between both i's and j's terminal node and their lowest common node: therefore, $S(\mathbf{x}_i,\mathbf{x}_j) = 2$ because it is the sum of splits to each terminal node. Also, $R(\mathbf{x}_i,\mathbf{x}_j) = 3+3 = 6$, because it is the sum for both nodes. The resulting distance is $D_F(\mathbf{x}_i,\mathbf{x}_j) = 2/6$. Relatively speaking, i and j in the left figure are considered much more similar to each other than in the second case (right figure) where the lowest common node is the root node itself. In the right figure, $S(\mathbf{x}_i,\mathbf{x}_j) = R(\mathbf{x}_i,\mathbf{x}_j) = 6$ and $D_F(\mathbf{x}_i,\mathbf{x}_j) = 1$. From this, we can see how the new measure can discern between the two cases, while the traditional proximity measure would have assigned both scenarios a value of zero since, in neither case, are observations i and j in the same terminal node.

As a more detailed example, consider Figure 16.8, which plots RF proximity versus the new RF distance for the Iris data. Here we have selected a reference point denoted by "X" to observe how distance and proximity differ. The size of points has been scaled by their proximity (or their distance) to "X" so that a small size reflects closeness. If proximity is zero or distance is one, then a point is color coded by cyan (for visibility these cyan points are depicted using small circles).

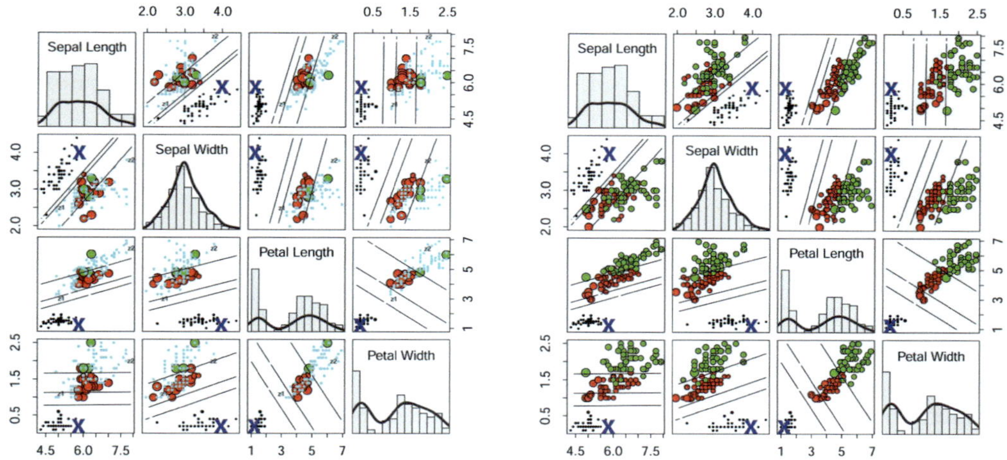

(a) *Iris data where size of points are scaled by* $1-\log(1-$ `proximity`) *to the reference point "X" (lines depict LDA decision boundary between clusters). Points are color coded by cluster except if the proximity is zero, then it is cyan (for visibility, cyan points are always depicted using small points).*

(b) *Same as (a), except values are calculated using the new random forest distance and size of points are scaled by* $1-\log($`distance`$)$. *There are no cyan points because there are no points with distance of one to the reference point "X".*

FIGURE 16.8
Random forest distance versus proximity on Iris data.

Notice for proximity (subpanel (a)) that many points are cyan, meaning these points are equidistant and far from the reference point of interest. This is the result of the very stringent cutoff that random forest proximity has for counting the proximity between two points.

In contrast, distance (subpanel (b)) can result in the value of one only under the extreme case that data points diverge after the root node split. The new distance allows for even small differences in the relatedness between points to be observed. Thus we see no cyan points in subpanel (b). Another important point is that distance appears to be smooth throughout much of the region and it is possible to observe where observations from different clusters are more similar to the point of interest, and where the points become different irrespective of their true cluster membership.

16.11 Potential Nearest Neighbors

Recall that Breiman's definition of a RF classifier called for trees split to purity. Due to the success of this method, the idea of using a small nodesize has been widely adopted in RF applications. The rationale is to create deep trees to encourage low bias, which, when combined with variance reduction due to averaging, reduces generalization error.

On the other hand, we saw in Figure 16.3 the first-hand effects of overfitting due to a small nodesize. Optimizing nodesize in Figure 16.4 revealed that the best fitting forest required a nodesize in excess of 40, a big increase from the nodesize = 1 used in the initial analysis.

This puzzle of why small nodesizes sometimes work was studied by Lin and Jeon [20]. They examined RF regression under non-adaptive splitting, using pure random splitting where each cell is cut randomly by choosing a variable and its split value randomly. This means the tree partition depends only on the feature space and is independent of the response Y. In addition to this, we will consider RF constructed without bootstrapping. We assume $(\mathbf{X}, Y), (\mathbf{X}_1, Y_1), \ldots, (\mathbf{X}_n, Y_n) \in \mathbb{R}^p \times \mathbb{R}$ are i.i.d. with distribution $\mathbb{P}$. The RF ensemble estimator for $F(\mathbf{x}) = \mathbb{E}[Y|\mathbf{X} = \mathbf{x}]$ is given by (16.3),

$$\hat{F}_{\text{ens}}(\mathbf{x}) = \sum_{i=1}^{n} W_i(\mathbf{x}) Y_i,$$

where $\{W_i(\mathbf{x})\}_1^n$ are the (full-sample) forest weights using non-adaptive RF.

Lin and Jeon [20] proved a lower bound for the MSE rate of convergence for RF by drawing a connection to what they termed *potential nearest neighbors* (PNN). The RF ensemble $\hat{F}_{\text{ens}}(\mathbf{x})$ is a weighted average but only a subset of the weights contribute to the prediction; the responses for these weights are called the *voting points*. For notational ease denote N for nodesize. For a RF constructed where each terminal node has nodesize N, the $K(\mathbf{x})$-PNN for a feature $\mathbf{x}$ are the $K := K(\mathbf{x})$ voting points $Y_{(1)}, \ldots, Y_{(K)}$ with non-zero weights $W_{(1)}(\mathbf{x}), \ldots, W_{(K)}(\mathbf{x}) > 0$. Note importantly that in addition to $\mathbf{x}$, K depends on the features and N. The crux of the argument is that the expected value for K is often much larger than N.

Consider a simulation with covariates $\mathbf{x}_i = (x_{i1}, x_{i2})^T$ drawn uniformly from $[0, 1]^2$ such that $y_i = x_{i2}^2 + \varepsilon_i$ where ε_i are i.i.d. $\text{N}(0, .2^2)$ for $i = 1, \ldots, n$. A sample size of $n = 1000$ was drawn and RF regression run on the data using $N = 5$ for the nodesize. The forest weights for $\mathbf{x} = (0.5, 0.5)^T$ are shown in Figure 16.9 as circles with radius scaled by weight size. The left and right side plots are RF using adaptive splitting and pure random splitting, respectively. Adaptive splitting RF is better since it recognizes that coordinate x_1 is noise and therefore spreads the weights out more evenly along the x_1 coordinate axis. However, pure random splitting is still relatively good as most of the forest weights are concentrated near the target $\mathbf{x} = (0.5, 0.5)^T$ value. Also notice that in both cases the number of voting points is much larger than N, which is 5 here.

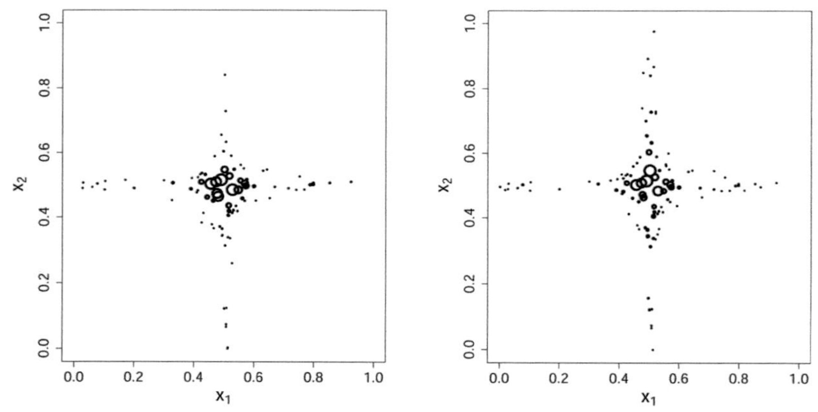

FIGURE 16.9
Forest weights for $\mathbf{x} = (0.5, 0.5)^T$ from regression simulation where only covariate x_2 is signal. Left plot is RF using adaptive splitting; right plot is RF using pure random splitting.

The following assumptions were used by Lin and Jeon [20].

A1. $\mathbf{X}$ is distributed on $[0, 1]^p$ with density bounded away from zero and infinity.

A2. The data is i.i.d. such that $Y = F(\mathbf{X}) + \varepsilon$ where $\mathbb{E}(\varepsilon) = 0$ and $\mathbb{E}(\varepsilon^2) = \sigma^2 < \infty$.

Theorem 16.2. *Under the above assumptions, for the non-adaptive RF with terminal nodesize N, there exists a finite constant $C > 0$, such that for any n*

$$\mathbb{E}\left[\left(\hat{F}_{ens}(\mathbf{X}) - F(\mathbf{X})\right)^2\right] \geq CN^{-1}(\log n)^{-(p-1)}.$$

Keep in mind that Theorem 16.2 provides a lower bound for the rate of convergence, and there is no guarantee that this bound is achievable. However, it shows that the fastest RF can converge is $N^{-1}(\log n)^{-(p-1)}$. This implies a very slow rate if N is small, specifically slower than the known optimal convergence rate of $n^{-2q/(2q+p)}$ for nonparametric regression (q is a measure of smoothness of the underlying regression function; see Stone [35]).

This result pertains to non-adaptive RF, and we can argue that adaptive RF would perform better. Nevertheless, it suggests that N should not be fixed and that letting $N \to \infty$ can improve convergence. Therefore, contrary to popular use, it seems that nodesize should be large rather than small.

However, the popularity of a small nodesize might not be a contradiction, since there is at least one case that suggests a smaller N is preferable, and this occurs when p is very large. For a fixed n if p is large enough we can have $(\log n)^{-(p-1)} < n^{-2q/(2q+p)}$. Therefore in high dimensions, growing deep trees might give the best performance. This is certainly consistent with empirical evidence of performance of RF in high dimensions.

Since the proof of Theorem 16.2 is relatively simple we provide a brief sketch. The proof uses two key facts. The first is that since tree splits are non-adaptive by assumption, the forest weights are conditionally independent of $\{Y_i\}_1^n$. By definition, the forest weights $\{W_i(\mathbf{X})\}_1^n$ are defined by the tree partition $\mathcal{T}$ which for a non-adaptive tree depends only only the features, $\mathcal{T} := \mathcal{T}(\mathbf{X}_1, \ldots, \mathbf{X}_n)$. Therefore, the following holds:

$$(Y_1, \ldots, Y_n | \mathbf{X}_1, \ldots, \mathbf{X}_n) \perp (W_1(\mathbf{X}), \ldots, W_n(\mathbf{X}) | \mathbf{X}_1, \ldots, \mathbf{X}_n).$$

The second fact involves the expected number of PNN. Let $K(\mathbf{x}; \{\mathbf{X}_i\}_1^n, N)$ be the number of K-PNN of a given data point $\mathbf{x}$ for a tree with nodesize N. Lin and Jeon [20] showed under Assumption (A1) that

$$C_1 N(\log n)^{p-1} \leq \mathbb{E}\left[K(\mathbf{X}; (\{\mathbf{X}_i\}_1^n, N))\right] \leq C_2 N(\log n)^{p-1} \tag{16.13}$$

for finite constants $C_1, C_2 > 0$. (This bound is the tricky part and we refer readers to the paper for details.)

The proof is as follows. For any $\mathbf{x} \in [0,1]^p$, write the ensemble in terms of the N voting points, $\hat{F}_{\text{ens}}(\mathbf{x}) = \sum_{l=1}^{K(\mathbf{x})} W_{(l)} Y_{(l)}$. Bounding the conditional MSE below by the conditional variance (see Chapter 7.3) and because the weights are independent of the responses given the features,

$$\begin{aligned}
&\mathbb{E}\left[\left(\hat{F}_{\text{ens}}(\mathbf{x}) - F(\mathbf{x})\right)^2 \middle| \mathbf{X}_1, \ldots, \mathbf{X}_n, W_{(1)}, \ldots, W_{(K)}\right] \\
&\quad \geq \text{Var}\left[\hat{F}_{\text{ens}}(\mathbf{x}) \middle| \mathbf{X}_1, \ldots, \mathbf{X}_n, W_{(1)}, \ldots, W_{(K)}\right] \\
&\quad = \text{Var}\left[\sum_{l=1}^{K(\mathbf{x})} W_{(l)} Y_{(l)} \middle| \mathbf{X}_1, \ldots, \mathbf{X}_n, W_{(1)}, \ldots, W_{(K)}\right] \\
&\quad = \sum_{l=1}^{K(\mathbf{x})} W_{(l)}^2 \text{Var}\left[Y_{(l)} \middle| \mathbf{X}_1, \ldots, \mathbf{X}_n, W_{(1)}, \ldots, W_{(K)}\right] \\
&\quad = \sigma^2 \sum_{l=1}^{K(\mathbf{x})} W_{(l)}^2 \geq \frac{\sigma^2}{K(\mathbf{x})}.
\end{aligned}$$

The last line uses Cauchy-Schwarz and $\sum_{l=1}^{K(\mathbf{x})} W_{(l)} = 1$. Therefore, we have by (16.13), and Jensen's inequality,

$$\mathbb{E}\left[\left(\hat{F}_{\text{ens}}(\mathbf{X}) - F(\mathbf{X})\right)^2\right] \geq \frac{\sigma^2}{\mathbb{E}(K(\mathbf{X}))} \geq C_2^{-1} N^{-1} (\log n)^{-(p-1)}.$$

Thus, completing the proof.

16.12 Random Forests Quantile Classifier (RFQ)

We have considered the two-class classification problem several times now, but an important special setting we have not addressed is the class imbalance problem. This occurs when the training data's class distribution is skewed, meaning there is a significant imbalance in the relative frequency of the observed class labels. This imbalance poses a major challenge for machine learning methods.

Class imbalance is more prevalent in the real world than one might expect. For instance, in biomedical applications, events like hospitalization or death occur far less frequently than control events. Positive diagnoses of diseases are rarer than negative ones, such as in the case of diagnostic testing for cancer versus benign tumors. Similarly, class imbalance is common in software and text recognition for tasks like spam filtering and code error detection. In industrial systems, it appears in fault detection and quality control. In business analytics, it is frequently encountered in customer churn data.

As an illustration, a study examining the value of adjuvant treatment versus surgery in esophageal cancer patients applied causal techniques, requiring a classification analysis to predict the treatment group based on patient features for a treatment overlap analysis. The dataset included $N_0 = 6649$ surgery patients and $N_1 = 988$ adjuvant treatment patients, resulting in an imbalanced ratio of $6649/988 \approx 6.8$, indicating a clear imbalance.

A random forest (RF) classifier was applied to the data, and the following confusion matrix (calculated using the OOB ensemble) was obtained, as shown in Table 16.1.

TABLE 16.1
Confusion matrix from RF classification analysis of esophageal cancer patients.

	predicted		
	surgery	adjuvant	error
surgery	6549	100	.015
adjuvant	694	294	.702

Overall error rate: 10.4%

The OOB estimated misclassification error is 10.4%, which initially appears to be a good result. However, a closer examination reveals a significant issue: the conditional error rates, which are conditioned on the true class labels, are markedly different between the classes. For the majority class (surgery group, labeled as "0"), the conditional error rate is nearly zero, at 0.015. In stark contrast, the minority class (adjuvant group, labeled as "1") has a very high conditional error rate of 0.702.

The imbalance ratio highlights the challenge in this type of analysis. This disparity causes the classifier to become biased toward the majority class, leading to poor predictive accuracy for the minority class. The root of this issue lies in the RF classifier's reliance on the Bayes decision rule for classification. According to this rule, patients are classified into the minority class if their predicted probabilities are greater than or equal to 0.5. However, due to the nature of the imbalanced data, the probability of a patient belonging to the minority class is almost always less than 0.5. Consequently, the classifier rarely assigns patients to the minority class.

16.12.1 Metrics suitable for imbalanced data

A 10.4% error rate is misleading, and focusing on this value would lead to incorrect conclusions in this example. This becomes clear when considering more appropriate metrics. The conditional error rates in the confusion matrix are closely related to the *true negative rate* (TNR) and *true positive rate* (TPR), which are commonly used in class imbalanced analyses to provide a more accurate assessment of a classifier's performance. For a classifier $\hat{\delta} : \mathscr{X} \to \{0,1\}$, these values are defined as

$$\mathrm{TNR} = \mathbb{P}\{\hat{\delta} = 0 | Y = 0\}, \quad \mathrm{TPR} = \mathbb{P}\{\hat{\delta} = 1 | Y = 1\}.$$

In the previous analysis, TNR = 0.985 and TPR = 0.298. To determine the effectiveness of the classifier, the TNR and TPR can be combined to form the geometric mean, defined as gmean = $(\mathrm{TNR} \times \mathrm{TPR})^{1/2}$. This metric measures the balance between classification performance on both the majority and minority classes [19]. In this example, the gmean = 0.542, which is very low and indicates poor performance. For contrast, pure random guessing results in a gmean of 0.5, while a gmean close to 1, which is highly desirable, occurs only when both the TNR and TPR are close to 1, with a small difference between the two.

In addition to the gmean, another metric for evaluating performance on class imbalanced data is the precision-recall area under the curve (PR-AUC). As discussed in Chapter 15.6, the receiver operating characteristic area under the curve (ROC-AUC) is commonly used to evaluate a classifier's performance. The precision-recall curve is created by plotting precision (positive predictive value) on the y-axis against recall (sensitivity, true positive rate) on the x-axis at various threshold levels. This differs from the ROC curve, which plots sensitivity against the false positive rate (1−specificity).

PR-AUC can be a more effective evaluation tool for imbalanced data [31]. Unlike ROC-AUC, which is insensitive to the imbalance ratio [25], PR-AUC provides a better reflection of performance when dealing with rare cases, which are usually associated with higher costs. Proper performance metrics should decrease monotonically with increasing imbalance, which is the case for PR-AUC, but not ROC-AUC.

Figure 16.10 shows the ROC and PR-AUC curves and their respective areas from the previous analysis. The ROC-AUC is extremely high, over 90%, which incorrectly suggests that the classifier is performing well. In contrast, the PR-AUC is much lower, at 63%, which more accurately reflects the challenges posed by the imbalanced data.

16.12.2 Definitions

Our goal is to build an accurate RF classifier for Y given $\mathbf{X} = \mathbf{x}$ when the learning data is imbalanced. To help quantify what is meant by "imbalancedness", we start by first defining the imbalance ratio (IR). Following the convention in the literature, we assume that the majority class labels are 0, and outnumber the minority class labels, 1.

Definition 16.2. The imbalance ratio (IR) is defined as IR = N_0/N_1 where N_0 and N_1 denote the cardinality of the majority and minority samples, respectively. A data set is imbalanced if IR is much greater than 1. □

Now we define some formal notions of imbalancedness. Following Fithian and Hastie [13], we distinguish between *marginally imbalanced* and *conditionally imbalanced* data.

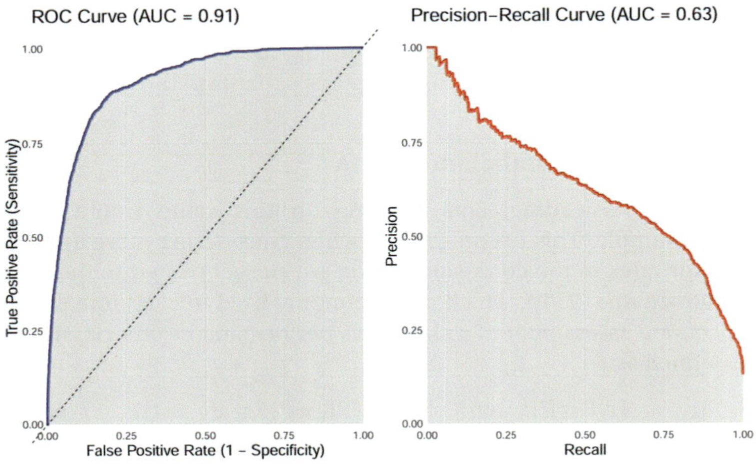

FIGURE 16.10
ROC curve and PR-AUC curve for RF classifier from esophageal study of adjuvant treatment versus surgery.

Definition 16.3. The data is marginally imbalanced if $F(\mathbf{x}) \ll 1/2$ for all $\mathbf{x} \in \mathscr{X}$ where $F(\mathbf{x}) = \mathbb{P}\{Y = 1|\mathbf{X} = \mathbf{x}\}$. □

Thus, marginally imbalanced data is data for which the probability of the minority class is close to zero throughout the feature space.

Definition 16.4. The data is conditionally imbalanced if there exists a set $A \subset \mathscr{X}$ with nonzero probability, $\mathbb{P}\{\mathbf{X} \in A\} > 0$, such that $\mathbb{P}\{Y = 1|\mathbf{X} \in A\} \approx 1$ and $F(\mathbf{x}) \ll 1/2$ for $\mathbf{x} \notin A$. □

In contrast to marginally imbalanced data, conditional imbalancedness occurs when the probability of the minority class is close to 1 given the features lie in a certain set, and approximately zero otherwise. In both cases, it is assumed that the minority class is rare.

16.12.3 Balanced random forests

A commonly used strategy in machine learning for addressing class imbalanced data is undersampling the majority class so that its cardinality matches that of the minority class. In the context of RF, undersampling the majority class improves classification performance for the minority class [6] and is the most common approach when using RF to learn from imbalanced data. This method is called *balanced random forests* (BRF) and is an example of what has been referred to in the literature [34] as a data-level method. Such methods transform the distributions of the classes in the training data. We will show that BRF has an important connection to a new approach called RFQ [29], which stands for *random forests quantile* (RFQ) classifier, even though surprisingly RFQ is not an example of a data-level method.

We will describe the RFQ method shortly, but for now, Figure 16.11 provides an empirical comparison of BRF to RFQ using a simulated high-dimensional class imbalanced dataset. Only the first two variables, X_1 and X_2, are signal variables. The top left panel gives a scatter plot for the data with true class labels colored in red and blue ($Y = 0$ and $Y = 1$, respectively). To the right of this is the RF classifier. As seen, due to the use of the Bayes rule, RF classifies all cases as majority class labels. The bottom left panel shows classified values using BRF, and the results are improved. However, the best results are seen for RFQ in the bottom right panel. RFQ is much more aggressive in identifying the minority class labels. This is particularly interesting since, as we stated, BRF and RFQ are related to one another.

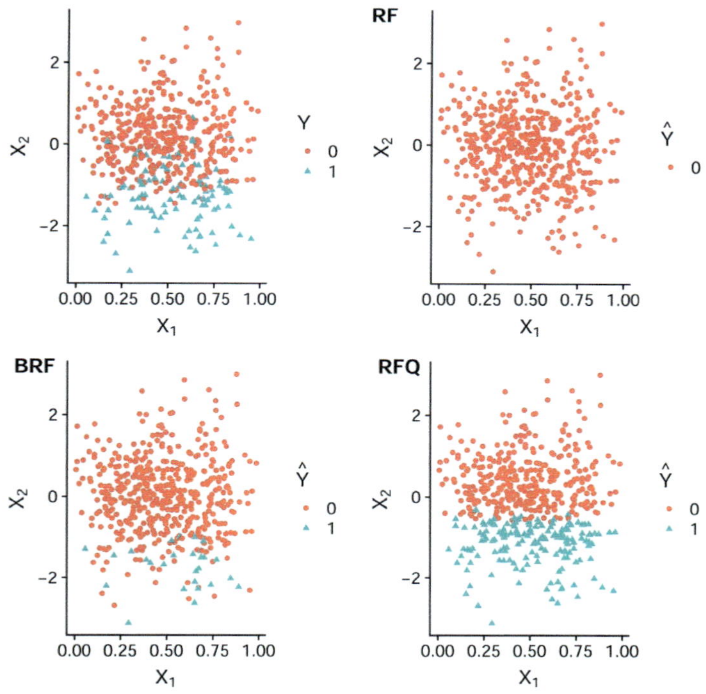

FIGURE 16.11
High-dimensional simulated class imbalanced data. The dimension is $p = 500$, but only variables X_1 and X_2 are relevant. The top left panel displays the data where red represents class label $Y = 0$ and blue represents class label $Y = 1$. The top right panel shows the RF classifier, which classifies all cases as majority labels. The bottom left panel shows the BRF classifier, which uses undersampling of the majority labels to balance the data. The performance of BRF is improved. However, the bottom right panel for the RFQ classifier shows that while BRF improves performance, it is not as aggressive as RFQ in identifying minority class labels.

As another illustration, we applied RFQ to our previous analysis involving esophageal cancer patients. Running a RF classifier but using the RFQ decision rule produced the confusion matrix shown in Table 16.2. This table indicates that the overall misclassification error has increased due to a higher number of misclassified majority cases (surgery patients), which is a result of RFQ's more aggressive classification approach. However, we observe a much more balanced value for the conditional error rates, 25.6% for the surgery group and 10.6% for the adjuvant group, resulting in a gmean of 0.813. This represents a significant improvement from the previous analysis.

16.12.4 Quantile classifiers and optimality goal

To define RFQ, we first introduce the definition of a quantile classifier. Following Mease et al. [26], a quantile classifier or q-classifier is defined as

$$\delta_{\text{q}}(\mathbf{x}) = 1_{\{F(\mathbf{x}) \geq q\}},$$

TABLE 16.2
Confusion matrix from RFQ analysis of esophageal cancer patients.

	predicted		
	surgery	adjuvant	error
surgery	4946	1703	.256
adjuvant	105	883	.106

Overall error rate: 23.6%
gmean: .813

where $0 < q < 1$ is a prespecified quantile threshold. If $q = 1/2$, this results in the *median classifier*, which is the familiar Bayes classifier, δ_{B}.

More generally, we can recognize that the cost-weighted Bayes classifier is an example of a quantile classifier. From Chapter 6.6, when the costs are $\ell_0 > 0$ for misclassifying a class 0 label and $\ell_1 > 0$ for misclassifying a class 1 label, the cost-weighted risk for a classifier $\hat{\delta}$ is defined as (cf. (6.5)):

$$r(\hat{\delta}, \ell_0, \ell_1) = \mathbb{E}\left[\ell_0 1_{\{\hat{\delta}(\mathbf{X})=1, Y=0\}} + \ell_1 1_{\{\hat{\delta}(\mathbf{X})=0, Y=1\}}\right].$$

Under cost-weighted risk the optimal classifier was shown to be the cost-weighted Bayes classifier, defined as

$$\delta_{\mathrm{WB}}(\mathbf{x}) = 1_{\{F(\mathbf{x}) \geq \ell_0/(\ell_0+\ell_1)\}}, \quad F(\mathbf{x}) = \mathbb{P}\{Y = 1|\mathbf{x}\}. \tag{16.14}$$

From this, we can observe that the cost-weighted Bayes classifier δ_{WB} is a quantile classifier with $q = \ell_0/(\ell_0 + \ell_1)$. This also implies an optimality property for quantile classifiers.

The following result summarizes this optimality.

Theorem 16.3. *The cost-weighted Bayes classifier δ_{WB} is optimal in that its risk satisfies $r(\delta_{WB}, \ell_0, \ell_1) \leq r(\hat{\delta}, \ell_0, \ell_1)$ for any classifier $\hat{\delta} : \mathscr{X} \to \{0, 1\}$. Its risk equals*

$$r(\delta_{WB}, \ell_0, \ell_1) = \mathbb{E}\left[\min\{\ell_1 F(\mathbf{X}), \ell_0(1 - F(\mathbf{X}))\}\right]. \tag{16.15}$$

Thus, (16.15) is the smallest weighted risk achievable under misclassification costs ℓ_0, ℓ_1. Moreover, this shows that the quantile classifier δ_q with $q = \ell_0/(\ell_0 + \ell_1)$ is the optimal cost-weighted classifier under costs ℓ_0, ℓ_1.

For the Bayes classifier, the costs are $\ell_0 = \ell_1 = 1$, which results in a cost-weighted risk

$$r(\hat{\delta}) = \mathbb{P}\{\hat{\delta}(\mathbf{X}) = 1, Y = 0\} + \mathbb{P}\{\hat{\delta}(\mathbf{X}) = 0, Y = 1\}.$$

Assuming marginal imbalance, i.e., $F \ll 0.5$, the cost-weighted risk, which in this case is the misclassification error, becomes

$$r(\delta_{\mathrm{B}}, 1, 1) = \mathbb{E}[\min\{F(\mathbf{X}), 1 - F(\mathbf{X})\}] \approx \mathbb{E}[F(\mathbf{X})] \approx 0.$$

This shows that equal misclassification costs provides a strong incentive to correctly classify majority class samples.

However, this approach of using equal costs is not well-balanced. Instead, it is preferable to achieve good performance under both types of classification errors. Therefore, our goal is to find a quantile classifier that achieves high values for both TNR and TPR.

Definition 16.5. A classifier $\hat{\delta} : \mathscr{X} \to \{0, 1\}$ is said to be TNR+TPR-optimal if it maximizes the sum of the rates, TNR + TPR. □

16.12.5 The q^*-classifier

To achieve the goal of TNR+TPR optimality, we introduce the following classifier motivated from a density-based approach. The classifier, denoted by $\delta_{\text{D}}(x)$, assigns $\mathbf{x}$ to the minority class if the minority class data density $f_{\mathbf{X}|1}(\mathbf{x}|1)$ is larger than majority class data density $f_{\mathbf{X}|0}(\mathbf{x}|0)$:

$$\delta_{\text{D}}(\mathbf{x}) = 1_{\{f_{\mathbf{X}|Y}(\mathbf{x}|1) \geq f_{\mathbf{X}|Y}(\mathbf{x}|0)\}}.$$

Basing the classifier on the conditional density of the features, $f_{\mathbf{X}|Y}$, rather than the conditional density of the response, F, removes the effect of the prevalence of the minority class labels. This is one way to see how δ_{D} is able to handle imbalancedness. More directly we can show that δ_{D} is TNR+TPR-optimal. Conditioning on the data, the sum of the TNR and TPR for a classifier $\hat{\delta}$ equals

$$\begin{aligned}
&\int_{\hat{\delta}(\mathbf{x})=0} f_{\mathbf{X}|Y}(\mathbf{x}|0)\, d\mathbf{x} + \int_{\hat{\delta}(\mathbf{x})=1} f_{\mathbf{X}|Y}(\mathbf{x}|1)\, d\mathbf{x} \\
&\quad \leq \int_{f_{\mathbf{X}|Y}(\mathbf{x}|0) > f_{\mathbf{X}|Y}(\mathbf{x}|1)} f_{\mathbf{X}|Y}(\mathbf{x}|0)\, d\mathbf{x} + \int_{f_{\mathbf{X}|Y}(\mathbf{x}|0) \leq f_{\mathbf{X}|Y}(\mathbf{x}|1)} f_{\mathbf{X}|Y}(\mathbf{x}|1)\, d\mathbf{x}.
\end{aligned}$$

The right-hand side is the sum of TNR and TPR for δ_{D}. Since this value does not depend on the data, we can conclude δ_{D} is the optimal TNR+TPR-rule.

As it stands, the classifier seems theoretical since it requires calculating the data density $f_{\mathbf{X}|Y}$ which can be difficult, especially if the feature dimension is high. However, using Bayes theorem, with a little bit of rearrangement, we have

$$\delta_{\text{D}}(\mathbf{x}) = 1_{\{\Delta_{\text{D}}(\mathbf{x}) \geq 1\}}, \quad \text{where } \Delta_{\text{D}}(\mathbf{x}) = \frac{f_{\mathbf{X}|Y}(\mathbf{x}|1)}{f_{\mathbf{X}|Y}(\mathbf{x}|0)} = \frac{F(\mathbf{x})(1-\pi)}{(1-F(\mathbf{x}))\pi} \tag{16.16}$$

for $\pi = \mathbb{P}\{Y = 1\}$. With some further rearrangement, we find that (16.16) is a q-classifier with $q = \pi$. Also notice when $\pi = 1/2$, we obtain the Bayes classifier

$$\delta_{\text{B}}(\mathbf{x}) = 1_{\{\Delta_{\text{B}}(\mathbf{x}) \geq 1\}}, \quad \text{where } \Delta_{\text{B}}(\mathbf{x}) = \frac{F(\mathbf{x})}{1 - F(\mathbf{x})}.$$

We have the following alternate definition of (16.16).

Definition 16.6. Call $\delta_{\text{q}^*}(\mathbf{x}) = 1_{\{F(\mathbf{x}) \geq \pi\}}$ the q^*-classifier (and keep in mind $\delta_{\text{q}^*} = \delta_{\text{D}}$) which is a quantile classifier with $q = \pi = \mathbb{P}\{Y = 1\}$. □

We can summarize our previous discussions as the following result [29].

Theorem 16.4. *The q^*-classifier where $q^* = \pi$ is TNR+TPR-optimal. Furthermore, it is the cost-weighted Bayes rule (16.14) under misclassification costs $\ell_0 = \pi$ and $\ell_1 = (1 - \pi)$.*

Theorem 16.4 shows that the density estimator we started with is actually a quantile classifier with threshold π, corresponding to the weighted Bayes rule with costs $\ell_0 = \pi$ and $\ell_1 = (1 - \pi)$. These costs match the optimal cost values needed to achieve TNR+TPR optimality. Therefore, this estimator, called the q^*-classifier, where $q^* = \pi$, achieves both TNR+TPR optimality *and* cost-weighted Bayes optimality. We can see that the threshold $q^* = \pi$ used for determining classification is very special. It is the value that leads to both classical Bayes optimality and the additional requirement of balancing TNR and TPR performance. Notice that equal costs $\ell_0 = \ell_1$ occur when $\pi = 1/2$, which is the only setting where the traditional Bayes rule is equal to the q^*-classifier.

16.12.6 BRF is a q^*-classifier

We now show, surprisingly, that BRF is actually a q^*-classifier, which means that it also has the optimality properties of Theorem 16.4. This might come as a surprise since BRF is an example of a data-level method that transforms the distributions of the training classes. Recall that BRF undersamples the majority class to evenly balance the data.

To make this connection we introduce the idea of response-based sampling. Define auxiliary variables $S_i \in \{0,1\}$ for $i = 1,\dots,n$ where values $S_i = 1$ are the subsampled cases from the learning data $\mathscr{L}$. The subsampled data is defined as $\mathscr{L}_S = \{(\mathbf{X}_i, Y_i) : S_i = 1, i = 1,\dots,n\}$. Response-based sampling selects the data values with probability that depends only on the value of Y_i and not $\mathbf{X}_i$. In particular, the probability a case is sampled is defined by

$$\mathbb{P}\{S=1|Y\} = \begin{cases} \pi_S(1), & \text{if } Y=1 \\ \pi_S(0), & \text{if } Y=0, \end{cases} \tag{16.17}$$

where $0 < \pi_S(Y) < 1$ is the auxiliary sampling distribution.

By (16.17), the probability a randomly selected Y from $\mathscr{L}_S$ equals $Y = 1$ is

$$\pi^S := \mathbb{P}\{Y=1|S=1\} = \frac{\mathbb{P}\{S=1|Y=1\}\mathbb{P}\{Y=1\}}{\mathbb{P}\{S=1\}} = \frac{\pi_S(1)\pi}{\mathbb{P}\{S=1\}}. \tag{16.18}$$

Likewise, $1-\pi^S = \mathbb{P}\{Y=0|S=1\} = \pi_S(0)(1-\pi)/\mathbb{P}\{S=1\}$. In order to have balanced labels we must have $\pi^S = 1/2$, or equivalently $\pi^S = 1-\pi^S$, which implies by (16.18)

$$\frac{\pi_S(1)}{\pi_S(0)} = \frac{1-\pi}{\pi}. \tag{16.19}$$

To balance the data, BRF uses (16.19) and applies response-based sampling with

$$\frac{\pi_S(1)}{\pi_S(0)} = \frac{1-\dfrac{N_1}{N_0+N_1}}{\dfrac{N_1}{N_0+N_1}} = \frac{N_0+N_1-N_1}{N_1} = \frac{N_0}{N_1}.$$

In other words, BRF undersamples the majority cases by a ratio of N_1/N_0.

Let δ_{B}^S be the Bayes rule constructed using $\mathscr{L}_S$ (call this the subsampled Bayes rule). For a given $\mathbf{x}$,

$$\delta_{\mathrm{B}}^S(\mathbf{x}) = 1 \quad \text{if} \quad \frac{F^S(\mathbf{x})}{(1-F^S(\mathbf{x}))} \geq 1,$$

where by definition $F^S(\mathbf{x}) = \mathbb{P}\{Y=1|\mathbf{X}=\mathbf{x}, S=1\}$. Using Bayes theorem, it can be shown that the subsampled Bayes rule can be rewritten as [29]

$$\delta_{\mathrm{B}}^S(\mathbf{x}) = 1 \quad \text{if} \quad \frac{F(\mathbf{x})}{(1-F(\mathbf{x}))} \geq \frac{\pi_S(0)}{\pi_S(1)}.$$

Under balanced sampling (16.19), the right-hand side equals $\pi/(1-\pi)$, and hence $\delta_{\mathrm{B}}^S = \delta_{\mathrm{D}}$ by (16.16). This implies that by using balanced sampling, the subsampled Bayes rule equals the density-based classifier, which implies that it is TNR+TPR-optimal.

Theorem 16.5. *Under balanced sampling (16.19), the subsampled Bayes rule δ_B^S is TNR+TPR-optimal.*

We can conclude that BRF is a response-based sampling method that by balancing the data, becomes a subsampled Bayes classifier. BRF uses the subsampled probability estimator $F^S(\mathbf{x}) = \mathbb{P}\{Y=1|\mathbf{X}=\mathbf{x}, S=1\}$, which although not a correct estimator for $F(\mathbf{x})$, nevertheless achieves TNR+TPR-optimality since cases are classified according to $F^S(\mathbf{x}) \geq 1/2$, which is equivalent to classification according to $F(\mathbf{x}) \geq \pi$ which is the q^*-classifier.

16.12.7 The q^*-classifier is invariant to response-based sampling

In contrast to BRF, the q^*-classifier is unaffected by response-based sampling and retains TNR+TPR-optimality no matter what the target balance ratio is. Let $\delta^S_{q^*}$ be the q^*-classifier constructed using $\mathscr{L}_S$. By definition, $\delta^S_{q^*}(\mathbf{x}) = 1_{\{F^S(\mathbf{x}) \geq \pi^S\}}$, or equivalently,

$$\delta^S_{q^*}(\mathbf{x}) = 1 \quad \text{if} \quad \frac{F^S(\mathbf{x})(1-\pi^S)}{(1-F^S(\mathbf{x}))\pi^S} \geq 1.$$

It can be shown this implies [29]

$$\delta^S_{q^*}(\mathbf{x}) = 1 \quad \text{if} \quad \frac{F(\mathbf{x})/\pi}{(1-F(\mathbf{x}))/(1-\pi)} \geq 1. \tag{16.20}$$

In other words, $\delta^S_{q^*} = \delta_{q^*}$ (compare (16.20) to (16.16)). We can therefore conclude that δ_{q^*} remains TNR+TPR-optimal. Combined with Theorem 16.5, we have therefore established the following.

Theorem 16.6. *Under response-based sampling (16.17), $\delta^S_{q^*} = \delta_{q^*}$, and therefore $\delta^S_{q^*}$ is TNR+TPR-optimal. Moreover, under balanced sampling (16.19), all three methods are equivalent:*

$$\delta^S_B = \delta^S_{q^*} = \delta_{q^*},$$

and all three methods are TNR+TPR-optimal.

16.12.8 Empirical results

In practice, F is unknown and must be estimated. In this scenario, when we refer to q^*-classification, we mean classification using an estimated value $\hat{F}$ for F. When we use this classifier with RF, we call the resulting procedure RFQ. So to be clear: RFQ runs RF as before, but the difference is that the classifier is no longer based on the Bayes decision rule but uses the q^*-classification rule. Therefore, if $\hat{F}^*_{\text{ens}}$ is the in-bag RF ensemble estimator for F, classification is decided using the rule: $\hat{F}^*_{\text{ens}}(\mathbf{x}) \geq \pi$. In a likewise fashion we obtain the OOB RFQ classifier.

When we use q^*-classification with RF we call the resulting procedure RFQ. To be clear, RFQ is nearly identical to RF and uses the same probability estimates as RF but it replaces the Bayes classification rule used by RF with the q^*-classification rule.

To apply BRF we make the following adjustment to the bootstrap. Instead of sampling n i.i.d. values from the empirical distribution $\mathbb{P}_n$ (i.e. a bootstrap draw), a sample of size $2N_1$ is used, where the probabilities for selecting minority and majority class instances are $\pi_S(1) = (N_0/N_1)\pi_S(0)$; thus satisfying the balancing condition (16.19). Keep in mind that BRF uses the subsampled Bayes rule for classification $F^S \geq 1/2$. Therefore denoting $\hat{F}_{\text{BRF}}$ for the subsampled RF estimator of F, classification is decided using the rule: $\hat{F}_{\text{BRF}}(\mathbf{x}) \geq 1/2$.

In theory, both RFQ and BRF are TNR+TPR-optimal due to Theorem 16.6, but in practice since both methods must estimate F, the advantage will go to RFQ because it uses a bootstrap sample size of $.632n$ which is much larger than BRF which only uses $2N_1$ cases. The scarcity of minority cases in imbalanced data makes BRF's sample size much smaller.

To evaluate performance, the procedures were tested using a diverse collection of 143 imbalanced data sets (for details see O'Brien and Ishwaran [29]). The gmean performance estimated by 5-fold cross-validation is reported in Figure 16.12. From the figure, it is clear that

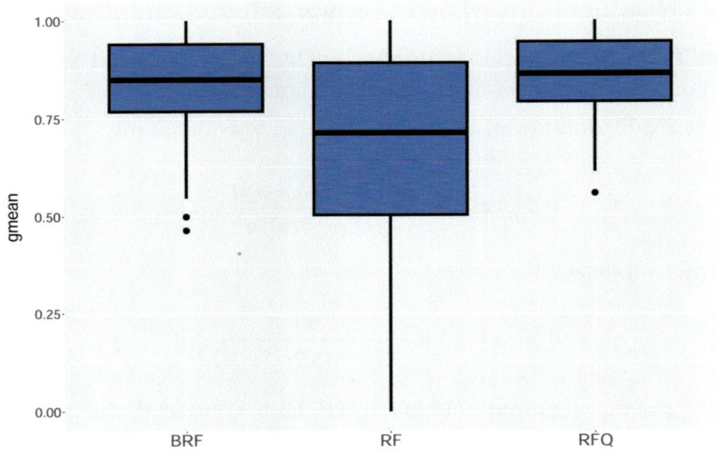

FIGURE 16.12
Five-fold cross-validation gmean performance of RF, BRF and RFQ for 143 benchmark imbalanced data sets.

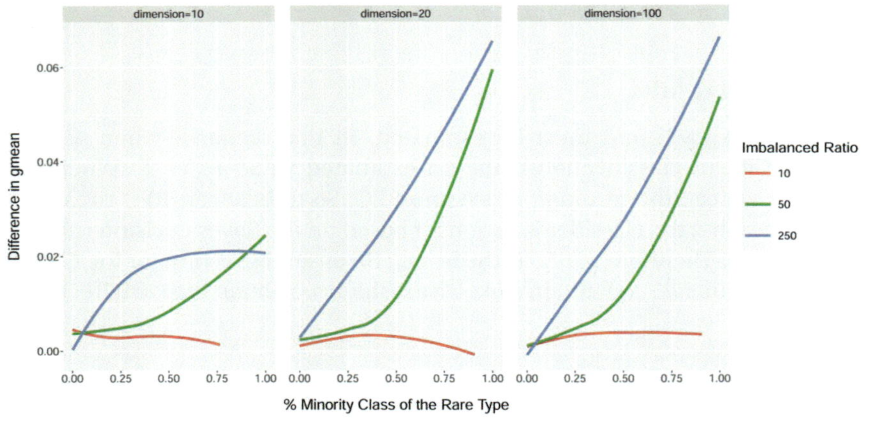

FIGURE 16.13
Difference in gmean for RFQ and BRF from Figure 16.12 where positive means RFQ is outperforming BRF. Values are given as a function of rareness of minority cases and are conditioned on imbalanced ratio and dimension p of the data. As seen, performance of RFQ improves with increasing difficulty of the data.

RFQ and BRF have superior performance to RF, however, it appears they have roughly similar performance to each other overall.

More careful inspection (Figure 16.13) reveals, however, a systematic difference as the difficulty of the data increases. The difference in gmean between RFQ and BRF is shown on the vertical axis and has been conditioned on imbalanced ratio and dimension, p. On the horizontal axis is rarity of minority classes (defined as in Blaszczynski and Stefanowski [2] and Napierala and Stefanowski [28]). The gmean difference increases as complexity increases if the imbalanced ratio is moderate to high. Dimension also plays a role. For low dimensional datasets the effect is modest, otherwise the effect is pronounced. We can conclude that as difficulty of data increases, the advantage of using the larger sample size begins to favor RFQ over BRF and this improvement can become quite significant.

16.12.9 BRF calibration

In addition to using less data than RFQ, another disadvantage of BRF is that the estimated probability will not accurately reflect the true probability. This is because the response-based sampling used by BRF alters the likelihood, resulting in a subsampled probability estimator $F^S(\mathbf{x}) = \mathbb{P}\{Y = 1|\mathbf{X} = \mathbf{x}, S = 1\}$ for $F(\mathbf{x})$ which is biased. To produce correct probability estimates, we need to recalibrate the estimator.

This can be accomplished as follows. First observe we have the following quantities, where recall that $S \in \{0,1\}$ is the binary response auxiliary variable sampled according to $S|Y \sim \pi_S(Y)$ where $S = 1$ indicates the data point is subsampled:

$$\begin{aligned}
F^S(\mathbf{x}) &= \mathbb{P}\{Y = 1|\mathbf{X} = \mathbf{x}, S = 1\} \quad \text{(BRF estimator)}\\
F(\mathbf{x}) &= \mathbb{P}\{Y = 1|\mathbf{X} = \mathbf{x}\} \quad \text{(Target function)}\\
\pi_S(0) &= \mathbb{P}\{S = 1|Y = 0\}\\
\pi_S(1) &= \mathbb{P}\{S = 1|Y = 1\}\\
\pi &= \mathbb{P}\{Y = 1\}\\
N_0 &= \text{Observed frequency of class 0 labels}\\
N_1 &= \text{Observed frequency of class 1 labels.}
\end{aligned}$$

For notational ease, define

$$\begin{aligned}
F(0|\mathbf{x}) &= 1 - F(\mathbf{x}) \qquad & F(1|\mathbf{x}) &= F(\mathbf{x})\\
F^S(0|\mathbf{x}) &= 1 - F^S(\mathbf{x}) \qquad & F^S(1|\mathbf{x}) &= F^S(\mathbf{x}).
\end{aligned}$$

The values $F^S(j|\mathbf{x})$ are the BRF estimated values which we would like to use to obtain a correct estimator for $F(j|\mathbf{x})$, $j = 0,1$. To obtain the correct conditional probabilities, we must solve for $F(\cdot|\mathbf{x})$ from

$$F^S(0|\mathbf{x}) = \frac{\pi_S(0)F(0|\mathbf{x})}{\sum_{j=0}^1 \pi_S(j)F(j|\mathbf{x})}, \qquad F^S(1|\mathbf{x}) = \frac{\pi_S(1)F(1|\mathbf{x})}{\sum_{j=0}^1 \pi_S(j)F(j|\mathbf{x})}$$

subject to the constraint $F(0|\mathbf{x}) + F(1|\mathbf{x})$ and $0 \le F(0|\mathbf{x}), F(1|\mathbf{x}) \le 1$.

To go about solving this, set $w = \pi_S(0)/\pi_S(1)$, to obtain

$$F^S(1|\mathbf{x}) = \frac{F(1|\mathbf{x})}{w(1 - F(1|\mathbf{x})) + F(1,\mathbf{x})},$$

which after a little manipulation yields

$$F(1|\mathbf{x}) = \frac{wF^S(1|\mathbf{x})}{1 - F^S(1|\mathbf{x}) + wF^S(1|\mathbf{x})}. \tag{16.21}$$

To check that (16.21) is a legitimate solution, it suffices to check $0 \le F(1|\mathbf{x}) \le 1$. First notice that $F(1|\mathbf{x}) \ge 0$ if and only if

$$1 \ge F^S(1|\mathbf{x})(1 - w)$$

which holds if $w \ge 0$ because $F^S(1|\mathbf{x}) \le 1$ by definition. Meanwhile, $F(1|\mathbf{x}) \le 1$ if and only if

$$wF^S(1|\mathbf{x}) \le 1 - F^S(1|\mathbf{x}) + wF^S(1|\mathbf{x})$$

which holds from the fact that $F^S(1|\mathbf{x}) \le 1$. Therefore, this verifies that (16.21) is the calibration solution.

To apply this to BRF, we have from (16.19)

$$w = \frac{\pi_S(0)}{\pi_S(1)} = \frac{\pi}{1-\pi}.$$

Substituting into (16.21), we obtain

$$F(\mathbf{x}) := F(1|\mathbf{x}) = \frac{\left(\frac{\pi}{1-\pi}\right) F^S(\mathbf{x})}{1 - F^S(\mathbf{x}) + \left(\frac{\pi}{1-\pi}\right) F^S(\mathbf{x})} = \frac{\pi F^S(1|\mathbf{x})}{(1-\pi)(1-F^S(\mathbf{x})) + \pi F^S(\mathbf{x})}$$

which yields the calibrated BRF estimator

$$\hat{F}(\mathbf{x}) = \frac{N_1 F^S(\mathbf{x})}{N_0(1-F^S(\mathbf{x})) + N_1 F^S(\mathbf{x})}.$$

16.12.10 AUC splitting

RFQ and BRF use the AUC splitting rule of Chapter 15.6 and not the Gini splitting rule used by CART. We prefer AUC due to a connection between maximizing the AUC and the goal of maximizing TNR+TPR.

To explain this connection, recall the notation from Chapter 15.6. Let $Z \in [0,1]$ be the elementary predictor used for classification and we wish to identify the best split for a node into daughters L and R so as to maximize AUC. For simplicity suppose the node to split is the root node of sample size n. Then n_L and n_R are the sample sizes for L and R where $n = n_L + n_R$. This yields a 2×2 frequency table (Table 16.3).

TABLE 16.3
Frequency of cases in left and right daughter nodes by splitting the parent node.

	Left	Right	
$Y=0$	n_{0L}	n_{0R}	n_L
$Y=1$	n_{1L}	n_{1R}	n_R

To determine the AUC for Z we vary the cutoff $\tau \in [0,1]$ to determine the ROC curve. There are at most two distinct values for τ, the left daughter relative frequency of class 1 labels, n_{1L}/n, and the right daughter relative frequency, n_{1R}/n. Chapter 15.6 showed that the AUC for Z equals

$$\max\left(\frac{1}{2}(1-\alpha+\beta),\ \frac{1}{2}(1-\beta+\alpha)\right), \quad (\alpha,\beta) = \left(\frac{n_{0R}}{n_{0L}+n_{0R}},\ \frac{n_{1R}}{n_{1L}+n_{1R}}\right).$$

Therefore to maximize the AUC, the split-statistic to maximize is

$$\hat{\gamma}(\alpha,\beta) = \max\left(\frac{1}{2}(1-\alpha+\beta),\ \frac{1}{2}(1-\beta+\alpha)\right).$$

By symmetry, we can consider the first term $(1-\alpha+\beta)$ appearing in $\hat{\gamma}(\alpha,\beta)$. Let $\mathbb{P}_n$ be the empirical measure for (Z,Y),

$$\begin{aligned}
(1-\alpha+\beta) &:= 1 - \mathbb{P}_n\{Z>\tau|Y=0\} + \mathbb{P}_n\{Z>\tau|Y=1\} \\
&= \text{Specificity}(\tau) + \text{Sensitivity}(\tau) \\
&= 1 - \text{False Positive Rate}(\tau) + \text{True Positive Rate}(\tau) \\
&= \text{True Negative Rate}(\tau) + \text{True Positive Rate}(\tau).
\end{aligned}$$

We can conclude that maximizing the AUC splitting rule is equivalent to the goal of maximizing TNR+TPR.

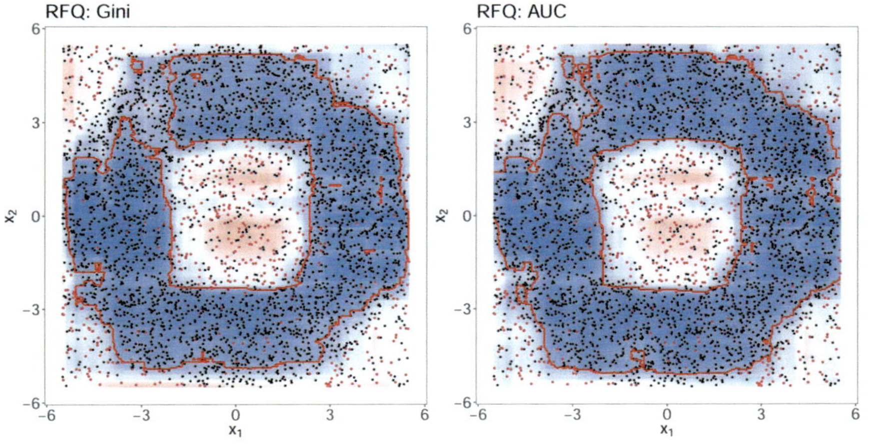

FIGURE 16.14
Decision boundary for RFQ for circle in a square class imbalanced simulation using Gini and AUC splitting.

To illustrate the difference between Gini and AUC splitting, data was simulated from circle-in-a-square simulation (Figure 16.14). The majority class 0 cases are inside a two-dimensional circle, while the minority class 1 cases are between the circle and the square. The left panel shows the decision boundary from RFQ using Gini splitting, and the right panel shows the boundary from AUC splitting. Although the differences are subtle, AUC splitting appears to perform better. Specifically, the decision boundary (red lines) for AUC splitting is more stable and exhibits lower variance.

16.13 Quantile Regression Forests

Quantile regression is a more general regression which considers the distribution of Y. Unlike regression analysis where the goal is to estimate the conditional mean $\mathbb{E}[Y|\mathbf{X}=\mathbf{x}]$, in quantile regression analysis the goal is to determine the conditional distribution of the response

$$F(y|\mathbf{x}) = \mathbb{P}\{Y \leq y|\mathbf{X}=\mathbf{x}\}.$$

For a target quantile value $0 < \alpha < 1$ and a given covariate $\mathbf{x}$ the task is to estimate the value $q_\alpha(\mathbf{x})$ defined by

$$\mathbb{P}\{Y \leq q_\alpha(\mathbf{x})|\mathbf{X}=\mathbf{x}\} = \alpha.$$

More technically, $q_\alpha(\mathbf{x})$ can be written as

$$q_\alpha(\mathbf{x}) = \inf\{y : F(y|\mathbf{x}) \geq \alpha\}. \tag{16.22}$$

With this knowledge, we can construct a prediction interval for the conditional distribution. Thus, a $1-\alpha$ prediction region for the conditional distribution is $I_\alpha(\mathbf{x}) = [q_{\alpha/2}(\mathbf{x}), q_{1-\alpha/2}(\mathbf{x})]$, since

$$\mathbb{P}\{Y \in I_\alpha(\mathbf{x})|\mathbf{X}=\mathbf{x}\} = F(q_{1-\alpha/2}|\mathbf{x}) - F(q_{\alpha/2}|\mathbf{x}) = 1 - \alpha/2 - \alpha/2 = 1 - \alpha.$$

Quantile regression forests is a RF procedure for estimating $q_\alpha(\mathbf{x})$ [27]. We can estimate this value by noticing that the conditional distribution function is the expectation of the function $g(y) = 1_{\{Y \le y\}}$,

$$F(y|\mathbf{x}) = \mathbb{P}\{Y \le y|\mathbf{X} = \mathbf{x}\} = \mathbb{E}\left[1_{\{Y \le y\}}|\mathbf{X} = \mathbf{x}\right] = \mathbb{E}\left[g(y)|\mathbf{X} = \mathbf{x}\right].$$

As was show in Chapter 16.4, this suggests using the RF ensemble estimator defined by

$$\hat{F}_n(y|\mathbf{x}) = \sum_{i=1}^{n} W_i(\mathbf{x}) g(Y_i) = \sum_{i=1}^{n} W_i(\mathbf{x}) 1_{\{Y_i \le y\}} \tag{16.23}$$

where $\{W_i(\mathbf{x})\}_1^n$ are the forest weights (16.4). The RF quantile estimates $\hat{q}_{\alpha,n}$ are obtained by substituting the RF estimator $\hat{F}_n(y|\mathbf{x})$ for $F(y|\mathbf{x})$ in (16.22):

$$\hat{q}_{n,\alpha}(\mathbf{x}) = \inf\{y : \hat{F}_n(y|\mathbf{x}) \ge \alpha\}.$$

To solve this in practice, we can evaluate $\hat{F}_n(\cdot|\mathbf{x})$ over a grid of y values, $y_1^o, \ldots, y_m^o$, and selecting the best approximation

$$\hat{q}_{n,\alpha}(\mathbf{x}) = \underset{y \in \{y_1^o, \ldots, y_m^o\}}{\arg\min} |\hat{F}_n(y|\mathbf{x}) - \alpha|.$$

This gives the RF prediction interval

$$\hat{I}_{n,\alpha}(\mathbf{x}) = [\hat{q}_{n,\alpha/2}(\mathbf{x}), \hat{q}_{n,1-\alpha/2}(\mathbf{x})]. \tag{16.24}$$

The validity of quantile regression forests was shown in Meinhausen [27] who proved that (16.23) is consistent for $F(y|\mathbf{x})$. The proof assumes the estimator is constructed using the full-sample data without bootstrapping. In addition, the following assumptions are required (for simplicity, we state these for a single tree):

1. The data are i.i.d. and the distribution of $\mathbf{X}$ is uniform on $\mathscr{X} = [0,1]^p$.
2. Let $\mathcal{T}_n$ be the partition of the tree and $N_n(\mathbf{X}) = |\mathcal{T}_n(\mathbf{X})|$ the number of data points falling in the terminal node of a random $\mathbf{X}$. Then there exists sequences $m_{n,U} \ge m_{n,L}$ such that

$$o(n) = m_{n,U} \ge N_n(\mathbf{X}) \ge m_{n,L} \to \infty, \quad \text{as } n \to \infty.$$

3. When constructing the tree, each variable is selected with probability no smaller than some $\pi > 0$ for every split in the tree and each split in the tree yields left and right daughters each having at least $\pi' > 0$ of the parent node cases for some π' satisfying $0.5 \ge \pi' > 0$.
4. F satisfies the Lipschitz condition

$$|F(y|\mathbf{x}) - F(y|\mathbf{x}')| \le C||\mathbf{x} - \mathbf{x}'||$$

   for some finite constant C for all $\mathbf{x}, \mathbf{x}' \in \mathscr{X}$.
5. $F(y|\mathbf{x})$ is strictly monotonically increasing in y for each $\mathbf{x} \in \mathscr{X}$.

Theorem 16.7. *Under the above assumptions, for each* $\boldsymbol{x} \in \mathscr{X}$,

$$\sup_{y \in \mathbb{R}} |\hat{F}_n(y|\boldsymbol{x}) - F(y|\boldsymbol{x})| \xrightarrow{\text{p}} 0, \quad \textit{for } n \to \infty.$$

16.13.1 Another approach using OOB data

Several authors have proposed ways to improve quantile regression forests. One such approach, introduced by Zhang et al. [39], has reported notable success. This method uses OOB residuals, similar to the technique discussed in Chapter 16.9.

Using the same notation as Chapter 16.9, let $\hat{Y}_i^{(i)}$ be the OOB estimator for $F_i := \mathbb{E}[Y_i|\mathbf{X}_i]$ and $\tilde{\varepsilon}_i = Y_i - \hat{Y}_i^{(i)}$ be the OOB residual, $i = 1,\dots,n$. Denote $\mathbb{P}_{n,\tilde{\varepsilon}}$ as the empirical measure for $\tilde{\varepsilon}_1,\dots,\tilde{\varepsilon}_n$. Then Zhang et al. [39] suggest using $\mathbb{P}_{n,\tilde{\varepsilon}}$ to approximate the distribution of ε under the assumption of homogeneity of variance. Assuming

$$Y_i \stackrel{d}{=} F_i + \tilde{\varepsilon}_i,$$

we obtain

$$\mathbb{P}\{F_i + \tilde{q}_{n,\alpha/2} \le Y_i \le F_i + \tilde{q}_{n,1-\alpha/2}\} = 1-\alpha$$

where $\tilde{q}_{n,\alpha/2}, \tilde{q}_{n,1-\alpha/2}$ are the $\alpha/2$ and $(1-\alpha/2)$-quantiles of $\mathbb{P}_{n,\tilde{\varepsilon}}$. The OOB prediction interval for Y_i given $\mathbf{X}_i$ is obtained by estimating F_i using $\hat{Y}^{(i)}$:

$$\tilde{I}_{n,\alpha}(\mathbf{X}_i) = [\hat{Y}^{(i)} + \tilde{q}_{n,\alpha/2}, \hat{Y}^{(i)} + \tilde{q}_{n,1-\alpha/2}].$$

This same strategy can be used to obtain the prediction interval for Y for an arbitrary feature $\mathbf{X} = \mathbf{x}$ using

$$\tilde{I}_{n,\alpha}(\mathbf{x}) = [\hat{F}_n(\mathbf{x}) + \tilde{q}_{n,\alpha/2}, \hat{F}_n(\mathbf{x}) + \tilde{q}_{n,1-\alpha/2}] \tag{16.25}$$

wheree $\hat{F}_n(\mathbf{x})$ is the RF ensemble estimator for $F(\mathbf{x}) = \mathbb{E}[Y|\mathbf{X} = \mathbf{x}]$

16.13.2 Quantile splitting rule

Meinhausen [27] used standard weighted variance splitting for quantile regression forests. However, we describe a *quantile regression splitting rule* designed to optimize quantile estimation that may be more suitable. The motivation for the new splitting rule is that it directly minimizes the risk for quantile regression using gradient descent. To explain this, it is helpful to state the loss function for quantile regression.

Theorem 16.8. *Let Y be a continuous random variable with distribution function F. Let $\phi_\alpha(u) = u(\alpha - 1_{\{u<0\}})$. The α quantile for Y equals*

$$q_\alpha = \arg\min_u \mathbb{E}\left[\phi_\alpha(Y-u)\right]. \tag{16.26}$$

Consequently $L_\alpha(Y,u) = \phi_\alpha(Y-u)$ is the quantile loss function (pinball loss). Moreover, the derivative of the risk function $R(Y,u) = \mathbb{E}[L_\alpha(Y,u)]$ is $\phi'_\alpha(u) = F(u) - \alpha$.

Proof. By the definition of $\phi_\alpha(u)$, we have

$$\mathbb{E}\left[\phi_\alpha(Y-u)\right] = (\alpha-1)\int_{-\infty}^{u}(y-u)dF(y) + \alpha\int_{u}^{\infty}(y-u)dF(y).$$

To find the minimum, take the derivative with respect to u and set to zero. It can be shown that the derivative equals

$$(1-\alpha)\int_{-\infty}^{u} dF(y) - \alpha\int_{u}^{\infty} dF(y) = F(u) - \alpha.$$

Set the derivative to zero. If q_α is the solution, we have $F(q_\alpha) = \alpha$ which verifies (16.26). □

We use a gradient descent approach to develop the split-rule. The loss function is $L_\alpha(Y,u)$. Given a cell A, let F_A be the distribution function of Y given A. Then conditioning on A, by Theorem 16.8, the value u minimizing the risk under the loss $L_\alpha(Y,u)$ is the solution to

$$\phi'_{\alpha,A} = F_A(u) - \alpha = 0,$$

where $\phi'_{\alpha,A}$ is the gradient conditional on A. The solution is $F_A(u) = \alpha$, but since we don't know F_A, we replace the optimization using the empirical gradient

$$\hat{\phi}'_{\alpha,A} = \frac{1}{|A|}\sum_{i\in A} 1_{\{Y_i\le u\}} - \alpha = \frac{1}{|A|}\left(\sum_{i\in A} 1_{\{Y_i\le u\}} - \alpha\right). \tag{16.27}$$

Setting this to zero, the solution is the α quantile for A which is denoted by $\hat{q}_{\alpha,A}$.

From (16.27), we can interpret

$$g_i(\hat{F}_A) := \alpha - 1_{\{Y_i\le \hat{q}_{\alpha,A}\}}$$

as negative gradient values for the risk for the cell estimator $\hat{F}_A := \hat{q}_{\alpha,A}$, which we can use to cut A by gradient descent. Similar to gradient boosting, which applies matching pursuit to find the base learner closest to the negative gradient value by fitting a weak regression tree, we apply matching pursuit by fitting a one-split tree to the values $\{g_i(\hat{F}_A)\}_{i\in A}$.

Consider updating the cell estimator $\hat{F}_A$ by $\hat{F}_{A_L} + \hat{F}_{A_R}$ which are constant piecewise over cells A_L, A_R where $A = A_L \cup A_R$. Therefore, approximating the gradient using L_2-distance, the goal is to solve for g_{A_L} and g_{A_R} minimizing

$$\begin{aligned}&\sum_{i\in A}\left[g_i(\hat{F}_A) - (g_{A_L}1_{\{\mathbf{X}_i\in A_L\}} + g_{A_R}1_{\{\mathbf{X}_i\in A_R\}})\right]^2\\ &= \sum_{i\in A}\left[\left(1_{\{Y_i\le g_{\alpha,A}\}} - \alpha\right) - (g_{A_L}1_{\{\mathbf{X}_i\in A_L\}} + g_{A_R}1_{\{\mathbf{X}_i\in A_R\}})\right]^2.\end{aligned}$$

This leads to the following definition of the new splitting rule.

Definition 16.7. The quantile regression split-statistic for cutting a cell A into A_L and A_R is

$$\hat{\gamma}(A, A_L, A_R) = \sum_{i\in A_L}(g_{i,\alpha} - \overline{g}_{L,\alpha})^2 + \sum_{i\in A_R}(g_{i,\alpha} - \overline{g}_{R,\alpha})^2,$$

where $\overline{g}_{L,\alpha}$ and $\overline{g}_{R,\alpha}$ are the mean of $g_{i,\alpha} = 1_{\{Y_i\le \hat{q}_{\alpha,A}\}}$ in the left and right daughters, A_L, A_R, respectively. The best split is the value minimizing $\hat{\gamma}(A, A_L, A_R)$ over all possible cuts A_L, A_R. □

We make the following remarks.

1. The splitting rule replaces the negative gradient values $\alpha - 1_{\{Y_i\le \hat{q}_{\alpha,A}\}}$ with $1_{\{Y_i\le \hat{q}_{\alpha,A}\}}$ without loss of generality.
2. Quantile splitting is identical to weighted variance splitting but where pseudo-response values $g_{i,\alpha}$ are used in place of Y_i.
3. Since $g_{i,\alpha}$ are binary values, the new splitting rule is equivalent to Gini splitting.

So far we have only considered the case where there is one target α value. However, in the case where we might be interested simultaneously in J target α values

$$0 < \alpha_1 < \alpha_2 < \cdots < \alpha_J < 1,$$

we extend our approach to avoid having to run a forest for each quantile α_j.

The following modification is used. For the parent cell A, calculate its quantiles

$$\hat{q}_{\alpha_1,A} \le \hat{q}_{\alpha_2,A} \le \cdots \le \hat{q}_{\alpha_J,A}.$$

Define pseudo-response values

$$g_i = \sum_{j=1}^{J+1} j 1_{\{\hat{q}_{\alpha_{j-1},A} < Y_i \le \hat{q}_{\alpha_j,A}\}}, \ i \in A, \quad \text{where } \hat{q}_{\alpha_0,A} = -\infty, \quad \hat{q}_{\alpha_{J+1},A} = +\infty. \tag{16.28}$$

Note that $g_i \in \{1,\dots,J+1\}$ and therefore can be treated as a multiclass (factor) outcome. Therefore in order to split A, we apply Gini splitting using pseudo-responses g_i for the outcome. When $J = 1$ this reverts back to the quantile splitting rule given in Definition 16.7.

Remark 16.2. A locally adaptive modification is used to further improve the quantile splitting rule. This requires that for a proposed split we center the observations by their means within the proposed daughters. Then instead of defining g_i within the parent node A as in (16.28), we define values g_{i,A_L} and g_{i',A_R} for the proposed daughters as follows:

$$\begin{aligned} g_{i,A_L} &= \sum_{j=1}^{J+1} j 1_{\{\hat{q}_{\alpha_{j-1},A_L} < Y_{i,A_L} \le \hat{q}_{\alpha_j,A_L}\}}, \quad i \in A_L \\ g_{i',A_R} &= \sum_{j=1}^{J+1} j 1_{\{\hat{q}_{\alpha_{j-1},A_R} < Y_{i',A_R} \le \hat{q}_{\alpha_j,A_R}\}}, \quad i' \in A_L \end{aligned}$$

where $Y_{i,A_L} = Y_i - \overline{Y}_{A_L}$ and $Y_{i',A_R} = Y_{i'} - \overline{Y}_{A_R}$ are the daughter centered responses. The g_{i,A_L} and $g_{',A_R}$ are multiclass outcomes with class labels $\{1,\dots,J+1\}$. Now instead of using Gini splitting applied to $\{g_i\}_{i\in A}$ we use Gini splitting applied to $\{g_{i,A_L}, g_{i',A_R}\}_{i\in A_L, i'\in A_R}$. This will be more computationally expensive because it requires recalculating the multiclass outcomes g_{i,A_L}, g_{i',A_R} for each proposed split, however, the benefit from local adaptivity offsets this. □

16.13.3 Boston Housing illustration

Meinhausen [27] used the Boston Housing data for illustration of quantile regression. We have reproduced his analysis here. Figure 16.15 displays the 95% prediction intervals (16.24). The right side plot displays the centered prediction intervals sorted by length of interval for better visualization.

Figure 16.16 shows the 95% prediction intervals using the OOB method (16.25). Locally adaptive quantile regression splitting was used. Two things immediately stand out: (1) The prediction intervals are shorter; (2) The length of a prediction interval is roughly the same across the data. The second point occurs due to exchangeability, which is a consequence of the equal variance assumption. Whether this assumption is always valid needs to be assessed on a case-by-case basis, and there is strong evidence of heterogeneity in the Boston Housing data. On the other hand, Zhang et al. [39] showed that the OOB method is surprisingly robust to departures from equal variance. The first point has been observed by other authors. The original Meinhausen [27] method tends to produce prediction intervals that are too wide. Overall, we recommend the OOB method unless there is serious doubt regarding the exchangeability assumption.

16.14 Random Forests Imputation for Missing Data

Missing data is a common problem in applied settings. Statistical analyses almost always require complete data, so researchers must either impute data or discard missing values

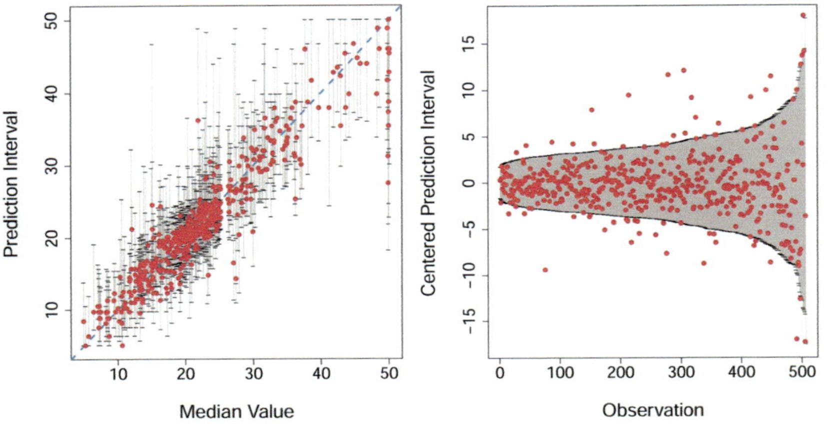

FIGURE 16.15
Quantile regression forests applied to Boston Housing data. Left side are 95% prediction intervals for each data point where prediction intervals are shown by grey bars. Red points are median values. Right side are centered prediction intervals where data have been sorted by length of prediction interval. Prediction intervals and median values for each case have been subtracted by the mean value of the upper and lower end of that case's prediction intervals.

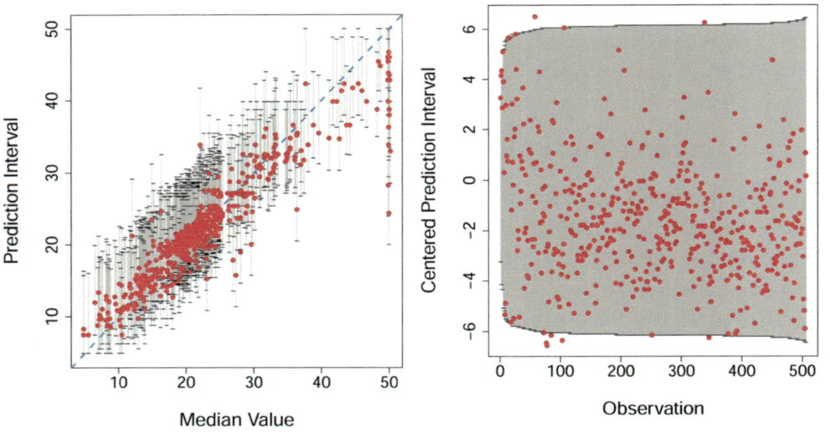

FIGURE 16.16
Same as Figure 16.15 but where prediction intervals have been calculated using the OOB residual method of Chapter 16.13.1.

when encountered. However, discarding missing values (complete case analysis) leads to lost information and compromises inferential power [12]. This can cause selection bias and may result in removing most of the data. Thus, imputing missing data is usually a more reasonable and practical approach.

There are many imputation methods for missing data, but most standard procedures perform poorly in high-dimensional and large-scale settings, such as genomics, proteomics, neuroimaging, and other high-throughput fields. Including all variables in multiple imputation is recommended to avoid bias in correlation estimates [30], but this can cause overparameterization with many variables and a moderate sample size, which is common for genomic data.

Additionally, computational issues like non-convexity in log-likelihood maximization occur with traditional methods like the EM algorithm in high dimensions. Standard imputation approaches also struggle with complex interactions and nonlinearity, often failing to incorporate interaction effects, leading to biased parameter estimates when interactions are present [11].

For these reasons, there has been much interest in using machine learning methods for missing data imputation. RF has proven to be very competitive because it can handle mixed types of missing data, manage interactions and nonlinearity, and scale to high dimensions. One way to use RF is to embed it within a pre-existing imputation method. For instance, Doove et al. [11] proposed using RF with multivariate imputation by chained equations (MICE [38]). See also Shah et al. [32].

Another strategy is to incorporate missing data methods directly into RF. These can be grouped into three general strategies with algorithms described as follows [36].

1. Preimpute the data; grow the forest; update the original missing values using proximity of the data. Iterate for improved results.
2. Simultaneously impute data while growing the forest; iterate for improved results.
3. Preimpute the data; grow a forest using in turn each variable that has missing values; predict the missing values using the grown forest. Iterate for improved results.

Each of these are discussed in detail next.

16.14.1 Proximity imputation

In *proximity imputation* [5], the data are first roughly imputed using a quick imputation (this is typically very crude; for example using mean imputation for continuous variables and maximal class imputation for categorical variables). A RF is fit using the imputed data. Using the resulting forest, the proximity matrix is determined where the (i, j) entry records the in-bag frequency that cases i and j share the same terminal node. The proximity matrix is used to impute the original missing values. For continuous variables, the proximity weighted average of non-missing data is used; for categorical variables, the largest average proximity over non-missing data is used. The updated data are used to grow a new RF and the procedure is iterated.

16.14.2 On the fly imputation

Methods where the data is imputed as the tree is grown are referred to as *on the fly imputation (OTFI)*. A disadvantage of proximity imputation is that it leads to biased OOB estimates for prediction error. The method is also awkward to implement on test data with missing values. The OTFI method was devised to address these issues. The procedure is outlined below (see Tang and Ishwaran [36] for further details).

1. Only non-missing data are used for split-statistics for tree splitting.
2. When assigning left and right daughter node membership if the variable used to split the node has missing data, missing data for that variable are imputed by drawing a random value from the in-bag non-missing data.
3. Following a node split, imputed data are reset to missing and the process is repeated until terminal nodes are reached. After terminal node assignment, imputed data are reset back to missing.
4. Missing data in terminal nodes are imputed using OOB non-missing terminal node data from all the trees. For integer-valued variables, a maximal class rule is used; a mean rule is used for continuous variables.

The purpose of the imputed data in Step 2 is to make it possible to assign cases to daughter nodes. Imputed data are not used to calculate the split-statistic and are reset to missing after node assignment. Thus, at the completion of growing the forest, the resulting forest contains missing values in its terminal nodes and no imputation has been done up to this point. Step 4 is included as a means for imputing the data, but this step is skipped when the goal is to use the forest solely for the purpose of prediction.

OTFI can also be used for test data with missing values. In such a scenario, test data assignment works in the same way as in step 2. That is, for missing test values, values are imputed as in step 2 using the original grow distribution from the training forest, and the test case assigned to its daughter node. Following this, the missing test data are reset back to missing as in step 3. Step 4 imputes the test missing data in a terminal node using the training data.

The above assumes a supervised setting where OTFI is performed simultaneously with the task of creating a supervised predictor. However, OTFI can be run in an unsupervised framework without a response variable. For this we use a multivariate splitting rule Chapter 15.10. As in the original RF algorithm, at each tree node, a set of mtry variables are selected as potential splitting variables. However, as there is no outcome, a random set of "ytry" variables is selected from the remaining variables. These are used to define the multivariate response (called the pseudo-outcomes). The multivariate normalized composite splitting rule Chapter 15.10 is applied to the pseudo-outcomes and the best split determined from the randomly selected mtry variables. We note that missing values in the response are excluded when evaluating the composite multivariate splitting rule [36].

16.14.3 missForest and mForest imputation

In the third strategy, which is called *missForest* [33], the approach turns from imputation to prediction. This approach is different as data are imputed by regressing each variable in turn against all other variables and then predicting missing data for the dependent variable using the fitted forest. Thus, missForest recasts the missing data problem as a prediction problem.

One issue is that with p variables, this means that p forests must be fit for each iteration. This makes the algorithm computationally expensive for high dimensional problems. A multivariate missForest called *mForest* [36] can be used to reduce computational burden. The algorithm is described as follows. First a quick strawman imputation of the data is performed. The p variables in the data set are then randomly assigned into mutually exclusive groups of approximate size αp where $0 < \alpha \leq 1$ (when $\alpha = 1$ the algorithm reverts to the original missForest). Each group in turn acts as the multivariate response to be regressed on the remaining variables. Cuts use a composite multivariate split-rule, where missing values in the multivariate responses are excluded and the split-rule averaged over non-missing responses. Upon completion of the forest, the missing response values are imputed using prediction. The $\approx 1/\alpha$ multivariate regressions are cycled in turn, completing one iteration. The algorithm is repeated until the imputation accuracy of the current imputed data relative to the previously imputed data no longer increases beyond an ε-tolerance value. See Tang and Ishwaran [36] for more details.

16.14.4 Different types of missingness

Depending on whether missingness is related to the data values, three types of missingness are distinguished: (1) missing completely at random (MCAR); (2) missing at random (MAR); and (3) missing not at random (MNAR) [22].

Formally, let $\mathbf{Z} = (Z_{ij})$ denote the complete data matrix that we would have observed in the absence of missing values; thus, $\mathbf{Z}$ includes observed values $\mathbf{Z}_{\text{obs}}$ and potentially unobserved missing data values $\mathbf{Z}_{\text{mis}}$. Let $\mathbf{R}$ be the missing data indicator matrix $\mathbf{R}$, with (i, j)th element $R_{ij} = 1$ if Z_{ij} is observed and $R_{ij} = 0$ if Z_{ij} is missing. The notion of the missing data mechanism is formalized in terms of the conditional distribution of **R** given the complete data **Z**. MCAR implies the distribution of **R** does not depend on **Z**. For MAR, **R** depends on $\mathbf{Z}_{\text{obs}}$ but not on $\mathbf{Z}_{\text{mis}}$. MNAR means **R** does not depend on $\mathbf{Z}_{\text{obs}}$ alone. Thus, we can summarize the three types of missingness as follows:

$$\begin{aligned} \text{MCAR:} &\qquad \mathbf{R}|\mathbf{Z} \stackrel{d}{=} \mathbf{R} \\ \text{MAR:} &\qquad \mathbf{R}|\mathbf{Z} \stackrel{d}{=} \mathbf{R}|\mathbf{Z}_{\text{obs}} \\ \text{MNAR:} &\qquad \mathbf{R}|\mathbf{Z} \stackrel{d}{\neq} \mathbf{R}|\mathbf{Z}_{\text{obs}}. \end{aligned}$$

A simple regression illustration of the three missingness mechanisms was given by Little [21]. The data consists of covariates $\mathbf{X} = (X_1, \ldots, X_p)^T$ and a response value Y and only X_1 has missing values. The probability that X_1 is missing for a case may be (1) independent of data values (thus yielding MCAR); (2) depend on the values of $X_2, \ldots, X_p$ and Y for that case (thus yielding MAR); (3) depends on the value of X_1 for that case (thus yielding MNAR).

16.14.5 Performance

Tang and Ishwaran [36] performed a detailed comparison of the different RF missing data procedures and came to the following conclusions:

1. Correlation plays a strong role in the performance of all RF procedures. Imputation generally improves with increasing correlation of features, even in the presence of high levels of missingness. RF procedures perform well on MCAR and MAR problems.
2. In high correlation settings, missForest generally performs the best.
3. Whenever correlation is present, even moderate amounts, RF algorithms are found to perform better than other procedures.

16.15 Isolation Forest

Isolation forest is an unsupervised tree-based method for detecting anomalous data [23]. "Isolation" refers to separating a case from the rest of the data, while "anomalous" cases are those that are few and different, making them more susceptible to isolation. The method uses random trees constructed through random splitting and subsampling, typically with a subsample size much smaller than 0.632 (the rate used for RF). Under random splitting and extreme subsampling, anomalous data tends to split early and become isolated quickly. Consequently, the depth of an observation (path length) equal to the number of edges from the root node to the data point can be used to identify anomalous observations. To reduce variability, the entire procedure is repeated many times to obtain a forest anomalous score. The resulting forest is referred to as an iForest.

Despite its success, iForest has limitations [8]. Anomalous data described as "few and different" doesn't cover all scenarios. For instance, *swamping* occurs when normal cases are close to anomalous cases, requiring more refined partitions to isolate anomalies, which lengthens path lengths and reduces effectiveness. Another issue is *masking*, where anomalies cluster together, forming a separate cluster that may appear normal, especially with multimodal normal data. This makes detection difficult when anomalous clusters are dense and proximate.

Another concern with the original method is the use of pure random splitting, which struggles with clustered anomalies. To address this, Liu, Kai and Zhou [24] introduced SCIforest, which uses random half-space cuts and a splitting rule based on gain in standard deviation. Related methods are discussed in Hariri, Kind and Brunner [14]. A review of existing methods is provided by Cortes [8].

The following example illustrates some of the limitations with random splitting and the issues with masking for anomaly detection. We use a simulation where there are two clusters, where cluster 1 are $(X_1, X_2)^T$ values

$$X_1 \sim \mathrm{N}(-1, .4), \quad X_2 \sim \mathrm{N}(-1, .2)$$

and cluster 2 are $(X_1, X_2)^T$ values

$$X_1 \sim \mathrm{N}(1, .2), \quad X_2 \sim \mathrm{N}(1, .4).$$

We draw $n = 1000$ independent values from each cluster. To the data we add n' independently drawn anomalous values

$$X_1 \sim \mathrm{N}(-1, \sigma^2), \quad X_2 \sim \mathrm{N}(1, \sigma^2).$$

The subsample size was set to $2^6 = 64$, which is a typical value used by iForest. A total of 500 trees were employed, each split to purity. The use of deep trees is also typical with this approach. Otherwise, due to extreme subsampling, the tree would have only a few terminal nodes, which would hinder anomaly detection. Three different splitting rules were used.

1. Pure random splitting (the original method used by iForest).
2. Unsupervised splitting. This type of splitting was explained in Chapter 16.14 for unsupervised OTF imputation. Briefly, this uses multivariate splitting where a subset of ytry variables are selected to become the pseudo-responses for splitting a node. The ytry pseudo-responses are selected randomly for each node alongside the random mtry set of features.
3. Multivariate splitting where the features are used for both the **Y** response and **X** predictors. We call this *autoencoder* splitting.

The number of anomalies n' was varied using $n' = 1, 5$ and σ was varied using $\sigma = .1, .5$, where larger σ encourage more anomalous data. The anomaly score for each data point was converted to a percentile where a small percentile corresponds to a data point with a short path to the root node. The experiment was repeated 250 times. The percentile values for the anomaly scores are given in Figure 16.17 for each of the 4 experiments. We draw the following conclusions.

1. Pure random splitting is the least effective of the three methods.
2. The autoencoder is significantly better than all other methods when there is only one outlier, $n' = 1$.
3. When there are several outliers, $n' = 5$, unsupervised splitting is slightly better than autoencoder, especially in the more challenging simulation (C).

Figure 16.18 helps to provide insight into these findings. Simulations (A)–(D) are shown from left to right, where density contour plots for the anomaly scores are given for the three different methods. The middle panel figures are random splitting. Notice here that density contours are distorted and enlarged and include the anomalous data points within the two normal clusters. This is because pure random splitting tends to create splits along the x_2 direction from the first cluster center $(-1, -1)$ and along the x_1 direction from the second cluster center $(1, 1)$, which causes the anomalous cases to be contained within one of the two clusters instead of becoming isolated. The top and bottom panels are for autoencoder and unsupervised methods, respectively. Compared to random splitting, these methods are far better at isolating the anomalous cases. In simulation (C), the unsupervised rule even exhibits a density contour

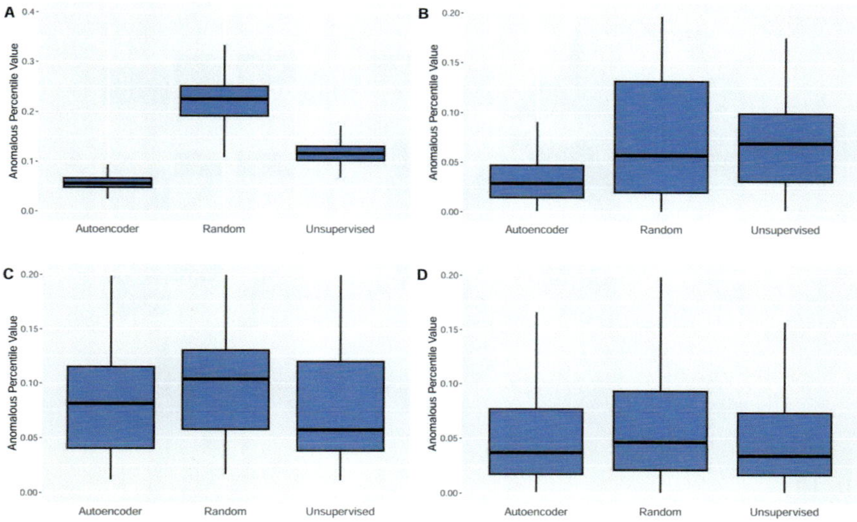

FIGURE 16.17
Simulation experiment to assess which splitting rules work best with isolation forest. The number of outliers n' and noise σ used in the simulation were varied: (A) $(n',\sigma) = (1,.1)$; (B) $(n',\sigma) = (1,.5)$; (C) $(n',\sigma) = (5,.1)$; (D) $(n',\sigma) = (5,.5)$.

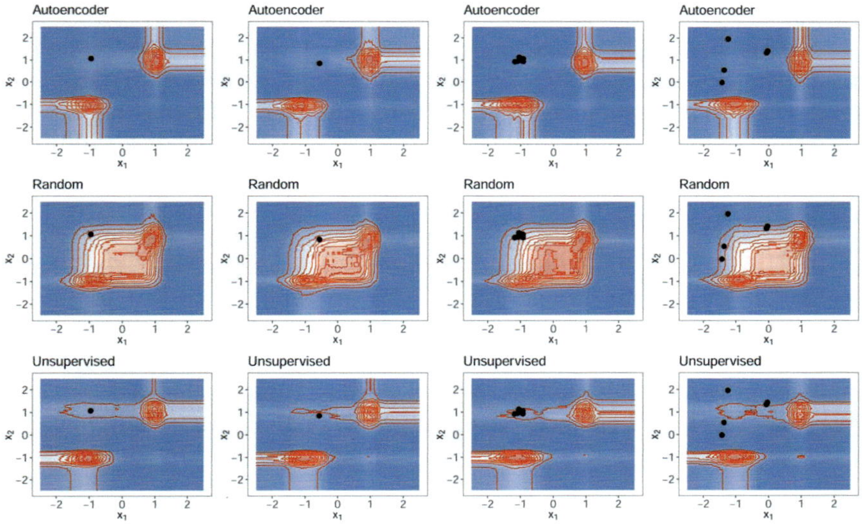

FIGURE 16.18
Contour density plot for anomaly scores from simulations (A)–(D) going from left to right for each of the three splitting rules going from top to bottom.

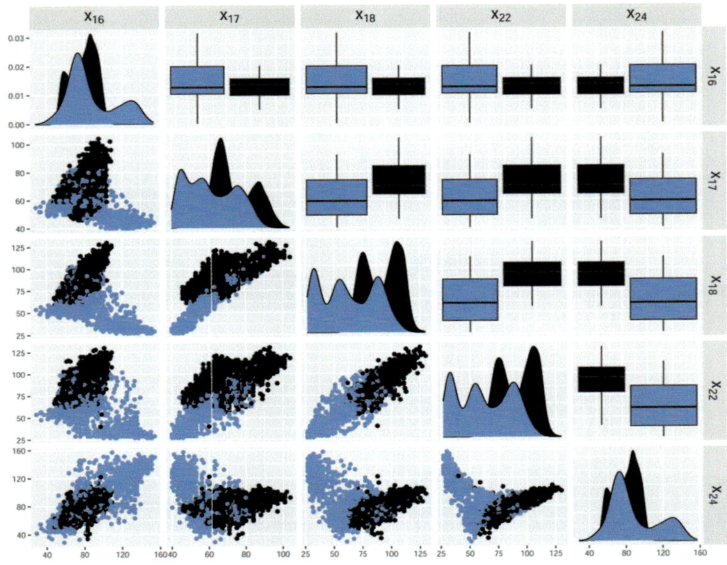

FIGURE 16.19
Satellite data for normal and anomalous data (black are normal cases; blue are anomalous data defined as cases with classification labels 4, 5, and 6). To aid in visualization, data is shown for some but not all of the features.

region for the tightly clustered anomaly points separate from the main data. Presumably, this is why it performed better than the autoencoder for this simulation.

As a second illustration, we use the satellite data available from the R-package `mlbench`, which is a frequently used dataset for illustrating anomaly detection. This is a classification problem where the task is to identify different terrain types from a satellite image using multi-spectral pixel values. There are six class labels corresponding to different terrain regions: (1) red soil; (2) cotton crop; (3) grey soil; (4) damp grey soil; (5) soil with vegetation stubble; (6) very damp grey soil. The class labels are removed from the dataset, making it an unsupervised problem, and we define the anomalous data as class labels 4, 5, and 6. The sample size for the entire data is $n = 6435$, and the number of anomalous data values is $n' = 2036$. Therefore, this is an example of swamping, where the sample size for the anomalous cluster is very large, and the cluster is potentially close to the normal cases; see Figure 16.19.

The anomaly scores (converted to percentiles) are given in Figure 16.20. Below this are the ROC curves and the AUC for each method where ROC curves are calculated using outlier status as the response. The best method is unsupervised followed by autoencoder while random splitting is substantially worse. These findings are in agreement with our previous example.

16.16 Super Greedy Forests

We have discussed tree cuts and the advantage of flexible multivariate cuts used by super greedy trees (SGT; see Chapter 15.11). In this section, we investigate how ensembles of SGT's perform.

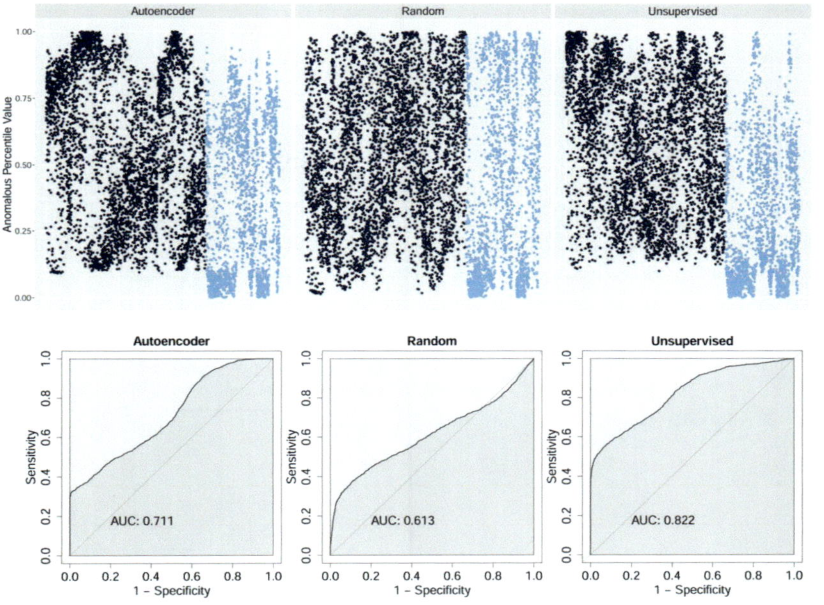

FIGURE 16.20
Top: anomaly percentiles for satellite data. Bottom: using outlier status as the response, performance of a procedure evaluated by AUC.

An ensemble of SGTs is referred to as a super greedy forest (SGF). The performance of SGF was compared in Ishwaran [18] to RF, and it was found that in many datasets, SGF had superior prediction performance. We revisit three of the simulations used in that paper. The regression model is assumed to be $Y = F(\mathbf{X}) + \varepsilon$ and the models considered are as follows.

1. *friedman1*: $F(\mathbf{X}) = 10\sin(\pi X_1 X_2) + 20(X_3 - 0.5)^2 + 10X_4 + 5X_5$, $X_j \sim \mathrm{U}(0,1)$, $\varepsilon \sim \mathrm{N}(0,1)$.
2. *peak*: $F(\mathbf{X}) = 25\exp(-0.5R^2)$, $R \sim \mathrm{U}(0,3)$, $Z_j \overset{\text{iid}}{\sim} \mathrm{N}(0,1)$, $X_j = \dfrac{RZ_j}{\sqrt{\sum_{j=1}^{p} Z_j^2}}$ for $j = 1,\ldots,20$.
3. *lmi2*: $F(\mathbf{X}) = 3(\sum_{j=1}^{15} X_j)^2$, $X_j \overset{\text{iid}}{\sim} \mathrm{N}(0,1)$, $\varepsilon \sim N(0,1)$.

Simulations *friedman1* and *peak* are well known experiments and are available in the `mlbench` R-package. Simulation *lmi2* is a linear model with main effects, all pairwise interactions and all quadratic terms for the first 15 features. For all three simulations the dimension is $p = 40$ to introduce noise variables. The sample size is $n = 1000$ and a test set of size $n' = 25{,}000$ is used for determining standardized prediction error.

Each SGF used 100 trees where the parametric model (15.24) was chosen with `hcut` = 3 (this corresponds to all main effects, quadratic effects and all pairwise interactions between variables). The value `hcut` = 0 was used for calculating RF. The prediction error for the first 40 trees (Figure 16.21) shows SGF's superior prediction performance. In all three experiments, SGF's prediction error is nearly zero with only a few trees, while RF requires at least 40 trees for error rate stabilization, with error rates far higher than SGF's.

Figure 16.22 displays the ensemble estimated β coefficients for each of the experiments. The β values were obtained as follows. Let Π_b be the bth tree partition, $b = 1,\ldots,B$, and adopting our previous notation, let $\Pi_b(\mathbf{x})$ the function that for a given case $\mathbf{x}$, identifies $\mathbf{x}$'s terminal in Π_b. Recall from Chapter 15.12 that the tree predictor is a piecewise combination of lasso fitted models. For

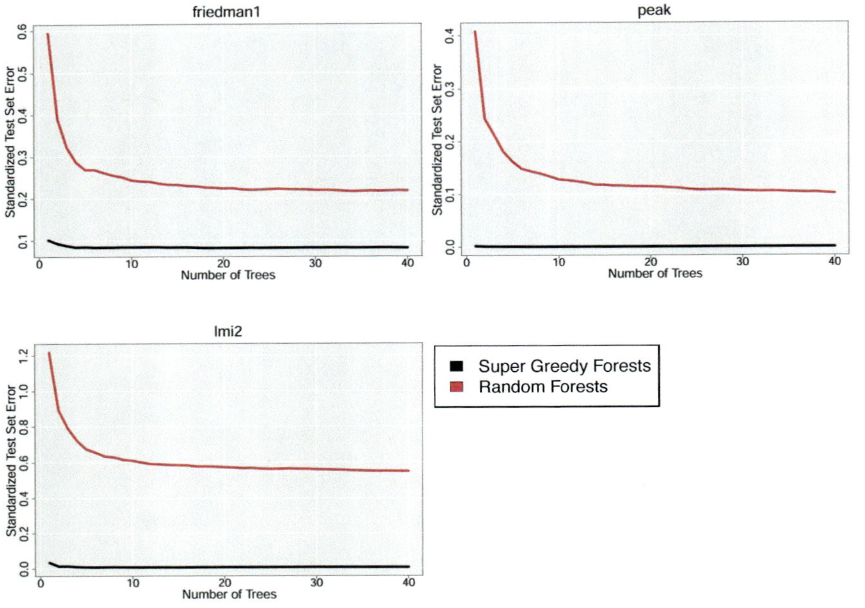

FIGURE 16.21
Test set prediction error for synthetic experiments using SGF.

the general model (15.24), we can write the tree fitted model as

$$\begin{aligned}\hat{F}_{\Pi_b}(\mathbf{x}) &= \hat{\beta}_{\Pi_b(\mathbf{x}),0} + \sum_{l=1}^{p} \hat{\beta}_{\Pi_b(\mathbf{x}),l} x_l + \sum_{l=1}^{p} \hat{\beta}_{\Pi_b(\mathbf{x}),ll} x_l^2 \\ &+ \sum_{1 \le l_1 < l_2 \le p} \hat{\beta}_{\Pi_b(\mathbf{x}),l_1 l_2} x_{l_1} x_{l_2} + \sum_{1 \le l_1 \le l_2 \le l_3 \le p} \hat{\beta}_{\Pi_b(\mathbf{x}),l_1 l_2 l_3} x_{l_1} x_{l_2} x_{l_3}.\end{aligned}$$

Averaging over trees, the ensemble predictor is

$$\begin{aligned}\hat{F}(\mathbf{x}) &= \hat{\beta}_0(\mathbf{x}) + \sum_{l=1}^{p} \hat{\beta}_l(\mathbf{x}) x_l + \sum_{l=1}^{p} \hat{\beta}_{ll}(\mathbf{x}) x_l^2 \\ &+ \sum_{1 \le l_1 < l_2 \le p} \hat{\beta}_{l_1 l_2}(\mathbf{x}) x_{l_1} x_{l_2} + \sum_{1 \le l_1 \le l_2 \le l_3 \le p} \hat{\beta}_{l_1 l_2 l_3}(\mathbf{x}) x_{l_1} x_{l_2} x_{l_3}\end{aligned}$$

where $\hat{\beta}$ are the tree averaged coefficients. For example, $\hat{\beta}_0(\mathbf{x}) = \sum_{b=1}^{B} \hat{\beta}_{\Pi_b(\mathbf{x}),0}/B$ equals the intercept function and $\hat{\beta}_l(\mathbf{x}) = \sum_{b=1}^{B} \hat{\beta}_{\Pi_b(\mathbf{x}),l}/B$ is the gradient function for coordinate l.

Only non-zero coefficients are shown in Figure 16.22. Nearly all noisy variables have zero coefficients, as evidenced by the fact that most of the shown coefficients correspond to signal variables. The coefficients for *friedman1* (top panel of Figure 16.22) highlight the strong quadratic effect for variable 3 and the significant interaction between variables 1 and 2. The coefficients for *peak* (middle panel) are nearly perfect, identifying all 20 quadratic terms. Similarly, for the linear model *lmi2* (bottom panel), all 15 quadratic terms are identified. However, the main effects and pairwise interactions have either zero or near-zero coefficient estimates. This does not seem to affect the nearly perfect prediction error performance shown in Figure 16.21. In conclusion, we find that SGF is a very promising new adaptation of RF using lasso.

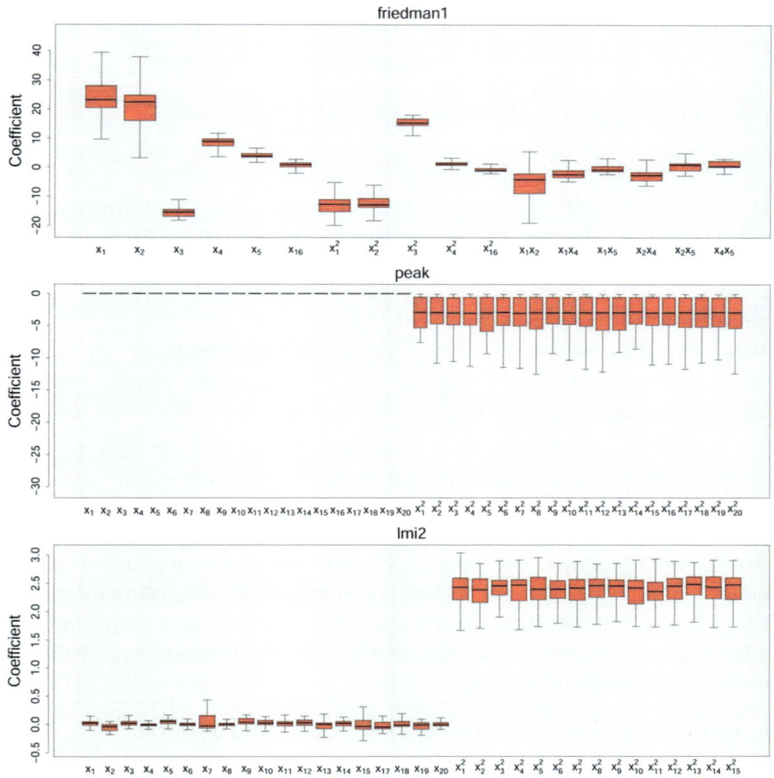

FIGURE 16.22
SGF estimated β coefficient values from synthetic experiments.

Exercises

Exercise 16.1. Train and evaluate a random forest (RF) regressor on a standard dataset.

1. Load the `BostonHousing` dataset.
2. Split the data into training and testing sets.
3. Train a random forest regressor on the training set.
4. Evaluate the model's performance on the testing set using mean squared error.

Exercise 16.2. Let $h_1, \ldots, h_B$ be classifiers which vote 0 or 1 in a two-class problem. Assume that the classifiers are independent and each classifier has a misclassification error $0 < p < 1$,

$$\mathbb{P}\{h_b(\mathbf{X}) \neq Y\} = p.$$

Let $\hat{\delta}_{\text{ens},B}(\mathbf{x})$ be the ensemble classifier defined by a majority vote among the B classifiers and let p_B be the probability $\hat{\delta}_{\text{ens},B}$ is in error.

1. Show that $p_B \downarrow 0$ if $p < 1/2$ and $p_B \uparrow 1$ if $p > 1/2$ as $B \to \infty$. Hint, notice that by independence,

$$\binom{B}{k} p^k (1-p)^{B-k}$$

   equals the probability of exactly k errors among the B classifiers.

2. Part 1 suggests that if $p < 1/2$, then the ensemble classifier

$$1_{\{\sum_{b=1}^{B} h_b(\mathbf{X}) > B/2\}}$$

achieves near perfect accuracy. What is wrong with this argument?

Exercise 16.3. Prove Theorem 16.1.

Exercise 16.4. In this example you will devise a regression simulation and construct an ensemble of trees to study Theorem 16.1. Specifically, simulate data from your experiment and do the following:

1. Estimate the squared bias $\mathbb{E}_{\mathbf{X}}(\text{Bias}\{h|\mathbf{X}\}^2)$ in Theorem 16.1.
2. Estimate the covariance between trees $\mathbb{E}_{\mathbf{X}}(\overline{\text{Cov}}(\mathbf{X}))$ in Theorem 16.1.
3. Estimate the generalization error using test data. Compare this value to Theorem 16.1 using your estimated values from parts 1 and 2.
4. Discuss how to improve your forest to improve the generalization error based on the results of parts 1 and 2. Implement this solution and report its generalization error. Compare this to part 3 and comment.

Exercise 16.5. In this exercise we return to the synthetic logistic regression example of Figure 16.4 illustrating the difference between in-bag and OOB classification. In that example the feature space is discrete so that the Bayes error rate can be calculated exactly. Explain (mathematically) how to calculate the Bayes error rate in this example and then write a program to calculate this value. Show your results.

Exercise 16.6. Train and evaluate a RF classifier on a standard dataset.

1. Load an interesting classification dataset.
2. Split the data into training and testing sets.
3. Train a random forest classifier on the training set.
4. Evaluate the model's performance on the testing set using accuracy and confusion matrix.

Exercise 16.7. Find a suitable data set for a RF analysis. Determine the proximity between observations. Do the same for the forest distance. Systematically compare the two and summarize your findings.

Exercise 16.8. Construct a regression simulation that will be used for studying the effect of nodesize on RF performance.

1. Simulate the data and run a RF regression analysis under different values of nodesize. Determine the optimal nodesize value by minimizing the estimated generalization error of the OOB ensemble.
2. Using simulated data sets repeat part 1 but use test set estimated generalization error to obtain the optimal nodesize.
3. Compare your results of parts 2 and 3 and discuss your findings.

Exercise 16.9. Optimize the hyperparameters of a RF regressor.

1. Load the `California Housing` dataset.
2. Split the data into training and testing sets.
3. Use grid search or random search to find the optimal hyperparameters for the RF regressor using the `caret` R-package.
4. Train the model with the best hyperparameters.
5. Evaluate the performance on the testing set and compare it to a default RF regressor.

Exercise 16.10. Find 5-10 classification datasets, or use simulated data, and run RF on the data for different values of mtry. Grouping the data by signal to noise ratio, or some other criteria, identify a pattern between prediction performance and mtry. Summarize your results clearly.

Exercise 16.11. In this example we study the non-adaptive RF of Chapter 16.11. Recall that $(\mathbf{X}, Y), (\mathbf{X}_1, Y_1), \ldots, (\mathbf{X}_n, Y_n)$ are i.i.d. and $Y = F(\mathbf{X}) + \varepsilon$ where $\mathbf{X}$ is independent of ε which has mean 0 and variance $\sigma^2 > 0$. The non-adaptive RF estimator for $F(\mathbf{X})$ is $\hat{F}_{\text{ens}}(\mathbf{X}) = \sum_{i=1}^n W_i(\mathbf{X}) Y_i$ where the following conditional assumption holds

$$(Y_1, \ldots, Y_n | \mathbf{X}_1, \ldots, \mathbf{X}_n) \perp (W_1(\mathbf{X}), \ldots, W_n(\mathbf{X}) | \mathbf{X}_1, \ldots, \mathbf{X}_n).$$

We will work out the variance for $\hat{F}_{\text{ens}}(\mathbf{X})$ by breaking the calculation up into two steps.

1. Stating all your assumptions carefully, prove that $\mathbb{E}(\hat{F}_{\text{ens}}(\mathbf{X})) = \mathbb{E}(Y_i)$.
2. Now show that

$$\text{Var}\left(\hat{F}_{\text{ens}}(\mathbf{X}) | \mathbf{X}, W_1(\mathbf{X}), \ldots, W_n(\mathbf{X}), \mathbf{X}_1, \ldots, \mathbf{X}_n\right) = \sigma^2 \sum_{i=1}^n W_i^2(\mathbf{X}) := \sigma^2 \Theta(\mathbf{X}).$$

3. Now use Tower's property (see (2.3)) to show that $\text{Var}(\hat{F}_{\text{ens}}(\mathbf{X})) = \sigma^2 \Theta + \text{Var}(F(\mathbf{X}))$ where $\Theta = \mathbb{E}(\Theta(\mathbf{X}))$.

Exercise 16.12. Prove Theorem 16.3. Hint: use Chapter 6.6.

Exercise 16.13. Prove Theorem 16.4. You are allowed to use Theorem 16.3 in your proof.

Exercise 16.14. Using simulations compare the coverage probability of the quantile regression confidence region using (16.24) and the OOB method (16.25). Which performs best?

Exercise 16.15. Train and evaluate a quantile RF model.

1. Load the `BostonHousing` dataset.
2. Split the data into training and testing sets.
3. Train a quantile RF to predict the 5th, 50th, and 95th percentiles of the target variable.
4. Evaluate the performance of the model over the test data in terms of coverage probability.

Exercise 16.16. Simulate a class imbalanced data set. Run RF and RFQ and compare their gmean importance. Repeat the analysis under different imbalanced ratio (IR) settings.

Exercise 16.17. Address class imbalance using different techniques in RF classification.

1. Load the `GermanCredit` dataset from the `caret` package.
2. Split the data into training and testing sets.
3. Train a default RF classifier and evaluate its performance.
4. Apply techniques such as RFQ, class weighting, SMOTE, or undersampling to address class imbalance.
5. Compare the performance of the RF classifier with and without the imbalance handling techniques. For performance metrics use gmean, PR-AUC and pinball loss.

Exercise 16.18. Train and evaluate a RF multivariate regression forest.

1. Find an appropriate and interesting multivariate regression dataset.
2. Split the data into training and testing sets.
3. Train a RF multivariate regression forest using the `randomForestSRC` package.
4. Evaluate the model's performance on the testing set using appropriate metrics for multivariate regression.

Exercise 16.19. Optimize the hyperparameters of an isolation forest to detect anomalies in a high-dimensional dataset and evaluate its performance.

1. Load the `Fashion-MNIST` dataset, which consists of $n = 60,000$ training cases and $n' = 10,000$ test cases of 28×28 grayscale images of 10 different categories of clothing items.
2. Preprocess the data by flattening the images into a high-dimensional feature space ($p = 784$ features per image).
3. Use grid search or random search to find the optimal hyperparameters for the isolation forest.
4. Train the isolation forest with the best hyperparameters on the training set.
5. Use the trained model to detect anomalies in the testing set.
6. Evaluate the performance of the model using gmean, PR-AUC, precision, recall and the F1-score.
7. Visualize the detected anomalies by reshaping the flattened images back to their original dimensions and plotting them.

Exercise 16.20. Apply isolation forests to detect anomalies in time series data and evaluate their effectiveness.

1. Load the `Yahoo S5` time series anomaly detection dataset, which contains real and synthetic time series with labeled anomalies.
2. Preprocess the data by creating rolling windows to convert the time series data into a format suitable for isolation forests.
3. Split the data into training and testing sets, ensuring that the testing set includes labeled anomalies for evaluation.
4. Train an isolation forest on the training set.
5. Use the trained model to detect anomalies in the testing set.
6. Evaluate the performance of the model using gmean, PR-AUC, precision, recall and the F1-score.
7. Compare the performance of the isolation forest with other anomaly detection methods, such as Local Outlier Factor (LOF) and One-Class SVM.
8. Visualize the detected anomalies on the time series plots and discuss the results.

Bibliography

[1] Amit Y. and Geman D. Shape quantization and recognition with randomized trees. *Neural Computation*, 9:1545–1588, 1997.

[2] Blaszczynski J. and Stefanowski J. Neighbourhood sampling in bagging for imbalanced data. *Neurocomputing*, 150:529–542, 2015.

[3] Breiman L. Bagging predictors. *Machine Learning*, 26:123–140, 1996.

[4] Breiman L. Random forests. *Machine Learning*, 45:5–32, 2001.

[5] Breiman L. Manual on setting up, using, and understanding random forests. Version 4.0, 2003.

[6] Breiman L., Chen C. and Liaw A. Using random forest to learn imbalanced data *Technical Report, University of California, Berkeley*, 2004.

[7] Biau G., Devroye L. and Lugosi G. Consistency of random forests and other classifiers. *Journal of Machine Learning Research*, 9:2039–2057, 2008.

[8] Cortes D. Revisiting randomized choices in isolation forests. *arXiv preprint arXiv:2110.13402*, 2021.

[9] Cutler A. and Zhao G. Pert - perfect random tree ensembles. *Computing Science and Statistics*, 33:490–497, 2001.

[10] Dietterich T.G. An experimental comparison of three methods for constructing ensembles of decision trees: bagging, boosting, and randomization. *Machine Learning*, 40:139–157, 2000.

[11] Doove, L.L. and Van Buuren, S. and Dusseldorp E. Recursive partitioning for missing data imputation in the presence of interaction effects. *Computational Statistics & Data Analysis*, 72, 92–104, 2014.

[12] Enders C.K., *Applied Missing Data Analysis.* Guilford Publications, 2022.

[13] Fithian W. and Hastie T. Local case-control sampling: Efficient subsampling in imbalanced data sets. *Quality Control and Applied Statistics*, 60:187–190, 2015.

[14] Hariri S., Kind M.C. and Brunner R.J. Extended isolation forest. *IEEE Transactions on Knowledge and Data Engineering*, 33(4):1479–1489, 2019.

[15] Harrison D. and Rubinfeld D.L. Hedonic prices and the demand for clean air. *Journal of Environmental Economics and Management*, (5), 81–102, 1978.

[16] Ho T.K. The random subspace method for constructing decision forests. *IEEE Trans. on Pattern Analysis and Machine Intelligence*, 20(8), 832–844, 1998.

[17] Ishwaran H. and Kogalur U.B. randomForestSRC: Random Forests for Survival, Regression and Classification. R package version 3.2.3. http://cran.r-project.org, 2024.

[18] Ishwaran H. Super greedy trees *Technical Report, University of Miami*, 2024.

[19] Kubat, M., Holte, R. and Matwin, S. (1997) Learning when negative examples abound. In *European Conference on Machine Learning*, 146–153. Springer.

[20] Lin Y. and Jeon Y. Random forests and adaptive nearest neighbors. *J. Amer. Statist. Assoc.*, 101: 578–590, 2006.

[21] Little R.J. Regression with missing X's: a review. *Journal of the American Statistical Association*, 87(420), 1227–1237, 1992.

[22] Little R.J. and Rubin D.B. *Statistical analysis with missing data*, Volume 793, John Wiley & Sons.

[23] Liu F.T., Ting K.M. and Zhou Z-H. Isolation forest. In *2008 Eighth IEEE International Conference on Data Mining*, 413–422, 2008.

[24] Liu F.T., Ting K.M. and Zhou Z-H. On detecting clustered anomalies using sciforest. In *Machine Learning and Knowledge Discovery in Databases: European Conference, Barcelona, Spain*, II(21):274–290, 2010.

[25] Lyu J. and Ishwaran H. Commentary: To classify means to means to choose a threshold. *The Journal of Thoracic and Cardiovascular Surgery*, 165(4), 1443–1445, 2023.

[26] Mease D., Wyner A.J., Buja A. Boosted classification trees and class probability/quantile estimation. *Journal of Machine Learning Research*, 8:409–439, 2007.

[27] Meinshausen N. Quantile regression forests. *Journal of Machine Learning Research*, 7(6):983-999, 2006.

[28] Napierala K. and Stefanowski J. Types of minority class examples and their influence on learning classifiers from imbalanced data. *Intelligent Information Systems*, 46(3):563–597, 2016.

[29] O'Brien R. and Ishwaran H. A random forests quantile classifier for class imbalanced data. *Pattern Recognition*, 90:232–249, 2019.

[30] Rubin D.B., Multiple imputation after 18+ years. *Journal of the American Statistical Association*, 91(434), 473–489, 1996.

[31] Saito T and Rehmsmeier M. The precision-recall plot is more informative than the ROC plot when evaluating binary classifiers on imbalanced datasets. *PLoS One*,10:e0118432, 2015.

[32] Shah A.D. et al. Comparison of random forest and parametric imputation models for imputing missing data using MICE: a CALIBER study. *American Journal of Epidemiology*, 179(6), 764–774, 2014.

[33] Stekhoven D.J. and Bühlmann P., MissForest-non-parametric missing value imputation for mixed-type data. *Bioinformatics*, 28(1), 112–118, 2012.

[34] Stefanowski J. Dealing with data difficulty factors while learning from imbalanced data. *In: Challenges in Computational Statistics and Data Mining*, Springer, 333–363, 2016.

[35] Stone C.J. Optimal rates of convergence for nonparametric estimators. *Ann. Stat.*, 8(6): 1348–1360, 1980.

[36] Tang F. and Ishwaran H. Random forest missing data algorithms. *Statistical Analysis and Data Mining*, 10(6), 363–377, 2017.

[37] Ueda N and Nakano R. Generalization error of ensemble estimators. In: *Proceedings of International Conference on Neural Networks, IEEE*, 90–95, 1996.

[38] Van Buuren S. Multiple imputation of discrete and continuous data by fully conditional specification. *Statistical Methods in Medical Research*, 16(3), 219–242, 2007.

[39] Zhang H. and Zimmerman J. and Nettleton D. and Nordman D.J. Random forest prediction intervals. *The American Statistician*, 74(4), 392–406, 2019.

17

Random Forests Variable Selection

17.1 Introduction

Although RF is capable of modeling a large number of variables and achieving high prediction accuracy, finding a small number of features with equivalent or near equivalent prediction performance is highly desirable. This allows the researcher to identify which variables play a key role in their problem setting, thus providing insight into the underlying mechanism for what otherwise might be considered a black box.

An important feature of RF is that it provides a rapidly computable internal measure of variable importance (VIMP) that can be used for ranking variables and for variable selection. This is especially useful for high-dimensional data, but can be used in low-dimensions as well. Two commonly evaluated importance measures are node impurity indices (such as the Gini index) and permutation importance. As an example of the former, the importance of a variable in classification is defined as the Gini index reduction for the variable summed over all nodes for each tree in the forest, normalized by the number of trees.

Permutation importance, also known as "Breiman-Cutler" importance [1], is the most commonly used measure of importance for Random Forests (RF) and will be our primary focus. In this chapter, we will review this method and provide some illustrations. One issue is that we need a means to threshold this value to determine statistically significant non-zero importance values. We describe a method based on subsampling to estimate the standard error and construct confidence intervals for this purpose.

In addition to "Breiman-Cutler" importance, we will review other methods for variable selection, including minimal depth, and a new rule-based procedure that can be used not only for RF but for any other machine learning procedure that produces suitable tree rules. Examples illustrating these methods will be given.

17.2 Permutation Variable Importance

For convenience, we will refer to Breiman-Cutler importance as VIMP or permutation VIMP. To calculate a variable's permutation importance, the variable is randomly permuted in the out-of-bag (OOB) data, and the permuted OOB data is then passed through the tree. The OOB error is calculated, and the difference between this error and the OOB error without permutation (i.e., from the original tree), averaged over all trees, is defined as the VIMP. The larger this value, the more predictive the variable [1]. Negative values can also occur, indicating that the variable is a non-signal variable since performance improves without it. Thus, VIMP relies on prediction error for its definition, with OOB data serving as the test data to provide a cross-validated estimate.

DOI: 10.1201/9781003493679-17

> *In the OOB cases for a tree, randomly permute all values of the jth variable. Put these new covariate values down the tree and compute a new internal error rate. The amount by which this new error exceeds the original OOB error is defined as the importance of the jth variable for the tree. Averaging over the forest yields VIMP.*
>
> — *Measure 1 (Manual On Setting Up, Using, And Understanding Random Forests V3.1)*

A mathematical description of VIMP is as follows. Let $\mathscr{L}_n = \{(\mathbf{X}_i, Y_i) : i = 1,\dots,n\}$ be the learning data. Let $h^*(\cdot,\mathbb{T}_b,\mathscr{L}_n) := h^*(\cdot,\mathbb{T}_b)$ be the bth in-bag tree learner with partition $\mathbb{T}_b$ and let $I_{i,b}$ be the binary value that is 1 if i is OOB for tree b. If $L(y,h)$ is the loss function, then the estimated generalization error for the tree learner is

$$\frac{1}{\sum_{i=1}^n I_{i,b}} \sum_{i=1}^n I_{i,b} L\big(Y_i, h^*(\mathbf{X}_i,\mathbb{T}_b)\big). \tag{17.1}$$

This equals the average loss of the in-bag learner over the OOB cases (equivalently, (17.1) equals the empirical risk for the OOB tree estimator).

Averaging (17.1) over all B trees provides the estimated generalization error for the forest (cf. (16.10)). However, we are specifically interested in the error rate for each individual tree, (17.1). This baseline tree error rate is compared to the tree error rate with permuted data. To determine this second quantity, let v be the variable of interest. Represent $\mathbf{X}$ as $(X^{(1)},\dots,X^{(p)})^T$, where $X^{(v)}$ is the vth feature coordinate. The permuted value of $X^{(v)}$ is denoted by $\tilde{X}^{(v)}$. Substituting this into the vth coordinate of $\mathbf{X}$ results in $\tilde{\mathbf{X}}^{(v)}$:

$$\tilde{\mathbf{X}}^{(v)} = (X^{(1)},\dots,X^{(v-1)},\tilde{X}^{(v)},X^{(v+1)},\dots,X^{(p)})^T.$$

Using $\mathbf{X}_1,\dots,\mathbf{X}_n$, randomly permute the vth coordinate. Let $\tilde{\mathbf{X}}_{1,b}^{(v)},\dots,\tilde{\mathbf{X}}_{n,b}^{(v)}$ denote the resulting sample. Then the estimated generalization error for the *noised-up tree learner* is

$$\frac{1}{\sum_{i=1}^n I_{i,b}} \sum_{i=1}^n I_{i,b} L\Big(Y_i, h^*(\tilde{\mathbf{X}}_{i,b}^{(v)},\mathbb{T}_b)\Big). \tag{17.2}$$

We might think of (17.2) informally as the error rate for the tree learner without the advantage of variable v.

The difference between (17.1) and (17.2) yields the tree-VIMP for v defined by

$$I(X^{(v)},\mathbb{T}_b) = \frac{\sum_{i=1}^n I_{i,b} L(Y_i, h^*(\tilde{\mathbf{X}}_{i,b}^{(v)},\mathbb{T}_b))}{\sum_{i=1}^n I_{i,b}} - \frac{\sum_{i=1}^n I_{i,b} L(Y_i, h^*(\mathbf{X}_i,\mathbb{T}_b))}{\sum_{i=1}^n I_{i,b}}. \tag{17.3}$$

The left-hand term will be larger than the right-hand term if v is a signal variable in the true model, thus resulting in a large tree importance value. The VIMP for v is obtained by tree averaging (17.3):

$$I(X^{(v)}) = \frac{1}{B}\sum_{b=1}^B I(X^{(v)},\mathbb{T}_b). \tag{17.4}$$

Large positive values identify signal variables. Figure 17.1 illustrates how the strength of a variable is tied to how large the noised up prediction error for a tree becomes.

As a simple illustration, VIMP was obtained from a RF regression of the Boston Housing data with median house price used for the outcome (left panel of Figure 17.2). Variables lower status and room have the two largest values.

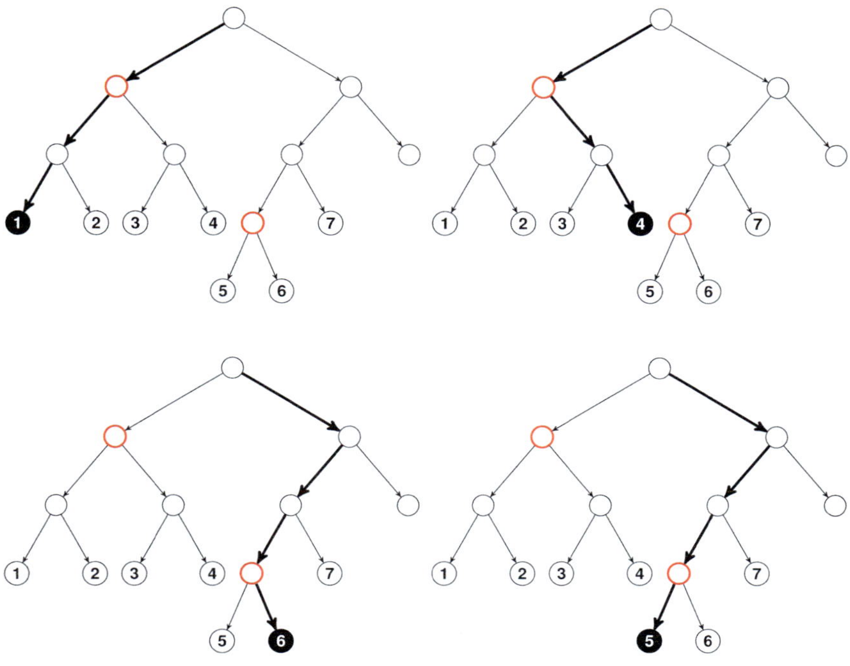

FIGURE 17.1
Illustration of how randomly permuting a variable leads to different terminal node assignments, which is at the heart of why VIMP works. Nodes colored red are tree nodes that split on the target variable v. In the top panel on the left, the bold arrows show the path that a data point $\mathbf{x}$ takes as it traverses through the tree to its terminal node assignment "1". On the right, the v coordinate of $\mathbf{x}$ has been randomly permuted and its new terminal node assignment is now "4". In the bottom panel, another $\mathbf{x}$ value follows its path to terminal node assignment "6". However, when v is permuted, its new terminal assignment becomes "5". Importantly, notice that in the top panel, the terminal node assignment after randomly permuting v is much farther from its original terminal node assignment than in the bottom panel. This shows that the higher v splits in the tree, the more an effect permutation has on final terminal node membership, and hence on prediction error.

In this regression setting, we can express tree-VIMP in a more direct way. Let $\hat{Y}_b^{(i)} = h^*(\mathbf{X}_i, \mathcal{T}_b)$ be the OOB regression estimator and $\hat{Y}_b^{(v,i)} = h^*(\tilde{\mathbf{X}}_{i,b}^{(v)}, \mathcal{T}_b)$ be the noised up estimator for the OOB case i. Then

$$I(X^{(v)}, \mathcal{T}_b) = \frac{\sum_{i=1}^{n} I_{i,b}\left(Y_i - \hat{Y}_b^{(v,i)}\right)^2}{\sum_{i=1}^{n} I_{i,b}} - \frac{\sum_{i=1}^{n} I_{i,b}\left(Y_i - \hat{Y}_b^{(i)}\right)^2}{\sum_{i=1}^{n} I_{i,b}}.$$

This equals the difference in OOB error under L_2-loss. The tree-averaged value of this is what is plotted in the left panel of Figure 17.2.

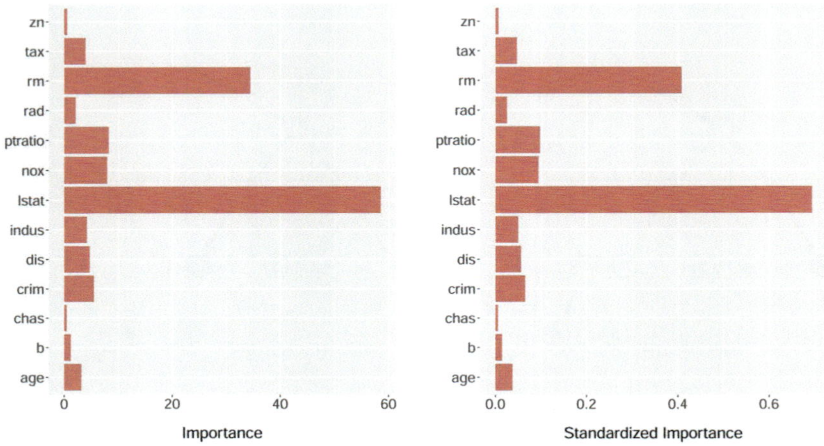

FIGURE 17.2
Breiman-Cutler permutation importance (VIMP) for Boston Housing data. Left and right are unstandardized and standardized values.

Examining this value, we see that the metric is calibrated to the scale of Y, making it less interpretable. To obtain a more understandable value, we can standardize it by dividing by the sample variance, similar to what we did for the RF regression benchmark experiment in Figure 16.1:

$$\frac{I(X^{(v)}, \mathfrak{T}_b)}{\sum_{i=1}^{n}(Y_i - \overline{Y})^2/n}.$$

This standardizes VIMP relative to the baseline predictor $\overline{F} = n^{-1}\sum_{i=1}^{n} F(\mathbf{X}_i)$ where $F(\mathbf{X}_i) = \mathbb{E}[Y_i|\mathbf{X}_i]$. Standardized VIMP is displayed in the right panel of Figure 17.2.

17.3 Subsampling for Standard Errors and Confidence Intervals for Variable Importance

One of the advantages of VIMP is that it removes the arbitrariness of having to select a cutoff value when determining the effectiveness of a variable. Regardless of the problem, a VIMP of zero is meant to reflect the point at which a variable no longer contributes predictive power to the model. However, in practice one may observe values close to zero, and the meaning of what constitutes being zero becomes unclear. In fact, we can already see this for the Boston Housing example of Figure 17.2. Therefore, it is not always straightforward to use VIMP to select the final model.

One attempt to resolve this are stepwise procedures which systematically remove features by using VIMP ordering. For example, Diaz-Uriarte and Alvares [5] described a backward elimination procedure using RF for selecting genes from microarray data. This method consists of the following steps: (1) rank genes according to permutation VIMP; (2) iteratively fit RF, and at each iteration remove a fixed portion of the lowest ranked genes; (3) select the group of genes achieving the smallest OOB error rate; (4) estimate prediction error for the selected model using the .632+ bootstrap method [7] to mitigate selection bias.

However, these are only indirect methods and do not address the underlying issue of quantifying the uncertainty in VIMP. A more direct solution is to approximate the distribution of VIMP through resampling. We describe an approach developed by Ishwaran and Lu [9] using subsampling [13]. The technique of subsampling is a general methodology for approximating the distribution of a statistic. In the following sections, we review subsampling and describe the subsampling estimator for the variance and confidence regions for VIMP used by Ishwaran and Lu [9]. We also discuss a delete-d jackknife variance estimator [15], which is closely related to the subsampling variance estimator.

17.3.1 Subsampling theory

We begin by reviewing some basic theory of subsampling. Let $X_1, \ldots, X_n$ be i.i.d. random values with common distribution $\mathbb{P}$. Let $\hat{\theta}_n = \hat{\theta}(X_1, \ldots, X_n)$ be some estimator for $\theta(\mathbb{P})$, an unknown real-valued parameter we wish to estimate. The bootstrap estimator [6] for the variance of $\hat{\theta}_n$ uses the following simple idea. Let $\mathbb{P}_n$ be the empirical measure and let $X_1^*, \ldots, X_n^*$ be a bootstrap sample obtained by independently sampling n points from $\mathbb{P}_n$. Because $\mathbb{P}_n$ converges to $\mathbb{P}$, we should expect the moments of the bootstrap estimator $\hat{\theta}_n^* = \hat{\theta}(X_1^*, \ldots, X_n^*)$ to closely approximate those of $\hat{\theta}_n$. In particular, the bootstrap variance estimator

$$\mathrm{Var}_{\mathbb{P}_n}(\hat{\theta}_n^*) = \mathbb{E}_{\mathbb{P}_n}\left(\hat{\theta}_n^* - \mathbb{E}_{\mathbb{P}_n}(\hat{\theta}_n^*)\right)^2$$

should closely approximate the population variance $\mathrm{Var}(\hat{\theta}_n) := \mathrm{Var}_{\mathbb{P}}(\hat{\theta}_n)$.

Subsampling [13] employs the same strategy as the bootstrap but is based on sampling *without replacement*. For $m := m(n)$ such that $m \ll n$, let S_m be the entire collection of subsets of $\{1, \ldots, n\}$ of size m. For each $s = \{i_1, \ldots, i_m\} \in S_m$, let $\hat{\theta}_{n,m,s} = \hat{\theta}_n(X_{i_1}, \ldots, X_{i_m})$ be the estimator evaluated using s. The goal is to estimate the distribution of $\hat{J}_n = n^{1/2}(\hat{\theta}_n - \theta(\mathbb{P}))$ using the subsampled statistics $\hat{\theta}_{n,m,s}$. This is advantageous computationally, since these statistics are calculated using a very small sample size m relative to n. In practice this can make a significant difference in compute time, especially for an expensive calculation like VIMP.

Subsampling provides a consistent estimate of the distribution of $\hat{J}_n$ under certain conditions. These conditions are simpler than those for the bootstrap, and there are cases where subsampling works but the bootstrap does not [13]. Let $\mathbb{Q}_n$ denote the distribution of $\hat{J}_n$ and define the subsampling distribution function

$$\widetilde{U}_{n,m}(x) = \frac{1}{C_m} \sum_{s \in S_m} 1_{\{m^{1/2}(\hat{\theta}_{n,m,s} - \hat{\theta}_n) \le x\}}, \quad \text{where } C_m = \binom{n}{m} = |S_m|. \tag{17.5}$$

Assuming $\mathbb{Q}_n$ converges weakly to a proper limiting distribution $\mathbb{Q}$,

$$\mathbb{Q}_n \overset{d}{\rightsquigarrow} \mathbb{Q}, \tag{17.6}$$

and $m/n \to 0$ as $m \to \infty$, then

$$\widetilde{U}_{n,m}(x) \xrightarrow{p} F(x) = \mathbb{Q}[-\infty, x] \tag{17.7}$$

for each x that is a continuity point of the limiting cumulative distribution function F.

What (17.5) and (17.7) indicate is that we can use the subsampled test statistic values $\tilde{J}_{n,m,s} = m^{1/2}(\hat{\theta}_{n,m,s} - \hat{\theta}_n)$ to approximate the distribution of $\mathbb{Q}$, which is the limit of the distribution of $\hat{J}_n = n^{1/2}(\hat{\theta}_n - \theta(\mathbb{P}))$. Therefore, we can use $\{\tilde{J}_{n,m,s}\}$ to estimate the variance of $\hat{\theta}_n$.

The key to this argument is to recognize that due to (17.6) and $m \ll n$, $\widetilde{U}_{n,m}$ closely approximates

$$U_{n,m}(x) = \frac{1}{C_m} \sum_{s \in S_m} 1_{\{m^{1/2}(\hat{\theta}_{n,m,s} - \theta(\mathbb{P})) \le x\}},$$

which is a U-statistic of order m (see Chapter 2.19).

There are three steps to the proof (details can be found in Politis and Romano [13]).

1. If $x \pm \varepsilon$ are continuity points of F, then with probability tending to one,
$$U_{n,m}(x-\varepsilon) \le \widetilde{U}_{n,m}(x) \le U_{n,m}(x+\varepsilon)$$
for any $\varepsilon > 0$, and hence it suffices to show $U_{n,m}(x) \xrightarrow{\mathrm{p}} F(x)$ for every continuity point.
2. Let $F_{n,m}$ be the distribution function for $\hat{J}_{n,m} = m^{1/2}(\hat{\theta}_{n,m} - \theta(\mathbb{P}))$. Then because $U_{n,m}$ is U-statistic of order m, and $0 \le U_{n,m}(x) \le 1$ and $\mathbb{E}[U_{n,m}(x)] = F_{n,m}(x)$, we have by Theorem 2.45 that for any $t > 0$
$$\mathbb{P}\{|U_{n,m}(x) - F_{n,m}(x)| \ge t\} \le 2\exp\left(-\frac{2nt^2}{m^2}\right).$$
Therefore $U_{n,m}(x) - F_{n,m}(x) \xrightarrow{\mathrm{p}} 0$ (a standard Borel-Cantelli argument; see Chapter 2.14).
3. Finally by (17.6), we have $\mathbb{Q}_{n,m} \overset{\mathrm{d}}{\rightsquigarrow} \mathbb{Q}$, or equivalently $F_{n,m}(x) \xrightarrow{\mathrm{p}} F(x)$ for all continuity points, thus completing the proof.

17.3.2 Subsampling estimator for the variance

The ability to approximate the distribution of $\hat{\theta}_n$ suggests, similar to the bootstrap, that we can approximate moments of $\hat{\theta}_n$ with those from the subsampled estimator: in particular, we should be able to approximate the variance. Unlike the bootstrap, however, subsampled statistics are calculated using a sample size m and not n. Therefore to estimate the variance of $\hat{\theta}_n$ we must apply a scaling factor to correct for sample size. The subsampled estimator for the variance is (see Radulović [14] and Section 3.3.1 from Politis and Romano [13])

$$\hat{\upsilon}_m^2 = \frac{m/n}{C_m} \sum_{s \in S_m} \left(\hat{\theta}_{n,m,s} - \frac{1}{C_m} \sum_{s' \in S_m} \hat{\theta}_{n,m,s'}\right)^2. \tag{17.8}$$

The estimator (17.8) is closely related to the delete-d jackknife [15]. The delete-d estimator works on subsets of size $r = n - d$ and is defined as

$$\hat{\upsilon}_{J(d)}^2 = \frac{r/d}{C_r} \sum_{s \in S_r} (\hat{\theta}_{n,r,s} - \hat{\theta}_n)^2.$$

With a little bit or rearrangement, this can be rewritten as

$$\hat{\upsilon}_{J(d)}^2 = \frac{r/d}{C_r} \sum_{s \in S_r} \left(\hat{\theta}_{n,r,s} - \frac{1}{C_r} \sum_{s' \in S_r} \hat{\theta}_{n,r,s'}\right)^2 + \frac{r}{d}\left(\frac{1}{C_r} \sum_{s \in S_r} \hat{\theta}_{n,r,s} - \hat{\theta}_n\right)^2.$$

Setting $d = n - m$, we obtain

$$\hat{\upsilon}_{J(d)}^2 = \frac{m/(n-m)}{C_m} \sum_{s \in S_m} \left(\hat{\theta}_{n,m,s} - \frac{1}{C_m} \sum_{s' \in S_m} \hat{\theta}_{n,m,s'}\right)^2 + \underbrace{\frac{m}{n-m}\left(\frac{1}{C_m} \sum_{s \in S_m} \hat{\theta}_{n,m,s} - \hat{\theta}_n\right)^2}_{\text{bias}}. \tag{17.9}$$

The first term closely approximates (17.8) since $m/n \to 0$, while the second term is a bias estimate of the subsampled estimator. Thus, the delete-d estimator (17.9) can be seen to be a bias corrected version of (17.8). Furthermore this correction is always upwards because the bias term is squared and always positive.

17.3.3 Subsampling and delete-d jackknife algorithms for VIMP

We now describe how to apply subsampling for the variance of VIMP. To map our problem back to the notation of the previous section we introduce the following estimators:

$$\hat{\theta}^{(v)}_{n,B} = \frac{1}{B}\sum_{b=1}^{B} I(X^{(j)}, \Pi_b, \mathcal{L}_n) \qquad \text{Monte Carlo VIMP}$$

$$\hat{\theta}^{(v)}_{n} = \mathbb{E}_{\Theta}\left\{I(X^{(j)}, \Pi(\Theta), \mathcal{L}_n)\right\} \qquad \text{VIMP}$$

$$\theta^{(v)} = \lim_{n\to\infty} \mathbb{E}\{\hat{\theta}^{(v)}_n\} \qquad \text{Population VIMP.}$$

The first line is the forest VIMP, $I(X^{(v)})$, defined earlier (17.4). This can be called Monte Carlo VIMP because the estimator depends on the number of trees, B. As $B \to \infty$, due to the law of large numbers, we obtain $\hat{\theta}^{(v)}_n$ of line 2 and this is called VIMP. For theoretical reasons we will work with $\hat{\theta}^{(v)}_n$ instead of the actual $\hat{\theta}^{(v)}_{n,B}$ used in practice. Finally, as $n \to \infty$ and $B \to \infty$, we obtain the population VIMP of line 3.

In the previous notation, the parameter of interest is $\theta^{(v)}$ and its estimator is $\hat{\theta}^{(v)}_n$. Thus the goal is to approximate the distribution of $\hat{J}^{(v)}_n = n^{1/2}(\hat{\theta}^{(v)}_n - \theta^{(v)})$ using subsampled $\hat{\theta}^{(v)}_{n,m,s}$ values for VIMP so as to approximate the standard error of $\hat{\theta}^{(v)}_n$:

$$\sigma^{(v)}_n = \sqrt{\mathrm{Var}(\hat{\theta}^{(v)}_n)}.$$

Algorithm 17.1 describes the subsampling algorithm used to obtain $\sigma^{(v)}_n$. It is assumed that the subsampling size $m := m(n)$ is chosen so that $m \ll n$.

Algorithm 17.1 *m-subsampling estimator for $\sigma^{(v)}_n$*

1: Draw a random set $s \in S_m$. Let $\mathcal{L}_{n,s} = \{(\mathbf{X}_i, Y_i) : i \in s\}$ be $\mathcal{L}_n$ restricted to s.
2: Calculate $\hat{\theta}^{(v)}_{n,B}$ using $\mathcal{L}_{n,s}$. Let $\hat{\theta}^{(v)}_{n,m,s}$ denote this value.
3: Repeat $K > 1$ times obtaining $\hat{\theta}^{(v)}_{n,m,s_1}, \ldots, \hat{\theta}^{(v)}_{n,m,s_K}$. Estimate $\sigma^{(v)}_n$ using

$$\hat{\sigma}^{(v)}_n = \sqrt{\frac{m}{nK}\sum_{k=1}^{K}\left(\hat{\theta}^{(v)}_{n,m,s_k} - \frac{1}{K}\sum_{k'=1}^{K}\hat{\theta}^{(v)}_{n,m,s_{k'}}\right)^2}.$$

The delete-d jackknife estimator is obtained by a slight modification; see Algorithm 17.2. Notice that this uses Monte Carlo forest VIMP calculated from the full data set in place of averaged subsampled VIMP as used in Algorithm 17.1.

Algorithm 17.2 *delete-d jackknife estimator ($d = n - m$) for $\sigma^{(v)}_n$*

1: Using the entire learning set $\mathcal{L}_n$, calculate the Monte Carlo VIMP $\hat{\theta}^{(v)}_{n,B}$.
2: Run the m-subsampling estimator as above, but replace the estimator in step 3 with

$$\hat{\sigma}^{(v)}_{J(d),n} = \sqrt{\frac{m}{(n-m)K}\sum_{k=1}^{K}\left(\hat{\theta}^{(v)}_{n,m,s_k} - \hat{\theta}^{(v)}_{n,B}\right)^2}.$$

Ishwaran and Lu [9] investigated the performance of the two estimators. We reproduce the results from their experiments involving RF regression. The simulation models used are listed below for convenience.

1. $Y = 10\sin(\pi X_1 X_2) + 20(X_3 - 0.5)^2 + 10X_4 + 5X_5 + \varepsilon$; $\{X_j\} \sim \mathrm{U}(0,1)$; $\varepsilon \sim \mathrm{N}(0,1)$.
2. $Y = (X_1^2 + [X_2X_3 - (X_2X_4)^{-1}]^2)^{1/2} + \varepsilon$; $X_1 \sim \mathrm{U}(0,100), X_2 \sim \mathrm{U}(40\pi, 560\pi), X_3 \sim \mathrm{U}(0,1), X_4 \sim \mathrm{U}(1,11)$; $\varepsilon \sim \mathrm{N}(0,125^2)$.
3. $Y = \tan^{-1}([X_2X_3 - (X_2X_4)^{-1}]/X_1) + \varepsilon$; $X_1 \sim \mathrm{U}(0,100), X_2 \sim \mathrm{U}(40\pi, 560\pi), X_3 \sim \mathrm{U}(0,1), X_4 \sim \mathrm{U}(1,11)$; $\varepsilon \sim \mathrm{N}(0,.1^2)$.
4. $Y = X_1X_2 + X_3^2 - X_4X_7 + X_8X_{10} - X_6^2 + \varepsilon$; $\{X_j\} \sim \mathrm{U}(-1,1)$; $\varepsilon \sim \mathrm{N}(0,.1^2)$.
5. $Y = 1\{X_1 > 0\} + X_2^3 + 1\{X_4 + X_6 - X_8 - X_9 > 1 + X_{10}\} + \exp(-X_2^2) + \varepsilon$; $\{X_j\} \sim \mathrm{U}(-1,1)$; $\varepsilon \sim \mathrm{N}(0,.1^2)$.
6. $Y = X_1^2 + 3X_2^2X_3\exp(-|X_4|) + X_6 - X_8 + \varepsilon$, $\{X_j\} \sim \mathrm{U}(-1,1)$; $\varepsilon \sim \mathrm{N}(0,.1^2)$.
7. $Y = 1\{X_1 + X_4^3 + X_9 + \sin(X_2X_8) + \varepsilon > 0.38\}$; $\{X_j\} \sim \mathrm{U}(-1,1)$; $\varepsilon \sim \mathrm{N}(0,.1^2)$.
8. $Y = \log(X_1 + X_2X_3) - \exp(X_4X_5^{-1} - X_6) + \varepsilon$; $\{X_j\} \sim \mathrm{U}(0.5,1)$; $\varepsilon \sim \mathrm{N}(0,.1^2)$.
9. $Y = X_1X_2^2|X_3|^{1/2} + \lfloor X_4 - X_5X_6 \rfloor + \varepsilon$; $\{X_j\} \sim \mathrm{U}(-1,1)$; $\varepsilon \sim \mathrm{N}(0,.1^2)$.
10. $Y = X_3(X_1+1)^{|X_2|} - (X_5^2[|X_4| + |X_5| + |X_6|]^{-1})^{1/2} + \varepsilon$; $\{X_j\} \sim \mathrm{U}(-1,1)$; $\varepsilon \sim \mathrm{N}(0,.1^2)$.
11. $Y = \cos(X_1 - X_2) + \sin^{-1}(X_1X_3) - \tan^{-1}(X_2 - X_3^2) + \varepsilon$; $\{X_j\} \sim \mathrm{U}(-1,1)$; $\varepsilon \sim \mathrm{N}(0,.1^2)$.
12. $Y = \varepsilon$; $\varepsilon \sim \mathrm{N}(0,1)$.

The dimension was set to $p = 20$ to add noise variables. Noise variables where $\mathrm{U}(0,1)$ for simulations 1–3; $\mathrm{U}(-1,1)$ for simulations 4–11; $\mathrm{N}(0,1)$ for simulation 12. Simulation 12 is a pure noise model. All features were simulated independently.

The sample size was $n = 250$. The subsample size was $m = n^{1/2}$, which in this case is $m = 16$. Practically speaking this is a very small sample size for VIMP but remember our goal is not estimation of VIMP but estimation of its standard error. The small m is advantageous and allows VIMP to be rapidly computed.

The bias and standardized mean-squared-error (SMSE) for each procedure was obtained using 250 replications. Bias was calculated by averaging the estimated value for $\sigma_n^{(v)}$ and taking the difference between this and the "true" value $\sigma_n^{(v)}$ approximated by Monte Carlo. MSE was estimated by averaging the squared difference between estimated and true value. SMSE equals MSE divided by the true value.

See Figure 17.3 for the results. To better understand performance differences, results are stratified by size of a variable's true VIMP (approximated by Monte Carlo). There are 6 groups for 10, 25, 50, 75, and 90th percentiles of true VIMP (standardized by the Y variance to make VIMP comparable across simulations).

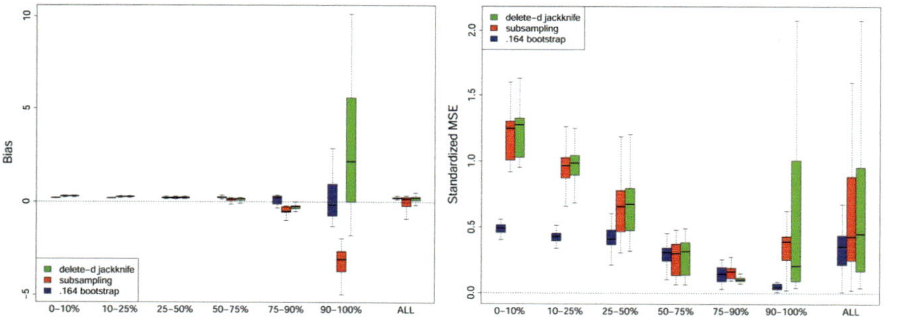

FIGURE 17.3
Bias and standardized mean-squared-error (SMSE) performance for estimating the VIMP standard error $\sigma_n^{(v)}$ from RF regression simulations. In total there are 240 variables (12 simulations, $p = 20$ variables in each simulation). These 240 variables have been stratified into 6 groups using 10, 25, 50, 75, and 90th percentiles of true standardized VIMP. Extreme right boxplots labeled "ALL" display performance for all 240 variables simultaneously.

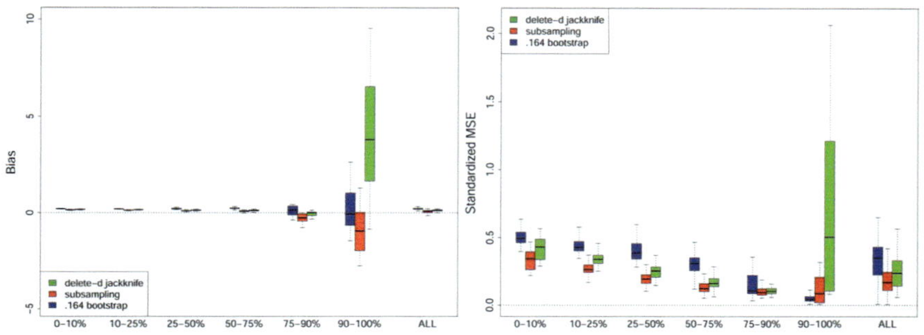

FIGURE 17.4
Experiments as in Figure 17.3 but with increased subsampling size $m = n^{3/4}$. Notice the improvement in bias and SMSE for the subsampling estimator relative to the delete-d estimator.

All methods exhibit a positive but small bias for small VIMP (i.e. weak signal or noisy variables). As VIMP increases (i.e. variables become more important), bias for the subsampling estimator $\hat{\sigma}_n^{(v)}$ increases and becomes negative, reflecting that standard error is underestimated for signal variables. The delete-d estimator $\hat{\sigma}_{J(d),n}^{(v)}$ does much better. This is due to the bias correction factor (17.9) which kicks in when signal increases. The pattern seen for bias is reflected in the results for SMSE: the delete-d is similar to the subsampling estimator except for large VIMP where it does better. In the figure we have also included results from a .164 bootstrap estimator introduced in Ishwaran and Lu [9]. It is more computationally demanding which is its main disadvantage. However, in this example the .164 estimator has the best performance.

In Figure 17.4 the previous experiments were repeated but with subsampling size increased from $m = n^{1/2}$ to $m = n^{3/4}$ ($m = 16$ to $m = 63$). Both $\hat{\sigma}_n^{(v)}$ and $\hat{\sigma}_{J(d),n}^{(v)}$ are improved. Also notice the improved bias and SMSE for the subsampling estimator, $\hat{\sigma}_n^{(v)}$. It performs the best of all three procedures. We observe that the .164 bootstrap estimator now has the worst performance.

17.3.4 Confidence intervals for VIMP

Subsampling can be used to obtain nonparametric confidence intervals for the target parameter $\theta(\mathbb{P})$ [13]. Define the $1-\alpha$ quantile for the subsampling distribution as $c_{n,m}(1-\alpha) = \inf\{x : \widetilde{U}_{n,m}(x) \geq 1-\alpha\}$. Similarly, define the $1-\alpha$ quantile for the limiting distribution $\mathbb{Q}$ of $n^{1/2}(\hat{\theta}_n - \theta(\mathbb{P}))$ as $c(1-\alpha) = \inf\{t : F(x) = \mathbb{Q}[-\infty, x] \geq 1-\alpha\}$. Then, assuming (17.6) and $m \ll n$, the interval

$$\left[\hat{\theta}_n - n^{-1/2} c_{n,m}(1-\alpha), \infty\right)$$

contains $\theta(\mathbb{P})$ with asymptotic probability $1-\alpha$ if $c(1-\alpha)$ is a continuity point of F.

Ishwaran and Lu [9] considered this procedure but found that a more stable solution could be obtained under the assumption of asymptotic normality. As before, let $\hat{\theta}_n^{(v)}$ be the limiting forest VIMP and let $\theta^{(v)}$ be the limiting population VIMP. Assuming asymptotic normality,

$$\frac{\hat{\theta}_n^{(v)} - \theta^{(v)}}{\sigma_n^{(v)}} \overset{d}{\rightsquigarrow} \mathrm{N}(0,1),$$

an asymptotic $100(1-\alpha)$ confidence region for $\theta^{(v)}$ is defined as

$$\hat{\theta}_n^{(v)} \pm z_{\alpha/2} \sigma_n^{(v)}$$

where $z_{\alpha/2}$ is the $1-\alpha/2$-quantile from a standard normal, $\mathbb{P}\{\mathrm{N}(0,1) \leq z_{\alpha/2}\} = 1-\alpha/2$.

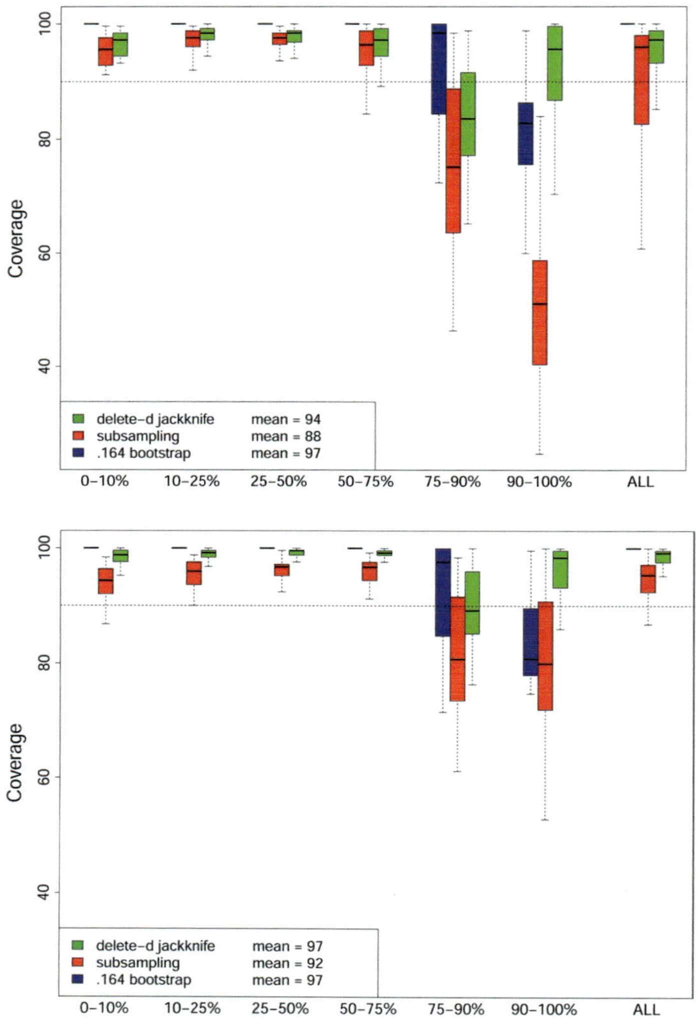

FIGURE 17.5

Coverage of 90% asymptotic normal confidence intervals for VIMP from RF regression simulations. Top and bottom figures use subsampling rates $m = n^{1/2}$ and $m = n^{3/4}$ respectively. Confidence regions for the 240 variables from the 12 simulation experiments have been stratified into 6 groups using 10, 25, 50, 75, and 90th percentiles of true VIMP values.

Coverage probabilities for 90% confidence intervals are provided for the previous simulations in Figure 17.5 (top and bottom panels correspond to Figure 17.3 and Figure 17.4 for subsample sizes $m = n^{1/2}$ and $m = n^{3/4}$). All procedures give confidence intervals that are too large when VIMP is small (reflected by coverage probabilities exceeding 90%). But this can be considered a useful property since over-estimating confidence intervals for noisy variables makes it less likely to select them. For larger VIMP, the subsampling estimator tends to produce intervals that are too small. As found earlier, this is because it tends to underestimate the standard error. The delete-d estimator performs much better. However, when the subsampling rate is increased (bottom panel), the subsampling estimator is generally superior to the delete-d estimator. Its overall mean coverage rate is 92 which is much better than the delete-d which achieves a coverage rate of 97. Overall it is also better than the .164 bootstrap estimator.

17.3.5 Illustration using Iowa housing data

To close out this section, we present an example using a real-world dataset. We use the "Iowa housing data," which consists of data collected from residential properties sold in Ames, Iowa, from 2006 to 2010 [4]. Similar to the famous Boston housing dataset, this is a regression problem where the goal is to predict the sales price of the home. However, in this problem, there are many more variables, and the dataset is much larger. In total, there are $p = 79$ variables measuring various aspects of the house and residential area. The data consists of $n = 2930$ homes. There are many missing values in the data, which also makes the analysis more challenging. To address this, we pre-imputed the data using the missForest procedure described in Chapter 16.14.

Figure 17.6 displays 95% confidence regions for permutation VIMP obtained from the RF analysis. We observe six variables with significantly large VIMP: overall quality of the home (measure of material and finish), living area (square feet), garage size (car capacity), basement area (total square feet), year built, and first-floor size (square feet). There are some intermediate values following these top six, where the confidence region excludes zero. Finally, there are many variables with non-significant VIMP. From this analysis, we can quickly understand which variables affect prediction accuracy: primarily the size of the house as measured by various attributes and the overall quality of the house in terms of its finishing.

17.4 VIMP for Class Imbalanced Data

Chapter 16.12 discussed the two-class imbalanced problem. We saw that imbalanced class frequencies favoring the majority class label have a very negative effect on prediction performance of machine learning methods using the Bayes decision rule. These same problems also manifest themselves for VIMP.

As an example, consider Figure 17.7 which displays VIMP from a RF classification analysis. The data used is a breast cancer dataset [16] where the goal is to classify patients as recurrent (R) and non-recurrent (N) using biopsy attributes (note this is a different subset of the data than Figure 15.1 from Chapter 15.2). The sample size is $n = 194$ with majority class frequency $N_0 = 148$ (non-recurrent class) and minority class frequency $N_1 = 46$ (recurrent class), an imbalanced ratio of 3.2. The OOB misclassification error is 0.24 but the RF classifier mostly favors classification into the majority class. Here the gmean is 0.21, where recall the gmean is the geometric mean of the TNR (true negative rate) and the TPR (true positive rate). A value of 0.21 is very low and typical of imbalanced problems using Bayes rule.

The problem also appears in VIMP since these values are obtained from 0-1 loss using Bayes rule. Using the notation of Chapter 17.2, Breiman-Cutler VIMP is

$$I(X^{(v)}, \mathcal{T}_b) = \frac{\sum_{i=1}^{n} I_{i,b} 1_{\{Y_i \neq \hat{\delta}_b^{(v,i)}\}}}{\sum_{i=1}^{n} I_{i,b}} - \frac{\sum_{i=1}^{n} I_{i,b} 1_{\{Y_i \neq \hat{\delta}_b^{(i)}\}}}{\sum_{i=1}^{n} I_{i,b}}$$

where $\hat{\delta}_b^{(i)} = \hat{\delta}_b^*(\mathbf{X}_i)$, $\hat{\delta}_b^{(v,i)} = \hat{\delta}_b^*(\tilde{\mathbf{X}}_{i,b}^{(v)})$ and

$$\hat{\delta}_b^*(\mathbf{x}) = \arg\max_{1 \le j \le J} \left\{ h_j^*(\mathbf{x}, \mathcal{T}_b) \right\}$$

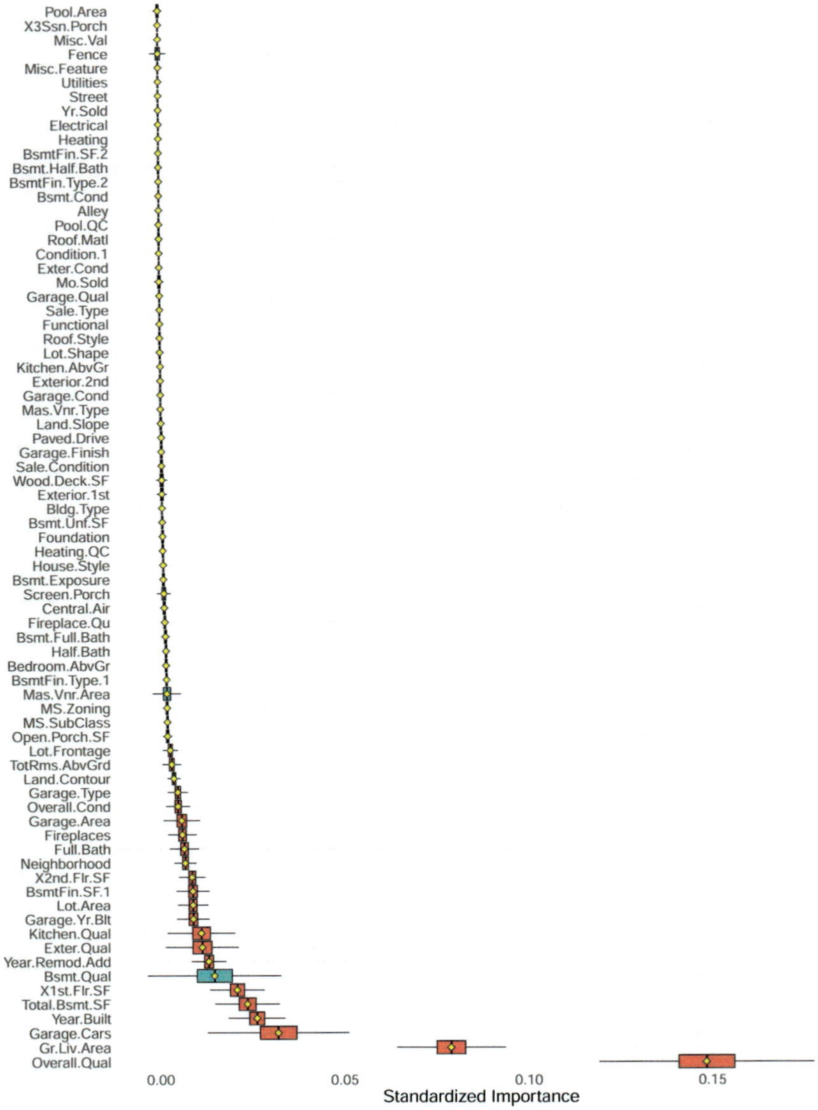

FIGURE 17.6
Confidence intervals for permutation VIMP from RF regression analysis of Iowa Housing data.

is the Bayes classifier for the in-bag estimator $h_j^*(\mathbf{x}, \mathcal{I}_b)$ of $\mathbb{P}\{Y = j|\mathbf{X} = \mathbf{x}\}$ (for generality, the definition provided here applies for $J \geq 2$).

The barplots at the top of Figure 17.7 labelled "all" give Breiman-Cutler importance. Notice how nearly all variables have non-zero importance values which suggests an underlying non-sparse model. Below this are conditional importance values for the non-recurrent and recurrent classes. These are calculated as before, but are conditioned so as to use only data from the respective class. Since in this example the RF classifier favors the majority class and classifies nearly all cases as from the majority class, this results in a positive VIMP for almost all variables for the non-recurrent cases, mirroring the unconditional VIMP, and negative values for the recurrent cases.

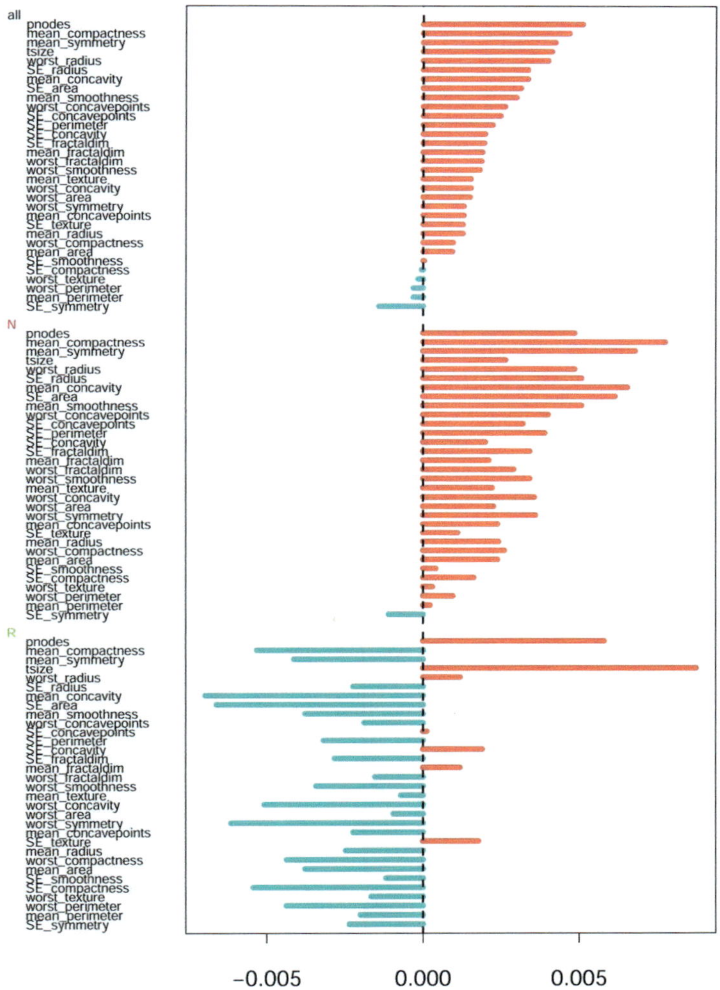

FIGURE 17.7
Breiman-Cutler permutation importance (VIMP) for Wisconsin breast cancer dataset (labeled "all" and given in the top panel). Values that have been conditioned on data for each class (N: non-recurrent, R: recurrent) are given in middle and bottom panels.

O'Brien and Ishwaran [11] suggested a different method for class imbalanced data by using gmean importance calculated using the q^*-classifier. Instead of using the in-bag Bayes classifier $\hat{\delta}_b^{(i)}$ with performance measured using misclassification error, they proposed using the in-bag q^*-classifier $\hat{\delta}_{q^*,b}^{(i)}$ with performance measured using the gmean. The proposed gmean VIMP is defined as

$$I(X^{(v)},\mathcal{T}_b) = \sqrt{\frac{\sum_{i=1}^{n} I_{i,b}1_{\{Y_i=\hat{\delta}^{(i)}_{q^*,b}=0\}}}{\sum_{i=1}^{n} I_{i,b}1_{\{Y_i=0\}}} \times \frac{\sum_{i=1}^{n} I_{i,b}1_{\{Y_i=\hat{\delta}^{(i)}_{q^*,b}=1\}}}{\sum_{i=1}^{n} I_{i,b}1_{\{Y_i=1\}}}} - \sqrt{\frac{\sum_{i=1}^{n} I_{i,b}1_{\{Y_i=\hat{\delta}^{(v,i)}_{q^*,b}=0\}}}{\sum_{i=1}^{n} I_{i,b}1_{\{Y_i=0\}}} \times \frac{\sum_{i=1}^{n} I_{i,b}1_{\{Y_i=\hat{\delta}^{(v,i)}_{q^*,b}=1\}}}{\sum_{i=1}^{n} I_{i,b}1_{\{Y_i=1\}}}}.$$

O'Brien and Ishwaran [11] also suggested a *blocking technique* where rather than using one tree for the calculation, a small group (say 10 trees) are used to calculate the q^*-classifier and its OOB gmean performance calculated analogously as above. See O'Brien and Ishwaran [11] for details.

In Figure 17.8 we have redone the previous analysis to compare Breiman-Cutler VIMP to gmean VIMP (using a block size of 10). There are clear differences in the two methods. By far, gmean concentrates on fewer variables and there are several variables with negative VIMP which is what we would expect since it is unreasonable to assume that all variables are significant as suggested by Breiman-Cutler VIMP. In our opinion, gmean VIMP performs much better. This general conclusion was also reached by O'Brien and Ishwaran [11].

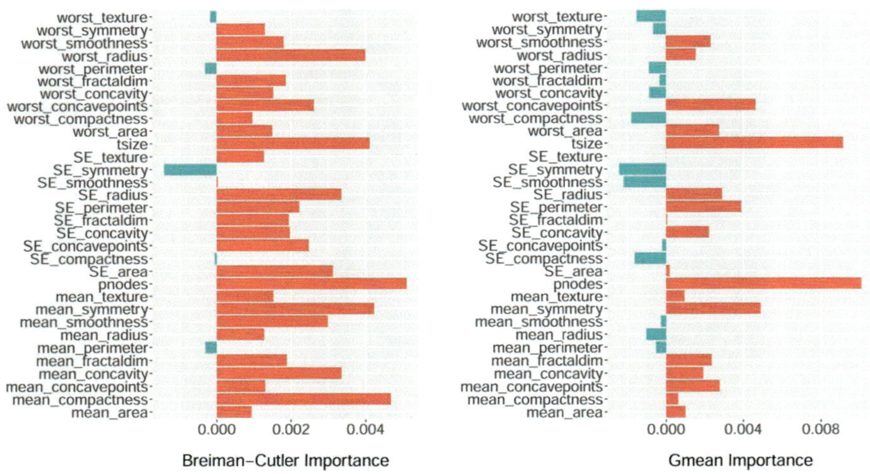

FIGURE 17.8
Gmean importance using q^*-classifier (right) versus Breiman-Cutler importance (left) using Bayes rule.

17.5 Minimal Depth

Ishwaran et al. [8] described a high-dimensional variable selection method using *minimal depth*, which measures the importance of a variable based on its split depth relative to the root node.

This differs from VIMP, as it does not use prediction error and is a simpler, non-randomized procedure using only generic tree-based concepts, making minimal depth easier to study. At the same time, this approach captures the essence of VIMP. Variables splitting close to the root node with a small depth value strongly affect prediction accuracy and thus VIMP. Conversely, variables splitting deeper in the tree with a larger depth value have a limited effect on prediction error, as terminal node assignments are not as perturbed. To formally define minimal depth, we introduce the notion of a *maximal subtree.*

Definition 17.1. The maximal subtree for a variable v is the largest subtree whose root node is split using v (i.e., no parent node of the subtree is split using v). The shortest distance from the root of the tree to the root of the closest maximal subtree of v is the minimal depth of v. This equals the distance at which v first splits the tree. A smaller minimal depth identifies a more predictive variable. □

Figure 17.9 shows a tree from an analysis involving cardiovascular disease. There are three variables: peak VO2 (red), BUN (green), and exercise time (orange). Maximal subtrees are indicated by color; node depth is indicated by an integer located in the center of a tree node. For example, the root node is split by exercise time; thus its maximal subtree is the entire tree and its minimal depth is 0. For BUN and peak VO2, there are two maximal subtrees on each side of the tree. The closest to the root node is on the left side for peak VO2 with minimal depth 1. For BUN, both subtrees have depth 2; its minimal depth is 2.

FIGURE 17.9
Illustration of minimal depth. The depth of a node, d, is the distance to the root node (depicted here at the bottom of the tree). Therefore, $d \in \{0, 1, \ldots, D\}$, where D is the depth of the tree, defined as the distance from the root node to the farthest terminal node. Here $D = 8$ and the first split at depth $d = 0$ (the root node) is on exercise time.

Denote the minimal depth for a variable v by D_v. We will derive the distribution of D_v to determine a threshold for selecting variables using minimal depth. Let $\pi_{v,j}(t)$ be the probability that v is one of the mtry candidate variables for splitting a node t at depth j, assuming v has not split the tree earlier. Let $\theta_{v,j}(t)$ be the probability that v splits a node t at depth j, given that v is a candidate for splitting t, and has not split the tree earlier. We assume the tree depth $D \geq 1$ is fixed and that the tree is balanced, i.e., $\ell_d = 2^d$, where ℓ_d is the number of non-terminal nodes at depth d.

Theorem 17.1. *Assume $\pi_{v,j}(t)$ and $\theta_{v,j}(t)$ depend only on the depth of the node t and not the node t itself. Then, for a balanced tree*

$$\mathbb{P}\left\{D_v = d \,\middle|\, \ell_0,\ldots,\ell_D\right\} = \left[\prod_{j=0}^{d-1}\left(1-\pi_{v,j}\theta_{v,j}\right)^{\ell_j}\right]\left[1-\left(1-\pi_{v,d}\theta_{v,d}\right)^{\ell_d}\right], \tag{17.10}$$

for $0 \leq d \leq D-1$, where $\pi_{v,j} := \pi_{v,j}(t)$, and $\theta_{v,j} := \theta_{v,j}(t)$.

In high-dimensional sparse settings, assuming v is noisy, we can derive a more straightforward distribution of minimal depth. We start with the following assumptions, which we will later explain.

Assumption 17.1. If v is a noisy variable assume that for each internal node t

(i) $\pi_{v,j}(t) = \pi_{v,j} = \text{mtry}/p$.

(ii) $\theta_{v,j}(t) = \theta_{v,j} = 1/\text{mtry}$.

□

Under Assumption 17.1, we have

$$\pi_{v,j}\theta_{v,j} = \frac{\text{mtry}}{p} \times \frac{1}{\text{mtry}} = \frac{1}{p}.$$

Substituting this into (17.10) yields

$$\mathbb{P}\left\{D_v = d \,\middle|\, \ell_0,\ldots,\ell_D, v \text{ is noisy}\right\} = \left[1-\left(1-\frac{1}{p}\right)^{\ell_d}\right]\prod_{j=0}^{d-1}\left(1-\frac{1}{p}\right)^{\ell_j}. \tag{17.11}$$

Note that the sum of probabilities (17.11) over $0 \leq d \leq D-1$ can be less than 1, so we must normalize them by

$$\begin{aligned}&\mathbb{P}\left\{D_v = D \,\middle|\, \ell_0,\ldots,\ell_D, v \text{ is noisy}\right\}\\ &\quad = 1 - \sum_{d=0}^{D-1}\mathbb{P}\left\{D_v = d \,\middle|\, \ell_0,\ldots,\ell_D, v \text{ is noisy}\right\}.\end{aligned}$$

The normalization is required to address the case when v is never split. When this happens, v is assigned the maximum depth possible, D.

The rationale for Assumption 17.1 can be explained as follows. Consider first $\pi_{v,j}(t)$. This equals 1 minus the probability that v is not one of the mtry candidate variables for splitting t, given that v has not been split on yet. Since a RF tree uses random feature selection by selecting mtry variables at random for each node, if p is large and $\text{mtry}/p = o(1)$, $\pi_{v,j}(t)$ can be approximated by

$$\pi_{v,j} = 1 - \prod_{k=0}^{\text{mtry}-1}\left(1-\frac{1}{p-k}\right) \asymp \frac{\text{mtry}}{p}.$$

This holds regardless of whether v is signal or noisy and is independent of the depth j and the node t. Thus, we arrive at Assumption 17.1(i).

Now consider $\theta_{v,j}(t)$. Let $0 < \tau_0 < 1$ be the fraction of signal variables. We assume that the probability t is split by a signal variable is $S \geq 1$ times larger than the probability that a noisy variable splits t,

$$\begin{aligned} \theta_0 &= \mathbb{P}\{v \text{ splits } t | v \text{ is noisy, } v \text{ is a candidate}\} \\ S\theta_0 &= \mathbb{P}\{v \text{ splits } t | v \text{ is signal, } v \text{ is a candidate}\}. \end{aligned}$$

Let $\mathscr{V}(t)$ be the set of mtry variables for splitting t. Using a conditioning argument, we can deduce that

$$1 = \sum_{v \in \mathscr{V}(t)} \mathbb{P}\{v \text{ splits } t\} = \sum_{v \in \mathscr{V}(t)} \Big[\theta_0(1-\tau_0) + S\theta_0\tau_0\Big].$$

By exchangeability, we can assume that $\theta_{v,j} = \theta_0$. Thus,

$$\theta_{v,j}\Big[(1-\tau_0) + S\,\tau_0\Big] = \frac{1}{\text{mtry}}.$$

Assuming a sparse setting in which $\tau_0 \ll 1$, we obtain the approximation

$$\theta_{v,j} = \frac{1}{\text{mtry}(1-\tau_0) + S\,\text{mtry}\tau_0} = \frac{1}{\text{mtry}}\Big(1 - \tau_0(S-1) + o(S\tau_0)\Big) \asymp \frac{1}{\text{mtry}}.$$

This is Assumption 17.1(ii).

Thus, from the above, Assumption 17.1 can reasonably be expected to hold in high-dimensional sparse settings for noisy variables. Therefore, (17.11) can be used to determine a threshold for selecting variables in high dimensions. Ishwaran et al. [8] proposed using the mean of this distribution to identify weak and signal variables, terming the method *minimal depth thresholding.*

Select v if its tree averaged minimal depth is less than or equal to the mean minimal depth $\mathbb{E}(D_v)$ under the null distribution (17.11). Call this variable selection method, minimal depth thresholding.

Proof of Theorem 17.1. By the definition of $\pi_{v,j}$ and $\theta_{v,j}$, if t is a node of depth $j \leq d$, then

$$\mathbb{P}\{v \text{ does not split } t | D_v \geq j\} = 1 - \pi_{v,j}\theta_{v,j}.$$

Furthermore, $\mathbb{P}\{D_v = 0\} = \pi_{v,0}\theta_{v,0}$. Therefore, the probability that no maximal v-subtree exists at depth less than $d \geq 1$ is

$$\begin{aligned} \mathbb{P}\{D_v \geq d\} &= \mathbb{P}\{D_v \geq 1, D_v \geq 2, \ldots, D_v \geq d\} \\ &= \mathbb{P}\{D_v \geq 1\} \prod_{j=1}^{d-1} \mathbb{P}\{D_v \geq j+1 | D_v \geq j\} \\ &= \mathbb{P}\{D_v \geq 1\} \prod_{j=1}^{d-1} \mathbb{P}\{v \text{ is not split at depth } j | D_v \geq j\} \\ &= \Big[1 - \mathbb{P}\{D_v = 0\}\Big] \prod_{j=1}^{d-1} \mathbb{P}\{v \text{ is not split at depth } j | D_v \geq j\} \\ &= \prod_{j=0}^{d-1} \big(1 - \pi_{v,j}\theta_{v,j}\big)^{\ell_j}, \end{aligned} \tag{17.12}$$

where $\ell_j = 2^j$ equals the total number of nodes of depth j. Given $D_v \geq d$, the probability that v splits a node of depth d is one minus the probability that each node of depth d is split on some other variable than v. This probability is

$$1-\left(1-\pi_{v,d}\theta_{v,d}\right)^{\ell_d}. \tag{17.13}$$

Using

$$\mathbb{P}\{D_v = d\} = \mathbb{P}\{v \text{ is split at depth } d|D_v \geq d\} \times \mathbb{P}\{D_v \geq d\},$$

the result for $d \geq 1$ follows upon multiplying (17.12) and (17.13) (the case $d = 0$ holds from $\mathbb{P}\{D_v = 0\} = \pi_{v,0}\theta_{v,0}$). □

17.5.1 Estimating the minimal depth threshold

The minimal depth distribution (17.11) depends on unknown parameters ℓ_d and D which must be estimated. There are two simple methods for doing so. The first method uses tree-averaged estimates for the unknown parameters. Thus, if $\overline{\ell}_d$ and $\overline{D}$ are the tree-averaged estimates of ℓ_d and D, the tree averaged minimal depth threshold is

$$\mathbb{E}\left[D_v \middle| \overline{\ell}_0, \ldots, \overline{\ell}_{\overline{D}}, v \text{ is noisy}\right]. \tag{17.14}$$

A second approach is to take a tree average, where for each tree, the mean of (17.11) is calculated by substituting tree values for ℓ_d and D. Thus, using a subscript of b to indicate dependence on the tree, the tree averaged minimal depth threshold is

$$\frac{1}{B}\sum_{b=1}^{B}\mathbb{E}\left[D_v \middle| \ell_{0,b}, \ell_{1,b} \ldots, v \text{ is noisy}\right]. \tag{17.15}$$

Generally (17.14) and (17.15) are roughly the same, although (17.15) tends to be slightly larger.

Minimal depth thresholding can sometimes be conservative. For a more liberal threshold, and larger sized models, a "one standard error" threshold rule can be used. Let $\mathbb{E}_{v,B}$ denote the expectation under the conditional distribution for the forest and let $\sigma^2_{v,B}$ be the variance. The one standard error threshold is defined as

$$\mathbb{E}_{v,B}(D_v) + \frac{\sigma_{v,B}}{\sqrt{n}}. \tag{17.16}$$

17.5.2 Benchmark analysis of minimal depth

A large benchmark analysis was used to study the performance of minimal depth. In total, 37 regression datasets of differing dimensions p and sample sizes n were used (Table 17.1), which included both real data (in capitals) and synthetic data (in lower case). Many of the synthetic datasets were obtained from the `mlbench` R-package (e.g., datasets listed in Table 17.1 starting with "frdmn" are the class of Friedman simulations included in the package). For each dataset, we ran a RF-R analysis using parameters $(\text{mtry}, \text{nodesize}, B) = (p/3, 5, 1000)$. This was followed by an additional RF-R analysis using only variables found significant under minimal depth thresholding. Two separate analyses were run using threshold values (17.15) and (17.16). We refer to the first (standard) analysis as RF-R and the second and third (using minimal depth) by RF-MD and RF-1se. Test set prediction error was estimated by 10-fold cross-validation.

From Table 17.1 we can draw the following general conclusions:

1. RF-MD and RF-1se reduce dimension while maintaining low error rates.
2. RF-MD tends to be sparser, especially in the higher-dimensional problems.

TABLE 17.1
Test set prediction error for regression benchmark data. RF-MD and RF-1se are RF-R using minimal depth selection under thresholds (17.15) and (17.16). Model sizes for RF-MD and RF-1se are listed under $\hat{p}_{md}$ and $\hat{p}_{1se}$.

	n	p	$\hat{p}_{\text{md}}$	$\hat{p}_{\text{1se}}$	RF-R	RF-MD	RF-1se
Air	111	5	3	3	0.21	0.21	0.21
Automobile	193	24	9	10	5.19	5.54	5.36
Bodyfat	252	13	4	6	22.54	23.44	22.97
Boston	506	13	6	6	12.39	13.55	13.19
CMB	899	4	3	3	11.79	11.58	11.61
Crime	47	15	4	5	88634.02	99240.79	93801.49
Diabetes	442	10	4	4	3157.69	3334.62	3344.74
DiabetesI	442	64	22	49	3300.07	3300.21	3320.76
Fitness	31	6	1	2	18.44	11.50	16.28
Highway	39	11	3	6	0.20	0.23	0.20
Iowa	33	9	3	6	111.49	105.61	107.60
Ozone	203	12	6	8	17.41	17.42	17.72
Pollute	60	15	4	8	1955.87	2082.34	1941.90
Prostate	97	8	2	5	0.59	0.65	0.62
Servo	167	19	10	13	0.44	0.51	0.52
Servofactor	167	4	2	3	0.43	1.04	0.75
Tecator	215	22	7	12	27.78	33.54	27.08
Tecator2	215	100	30	44	59.15	53.43	54.14
Windmill	1114	12	5	5	4.44	4.82	4.76
expon	250	2	1	2	5.70	11.96	7.74
expon.noise	250	17	4	10	6.55	5.53	6.44
frdmn1	250	10	4	4	6.25	5.77	5.55
frdmn1.bigp	250	250	36	157	11.41	9.80	11.06
frdmn2	250	4	2	2	24235.94	21379.75	21025.05
frdmn2.bigp	250	254	8	48	56929.30	42400.25	52433.19
frdmn3	250	4	2	3	0.02	0.03	0.03
frdmn3.bigp	250	254	52	237	0.06	0.05	0.06
peak	250	20	6	19	15.37	17.87	15.62
peak.bigp	250	220	20	20	15.00	14.36	14.32
noise	250	500	29	86	1.20	1.28	1.24
sine	250	2	1	2	14.00	35.72	14.00
sine.noise	250	5	2	3	25.25	39.42	38.69
AML	116	629	14	19	71.43	83.34	76.79
DLBCL	240	740	211	468	70.14	69.67	70.29
Lung	86	713	9	11	4.45	4.42	4.21
MCL	92	881	10	10	51.80	58.90	57.39
VandeVijver	78	475	9	10	7.11	8.95	8.48

17.6 VarPro a New Rule-Based Approach to Variable Selection

So far we have focused on variable selection using RF, but in this section we describe a new approach based on the work of Lu and Ishwaran [10] that can be used with any rule-based learning method.

The notational framework is as follows. Let $Y \in \mathscr{Y}$ be the response variable and $\mathbf{X} = (X^{(1)},\dots,X^{(p)})^T \in \mathscr{X}$ the covariate vector. We consider the setting where the analyst is interested in some prechosen function $\psi(\mathbf{X}) = \mathbb{E}[g(Y)|\mathbf{X}]$ equal to the conditional expectation of a function g of Y. For example, in regression, $\psi(\mathbf{X}) = \mathbb{E}[Y|\mathbf{X}]$ and $g(Y) = Y$. For classification, $\psi(\mathbf{X}) = \mathbb{P}\{Y = j|\mathbf{X}\}$ for $j = 1,\dots,J$ and $g(Y) = 1_{\{Y=j\}}$.

The goal is to identify variables of importance for the functional of interest ψ. We call the minimal set of variables $\mathscr{S} \subset \{1,\dots,p\}$ that ψ depends on the "signal variables" and the task is to identify this set. A more formal definition is as follows. Write $\mathbf{X}^{(S)} = \{X^{(j)}\}_{j\in S}$ for $\mathbf{X}$ restricted to coordinates $j \in S$ and $\mathbf{X}^{\setminus(S)} = \{X^{(j)}\}_{j\notin S}$ for coordinates not in S.

Definition 17.2. *$\mathscr{S} \subset \{1,\dots,p\}$ is the set of signal variables if $\mathscr{S}$ is the minimal set of coordinates that ψ depends on. Thus, $\mathscr{S}$ is the smallest subset of coordinates satisfying $\mathbb{P}\{\psi(\mathbf{X}) = \psi(\mathbf{X}^{(\mathscr{S})})\} = 1$ such that $\mathbb{P}\{\psi(\mathbf{X}^{(\mathscr{S})}) = \psi(\mathbf{X}^{(S')})\} = 0$ for every S' where $\mathscr{S} \not\subseteq S'$. The complementary set $\mathscr{N} = \{1,\dots,p\} \setminus \mathscr{S}$ is unrelated to ψ and therefore contains the noisy variables. If $\mathscr{S} = \emptyset$, then $\psi = \mathbb{E}(g(Y))$ is constant and $\mathscr{N} = \{1,\dots,p\}$. However, we rule this trivial case out and always assume $\mathscr{S} \neq \emptyset$.* □

The definition of $\mathscr{S}$ specifies that it is the smallest set of coordinates such that $\psi(\mathbf{x}) = \psi(\mathbf{x}^{(\mathscr{S})})$ for almost all $\mathbf{x}$. Furthermore, there is no other set $S' \neq \mathscr{S}$ (where $\mathscr{S} \not\subseteq S'$) for which $\psi(\mathbf{X}) = \psi(\mathbf{X}^{(S')})$ holds with non-zero probability. Another way to view Definition 17.2 is that it implies a type of conditional independence for noise variables. For example, if $p = 2$ and $\mathscr{S} = \{1\}$, then since $X^{(2)}$ is a noise variable

$$\mathbb{E}(g(Y)|X^{(1)}, X^{(2)}) = \mathbb{E}(g(Y)|X^{(1)}). \tag{17.17}$$

The relationship (17.17) is weaker than the usual conditional independence assumption used for variable selection (see (17.18) below) since it depends upon the choice of g and ψ. For example, in classification, if the analysts chooses $g(Y) = I\{Y = c_l\}$ for a specific class label c_l of interest, (17.17) implies a conditional independence of $X^{(2)}$ for class label c_l, but not necessarily for other class labels. Thus, variables affecting relapse of cancer may be different than those affecting absence or death due to cancer.

The rule-based approach to be discussed works with the functional ψ, which is obtained by integration with respect to the conditional distribution of Y given $\mathbf{X}$, but an often used method for variable selection is to directly use the conditional distribution. This assumes a conditional independence between Y and the noise variables given the signal variables:

$$Y \perp \mathbf{X}^{\setminus(\mathscr{S})}|\mathbf{X}^{(\mathscr{S})}. \tag{17.18}$$

Let $S \subset \{1,\dots,p\}$ be a set of variables of interest where the goal is to determine if S contains signal variables. The strategy used is to construct a test statistic $\hat{\theta}_n(S)$ using augmented features $(\mathbf{X}^{(S)},\mathbf{X}^{\setminus(S)},\tilde{\mathbf{X}}^{(S)},\tilde{\mathbf{X}}^{\setminus(S)})$, where $\tilde{\mathbf{X}}^{(S)}$ and $\tilde{\mathbf{X}}^{\setminus(S)}$ are new artificial features. The test is designed so that, due to conditional independence, $\hat{\theta}_n(S)$ is statistically non-significant if S contains only noise variables.

Two popular approaches using this idea include permutation variable importance, previously discussed in Chapter 17.2, and simulation according to the distribution of $\mathbf{X}$ to obtain $(\tilde{\mathbf{X}}^{(S)},\tilde{\mathbf{X}}^{\setminus(S)})$. This latter method is called *knockoffs* [3]. One of the problems with VIMP (to be illustrated) is that the permuted sample does not have the same distribution as **X** which can lead to flawed variable selection in some settings. Knockoffs [3] innovatively takes a different approach by simulating the artificial data to satisfy

$$(\mathbf{X}^{(S)}, \mathbf{X}^{\setminus(S)}, \tilde{\mathbf{X}}^{(S)}, \tilde{\mathbf{X}}^{\setminus(S)}) \stackrel{d}{=} (\mathbf{X}^{(S)}, \tilde{\mathbf{X}}^{\setminus(S)}, \tilde{\mathbf{X}}^{(S)}, \mathbf{X}^{\setminus(S)}). \tag{17.19}$$

This is used to compute a knockoff statistic for filtering variables. Using (17.18), the knockoff test statistic can achieve a desired false discovery level. This avoids the problems of permutation importance, however, this relies on strong assumptions about the parent distribution and achieving (17.19) may be difficult.

17.6.1 A new rule-based approach

Lu and Ishwaran [10] approached the problem from a different perspective, introducing a new procedure called Variable Priority (VarPro) that performs variable selection by using rules. This works with any machine learning method that produces decision boundaries, or which cuts the space into identifiable regions. Let S represent a set of variables the analyst wants to test. Given a rule ζ, VarPro calculates a sample-averaged estimator $\hat{\theta}_n(\zeta)$ for the target functional ψ of interest using the data in ζ's *region*. Then, a *released rule* ζ^S is constructed by removing any constraints on the indices in S, and its sample-averaged estimator $\hat{\theta}_n(\zeta^S)$ is calculated over the *released region*. $\hat{\theta}_n(\zeta^S)$ is then contrasted with $\hat{\theta}_n(\zeta)$ to measure the importance of S.

Many existing methodologies for variable importance rely on resampling or refitting models, which can introduce finite sample bias, or they use artificial data. The merit of VarPro is that it avoids these issues by calculating its estimators directly from the original data. By releasing the coordinates on S with the released rule, under certain conditions, it follows that $\lim_{n\to\infty} |\hat{\theta}_n(\zeta) - \hat{\theta}_n(\zeta^S)| \xrightarrow{\text{p}} 0$ for noisy variables S. This makes it possible to consistently select variables using the VarPro importance estimator.

A key aspect of VarPro is the idea of the region of a rule ζ. A region is a subset of the feature space obtained by a function R that maps $\zeta \mapsto \mathscr{X}$. In many cases, ζ can be associated with a series of univariate rules. The region mapped by R in this case is denoted by $R(\zeta) = \{\mathbf{x} \in \mathscr{X} : x^{(1)} \in R(\zeta^{X^{(1)}}), \dots, x^{(p)} \in R(\zeta^{X^{(p)}})\}$, where $\zeta^{X^{(j)}}$ denotes the univariate rule for feature j. For example, if all the features are continuous, we can imagine a rule with region $R(\zeta) = \{\mathbf{x} \in \mathbb{R}^p : a_1 \le x^{(1)} \le b_1, \dots, a_p \le x^{(p)} \le b_p\}$. Boundaries like this naturally arise in methods such as trees.

In order to construct the VarPro importance estimator we define a released rule and released region.

Definition 17.3. *Let $R(\zeta) \subset \mathscr{X}$ be a region of the feature space. To check the importance of the variables $\mathbf{X}^{(S)}$ to the rule ζ, we introduce the concept of a released region $R(\zeta^S)$ for the released rule ζ^S obtained by removing the dependence of ζ on the coordinates $\mathbf{X}^{(S)}$:*

$$R(\zeta^S) = \{\boldsymbol{x} \in \mathscr{X} : \boldsymbol{x}^{\setminus(S)} \in R(\zeta)\}.$$

In other words, the released region $R(\zeta^S)$ is the set of all $\boldsymbol{x}$ whose S-coordinate values are unconstrained but with non-S-coordinate values that lie in $R(\zeta)$. In particular, $R(\zeta) \subseteq R(\zeta^S)$. □

Example 17.1. In this example, shown in Figure 17.10, the feature space is $\mathscr{X} \subseteq \mathbb{R}^2$ and the rule ζ corresponds to the branch of a tree using splits $x^{(1)} \le -0.7$, $x^{(2)} \ge -0.8$, $x^{(2)} \le 0.7$ and $x^{(1)} \ge -1.95$. We can think of ζ as a product of indicator functions

$$\zeta = I\{x^{(1)} \le -0.7\} I\{x^{(2)} \ge -0.8\} I\{x^{(2)} \le 0.7\} I\{x^{(1)} \ge -1.95\}.$$

The region $R(\zeta)$ for ζ is the blue rectangle in Figure 17.10 (A):

$$R(\zeta) = \{(x^{(1)}, x^{(2)}) : -1.95 \le x^{(1)} \le -0.7, -0.8 \le x^{(2)} \le 0.7\}.$$

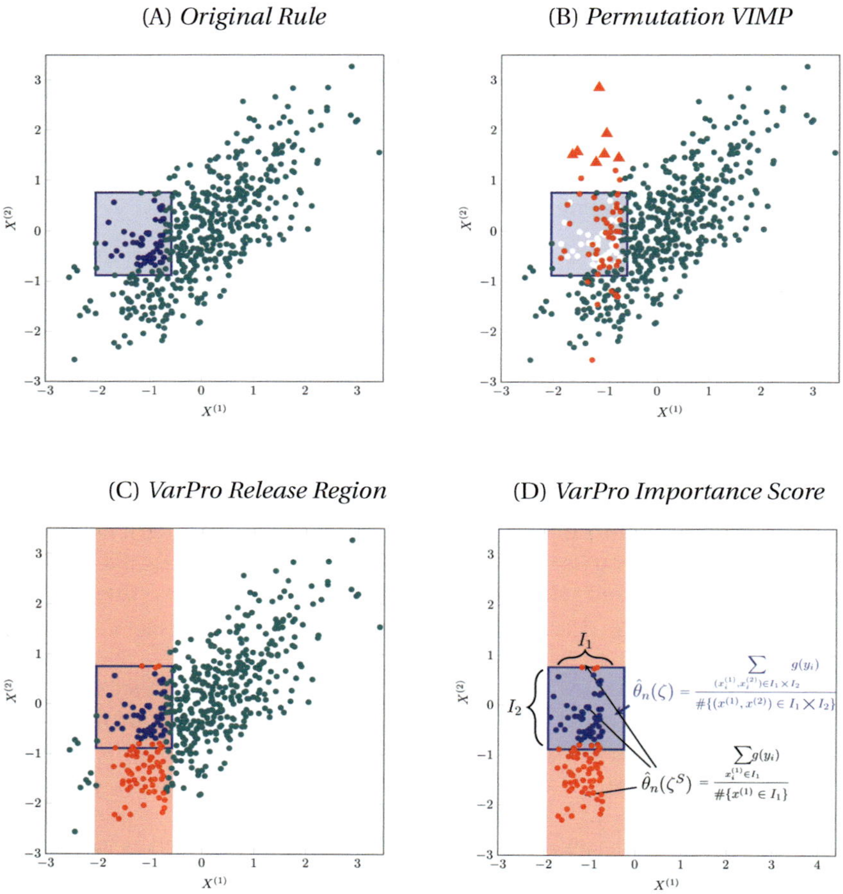

FIGURE 17.10

Two-dimensional illustration of how VarPro differs from artificial data methods. (A) The two-dimensional region for ζ is a rectangle. The data of interest are marked in blue. (B) Permutation variable importance (VIMP) for $X^{(2)}$. The data has been permuted along $X^{(2)}$ and data marked in red with triangles identify values that do not match the joint distribution of $\mathbf{X}$. The model-based predicted values $\tilde{y} := \tilde{y}(x^{(1)}, \tilde{x}^{(2)})$ for these artificial points are extrapolated from a region of the feature space that could be from potentially different responses. (C) VarPro release region for $X^{(2)}$. The original rule is modified to the S-released rule ζ^S (where $S = \{2\}$) shown using a pink background color. (D) VarPro importance score is defined using the estimator calculated using observed data values in blue compared to the estimator where the new released values in red are additionally used. No artificial data is created.

Consider testing whether $X^{(2)}$ is a noisy variable. Then by Definition 17.3, we set $S = \{2\}$ and obtain the released region by releasing the dependence on coordinate $X^{(2)}$: $R(\zeta^S) = \{(x^{(1)}, x^{(2)}) : -1.95 \le x^{(1)} \le -0.7\}$. The released region is therefore the rectangle with sides for the release coordinates removed (see the pink region in Figure 17.10 (C)). A unique aspect of working with trees, is that the released rule is itself a tree branch equal to the original rule altered such that whenever a binary decision is made on a variable in S, the decision is always 1. In this example, $\zeta^S = I\{x^{(1)} \le -0.7\}I\{x^{(1)} \ge -1.95\}$. □

Example 17.2. The following example shows that Definition 17.3 applies to rules other than from trees using axis-aligned cuts. Shown in Figure 17.11 (left panel) is an ellipse which corresponds to the region $R(\zeta)$ for a rule ζ. According to Definition 17.3, in order to release the region along $X^{(2)}$, we take the set of all $\mathbf{x}$ with coordinates $X^{(1)}$ inside the ellipse while $X^{(2)}$ is left unconstrained. Therefore, the released region $R(\zeta^S)$ is a box that touches the left and right sides of the ellipse (see right side of Figure 17.11). Note that unlike the previous example, the released rule ζ^S may not correspond to a rule generated by the original machine learning procedure. However, this is not necessary for the theory to work. □

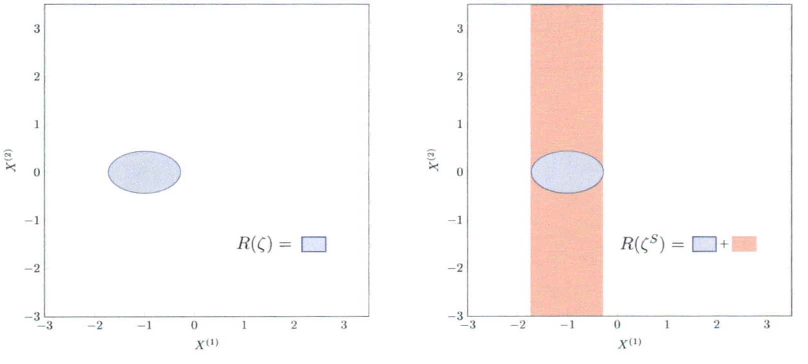

FIGURE 17.11
Left panel displays the elliptical region $R(\zeta)$ for a rule ζ. Right panel is the released region $R(\zeta^S)$ for $S = \{2\}$ from releasing $R(\zeta)$ along coordinate $X^{(2)}$.

17.6.2 VarPro estimator

The VarPro estimator can be described as follows. Let $\{(\mathbf{X}_i, Y_i)\}_{i=1}^n \in \mathscr{X} \times \mathscr{Y}$ be the i.i.d. learning data sampled from $\mathbb{P}$. Observations Y_i for $\mathbf{X}_i \in R(\zeta)$ are used to estimate the conditional mean of $g(Y)$ in $R(\zeta)$ via

$$\hat{\theta}_n(\zeta) = \frac{1}{\#\{\mathbf{X}_i \in R(\zeta)\}} \sum_{\mathbf{X}_i \in R(\zeta)} g(Y_i). \tag{17.20}$$

The released region $R(\zeta^S)$ is used to obtain the released estimator

$$\hat{\theta}_n(\zeta^S) = \frac{1}{\#\{\mathbf{X}_i \in R(\zeta^S)\}} \sum_{\mathbf{X}_i \in R(\zeta^S)} g(Y_i). \tag{17.21}$$

Then, the VarPro importance of S is defined to be the absolute difference between (17.20) and (17.21), $|\hat{\theta}_n(\zeta^S) - \hat{\theta}_n(\zeta)|$.

We return to our previous example in Figure 17.10 to give some intuition for the VarPro estimator by contrasting it with permutation importance (VIMP). Recall that $R(\zeta)$ is the blue rectangle in (A) and the released region $R(\zeta^S)$ is the pink area in (C). Panel (B) displays the data after permutation on coordinate $X^{(2)}$. Permuted data $(x^{(1)}, \tilde{x}^{(2)}) \in R(\zeta^S)$ are marked in red. Notice that in some cases (indicated by triangles), these data deviate strongly from the true distribution of $(X^{(1)}, X^{(2)})$. This creates problems for VIMP because VIMP calculates the importance of $X^{(2)}$ by comparing the test statistic calculated using the observed data in $R(\zeta)$ (blue points in (A)) to that calculated using the permuted data in $R(\zeta^S)$ (red points in (B)).

In a regression setting, where the goal is to estimate the conditional mean, this corresponds to averaging observations y_i for $i \in R(\zeta)$ and comparing them to averaged estimated values $\tilde{y}_i := \tilde{y}_i(x_i^{(1)}, \tilde{x}_i^{(2)})$. Unfortunately, since $\tilde{y}_i$ has to be model estimated (because it has to estimate ψ for features not in the training data), this produces atypical $\tilde{y}_i$ if $\tilde{x}_i^{(2)}$ is from a different region of the data space than the original data (like the red triangular points). This can result in large VIMP even when $X^{(2)}$ is a noisy variable.

The problem with VIMP is that the learner is applied to artificially created data that can deviate from the feature distribution. To avoid this, VarPro takes a direct approach by using only the original training data. Estimates are obtained for the conditional mean using observed values $g(y_i)$ via (17.20) and (17.21), where for the example above, $g(y) = y$. As shown in figure (D), $\hat{\theta}_n(\zeta)$ is calculated from observations in the blue box (blue points), which is compared to $\hat{\theta}_n(\zeta^S)$ calculated using the data in the pink rectangle (blue and red points), yielding the absolute difference

$$\left|\hat{\theta}_n(\zeta^S) - \hat{\theta}_n(\zeta)\right| = \left|\frac{\sum_{x_i^{(1)} \in I_1} y_i}{\#\{x^{(1)} \in I_1\}} - \frac{\sum_{(x_i^{(1)}, x_i^{(2)}) \in I_1 \times I_2} y_i}{\#\{(x^{(1)}, x^{(2)}) \in I_1 \times I_2\}}\right|,$$

where $I_1 = [-1.95, -0.7]$ and $I_2 = [-.8, 0.7]$. If $X^{(1)}$ is a signal feature, but $X^{(2)}$ is a noisy variable, then the two averages should converge to nearly the same value because $\psi(\mathbf{X}) = \mathbb{E}(Y|\mathbf{X}) = \psi(X^{(1)})$ depends only on $X^{(1)}$. On the other hand, if $X^{(2)}$ is a signal feature, then the difference will not necessarily be zero.

Releasing a rule ζ along noisy coordinates does not change ψ, therefore the difference between the average of g in the rule's region and its released region will be asymptotically the same, and subsequently we can expect a zero importance. For signal variables, the opposite happens, and since ψ changes along the released direction, we can expect a difference in average g values in the released and non-released regions, and therefore we can expect a non-zero importance value asymptotically.

17.6.3 Connection of VarPro to permutation tests

To further illustrate the dangers of using artificial data and how VarPro avoids this problem, we show that the VarPro estimator can be written as a permutation test. We contrast this with model-based importance to explain the differences.

Without loss of generality, let us assume that the coordinates of $\mathbf{x}$ have been reordered so that $\mathbf{x} = (\mathbf{x}^{(S)}, \mathbf{x}^{\setminus(S)})$. Consider all possible permutations of the training data $\{(y_i, \mathbf{x}_j^{(S)}, \mathbf{x}_i^{\setminus(S)})\}_{1 \le i,j \le n}$.

Then it can be shown [10] that

$$\frac{\sum_{i=1}^n g(y_i) 1_{\{\mathbf{x}_i \in R(\zeta^S)\}}}{\sum_{i=1}^n 1_{\{\mathbf{x}_i \in R(\zeta^S)\}}} = \frac{\sum_{i=1}^n \sum_{j=1}^n g(y_i) 1_{\{(\mathbf{x}_j^{(S)}, \mathbf{x}_i^{\setminus(S)}) \in R(\zeta)\}}}{\sum_{i=1}^n \sum_{j=1}^n 1_{\{(\mathbf{x}_j^{(S)}, \mathbf{x}_i^{\setminus(S)}) \in R(\zeta)\}}}. \tag{17.22}$$

The left-hand side is the VarPro released estimator $\hat{\theta}_n(\zeta^S)$ described in (17.21), whereas the right-hand side is an estimator using the permuted data.

Obviously the permuted estimator on the right side of (17.22) is impractical since it involves $O(n^2)$ calculations and would be an inefficient means for calculating $\hat{\theta}_n(\zeta^S)$. However, its use will be in describing why permutation VIMP generally does not satisfy an identity like (17.22) which will then highlight the main problem.

(17.22) is due to the following identity (see [10] for further details). Let $\zeta^{\setminus(S)}$ be the rule that relaxes the constraints of ζ on the indices not in S. If ζ is a rule expressible as $\zeta = \prod_{s=1}^p 1_{\{x^{(s)} \in I_s\}}$ where $I_s \subseteq \mathbb{R}$ are real-valued intervals (as in the example of Figure 17.10), then

$$1_{\{\mathbf{x} \in R(\zeta)\}} = 1_{\{\mathbf{x}^{\setminus(S)} \in R(\zeta)\}} 1_{\{\mathbf{x}^{(S)} \in R(\zeta)\}} = 1_{\{\mathbf{x} \in R(\zeta^S)\}} 1_{\{\mathbf{x} \in R(\zeta^{\setminus(S)})\}}. \tag{17.23}$$

The right side follows by Definition 17.3 since $\mathbf{x} \in R(\zeta^S)$ implies $\mathbf{x}^{\setminus(S)} \in R(\zeta)$ and $\mathbf{x} \in R(\zeta^{\setminus(S)})$ implies $\mathbf{x}^{(S)} \in R(\zeta)$.

Now consider how VIMP would be applied. Recall that unlike VarPro, VIMP uses a model-based estimator to estimate ψ. Let ψ_n be this estimator. For example, this could be the ensemble estimator from a random forest analysis or a boosted tree estimator using gradient boosting. Then using the permuted data as above, the VIMP estimator is

$$\begin{aligned}
\tilde{\theta}_{\text{VIMP}}(\zeta^S) &= \frac{\sum_{i=1}^n \sum_{j=1}^n \psi_n(\mathbf{x}_j^{(S)}, \mathbf{x}_i^{\setminus(S)}) 1_{\{(\mathbf{x}_j^{(S)}, \mathbf{x}_i^{\setminus(S)}) \in R(\zeta)\}}}{\sum_{i=1}^n \sum_{j=1}^n 1_{\{(\mathbf{x}_j^{(S)}, \mathbf{x}_i^{\setminus(S)}) \in R(\zeta)\}}} \\
&= \frac{\sum_{i=1}^n \sum_{j=1}^n \psi_n(\mathbf{x}_j^{(S)}, \mathbf{x}_i^{\setminus(S)}) 1_{\{\mathbf{x}_i \in R(\zeta^S)\}} 1_{\{\mathbf{x}_j \in R(\zeta^{\setminus(S)})\}}}{\sum_{i=1}^n 1_{\{\mathbf{x}_i \in R(\zeta^S)\}} \sum_{j=1}^n 1_{\{\mathbf{x}_j \in R(\zeta^{\setminus(S)})\}}},
\end{aligned}$$

where the last identity is due to (17.23). However, unlike the previous identity, it not true that $\tilde{\theta}_{\text{VIMP}}(\zeta^S)$ is the same as the non-permuted estimator

$$\tilde{\theta}_n(\zeta^S) = \frac{\sum_{i=1}^n \psi_n(\mathbf{x}_i) 1_{\{\mathbf{x}_i \in R(\zeta^S)\}}}{\sum_{i=1}^n 1_{\{\mathbf{x}_i \in R(\zeta^S)\}}}$$

which is the analog to $\hat{\theta}_n(\zeta^S)$ since it replaces the observed value $g(y_i)$ with the model estimated value $\psi_n(\mathbf{x}_i)$. Just like $\hat{\theta}_n(\zeta^S)$, the estimator $\tilde{\theta}_n(\zeta^S)$ leads to consistent variable selection [10] so the equality would show that the model-based importance has good properties.

Unfortunately, the two estimators become the same only if

$$\psi_n(\mathbf{x}) = \psi_n(\mathbf{x}^{\setminus(S)}) \tag{17.24}$$

because then

$$\begin{aligned}
\tilde{\theta}_{\text{VIMP}}(\zeta^S) &= \frac{\sum_{i=1}^n \psi_n(\mathbf{x}_i^{\setminus(S)}) 1_{\{\mathbf{x}_i \in R(\zeta^S)\}} \sum_{j=1}^n 1_{\{\mathbf{x}_j \in R(\zeta^{\setminus(S)})\}}}{\sum_{i=1}^n 1_{\{\mathbf{x}_i \in R(\zeta^S)\}} \sum_{j=1}^n 1_{\{\mathbf{x}_j \in R(\zeta^{\setminus(S)})\}}} \\
&= \frac{\sum_{i=1}^n \psi_n(\mathbf{x}_i^{\setminus(S)}) 1_{\{\mathbf{x}_i \in R(\zeta^S)\}}}{\sum_{i=1}^n 1_{\{\mathbf{x}_i \in R(\zeta^S)\}}} := \tilde{\theta}_n(\zeta^S).
\end{aligned}$$

If S consists entirely of noisy variables, then $\psi(\mathbf{x}) = \psi(\mathbf{x}^{\setminus(S)})$, so (17.24) is asserting that the model-based estimator has correctly eliminated all noisy variables. This is a very strong requirement and is asking a lot from the underlying procedure since after all this would mean that the procedure has achieved perfect variable selection on its own.

So since it is not reasonable to expect (17.24) to hold, this explains why $\tilde{\theta}_{\mathrm{VIMP}}(\zeta^S)$ calculated using $\psi_n(\mathbf{x}_j^{(S)}, \mathbf{x}_i^{\setminus(S)})$ from permuted data and model estimated values can have degraded performance in some settings.

17.6.4 Theory

To state theoretical properties for VarPro, we will assume that the analyst has already devised a rule, or more generally, a collection of suitable rules for variable selection. These rules are assumed to have been constructed independently of the data used by VarPro, and thus, without loss of generality, the rules are assumed to be deterministic.

For each n, let $\zeta_{n,1}, \ldots, \zeta_{n,K_n}$ denote these rules. Note that the number of rules K_n can vary with n and that the rules themselves are allowed to change with n. For a given rule ζ, the rule-based estimator is

$$\hat{\theta}_n(\zeta) = m_n(\zeta)^{-1} \sum_{i=1}^{n} g(Y_i) 1_{\{\mathbf{X}_i \in \mathscr{R}(\zeta)\}}, \quad m_n(\zeta) = \sum_{i=1}^{n} 1_{\{\mathbf{X}_i \in \mathscr{R}(\zeta)\}},$$

which is a slightly more formal way of writing $\hat{\theta}_n$ than (17.20). In a likewise fashion, we can define $\hat{\theta}_n(\zeta^S)$. For notational ease, let $m_{n,k} = m_n(\zeta_{n,k})$, $m_{n,k}^S = m_n(\zeta_{n,k}^S)$ and $\mathscr{R}_{n,k} = \mathscr{R}(\zeta_{n,k})$, $\mathscr{R}_{n,k}^S = \mathscr{R}(\zeta_{n,k}^S)$. The importance for S is defined as the weighted averaged difference

$$\begin{aligned} \Delta_n(S) &= \sum_{k=1}^{K_n} W_{n,k} |\hat{\theta}_n(\zeta_{n,k}^S) - \hat{\theta}_n(\zeta_{n,k})| \\ &= \sum_{k=1}^{K_n} W_{n,k} \left| \frac{1}{m_{n,k}^S} \sum_{i=1}^{n} g(Y_i) 1_{\{\mathbf{X}_i \in \mathscr{R}_{n,k}^S\}} - \frac{1}{m_{n,k}} \sum_{i=1}^{n} g(Y_i) 1_{\{\mathbf{X}_i \in \mathscr{R}_{n,k}\}} \right| \end{aligned} \tag{17.25}$$

where $0 \le W_{n,k} \le 1$ are any weights satisfying $\sum_{k=1}^{K_n} W_{n,k} = 1$.

The following theorem shows, that under mild assumptions, VarPro is consistent for noisy variables (see [10] for details).

Theorem 17.2. *Assume conditions (A1), (A2), (A3) and (A4) from Lu and Ishwaran [10]. If $K_n \le O(\log n)$ and $m_{n,k} \ge m_n = n^{1/2}\gamma_n$ where $\gamma_n \uparrow \infty$ at a rate faster than $\log n$, then $\Delta_n(S) \xrightarrow{\mathrm{p}} 0$ if $S \subseteq \mathscr{N}$.*

Assumption (A1) is an integrability condition needed for estimators to converge. Assumptions (A2) and (A4) are smoothness and boundedness conditions for ψ and its derivative. Roughly speaking, since the sample size of a region $m_n \gg n^{1/2} \to \infty$, then due to the large sample properties of averages, $\hat{\theta}_n(\zeta_{n,k}^S) - \hat{\theta}_n(\zeta_{n,k})$ asymptotically equals $\mathbb{E}(\psi(\mathbf{X})|\mathbf{X} \in R_{n,k}^S) - \mathbb{E}(\psi(\mathbf{X})|\mathbf{X} \in R_{n,k})$, and in order for this to provide useful information about a variable's importance, it is necessary for ψ to have certain smoothness properties. Assumptions (A2) and (A4) ensure that this happens. Capping the number of rules by $K_n \le O(\log n)$ ensures that the approximation holds uniformly. Assumption (A3) is a condition that regulates the region of a rule. It requires that the diameter of a region shrinks to zero uniformly. However, the shrinkage rate is left unspecified, and the shrinkage only has to hold along the signal feature coordinates.

The assumptions on shrinkage and sample size of a region are expected to hold for any reasonable procedure that produces rules. Similar assumptions are fairly standard in the asymptotic analysis of trees. Actually we have already seen similar conditions used for proving tree consistency in Chapter 15. Theorem 15.1 and Theorem 15.6 for partitioning classifiers and partitioning histogram estimators by assuming the diameter of a cell converges to 0 and the number of points in the cell converges to ∞ in probability.

17.6.5 Regression and classification

The following computational procedure for VarPro was described by Lu and Ishwaran [10] for regression and classification problems. First, before implementing VarPro, a lasso model is fit, and the absolute coefficient value for each standardized variable is recorded (for classification, the mean absolute value across class labels is used). This value is used as the split-weight for a variable in the rule generation step (this guides the rule generation to use signal features as much as possible to satisfy assumption (A3), which requires rules to shrink along the signal feature directions). It was observed that in certain cases (such as nonparametric models), lasso could become overly sparse due to model misspecification, sometimes resulting in all coefficients being zero. To address this, 500 trees are additionally constructed, and the relative split frequency of a variable is recorded and added to the lasso coefficient estimate for the finalized split-weight.

Following this, a tree is constructed using $0.632n$ of the data using weighted feature selection. At each node, the tree is split by selecting a random subset of covariates, with covariates chosen randomly with probability proportional to their split-weight. From the finalized tree, a random K_n of tree branches are selected, yielding rules $\mathscr{R}_{n,1},\ldots,\mathscr{R}_{n,K_n}$. The remaining data ($0.368n$) is used for calculating $\Delta_n(s)$ using (17.25).

For a split-sample measure of variability, the process of rule generation and VarPro estimation is repeated $B > 1$ times. This yields an averaged $\Delta_n^*(s)$ and sample variance $\mathrm{Var}_n^*(\Delta_n(s))$, which is converted to a standardized estimate:

$$\text{Standardized importance for variable } X^{(s)} = \frac{\Delta_n^*(s)}{\sqrt{\mathrm{Var}_n^*(\Delta_n(s))}}. \tag{17.26}$$

Variable $X^{(s)}$ is selected if the standardized importance (17.26) exceeds a preset cutoff value, such as $Z_0 = 2$. It is also possible to select Z_0 using out-of-sample validation (see Lu and Ishwaran [10] for details). Algorithm 17.3 provides a formal description of the procedure.

Algorithm 17.3 VarPro standardized importance

1: Calculate split-weights for each of the p variables.
2: **while** $1 \le b \le B$ **do**
3: Split the data randomly into $0.632n$ and $0.368n$ sizes. Using the larger portion, $0.632n$, construct a tree using weighted feature selection with selection probability proportional to the split-weight.
4: Randomly select K_n branches (rules) from the tree. Over the smaller portion of the data, $0.368n$:
5: **for** $1 \le s \le p$ **do**
6: Calculate $\Delta_n(s)$ (17.25) for variable $X^{(s)}$ using the K_n rules.
7: **end for**
8: **end while**
9: Using the B importance values, calculate $\Delta_n^*(s)$, $\mathrm{Var}_n^*(\Delta_n(s))$ and the standardized importance (17.26) for each s: select $X^{(s)}$ if (17.26) exceeds Z_0.

17.6.5.1 Boston Housing

For a first illustration we re-analyzed the Boston Housing data. The importance values $\Delta_n(s)$ are given in the left panel of Figure 17.12; the right panel displays the standardized importance (17.26). These latter estimates really help to separate noisy from signal variables.

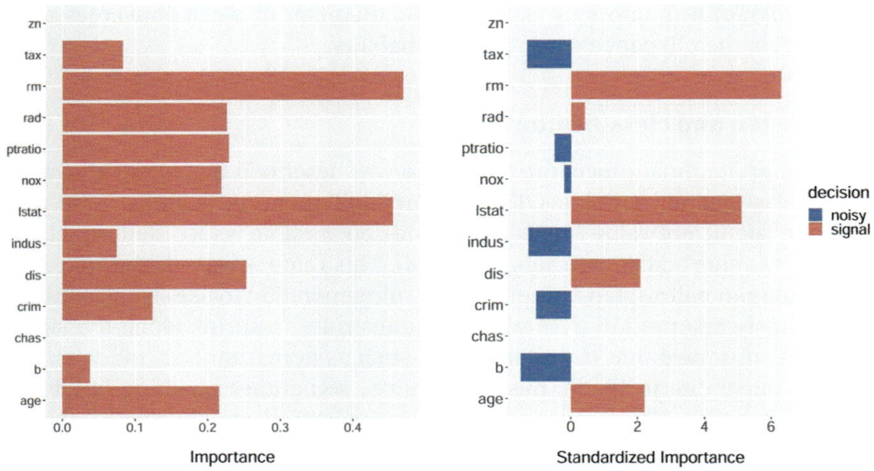

FIGURE 17.12
VarPro importance values for Boston Housing regression. Left panel shows unstandardized values; right panel are standardized such that positive values are identified as being signal.

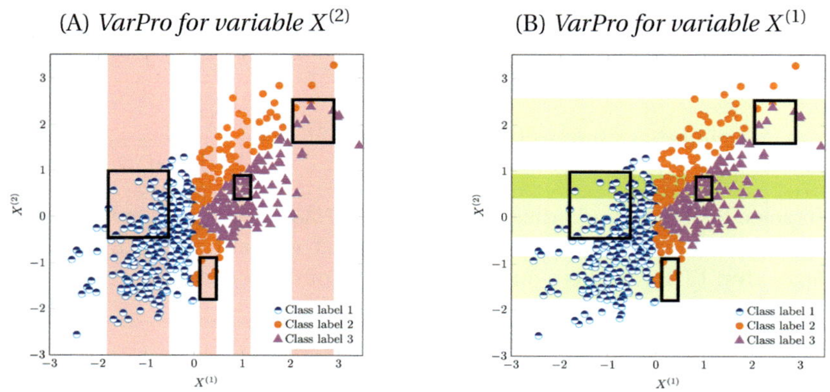

FIGURE 17.13
Illustration of VarPro for $L = 3$ classification (blue, orange and purple are class labels 1, 2 and 3). The classification boundary for class label 1 is vertical making $X^{(2)}$ a noisy variable for it but not for other class labels.

Room and lower status are the top two variables which is in agreement with our previous analysis of Figure 17.2. However, using VarPro, radius, distance and age are found to be significant, while all remaining variables are not. This is different than Figure 17.2. In that analysis nearly all variables have roughly the same importance values. Thus, VarPro is much sparser.

17.6.5.2 Multiclass illustration

Next we illustrate VarPro in multiclass problems. We first mention some interesting differences from regression. This is due to the definition of noisy and strong features via conditional independence (17.17) used by Definition 17.2, which depends on the choice of g and ψ.

Consider Figure 17.13 which displays data from an $L = 3$ multiclass problem with two variables. In this example there are three functions, ψ_1, ψ_2, ψ_3, where $\psi_l(\mathbf{x}) = \mathbb{P}\{Y = l|\mathbf{X} = \mathbf{x}\}$ and

$\mathbf{X} = (X^{(1)}, X^{(2)})^T$, for $l = 1,2,3$. In this setting the target g function is $g(Y) = I\{Y = l\}$ for some l. Shown in the figure are the regions of 4 different rules (given by black rectangles). In (A), the rules are released along $X^{(2)}$ shown in pink which are the original regions but now unconstrained along $X^{(2)}$. The first box on the left lies within the classification boundary for class 1, which is vertical. This vertical boundary implies that ψ_1 does not depend on $X^{(2)}$, and therefore $X^{(2)}$ is a noisy variable for this class. But this is not the case for the three boxes on the right where ψ_2 and ψ_3 depend on $X^{(2)}$. Therefore $X^{(2)}$ is a signal variable for class labels 2 and 3. In (B), the rules are released along $X^{(1)}$ shown by green regions. All regions vary with $X^{(1)}$ which shows that $X^{(1)}$ is a signal variable for all three classes. This illustrates that in multiclassification the definition of a noisy and signal variable can depend on the choice of g.

For an empirical demonstration of VarPro, data was simulated from a $J = 3$ classification model specified by

$$
\begin{aligned}
y &= \arg\max_{l}\{\psi_l(\mathbf{x})\}, \quad \text{where } \psi_l(\mathbf{x}) = \frac{\phi_l(\mathbf{x})}{\sum_{l=1}^{L}\phi_l(\mathbf{x})}, \ \phi_l(\mathbf{x}) = \exp\left(\sum_{j=1}^{p}\beta_{j,l}x^{(j)}\right) \\
& \beta_{j,1} = 1,\ j = 1,2,3, \text{otherwise}\, \beta_{j,1} = 0 \\
& \beta_{j,2} = 1,\ j = 4,5,6, \text{otherwise}\, \beta_{j,2} = 0 \\
& \beta_{j,3} = 1,\ j = 7,8,9, \text{otherwise}\, \beta_{j,3} = 0.
\end{aligned}
$$

Due to the constraint $\sum_j \psi_j = 1$, all 9 covariates $X^{(1)},\ldots,X^{(9)}$ are signal variables for all three classes. However, coordinates 1,2,3 are especially informative for class 1, coordinates 4,5,6 for class 2, and coordinates 7,8,9 for class 3. The covariates were drawn so that each coordinate had a marginal uniform distribution $X^{(j)} \sim \mathrm{U}(0,1)$ independent of other coordinates except for pairs $\{X^{(3)}, X^{(10)}\}$, $\{X^{(6)}, X^{(15)}\}$ and $\{X^{(9)}, X^{(20)}\}$ which were transformed to have correlation $\rho = .9$. Simulations used $N = 2000$ and $p = 20$.

VarPro importance values $\Delta_{n,j}$ were obtained for $j = 1,2,3$. The results from 250 independent runs are displayed in the top panel of Figure 17.14. VarPro's performance is very good, as seen by comparison with the bottom panel, which shows permutation VIMP from RF. It is clear that VIMP performs poorly in this simulation. The poor performance is due to the correlation between the signal variables $X^{(3)}, X^{(6)}, X^{(9)}$ and noise variables $X^{(10)}, X^{(15)}, X^{(20)}$, which causes permutation importance for signal variables to be degraded while artificially increasing the importance of noise variables. This contrasts sharply with VarPro, where the importance of signal variables is not degraded, and noisy variables have substantially smaller, non-significant importance values. Additionally, the group structure is clear (for example, covariates $X^{(1)}, X^{(2)}, X^{(3)}$ are highly informative for class 1), and it is evident that each conditional probability depends on all 9 variables $X^{(1)},\ldots,X^{(9)}$.

Exercises

Exercise 17.1. Using the `METABRIC` (Breast Cancer Gene Expression Profiles) dataset available on Kaggle, your goal is to predict the mutation count of breast cancer patients using clinical attributes, mRNA levels, and gene mutations. Apply RF to identify the most important predictors.

1. Preprocess the data by imputing missing values (Chapter 16.14).
2. Split the data into training and testing sets.
3. Fit a RF regression model using the training data to predict the mutation count (treat mutation count as a continuous outcome).
4. Use cross-validation to tune the hyperparameters of the RF model.

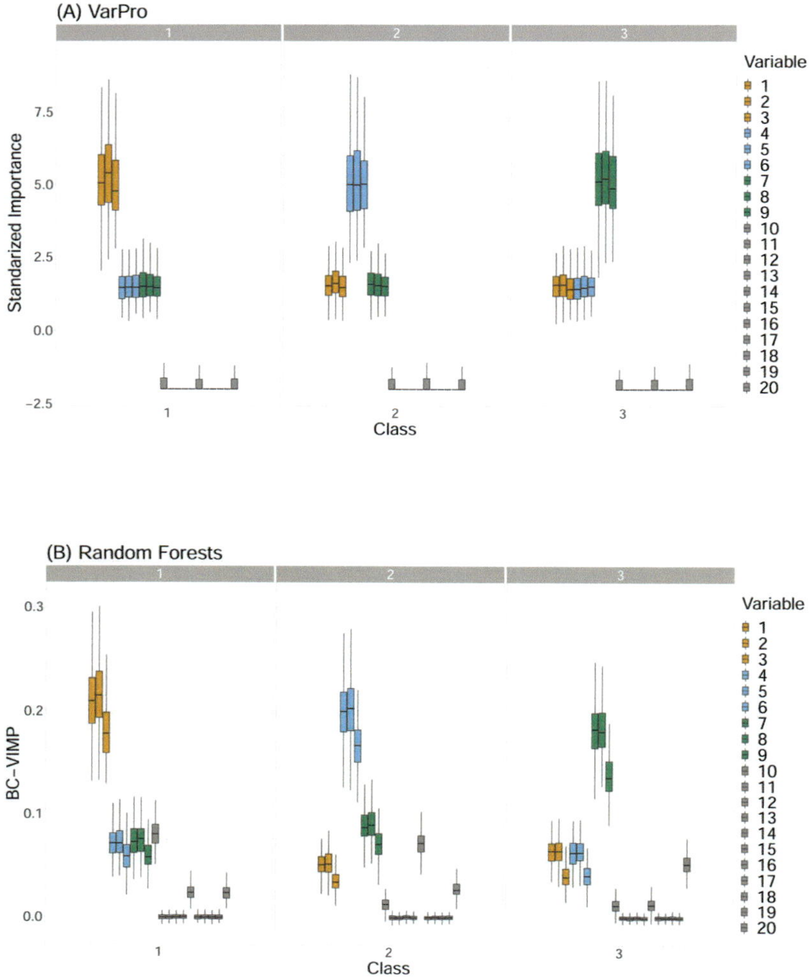

FIGURE 17.14
Results from a $J = 3$ multiclass simulation where variables 1–3 are most informative for class 1; variables 4–6 for class 2; variables 7–9 for class 3. For all classes, variables 10–20 are noisy variables, also variables 3 and 10, 6 and 15, 9 and 20 are strongly correlated: thus there is correlation between signal and noisy features. (A) VarPro standardized importance performs very well and identifies the group structure and is not influenced by correlation. (B) Permutation VIMP from RF is influenced by correlation which degrades its performance.

5. Calculate permutation VIMP and minimal depth importance for each predictor. Compare and visualize the top predictors identified by the two methods. Interpret the most important predictors identified by each method in the context of biological and clinical relevance.
6. Build reduced models using the top predictors identified by each method and compare their predictive performance over the test data.
7. Provide a detailed discussion on the advantages and limitations of the two different methods.

Exercise 17.2. Create a simulation for testing variable selection performance. Compare the performance of permutation VIMP, minimal depth and VarPro on your simulation. Repeat the previous analysis but using correlated features. How have your conclusions changed and if so in which way?

Exercise 17.3. Use the `METABRIC` dataset to perform a classification analysis. For the outcome select a binary class outcome with a reasonably large imbalanced class ratio. Build a RF classifier and a RFQ classifier. Compare the VIMP from the RF classifier with the gmean VIMP from the RFQ classifier. Discuss your findings, focusing on their biological and clinical relevance.

Exercise 17.4. In this exercise the training set consists of n points uniformly distributed in $[0,1]^p$. We will use subsampling to estimate the variance of the 1-NN (nearest neighbor) distance D from a query point $\mathbf{X}_0$ to its nearest neighbor.

1. Generate a uniformly distributed dataset in the unit cube $[0,1]^p$. Sample a random query point $\mathbf{X}_0$ from this space.
2. Subsample the dataset multiple times without replacement, compute the 1-NN distance for each subsample. Use the distances from subsampling to estimate the variance, applying the proper sample size correction:

$$\hat{\sigma}_n^2 = \text{Var(1-NN distances)} \times \sqrt{\frac{\text{subsample size}}{n}}.$$

3. Perform multiple simulations of generating new datasets. Calculate the variance of the distances from these simulations.
4. Compare the variance obtained from subsampling with the empirical variance from replicated simulations.

   (a) How close is the variance from subsampling to the empirical true variance?
   (b) How does the subsample size affect the variance estimate from subsampling?
   (c) How does changing n and p alter the results?

Exercise 17.5. Consider a time series dataset $\{Y_t\}_{t=1}^{1000}$ representing daily returns of a financial asset. The objective is to estimate the distribution of the sample mean return $\overline{Y} = \sum_{t=1}^{1000} Y_t/1000$ using subsampling methods.

1. Generate a time series $\{Y_t\}_{t=1}^{1000}$ using an AR(1) process:

$$Y_t = \phi Y_{t-1} + \varepsilon_t$$

   where $\varepsilon_t \sim \text{N}(0,\sigma^2)$, $\phi = 0.5$, and $\sigma = 1$. Set $Y_0 = 0$ to start the process.
2. Compute the subsample means $\overline{Y}_{s_k,b} = b^{-1}\sum_{t\in s_k} Y_t$, where s_k is the k-th subsample, $k = 1,\dots,K$, using a subsample size $b \lll 1000$.
3. Use the subsampled means to construct a 95% confidence interval for the mean.
4. Discuss your findings, focusion on: (i) the selection of subsample size b; (b) the computational complexity and efficiency; (iii) the impact of dependence in the time series on the subsampling results.
5. Compare the subsampling results to other resampling methods like the bootstrap.

Exercise 17.6. In this exercise, you will use the `METABRIC` dataset to obtain confidence regions for VIMP using subsampling. Use the `subsample` function from the `randomForestSRC` for this purpose.

1. Train a RF classifier model. Use subsampling to obtain confidence intervals for permutation VIMP. Repeat this step for conditional class label VIMP for each class.
2. Interpret the permutation importance values and their confidence intervals, focusing on biological and clinical relevance of your findings.

Exercise 17.7. The VarPro variable selection method (Chapter 17.6) estimates importance values using a rule-based approach.

1. Explain exactly how VarPro calculates its importance value. For your explanation, describe a rule and then using simulated data illustrate how the VarPro importance value is obtained.
2. Explain how VarPro differs from permutation VIMP and knockoffs. Use the rule and simulated data from part 1 to aid in this.
3. The mean decrease impurity (MDI) score [2] is a tree-based procedure for calculating importance values. For a given variable X, the MDI score is obtained by summing the weighted impurity decrease across all tree nodes that involve a split using X. Explain why MDI shares similarities to VarPro but also indicate important differences between the methods.
4. Explain the concept of guided tree-splitting in the VarPro method.
5. Outline the steps of the VarPro algorithm for variable importance estimation.
6. Discuss how VarPro could be extended to applications outside of regression and multiclass data. Give a concrete example using simulated data.
7. Describe the out-of-sample method used in Lu and Ishwaran [10] to automate the VarPro cutoff value. Come up with an alternative strategy and discuss its pros and cons.
8. Apply a VarPro analysis to a real-world high-dimensional dataset. Discuss your findings and any practical challenges that you encountered.

Bibliography

[1] Breiman L. Random forests. *Machine Learning,* 45:5–32, 2001.

[2] Breiman L., Friedman J.H., Olshen R.A., and Stone C.J. *Classification and Regression Trees,* Belmont, California, 1984.

[3] Candes E. and Fan, Y. and Janson L. and Lv J. Panning for gold:'model-x' knockoffs for high dimensional controlled variable selection. *Journal of the Royal Statistical Society: Series B (Statistical Methodology),* 80, 551–577, 2018.

[4] De Cock D. Ames, Iowa: alternative to the Boston housing data as an end of semester regression project. *Journal of Statistics Education,* 19(3), 1-14, 2011.

[5] Diaz-Uriarte R. and Alvarez de Andres S. Gene selection and classification of microarray data using random forest. *BMC Bioinformatics,* 7:3, 2006.

[6] Efron B. Bootstrap methods: another look at the jackknife. *Annals of Statistics,* 7(1), 1–26, 1979.

[7] Efron B. and Tibshirani R. Improvements on cross-validation: The .632+ bootstrap method. *J. Amer. Stat. Assoc.,* 92:548–560, 1997.

[8] Ishwaran H., Kogalur U.B., Gorodeski E.Z., Minn A.J., and Lauer M.S. High-dimensional variable selection for survival data. *Journal of the American Statistical Association,* 105:205-217, 2010.

[9] Ishwaran H. and Lu M. Standard errors and confidence intervals for variable importance in random forest regression, classification, and survival. *Statistics in Medicine,* 38(4):558–582, 2019.

[10] Lu M. and Ishwaran H. Model-independent variable selection via the rule-based variable priority. *Techincal Report*, 2022.

[11] O'Brien R. and Ishwaran H. A random forests quantile classifier for class imbalanced data. *Pattern Recognition*, 90:232–249, 2019.

[12] Politis D. N. and Romano J. P. and Wolf M. Subsampling. *Springer Series in Statistics*. Springer-Verlag, New York, 1999.

[13] Politis D. N. and Romano J. P. Large sample confidence regions based on subsamples under minimal assumptions. *The Annals of Statistics*, 22(4):2031–2050, 1994.

[14] Radulović D. On the subsample bootstrap variance estimation. *Test*, 7(2):295–306, 1998.

[15] Shao J. and Wu C. J. A general theory for jackknife variance estimation. *The Annals of Statistics*, 17(3):1176–1197, 1989.

[16] Wolberg W.H., Street W.N., Heisey D.M., and Mangasarian O.L. Computerized breast cancer diagnosis and prognosis from fine needle aspirates. *Archives of Surgery*, 130:511-516, 1995.

18

The Effect of Splitting Rules on Random Forests

18.1 Introduction

In our treatment of trees, we have examined various splitting rules, starting with CART impurity and extending to specialized settings like quantile regression, AUC splitting, and lasso splitting for SGTs. In this chapter, we return to classical CART splitting rules to closely examine their properties. We investigate how the split-point responds to the underlying function being estimated. For example, what is the optimal split-point for a wiggly function? What about a smooth function? How does the split-point track geometric features of the function? We derive a theoretical result for the population limiting case, providing insight into these questions.

Our split-point analysis considers when the signal is present, but an interesting phenomenon arises when the split is on a noise variable. In such cases, the split tends towards the edge of the available split values, known as the *end-cut preference* (ECP) splitting property [9]. This edge-splitting behavior for noisy variables is often seen as a nuisance for CART. According to Breiman et al. [1, Chapter 11.8], the delta splitting rule used by THAID was introduced to suppress ECP splitting. An illustration of ECP was shown in Chapter 15.5.2 for a prostate data set, where the splitting behavior of each variable was shown after the variable was randomly permuted. We observed that cuts were often near the edge (see Figure 15.5).

Definition 18.1. A splitting rule has the ECP property if it tends to split near the edge for noisy variables. □

Even though this behavior is seen as a nuisance for CART, our results suggest that ECP splitting is beneficial for random forests (RF). By splitting towards the edge for a noisy variable, the resulting split maximizes the tree node sample size, helping the tree recover from a bad split downstream. This advantage is significant for the deep trees used in RF. Additionally, since RF employs random feature selection, it's likely that a node may split on a noisy variable, especially in high-dimensional settings. ECP is advantageous in these cases as it maintains the sample size as much as possible.

We will consider a class of weighted splitting rules, including the CART weighted variance splitting rule and Gini index, and identify the properties needed for the ECP property. We will also examine random splitting and whether ECP is preserved in these cases. Random split-point selection is an active area of research in various RF applications. Dietterich [3] studied bagged trees where the split-point for a variable is randomly selected from the top 20 split-points based on CART splitting. Perfect random trees for ensemble classification [2] randomly choose a variable and then randomly select the split-point for this variable from the observed values of two randomly chosen points from different classes. Ishwaran et al. [5, 6] proposed a partially randomized splitting rule for survival forests (Chapter 19), where a fixed number of randomly selected split-points are chosen for each variable and the top split-point based on a survival splitting rule is selected. Related to this is Geurts et al. [4], who investigated extremely

DOI: 10.1201/9781003493679-18

randomized trees, selecting a single random split-point for each variable and then choosing the top split-point.

Much of the material presented here is condensed from the work of Ishwaran [7]. For brevity our analysis will be confined to the regression problem, readers can consult that paper for details regarding classification.

18.2 Theoretical Derivation of the Split-Point

The learning data is $(\mathbf{X}_1, Y_1), \ldots, (\mathbf{X}_n, Y_n) \in \mathbb{R}^p \times \mathbb{R}$ which are i.i.d. from $\mathbb{P}$. A generic data point from $\mathbb{P}$ is denoted by $(\mathbf{X}, Y)$. Given $\mathbf{X}_i = \mathbf{x}_i$, we assume that

$$Y_i = F(\mathbf{x}_i) + \varepsilon_i, \quad i = 1 \ldots, n, \tag{18.1}$$

where F is the unknown regression function assumed to be continuous and $(\varepsilon_i)_{i=1}^n$ are i.i.d., independent of $(\mathbf{X}_i)_{i=1}^n$, such that $\mathbb{E}(\varepsilon_i) = 0$, $\mathbb{E}(\varepsilon_i^2) = \sigma^2 > 0$.

Recall that the tree split-point is found by optimizing the CART splitting rule, i.e., by reducing impurity the most. But how does the optimized split-point depend on the underlying regression function F? What are its properties when F is flat, linear, or wiggly?

We answer this by theoretically determining the split-point for a given F using an infinite sample paradigm. Let t be a node of a tree to be split. Recall from Chapter 15.7, that the impurity splitting rule for splitting t using the variable X at split-point s is

$$\hat{\gamma}(s, t) = \hat{P}(t_L)\hat{\Gamma}(t_L) + \hat{P}(t_R)\hat{\Gamma}(t_R)$$

where $t_L = \{\mathbf{X}_i \in t : X \le s\}$, $t_R = \{\mathbf{X}_i \in t : X > s\}$ and $\hat{P}(\cdot)$ is the proportion of cases. The goal is to minimize $\hat{\gamma}(s, t)$. When $\hat{\Gamma}(\cdot)$ is within node variance this gives the weighted variance splitting rule

$$\hat{\gamma}(s, t) = \frac{N_L}{N}\left[\frac{1}{N_L}\sum_{i \in t_L}\left(Y_i - \overline{Y}_{t_L}\right)^2\right] + \frac{N_R}{N}\left[\frac{1}{N_R}\sum_{i \in t_R}\left(Y_i - \overline{Y}_{t_R}\right)^2\right]$$

where $N_L = |t_L|$, $N_R = |t_R|$, $N = |t| = N_L + N_R$ and $\overline{Y}_{t_L}, \overline{Y}_{t_R}$ are the response averages in t_L, t_R. Equivalently, the minimizer of $\hat{\gamma}(s, t)$ is the same as the maximizer of

$$\hat{\Psi}_t(s) = \frac{N_L}{N}\left(\overline{Y}_{t_L}\right)^2 + \frac{N_R}{N}\left(\overline{Y}_{t_R}\right)^2. \tag{18.2}$$

Let $\hat{s}_N$ denote the optimized split-point for (18.2).

To study $\hat{s}_N$, we replace $\hat{\gamma}(s, t)$ with its analog based on population parameters:

$$\gamma(s, t) = P(t_L)\Gamma(t_L) + P(t_R)\Gamma(t_R) \tag{18.3}$$

where $\Gamma(t_L)$ and $\Gamma(t_R)$ are the population conditional variances

$$\Gamma(t_L) = \mathrm{Var}(Y|X \le s, \mathbf{X} \in t), \quad \Gamma(t_R) = \mathrm{Var}(Y|X > s, \mathbf{X} \in t),$$

and $P(t_L)$ and $P(t_R)$ are the population conditional probabilities

$$P(t_L) = \mathbb{P}\{X \le s|\mathbf{X} \in t\}, \quad P(t_R) = \mathbb{P}\{X > s|\mathbf{X} \in t\}.$$

We want to minimize $\gamma(s, t)$ over s to determine the theoretical split-point, s_∞. Surprisingly, there is a closed form solution in the one-dimensional case, $p = 1$. We state this formally in the following result.

Theorem 18.1. *Let $\mathbb{P}_t$ denote the conditional distribution for X given that $X \in t$. Let $\mathbb{P}_{t_L}(\cdot)$ and $\mathbb{P}_{t_R}(\cdot)$ denote the conditional distribution of X given that $X \in t_L$ and $X \in t_R$, respectively. The minimizer of (18.3) is the value for s maximizing*

$$\Psi_t(s) = \mathbb{P}_t\{X \le s\}\left(\int_a^s F(x)\, d\mathbb{P}_{t_L}(x)\right)^2 + \mathbb{P}_t\{X > s\}\left(\int_s^b F(x)\, d\mathbb{P}_{t_R}(x)\right)^2 \tag{18.4}$$

where $\Psi_t(s)$ can be considered as the population analog of $\hat{\Psi}_t(s)$. Moreover, if $t = [a,b]$ and $d\mathbb{P}_t(x) > 0$ is continuous, then the maximizer of (18.4) satisfies

$$2F(s) = \int_a^s F(x)\, d\mathbb{P}_{t_L}(x) + \int_s^b F(x)\, d\mathbb{P}_{t_R}(x).$$

Note that a solution is permissible only if $a \le s \le b$.

An explicit solution exists if X is assumed to be equispaced and F is a polynomial function.

Theorem 18.2. *Suppose that $F(x) = c_0 + \sum_{j=1}^q c_j x^j$. If $\mathbb{P}_t$ is the uniform distribution on $t = [a,b]$, then the minimizer of (18.3) is a solution to*

$$\sum_{j=0}^{q} \left(U_j + V_j - 2c_j\right) s^j = 0, \tag{18.5}$$

where $U_j = c_j/(j+1) + ac_{j+1}/(j+2) + \cdots + a^{q-j} c_q/(q+1)$ and $V_j = c_j/(j+1) + bc_{j+1}/(j+2) + \cdots + b^{q-j} c_q/(q+1)$. To determine which value is the true minimizer, discard all solutions not in t (including imaginary values) and choose that value which maximizes

$$\begin{aligned} \Psi_t(s) &= \frac{1}{(b-a)(s-a)}\left(\sum_{j=0}^{q} \frac{c_j}{j+1}\left(s^{j+1} - a^{j+1}\right)\right)^2 \\ &\quad + \frac{1}{(b-a)(b-s)}\left(\sum_{j=0}^{q} \frac{c_j}{j+1}\left(b^{j+1} - s^{j+1}\right)\right)^2. \end{aligned} \tag{18.6}$$

Example 18.1. As a first illustration, suppose that $F(x) = c_0 + c_1 x$ for $x \in [a,b]$. Then, $U_0 = c_0 + ac_1/2$, $V_0 = c_0 + bc_1/2$, and $U_1 = V_1 = c_1/2$. Hence, (18.5) equals

$$\frac{c_1}{2}(a+b) - c_1 s = 0.$$

If $c_1 \ne 0$, then $s = (a+b)/2$; which is a permissible solution. Thus, for simple slope-intercept functions, node-splits are always at the midpoint. □

Example 18.2. Now consider a more complicated polynomial $F(x) = 2x^3 - 2x^2 - x$, where $x \in [-3,3]$. We numerically solve (18.5) and (18.6). The solutions are displayed recursively in Figure 18.1. The first figure is the optimal split over the root node $[-3,3]$. There is one distinct solution $s = -1.924$. The second figure is the optimal split over the daughters arising from the first figure. There are 3 solutions over $(-1.924, 3]$ and 1 solution over $[-3, -1.924]$. The third figure are the optimal splits arising from the second figure, and so forth.

The derivative of F is $F'(x) = 6x^2 - 4x - 1$. Inspection of the derivative shows that F is increasing most rapidly for $-3 \le x \le -2$, followed by $2 \le x \le 3$, and then $-2 < x < 2$. Interestingly, the order of splits in Figure 18.1 follows this pattern, showing that node splitting tracks the curvature of F, with splits occurring first in regions where F is most steep, and last in places where F is flattest.

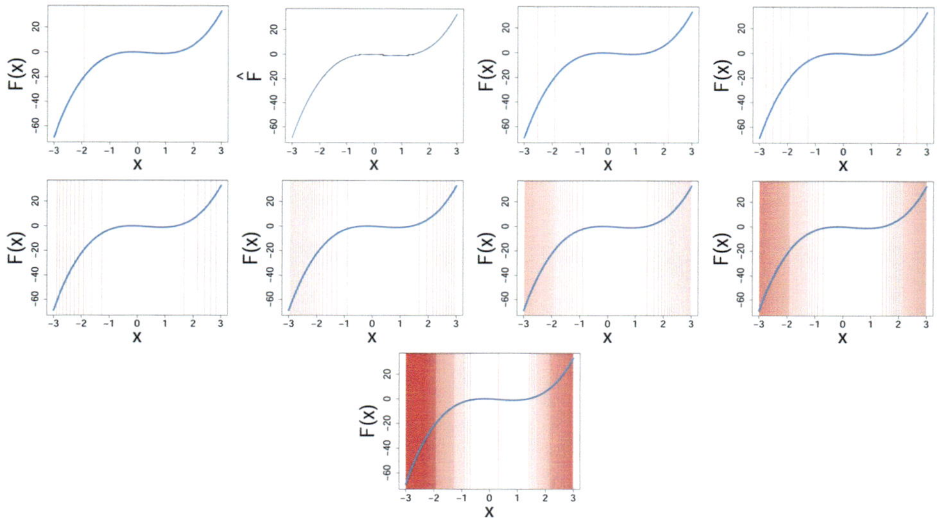

FIGURE 18.1
Theoretical split-points for X under weighted variance splitting (displayed using vertical red lines) for $F(x) = 2x^3 - 2x^2 - x$ assuming a uniform $[-3,3]$ distribution for X.

These results assume a $p = 1$ one-dimensional scenario. To test how well they extrapolate to higher dimensions we simulated $n = 10,000$ values from

$$Y_i = F(X_i) + C_1 \sum_{k=1}^{d} U_{i,k} + C_2 \sum_{k=d+1}^{D} U_{i,k} + \varepsilon_i, \quad i = 1\ldots, n, \tag{18.7}$$

for F defined as above, $\varepsilon_i \overset{\text{iid}}{\sim} \mathrm{N}(0,\sigma^2)$ for $\sigma = 2$, $X_i \overset{\text{iid}}{\sim} \mathrm{U}[-3,3]$ and $U_{i,k} \overset{\text{iid}}{\sim} \mathrm{U}[-3,3]$. We set $d = 10, D = 13$ where the first $1 \le k \le d$ variables $U_{i,k}$ are signal variables with signal $C_1 = 3$ while the remaining $U_{i,k}$ variables, $d+1 \le k \le D$, are noise variables with $C_2 = 0$. Weighted variance splitting was used to split X. The data-optimized split-points $\hat{s}_N$ for the splits on X are shown in Figure 18.2. Notice how closely these track the theoretical splits of Figure 18.1. □

The near-equivalence of the split-points of Figure 18.1 and Figure 18.2 is not a coincidence and is a direct consequence of the following limiting result. If $\Psi_t(s)$ has a unique global maximum at an interior point of t, then

$$\hat{s}_N \xrightarrow{\text{p}} s_\infty = \underset{a \le s \le b}{\arg\max}\, \Psi_t(s), \quad \text{as } N \to \infty$$

where s_∞ is the unique population maximizer. With some rearrangement, (18.6) becomes

$$\Psi_t(s) = (s-a)\left(\sum_{j=0}^{q} A_j s^j\right)^2 + (b-s)\left(\sum_{j=0}^{q} B_j s^j\right)^2,$$

where A_j, B_j are constants that depend on a and b. Therefore Ψ_t is a polynomial and achieves a global maximum over t or over a sufficiently small subregion t'.

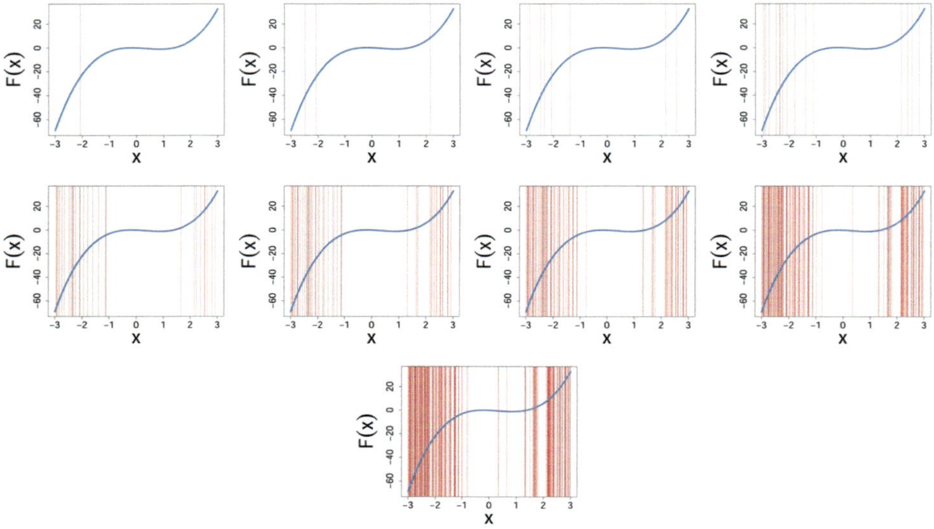

FIGURE 18.2
Data optimized split-points $\hat{s}_N$ for X from a regression tree using weighted variance splitting from simulated data (18.7).

Figure 18.3 illustrates how $\Psi_{t'}(s)$ depends on t' for $F(x)$ of Example 18.2. The first subpanel displays $\Psi_t(s)$ over the entire range $t = [-3,3]$. Clearly, it achieves a global maximum. Furthermore, when $[-3,3]$ is broken up into contiguous subregions t', $\Psi_{t'}(s)$ becomes nearly concave (last six panels) and its maximum becomes more pronounced. Therefore, $\hat{s}_N$ approximates s_∞ over these regions.

It can be shown in general, regardless of whether F is a polynomial, that the maximum of Ψ_t depends on the curvature of F (see Ishwaran [7]). Thus over a small enough region so that Ψ_t achieves a global maximum, $\hat{s}_N$ will approximate s_∞ and the data optimized split-point will track the curvature of F.

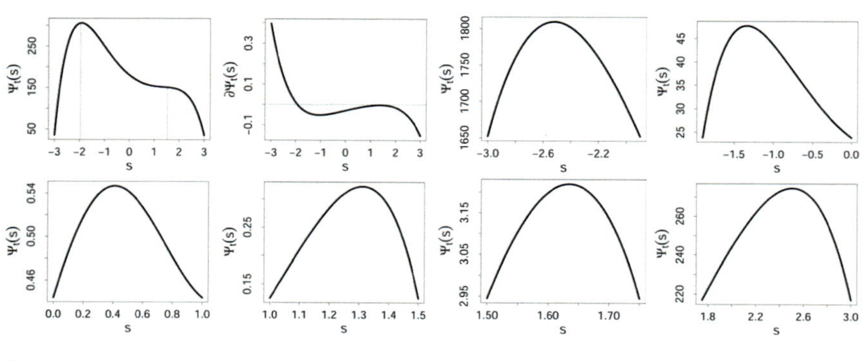

FIGURE 18.3
The first two panels are $\Psi_t(s)$ and its derivative $\Psi'_t(s)$ for $F(s) = 2s^3 - 2s^2 - s$ where $t = [-3,3]$. Remaining panels are $\Psi_{t'}(s)$ for $t' = [-3,-1.9]$, $t' = [-1.9,0]$, $t' = [0,1]$, $t' = [1,1.5]$, $t' = [1.5,1.75]$ and $t' = [1.75,3]$. Vertical lines in first subpanel identify stationary points of $\Psi_t(s)$.

18.3 Weighted Variance Splitting has the ECP Property

Example 18.1 showed that the population weighted variance splitting rule splits at the mid-point for simple linear functions $F(x) = c_0 + c_1 x$. This applies for any $c_1 \neq 0$, and in particular, it holds for arbitrarily small $c_1 \neq 0$. Thus, weighted variance will split near the center even when X has near zero strength. However, when X is a noisy variable, i.e. $c_1 = 0$, weighted variance splits tend to be near the edge and not near the middle. As discussed, this behavior is referred to as the ECP property.

The following result shows that weighted splitting has the ECP property asymptotically. The proof uses two key facts: Kolmogorov's inequality (Theorem 2.42) and the law of the iterated logarithm (Theorem 2.19).

Theorem 18.3. *[Theorem 11.1; 1]. Let $Z_1, \ldots, Z_N$ be i.i.d. with finite variance $\sigma^2 > 0$. Let*

$$\xi_{N,m} = \frac{1}{m}\left(\sum_{i=1}^{m} Z_i\right)^2 + \frac{1}{N-m}\left(\sum_{i=m+1}^{N} Z_i\right)^2, \qquad 1 \le m \le N-1. \tag{18.8}$$

Then for any $0 < \delta < 1/2$ and any $0 < \tau < \infty$

$$\lim_{N\to\infty} \mathbb{P}\left\{\max_{1\le m\le N\delta} \xi_{N,m} > \max_{N\delta<m<N(1-\delta)} \tau\xi_{N,m}\right\} = 1$$

and

$$\lim_{N\to\infty} \mathbb{P}\left\{\max_{N(1-\delta)\le m\le N} \xi_{N,m} > \max_{N\delta<m<N(1-\delta)} \tau\xi_{N,m}\right\} = 1.$$

We refer to rules (18.8) as weighted splitting rules. Substituting Y_i for Z_i in (18.8) yields the weighted variance splitting rule, since (18.2) can be rewritten as

$$\frac{1}{N_L}\left(\sum_{i\in t_L} Y_i\right)^2 + \frac{1}{N_R}\left(\sum_{i\in t_R} Y_i\right)^2.$$

Theorem 18.3 applies to i.i.d. data, which by our regression model (18.1) applies when $F(x)$ is a constant, i.e. when X is a noisy variable. Therefore, Theorem 18.3 shows that weighted variance splitting has the ECP property.

Combining this result with Theorem 18.1, we now see that weighted variance splitting possesses an optimality property. If the variable has a non-zero effect (no matter how small), the split is at the point of greatest curvature, which gives the greatest power to uncover signal, whereas if the variable is noisy, the split is towards the edge, which preserves the sample size as much as possible, and maximize power.

18.4 Other Weighted Splitting Rules

Weighted variance splitting uses weights equal to the daughter sample sizes. Here we introduce two new classes of splitting rules using different weighting schemes and we explore their properties.

The first new class uses equal weights for the daughter node variances. This rule is referred to as unweighted variance splitting. Formally, let

$$\hat{\gamma}_U(s,t) = \hat{\Gamma}(t_L) + \hat{\Gamma}(t_R). \tag{18.9}$$

Unweighted variance splitting finds the best split s minimizing $\hat{\gamma}_U(s,t)$. Notice that (18.9) can be rewritten as

$$\hat{\gamma}_U(s,t) = \frac{1}{N_L}\sum_{i\in t_L} Y_i^2 + \frac{1}{N_R}\sum_{i\in t_R} Y_i^2 - \frac{1}{N_L^2}\left(\sum_{i\in t_L} Y_i\right)^2 - \frac{1}{N_R^2}\left(\sum_{i\in t_R} Y_i\right)^2.$$

The following theorem shows these rules possess the ECP property.

Theorem 18.4. *Let $Z_1,\ldots,Z_N$ be i.i.d. such that $\mathbb{E}(Z_1^4)<\infty$. Let*

$$\begin{aligned}\zeta_{N,m} &= \frac{1}{m}\sum_{i=1}^{m} Z_i^2 + \frac{1}{N-m}\sum_{i=m+1}^{N} Z_i^2 \\ &\quad - \frac{1}{m^2}\left(\sum_{i=1}^{m} Z_i\right)^2 - \frac{1}{(N-m)^2}\left(\sum_{i=m+1}^{N} Z_i\right)^2, \quad 1\le m\le N-1.\end{aligned} \tag{18.10}$$

Then for any $0<\delta<1/2$

$$\lim_{N\to\infty}\mathbb{P}\left\{\min_{1\le m\le N\delta}\zeta_{N,m} < \min_{N\delta<m<N(1-\delta)}\zeta_{N,m}\right\} = 1$$

and

$$\lim_{N\to\infty}\mathbb{P}\left\{\min_{N(1-\delta)\le m\le N}\zeta_{N,m} < \min_{N\delta<m<N(1-\delta)}\zeta_{N,m}\right\} = 1.$$

Our second new class of rules use heavy weighted splitting by weighting the daughter node sample sizes using a more aggressive weight. The heavy weighted variance splitting rule is

$$\hat{\gamma}_H(s,t) = \hat{P}(t_L)^2\hat{\Gamma}(t_h) + \hat{P}(t_R)^2\hat{\Gamma}(t_R). \tag{18.11}$$

The best split s is found by minimizing $\hat{\gamma}_H(s,t)$. Notice that $\hat{\gamma}_H(s,t)$ weights the variance by using the squared daughter node size, which is a power larger than the weights used by weighted variance splitting. It is convenient to rewrite (18.11) as

$$\hat{\gamma}_H(s,t) = N_L\sum_{i\in t_L} Y_i^2 + N_R\sum_{i\in t_R} Y_i^2 - \frac{1}{N_L}\left(\sum_{i\in t_L} Y_i\right)^2 - \frac{1}{N_R}\left(\sum_{i\in t_R} Y_i\right)^2.$$

Unlike weighted and unweighted variance splitting, heavy weighted variance splitting does not have the ECP property. The following result shows that heavy weighted splitting splits near the center for noisy variables. Therefore, it is the greediest in terms of using up the sample size.

Theorem 18.5. *Let $Z_1,\ldots,Z_N$ be i.i.d. such that $\mathbb{E}(Z_1^4)<\infty$. Let*

$$\begin{aligned}\varphi_{N,m} &= m\sum_{i=1}^{m} Z_i^2 + (N-m)\sum_{i=m+1}^{N} Z_i^2 \\ &\quad - \frac{1}{m}\left(\sum_{i=1}^{m} Z_i\right)^2 - \frac{1}{N-m}\left(\sum_{i=m+1}^{N} Z_i\right)^2, \quad 1\le m\le N-1.\end{aligned} \tag{18.12}$$

Then for any $0<\delta<1/2$,

$$\lim_{N\to\infty}\mathbb{P}\left\{\min_{1\le m<N\delta}\varphi_{N,m} > \min_{N\delta\le m\le N(1-\delta)}\varphi_{N,m}\right\} = 1$$

and

$$\lim_{N\to\infty}\mathbb{P}\left\{\min_{N(1-\delta)<m\le N}\varphi_{N,m} > \min_{N\delta\le m\le N(1-\delta)}\varphi_{N,m}\right\} = 1.$$

We observe that the ECP property only holds for weighted and unweighted splitting rules, but not heavy weighted splitting rules. For convenience, we summarize the three splitting rules below.

> *Splitting rules (18.8) are called weighted splitting rules, (18.10) are called unweighted splitting rules, while (18.12) are called heavy weighted rules.*

18.5 Comparison of Split-Rules in the One-Dimensional Case

To investigate the differences between our three splitting rules we simulated $n = 100$ observations from the $p = 1$ dimensional model

$$Y_i = c_0 + c_1 X_i + \varepsilon_i, \qquad i = 1, \ldots, n,$$

where X_i was drawn independently from a uniform distribution on $[-3, 3]$ and ε_i was drawn independently from a standard normal. We considered three scenarios: (a) noisy $(c_0 = 1, c_1 = 0)$; (b) moderate signal $(c_0 = 1, c_1 = 0.5)$; and (c) strong signal $(c_0 = 1, c_1 = 2)$.

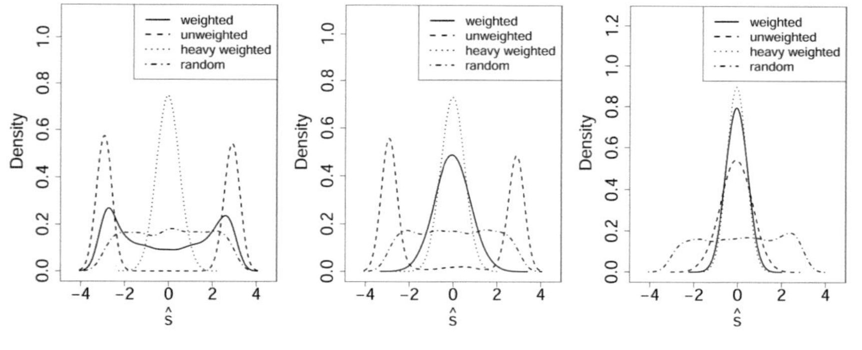

FIGURE 18.4
Density for split-point $\hat{s}$ under weighted variance, unweighted variance, heavy weighted variance, and random splitting where $F(x) = c_0 + c_1 x$ for (a) $c_0 = 1, c_1 = 0$ (left); (b) $c_0 = 1, c_1 = 0.5$ (middle); (c) $c_0 = 1, c_1 = 2$ (right).

The simulation was repeated 10,000 times independently. The optimal split-point $\hat{s}$ under each rule was recorded. In addition to this we also recorded the split-point behavior under pure random splitting in which the split-point was randomly selected. Figure 18.4 displays the density for $\hat{s}$ for each of the four splitting rules. In the noisy variable setting (a), only weighted and unweighted splitting exhibit ECP behavior. When the signal increases (b), weighted and heavy weighted splitting tends to split in the middle, which is optimal for uncovering the signal, whereas unweighted splitting continues to exhibit ECP behavior. In the strong signal setting (c), unweighted splitting finally recognizes the signal and accordingly splits near the middle. In all three scenarios, heavy weighted splitting splits near the middle, random splitting is uniform, while weighted splitting performs correctly in all cases.

This confirms our earlier hypothesis: weighted splitting is the most adaptive. In noisy scenarios it exhibits ECP tendencies but even in moderate signal scenarios it is able to shut off ECP splitting enabling it to recover signal.

18.6 The ECP Statistic

We return to the $p > 1$ simulation (18.7) in order to see if our previous results hold in higher dimensions. To facilitate this analysis, we use the ECP statistic measuring the closeness of a split to an edge. Let $\hat{s}$ be the optimized split for a continuous variable X with values $x_1 < x_2 < \cdots < x_N$ in the node t to be split. Then, $\hat{s} = x_j$ for some $1 \le j \le N-1$. Let $j(\hat{s})$ denote this j. The ECP statistic is defined as

$$\text{ecp}(\hat{s}) = \frac{1}{2} - \frac{\min\left\{N-1-j(\hat{s}),\ j(\hat{s})-1\right\}}{N-1}.$$

The ECP statistic is motivated by the following observations. The closest that $\hat{s}$ can be to the right and left most splits is $j(\hat{s}) = N-1$ and $j(\hat{s}) = 1$. The second term on the right chooses the smallest of these two values and divides by the total number of available splits, $N-1$. This ratio is bounded by $1/2$. Subtracting this from $1/2$ yields a statistic between 0 and $1/2$ that is largest when the split is nearest an edge and smallest when the split is away from an edge.

We simulated $n = 1000$ observations from (18.7). All parameters were kept the same as before, except now 25 noise variables were used in place of only 3 noisy variables. Figure 18.5 displays the ecp($\hat{s}$) statistic for variable X (strong non-linear variable), variable U_1 (linear variable with moderate signal) and U_{d+1} (noisy variable). Large points in red indicate high ECP values, smaller points in blue are moderate ECP values, and small black points are small ECP values.

For weighted splitting (top panel), ECP values are high near -1 and 1.5 and low between $[-1, 1.5]$. This is because the observed values of Y are relatively constant in the range $[-1, 1.5]$ which causes splits to occur relatively infrequently in this region, similar to Figure 18.2 and end-cut splits to occur at its edges. Most of the splits occur in $[-3, -1)$ and $(1.5, 3]$ where Y is non-linear in X, and many of these splits occur at relatively small depths, reflecting a strong signal in X in these regions. For U_1, ECP behavior is generally uniform. This is expected because U_1 contributes a linear term to Y so that splits will be at the mid-point of each node. For the noisy variable, ECP behavior is confined to the edges -3 and 3.

Unweighted splitting (middle panel) exhibits aggressive ECP behavior for X across much of its range of values (but excluding $[-1, 1.5]$, where again splits of any kind are infrequent). For U_1, ECP behavior is more apparent at the edges. For the noisy variable, aggressive ECP behavior is seen at the edges -3 and 3. Finally, heavy weighted splitting (bottom panel) responds well to the signal variable X but does not have good ECP behavior for the noise variable.

18.7 Regression Benchmark Results

We previously argued that the ECP property, while often viewed as undesirable for individual trees, can be advantageous for a procedure like Random Forests (RF). To investigate this, Ishwaran [7] conducted an extensive benchmark analysis of RF regression. They employed various splitting criteria, including weighted variance, unweighted variance, heavily weighted variance, and pure random splitting, across multiple datasets. The prediction error on the test

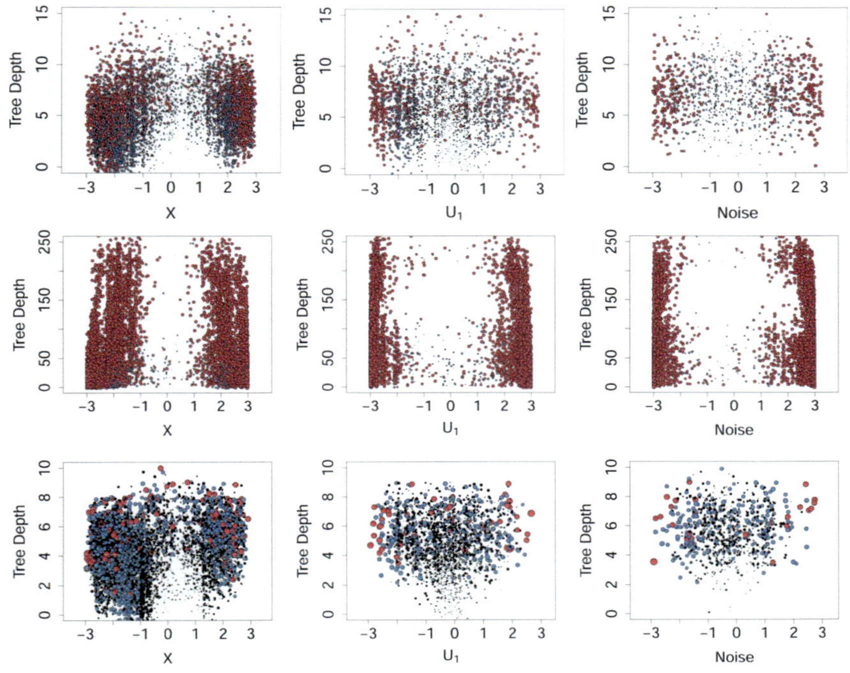

FIGURE 18.5

ECP statistic, ecp($\hat{s}$), for variable X (strong, non-linear), U_1 (moderate, linear), and U_{d+1} (noisy, labeled as "Noise") from simulation (18.7) with $n = 1000$, and $p = 36$ ($d = 10$, $D = 35$). Top, middle, and bottom panels are weighted, unweighted, and heavy weighted variance splitting. Circles are proportional to their ecp($\hat{s}$) statistic. Black, blue, and red indicate low, medium, and high ecp($\hat{s}$) values. Values are jittered to aid visualization.

set was assessed using 10-fold cross-validation. The datasets used for this benchmark are the same as those discussed in Chapter 17.5.

We summarize the results of Table 18.1.

1. Pure random splitting is consistently the worst especially over the synthetic examples with highly nonlinear functions (friedman and peak) and for the high-dimensional examples. The only instance where it performs well is the "CMB" dataset with a large n and small p.
2. Heavy weighted splitting is the second worst.
3. The best is weighted splitting, followed by unweighted splitting.

Regarding the last point, we note that there is one example where unweighted splitting performs exceptionally well: the "peak" dataset. This result is intriguing and not easily explained, likely it is due to the geometric shape of the underlying function, which seems to favor this type of split. To investigate further, we calculated the ecp($\hat{s}$) statistics for a subset of the variables from the data, which are displayed in Figure 18.6. Unweighted splitting exhibits aggressive ECP behavior, consistent with previous observations. However, split-points are distributed across a wide range of x values $[-1.5, 1.5]$, whereas for weighted variance splitting, the split-points are concentrated over a narrower range $[-1, 1]$. Additionally, the tree depth for unweighted splitting is significantly greater than that for weighted variance splitting, as also noted in previous figures. These two features may provide unweighted splitting with an advantage in this particular problem.

TABLE 18.1
RF regression test set prediction error.

	n	p	weighted	unweighted	heavy weighted	random
Air	111	5	0.21	0.20	0.24	0.33
Automobile	193	24	5.54	5.46	5.91	16.92
Bodyfat	252	13	22.38	22.76	23.75	31.94
Boston	506	13	12.82	15.13	13.52	26.07
CMB	899	4	11.12	10.45	10.76	9.73
Crime	47	15	89194.67	90463.13	88629.05	113937.42
Diabetes	442	10	3242.19	3531.40	3283.92	3520.65
DiabetesI	442	64	3248.11	3980.96	3288.99	4630.17
Fitness	31	6	18.30	18.00	19.14	22.86
Highway	39	11	0.17	0.17	0.19	0.30
Iowa	33	9	107.60	108.08	116.75	149.24
Ozone	203	12	17.35	15.42	18.24	21.34
Pollute	60	15	1884.61	1760.53	1956.96	2611.61
Prostate	97	8	0.61	0.62	0.63	0.66
Servo	167	19	0.53	0.42	0.73	1.04
Servo_factor	167	4	0.70	0.70	0.79	1.19
Tecator	215	22	28.99	32.74	31.32	84.07
Tecator2	215	100	55.66	58.89	56.73	88.88
Windmill	1114	12	4.46	5.00	4.55	5.19
expon	250	2	4.02	3.97	4.65	5.38
expon.noise	250	17	6.20	8.27	9.43	10.68
friedman1	250	10	7.28	8.73	8.96	15.42
friedman1.bigp	250	250	11.75	21.71	13.95	24.54
friedman2	250	4	44286.80	45130.04	47909.32	67308.38
friedman2.bigp	250	254	49781.22	86684.01	55236.97	159378.46
friedman3	250	4	0.04	0.05	0.04	0.07
friedman3.bigp	250	254	0.05	0.08	0.05	0.08
peak	250	20	15.77	5.39	26.28	28.51
peak.bigp	250	220	13.24	5.70	18.63	28.73
noise	250	500	0.88	0.92	0.88	0.88
sine	250	2	13.70	17.08	13.05	19.71
sine.noise	250	5	25.66	36.37	22.11	30.78
AML	116	629	73.93	77.73	77.29	118.51
DLBCL	240	740	70.09	72.27	79.29	126.88
Lung	86	713	4.42	5.15	4.74	9.85
MCL	92	881	51.83	49.58	54.08	133.45
VandeVijver	78	475	7.45	7.56	7.66	14.52

18.8 ECP Behavior for Partially Randomized Splitting Rules

We found that pure random splitting is rarely effective. It does not possess the ECP property and it does not adapt to signal. However, randomized rules are desirable because of their

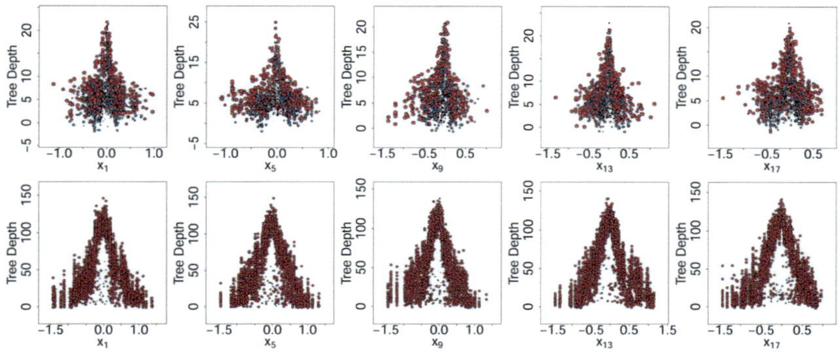

FIGURE 18.6
ECP statistic, ecp($\hat{s}$), for peak data using weighted variance splitting (top) and unweighted variance splitting (bottom).

computational efficiency, therefore to take advantage of this while achieving adaptivity, we consider *partially randomized* splitting. In place of deterministic splitting in which the splitting rule is calculated for the entire set of N available split-points for a variable, the partially randomized rule is confined to a set of split-points indexed by $I_N \subseteq \{1,\ldots,N\}$, where $|I_N|$ is typically much smaller than N. This reduces the search for the optimal split-point from a maximum of N split-points to the much smaller $|I_N|$.

To study the ECP behavior for these rules, we confine our analysis to weighted splitting. Thus, in place of maximizing (18.8) over the available split values for $1 \le m \le N-1$, we restrict the maximization to I_N. The optimal split-point is determined by maximizing the restricted splitting rule:

$$\xi^r_{N,m} = \frac{1}{m}\left(\sum_{i=1}^{m} Z_{N,i}\right)^2 + \frac{1}{R_N - m}\left(\sum_{i=m+1}^{R_N} Z_{N,i}\right)^2, \qquad 1 \le m \le R_N - 1, \tag{18.13}$$

where $R_N = |I_N|$ and $(Z_{N,i})_{i=1}^{R_N}$ denotes the sequence of values $\{Z_i : i \in I_N\}$.

To select I_N, we choose nsplit candidate split-points at random. This corresponds to sampling R_N-out-of-N values from $\{1,\ldots,N\}$ without replacement, where R_N = nsplit. The following result shows that (18.13) has the ECP property under any sequence $(I_N)_{N\ge1}$ where $R_N \to \infty$, which covers the nsplit method for selection just described. The result requires only a slightly stronger moment assumption than Theorem 18.3. The proof makes use of a law of iterated logarithms for triangular arrays [8].

Theorem 18.6. *Let $(Z_i)_{i=1}^N$ be independent with a common mean and variance and assume $\sup_i \mathbb{E}(|Z_i|^q) < \infty$ for some $q > 2$. Let $(I_N)_{N\ge1}$ be a sequence of index sets such that $R_N \to \infty$. Then for any $0 < \delta < 1/2$ and any $0 < \tau < \infty$:*

$$\lim_{N\to\infty} \mathbb{P}\left\{\max_{1\le m\le R_N\delta} \xi^r_{N,m} > \max_{R_N\delta < m < R_N(1-\delta)} \tau\xi^r_{N,m}\right\} = 1$$

and

$$\lim_{N\to\infty} \mathbb{P}\left\{\max_{R_N(1-\delta)\le m\le R_N} \xi^r_{N,m} > \max_{R_N\delta < m < R_N(1-\delta)} \tau\xi^r_{N,m}\right\} = 1.$$

Remark 18.1. As a special case, Theorem 18.6 yields Theorem 18.3 for the sequence $I_N = \{1,\ldots,N\}$. Note that while the moment condition is somewhat stronger, Theorem 18.6 does not require $(Z_i)_{i=1}^N$ to be i.i.d. but only independent. □

Theorem 18.6 shows that the ECP property holds if nsplit $\to \infty$. Because any rate is possible, the condition is mild and provides justification for partially randomized rules which are commonly used with RF. However, it is interesting to note that the choice nsplit $= 1$, which corresponding to the popular extremely randomized tree method of Geurts et al. [4], does not satisfy the rate condition.

18.9 Conclusion

In the presence of signal, weighted splitting has the desirable property of splitting at points of curvature of the underlying target function. For noisy variables, weighted splitting favors end-cut splits. This adaptivity to both noise and signal does not hold for unweighted and heavy weighted splitting. These latter procedures are either too greedy, making them poor at recognizing noisy scenarios, or they are overly ECP aggressive, making them poor at recognizing signal. The least effective procedure is pure random splitting which does not adapt to signal and does not have the ECP behavior. On the other hand, partially randomized splitting employing random split-point selection has the ECP property if nsplit is not too small relative to the sample size. Ishwaran [7] studied their empirical properties and found the procedure to be competitive when used in combination with weighted splitting, especially with increasing nsplit.

Exercises

The objective in these exercises is to review and understand the concept of the parameter r in the weighted splitting criteria for regression and extend this to classification. We have considered split-statistics

$$\hat{\gamma}(s,t) = \hat{P}(t_L)^r\,\hat{\Gamma}(t_L) + \hat{P}(t_R)^r\,\hat{\Gamma}(t_R), \qquad r = 0,1,2.$$

The case $r = 1$ is the classical CART split-statistic (weighted variance for regression). Values $r = 0,2$ are unweighted and heavy weighted splitting rules, respectively.

Exercise 18.1. Using the above notation:

1. Define the Gini impurity split-statistic and the weighted and unweighted Gini split-statistic for classification.
2. Explain why the theory we developed in this chapter for regression also holds for the classification case.

Exercise 18.2. Building on the previous example, let us consider the classification setting.

1. Design a custom weighted variance splitting criterion for a classification task using weighted Gini impurity with $r = 0,1,2$.
2. Implement the custom splitting criterion in a decision tree algorithm.
3. Train the modified decision tree on a classification dataset.
4. Evaluate the performance of the custom criterion compared to the standard criterion ($r = 1$).

Exercise 18.3. Analyze the splitting behavior of random forests (RF) using weighted variance and its impact on model performance for different values of r in classification:

1. Train a RF model on a classification dataset and a regression dataset using $r = 0,1,2$.

2. Extract and analyze the splitting rules used in the first few trees of the forest. Identify the most frequently used features and splitting points for each value of r.
3. Discuss how the choice of r impacts the diversity and performance of the ensemble model.

Exercise 18.4. Explore how different values of r in the weighted variance splitting criteria affect the tendency of a decision tree or RF to overfit:

1. Train decision trees and random forests with different values of r in weighted variance splitting criteria (e.g., $r = 0, 1, 2$) on a synthetic dataset with known patterns and noise.
2. Evaluate the performance of each model on training and testing datasets.
3. Analyze the extent of overfitting for each value of r and discuss the results.

Exercise 18.5. Investigate the effect of the maximum depth parameter on the splitting behavior and performance of RF with weighted variance splitting criteria for different values of r:

1. Train random forest models with varying maximum depths on classification and regression datasets using $r = 0, 1, 2$.
2. Analyze how the maximum depth affects the number of splits, the complexity of the trees, and the overall model performance for each value of r.
3. Discuss the trade-offs between model complexity, interpretability, and performance when adjusting the maximum depth parameter for different values of r.

Exercise 18.6. Explore the impact of weighted splitting on the bias and variance of RF.

1. Discuss how the choice of different weights $\hat{P}(t_L)^r$ and $\hat{P}(t_R)^r$ for $r = 0, 1, 2$ influences the bias and variance of the individual trees in the forest.
2. Provide a theoretical analysis of how weighted splitting affects the overall bias-variance tradeoff in a random forest ensemble. For this analysis you are allowed to simplify the forest construction in any way you deem helpful.

Exercise 18.7. Examine how weighted splitting affects feature importance in random forests. Specifically, using carefully designed experiments, discuss how the use of different weights $\hat{P}(t_L)^r$ and $\hat{P}(t_R)^r$ for $r = 0, 1, 2$ changes the interpretability and reliability of the feature importance measures.

Bibliography

[1] Breiman L., Friedman J.H., Olshen R.A., and Stone C.J. *Classification and Regression Trees*, Belmont, California, 1984.

[2] Cutler A. and Zhao G. Pert - perfect random tree ensembles. *Computing Science and Statistics*, 33:490–497, 2001.

[3] Dietterich T.G. An experimental comparison of three methods for constructing ensembles of decision trees: bagging, boosting, and randomization. *Machine Learning*, 40:139–157, 2000.

[4] Geurts P., Ernst D. and Wehenkel L. Extremely Randomized Trees. *Machine Learning*, 63: 3–42, 2006.

[5] Ishwaran H., Kogalur U.B., Blackstone E.H., and Lauer M.S. Random survival forests. *Annals of Applied Statistics*, 2:841–860, 2008

[6] Ishwaran H., Kogalur U.B., Gorodeski E.Z., Minn A.J., and Lauer M.S. High-dimensional variable selection for survival data. *Journal of the American Statistical Association*, 105:205–217, 2010.

[7] Ishwaran H. The effect of splitting on random forests. *Machine Learning*, 99(1):75–118, 2015.

[8] Lai T. L. and Wei C.Z. A law of the iterated logarithm for double arrays of independent random variables with applications to regression and time series models. *Annals of Probability*, 19:320–335, 1982.

[9] Morgan J.N. and Messenger R.C. *THAID: a Sequential Search Program for the Analysis of Nominal Scale Dependent Variables*, Survey Research Center, Institute for Social Research, University of Michigan", 1973.

19

Random Survival Forests

19.1 Introduction

In this chapter, we explore random survival forests (RSF) [17], an extension of random forests (RF) designed for right-censored survival data. In Chapter 16, we discussed RF in the context of regression and classification, along with some extensions such as multivariate forests and quantile regression forests. The core principles of RF can be adapted to more complex tasks like survival analysis. We begin with a high-level definition of RSF.

Definition 19.1. A random survival forests (RSF) consists of a collection of survival-tree structured learners $\{h(\cdot, \mathcal{T}_b(\Theta_b)), b = 1, \ldots, B\}$ where $\{\Theta_b\}$ are i.i.d. elements encoding the bootstrapping and random feature instructions for constructing the survival tree partition $\mathcal{T}_b := \mathcal{T}_b(\Theta_b)$. The RSF ensemble is the tree-averaged estimator $\sum_{b=1}^{B} h(\cdot, \mathcal{T}_b)/B$. □

Although Definition 19.1 resembles the definition of RF, survival analysis differs significantly from regression and classification, necessitating certain modifications to RF. One of the primary differences is the way the data is collected, as survival analysis involves some form of censoring. This means the data used for training is only partially observed, which is quite different from classical RF settings. In the following sections, we will highlight these differences and introduce the Kaplan-Meier estimator to address them, before delving into the details of RSF, its extensions, and practical applications.

19.2 The Difference Between Survival Data and Usual Settings

We use the following survival notation to clarify the difference between survival data and the usual regression or classification setting. Let $(\mathbf{X}, T, \delta)$ be the data with distribution $\mathbb{P}$. As usual, $\mathbf{X}$ represents the p-dimensional feature vector from the feature space $\mathscr{X}$. However, the response outcome differs from typical settings and is composed of two values: (T, δ), where $T = \min(T^o, C) \geq 0$ denotes the observed time, and $\delta = 1_{\{T^o \leq C\}} \in \{0, 1\}$ is the binary censoring indicator.

The value T is what we observe, but the true lifetime is T^o, which may be unobserved depending on whether T^o is greater than C. The variable C represents the censoring time, meaning we observe the true lifetime T^o only if $T^o \leq C$. In survival analysis, when this condition holds, $\delta = 1$, indicating that the individual experienced the event at time $T = T^o$. Otherwise, the individual is right-censored at time $T = C$, and $\delta = 0$. Thus, because of censoring, we do not know exactly when the individual experiences the event of interest, but only that they have survived up to time $T = C$.

DOI: 10.1201/9781003493679-19

$$
\begin{array}{rcl}
(T,\delta) & = & \textit{Observed time } T \geq 0 \textit{ and censoring indicator, } \delta \in \{0,1\} \\
\delta = 1 & \rightarrow & \textit{Event has occured at time } T \\
\delta = 0 & \rightarrow & \textit{Censoring has occurred at } T \textit{ and the true event time is} > T \\
T^o & = & \textit{True event time which can potentially be unobserved} \\
C & = & \textit{Censoring time which can potentially be unobserved}
\end{array}
$$

This type of data is called *right-censored*, and it arises in studies utilizing a *Type-I censoring* experimental design. These designs terminate a study after a specified period, resulting in censored data where the exact event time may be unknown. For instance, in a study investigating the effect of a treatment on cancer, researchers may aim to determine whether the treatment extends the time until cancer recurrence, which is the event of interest. In this context, the goal is to assess whether the treatment improves a patient's survival. However, since the study cannot continue indefinitely, patients may only be observed for a limited time.

In the strictest form of Type-I censoring, the study begins at a specific time point and concludes at a predetermined time. As a result, by the end of the study, all patients will either have experienced the event, with the event time recorded as the time of recurrence (the event of interest), or they will be right-censored at the study's conclusion. In the former case, the recorded value T represents the time from the study's start until the event of interest. In the latter case, where a patient is censored, the recorded value T is the duration of the study minus the start time.

A more flexible approach is *generalized Type-I censoring* experimental designs, which are more commonly used in medicine and clinical trials. In these designs, patients enter the study at different times and are followed until the study concludes. As a result, the follow-up time differs for each patient. For example, in a cancer study that begins recruiting patients in 2000 and ends in 2005, patients can enroll at any point during this period. The recorded value T for an event is calculated as the time from the occurrence of the event to the patient's entry into the study. For right-censored observations, where the patient does not experience the event before the study ends in 2005, T is the difference between the study's end time (2005) and the patient's entry time.

Figure 19.1 illustrates an example of this type of experimental design with 5 subjects entering at different times. Each subject's time is set to zero upon entry, and they are followed until either an event occurs or the study concludes. The blue line represents the true event time, while the red line represents the censoring time. The observed time is the smaller of these two values, indicated by "x" for an event and "o" for a right-censored observation. For instance, Subject 5 has a true event time of 15 and a censoring time of 13, meaning this subject is right-censored with an observed time of 13. Subject 5 also has the longest observed time of 13, indicating that the study continued for at least 13 time units.

Other types of censoring can also occur in survival analysis, such as *left-censoring*. Left-censoring occurs when the event of interest may have happened before the study begins, but the exact time is unknown. For example, in a study investigating the time to onset of diabetes among adults, participants may be enrolled at a specific point in time, but some may have already developed diabetes prior to joining the study. While the exact time of diabetes onset is unknown, it is known that these participants had diabetes upon enrollment. This is an example of left-censoring.

Mathematically, the data from left-censoring is represented by a pair (T,ε), where ε indicates whether the exact event time is observed ($\varepsilon = 1$) or not ($\varepsilon = 0$). If an individual's true event time

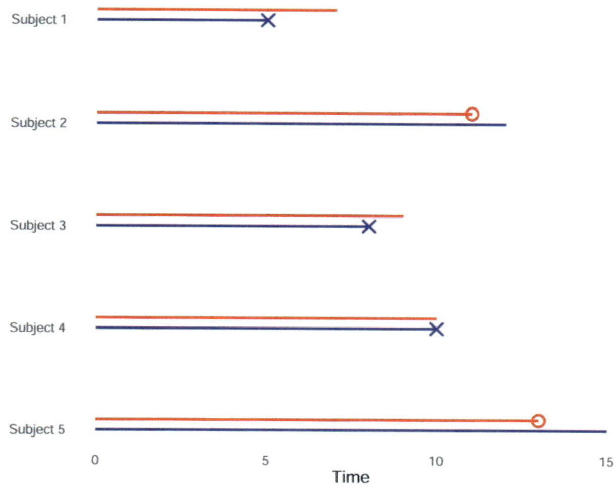

FIGURE 19.1
Illustration of generalized Type-I censoring. Blue and red lines show the true event and censoring time. The observed time is the smaller of these and is shown by a symbol where "x" is an event and "o" is a censored case.

T^o is less than or equal to the observed time T, the event is left-censored, and the censoring indicator is $\varepsilon = 0$ because the exact event time is unknown, though it occurred before C_l, the left-censoring time, i.e., the time the individual enters the study (in the previous example, this would be the case if the individual developed diabetes prior to the start of the study). If the observed time equals the true event time, then $\varepsilon = 1$, meaning the event time is observed (in this case, the onset of diabetes is recorded). Therefore, in left-censoring, $T = \max(T^o, C_l)$, and the event time is observed when $\varepsilon = 1$, which occurs if and only if $T^o \geq C_l$.

Another complication in survival analysis is *truncation*, which differs from censoring. Truncation occurs when only patients with event times within an observational interval (τ_L, τ_R) are observed. If the event time falls outside this interval, no information is available. Unlike censoring, no partial information is available in this case. A common special case is left-truncation, where $\tau_R = \infty$. In this scenario, we only observe information if the event time exceeds the truncation time τ_L. In medical studies, left-truncation may arise when examining intermediate events. For instance, in a study of death following a heart transplant, patients must first survive while waiting for a heart; those who do not survive are truncated from the study. Censoring can also occur with truncated data. For example, if a heart transplant patient is still alive at the end of the study period, they are both right-censored and left-truncated.

In this chapter we will only be interested in right-censoring without truncation. In order to develop proper inference in the setting of right-censored survival analysis it is required that some form of independence of T^o and C must hold. A common assumption is the random censorship model, which assumes the two values are independent, $T^o \perp C$. A weaker and more flexible assumption is conditional independence

$$T^o \perp C \mid \mathbf{X} \tag{19.1}$$

which states the two quantities are independent, conditionally on the feature $\mathbf{X}$.

Under the random censorship assumption, the joint density for (T^o, C) factorizes. Therefore, the joint density for (T, δ) can be described by

$$\begin{aligned} f_{T,\delta}(t,\delta) &= 1_{\{\delta=0\}}\mathbb{P}\{T=t,\delta=0\} + 1_{\{\delta=1\}}\mathbb{P}\{T=t,\delta=1\} \\ &= 1_{\{\delta=0\}}\mathbb{P}\{T^o>t, C=t\} + 1_{\{\delta=1\}}\mathbb{P}\{T^o=t, C>t\} \\ &= 1_{\{\delta=0\}}\mathbb{P}\{T^o>t\}f_C(t) + 1_{\{\delta=1\}}f_{T^o}(t)\mathbb{P}\{C>t\} \\ &= 1_{\{\delta=0\}}S(t)f_C(t) + 1_{\{\delta=1\}}f_{T^o}(t)G(t) \end{aligned}$$

where S and G are the survival and censoring functions defined by

$$S(t) = \mathbb{P}\{T^o > t\}, \quad G(t) = \mathbb{P}\{C > t\}$$

where we have assumed that all random variables are continuous. The densities for T^o and C are denoted by f_{T^o} and f_C.

Therefore, if $(T_1, \delta_1), \ldots, (T_n, \delta_n)$ are i.i.d. values, under the random censorship model, the likelihood for the observed data is

$$\begin{aligned} \prod_{i=1}^{n} f_{T,\delta}(t_i,\delta_i) &= \left[\prod_{\delta_i=0} S(t_i)f_C(t_i)\right]\left[\prod_{\delta_i=1} f_{T^o}(t_i)G(t_i)\right] \\ &= \left[\prod_{i=1}^{n} f_C(t_i)^{1-\delta_i}G(t_i)^{\delta_i}\right]\left[\prod_{i=1}^{n} f_{T^o}(t_i)^{\delta_i}S(t_i)^{1-\delta_i}\right]. \end{aligned}$$

Since T^o and C are assumed to be independent, the first term in the likelihood will involve parameters independent of the second term, therefore we can work strictly with the latter term, which is the likelihood for the observed data expressed in terms of the lifetime density and the lifetime survival function

$$\prod_{i=1}^{n} f_{T^o}(t_i)^{\delta_i}S(t_i)^{1-\delta_i}.$$

This makes it apparent that the survival function is an important quantity for inference.

More generally, if the data $(\mathbf{X}_1, T_1, \delta_1), \ldots, (\mathbf{X}_n, T_n, \delta_n)$ are assumed to be i.i.d., then under the conditional independence assumption (19.1), using the same argument, the relevant quantity from the likelihood is

$$\prod_{i=1}^{n} f_{T^o|\mathbf{X}}(t_i|\mathbf{x}_i)^{\delta_i}S(t_i|\mathbf{x}_i)^{1-\delta_i} \tag{19.2}$$

where $f_{T^o|\mathbf{X}}$ is the conditional density of T^o given $\mathbf{X}$ and $S(t|\mathbf{X}) = \mathbb{P}\{T^o > t|\mathbf{X}\}$ is the survival function conditioned on $\mathbf{X}$.

From the preceding discussions, it is clear that the quantities involved in the likelihood for survival analysis differ significantly from those in standard applications of RF, such as regression and classification. One particularly important quantity is the survival function $S(t) = \mathbb{P}\{T^o > t\}$, or more generally, the conditional survival function $S(t|\mathbf{X}) = \mathbb{P}\{T^o > t|\mathbf{X}\}$, which summarizes lifetime behavior and represents the probability that an individual survives beyond a given time t.

19.3 Kaplan-Meier Estimator

There are several key aspects to consider when constructing RSF, but one of the most important is the survival function, which is central to survival analysis. To address this, we begin by discussing the Kaplan-Meier estimator (KM estimator) [20], a nonparametric estimator of the survival function that serves as the terminal node estimator in RSF. The KM estimator was briefly introduced in Chapter 15.9. A more formal description is provided below.

Let $\mathrm{t}_1 < \mathrm{t}_2 < \cdots < \mathrm{t}_m$ denote the $m \le n$ distinct ordered event times from $T_1, \ldots, T_n$. Let $d(\mathrm{t}_k) = \sum_{i=1}^n 1_{\{T_i = \mathrm{t}_k, \delta_i = 1\}}$ be the number of events at t_k, $N(t) = \sum_{i=1}^n 1_{\{T_i \le t, \delta_i = 1\}}$ be the number of events in $[0, t]$ and $Y(t) = \sum_{i=1}^n 1_{\{T_i \ge t\}}$ be the number of individuals at risk (event-free and uncensored) just prior to t. The KM estimator for the survival function is given by

$$\hat{S}(t) = \prod_{s \le t} \left(1 - \frac{N(ds)}{Y(s)}\right) = \prod_{k=1}^{m(t)} \left(1 - \frac{d(\mathrm{t}_k)}{Y(\mathrm{t}_k)}\right) \tag{19.3}$$

where $m(t) = \max\{k : \mathrm{t}_k \le t\}$. The KM estimator uses all of the data, including the censored observations, to estimate the survival function. A slightly easier representation can be given if there are no ties. Then

$$\hat{S}(t) = \begin{cases} \prod^* \left(1 - \dfrac{1}{n - i + 1}\right) & \text{for } t \le T_{(n)} \\ 0 & \text{if } t > T_{(n)} \text{ and } \delta_{(n)} = 1 \\ \text{undefined} & \text{if } t > T_{(n)} \text{ and } \delta_{(n)} = 0 \end{cases}$$

where $T_{(1)} \le \cdots \le T_{(n)}$ are the ordered observed times and $\prod^*$ runs over all i such that $T_{(i)} \le t$ and $\delta_{(i)} = 1$.

The KM estimator is also referred to as the *product limit (PL) estimator*. This an efficient technique for using right-censored data for calculating conditional probabilities related to survival. The advantage of the PL method is explained using the following example which is taken from the original paper by Kaplan and Meier [20]. A random sample of 100 items are tested at the beginning of the year. Of these 70 die and 30 survive. In year two, a larger sample of 1000 identical items are available and these are tested. During year two, 15 items from the first sample die and 750 from the second sample die. See Table 19.1.

TABLE 19.1
The advantage of the Product Limit (PL) method.

Samples	I	II
Initial numbers	100	1000
Deaths in first year of age	70	750
One year survivors	30	250
Deaths in second year of age	15	
Two year survivors	15	

Consider estimating the proportion $S(2) = \mathbb{P}\{T^o > 2\}$, representing the fraction of items in the population surviving for more than two years. To demonstrate the efficiency of the PL estimator, we compare it to the *reduced sample estimator* $S^*(t)$, which naively discards information from sample II and ignores the data from year $t = 1$ for sample I. The reduced estimator compares the number alive at the end of the year to the number available at the start for sample I,

$$\hat{S}^*(2) = \frac{15}{100} = 0.15.$$

The PL estimator, on the other hand, makes use of all available information. First, note that the probability of surviving more than one year is

$$\overbrace{\hat{S}(1)}^{\text{PL}} = \overbrace{\hat{S}^*(1)}^{\text{Reduced}} = \frac{30 + 250}{100 + 1000} = 0.255,$$

where both estimators are in agreement. However, for year $t = 2$, the PL estimator uses the conditional probability of surviving more than two years, given survival beyond one year,

$$\hat{S}(2|1) = \frac{15}{30} = 0.5.$$

Thus,

$$\hat{S}(2) = \hat{S}(1) \times \hat{S}(2|1) = 0.255 \times 0.5 = 0.127,$$

which differs from $\hat{S}^*(2) = 0.15$, highlighting the inefficiency and poor performance of the reduced estimator.

In general, we can think of the PL estimator as a conditioning estimator with the general estimation strategy

$$\hat{S}(t) = \hat{S}(1) \times \hat{S}(2|1) \times \cdots \times \hat{S}(t|t-1)$$

with

$$\hat{S}(j|j-1) = \frac{\text{number surviving beyond } t = j}{\text{number surviving beyond } t = j-1}.$$

We can compare this to the KM estimator

$$\hat{S}(t) = \prod_{k=1}^{m(t)} \left(\frac{Y(\mathrm{t}_k) - d(\mathrm{t}_k)}{Y(\mathrm{t}_k)} \right) = \prod_{k=1}^{m(t)} \left(\frac{\text{number surviving beyond } \mathrm{t}_k}{\text{number surviving beyond } \mathrm{t}_{k-1}} \right).$$

19.4 The RSF Ensemble, Prediction and Performance

In order to construct RSF there are four aspects that need to be addressed: (1) Splitting rules; (2) Tree estimators and ensembles; (3) Prediction measure; (4) Performance metrics. We address each of these issues separately below.

19.4.1 Splitting rules

Due to working with survival data, the splitting rule for the random survival tree must take into account right-censoring. The most commonly used rule for this purpose is the log-rank rule (Chapter 15.9). Another example is log-rank score splitting [12].

19.4.2 Survival function tree estimator and ensemble

The RSF tree estimator for the survival function is defined by assigning all cases within a terminal node to the terminal node KM estimator. Let $\mathcal{T} = \{A_1, \ldots, A_K\}$ represent the tree survival partition. The survival estimator for the k-th cell is denoted as $\hat{S}^*(t|A_k)$, which is the KM estimator calculated using in-bag (bootstrap) cases within A_k (as before, we use * to indicate a calculation based on bootstrap data). The tree estimator for the survival function $S(t|\mathbf{x}) = \mathbb{P}\{T^o > t|\mathbf{x}\}$ is

$$\hat{S}^*(t|\mathbf{x}, \mathcal{T}) = \sum_{k=1}^{K} \hat{S}^*(t|A_k) 1_{\{\mathbf{x} \in A_k\}}.$$

Given in-bag survival trees $\hat{S}_1^*, \ldots, \hat{S}_B^*$ with partitions $\mathcal{T}_1, \ldots, \mathcal{T}_B$, the in-bag ensemble survival function estimator is

$$\hat{S}^*(t|\mathbf{x}) = \frac{1}{B} \sum_{b=1}^{B} \hat{S}^*(t|\mathbf{x}, \mathcal{T}_b).$$

Let $\mathcal{O}_i$ represent the out-of-bag (OOB) trees for $\mathbf{X}_i$ (recall that to be OOB, the case is not included in the bootstrap sample used to construct the tree learner). Then, the OOB ensemble for the i-th case is

$$\hat{S}^{(i)}(t) = \frac{1}{|\mathcal{O}_i|} \sum_{b \in \mathcal{O}_i} \hat{S}^*(t|\mathbf{X}_i, \mathcal{T}_b), \quad i = 1, \ldots, n.$$

As an illustration, estimated OOB survival curves from an RSF analysis are shown in Figure 19.2. In this analysis, we revisit the WECC esophageal cancer data described in Chapter 1 (readers may find it helpful to review that material). Recall that one of the goals of WECC was to develop a data-driven cancer staging system to refine and improve upon previous stage groupings. These previous groupings had been criticized for failing to adequately capture the complex interactions of cancer characteristics specific to the esophagus.

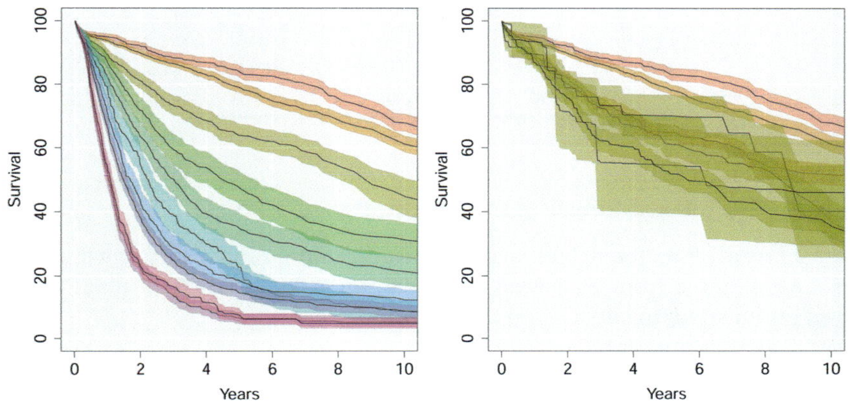

FIGURE 19.2
OOB survival curves for WECC esophageal cancer data where patients have been grouped by data-driven staging. Colored regions reflect ±1 standard error confidence limits where standard error is estimated by the bootstrap. Right figure shows survival curves for first 3 groups where group 3 has been expanded to show survival curves for its 4 patient subgroups.

For this analysis, we focused on adenocarcinoma patients (one of the findings of WECC was that survival depended significantly on the distinction between adenocarcinoma and squamous cell types). Cancer characteristics used in the RSF analysis included TNM classifications from pathology (pTNM), histologic grade (pG), which is a biological indicator of tumor growth, and the number of positive cancer lymph nodes. Unlike previous stage groupings that only considered the presence or absence of nodal involvement (pN0, pN1) and ignored grade, we included both pG and the number of lymph nodes for a more refined understanding of cancer involvement.

The left figure shows OOB survival curves for the first 10 stage groups, based on groupings suggested by the earlier analysis of Rice et al. [27]. The survival curves are distinct and decrease with increasing stage, as is expected from an effective stage grouping. Within each survival group, patients exhibit various cancer characteristics. Table 19.2 lists the first 3 stage groups and the associated cancer variables. For example, group 2 consists of patients with pT1 cancer depth and no nodal involvement (pN0), all having the same anatomic classification, pT1N0. However, histologic grade (pG) plays a role, with group 2 comprising both pG1 and pG2 grades. Thus, group 2 contains pT1N0G1 and pT1N0G2 patients.

The right figure expands on group 3, showing survival curves for its 4 patient subgroups: (i) pT1N0G3, (ii) pT2N0G1, (iii) pT2N0G2, and (iv) pT1N1G2 (with 1 regional node). This is a diverse collection, reflecting interactions between depth of invasion, tumor grade, and regional node activity. The survival curves capture this variability, though the 4 subgroups can be statistically merged into the larger stage group 3, as indicated by the overlap in the confidence limits of the survival curves. More discussion on merging stage groups can be found in Ishwaran et al. [15].

TABLE 19.2
Data-driven stage groups for WECC adenocarcinoma patients. All patients are pM0 without metastatic involvement.

group	n	pT	pN	pG	nodes
1	409	pTis	pN0	1	[0,1)
2	950	pT1	pN0	1	[0,1)
2	727	pT1	pN0	2	[0,1)
3	357	pT1	pN0	3	[0,1)
3	58	pT2	pN0	1	[0,1)
3	223	pT2	pN0	2	[0,1)
3	60	pT1	pN1	2	[1,2)

19.4.3 Cumulative hazard function tree estimator

The *cumulative hazard function* (CHF) is another important quantity in survival analysis. This quantity can be estimated from a survival tree partition as well. The CHF plays a role in RSF prediction and performance evaluation to be described later.

The Nelson-Aalen estimator [1] for the CHF is

$$\hat{H}(t) = \int_0^t \frac{N(ds)}{Y(s)} = \sum_{k=1}^{m(t)} \frac{d(\mathrm{t}_k)}{Y(\mathrm{t}_k)}, \qquad m(t) = \max\{k : \mathrm{t}_k \leq t\}. \tag{19.4}$$

The in-bag CHF tree estimator is

$$\hat{H}^*(t|\mathbf{x}, \mathcal{T}) = \sum_{k=1}^{K} \hat{H}^*(t|A_k) 1_{\{\mathbf{x} \in A_k\}}$$

where $\hat{H}^*(t|A_k)$ is the Nelson-Aalen estimator for cell A_k constructed using the bootstrap cases in A_k.

Given tree CHF estimators $\hat{H}_1^*, \ldots, \hat{H}_B^*$, the in-bag ensemble CHF estimator is

$$\hat{H}^*(t|\mathbf{x}) = \frac{1}{B} \sum_{b=1}^{B} \hat{H}^*(t|\mathbf{x}, \mathcal{T}_b). \tag{19.5}$$

The OOB ensemble estimator for i is

$$\hat{H}^{(i)}(t) = \frac{1}{|\mathcal{O}_i|} \sum_{b \in \mathcal{O}_i} \hat{H}^*(t|\mathbf{X}_i, \mathcal{T}_b), \qquad i = 1, \ldots, n. \tag{19.6}$$

19.4.4 Why two estimators?

Why does RSF estimate both the CHF and the survival function? There is a well known identity in survival analysis that shows that these two values are related, therefore it seems redundant to estimate both values. Let H and S be the CHF and survival function. Then assuming continuous random variables, we have the following well known identity

$$H(t) = -\log(S(t)). \tag{19.7}$$

A proof of (19.7) is as follows. Define the *hazard function* by

$$\lambda(t) = \lim_{\Delta t \to 0} \frac{\mathbb{P}\{t < T^o \leq t + \Delta t | T^o > t\}}{\Delta t} = \frac{f_{T^o}(t)}{S(t)} = -\frac{\partial \log(S(t))}{\partial t}. \tag{19.8}$$

The CHF is obtained by integrating the hazard function, $H(t) = \int_0^t \lambda(s)\,ds$. Then, $\int_0^t \lambda(s)\,ds = -\log(S(t))$ by the fundamental theorem of calculus because of (19.8).

Thus, given that H and S are monotonically related, why is it necessary to provide two separate estimates in RSF? The problem is that even if it were true that $H_b = -\log(S_b)$ for each tree $b = 1, \ldots, B$, the identity (19.7) will not hold for an ensemble. Let $\overline{S}$ and $\overline{H}$ denote the ensemble survival and ensemble CHF and let $\mathbb{E}_b$ denote ensemble expectation (i.e., $\overline{S} = \mathbb{E}_b[S_b]$ and $\overline{H} = \mathbb{E}_b[H_b]$). Then by Jensen's inequality for convex functions:

$$-\log(\overline{S}) = -\log(\mathbb{E}_b[S_b]) \leq \mathbb{E}_b[-\log(S_b)] = \mathbb{E}_b[H_b] = \overline{H}.$$

In other words, $-\log$ of the survival ensemble does not necessarily equal the ensemble of $-\log$ of the tree survival function. The inequality above also shows that taking $-\log$ of the ensemble survival will generally be smaller than the ensemble CHF. This is why RSF provides two separate values.

Remark 19.1. Identity (19.7) is part of a set of distributional relationships that hold between various survival quantities, which are summarized as follows (here U indicates a Uniform$[0,1]$ random variable, all variables are assumed to be continuous and $\stackrel{d}{=}$ means equal in distribution):

$$\begin{aligned}
T^o &\sim F := 1 - S \\
F(T^o) &\stackrel{d}{=} U \\
S(T^o) = 1 - F(T^o) &\stackrel{d}{=} U \\
\exp(-H(T^o)) &\stackrel{d}{=} U \\
H(T^o) &\stackrel{d}{=} -\log(U) \\
T^o &\stackrel{d}{=} H^{-1}(-\log U).
\end{aligned}$$

These identities are very helpful; for example, they are often used for simulating survival data (see Exercise 19.13). □

19.4.5 Prediction

It is useful to have a one-dimensional value that summarizes survival and lifetime behavior of an individual. This makes it convenient to define prediction and for evaluating the performance of the RSF ensemble.

RSF defines a one-dimensional predicted outcome by using the CHF. The approach rests on a *conservation-of-events* principle [25]. Under fairly general conditions, conservation of events asserts that the sum of the estimated CHF over observed times (both censored and uncensored) equals the total number of deaths. Let $\mathcal{T} = \{A_1, \ldots, A_K\}$ be the survival tree partition and let

$$\hat{H}(t|\mathbf{x}, \mathcal{T}) = \sum_{k=1}^{K} \hat{H}(t|A_k) 1_{\{\mathbf{x} \in A_k\}}$$

be the tree Nelson-Aalen CHF estimator where $\hat{H}(t|A_k)$ is the CHF estimator for cell A_k. For simplicity we work with the full sample estimator and not the in-bag estimator which is why we have dropped the * notation. Conservation of events asserts the following.

Lemma 19.1. $\sum_{i \in A_k} \hat{H}(T_i | A_k) = \sum_{i \in A_k} \delta_i$ *for each cell* $A_k \in \mathcal{T}$, $k = 1, \ldots, K$. *In other words, the total number of deaths is conserved within each cell of the partition.*

Lemma 19.1 shows that summing the CHF over observed survival times equals the total number of deaths within a terminal node: a type of conservation of events within the ends of a tree. An immediate corollary is that the total number of deaths in a tree is conserved.

Corollary 19.2. *For each tree grown from the original non-bootstrapped data,*

$$\sum_{i=1}^{n} \hat{H}(T_i|\mathbf{X}_i, \mathfrak{T}) = \sum_{k=1}^{K} \sum_{i \in A_k} \hat{H}(T_i|A_k) = \sum_{k=1}^{K} \sum_{i \in A_k} \delta_i = \sum_{i=1}^{n} \delta_i.$$

Note that the right-hand side equals the total number of deaths.

Corollary 19.2 motivates the concept of mortality; a measure of the expected number of deaths under a null hypothesis of similar survival behavior [17]. The mortality for case i equals the inbag CHF ensemble for i (19.5) summed over the observed times

$$\hat{M}_i = \sum_{j=1}^{n} \hat{H}^*(T_j|\mathbf{X}_i). \tag{19.9}$$

This value estimates the expected number of deaths assuming that all patients are similar to i. This can be compared to observed number of events as a way to gauge the mortality of a case.

OOB mortality is defined by replacing the inbag CHF by by its OOB counterpart (19.6). The OOB mortality for i is

$$\hat{M}^{(i)} = \sum_{j=1}^{n} \hat{H}^{(i)}(T_j). \tag{19.10}$$

We use this for estimating prediction error to be explained in the next subsection.

Proof of Lemma 19.1. With a slight abuse of notation, dropping the dependence on k, let d_i and Y_i denote the number of deaths and number of individuals at risk in cell A_k at time t_i where $\mathrm{t}_1, \ldots, \mathrm{t}_m$ are the unique event times in A_k. Let $N_i = \#\{j \in A_k : \mathrm{t}_i \le T_j < \mathrm{t}_{i+1}\}$ for $i = 1, \ldots, m$, where $\mathrm{t}_{m+1} = \infty$.

Conservation of events is equivalent to the assertion

$$\sum_{i=1}^{m} N_i \hat{H}(\mathrm{t}_i|A_k) - \sum_{i=1}^{m} d_i = 0.$$

Thus, the proof follows by showing that the above expression holds. First observe that $N_i = Y_i - Y_{i+1}$ where $Y_{m+1} = 0$. Define $d_0/Y_0 = 0/0 = 0$. We have,

$$\begin{aligned}
\sum_{i=1}^{m} N_i \hat{H}(\mathrm{t}_i|A_k) - \sum_{i=1}^{m} d_i &= \sum_{i=1}^{m} \left[(Y_i - Y_{i+1}) \sum_{s=1}^{i} \frac{d_s}{Y_s} - d_i \right] \\
&= \sum_{i=1}^{m} Y_i \sum_{s=1}^{i-1} \frac{d_s}{Y_s} - \sum_{i=1}^{m} Y_{i+1} \sum_{s=1}^{i} \frac{d_s}{Y_s} \\
&= \sum_{i=1}^{m-1} Y_{i+1} \sum_{s=1}^{i} \frac{d_s}{Y_s} - \sum_{i=1}^{m} Y_{i+1} \sum_{s=1}^{i} \frac{d_s}{Y_s} = -Y_{m+1} \sum_{s=1}^{m} \frac{d_s}{Y_s} = 0.
\end{aligned}$$

□

19.4.6 Prediction error

Harrell's concordance [8] is used for evaluating performance in right-censored settings. This estimates the probability that a predictor correctly ranks a randomly selected pair of data points, by ranking the case that fails first as having a worst predicted outcome. There are two primary advantages to this metric: (1) It is easily interpretable as a probability. (2) Unlike other measures of survival performance, it does not depend on fixed time points for its evaluation.

The concordance value is the fraction of concordant pairs from the total permissible pairs that are correctly ranked by the predictor. A random pair $(\mathbf{X}_i, T_i, \delta_i)$ and $(\mathbf{X}_j, T_j, \delta_j)$ is called permissible if the shorter survival time is an event. Specifically, if $T_i < T_j$, the pair (i, j) is permissible if $\delta_i = 1$. To use the concordance value with RSF, mortality values (19.9) $\hat{M}_i$ and $\hat{M}_j$ are used for predictors for i and j. A larger mortality means higher risk and worse survival, thus the pair (i, j) is concordant if $\hat{M}_i > \hat{M}_j$. The concordance for RSF is

$$\mathscr{C} = \frac{\sum_{i=1}^{n}\sum_{j=1}^{n} 1_{\{\hat{M}_i > \hat{M}_j\}} 1_{\{T_i < T_j, \delta_i = 1\}}}{\sum_{i=1}^{n}\sum_{j=1}^{n} 1_{\{T_i < T_j, \delta_i = 1\}}}. \tag{19.11}$$

The denominator of (19.11) equals the total number of permissible pairs, while the numerator equals the total number of such pairs that are concordant. The metric can be extended to address ties. See Ishwaran et al. [17] for details.

The prediction error, called the *C-error rate*, is defined as $\text{Err} = 1 - \mathscr{C}$. A value $\text{Err} = 0$ indicates perfect discrimination, while $\text{Err} = 0.5$ is equivalent to random guessing. The prediction error is estimated by the OOB estimator by replacing inbag mortality values $\hat{M}_j$ and $\hat{M}_j$ with OOB values (19.10) $\hat{M}^{(i)}$ and $\hat{M}^{(j)}$.

The *Brier score* is another measure used for assessing performance in survival analysis [6]. Let $S(t|\mathbf{X}) = \mathbb{P}\{T^o > t|\mathbf{X}\}$ be the true survival function. The Brier score is

$$\text{BS}(t, S) = \mathbb{E}\left(1_{\{T^o > t\}} - S(t|\mathbf{X})\right)^2.$$

Let $\hat{S}(\cdot|\mathbf{X})$ be an estimator of $S(\cdot|\mathbf{X})$ that we wish to determine the performance of. Then, the Brier score performance for $\hat{S}$ evaluated at time t is

$$\text{BS}(t, \hat{S}) = \mathbb{E}\left(1_{\{T^o > t\}} - \hat{S}(t|\mathbf{X})\right)^2.$$

Lower values for the Brier score indicate better prediction performance. The benchmark value for the Brier score is 0.25, since if $\hat{S}(t|\mathbf{X}) = 0.5$, representing a coin flip,

$$\text{BS}(t, \hat{S}) = \mathbb{E}\left(1_{\{T^o > t\}} - \frac{1}{2}\right)^2 = \mathbb{E}\left(\frac{1}{2}\right)^2 = \frac{1}{4} = 0.25.$$

Therefore, a good performing estimator should have values closer to zero and below 0.25. Note that the Brier score is minimized when $\hat{S}(t|\mathbf{X}) = S(t|\mathbf{X})$, since by conditional expectation, $\text{BS}(t, \hat{S})$ is minimized at the conditional mean $\hat{S}(t|\mathbf{X}) = \mathbb{E}(1_{\{T>t\}}|\mathbf{X})$.

Let $\hat{G}(t|\mathbf{X})$ be an estimator of the censoring function, $G(t|\mathbf{X}) = \mathbb{P}\{C > t|\mathbf{X}\}$. The Brier score for $\hat{S}$ is estimated by [4]

$$\hat{\text{BS}}(t, \hat{S}) = \frac{1}{n}\sum_{i=1}^{n}\left(1_{\{T_i > t\}} - \hat{S}(t|\mathbf{X}_i)\right)^2 w(t, \mathbf{X}_i, T_i, \delta_i, \hat{G}),$$

where

$$w(t, \mathbf{X}_i, T_i, \delta_i, \hat{G}) = \frac{1_{\{T_i \le t\}}\delta_i}{\hat{G}(T_i - |\mathbf{X}_i)} + \frac{1_{\{T_i > t\}}}{\hat{G}(t|\mathbf{X}_i)}.$$

The weights $w(\cdot, \mathbf{X}_i, T_i, \delta_i, \hat{G})$ are piecewise step functions that change only at event times. After some simplifications, the Brier score estimator can be more simply expressed as

$$\hat{\text{BS}}(t, \hat{S}) = \frac{1}{n}\sum_{i=1}^{n}\left\{\frac{\delta_i \hat{S}^2(t|\mathbf{X}_i) 1_{\{T_i \le t\}}}{\hat{G}(T_i - |\mathbf{X}_i)} + \frac{\left(1 - \hat{S}(t|\mathbf{X}_i)\right)^2 1_{\{T_i > t\}}}{\hat{G}(t|\mathbf{X}_i)}\right\}.$$

This is an unbiased estimator for $\mathrm{BS}(t, \hat{S})$ when $\hat{G}(t|\mathbf{X})$ is set to the true censoring function $G(T|\mathbf{X})$ (see Exercise 19.11). Therefore, by using a sensible estimator $\hat{G}(t|\mathbf{X})$ we can obtain a reliable estimator for the Brier score.

An overall summary of the Brier score is the *integrated Brier score* (IBS) at time τ defined by

$$\mathrm{IBS}(\tau) = \int_0^\tau \mathrm{BS}(t, S)\,dt.$$

The cumulative prediction error, or *continuous ranked probability score* (CRPS), is defined as the area under the prediction error curve; hence, the CRPS is an alternate name for the IBS. It can be helpful to standardize these values by dividing by time

$$\text{standardized CRPS} = \tau^{-1} \int_0^\tau \mathrm{BS}(t, S)\,dt.$$

To illustrate, we use survival data from Hsich et al. [11] of $n = 2231$ heart failure patients who underwent cardiopulmonary stress testing (more details about stress testing are discussed in the next section). During a mean follow-up of 5 years (maximum duration for survivors was 11 years), 742 patients died. The outcome is all-cause mortality and a total of $p = 39$ covariates were measured for each patient including demographic, cardiac and noncardiac comorbidity, and stress test information.

To calculate the Brier score, we use OOB survival for $\hat{S}$. The censoring distribution was estimated by two methods: (i) KM survival curves and (ii) RSF, where in the latter case the RSF nodesize was tuned to optimize the OOB C-error rate. The censoring distribution is estimated by repacing the censoring indicator δ_i with $1 - \delta_i$, thus effectively converting censoring to death and death to censoring.

The Brier score estimator is shown on the left in Figure 19.3. Its value increases with increasing time showing that the estimated survival curves becomes less predictive with increasing time, which is common in survival settings. The two censoring distribution methods produce nearly identical curves except for large time values where KM gives smaller values than the RSF censoring distribution estimator. This is probably due to the poor behavior in the tail of the KM estimator which is one of its weaknesses.

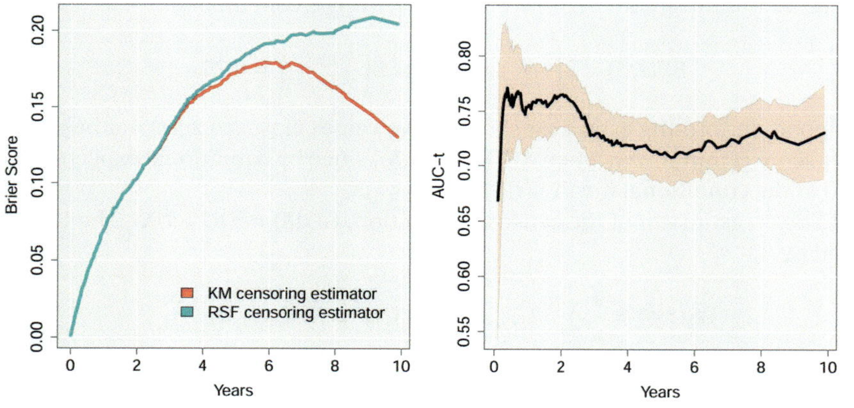

FIGURE 19.3
Brier score (left) and time-varying AUC (right) performance values using RSF for heart failure survival data.

For comparison, the right figure shows the *time-varying AUC* (AUC-t) [9]. Recall that the ROC curve (Chapter 15.6) displays the relationship between a predictor Z and a binary variable D by plotting the sensitivity, $\mathbb{P}\{Z > \tau | D = 1\}$, against one minus the specificity, $1 - \mathbb{P}\{Z \leq \tau | D = 0\}$, for all τ. The ROC curve at time t for survival is defined by rewriting the sensitivity and one minus the specificity using Bayes' theorem as:

$$\begin{aligned}\mathbb{P}\{Z > \tau | D(t) = 1\} &= \frac{\mathbb{P}\{Z > \tau, T^o \leq \tau\}}{1 - S(t)} = \frac{\mathbb{P}\{Z > \tau\} - S(\tau, t)}{1 - S(t)} = \frac{S(\tau, 0) - S(\tau, t)}{1 - S(-\infty, t)} \\ \mathbb{P}\{Z > \tau | D(t) = 0\} &= \frac{\mathbb{P}\{Z > \tau, T^o > \tau\}}{S(t)} = \frac{S(\tau, t)}{S(-\infty, t)},\end{aligned}$$

where

$$D(t) = \begin{cases} 1 & \text{if } T^o \leq t \\ 0 & \text{if } T^o > t \end{cases}$$

is a binary variable that records the failure status at time t. The quantity $S(\tau, t) = \mathbb{P}\{Z > \tau, T^o > t\}$ is the joint survival function for (Z, T^o).

The AUC-t curve in Figure 19.3 displays the AUC for the time-varying ROC curve (calculated using OOB mortality for Z). The bivariate distribution $S(\tau, t)$ needed to obtain the ROC curve was estimated nonparametrically using methods described in Heagerty, Lumley and Pepe [9] using the `riskRegression` R-package [5].

We observe that the AUC-t curve is highest for time less than two years and then decreases and remains generally flat thereafter except near the tails where the AUC-t increases slightly. However, this latter effect appears to be spurious as the increase is within tolerance of the 95% confidence regions obtained using the bootstrap.

As another performance comparison, the C-error rate using Harrell's concordance was 0.293, and one minus this value is 0.702, which is within the confidence limits for AUC-t in the right tail. Note that while Harrell's concordance is the same as AUC for binary classification problems without censored data, they are not the same in survival when censoring is present. Thus, comparison of AUC-t to the C-error should recognize that these values are not identical.

19.5 ECG and Long-Term Survival

In this and the next section, we provide in-depth examples of RSF applied to real-world survival data. In the first example, we use minimal depth to study whether ECG adds value over classical stress testing for predicting cardiovascular risk. In the second example, we assess the predictiveness of the well-known predictor heart rate recovery for risk prediction.

Let's begin with the first example related to ECG and stress testing. Electrocardiography (ECG) and exercise treadmill testing are commonly used diagnostic tests for evaluating cardiovascular risk. Both tests collect numerous variables, with over 500 variables typically recorded for ECG using digital technology. During the exercise test, patients walk on a treadmill with gradually increasing speed and grade until exhaustion, while detailed data on electrocardiographic changes, heart rate, heart rhythm, and symptoms are obtained.

Previous studies have shown that both ECG and exercise testing are powerful predictors of risk in patients with suspected cardiovascular disease. We aimed to test whether ECG variables add value for predicting long-term survival accounting for stress testing [18]. To do this, we ran RSF on a large dataset comprising both types of variables. The data included patients considered at risk for cardiovascular disease, all referred for ECG and exercise treadmill testing, with no gross abnormalities in their ECGs. This cohort was chosen because subtle quantitative differences in computerized measures may be prognostically important. The dataset contained $n = 19{,}530$ patients, of which 1,742 died. For each patient, $p = 346$ variables from stress testing and ECG were recorded. All-cause mortality was used as the survival outcome, and the data was right-censored with a mean follow-up time of 11 years.

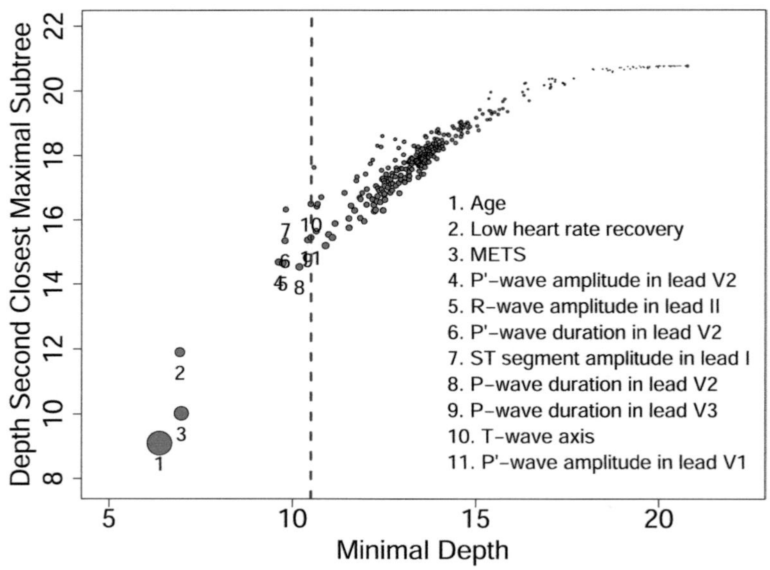

FIGURE 19.4
Minimal depth applied to large cohort of patients considered at high risk for cardiovascular disease ($n = 19,530$, $p = 346$).

Displayed in Figure 19.4 are the minimal depth values (Chapter 17.5) from the RSF analysis. On the vertical axis are the second order depth; defined as the minimal depth of the second closest maximal subtree to the root. A circle's diameter in the plot is proportional to the forest-averaged number of maximal subtrees for a variable. The three variables appearing in the extreme bottom left of the plot have the smallest minimal and second-order depths. These represent the most predictive variables. In order, they are age, low heart rate recovery (slow post-exercise attenuation of heart rate), and peak mets (the proportion of age- and gender-adjusted exercise capacity achieved). All are clinical variables, and all are known to be powerful predictors of cardiovascular disease.

The thick dashed blue line in Figure 19.4 is the mean minimal depth threshold value. In total, 11 variables have minimal depth less than this threshold: eight ECG variables in addition to the top three variables. The presence of the ECG variables are interesting and suggest ECG measurements play a role in long-term survival even after accounting for clinical and stress testing attributes.

19.6 Heart Rate Recovery

As discussed in the previous example, exercise stress testing is commonly used to assess patients at risk for heart disease. A useful predictor of mortality is the fall in heart rate immediately after exercise stress testing, or heart rate recovery. Heart rate recovery is formally defined as the heart rate at peak exercise minus the heart rate measured 1 minute later. The hypothesis that heart rate recovery predicts mortality has been tested and validated in a number of cohorts. Here we consider this issue by using variable selection using data from a study involving heart rate recovery [14].

The dataset has n = 23,701 patients referred to the Cleveland Clinic between 1990 and 1999. Each patient was stress tested and underwent an upright cool-down period for the first 2 minutes after recovery. Data regarding the reason for testing, symptoms, cardiac risk factors, other medical diagnoses, prior cardiac and noncardiac procedures, medications, resting electrocardiogram, resting heart rate, and blood pressure were recorded prospectively prior to testing. During each stage of exercise, and during the first 5 minutes of recovery, data were recorded regarding heart rate, blood pressure, ST-segment changes, symptoms, and arrhythmias. In total p = 85 variables were available for the analysis. All-cause mortality was used for the survival outcome. The data was right-censored with a mean follow-up for survivors of 5.7 years (range 0.75 to 10.1 years) during which 1,617 patients died.

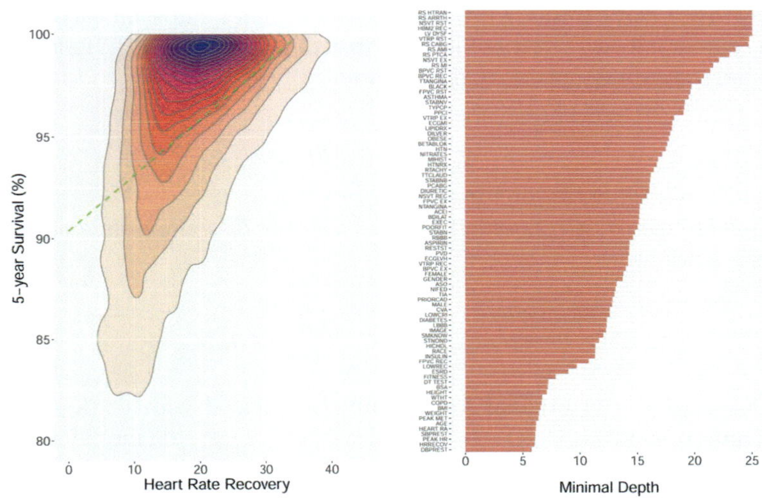

FIGURE 19.5
Left: 5-year OOB survival versus heart rate recovery. Right: Minimal depth for all variables.

On the left of Figure 19.5 is a contour plot from RSF. Five-year OOB survival for each patient is shown against heart rate recovery value. Survival clearly increases with increasing heart rate recovery. However, the distribution is diffuse, suggesting other variables might be involved. The right side of the figure gives the minimal depth for each variable, sorted from smallest (most important) to largest (least important). Heart rate recovery has the second smallest minimal depth, showing its importance, however, at the same time we observe many other variables with similar small values. Thus we cannot be sure to the extent of the influence of heart rate recovery using a simple method like minimal depth. In order to better assess the importance of heart rate recover we will turn to a VarPro analysis described next.

19.7 Using VarPro with RSF

Calculating variable importance (VIMP) using the C-error (19.11) is computationally demanding requiring $O(n^2)$ calculations which becomes prohibitive in large survival problems. For the heart recovery analysis just discussed this was already challenging, which is why we used minimal depth for the analysis and not permutation importance (VIMP). We have also seen problems with VIMP due to using artificial data. In Chapter 17.6 we discussed VarPro which is a new-ruled based procedure with observed excellent empirical performance. Here we discuss an extension of VarPro to survival analysis given in Lu and Ishwaran [23].

Recall from Chapter 17.6 that VarPro estimates $\psi(\mathbf{X}) = \mathbb{E}(g(Y)|\mathbf{X})$ by using a sample average of $g(Y)$ calculated locally within a region defined by a rule. For a rule $\zeta_{n,k}$ and its region $\mathscr{R}_{n,k} = \mathscr{R}(\zeta_{n,k})$ with sample size $m_{n,k}$, VarPro makes use of the estimator (17.20)

$$\hat{\theta}_n(\zeta_{n,k}) = \frac{1}{m_{n,k}} \sum_{i=1}^{n} g(Y_i) 1_{\{\mathbf{X}_i \in \mathscr{R}(\zeta_{n,k})\}}.$$

However, this may not always be suitable for all settings as it could be inefficient or difficult to estimate ψ locally this way.

The strategy to overcome this is to use an external estimator. Thus, for each n, let $\psi_n : \mathscr{X} \to \mathbb{R}$ be an external estimator for ψ. The modified VarPro estimator is defined as follows. For each rule $\zeta_{n,k}$, the modified procedure replaces $\hat{\theta}_n(\zeta_{n,k})$ with

$$\tilde{\theta}_n(\zeta_{n,k}) = \frac{1}{m_{n,k}} \sum_{i=1}^{n} \psi_n(\mathbf{X}_i) 1_{\{\mathbf{X}_i \in \mathscr{R}_{n,k}\}}.$$

Likewise, the release rule $\hat{\theta}_n(\zeta^S_{n,k})$, which releases the rule $\zeta_{n,k}$ along the coordinates of a set $S \subset \{1,\dots,p\}$, is replaced with

$$\tilde{\theta}_n(\zeta^S_{n,k}) = \frac{1}{m^S_{n,k}} \sum_{i=1}^{n} \psi_n(\mathbf{X}_i) 1_{\{\mathbf{X}_i \in \mathscr{R}^S_{n,k}\}}.$$

Therefore, $g(Y_i)$ used in the original formulation is replaced with $\psi_n(\mathbf{X}_i)$ which serves to estimate the conditional mean of $g(Y_i)$. Given rules $\zeta_{n,1},\dots,\zeta_{n,K_n}$, the modified VarPro measure of importance for S is

$$\tilde{\Delta}_n(S) = \sum_{k=1}^{K_n} W_{n,k} |\tilde{\theta}_n(\zeta^S_{n,k}) - \tilde{\theta}_n(\zeta_{n,k})|$$

for prechosen weights $0 \le W_{n,k} \le 1$, $\sum_{k=1}^{K_n} W_{n,k} = 1$.

The modified procedure is especially suitable for survival analysis. Because the survival function is $S(t|\mathbf{X}) = \mathbb{P}\{T^o > t|\mathbf{X}\}$ this corresponds to $\psi(\mathbf{X}) = \mathbb{E}(g(T^o)|\mathbf{X})$ where $g(T^o) = I\{T^o > t\}$. Let ζ be a rule and $\mathscr{R}(\zeta)$ its region. Because T^o is potentially unobserved, notice we cannot estimate $S(t|\mathbf{X})$ locally using an estimator like

$$\hat{\theta}_n(R(\zeta)) := \frac{\sum_{i=1}^n g(Y_i) 1_{\{\mathbf{X}_i \in \mathscr{R}(\zeta)\}}}{\sum_{i=1}^n 1_{\{\mathbf{X}_i \in \mathscr{R}(\zeta)\}}} = \frac{\sum_{i=1}^n 1_{\{T_i^o > t, \mathbf{X}_i \in \mathscr{R}(\zeta)\}}}{\sum_{i=1}^n 1_{\{\mathbf{X}_i \in \mathscr{R}(\zeta)\}}}.$$

If we want to use a sample average of the observed data, as in the original VarPro formulation, then we have to account for censoring. For example, under the assumption of conditional independence (19.1), one approach could be to use an IPCW (inverse of probability of censoring weighting) estimator such as

$$\hat{\theta}_n(\zeta) = \left(\sum_{i=1}^{n} 1_{\{\mathbf{X}_i \in \mathscr{R}(\zeta)\}} \right)^{-1} \sum_{i=1}^{n} \frac{\delta_i}{G(T_i)} \left[1_{\{T_i > t, \mathbf{X}_i \in \mathscr{R}(\zeta)\}} \right]$$

where $G(u) = \mathbb{P}\{C > u\}$ is the unknown censoring distribution. However, there are known issues with IPCW estimators, such as estimation for G, and problems with inverse weights becoming large [4, 6].

As another example, consider variable selection for the restricted mean survival time (RMST) [13]. The RMST is a useful quantity summarizing lifetime behavior and is defined as the survival function integrated from $[0, \tau]$ for some fixed time point $0 < \tau < \infty$:

$$\int_0^\tau S(t|\mathbf{X})\,dt = \mathbb{E}\left[\int_0^\tau 1_{\{T^o > t\}}\,dt \middle| \mathbf{X}\right] = \mathbb{E}\left[\int_0^{T^o \wedge \tau} dt \middle| \mathbf{X}\right] = \mathbb{E}\left[T^o \wedge \tau | \mathbf{X}\right]. \tag{19.12}$$

The time horizon τ is usually selected to represent a clinically meaningful endpoint, such as 1 year or 5 year survival. Notice that

$$T^o \wedge \tau = \tau - (\tau - T^o)1_{\{T^o \le \tau\}}.$$

Therefore taking expectations, the RMST can be written as $\psi(\mathbf{X}) = \mathbb{E}(g(T^o)|\mathbf{X})$ where $g(T^o) = \tau - (\tau - t)I\{T^o \le \tau\}$. Just like the survival function due to censoring, the RMST may be difficult to estimate locally.

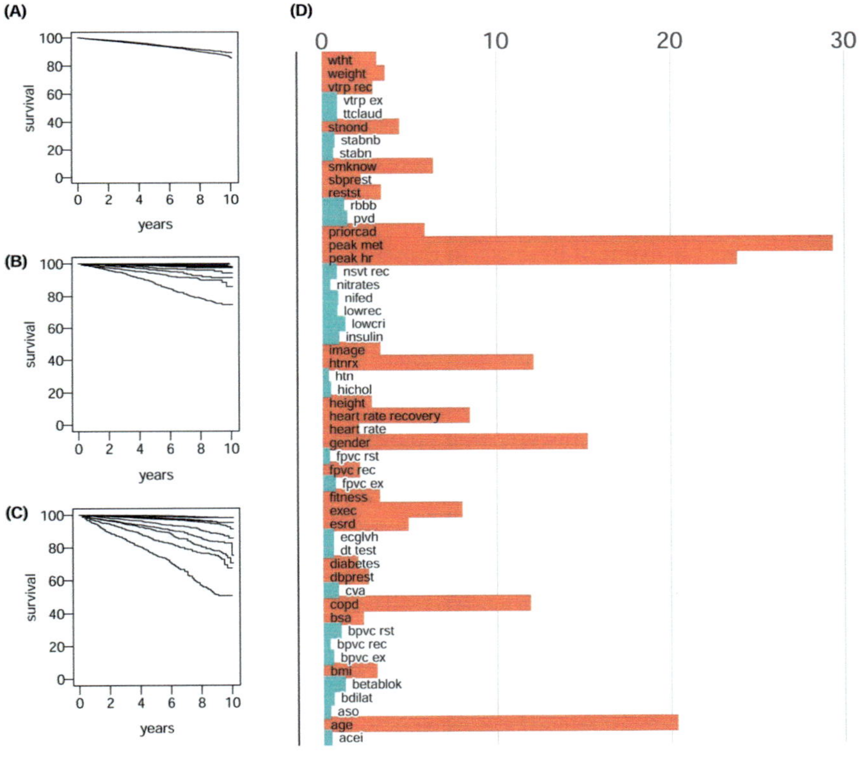

FIGURE 19.6
Results of heart rate recovery survival analysis using the modified VarPro procedure. (A) Survival for men and women; (B) Survival for women, stratified by peak met; (C) Survival for men, stratified by peak met; (D) Standardized importance for variables (red are signal, blue are noise variables).

To illustrate, we apply the modified VarPro importance to the previous heart rate recovery analysis where for ψ_n we used the RMST evaluated at $\tau = 3$ years. The RMST was calculated by using OOB survival.

The standardized importance values from VarPro are given in Figure 19.6 (D). Signal variables are indicated by red and noise variables by blue. Considering Figure 19.6 (D), we observe that heart rate recovery is identified as a signal variable. While this is not surprising in light of our previous discussions, at the same time there are several other significant variables found by the analysis, some with even larger standardized importance values, such as age, peak met, peak heart rate and gender. Therefore, while there is evidence that heart rate recovery is an important attribute for risk prediction, many other variables are also at play in the multivariable model.

An unexpected and interesting finding from this analysis is that gender has a high importance value, even larger than heart rate recovery. Sex is an important variable often overlooked in the heart failure literature. To further investigate this, we provide separate KM survival curves for men and women in Figure 19.6 (A), which shows minimal differences. However, when stratifying peak MET (a significant variable) into deciles and plotting survival curves for each category, a notable distinction emerges. For men Figure 19.6 (C), survival decreases more significantly at lower peak MET values compared to women Figure 19.6 (B) (survival curves descending from top to bottom). This suggests that men are more adversely affected by reduced peak MET, implying greater sensitivity to exercise capacity, as peak MET is an indicator of this. These observations are consistent with prior research by Hsich et al. [10], which indicated that women have a lower mortality risk than men at any given level of exercise or oxygen capacity.

19.8 Competing Risks

We have discussed survival analysis settings where the outcome is the time until an event of interest occurs. In medical studies, this might be the time until death or the time until the onset of a specific disease, such as cancer. However, there are situations where competing risks lead to more than one possible event of interest. In these cases, instead of focusing on the time to a *specific* event, the focus shifts to the time until the occurrence of *any* event, which is known as a competing risk analysis. In the classical competing risk setting, it is assumed that only one event can occur at a time, and the occurrence of one event precludes the occurrence of the other possible events. Type-I censoring and generalized Type-I censoring experimental designs also apply to competing risks. Thus, individuals subject to competing risks are observed from study entry until one of the events occurs, or, if no event is observed, their survival time is right-censored at the study's conclusion.

Formally, let T_i^o be the true event time for the ith subject, $i = 1, \dots, n$, and let δ_i^o be the event type, $\delta_i^o \in \{1, \dots, J\}$, where $J \geq 1$. Let C_i denote the censoring time for individual i such that the actual time of event T_i^o is potentially unobserved and we observe $T_i = \min(T_i^o, C_i)$ and the event indicator $\delta_i = \delta_i^o 1_{\{T_i^o \leq C_i\}}$. When $\delta_i = 0$, the individual is right-censored at T_i, otherwise if $\delta_i = j > 0$, the individual is said to have an event of type j at time T_i. The data are the set of values $(\mathbf{X}_i, T_i, \delta_i)_{1 \leq i \leq n}$.

In competing risks, the *cause-specific hazard* function [21, 26] for event j is

$$\lambda_j(t, \mathbf{x}) = \lim_{\Delta t \to 0} \frac{\mathbb{P}\{t < T^o \leq t + \Delta t, \delta^o = j \mid T^o > t, \mathbf{x}\}}{\Delta t} = \frac{f_j(t|\mathbf{x})}{S(t|\mathbf{x})}.$$

This describes the short term risk of event j for those still alive. Thus, factors $\mathbf{x}$ found to change the event risk are associated to the (biological) mechanism behind event j.

For long-term predictions of event j the situation is different. The probability that an event occurs in a specific time period, say $[0, t]$, depends on the cause-specific hazards of the other (fatal) events [7]. The probability of an event is determined by the *cumulative incidence function* (CIF) defined as the probability of experiencing an event of type j by time t, i.e. $F_j(t|\mathbf{x}) = \mathbb{P}\{T^o \leq t, \delta^o = j|\mathbf{x}\}$. The dependence of the CIF for a specific event j on all the competing risks can be seen in the following identity

$$F_j(t|\mathbf{x}) = \int_0^t \exp\left(-\int_0^s \sum_{l=1}^J \lambda_l(u, \mathbf{x})\, du\right) \lambda_j(s, \mathbf{x})\, ds = \int_0^t S(s - |\mathbf{x}) \lambda_j(s, \mathbf{x})\, ds.$$

The right hand side holds by the identity that relates the survival function to the CHF

$$S(t|\mathbf{x}) = \exp\left(-\int_0^t \lambda(u,\mathbf{x})du\right) \tag{19.13}$$

and because $\lambda(u,\mathbf{x}) = \sum_{j=1}^{J} \lambda_j(u,\mathbf{x})$ due to the assumption that competing risks are mutually exclusive. This means factors that change the long-term risk (cumulative incidence) of event j depend on the factors for all of the cause-specific hazard functions. This is because event j can only occur to those that survive the other risks.

We can summarize the differences in the two competing risk quantities as follows: when the aim is to assist decision making and for patient counseling we are interested in long-term predictions and factors that affect the cumulative incidence. For understanding and treating the intrisic risk for a specific event, e.g. the risk of relapse, we focus on the cause-specific hazard function of the event.

To apply RSF to the competing risk problems, we need to address four issues: (1) Splitting rules; (2) Tree and ensemble estimators; (3) Prediction measure; (4) Performance metrics. Each issue is treated separately below.

19.8.1 Splitting rules

Two splitting rules are used [16]. The first is a cause-j generalized log-rank test. This is a test of the null hypothesis that the cause-specific hazards for event j are equal in the proposed daughter nodes. Large values favor differences in cause-specific hazards and therefore encourage tree splitting where the goal is to identify risk factors associated with a specific event. The second splitting rule is Gray's test [7] which tests the null hypothesis that the CIF for event j is the same across the daughters.

Cause-j generalized log-rank splitting rule is useful if the main purpose is to detect variables that affect the cause-j specific hazard. It may not be optimal if the purpose is also prediction of cumulative event probabilities, which is the purpose of the second rule based on Gray's test. This latter rule is most appropriate when the goal is long term probability prediction.

19.8.2 Tree and ensemble estimators

We introduce notation to explain the tree and ensemble estimators. As before, $\mathrm{t}_1 < \mathrm{t}_2 < \cdots < \mathrm{t}_m$ denotes the $m \le n$ distinct ordered event times of $(T_i)_{1 \le i \le n}$. Let $d_j(\mathrm{t}_k) = \sum_{i=1}^{n} 1_{\{T_i = \mathrm{t}_k, \delta_i = j\}}$ be the number of type j events at t_k, and $N_j(t) = \sum_{i=1}^{n} 1_{\{T_i \le t, \delta_i = j\}}$ the number of type j events in $[0, \mathrm{t}_k]$. Define also $d(\mathrm{t}_k) = \sum_j \delta_j(\mathrm{t}_k)$, the total number of events occurring at time t_k, $N(t) = \sum_j N_j(t)$, the total number of events occurring in $[0, t]$ and $Y(t) = \sum_{i=1}^{n} 1_{\{T_i \ge t\}}$, the number of individuals at risk (event-free and uncensored) just prior to t.

A key value is the cumulative event-specific hazard function defined as

$$H_j(t) = \mathbb{E}[H_j(t|\mathbf{X})] = \mathbb{E}\left[\int_0^t \lambda_j(s,\mathbf{X})\, ds\right].$$

The Nelson-Aalen estimator for H_j is given by [1]

$$\hat{H}_j(t) = \int_0^t \frac{N_j(ds)}{Y(s)} = \sum_{k=1}^{m(t)} \frac{d_j(\mathrm{t}_k)}{Y(\mathrm{t}_k)}, \qquad m(t) = \max\{k : \mathrm{t}_k \le t\}.$$

A second function of interest is the cumulative incidence $F_j(t) = \mathbb{E}[F_j(t|\mathbf{X})]$. Let $\hat{S}(t)$ be the KM estimator (19.3) for the (event-free) survival function, then F_j is estimated by

$$\hat{F}_j(t) = \int_0^t \hat{S}(u-)\hat{H}_j(du) = \int_0^t \hat{S}(u-)Y(u)^{-1}N_j(du) = \sum_{k=1}^{m(t)} \hat{S}(t_{k-1})Y(t_k)^{-1}d_j(t_k).$$

Let $\mathcal{T} = \{A_1, \ldots, A_k\}$ be the survival tree partition. Using in-bag data, the tree estimators for $H_j(t|\mathbf{x})$ and $F_j(t|\mathbf{x})$ are given by

$$\hat{H}_j^*(t|\mathbf{x}, \mathcal{T}) = \sum_{k=1}^{K} \hat{H}_j^*(t|A_k) 1_{\{\mathbf{x} \in A_k\}}, \quad \hat{F}_j^*(t|\mathbf{x}, \mathcal{T}) = \sum_{k=1}^{K} \hat{F}_j^*(t|A_k) 1_{\{\mathbf{x} \in A_k\}}$$

where $\hat{H}_j^*(t|A_k)$ and $\hat{F}_j^*(t|A_k)$ are the estimators $\hat{H}_j(t)$ and $\hat{F}_j(t)$ for cell A_k estimated using bootstrap cases in A_k. The in-bag ensemble estimators are given by

$$\hat{H}_j^*(t|\mathbf{x}) = \frac{1}{B} \sum_{b=1}^{B} \hat{H}_j^*(t|\mathbf{x}, \mathcal{T}_b), \quad \hat{F}_j^*(t|\mathbf{x}) = \frac{1}{B} \sum_{b=1}^{B} \hat{F}_j^*(t|\mathbf{x}, \mathcal{T}_b).$$

OOB ensemble estimators $\hat{H}_j^{(i)}(t)$ and $\hat{F}_j^{(i)}(t)$ for a case i are obtained by restricting tree averaging to $\mathcal{O}_i$ equal to the set of trees over which i is OOB.

19.8.3 Predicted value

Similar to survival analysis, we can define a one-dimensional summary value for prediction and performance assessment in competing risks. This value summarizes the cumulative incidence predictions and is referred to as the expected number of life years lost due to cause j [2].

In right-censored data, obtaining a reliable estimate of the expected lifetime is not feasible. Therefore, the RMST, $\int_0^\tau S(t|\mathbf{x})\,dt$, is often used (see (19.12)). Based on the relation

$$S(t|\mathbf{x}) + \sum_{j=1}^{J} F_j(t|\mathbf{x}) = 1,$$

the expected number of years lost before time τ, which is the difference between τ and the RMST, can be written as

$$\tau - \int_0^\tau S(t|\mathbf{x})\,dt = \int_0^\tau \sum_{j=1}^{J} F_j(t|\mathbf{x})\,dt.$$

The cause-specific quantity $M_j(\tau|\mathbf{x}) = \int_0^\tau F_j(t|\mathbf{x})\,dt$ can thus be interpreted as the expected number of life years lost due to cause j before time τ. This value, $M_j(\tau|\mathbf{x})$, is also known as cause-j mortality.

Substituting the OOB estimator $\hat{F}_j^{(i)}(t)$ for $F_j(t|\mathbf{X}_i)$ results in

$$\hat{M}_j^{(i)}(\tau) = \int_0^\tau \hat{F}_j^{(i)}(t)\,dt,$$

which is used for performance assessment, such as via concordance-error or the integrated Brier score, similar to Chapter 19.4.5. An example is provided next.

19.8.4 Highly active antiretroviral therapy for HIV infection

We use the following example related to HIV to illustrate competing risks. The data comes from the Johns Hopkins HIV Clinical Cohort, a longitudinal study of HIV-infected patients receiving primary care through the Johns Hopkins AIDS Service [24]. We consider a subset of this data, consisting of $n = 2960$ individuals who initiated effective antiretroviral therapy (HAART) between 1996 and 2005. The events of interest are time to all-cause death and time to AIDS-defining illnesses after the start of effective treatment. Including death, there were 15 competing risk outcomes. The variables included laboratory measurements (CD4 nadir, HIV-RNA levels, total lymphocyte counts, hemoglobin, albumin, and creatinine levels) as well as patient variables (prior diagnosis of an AIDS-defining illness, prophylaxis for pneumonia, sex, race, history of injection drug use, history of heavy alcohol use, heroin use, cocaine use, and a medical history of personality disorder). In total, $p = 28$ variables were collected for each individual.

We conducted an event-specific analysis, running 15 separate forests where each forest used cause-j log-rank splitting. Recall that this splitting rule prioritizes splits that identify risk factors associated with the j-event. Permutation variable importance (VIMP, Chapter 17.2) was calculated for each of the $j = 1, \ldots, 15$ forests using the C-error rate as the prediction metric. The C-error rate evaluates concordance pairs using machine learning predictors. RSF competing risk predictors used were $\{\hat{M}_j^{(i)}(\tau)\}$, where $\hat{M}_j^{(i)}(\tau)$ and the event horizon τ was selected as the largest follow-up time (see Ishwaran et al. [16] for further details).

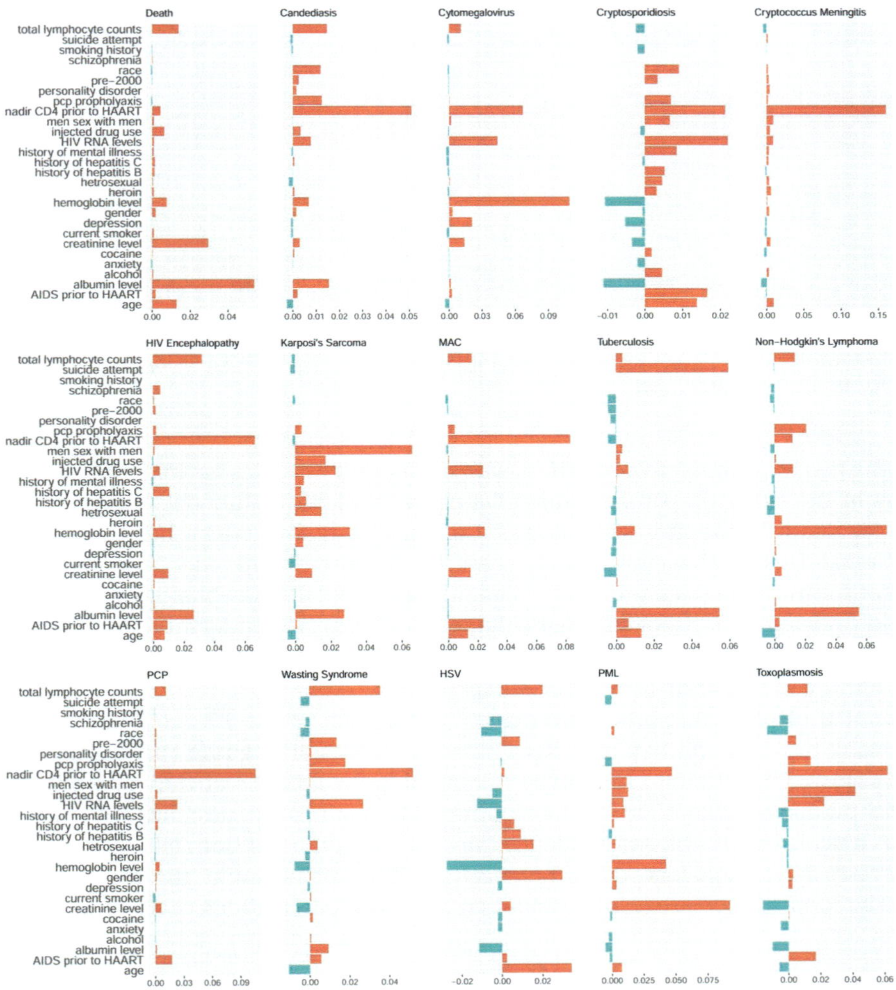

FIGURE 19.7
Permutation importance (VIMP) using RSF for each of the 15 competing risks from the HAART study. Red signifies VIMP that are positive.

VIMP for each of the $p = 28$ features is given in Figure 19.7. We observe that a feature's VIMP is highly dependent on the competing risk outcome. For all-cause death, creatinine and albumin

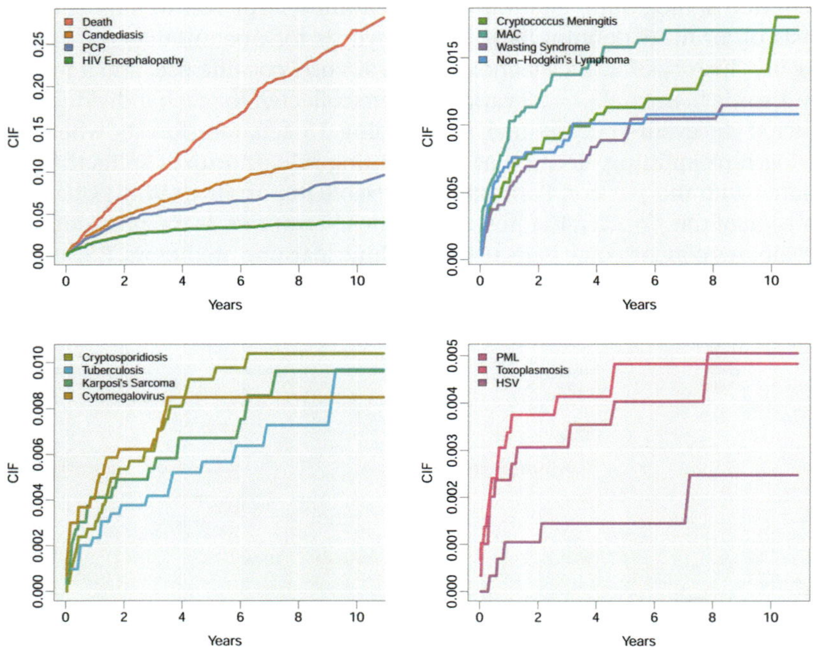

FIGURE 19.8
Patient averaged ensemble CIF for all 15 events from the HAART study using RSF. CIFs have been grouped for better visualization.

level are the two most predictive variables, with most other variables having small to zero VIMP. For many of the other outcomes, nadir CD4 count is highly influential. The importance of albumin for death is not surprising as it is a marker for many general health issues such as liver and renal disease. However, it is not a marker for immune function, whereas nadir CD4 is a marker for immune system damage. Candidiasis, PCP, HIV encephalopathy, and MAC are all infection-related, so it is not surprising that nadir CD4 is influential for these outcomes. For HIV encephalopathy, in addition to nadir CD4, total lymphocyte count is also highly influential, possibly acting as a marker for immune damage.

Figure 19.8 displays the patient-averaged CIF for each of the 15 outcomes from an RSF analysis using the Gray splitting rule. Recall that the Gray splitting rule encourages splits that separate cases with differing CIFs, making it suitable for analysis of the CIF and long-term probability prediction. The CIFs have been grouped with similar ranges for better visualization. Death has a near-uniform higher incidence rate, which is different from all other events. Some AIDS illnesses have incidence rates that peak rapidly. For example, the incidence of non-Hodgkins lymphoma increases rapidly and then begins to flatten after 4 years.

19.9 Random Hazard Forests

Lee et al. [22] described a gradient boosting tree procedure for estimating the hazard function with time varying variables. A key result was a smooth convex representation for the nonparametric log-likelihood functional and a closed-form expression for the gradient. The log-likelihood risk functional they described is especially amenable to trees, which is a key fact exploited by a new survival approach called *random hazard forests* (RHF).

RHF extends RSF in two important ways. The first is that it directly estimates the hazard function; the second is that it extends RSF to allow for time-dependent covariates. These ideas are interconnected.

To explain, recall by (19.13) that the conditional survival function is

$$S(t|\mathbf{x}) = \mathbb{P}\{T^o > t|\mathbf{X} = \mathbf{x}\} = \exp\left(-\int_0^t \lambda(s,\mathbf{x})\,ds\right), \tag{19.14}$$

where $\lambda(t,\mathbf{x})$ is the hazard function at time t given $\mathbf{X} = \mathbf{x}$ where $\mathbf{X} = (X^{(1)},\ldots,X^{(p)})^T$ is the p-dimensional time static covariate. As we have discussed, the survival function is a key quantity in survival analysis. However, in the general setting where the covariates $\mathbf{X}(t) = (X^{(1)}(t),\ldots,X^{(p)}(t))^T$ are allowed to be time varying, thus yielding what is called *functional survival data*, there is no meaningful analogue to the survival function since it involves integrating $\lambda(u,\mathbf{X}(s))$ along the unknown future trajectory of the covariates $\{\mathbf{X}(s)\}_{s\in(0,t]}$. Instead of the survival function, the interest shifts to estimating the hazard $\lambda(t,\mathbf{x})$ given $\mathbf{X}(t) = \mathbf{x}$, which, recall by (19.8), is the probability of the event happening in the near future given it has not happened yet:

$$\mathbb{P}\{T \in [t,t+dt)|T > t,\mathbf{X}(t) = \mathbf{x}\} = \lambda(t,\mathbf{x})dt. \tag{19.15}$$

In the time-static covariate setting, since $S(t|\mathbf{x})$ can be recovered from $\lambda(t,\mathbf{x})$ via (19.14), it follows that the hazard is the fundamental quantity unifying both settings. This is why RHF focuses on the hazard, as this unifies the problem.

19.9.1 Log-likelihood risk functional

The formal framework is as follows. In addition to the previous quantities, there is also an indicator $Y(t) \in \{0,1\}$ denoting whether the subject is at risk of experiencing the event during $[t,t+dt)$. Under this setting, the probability (19.15) of experiencing the event during $[t,t+dt)$ generalizes to $\lambda(t,\mathbf{x})Y(t)dt$. This allows for a variety of censoring schemes including right-censoring, corresponding to $Y(t) = 1_{\{T\geq t\}}$. Without loss of generality, we normalize the units of time so that $Y(t) = 0$ for $t > 1$, i.e. the subject is not a risk after time $t = 1$. This allows us to restrict attention to the time interval $(0,1]$ (a simple change of variables for the hazard can be used to map back to the original time scale). If the subject experiences the event in $(0,1]$ then $\delta = Y(T)$ equals 1, otherwise $\delta = 0$.

Given n functional data samples $\{(\mathbf{X}_i(\cdot), Y_i(\cdot), T_i)\}_{i=1}^n$, the evolution of subject i's event status can then be thought of as a sequence of coin flips at time increments $t = 0, dt, 2dt, \ldots$, with the probability of "heads" at each time point equal to $\lambda(t,\mathbf{x})Y(t)dt$. Therefore, subject i's contribution to the likelihood is

$$\begin{aligned}&\{1-\lambda(0,\mathbf{X}_i(0))Y_i(0)dt\}\times\{1-\lambda(dt,\mathbf{X}_i(dt))Y_i(dt)dt\}\times\cdots\times\lambda(T_i,\mathbf{X}_i(T_i))^{\delta_i}\\ &\qquad\xrightarrow[dt\downarrow 0]{}\left[e^{-\int_0^1 Y_i(t)\lambda(t,\mathbf{X}_i(t))\,dt}\right]\lambda(T_i,\mathbf{X}_i(T_i))^{\delta_i}.\end{aligned}$$

Define the log-hazard function $F(t,\mathbf{X}) = \log(\lambda(t,\mathbf{X}))$. This is the unknown nonparametric function we want to estimate. We define the empirical risk as the (scaled) negative log-likelihood functional,

$$R_n(F) = \frac{1}{n}\sum_{i=1}^n \int_0^1 Y_i(t)e^{F(t,\mathbf{X}_i(t))}dt - \frac{1}{n}\sum_{i=1}^n \delta_i F(T_i,\mathbf{X}_i(T_i)). \tag{19.16}$$

Observe that (19.16) is equivalent to the likelihood (19.2) in the time-static covariate case. Setting $\mathbf{X}_i(t) = \mathbf{X}_i$ and $Y_i(t) = 1_{\{T_i \geq t\}}$, and using

$$\lambda(t,\mathbf{x}) = \frac{f(t|\mathbf{x})}{S(t|\mathbf{x})}, \qquad \int_0^t \lambda(s,\mathbf{x})\, ds = -\log(S(t|\mathbf{x})),$$

(19.16) becomes

$$\frac{1}{n}\sum_{i=1}^{n}\int_0^{T_i} \lambda(t,\mathbf{X}_i)\, dt - \frac{1}{n}\sum_{i=1}^{n}\delta_i \log(\lambda(T_i,\mathbf{X}_i))$$
$$= -\frac{1}{n}\sum_{i=1}^{n}\log(S(T_i|\mathbf{X}_i)) - \frac{1}{n}\sum_{i=1}^{n}\delta_i \log\left(\frac{f(T_i|\mathbf{X}_i)}{S(T_i|\mathbf{X}_i)}\right).$$

This is the negative log-likelihood from (19.2).

19.9.2 Splitting a RHF tree

RHF uses best split first (BSF) where the tree splits are in the sequential order of largest decrease of empirical risk $R_n(F)$. The following notation is introduced to express the empirical risk in a useful form for explaining this. Define the empirical sub-probability measure on $(0,1] \times \mathscr{X}$ as

$$\tilde{\mathbb{P}}_n(B) = \frac{1}{n}\sum_{i=1}^{n}\int_0^1 Y_i(t)\cdot 1_{\{(t,\mathbf{X}_i(t))\in B\}}\, dt.$$

Intuitively, $\tilde{\mathbb{P}}_n$ measures the denseness of the sample time-covariate paths. To estimate F we consider the space of models $\mathscr{F} = \{\sum_j c_j 1_{\{(t,\mathbf{x})\in B_j\}} : c_j \in \mathbb{R}\}$ which are linear combinations of indicator functions over disjoint time-covariate cells $\{B_j\}$. The cells are constructed by using all possible observed time and covariate values; thus $\{B_j\}$ forms the most granular partition of the time-covariate space that can be built from the observed data.

We can write the tree estimator as $F(t,\mathbf{x}) = \sum_j c_j 1_{\{(t,\mathbf{x})\in B_j\}} \in \mathscr{F}$. The following recasts the log-likelihood risk as a smooth convex functional on $\mathscr{F}$ with inner product $(\cdot,\cdot)_{\tilde{\mathbb{P}}_n}$ and therefore provides a tool for optimizing the empirical risk functional.

Theorem 19.3. *The log-likelihood risk (19.16) for $F(t,\mathbf{x}) = \sum_j c_j 1_{\{(t,\mathbf{x})\in B_j\}} \in \mathscr{F}$ can be written as*

$$R_n(F) = \sum_{j:\tilde{\mathbb{P}}_n(B_j)>0}\left(e^{c_j}\cdot\tilde{\mathbb{P}}_n(B_j) - \frac{c_j\, \mathit{Fail}(B_j)}{n}\right), \tag{19.17}$$

where $\mathit{Fail}(B_j) = \sum_i \delta_i 1_{\{(T_i,\mathbf{X}_i(T_i))\in B_j\}}$ is the number of observed failures in the time-covariate region B_j.

Theorem 19.3 is used to determine the best split for a tree. For this argument and throughout we assume a right-censoring predictable function $Y_i(t) = 1_{\{T_i \geq t\}}$.

Begin by initializing the tree log-hazard function by the constant function

$$F_0(t,\mathbf{x}) = c_0 1_{\{(t,\mathbf{x})\in(0,1]\times\mathscr{X}\}}.$$

To determine c_0, rewrite F in terms of the disjoint time-covariate cells $\{B_j\}$,

$$F_0(t,\mathbf{x}) = c_0 \sum_j 1_{\{(t,\mathbf{x})\in B_j\}}.$$

Setting $F = F_0$ in (19.17), differentiating with respect to c_0, and setting the derivative to zero, we find the value of c_0 that minimizes $R_n(F_0)$ is

$$e^{c_0} = \frac{n^{-1}\sum_{\tilde{\mathbb{P}}_n(B_j)>0}\mathrm{Fail}(B_j)}{\sum_{\tilde{\mathbb{P}}_n(B_j)>0}\tilde{\mathbb{P}}_n(B_j)}.$$

This can be conveniently expressed as $c_0 = \log(V_0/U_0)$, where

$$\begin{aligned} V_0 &= \text{total number of deaths}/n \\ U_0 &= n^{-1}\sum_{i=1}^{n}\int_0^{T_i}\left[\sum_j 1_{\{(t,\mathbf{X}_i(t))\in B_j\}}\right]dt = n^{-1}\sum_{i=1}^{n} T_i = \text{total risk}/n. \end{aligned}$$

Now consider splitting the tree s times ($s = 0$ is the initialization step above). Let $F_s(t,\mathbf{x}) = \sum_{j=1}^{s+1} c_j 1_{\{(t,\mathbf{x})\in A_j\}}$ be the log-hazard estimator. To obtain the $s+1$ split, consider a cell A_j split into subregions $A_{j,1}$ and $A_{j,2}$, with F_s updated to F_{s+1} defined by

$$F_{s+1}(t,\mathbf{x}) = F_s(t,\mathbf{x}) - c_j 1_{\{(t,\mathbf{x})\in A_j\}} + c_{j,1}1_{\{(t,\mathbf{x})\in A_{j,1}\}} + c_{j,2}1_{\{(t,\mathbf{x})\in A_{j,2}\}}. \tag{19.18}$$

We want to find $\{A_{j,1}, A_{j,2}, c_{j,1}, c_{j,2}\}$ to maximize

$$R_n(F_s) - R_n(F_{s+1}) = R_n\left(c_j 1_{\{(t,\mathbf{x})\in A_j\}}\right) - R_n\left(\sum_{l=1}^{2} c_{j,l}1_{\{(t,\mathbf{x})\in A_{j,l}\}}\right).$$

Let $F_{j,l} = c_{j,l}1_{\{(t,\mathbf{x})\in A_{j,l}\}}$ for $l = 1,2$. Write $F_{j,l}$ in terms of $\{B_j\}$,

$$F_{j,l}(t,\mathbf{x}) = c_{j,l}\sum_{B_j\subseteq A_{j,l}} 1_{\{(t,\mathbf{x})\in B_j\}}.$$

Setting $F = F_{j,l}$ in Theorem 19.3 and solving for the gradient as before, the value $c_{j,l}$ that minimizes $R_n(F_{j,l})$ is $c_{j,l} = \log(V_{j,l}/U_{j,l})$, where

$$\begin{aligned} V_{j,l} &= n^{-1}\sum_{i=1}^{n}\delta_i\left[\sum_{B_j\subseteq A_{j,l}} 1_{\{(T_i,\mathbf{X}_i(T_i))\in B_j\}}\right] = n^{-1}\sum_{i=1}^{n}\delta_i 1_{\{(T_i,\mathbf{X}_i(T_i))\in A_{j,l}\}} \\ U_{j,l} &= n^{-1}\sum_{i=1}^{n}\int_0^{T_i}\left[\sum_{B_j\subseteq A_{j,l}} 1_{\{(t,\mathbf{X}_i(t))\in B_j\}}\right]dt = n^{-1}\sum_{i=1}^{n}\int_0^{T_i} 1_{\{(t,\mathbf{X}_i(t))\in A_{j,l}\}}\,dt. \end{aligned}$$

Informally, these values have the following interpretation:

$$\begin{aligned} V_{j,l} &= \text{number of deaths in } A_{j,l}/n \\ U_{j,l} &= \text{risk in } A_{j,l}/n. \end{aligned}$$

By induction, it follows that the value for c_j corresponding to the parent node A_j is $c_j = \log[(V_{j,1}+V_{j,2})/(U_{j,1}+U_{j,2})]$, so the empirical risk reduction by (19.17) equals

$$\begin{aligned} &R_n(F_s) - R_n(F_{s+1}) \\ &= \frac{1}{n}\sum_{i=1}^{n}\sum_{l=1}^{2}\left\{\int_0^{T_i}(e^{c_j} - e^{c_{j,l}})1_{\{(t,\mathbf{X}_i(t))\in A_{j,l}\}}\,dt - (c_j - c_{j,l})\delta_i 1_{\{(T_i,\mathbf{X}_i(T_i))\in A_{j,l}\}}\right\} \\ &= \sum_{l=1}^{2}\left\{(e^{c_j} - e^{c_{j,l}})U_{j,l} - (c_j - c_{j,l})V_{j,l}\right\} \\ &= V_{j,1}\log\left(\frac{V_{j,1}}{U_{j,1}}\right) + V_{j,2}\log\left(\frac{V_{j,2}}{U_{j,2}}\right) - (V_{j,1}+V_{j,2})\log\left(\frac{V_{j,1}+V_{j,2}}{U_{j,1}+U_{j,2}}\right) \\ &= V_{j,1}\log\left(\frac{V_{j,1}}{U_{j,1}}\right) + V_{j,2}\log\left(\frac{V_{j,2}}{U_{j,2}}\right) - V_j\log\left(\frac{V_j}{U_j}\right). \end{aligned} \tag{19.19}$$

Of all possible splits $A_{j,1}, A_{j,2}$, we choose the cut which maximizes (19.19). This leads to the following definition for the RHF split-statistic.

Definition 19.2. The RHF split-statistic for splitting a node A_j into regions $A_{j,1}, A_{j,2}$ is

$$V_{j,1}\log\left(\frac{V_{j,1}}{U_{j,1}}\right)+V_{j,2}\log\left(\frac{V_{j,2}}{U_{j,2}}\right) \tag{19.20}$$

and the empirical risk reduction from the split is

$$\Delta_n(A_{j,1},A_{j,2})=V_{j,1}\log\left(\frac{V_{j,1}}{U_{j,1}}\right)+V_{j,2}\log\left(\frac{V_{j,2}}{U_{j,2}}\right)-(V_{j,1}+V_{j,2})\log\left(\frac{V_{j,1}+V_{j,2}}{U_{j,1}+U_{j,2}}\right). \tag{19.21}$$

If $R_n(F_s)$ is the empirical risk for the current log-hazard, and $R_n(F_{s+1})$ is the risk for the updated log-hazard (19.18) from the split, then

$$R_n(F_{s+1})=R_n(F_s)-\Delta_n(A_{j,1},A_{j,2}).$$

The best split is the value that maximizes (19.20), or equivalently, that maximizes (19.21). □

Splitting the tree K times using Definition 19.2 results in the partition $\mathbb{\Pi}=\{A_1,\dots,A_{K+1}\}$, which is the tree partition of $(0,1]\times\mathscr{X}$. When constructing a RHF tree, only covariate splits are used. The reason time splits are not employed will be explained in the following section. In that section, we also define the hazard estimate for a terminal node, which will differ from the estimate at an intermediate node, where it is treated as constant in the above discussions.

19.9.3 Terminal node estimator

Due to the functional nature of the data, an individual's time history is separated into disjoint start-stop intervals. Case i in a terminal node A_s will have n_i records and each of these has a start-stop interval, with the last time point being either censored or an event. These n_i records can be considered as "pseudo-observations". These records looks something like:

id	start-stop	δ	$\mathbf{X}$
i	$(T_{i,1}^-, T_{i,1}]$	$\delta_i(T_{i,1})=0$	$\mathbf{X}_i(T_{i,1})$
i	$(T_{i,2}^-, T_{i,2}]$	$\delta_i(T_{i,2})=0$	$\mathbf{X}_i(T_{i,2})$
$\vdots$	$\vdots$	$\vdots$	
i	$(T_{i,n_i}^-, T_{i,n_i}]$	$\delta_i(T_{i,n_i})\in\{0,1\}$	$\mathbf{X}_i(T_{i,n_i})$

Here, $T_{i,j}^-$ denotes the left end point for the jth interval and satisfies $T_{i,j}^-\ge T_{i,j-1}$.

To define the terminal node estimator, let

$$0=T_0^s<T_1^s<\cdots<T_N^s<T_{N+1}^s=1$$

be the unique ordered time points in A_s (padded by 0 and 1) and let $\mathrm{t}_1<\cdots<\mathrm{t}_m$ denote the ordered event times from the full data set. Set $l(0)=0$ and define $\{l(k)\}$ to be the indices of $T_1^s,\dots,T_N^s$, in order, corresponding to the subset of the possible event times: $\{T_{l(k)}^s\}\subseteq\{\mathrm{t}_k\}$. For example, if $l(j)$ is the index corresponding to event time t_k, then $T_{l(j)}^s=\mathrm{t}_k$, and this would mean for example that $T_{l(j)-1}^s$ is the first time point in s left adjacent to t_k, but $T_{l(j)-1}^s$ is not necessarily an event time. The smallest event time in s less than $T_{l(j)}^s$ is $T_{l(j-1)}^s$. Thus we have

$$T_{l(j-1)}^s\le T_{l(j)-1}^s<T_{l(j)}^s=\mathrm{t}_k.$$

For a terminal node A_s, let $d_s(\mathrm{t}_k)=\sum_{i\in A_s}1_{\{T_{i,n_i}=\mathrm{t}_k,\,\delta_i(T_{i,n_i})=1\}}$ be the number of cases dead in A_s at time t_k and let $Y_s(\mathrm{t}_k)=\sum_{i\in A_s}1_{\{T_{i,n_i}\ge\mathrm{t}_k\}}$ be the number of pseudo-cases in A_s at risk prior to t_k. Using this we have the following definition for the RHF terminal node estimator.

Definition 19.3. The hazard estimator for a terminal node A_s is

$$\lambda_s(t) = \frac{d_s(T^s_{l(j)})}{(T^s_{l(j)} - T^s_{l(j-1)})Y_s(T_{l(j)})}, \quad T^s_{l(j-1)} < t \le T^s_{l(j)},$$

which equals number of deaths at $T^s_{l(j)}$ in A_s divided by number of pseudo-observations at risk prior to $T^s_{l(j)}$ in A_s, normalized by the length of time. □

With the above notation, integrating over time, we obtain the CHF

$$H_s(T^s_{l(k)}) = \sum_{j=1}^{k} \int_{T^s_{l(j)-1}}^{T^s_{l(j)}} \frac{d_s(T^s_{l(j)})}{(T^s_{l(j)} - T^s_{l(j)-1})Y_s(T_{l(j)})} dt = \sum_{j=1}^{k} \frac{d_s(T^s_{l(j)})}{Y_s(T_{l(j)})},$$

which equals the Nelson-Aalen CHF estimator (19.4) evaluated at the event time $T^s_{l(k)}$. Therefore, Definition 19.3 is chosen so that it matches Nelson-Aalen.

19.9.4 The RHF algorithm

In our earlier discussion of Definition 19.2, we noted that RHF splits only on covariates. As a result, using a constant hazard function for a terminal node, similar to what is done for intermediate nodes, would create a tree with a hazard function that remains unchanged over time, which poses significant issues. One potential solution to this problem is to allow splits based on time, thereby achieving a time-varying estimator. However, this approach introduces several challenges. Firstly, splitting on time can lead to an overfitted hazard estimator. Secondly, even when time splits are restricted to event times, the terminal node estimator might still fail to align with the Nelson-Aalen CHF.

To address these challenges, the terminal node estimator for RHF is defined as the time-varying hazard estimator described in Definition 19.3. This choice not only provides a more appropriate estimator but also simplifies the tree construction process by restricting splits to covariates.

Below is a formal description of RHF that incorporates the time-varying terminal node estimator. This algorithm can be applied in scenarios involving both time-static and time-dependent covariates.

Algorithm 19.1 *Random Hazard Forests (RHF) Algorithm*

1: Use the RHF split-statistic (19.20) to grow a tree using a pre-specified number of splits $K \ge 1$. Splits are applied only to covariates (which can be time-static or time-dependent). Use random feature selection when choosing covariates for splitting a node.

2: Let $\mathcal{T} = \{A_1, \ldots, A_{K+1}\}$ be the collection of terminal nodes of the resulting tree. The tree hazard estimator $\lambda(t, \mathbf{x}, \mathcal{T})$ is obtained by calculating each terminal node's hazard estimator using Definition 19.3.

3: Grow a collection of tree hazard learners $\{\lambda(t, \mathbf{x}, \mathcal{T}_b), b = 1, \ldots, B\}$ using steps 1 and 2. The RHF ensemble hazard estimator is the tree-averaged estimator $\sum_{b=1}^{B} \lambda(t, \mathbf{x}, \mathcal{T}_b)/B$.

Remark 19.2. Making use of pseudo-observations reveals a computationally efficient way to calculate the RHF split-statistic in Step 1 of Algorithm 19.1. It suffices to consider the relevant calculations for a cell A_s. The goal is to rapidly calculate

$$\begin{aligned} nV_s &= \text{number of deaths in } A_s \\ nU_s &= \text{total risk in } A_s. \end{aligned}$$

Observe that

$$nV_s = \sum_{i=1}^{n} \delta_i 1_{\{(T_i, \mathbf{X}_i(T_i)) \in A_s\}} = \sum_{i \in A_s} \sum_{j=1}^{n_i} \delta_i(T_{i,j}).$$

Therefore calculating V_s is straightforward since we just have to sum $\{\delta_i(T_{i,j})\}_{i,j}$ over all pseudo-observations. The more challenging calculation is U_s; however, this also simplifies by working with pseudo-observations. By definition,

$$nU_s = \sum_{i=1}^{n} \int_0^{T_i} 1_{\{(t, \mathbf{X}_i(t)) \in A_s\}} \, dt.$$

However, notice this can be simplified to

$$nU_s = \sum_{i \in A_s} \sum_{j=1}^{n_i} \int_0^{T_{i,j}} 1_{\{T_{i,j}^- < t \le T_{i,j}\}} \, dt = \sum_{i \in A_s} \sum_{j=1}^{n_i} (T_{i,j} - T_{i,j}^-).$$

This has a simple interpretation. The total risk equals the sum of the lengths of the time intervals over the pseudo-cases. □

Remark 19.3. Recall that RHF normalizes the units of time to the interval $(0,1]$. The final hazard estimator must therefore be transformed back to its original time scale. Suppose that T' denotes the failure time in the original units and let τ' be the longest observed follow up time. Then $T = T'/\tau'$ is the transformation used to process the data to unit time. The hazard function on the original scale $\lambda_{T'}$ can be recovered from $\lambda_T := \lambda$ using

$$\lambda_{T'}(t', \mathbf{x}(t')) = \frac{1}{\tau'} \lambda_T \left(\frac{t'}{\tau'}, \mathbf{x}\left(\frac{t'}{\tau'} \right) \right).$$

□

Exercises

Exercise 19.1. Give three real-world examples illustrating generalized Type-I censoring.

Exercise 19.2. Give a real-world example of left-censoring. Give the likelihood for this type of data, stating all your notation and assumptions clearly.

Exercise 19.3. A researcher conducted a trial to study whether a drug could shrink cancer tumors. At the end of the trial, survival curves were created to compare the patients with an early response to treatment (tumor shrinkage) to those without. It was found that responders had better survival curves. The researcher initially used Kaplan-Meier (KM) curves for the analysis but later decided to apply a random survival forest (RSF) analysis for the same comparison. Is this a better analysis? Are there problems with this approach?

Exercise 19.4. In this exercise we will learn how to estimate and interpret survival probabilities using the KM estimator.

1. Load an appropriate real-world survival dataset.
2. Use the KM estimator to estimate the survival function.
3. Plot the KM survival curves.
4. Compare the survival curves between two or more groups (e.g., based on treatment or demographic variables).

5. Interpret the results and draw conclusions about the survival probabilities for the different groups.

Exercise 19.5. In this exercise we will understand how to compare survival curves between groups using the log-rank test.

1. Load a survival dataset (e.g., the `lung` dataset from the `survival` R-package).
2. Use the KM estimator to estimate survival curves for two groups (e.g., smokers vs. non-smokers).
3. Perform a log-rank test to compare the survival distributions of the two groups.
4. Interpret the p-value and test statistic to determine if there is a significant difference between the groups.
5. Report your findings.

Exercise 19.6. Find a real-world survival dataset. Calculate the log-rank split-statistic for each feature using split-values over all possible observed values of the feature. Plot your results and comment on which features appear to be most informative for survival.

Exercise 19.7. Here we will fit parametric survival models and compare results to machine learning.

1. Load a real-world survival dataset.
2. Fit parametric survival models (e.g., exponential, Weibull, log-normal) to the data.
3. Compare the fit of the parametric models using criteria such as AIC or BIC.
4. Plot the survival functions from the different parametric models.
5. Interpret the parameters of the best-fitting model and report your conclusions.
6. Compare your results to the same analysis using RSF.

Exercise 19.8. Understand how to apply gradient boosting techniques to survival data.

1. Load a real-world survival dataset.
2. Train a gradient boosting survival model (e.g., using the `gbm` R-package).
3. Tune the hyperparameters of the model using cross-validation.
4. Evaluate the model's performance.
5. Use your model to understand your data and compare your result to RSF.
6. What are the strengths and weaknesses of the two methods?

Exercise 19.9. Use the `peakVO2` dataset from the `randomForestSRC` R-package [19] and perform a RSF analysis. Calculate the Brier score and the time-dependent AUC and plot these values as a function of time. Be sure to state all your assumptions. For example, state the estimator that was used for the censoring distribution. State your conclusions.

Exercise 19.10. Continuing from the previous exercise, using the same data, apply permutation importance VIMP, minimal depth and varPro variable selection. Summarize the results from the three different methods and state your conclusions about the data. State carefully any choices that were made in the analysis that might affect your results. For example, indicate the external estimator used for varPro. State what threshold techniques were used for each method in determining whether a variable was a signal variable.

Exercise 19.11. Using the notation from this chapter, show that

$$\widehat{\text{BS}}(t,\hat{S}) = \frac{1}{n}\sum_{i=1}^{n}\left\{\frac{\delta_i \hat{S}^2(t|\mathbf{X}_i)1_{\{T_i\le t\}}}{G(T_i-|\mathbf{X}_i)} + \frac{\left(1-\hat{S}(t|\mathbf{X}_i)\right)^2 1_{\{T_i>t\}}}{G(t|\mathbf{X}_i)}\right\}$$

is an unbiased estimator for the Brier score

$$\mathrm{BS}(t,\hat{S}) = \mathbb{E}\left(1_{\{T^o>t\}} - \hat{S}(t|\mathbf{X})\right)^2.$$

where $G(t|\mathbf{X}) = \mathbb{P}\{C > t|\mathbf{X}\}$ is the true censoring distribution and $\hat{S}(t|\mathbf{X})$ is the estimated survival function. Make sure to state all your assumptions clearly.

Exercise 19.12. Find a suitable dataset for a competing risk analysis and perform a RSF competing risk analysis. Run a VIMP analysis and use the results from this to draw conclusions about your data. State your results in a way that provide insight into how variables affect the different competing risks.

Exercise 19.13. In this exercise we will use simulations to study variable selection in survival analysis. Come up with a relatively interesting and complicated synthetic survival experiment (Remark 19.1 is helpful for this). Choose at least two different variable selection methods and study their performance over a replicated benchmark experiment using your synthetic data. What are your findings?

Exercise 19.14. Find a real-world survival dataset and test a hypothesis about one of the variables. Use any appropriate technique to draw conclusions regarding your hypothesis. Justify your approach.

Exercise 19.15. Write a program to use the bootstrap to calculate standard errors for the survival curves from a RSF. Use your program on a dataset to display some survival curves and add ± standard errors to the curves.

Exercise 19.16. Here we will learn how to tune hyperparameters of a RSF.

1. Load a real-world survival dataset.
2. Split the dataset into training and testing sets.
3. Train an RSF model using the `randomForestSRC` R-package in R, performing hyper-parameter tuning (e.g., node size, number of variables to try at each split).
4. Evaluate the model's performance on the testing data using appropriate metrics (e.g., concordance index, integrated Brier score).
5. Use variable importance measures to identify the most influential predictors and interpret the results.
6. Visualize the survival curves for different groups based on important variables.

Exercise 19.17. Here we will explore techniques for handling missing data in RSF and assess their impact on model performance.

1. Load a real-world survival dataset with missing values (e.g., the `pbc` dataset from the `randomForestSRC` R-package).
2. Split the dataset into training and testing sets.
3. Compare different strategies for handling missing data:
   - Ignoring missing data.
   - Imputing missing data using missForest (see Chapter 16.14).
   - Using the built-in imputation methods of `randomForestSRC`.
4. Train RSF models on the datasets prepared with different missing data handling strategies.
5. Evaluate and compare the models' performance using appropriate metrics.
6. Discuss the impact of different missing data handling techniques on the model results and performance.

Exercise 19.18. Stacking is an ensemble learning technique that combines multiple machine learning models to improve predictive performance. It involves training several base models (also known as level-0 models) and then training a meta-model (or level-1 model) to make the final predictions based on the outputs of the base models. Here we will implement and evaluate a stacked ensemble model combining RSF with other survival models.

1. Load a real-world survival dataset.
2. Split the dataset into training and testing sets.
3. Train several base models, including RSF, Cox proportional hazards model [3], and a gradient boosting survival model (e.g., using `gbm`).
4. Generate meta-features from the predictions of the base models.
5. Train a meta-model (e.g., a Cox model) using the meta-features.
6. Evaluate the performance of the stacked ensemble model on the testing data using appropriate metrics.
7. Compare the performance of the stacked ensemble model to the individual base models and discuss the results.

Bibliography

[1] Aalen O. and Johansen S. An empirical transition matrix for non-homogeneous Markov chains based on censored observations, *Scandinavian Journal of Statistics* 5:141–150, 1978.

[2] Andersen P.K. Decomposition of number of life years lost according to causes of death. *Statistics in Medicine*, 32(30), 5278–5285, 2013.

[3] Cox D. R. Regression models and life-tables. *Journal of the Royal Statistical Society. Series B (Methodological)*, 34(2), 187–220, 1972.

[4] Gerds T.A. and Schumacher, M. Consistent estimation of the expected Brier score in general survival models with right-censored event times. *Biometrical Journal*, 6:1029-1040, 2006.

[5] Gerds T.A. and Scheike T.H. and Andersen P.K. Absolute risk regression for competing risks: interpretation, link functions, and prediction. Statistics in Medicine,31(29):3921–3930, 2012.

[6] Graf, E., Schmoor, C., Sauerbrei, W. and Schumacher, M. Assessment and comparison of prognostic classification schemes for survival data. *Statistics in Medicine*, 18, 2529–2545, 1999.

[7] Gray R.J. A class of K-sample tests for comparing the cumulative incidence of a competing risk. *Annals of Statistics*, 16:1141–1154, 1988.

[8] Harrell F., Califf R. and Pryor D. and Lee K. and Rosati R. Evaluating the yield of medical tests. *Journal of American Medical Association*, 247:2543–2546, 1982.

[9] Heagerty P.J., Lumley T. and Pepe M.S. Time-dependent ROC curves for censored survival data and a diagnostic marker. *Biometrics*, 56(2): 337–344, 2000.

[10] Hsich E., Chadalavada S. and Krishnaswamy G. and Starling R. C. and Pothier C. E. and Blackstone, E. H. and Lauer M. S. Long-term prognostic value of peak oxygen consumption in women versus men with heart failure and severely impaired left ventricular systolic function. *The American Journal of Cardiology*, 100(2),291–295, 2007.

[11] Hsich E., Gorodeski E.Z. and Blackstone E.H. and Ishwaran H and Lauer M.S. Identifying important risk factors for survival in patient with systolic heart failure using random survival forests. *Circulation: Cardiovascular Quality and Outcomes*, 4:39–45, 2001.

[12] Hothorn T. and Lausen B. On the exact distribution of maximally selected rank statistics. *Computational Statistics & Data Analysis*, 43(2), 121–137, 2003.

[13] Irwin J. The standard error of an estimate of expectation of life, with special reference to expectation of tumourless life in experiments with mice. *Epidemiology & Infection*, 47, 188–189, 1949.

[14] Ishwaran H., Blackstone E.H., Pothier C.E., and Lauer M.S. Relative risk forests for exercise heart rate recovery as a predictor of mortality. *Journal of American Medical Association*, 99(467):591–600, 2004.

[15] Ishwaran H., Blackstone E.H., Hansen. C.A. and Rice T.W. A novel approach to cancer staging: application to esophageal cancer. *Biostatistics*, 10:603–620, 2009.

[16] Ishwaran H., Gerds T., Kogalur U.B., Moore R.D., Gange S.J., and Lau B.M. Random survival forests for competing risks. 2012.

[17] Ishwaran H., Kogalur U.B., Blackstone E.H. and Lauer M.S. Random survival forests. *Ann. Appl. Stat.*, 2:841–860, 2008.

[18] Ishwaran H., Kogalur U.B., Gorodeski E.Z., Minn A.J., and Lauer M.S. High-dimensional variable selection for survival data. *J. Amer. Stat. Assoc.*, 105:205-217, 2010.

[19] Ishwaran H. and Kogalur U.B. randomForestSRC: Random Forests for Survival, Regression and Classification. R package version 3.2.3. http://cran.r-project.org, 2024.

[20] Kaplan E.L. and Meier P. Nonparametric estimation from incomplete observations. *Journal of the American Statistical Association*, 53(282), 457–481, 1958.

[21] Larson M.G. Covariate analysis of competing-risks data with log-linear models. *Biometrics*, 40:459–469, 1978.

[22] Lee, D.K. and Chen N. and Ishwaran H. Boosted nonparametric hazards with time-dependent covariates. *Annals of Statistics*, 49, 21012128, 2021.

[23] Lu M. and Ishwaran H. Variable Priority for Variable Selection in Rule-based Models. *Techincal Report*, 2022.

[24] Moore R.D. Understanding the clinical and economic outcomes of HIV therapy: the Johns Hopkins HIV clinical practice cohort. *Journal of Acquired Immune Deficiency Syndrome and Human Retrovirology*, 17(Suppl.1):S38–S41, 1998.

[25] Naftel D., Blackstone E.H. and Turner M. Conservation of events. Unpublished notes, 1985.

[26] Prentice R.L, Kalbfleisch J.D.,Peterson A.V. JR., Flournoy N., Farewell V.T. and Breslow N.E. The analysis of failure times in the presence of competing risks. *Biometrics*, 34:541–554, 1978.

[27] Rice T.W., Rusch V.W., Ishwaran H. and Blackstone E.H. Cancer of the esophagus and esophagogastric junction: data-driven staging for the 7th edition of the AJCC/UICC cancer staging manuals. *Cancer*, 116(16):3763–3773, 2010.

20

Causal Estimates using Machine Learning

20.1 Introduction

Causal analysis is a method used to determine the cause-and-effect relationships between variables. It goes beyond mere correlation and aims to understand whether changes in one variable directly cause changes in another. This is fundamental in various fields such as medicine, social sciences, and economics, where understanding the impact of interventions, policies, or treatments is critical.

Traditional approaches to causal inference rely on randomized controlled trials (RCTs), which are considered the gold standard. In randomized experiments, subjects are randomly assigned to different groups, typically a treatment group and a control group. This randomization ensures that each group is statistically equivalent at the start of the experiment, controlling for both known and unknown confounders. This allows the observed treatment effect, defined as the difference in the effect size for the treatment compared to the control, to be directly interpreted as a causal effect.

However, RCTs are not always feasible due to ethical and logistical constraints. As a result, many real-life therapeutic decisions are based on observational data. In observational studies, the researcher observes the subjects without manipulating the assignment of treatments. The treatment and control groups are not randomly assigned but are naturally occurring or self-selected.

Because of this, several challenges arise when dealing with observational data. One issue is *confounding variables* which are variables that influence both treatment and outcome. These can create a spurious association between the treatment and the outcome if not properly adjusted for in the analysis. A related issue is *selection bias* which occurs when there are systematic differences in characteristics between treated and untreated groups. Finally, *partial overlap* is a problem which occurs when there is a limited overlap in characteristics of groups, thus making it infeasible to compare some groups of individuals.

In this chapter, we delve into the concept of causation and discuss various classical estimators used for the estimation of causal effects from observational data. Following this, we will explore advanced machine learning methods for estimating individual treatment effects (ITE), an extension of the population average treatment effect (ATE) used in classical analyses, using the concept of virtual twins (digital twins). Through this exploration, we aim to provide a comprehensive understanding of how these methodologies contribute to the accurate assessment of causal relationships and the assumptions necessary to achieve accurate estimates using observational data.

DOI: 10.1201/9781003493679-20

20.2 Treatment Effect in Randomized Studies

Formally, let (T, Y) denote the observed treatment assignment and the outcome measuring the treatment's effectiveness. Specifically, $T = 0$ represents the control group, and $T = 1$ represents the intervention group (generalizations to more than two groups will be addressed later). To accurately measure the effect of the treatment, we need two variables, $Y(0)$ and $Y(1)$, representing the potential outcomes under treatments $T = 0$ and $T = 1$. This is known as the counterfactual framework or potential outcomes model [8, 12].

In this framework, $Y(0)$ is the response if the individual had received treatment $T = 0$, and $Y(1)$ is the response if the individual had received treatment $T = 1$. The causal effect is the difference between these two values:

$$\text{Causal Effect} = Y(1) - Y(0). \tag{20.1}$$

However, the problem with this definition is that it is not possible to observe $Y(1)$ and $Y(0)$ simultaneously, which the potential outcomes framework implicitly requires. If these two values are not observed simultaneously, there is a potential for bias because the responses are collected at different times and under different conditions. This undermines the premise that all factors are the same, which is essential for determining the true causal effect. This issue leads to the fundamental problem of causal inference, as noted by Holland [5].

Fundamental Problem of Causal Inference: It is impossible to observe the value of $Y(0)$ and $Y(1)$ on the same unit and, therefore, it is impossible to observe the effect of T on Y.

This might make causal inference seem hopeless, but there is a solution. Instead of using the potential outcome response values to determine the causal effect size, we use the *average causal effect* by calculating the mean. This is a standard approach in statistical analyses. We rarely use the observed Y for inference because it can contain measurement errors and spurious effects. Instead, we use averages or regression models to capture the underlying effect, removing variability due to errors in measuring Y. This leads to the *average treatment effect* (ATE) defined as the difference between the mean potential outcomes

$$\tau = \mathbb{E}(Y(1)) - \mathbb{E}(Y(0)). \tag{20.2}$$

While estimating (20.1) is not possible due to the fundamental problem of causal inference, estimating (20.2) is feasible under certain conditions. Specifically, under the assumption of conditional independence, we can estimate the ATE from the observed data.

This independence assumption is met in randomized studies, which is why such studies are considered the gold standard for causal analysis. To understand this, we will consider two different studies with different types of randomization. As will be seen, both meet the criteria for unbiased ATE estimation.

1. The first study uses a completely randomized design. In this study, for a given population, individuals are selected randomly with probabilities of 30% and 70% to receive treatments $T = 1$ and $T = 0$, respectively.

2. For the second study, consider a population of individuals with cardiovascular disease. Among those predetermined to be in critical health, 85% are selected to receive surgical treatment, while the remaining 15% receive drug therapy with close follow-up monitoring. For individuals who are not in critical condition, 90% are randomly assigned to drug therapy, while the remaining 10% are assigned to surgical treatment. This second example is slightly more complicated and illustrates a randomized study using conditional randomization.

We start by looking at the first study which is more easier to understand. Unconditional randomization assumes that

$$\{Y(0), Y(1)\} \perp T.$$

That is, the potential outcomes are independent of the actual treatment assignment. This makes sense because individuals are randomly assigned to one of the two treatments, so the potential outcome of what would have been observed cannot depend on the treatment actually assigned. By conditional independence, it follows that

$$\begin{aligned} \tau &= \mathbb{E}[Y(1)] - \mathbb{E}[Y(0)] \\ &= \mathbb{E}[Y(1)|T=1] - \mathbb{E}[Y(0)|T=0] \quad \text{(by conditional independence)} \\ &= \mathbb{E}[Y|T=1] - \mathbb{E}[Y|T=0]. \end{aligned}$$

Notice that the last line conditions on the treatment assignment. For example, the first value, $\mathbb{E}[Y \mid T = 1]$, is the mean of the response given that the assignment was $T = 1$. Since this information is available from the observed data, the conditional mean is estimated directly by averaging the response values for treatment assignment $T = 1$. The same holds true for the second term, $\mathbb{E}[Y \mid T = 0]$. Therefore, the ATE, τ, can be estimated from the observable data.

It is important to note that the above argument makes a subtle assumption when going from the second to the third line. This step is only valid if the response to the treatment received is the same as the potential response to that treatment, meaning $Y = Y(T)$. This is called the consistency assumption, and we assume this is always the case.

Consistency: The response and the potential outcome response are related by $Y = \sum_{t=0}^{1} 1_{\{T=t\}} Y(t) = Y(T)$.

Now consider our second example, the conditionally randomized experiment. We extend the previous notation to allow for covariate information X and denote the observed data by (T, X, Y). In this example, randomization is based on a pre-determined variable recording health status of the patient. Call this variable X, where $X = 1$ indicates a critical condition and $X = 0$ is non-critical. Setting $T = 1$ for the surgical treatment, we have

$$\mathbb{P}\{T=1|X=1\} = .85, \quad \mathbb{P}\{T=1|X=0\} = .1,$$

and for the drug therapy, $T = 0$, we have

$$\mathbb{P}\{T=0|X=1\} = .15, \quad \mathbb{P}\{T=0|X=0\} = .9.$$

Even though this is not an unconditional randomized study, we can still achieve unbiased ATE estimation.

To see why, first observe that we have the following conditional independence

$$\{Y(0), Y(1)\} \perp T \mid X. \tag{20.3}$$

In our example, X is the critical health status of the individual. Because individuals are randomized to treatments conditional on this value, the potential outcomes are also conditionally independent of treatment assignment given X, and the assumption clearly holds. From (20.3), it follows that

$$\begin{aligned}\mathbb{E}[Y(1)|X]-\mathbb{E}[Y(0)|X] &= \mathbb{E}[Y(1)|T=1,X]-\mathbb{E}[Y(0)|T=0,X]\\ &= \mathbb{E}[Y|T=1,X]-\mathbb{E}[Y|T=0,X].\end{aligned}$$

Hence the ATE can be estimated from the observable data by taking the expectation over the distribution of X

$$\tau=\mathbb{E}\Big\{\mathbb{E}[Y(1)|X]-\mathbb{E}[Y(0)|X]\Big\}=\mathbb{E}\Big\{\mathbb{E}[Y|T=1,X]\Big\}-\mathbb{E}\Big\{\mathbb{E}[Y|T=0,X]\Big\}.$$

As an illustration, assume in our example that $Y\in\{0,1\}$ records whether an individual dies within 90 days of treatment. Then

$$\begin{aligned}\mathbb{E}[\mathbb{E}[Y|T=t,X]] &= \mathbb{E}[\mathbb{P}\{Y=1|T=t,X\}]\\ &= \sum_{x=0}^{1}\mathbb{P}\{Y=1|T=t,X=x\}\mathbb{P}\{X=x\}.\end{aligned}$$

This can be estimated from the observable data $(T_1,X_1,Y_1),\ldots,(T_n,X_n,Y_n)$ using empirical conditioning; i.e.,

$$\mathbb{P}\{Y=1|T=t,X=x\}\approx\frac{n^{-1}\sum_{i=1}^{n}1_{\{Y_i=1,T_i=t,X_i=x\}}}{n^{-1}\sum_{i=1}^{n}1_{\{T_i=t,X_i=x\}}},\quad \mathbb{P}\{X=x\}\approx\frac{1}{n}\sum_{i=1}^{n}1_{\{X_i=x\}}.$$

Therefore, we have the following ATE estimator

$$\begin{aligned}\hat{\tau} &= \sum_{x=0}^{1}\left\{\frac{\sum_{i=1}^{n}1_{\{Y_i=1,T_i=1,X_i=x\}}}{\sum_{i=1}^{n}1_{\{T_i=1,X_i=x\}}}\left(\frac{1}{n}\sum_{i=1}^{n}1_{\{X_i=x\}}\right)\right\}\\ &\quad-\sum_{x=0}^{1}\left\{\frac{\sum_{i=1}^{n}1_{\{Y_i=1,T_i=0,X_i=x\}}}{\sum_{i=1}^{n}1_{\{T_i=0,X_i=x\}}}\left(\frac{1}{n}\sum_{i=1}^{n}1_{\{X_i=x\}}\right)\right\}.\end{aligned}\tag{20.4}$$

20.3 Causal Estimates in Observational Studies

The randomization used in the last example was relatively simple and involved conditioning on only a single patient variable (health status). But in real life, achieving randomization can be much more difficult, and this is the scenario often encountered when dealing with observational data. Since observational data is collected without any formal randomization in place, it is not reasonable to expect that individuals are randomized to treatments. On the other hand, it can be reasonable to assume that individuals are conditionally randomized, if we condition on appropriate factors. If the feature information $\mathbf{X}$ collected for the study is rich enough, it can be reasonable to expect that potential outcomes are conditionally independent of treatment assignment given the feature information $\mathbf{X}$. In causal inference of observational data, this assumption is called *strong unconfoundedness* [6, 11].

Strong unconfoundedness is the assumption that potential outcomes are conditionally independent of treatment assignment given $\mathbf{X}$*:*

$$\{Y(0), Y(1)\} \perp T \mid \mathbf{X}. \tag{20.5}$$

Intuitively, this means that the feature space for the experiment is rich enough so that conditional on $\mathbf{X}$*, the response is unconfounded by the treatment.*

To illustrate that the ATE is estimable under this assumption, suppose as before that Y is the binary outcome but where now in place of a scalar X we have a feature vector $\mathbf{X}$. For simplicity, assume that $\mathbf{X}$ is discrete, then by (20.5) using a conditioning argument as in the preceding example, it follows that

$$\mathbb{E}(Y(t)) = \sum_{\mathbf{x}} \mathbb{P}\{Y = 1 | T = t, \mathbf{X} = \mathbf{x}\} \, \mathbb{P}\{\mathbf{X} = \mathbf{x}\}, \quad t = 0, 1. \tag{20.6}$$

Denote $\psi(t, \mathbf{x}) = \mathbb{P}\{Y = 1 | T = t, \mathbf{X} = \mathbf{x}\}$, therefore,

$$\tau = \sum_{\mathbf{x}} \psi(1, \mathbf{x}) \, \mathbb{P}\{\mathbf{X} = \mathbf{x}\} - \sum_{\mathbf{x}} \psi(0, \mathbf{x}) \, \mathbb{P}\{\mathbf{X} = \mathbf{x}\}.$$

The conditional probabilities $\psi(t, \mathbf{x})$ appearing in τ are values for the response conditional on treatment and factors. Therefore, this shows, as before, that the ATE is estimable from the data.

Now in order to estimate the ATE from the data, we cannot use an empirical conditioning estimator as in (20.4). This is because conditioning will lead to data too sparse for proper estimation, especially if the dimension of $\mathbf{X}$ is high. This is called the curse of dimensionality. To get around this, $\psi(t, \mathbf{x})$ is estimated by a regression model. For example, if $(T_1, \mathbf{X}_1, Y_1), \ldots, (T_n, \mathbf{X}_n, Y_n)$ is the data, $\psi(t, \mathbf{x})$ can be estimated by logistic regression by fitting the reponse values $\{Y_i\}_1^n$ using $\{(T_i, \mathbf{X}_i)\}_1^n$ for the features. Let $\hat{\psi}(t, \mathbf{X})$ be the trained model, then τ is estimated by

$$\hat{\tau} = \frac{1}{n} \sum_{i=1}^{n} \hat{\psi}(1, \mathbf{X}_i) - \frac{1}{n} \sum_{i=1}^{n} \hat{\psi}(0, \mathbf{X}_i).$$

In addition to strong unconfoundedness, there is one more assumption that is required if we want to achieve unbiased estimation. In randomized studies, every individual has a positive probability of receiving each treatment. However, in observational studies, this condition might not hold, as treatments are often limited by factors such as indication and appropriateness. Consequently, some individuals may not be eligible for certain treatments (for instance, heart failure patients over the age of 70 are generally ineligible for a heart transplant).

To deal with this, a treatment overlap assumption $0 < \mathbb{P}\{T = 1 | \mathbf{X} = \mathbf{x}\} < 1$ is required. Technically, the issue is that (20.6) is only well defined if $\psi(t, \mathbf{x})$ is well defined. By Bayes' rule

$$\psi(t, \mathbf{x}) = \frac{\mathbb{P}\{Y = 1, T = t, \mathbf{X} = \mathbf{x}\}}{\mathbb{P}\{T = t | \mathbf{X} = \mathbf{x}\} \, \mathbb{P}\{\mathbf{X} = \mathbf{x}\}}, \tag{20.7}$$

therefore

$$\mathbb{E}(Y(t)) = \sum_{\mathbf{x}} \left\{ \frac{\mathbb{P}\{Y = 1, T = t, \mathbf{X} = \mathbf{x}\}}{\mathbb{P}\{T = t | \mathbf{X} = \mathbf{x}\}} \times 1_{\{\mathbb{P}\{\mathbf{X} = \mathbf{x}\} > 0\}} \right\}.$$

Thus, (20.6) is well defined only if $\mathbb{P}\{T = t | \mathbf{X} = \mathbf{x}\} > 0$ for $t = 0, 1$ for all $\mathbf{x}$ in the support of the distribution for $\mathbf{X}$. This condition is known as the positivity condition or *treatment overlap* assumption. When combined with strong unconfoundness, this is called *strongly ignorable treatment assignment* (SITA).

The positivity condition, or treatment overlap assumption, requires that

$$0 < \mathbb{P}\{T = 1|\boldsymbol{X} = \boldsymbol{x}\} < 1 \tag{20.8}$$

for all $\mathbf{x}$ *in the support of the distribution of* $\mathbf{X}$*. Intuitively, satisfying treatment overlap means that we are only allowed to use* $\mathbf{X}$ *where it is plausible to assign* $\mathbf{X}$ *to both treatments. When strong unconfoundness (20.5) and treatment overlap (20.8) holds, we call this the strongly ignorable treatment assignment (SITA) assumption.*

20.3.1 Propensity score

In practice, the positivity condition is applied with a preset $0 < \alpha < 1$ using $\mathbf{X}$ satisfying $\alpha < \mathbb{P}\{T = 1|\mathbf{X} = \mathbf{x}\} < 1 - \alpha$. The quantity $\mathbb{P}\{T = 1|\mathbf{X} = \mathbf{x}\}$ is often estimated using logistic regression and is called the propensity score. For notational convenience, let $e(t,\mathbf{x}) = \mathbb{P}\{T = t|\mathbf{X} = \mathbf{x}\}$ and define the propensity score by $e(\mathbf{x}) := e(1,\mathbf{x}) = \mathbb{P}\{T = 1|\mathbf{X} = \mathbf{x}\}$.

The propensity score plays an important role in causal inference due to its two distinct properties [11]. First, it is a balancing score, which implies that $T \perp \mathbf{X} \,|\, e(\mathbf{X})$. This means that treatment assignment and features are conditionally independent given the propensity score. To achieve balanced samples for the treatment, we can condition on the propensity score instead of the covariates. This is particularly advantageous in high dimensions, as it overcomes the curse of dimensionality by using the one-dimensional propensity score in place of the multivariate $\mathbf{X}$.

Secondly, if treatment assignment is strongly ignorable for $\mathbf{X}$, it remains strongly ignorable for $e(\mathbf{X})$. Specifically, if $T \perp \{Y(0), Y(1)\} \,|\, \mathbf{X}$ and $0 < e(\mathbf{x}) < 1$ for all $\mathbf{x}$, then $T \perp \{Y(0), Y(1)\} \,|\, e(\mathbf{X})$ and $0 < \mathbb{P}\{T = 1|e(\mathbf{x})\} < 1$ for all $e(\mathbf{x})$. This property ensures that if strong ignorability of treatment assignment (SITA) holds, the ATE can be obtained by averaging the conditional treatment effect with respect to $e(\mathbf{X})$ (Theorem 4, Rosenbaum and Rubin [11]):

$$\tau = \mathbb{E}\Big\{\mathbb{E}\big[Y \,|\, T = 1, e(\mathbf{X})\big] - \mathbb{E}\big[Y \,|\, T = 0, e(\mathbf{X})\big]\Big\}.$$

Let $\psi(t,\mathbf{x}) = \mathbb{E}[Y \,|\, T = t, \mathbf{x}]$ for $t = 0, 1$ be the conditional regression model for the outcome. Then another way to write τ is

$$\tau = \mathbb{E}\Big\{\psi(1, e(\mathbf{X})) - \psi(0, e(\mathbf{X}))\Big\}.$$

This result is the foundation for ATE estimators based on stratification or matching of the data on propensity scores.

Another important application of the propensity score is inverse probability weighting, a method for achieving unbiased ATE estimation. Returning to the notation $e(T,\mathbf{X})$, and notice this value is random in both T and $\mathbf{X}$, the inverse probability weighted (IPW) estimator for τ is defined as

$$\tau_{\text{IPW}} = \frac{1_{\{T=1\}} Y}{e(T,\mathbf{X})} - \frac{1_{\{T=0\}} Y}{e(T,\mathbf{X})}. \tag{20.9}$$

The following calculation shows that (20.9) is unbiased. For simplicity we assume Y is binary and $\mathbf{X}$ is discrete:

$$\begin{aligned}
\mathbb{E}\left[\frac{1_{\{T=t\}}Y}{e(T,\mathbf{X})}\right] &= \sum_{\mathbf{x},y}\frac{1_{\{T=t\}}y}{e(t,\mathbf{x})}\mathbb{P}\{Y=y,T=t,\mathbf{X}=\mathbf{x}\} \\
&= \sum_{\mathbf{x}}\frac{1}{e(t,\mathbf{x})}\mathbb{P}\{Y=1,T=t,\mathbf{X}=\mathbf{x}\} \\
&= \sum_{\mathbf{x}}\frac{1}{e(t,\mathbf{x})}\psi(t,\mathbf{x})\,e(t,\mathbf{x})\,\mathbb{P}\{\mathbf{X}=\mathbf{x}\} \quad \text{(by (20.7))} \\
&= \sum_{\mathbf{x}}\psi(t,\mathbf{x})\,\mathbb{P}\{\mathbf{X}=\mathbf{x}\} = \mathbb{E}(Y(t)).
\end{aligned}$$

Thus, the IPW estimator is unbiased for the ATE:

$$\mathbb{E}\left[\tau_{\text{IPW}}\right] = \mathbb{E}\left[\frac{1_{\{T=1\}}Y}{e(T,\mathbf{X})}\right] - \mathbb{E}\left[\frac{1_{\{T=0\}}Y}{e(T,\mathbf{X})}\right] = \mathbb{E}(Y(1)) - \mathbb{E}(Y(0)) = \tau.$$

It is straightforward to modify the calculations for the general case of continuous and or discrete $\mathbf{X}$ and Y where $\psi(t,\mathbf{x}) = \mathbb{E}[Y|T=t,\mathbf{X}=\mathbf{x}]$. For convenience we state this result below but leave the proof as an exercise for readers (see Exercise 20.1).

Theorem 20.1. *Under the assumption of SITA and consistency,* $\mathbb{E}\left[\tau_{IPW}\right] = \tau$.

In practice, the IPW estimator is estimated from the observable data using an estimator $\hat{e}(\mathbf{X}_i)$ for the propensity score:

$$\hat{\tau}_{\text{IPW}} = \frac{1}{n}\sum_{i=1}^{n}\frac{T_iY_i}{\hat{e}(\mathbf{X}_i)} - \frac{1}{n}\sum_{i=1}^{n}\frac{(1-T_i)Y_i}{1-\hat{e}(\mathbf{X}_i)}.$$

A problem with this estimator is that it can become unstable if the propensity score estimator $\hat{e}(\mathbf{X}_i)$ becomes close to 0 or 1. In these instances, a stabilized version of the estimator can be used in its place

$$\hat{\tau}'_{\text{IPW}} = \left(\sum_{i=1}^{n}\frac{T_i}{\hat{e}(\mathbf{X}_i)}\right)^{-1}\sum_{i=1}^{n}\frac{T_iY_i}{\hat{e}(\mathbf{X}_i)} - \left(\sum_{i=1}^{n}\frac{1-T_i}{1-\hat{e}(\mathbf{X}_i)}\right)^{-1}\sum_{i=1}^{n}\frac{(1-T_i)Y_i}{1-\hat{e}(\mathbf{X}_i)}.$$

20.3.2 Doubly robust estimation

The IPW estimator can be further improved by combining the IPW estimator with an outcome model to derive a semiparametric doubly robust (DR) estimator [9]. This applies to the continuous regression case. Let $\hat{\psi}(t,\mathbf{x})$ be an estimator for the conditional mean $\psi(t,\mathbf{x}) = \mathbb{E}[Y|t,\mathbf{X}=\mathbf{x}]$, then the DR estimator is

$$\begin{aligned}
\hat{\tau}_{\text{DR}} &= \frac{1}{n}\sum_{i=1}^{n}\left[\frac{T_iY_i}{\hat{e}(\mathbf{X}_i)} - \frac{T_i-\hat{e}(\mathbf{X}_i)}{\hat{e}(\mathbf{X}_i)}\hat{\psi}(1,\mathbf{X}_i)\right] \\
&\quad -\frac{1}{n}\sum_{i=1}^{n}\left[\frac{(1-T_i)Y_i}{1-\hat{e}(\mathbf{X}_i)} - \frac{T_i-\hat{e}(\mathbf{X}_i)}{1-\hat{e}(\mathbf{X}_i)}\hat{\psi}(0,\mathbf{X}_i)\right]
\end{aligned}$$

which can also be written as

$$\hat{\tau}_{\text{DR}} = \frac{1}{n}\sum_{i=1}^{n}\left[\hat{\psi}(1,\mathbf{X}_i) + \frac{T_i\hat{R}_i}{\hat{e}(\mathbf{X}_i)}\right] - \frac{1}{n}\sum_{i=1}^{n}\left[\hat{\psi}(0,\mathbf{X}_i) + \frac{(1-T_i)\hat{R}_i}{1-\hat{e}(\mathbf{X}_i)}\right]$$

where $\hat{R}_i = Y_i - \hat{\psi}(T_i,\mathbf{X}_i)$ is the residual from the outcome model estimator $\hat{\psi}(t,\mathbf{x})$.

These estimators can achieve unbiasedness as can be seen by the following argument. Replacing the estimated $\hat{\psi}$ with the true model ψ, and $\hat{R}$ with the true model residual R, we have

$$\begin{aligned}\tau_{\text{DR}} &= \mathbb{E}\left\{\frac{TY}{e(\mathbf{X})} - \frac{T-e(\mathbf{X})}{e(\mathbf{X})}\psi(1,\mathbf{X})\right\} - \mathbb{E}\left\{\frac{(1-T)Y}{1-e(\mathbf{X})} - \frac{T-e(\mathbf{X})}{1-e(\mathbf{X})}\psi(0,\mathbf{X})\right\} \\ &= \mathbb{E}\left\{\psi(1,\mathbf{X}) + \frac{TR}{e(\mathbf{X})}\right\} - \mathbb{E}\left\{\psi(0,\mathbf{X}) + \frac{(1-T)R}{1-e(\mathbf{X})}\right\} \\ &= \mathbb{E}\left[\psi(1,\mathbf{X})\right] - \mathbb{E}\left[\psi(0,\mathbf{X})\right] = \tau.\end{aligned}$$

The DR estimator can be further improved by replacing the weights with values less sensitive to $\hat{e}(\mathbf{X}_i)$ becoming close to 0 or 1 [10]:

$$\begin{aligned}\hat{\tau}'_{\text{DR}} = &\left[\frac{1}{n}\sum_{i=1}^{n}\hat{\psi}(1,\mathbf{X}_i) + \left(\sum_{i=1}^{n}\frac{T_i}{\hat{e}(\mathbf{X}_i)}\right)^{-1}\sum_{i=1}^{n}\frac{T_i\hat{R}_i}{\hat{e}(\mathbf{X}_i)}\right] \\ &- \left[\frac{1}{n}\sum_{i=1}^{n}\hat{\psi}(0,\mathbf{X}_i) + \left(\sum_{i=1}^{n}\frac{1-T_i}{1-\hat{e}(\mathbf{X}_i)}\right)^{-1}\sum_{i=1}^{n}\frac{(1-T_i)\hat{R}_i}{1-\hat{e}(\mathbf{X}_i)}\right].\end{aligned}$$

20.3.3 Challenging simulation

Modern day observational data can be complicated, therefore we use a challenging simulation to test how well the previously described ATE estimators perform. There are two features X_1, X_2 in the simulation which are independent and uniformly distributed on a discrete grid over $[-1,1]$ with values $0, \pm 1/m, \pm 2/m, \ldots, \pm 1$ where $m = 50$. Therefore, the feature space is a discrete grid on $[-1,1] \times [-1,1]$ and there are $n = (2m+1)^2$ data values. The treatment variable T is simulated using the conditional probability

$$\mathbb{P}\{T = 1|\mathbf{X} = \mathbf{x}\} = \frac{1}{2}(x_1^2 + x_2^2).$$

The response is $Y = Y(T)$ where the potential outcomes are defined by

$$Y(0) = 1.5 + 1.5x_1x_2, \quad Y(1) = 2 + 2.5(x_1-1)^2(x_2+1)^2.$$

The mean values of the features are $\mathbb{E}(X_1) = \mathbb{E}(X_2) = 0$ and the second moments are

$$\mathbb{E}(X_1^2) = \mathbb{E}(X_2^2) = 2\sum_{j=1}^{m}\left(\frac{j}{m}\right)^2\frac{1}{2m+1} = \frac{2}{m^2(2m+1)}\sum_{j=1}^{m}j^2 = \frac{m+1}{3m} = 0.34 := \sigma^2.$$

Therefore, the true ATE in the simulation is

$$\tau = \mathbb{E}(Y(1)) - \mathbb{E}(Y(0)) = \left[2 + 2.5(\sigma^2+1)(\sigma^2+1)\right] - [1.5 + 1.5 \times 0] = 4.989.$$

Relevant simulation values are shown in Figure 20.1. Conditional probabilities $\mathbb{P}\{T = 1 \mid \mathbf{X} = \mathbf{x}\}$, where blue represents $T = 0$ and red represents $T = 1$, are given in (A). The response Y is shown in (B), and the potential outcomes in (C) and (D). We observe in (A) that the simulation model places low probability at the origin $(0,0)^T$, with probabilities increasing concentrically as the radius expands from this point. Therefore, values $T = 0$ occur abundantly near the center, whereas $T = 1$ values are more likely to occur at the edges. The observed Y in (B) is a mixture of the potential outcomes (C) and (D). Therefore, we tend to observe $Y = 1$ near the edges, which are generally small except for the top left, whereas $Y = 0$ occurs most often in the center, where its values are smaller when compared to the bottom left and top right. Therefore, we can see

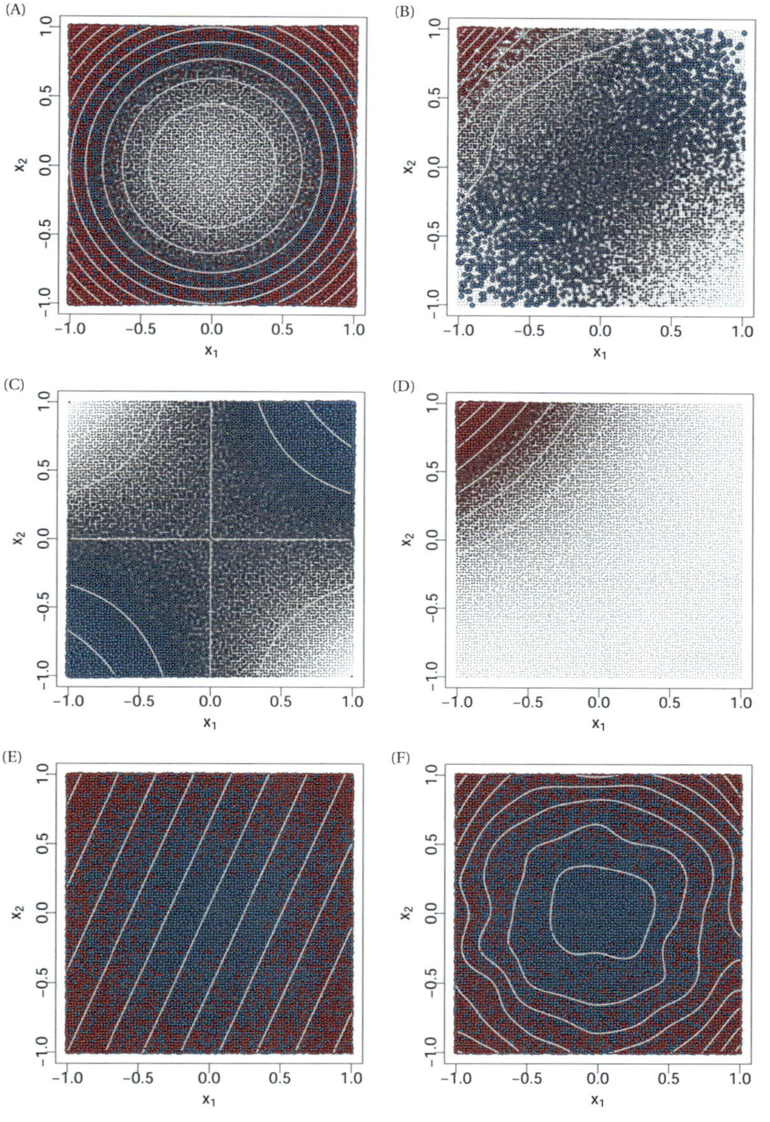

FIGURE 20.1
Challenging simulation used for testing ATE estimators. (A) Conditional probability $\mathbb{P}\{T = 1|\boldsymbol{X} = \boldsymbol{x}\}$ *where blue is* $T = 0$ *and red is* $T = 1$*; (B) Observed* Y*; (C) Potential outcome,* $Y(0)$ *(D) Potential outcome,* $Y(1)$*; (E) Estimated propensity score,* $\hat{e}(\boldsymbol{x})$*; (F) Inverse probability weights,* $1/\hat{e}(\boldsymbol{x}), 1/(1-\hat{e}(\boldsymbol{x}))$ *for each treatment group. The size of points in each figure has been scaled to the value being displayed. Gray lines are contours.*

from these plots that for both treatments, the effect size strongly depends on the covariates, and the probability of treatment assignment also strongly depends on those same covariates. This suggests that for the IPW estimators to work properly, the IP weights must be accurate. However, the propensity score in (E), estimated using logistic regression, does a poor job and fails to identify the center as having low probabilities. This will affect the accuracy of the inverse probability weights (F).

The results for the various ATE estimators are provided in the first row of Table 20.1. The table also includes the naive estimator, which compares the sample average of Y over the two treatment groups. All of the estimators perform poorly, with estimates far from the true value, $\tau = 4.99$. The IPW estimators (regular and robust) have nearly identical values of 5.94 and 5.93 and are hardly better than the naive estimator. DR estimators used linear models and random forests for the conditional mean estimator $\hat{\psi}(t,\mathbf{x})$ and have estimated values of 5.94 and 5.44, respectively. The linear model offers no improvement, while random forests is slightly better.

ATE estimators were then recalculated by using random forests (RF), super greedy forests (SGF) and gradient-boosted trees (GBT) in place of logistic regression for estimating the propensity score. The results, shown in the last three rows, are significantly improved over logistic regression (first row), especially for GBT. This demonstrates that with the proper use of machine learning, ATE estimation can be substantially improved. In the next section, we will discuss a general approach using machine learning for more effective causal inference.

TABLE 20.1
ATE estimators applied to challenging simulation.

$\hat{e}(\mathbf{X})$	τ (truth)	naive	$\hat{\tau}_{\text{IPW}}$	$\hat{\tau}'_{\text{IPW}}$	$\hat{\tau}_{\text{DR}}$ (linear)	$\hat{\tau}_{\text{DR}}$ (RF)
logistic	4.99	6.07	5.94	5.93	5.94	5.44
RF	4.99	6.07	7.97	4.90	4.33	4.50
SGF	4.99	6.07	2.72	5.64	5.78	5.39
GBT	4.99	6.07	4.78	5.06	5.12	5.00

20.4 Virtual (Digital) Twins and the Individual Treatment Effect

We have focused on the ATE which is a population-level measure of treatment effectiveness, however, there is a more granular measurement called the *individual treatment effect* (ITE) that plays an important role in causal inference. This value could be especially helpful to an individual who is interested in the best treatment decision for themselves. A population-level measure only tells the individual i what happens on average but does not inform them to the best option for their unique characteristics as measured by their feature vector $\mathbf{X}_i$. This makes the ITE valuable for a field like precision medicine.

We will make the following assumptions for case i with data $(T_i, \mathbf{X}_i, Y_i)$. For now we consider the scenario where the individual is confronted with two treatment options, $T_i \in \{0,1\}$, and we denote their potential outcomes by $\{Y_i(0), Y_i(1)\}$. The generalization to continuous and multivariate treatments is discussed in the next section.

1. *Consistency:* The outcome for i under the treatment received is the same as its potential outcome for that treatment, $Y_i = Y_i(T_i)$.
2. *Strong Unconfoundedness:* Potential outcomes for i are conditionally independent of the treatment assignment given the covariates, $\{Y_i(0), Y_i(1)\} \perp T_i \,|\, \mathbf{X}_i$.

3. *Treatment Overlap:* Both treatments have a positive probability of occurring, $0 < e(\mathbf{x}) < 1$ for $\mathbf{X}_i = \mathbf{x}$.

The ITE for i is defined as the conditional mean difference in potential outcomes

$$\tau(\mathbf{x}) = \mathbb{E}[Y_i(1)|\mathbf{X}_i = \mathbf{x}] - \mathbb{E}[Y_i(0)|\mathbf{X}_i = \mathbf{x}].$$

Notice that by the assumption of strong unconfoundedness and consistency

$$\mathbb{E}[Y_i(t)|\mathbf{X}_i = \mathbf{x}] = \mathbb{E}[Y_i(t)|T_i = t, \mathbf{X}_i = \mathbf{x}] = \mathbb{E}[Y_i|T_i = t, \mathbf{X}_i = \mathbf{x}], \quad \text{for } t = 0, 1,$$

therefore

$$\tau(\mathbf{x}) = \mathbb{E}[Y|T = 1, \mathbf{X} = \mathbf{x}] - \mathbb{E}[Y|T = 0, \mathbf{X} = \mathbf{x}]. \tag{20.10}$$

This value no longer depends on i since we assume that the data is independent and identically distributed. Also, because the conditioning on the response is in terms of the treatment and covariates, this shows that the ITE is estimable from the data. Therefore, once again we overcome the fundamental problem of causal inference by focusing on averages rather than differences in potential outcomes which may not be estimable. The only difference is now we are considering effects for a specific covariate feature of interest, $\mathbf{x}$.

Another important implication is that the ITE leads directly to the ATE. Since we have

$$\mathbb{E}(\tau(\mathbf{X})) = \mathbb{E}\left\{\mathbb{E}[Y|T = 1, \mathbf{X}] - \mathbb{E}[Y|T = 0, \mathbf{X}]\right\} = \mathbb{E}[Y|T = 1] - \mathbb{E}[Y|T = 0] = \tau,$$

this shows averaging $\tau(\mathbf{X})$ over the data provides a means for estimating the ATE. Therefore, the ITE provides a granular feature level effect and by the process of averaging it also provides us with a population effect size.

To estimate $\tau(\mathbf{x})$ we use the *virtual (digital) twin method* from machine learning [3]. In the virtual twin method, a prediction model $\hat{\psi}(t, \mathbf{x})$ is obtained by training a machine learning method on the data $(T_1, \mathbf{X}_1, Y_1), \ldots, (T_n, \mathbf{X}_n, Y_n)$. Denote the true conditional mean value by $\psi(t, \mathbf{x}) = \mathbb{E}[Y|T = t, \mathbf{X} = \mathbf{x}]$, then using (20.10) we estimate $\tau(\mathbf{x}) = \psi(1, \mathbf{x}) - \psi(0, \mathbf{x})$ by

$$\hat{\tau}_{\text{VT}}(\mathbf{x}) = \hat{\psi}(1, \mathbf{x}) - \hat{\psi}(0, \mathbf{x}).$$

Conceptually, this amounts to having virtual twins (Figure 20.2), where each twin is exactly the same and therefore has the same covariate $\mathbf{x}$. One twin is studied under the actual treatment T^a (the actual individual under the assigned treatment), and the other under the causal treatment T^c (the virtual twin under the hypothesized treatment). This is analogous to the potential outcomes framework, but the virtual twin analogy is more intuitive for researchers in machine learning. It also emphasizes the individual $\mathbf{x}$ feature under consideration. Therefore, we will use this analogy framework in our subsequent discussion.

Recall, a key assumption needed for correct inference is the positivity assumption. This requires that, for a given feature $\mathbf{x}$, the individual has a positive probability of being assigned to either treatment. In the virtual twin analogy, this means that a virtual twin under the alternate treatment assignment must be possible. However, this may not always be the case since some treatments may not be plausible or even ethically permissible for an individual. Therefore, we must ask, "What if the individual received the alternate treatment?" and be convinced that such a treatment assignment is plausible. If we pass the what-if test, then we can use the machine learning regression model applied to the virtual twin to estimate the true model value under the causal treatment assignment.

$$\underbrace{\hat{\psi}(T^c, \mathbf{x})}_{\substack{\text{regression model} \\ \text{estimator for what-if virtual twin}}} \quad \Longrightarrow \quad \underbrace{\psi(T^c, \mathbf{x})}_{\substack{\text{true value} \\ \text{for virtual twin under causal assignment}}}$$

FIGURE 20.2
Virtual (digital) twins where one is the causal twin under the causal treatment (right) and the other is the original twin under the assigned treatment (left).

20.5 Unlimited Virtual Twins and partialPro

There are two limitations to the virtual twin method. One is that it only applies to studies with binary treatment options. The second is the absence of a systematic means for assessing the positivity criteria. A solution that simultaneously addresses these two issues is called *partialPro* which is the subject of this section.

This general approach works as follows. We consider an extended potential outcomes framework with a potential outcome $Y(z)$ indexed by an abstract treatment value z. In this extension, the parameter of interest is abstracted from the individual treatment effect to a more general "dose-response function" defined as

$$F(z,\mathbf{x}) = \mathbb{E}\left[g(Y)|Z = z,\mathbf{X} = \mathbf{x}\right]$$

where g is a pre-chosen function. Integrating over $\mathbf{X}$ yields the average dose-response function

$$F(z) = \mathbb{E}\left[F(z,\mathbf{X})\right].$$

For regression, for example, we could be interested in studying the treatment effect for a binary treatment $z \in \{0,1\}$. Then this corresponds to the previous discussions where $\{Y(0), Y(1)\}$ are the continuous potential outcomes under the two treatment options and $F(z) = \mathbb{E}[Y(z)]$ is the mean for the potential outcome $Y(z)$ where $g(y) = y$. The response function is the conditional mean under treatment z, $F(z,\mathbf{x}) = \mathbb{E}[Y|Z = z,\mathbf{X} = \mathbf{x}]$. For survival, imagine now we are interested in studying a dose-level effect on patient survival, then $Y(z) = T^o(z)$ would be the potential event time under a dose-level of z. In other words, it is the event time for the patient if a dose of z was administered. The parameter of interest could be the survival function $F(z) = \mathbb{P}\{T^o(z) > t\}$ at time t corresponding to $g(y) = I\{y > t\}$; or it could be $F(z) = \int_0^{t_\infty} \mathbb{P}\{T^o(z) > t\}dt$ equal to the restricted mean survival time (RMST) at time horizon t_∞, which corresponds to $g(y) = \max(y, t_\infty)$.

20.5.1 The partial dependence function as an ATE estimator

The previous examples all consider Z to be a treatment variable separate from the feature vector $\mathbf{X} = (X^{(1)}, \ldots, X^{(p)})^T$, however, Z can be further abstracted by allowing it to be one of the feature coordinates. Thus, we are interested in causal effects without any specific "treatment" in mind.

This is very useful since often in multivariate analysis we are interested in understanding the effects of the features on the outcome. For example, in nonparametric regression, it is very helpful to understand how the conditional mean $\psi(\mathbf{X}) = \mathbb{E}[Y|\mathbf{X}]$ depends on a feature $X^{(j)}$. Then, attention could turn to the dose-response function defined as

$$\begin{aligned} F(z, \mathbf{X}^{\setminus(j)}) &= \mathbb{E}\left[Y \middle| X^{(j)} = z, \mathbf{X}^{\setminus(j)}\right] \\ &= \psi\left(X^{(1)}, \ldots, X^{(j-1)}, z, X^{(j+1)}, \ldots, X^{(p)}\right) \end{aligned} \tag{20.11}$$

where $Z = X^{(j)}$ and $\mathbf{X}^{\setminus(j)} = \{X^{(k)}\}_{k \neq j}$ is the feature vector with coordinate j removed.

Integrating (20.11) gives the average dose-response function, $F(z)$. Interestingly this is the same as the marginal partial dependence function used in machine learning [4]. This quantity is obtained by integrating $\psi(\mathbf{X})$ over $\mathbf{X}^{\setminus(j)}$ (the coordinates to be adjusted for) and is defined by

$$\psi_j(z) = \mathbb{E}\left[\psi\left(X^{(1)}, \ldots, X^{(j-1)}, z, X^{(j+1)}, \ldots, X^{(p)}\right)\right]. \tag{20.12}$$

Thus the average dose-response function for (20.11) and the partial dependence function are the same thing, $F(z) = \psi_j(z)$.

The connection between the partial dependence function and the mean dose-response function suggests that the partial dependence function has a causal interpretation. By integrating over $\mathbf{X}^{\setminus(j)}$, the function yields an estimate for the effect size for $X^{(j)}$ controlling for the other features. As an illustration of the importance of this type of integration, imagine that instead of $F(z)$ we use

$$\tilde{F}(z) = \mathbb{E}\left\{\psi\left(X^{(1)}, \ldots, X^{(j-1)}, z, X^{(j+1)}, \ldots, X^{(p)}\right) \middle| X^{(j)} = z\right\}. \tag{20.13}$$

Even though $\tilde{F}(z)$ looks almost identical to $F(z)$, they are not equal in general. $\tilde{F}$ integrates

$$\psi\left(X^{(1)}, \ldots, X^{(j-1)}, z, X^{(j+1)}, \ldots, X^{(p)}\right)$$

with respect to the conditional distribution of $\mathbf{X}$ given $X^{(j)}$, whereas F uses the joint distribution of $\mathbf{X}$. In particular, the two values are only equal if there is conditional independence between the coordinates, but otherwise $\tilde{F}$ will be inferior.

Example 20.1. As a concrete example, suppose that the true model is linear, $\mathbb{E}[Y|\mathbf{X}] = \psi(\mathbf{X}) = \sum_{k=1}^{p} \beta_k X^{(k)}$. Then

$$\tilde{F}(z) = \beta_j z + C(z)$$

where $C(z) = \sum_{k \neq j} \beta_k \mathbb{E}[X^{(k)}|X^{(j)} = z]$ and

$$F(z) = \beta_j z + C$$

where $C = \sum_{k \neq j} \beta_k \mathbb{E}[X^{(k)}]$. Further insight can be gained by considering the specialized case where the covariates have a multivariate normal distribution. Suppose that $\mathbf{X}$ is normal and $\mathbb{E}(X^{(k)}) = 0$, $\mathrm{Var}(X^{(k)}) = 1$ and $\mathrm{Cov}(X^{(k)}, X^{(k')}) = \rho$ for $k \neq k'$. Then $\mathbb{E}(X^{(k)}|X^{(j)} = z) = \rho z$, from which it follows that

$$\tilde{F}(z) = \beta_j z + \rho z \sum_{k \neq j} \beta_k = \left(\beta_j + \rho \sum_{k \neq j} \beta_k\right) z.$$

On the other hand, using the fact that the covariates have zero mean, we have

$$F(z) = \beta_j z.$$

Therefore, F correctly identifies the partial effect, whereas $\tilde{F}$ incorrectly adjusts the contribution of z by inflating the value depending on the feature correlation. □

The problem with $\tilde{F}$ is similar to what happens in observational data when the observed treatment effect is incorrectly used for causal estimation. For example, let $(Z_i, \mathbf{X}_i, Y_i)$ be the data, $i = 1, \ldots, n$, and let $\hat{F}(z, \mathbf{x})$ be a prediction model for $F(z, \mathbf{x})$ obtained using a machine learning procedure. Then if $z \in \{0, 1\}$ is binary, the ATE can be estimated as

$$\hat{\tau} = \hat{F}(1) - \hat{F}(0) = \frac{\sum_{i=1}^n \hat{F}(z = 1, \mathbf{X}_i)}{n} - \frac{\sum_{i=1}^n \hat{F}(z = 0, \mathbf{X}_i)}{n}. \tag{20.14}$$

Note that the difference between (20.14) and the observed mean difference

$$\hat{\tau}_{\text{naive}} = \frac{\sum_{i=1}^n 1_{\{Z_i=1\}} Y_i}{\sum_{i=1}^n 1_{\{Z_i=1\}}} - \frac{\sum_{i=1}^n 1_{\{Z_i=0\}} Y_i}{\sum_{i=1}^n 1_{\{Z_i=0\}}} \tag{20.15}$$

is that (20.14) is equivalent to integrating $\hat{F}(z, \mathbf{x})$ using the joint density $f_{\mathbf{X}}$, and is similar to what occurs for the partial dependence function (20.12), and yields a sensible causal estimate, whereas (20.15) uses the conditional density of $\mathbf{X}$ given Z, $f_{\mathbf{X}|Z}$, and is similar to the conditional expectation (20.13) and equals the naive observed treatment effect which is potentially biased.

Remark 20.1. An equally poor estimator with the same problems as (20.15) is

$$\hat{\tau}'_{\text{naive}} = \frac{\sum_{i=1}^n 1_{\{Z_i=1\}} \hat{F}(z = 1, \mathbf{X}_i)}{\sum_{i=1}^n 1_{\{Z_i=1\}}} - \frac{\sum_{i=1}^n 1_{\{Z_i=0\}} \hat{F}(z = 0, \mathbf{X}_i)}{\sum_{i=1}^n 1_{\{Z_i=0\}}}$$

which is equivalent to integrating $\hat{F}(z, \mathbf{x})$ with respect to the conditional density $f_{\mathbf{X}|Z}$. Just like (20.15) this will suffer from potential bias as it does not control for confounding. The next example illustrates this point. □

20.5.2 Simple illustration of partialPro

Figure 20.3 (A) demonstrates that integrating the marginal distribution of $\mathbf{X}$ using (20.12) to explore the relationship between Z and Y is of paramount importance for controlling the effect of $\mathbf{X}$. In this example, we use a continuous treatment variable Z and the true model is $\mathbb{E}(Y|z, x) = \psi(z, x) = -z^2 + \beta x$, where $Y_i = \psi(Z_i, X_i) + \varepsilon_i$, $\beta = 2.5$, $\varepsilon_i \overset{\text{iid}}{\sim} \text{N}(0, 0.5)$ and

$$\begin{pmatrix} Z_i \\ X_i \end{pmatrix} \overset{\text{iid}}{\sim} \text{N}(\mathbf{0}, \mathbf{\Sigma}), \quad \text{where } \mathbf{\Sigma} = \begin{pmatrix} 1.0 & 0.9 \\ 0.9 & 1.0 \end{pmatrix}.$$

Although the treatment variable Z has a quadratic effect on the outcome Y, it demonstrates a positive near linear effect on Y in the scatterplot because Z is strongly correlated with the confounding variable X. Therefore, the scatter plot is misleading since it does not control for the confounding variables. Integrating $F(z, x)$ over the marginal distribution of X (red triangles) reflects the true causal relationship for Z (in green), while integrating $F(z, x)$ over the conditional distribution of X given Z (in blue) is similar to checking the scatterplot between Z and the outcome and is incorrect.

Suppose $z_{(1)}, \ldots, z_{(M)}$ are the unique values of Z, then we create M virtual twins for each X_i obtaining $\hat{F}(z_{(1)}, X_i), \ldots, \hat{F}(z_{(M)}, X_i)$ where $\hat{F}(z, x)$ is the prediction model for $F(z, x)$. We estimate $F(z_j)$ in the same fashion as (20.14) using

$$\hat{F}(z_{(j)}) = \frac{1}{n} \sum_{i=1}^n \hat{F}(z = z_{(j)}, X_i). \tag{20.16}$$

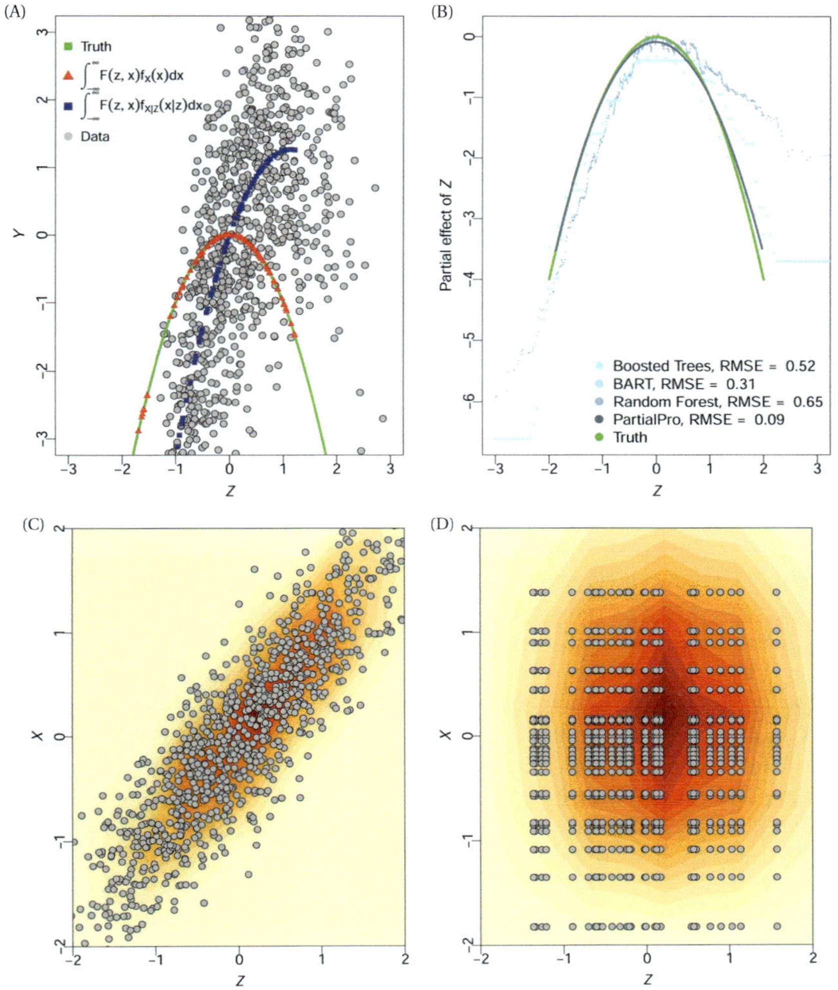

FIGURE 20.3
Illustration of partialPro in a two-dimensional example where the true causal relationship is quadratic but due to correlation the effect appears nearly linear in a naive analysis (blue line in A). Even powerful machine learning procedures (B) can be suboptimal due to using artificial data (D) which does not look like the original data (C). partialPro (dark blue points in B nearly equal to the green line) removes outliers and forms its estimate using unlimited virtual twins to mitigate the problem.

Plotting the M values of $\hat{F}(z_{(j)})$ against $z_{(j)}$ is similar to the partial dependence plot. Figure 20.3 (B) shows three partial plots from different models: boosted regression trees, random forests and Bayesian additive regression trees, BART [1]. Here we generate 100 training and testing samples to evaluate the RMSE (root mean-squared error).

The accuracy of a partial plot depends on the accuracy of the underlying learner $\hat{F}$. In this example, all the procedures perform similarly, with BART performing slightly better due to its improved behavior near the edges. However, regardless of the learner used, the performance of the partial plot is influenced by the distribution of the artificial data used for obtaining $\hat{F}(z_{(j)})$.

We can observe that the artificial data $(z_{(j)}, X_i)^T$ in (D), which is used for $\hat{F}(z_{(j)})$, has a different distribution from the original data $(Z_i, X_i)^T$ in (C).

This is similar to the problems with variable importance measures that use artificial data, such as permutation importance (Chapter 17.2). This was the basis for the variable selection procedure VarPro introduced in Chapter 17.6. The partialPro method is designed to solve this problem by recognizing that artificial data can create implausible virtual twins. Observe that the partialPro estimator, which is also given in (B), has the best performance (with near-zero RMSE and a curve almost identical to the truth shown in green). PartialPro identifies and discards outliers in the artificial predictor values. Then, using only the non-outliers, it extrapolates the shape between the predicted outcomes $\hat{F}$ and the treatment variable Z. This improves the performance of the partial plot, regardless of which prediction model is chosen. In Figure (B), partialPro was obtained using the BART predictor. So, in this example, it took the best-performing predictor and made it even better.

20.5.3 The partialPro method and unlimited virtual twins

The partialPro method illustrated in our previous example can be formally described as follows. To estimate $F(z_{(j)})$, instead of averaging $\hat{F}(z_{(j)},\mathbf{X}_i)$ for all the $\mathbf{X}_i$ as in (20.16), as done by partial plots, partialPro fits a parametric regression function $\sum_{k=0}^{K}(\beta_k \times z^k)$ of z for a given $\mathbf{X}_i$ using the predicted values $\{\hat{F}(z_{(j)},\mathbf{X}_i)\}$ as the y-values. Here, $K \geq 1$ represents the degrees of freedom (for example, $K \leq 3$ is a reasonable choice for many problems).

Since $\{\hat{F}(z_{(j)},\mathbf{X}_i)\}$ are predicted values from artificial data, to avoid the same issues encountered with partial plots, the regression model is fit using only a carefully selected subset $\mathcal{Y}_i \subseteq \{\hat{F}(z_{(j)},\mathbf{X}_i)\}_{1\leq j\leq M}$. To achieve this, partialPro utilizes isolation forests (Chapter 16.15). Given the $n \times M$ artificial data points $\{(z_{(j)},\mathbf{X}_i)\}_{1\leq j\leq M, 1\leq i\leq n}$, isolation forests are applied to efficiently filter and remove outliers from the data.

As demonstrated by Figure 20.3 (D), the issue arises because the $n \times M$ data points $\{(z_{(j)},\mathbf{X}_i)\}_{1\leq j\leq M, 1\leq i\leq n}$ used to obtain the predicted values may have a different distribution compared to the original n training data points (C), potentially violating the positivity assumption. Since isolation forests assess positivity across multiple virtual twins simultaneously, we refer to this procedure as *unlimited virtual twins*.

The estimator obtained by fitting the parametric function using $\mathcal{Y}_i$ for the outcomes is denoted as $\hat{F}_{\text{Pro}}(z,\mathbf{X}_i) = \sum_{k=0}^{K}(\hat{\beta}_{ik} \times z^k)$. The partialPro causal effects estimator is defined as the average:

$$\hat{F}_{\text{Pro}}(z_{(j)}) = \frac{1}{n}\sum_{i=1}^{n}\hat{F}_{\text{Pro}}(z_{(j)},\mathbf{X}_i) = \frac{1}{n}\sum_{i=1}^{n}\sum_{k=0}^{K}\hat{\beta}_{ik}z_{(j)}^{k}. \tag{20.17}$$

Besides filtering the artificial data to remove implausible twins, another key difference between (20.17) and (20.16) is that, instead of averaging $\hat{F}(z_{(j)},\mathbf{X}_i)$, partialPro averages the "shape" estimated from $\hat{F}(z_{(j)},\mathbf{X}_i)$. By averaging parametric functions, this allows inference of a causal effect for z not necessarily contained within the observed $\{z_{(j)}\}$.

The estimator (20.17) works best when the effect can be reasonably approximated by a polynomial. However, for greater flexibility, a local regression procedure can be employed. For each $z_{(j)}$, the values of $\hat{F}(z_{(j)},\mathbf{X}_i)$ are averaged over i indices for values $(z_{(j)},\mathbf{X}_i)$ not filtered by isolation forests. Let this value be denoted as $\overline{F}_{\text{Pro}}(z_{(j)})$. Each $z_{(j)}$ is assigned a weight $W_j \geq 0$, reflecting how frequently the value is associated with a rejected virtual twin (smaller weights are given to $z_{(j)}$ with many rejections). Then, loess (locally estimated scatterplot smoothing) is applied to the data $(z_{(1)},\overline{F}_{\text{Pro}}(z_{(1)})),\ldots,(z_{(M)},\overline{F}_{\text{Pro}}(z_{(M)}))$ using weights $W_1,\ldots,W_M$ to obtain the estimator.

20.5.4 Cirrhosis of the liver

The Primary Biliary Cirrhosis (PBC) data is a famous benchmark dataset involving cirrhosis of the liver [2]. Cirrhosis is a fibrosis (scarring) of the liver caused by liver disease which can occur for a variety of reasons such as hepatitis and chronic alcoholism. The goal of the PBC study was to determine the effectiveness of the drug D-penicillamine on improving survival for patients with liver disease. The data was collected from a randomized trial run by the Mayo Clinic between 1974 and 1984 involving a total of $n = 424$ patients. Patients who met eligibility criteria were randomized to placebo and the drug D-penicillamine. The first $n = 312$ cases in the data set participated in the randomized trial and contain largely complete data, which is the data that we use.

The outcome is the number of years between study entry and the earlier of death, transplantation, or study end. Therefore we perform a partialPro analysis for survival. For the dose-response function we consider the RMST using a time horizon of 3 years. The top panel of Figure 20.4 displays some of the data including the treatment variable.

Because the study was randomized on treatment, we would expect to observe no significant differences in patient treatment assignment. This was confirmed by running logistic regression using treatment for the response; no significant coefficients from the analysis were found (Table 20.2). Because the data should be balanced by randomization, we also should expect that the IP weights will have the same distribution under the two treatments. The bottom panel of Figure 20.4 records the density estimator for the IP weights $1/(1-\hat{e}(\mathbf{X}_i))$ and $1/\hat{e}(\mathbf{X}_i)$ for the placebo ($T_i = 0$) and D-penicillamine ($T_i = 1$) treatment groups. There is very little difference between the two groups.

Causal effect estimates $\hat{F}_{\text{Pro}}$ (20.17) are shown in Figure 20.5 for the top 6 features selected by VarPro. BART was used for the prediction model. The vertical axis equals the effect size for RMST at 3 years adjusted for all other variables. Several of the variables have noticeable effect sizes. For example, the presence or absence of ascites has an effect size of approximately half a year on expected survival.

The treatment was not selected by VarPro and has a zero effect size. From the perspective of the original trial, this would mean that the treatment under study proved non-beneficial. Generally, without any additional secondary endpoints of interest, this would signify that there was nothing more to be learned from the data. However, as shown by this analysis, valuable relationships can be uncovered by a partialPro causal analysis regardless if there is a "treatment" variable under consideration. The variables bilirubin, copper, ascites, and other variables in the data were not randomized, and therefore we employed causal inference techniques to extract their effect sizes. The unlimited virtual twin method easily handled this.

20.5.5 Challenging simulation with partialPro

To close out this chapter, we revisit the challenging simulation in Chapter 20.3.3, which proved difficult for classical ATE estimators.

We ran the same data through partialPro without indicating that it was a causal analysis or giving special handling to the treatment variable. Every variable was treated equally for determining causal effect sizes. The partialPro plots are shown in Figure 20.6. The treatment effect is approximately 4.9, close to the true ATE value, τ. This demonstrates that removing implausible twins and using local smoothing produces reliable causal estimates. The partial effect sizes for x_1 and x_2 are equally impressive, indicating a quadratic effect consistent with the simulated potential outcome models.

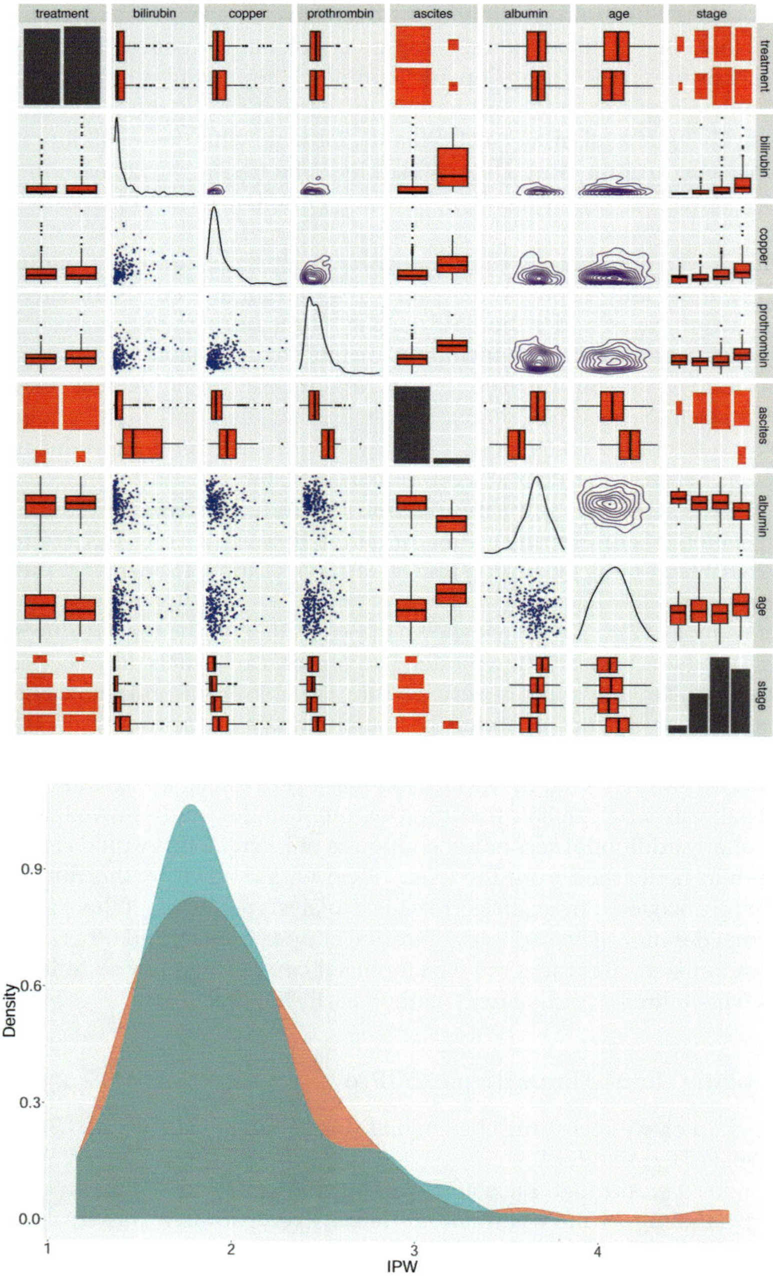

FIGURE 20.4
Top: Exploratory data plot of selected features from the PBC data and the treatment variable, D-penicillamine and placebo. Bottom: Inverse probability weights (IPW) for the two treatment groups.

TABLE 20.2
Logistic regression using treatment as the response for the PBC data.

	Estimate	Std. Error	Z-statistic	p-value
Intercept	-2.346	2.383	-0.985	0.325
age	-0.000	0.000	-1.873	0.061
sex	-0.330	0.434	-0.761	0.446
ascites	-0.436	0.724	-0.602	0.547
hepatomegaly	0.493	0.278	1.775	0.076
spiders	-0.262	0.316	-0.829	0.407
edema	-0.294	0.345	-0.851	0.395
bilirubin	0.073	0.046	1.584	0.113
cholesterol	-0.001	0.001	-1.024	0.306
albumin	0.232	0.370	0.627	0.531
copper	-0.001	0.002	-0.780	0.435
alkaline phosphatase	-0.000	0.000	-0.492	0.622
sgot	0.001	0.003	0.224	0.823
tryglicerides	0.000	0.002	0.006	0.995
platelets	0.001	0.001	0.669	0.503
prothrombin	0.275	0.157	1.755	0.079

Exercises

Exercise 20.1. Prove that the IPW estimator (20.9) is unbiased for continuous Y where $\mathbf{X}$ can be either continuous or discrete. State all your assumptions carefully.

Exercise 20.2. In this exercise, we will review propensity scores and their use in causal inference.

1. Define the propensity score and explain its role in causal inference.
2. Discuss the properties of the propensity score, including the balancing property.
3. Explain how propensity score matching, stratification, and weighting are used to estimate causal effects.
4. Discuss the limitations and potential biases that can arise when using propensity scores.

Exercise 20.3. In this exercise we learn how to estimate the Average Treatment Effect (ATE) using propensity score matching.

1. Load the `lalonde` dataset from the `MatchIt` R-package.
2. Estimate the propensity score for each individual using logistic regression.
3. Perform propensity score matching to pair treated and untreated individuals.
4. Estimate the ATE by comparing the outcomes of the matched pairs.
5. Assess the balance of covariates before and after matching and discuss the results.

Exercise 20.4. Here we look at estimating the ATE using IPW.

1. Load the `nhanes` dataset from the `survey` package.
2. Estimate the propensity score for each individual using logistic regression.
3. Calculate the inverse probability weights.

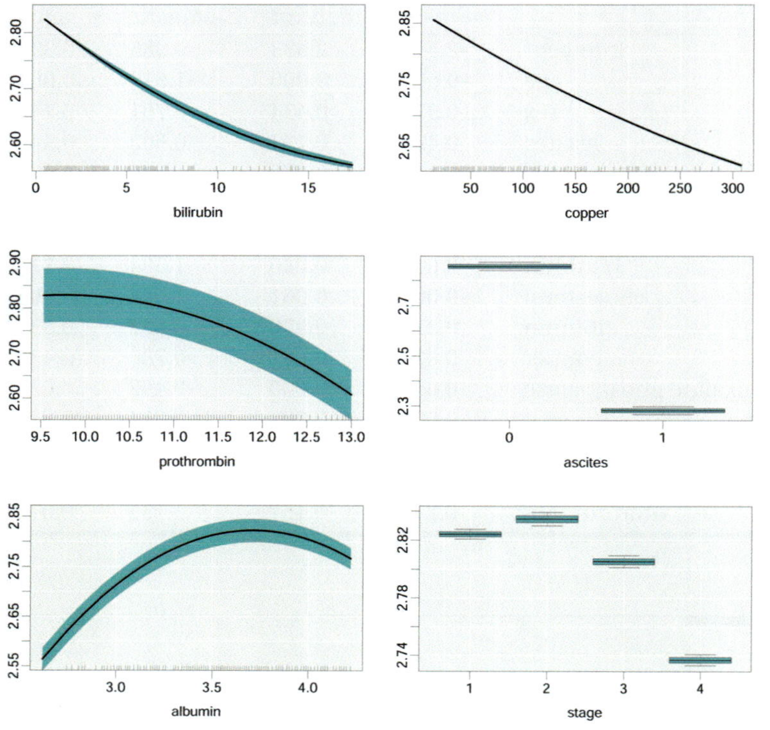

FIGURE 20.5
Causal effect size (RMST at three-year time horizon) for top features using partialPro for PBC data.

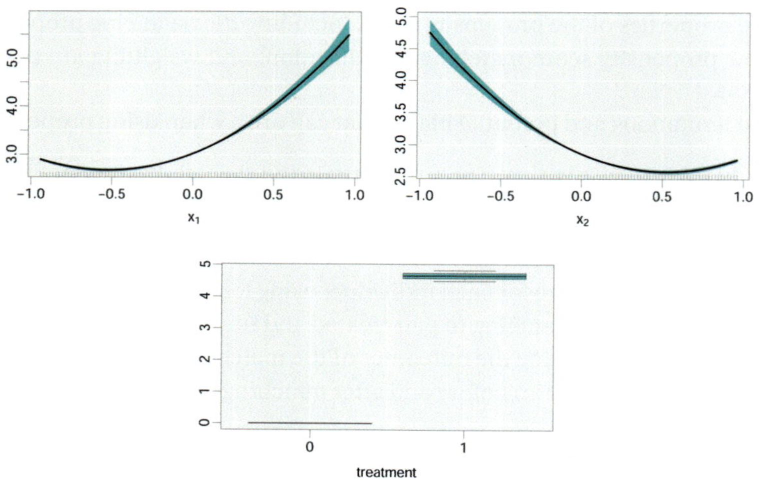

FIGURE 20.6
Causal effect sizes using partialPro for the challenging simulation used in Chapter 20.3.3.

4. Use the weights to estimate the ATE by fitting a weighted regression model.
5. Compare the IPW estimate of the ATE to the estimate obtained from propensity score matching and discuss the differences.

Exercise 20.5. Here we learn how to perform a doubly robust estimation of the ATE.

1. Load the `nsw74psid1` from the `twang` R-package.
2. Estimate the propensity score for each individual using logistic regression.
3. Calculate the inverse probability weights.
4. Fit a regression model to predict the outcome using the covariates.
5. Combine the IPW and the regression models to performs doubly robust estimation of the ATE.
6. Compare the doubly robust estimate of the ATE to the estimates obtained from propensity score matching and IPW, and discuss the robustness of the results.

Exercise 20.6. Here we explore the estimation of individual treatment effects (ITE) using the virtual twins method. We will use the `lalonde` dataset for this exercise.

1. Preprocess the data and split it into training and testing sets.
2. Implement the virtual twins method to estimate ITE for each individual on the training data.
3. Evaluate the performance of the virtual twin method on the testing data using appropriate metrics (e.g., mean squared error).
4. Interpret the estimated ITEs and identify subgroups of individuals who benefit most from the treatment.

Exercise 20.7. Here we conduct a sensitivity analysis to assess the robustness of causal estimates to unmeasured confounding.

1. Find a suitable dataset for a causal analysis.
2. Estimate the ATE using propensity score matching, IPW, and doubly robust methods.
3. Introduce a hypothetical unmeasured confounder and simulate its effect on the treatment assignment and outcome.
4. Re-estimate the ATE accounting for the simulated unmeasured confounder.
5. Conduct a sensitivity analysis to assess how the presence of the unmeasured confounder affects the causal estimates and discuss the implications.

Exercise 20.8. In this exercise, we will introduce and study another technique used for causal analysis. An instrumental variable (IV) is used in causal inference to estimate causal relationships when controlled experiments are not feasible, and there is a concern about the endogeneity of the explanatory variables. An IV is a variable that is correlated with the treatment variable but is not directly related to the outcome except through the treatment variable. Two-stage least squares (2SLS) is a common technique used to estimate causal effects using IVs. The method involves two-stages:

(i) Regress the treatment variable on the IV(s) to obtain predicted values of the treatment.
(ii) Regress the outcome variable on the predicted values of the treatment obtained from the first stage.

Complete the following exercises:

1. Using the `lalonde` dataset, suppose you want to estimate the effect of the treatment (`treat`) on the outcome (`re78`). Assume that the variable `nodegree` can serve as an instrumental variable. Explain why `nodegree` might be a valid IV in this context.
2. Implement 2SLS to estimate the effect of the treatment on the outcome. Interpret the results.

3. Evaluate the strength of the instrumental variable `nodegree` using an appropriate test statistic in the first stage regression. Discuss the implications of the results for the validity of the IV.
4. Discuss potential limitations and biases that might arise when using instrumental variables, particularly in the context of the `lalonde` dataset. How might weak instruments or violations of the exogeneity assumption affect your estimates (you will have to do a little research on your own to answer this)?

Bibliography

[1] Chipman, H.A. and George E.I. and McCulloch R.E. BART: Bayesian additive regression trees. *The Annals of Applied Statistics*, 4(1):266–298, 2010.

[2] Fleming T. and Harrington D. *Counting Processes and Survival Analysis*. Wiley, New York, 1991

[3] Foster J.C. and Taylor J.M. and Ruberg S.J. Subgroup identification from randomized clinical trial data. *Statistics in Medicine*, 30(24), pp.2867—2880, 2011.

[4] Friedman J.H. Greedy function approximation: A gradient boosting machine. *Annals of Statistics*, 29:1189–1232, 2001.

[5] Holland P.W. Statistics and causal inference. *J*ournal of the American statistical Association, 81(396):945–960, 1986.

[6] Imbens G. W. The role of the propensity score in estimating dose-response functions. *Biometrika*, 87(3):706–710, 2000.

[7] Imbens G. W. and Wooldridge J. M. Recent developments in the econometrics of program evaluation. *Journal of Economic Literature*, 47(1):5–86, 2009.

[8] Neyman, J. and Dabrowska D. and Speed T., et al. On the application of probability theory to agricultural experiments. *Statistical Science*, 5(4):465–472, 1990.

[9] Robins J.M. and Rotnitzky A. and Zhao L.P. Estimation of regression coefficients when some regressors are not always observed. *Journal of the American statistical Association*, 89(427):846–866, 1994.

[10] Robins J.M. and Sued M. and Lei-Gomez Q. and Rotnitzky A. Double-robust and efficient methods for estimating the causal effects of a binary treatment. *arXiv:2008.00507*, 2020.

[11] Rosenbaum P. R and Rubin D. B. The central role of the propensity score in observational studies for causal effects. *Biometrika*, 70(1):41–55, 1983.

[12] Rubin D. B. Estimating causal effects of treatments in randomized and nonrandomized studies. *Journal of Educational Psychology*, 66(5):688, 1974.

Index

For Product Safety Concerns and Information please contact our EU representative GPSR@taylorandfrancis.com Taylor & Francis Verlag GmbH, Kaufingerstraße 24, 80331 München, Germany